METAL IONS IN BIOLOGY AND MEDICINE

LES IONS MÉTALLIQUES EN BIOLOGIE ET EN MÉDECINE

Volume 7

Sponsored by

- Bashkir State Pedagogical University (Russia)
- Publishing House "Alpha Color" (Russia)
- Publishing House "Belaya Reka" (Russia)

The Editorial Committee wishes to thank
– the Conseil Régional de Champagne-Ardenne (France),
– the Institut International de Recherche sur les Ions Métalliques (Reims, France),
– the Faustus Forschungs Cie, Translational Cancer Research GmbH, Leipzig, Deutschland,
– the Nickel Producers Environmental Research Association (NIPERA) (Durham, USA),
– the Polyclinique Maymard, Bastia, France,
for their support in the realization of this book.

METAL IONS IN BIOLOGY AND MEDICINE

LES IONS MÉTALLIQUES EN BIOLOGIE ET EN MÉDECINE

Volume 7

Proceedings of
the seventh International Symposium
on Metal Ions in Biology and Medicine held in
Saint Petersburg State University,
Saint Petersburg, Russia, on May 5-9, 2002

Septième Symposium International sur les Ions
Métalliques en Biologie et en Médecine,
Saint Petersburg State University,
Saint Petersbourg, Russie, 5-9 mai 2002

Edited by
Lylia Khassanova
Philippe Collery
Ivan Maymard
Zilara Khassanova
Jean-Claude Étienne

John Libbey EUROTEXT

The first volume of this series was published in May 1990
Eds. Ph. Collery, L.A. Poirier, M. Manfait, J.-C. Étienne

The second volume of this series was published in May 1992
Eds. J. Anastassopoulou, Ph. Collery, J.-C. Étienne, T. Theophanides

The third volume of this series was published in May 1994
Eds. Ph. Collery, L.A. Poirier, N.A. Littlefield, J.-C.-Étienne

The fourth volume of this series was published in May 1996
Eds. Ph. Collery, J. Corbella, J.L. Domingo, J.-C. Étienne, J.M. Llobet

The fifth volume of this series was publishecd in May 1998
Eds. Ph. Collery, P. Brätter, V. Negretti de Brätter, L. Khassanova, J.-C. Étienne

The sixth volume of this session was published in May 2000
Eds. J.A. Centeno, Ph. Collery, G. Vernet, R.B. Finkelman, H. Gibb, J.-C. Étienne

Editor in chief : Ph. Collery
Service de cancérologie
Polyclinique Maymard
20200 Bastia, France
Philippe.Collery@wanadoo.fr

Éditions John Libbey Eurotext
127, avenue de la République, 92120 Montrouge, France
Tél. : (1) 46.73.06.60 – Fax : (1) 40.84.09.99
e-mail : contact@john-libbey-eurotext.fr
Website : http://www.john-libbey-eurotext.fr

John Libbey and Company Ltd
163-169 Brompton Road, Knightsbridge
London SW 3 IPY, England
Tél. : (44) (0) 23 80 65 02 08

John Libbey CIC
Corso Trieste 42
00198 Roma, Italia
Tél. : (39) 06 841 2673

ISBN 2-7420-0429-7

List and addresses of editors

Lylia Khassanova, Departement of Environmental Protection, Bashkir State University, Frunze Street, 32, 450074 Ufa, Bashkortostan, Russia.

Philippe Collery, Service de Cancérologie, Polyclinique Maymard, 20200 Bastia, France.

Ivan Maymard, Polyclinique Maymard, 20200 Bastia, France.

Zilara Khassanova, Department of Botany, Bashkir State Pedagogical University, October revolution Street, 3a, 450025 Ufa, Bashkortostan, Russia.

Jean-Claude Étienne, Centre Hospitalier Universitaire de Reims, 51092 Reims Cedex, France.

Domingo J. (Spain)
Durlach J. (France)
Ermakov V.V. (Russia)
Ershov Yu.A. (Russia)
Etcheverry S.B. (Argentina)
Evangelou A. (Greece)
Farzami B. (Iran)
Fields M. (USA)
Finkelman R.B. (USA)
Flores-Arce M.F. (Mexico)
Fomchenkov V.M. (Russia)
Fowler B. (USA)
Garfinkel D. (Israel)
Gibb H. (USA)
Gielen M. (Belgium)
Gogotov I.N. (Russia)
Golubkina N.A. (Russia)
Gromov B.V. (Russia)
Guillard O. (France)
Hackl E. (Ukraine)
Haroutiuniuan S. (Armenia)
Haywood S. (UK)
Hoel P.J. (Norway)
Hura C. (Romania)
Ilyaletdinov A.N. (Kazachstan)
Irgolic K. (Austria)
Ivanov A.Yu. (Russia)
Jimenez B. (Puerto Rico)
Kalfakakou V. (Greece)
Kamnev A.A. (Russia)
Kantola M. (Finland)
Karavaiko G.I. (Russia)
Kasprzak K.S. (USA)
Keppler B. (Austria)
Khassanova Z.M. (Russia)
Kisters K. (Germany)
Knekt P. (Finland)
Koudrine A.V. (Russia)
Kruse-Jarres J. (Germany)
Kumerova A. (Latvia)
Kvicala J. (Czechia)
Kylyvnyk K. (Ukraine)
Lando D. (Belarus)
Lebedev V.S. (Russia)
Limouris G. (Greece)
Littlefield N.A. (USA)
Llobet J. (Spain)
Longfellow D. (USA)
Lopes A.M. (Brazil)
Maximova I.A. (Russia)
Mazej D. (Slovenia)
Mazumder D.G.N. (India)
Meltzer H.M. (Norway)
Mendz G.L. (Australia)
Michalke B. (Germany)
Moore T. (New Zealand)
Mullick F.G. (USA)
Natochin Yu.V. (Russia)
Negretti de Bratter V. (Germany)
Nelson R. (USA)
Neve J. (Belgium)
Nielsen J.B. (Denmark)
Nowak G. (Poland)
Oller A.R. (USA)
Omu A.E. (Kuwait)
Pineau A. (France)
Pinevich A.V. (Russia)
Poirier L. (USA)
Prasad S.B. (India)
Rubikas J. (Lithuania)
Roussel A. M. (France)
Sadler P. (UK)
Sastre M. (Puerto Rico)

Organizing Committee

Bolshakov V.N. (Russia)
Maximova I.A. (Russia)
Stefanov A.V. (Russia)
Visulainen V.V. (Russia)

Coordination of the series of conferences:

Institut International de Recherche sur les Ions Métalliques
International Research Institute on Metal Ions

Address: Université de Reims Champagne Ardennes
UFR Sciences de Reims, Moulin de la Housse
BP 1039
51687 Reims Cedex
France

CONTENTS/SOMMAIRE

I CURRENT TOPICS ON SPECIATION

II MOLECULAR AND BIOLOGICAL EFFECTS

III CELL EFFECTS

IV DISTRIBUTION

V *METAL IONS AND MICROORGANISMS*

VI *METAL IONS AND PLANTS*

VII *TOXICOLOGY*

VIII *ENVIRONMENT*

IX EPIDEMIOLOGY

X NUTRITION

XI INFLAMMATION, IMMUNOLOGY

XIII ONCOLOGY

XIV *MISCELLANEOUS*

I CURRENT TOPICS ON SPECIATION

Metal Ions in Biology and Medicine: vol. 7. Eds L. Khassanova, Ph. Collery, I. Maymard, Z. Khassanova, J.-C. Étienne.
John Libbey Eurotext, Paris © 2002 pp. 3-10.

Element speciation, specifically in the bio-medical field

Bernhard Michalke

GSF National Research Center for Environment and Health, Ingolstaedter Landstr. 1, 85764 Neuherberg, Germany

INTRODUCTION

It is generally accepted that total concentrations of the elements mostly cannot give the required information about mobility, bioavailability and finally the impact of elements on ecological systems or biological organisms. Only the knowledge of the chemical species of the elements can provide an understanding of chemical and biochemical reactions and thus leading to more information about toxicity or essentiality. Combination and hyphenation of separation technologies and element selective or molecule selective detection systems are generally the basis for speciation analyses. Still many methodical developments are necessary, predominantly for hyphenation and generally quality control strategies. Investigations on quality control finally showed that during sampling, sample preparation and storage, separation and detection changes in "original species information" are easily occurring: the final result then can give only poor information about speciation in the original sample. The different results then are characterized by (partly not recognized) alterations caused from different methodologies. The need for quality control strategies is widely recognized now and practiced more and more. This contribution will first describe shortly the technical demands for element speciation and finally shows exemplarily some trends of speciation in the biomedical field. Especially recent investigations on Se-speciation and the various speciation approached aimed for different isoforms of metallothioneins will be focused on.

DEFINITIONS OF TERMS RELATED TO SPECIATION

There are now several definition in use about terms related to speciation. These definitions are derived from [1] and are listed shortly in *table 1*:

Table 1. Definitions ("short forms")

chemical species	Specific form of an element (specific molecule, complex, oxidation state).
speciation analysis	Analytical activity of identifying and measuring species in a real and representative sample / matrix. Identifying and measuring strictly includes quality control.
operationally defined characterization	Not chemical speciation analysis. No species identification provided. Characterization of molecule groups according to a given analytical procedure.
functionally defined characterization	Not chemical speciation analysis. No species identification provided. Characterization of molecule groups according to their impact on organisms ("essential", "toxic").
Speciation	Distribution of species in a particular sample or matrix.

SPECIATION ANALYSIS - BASIC PRINCIPLES

Speciation schemes generally consists of the sampling and sample preparation step and of the speciation analysis step (including separation and detection).

SAMPLING

The sampling step is one of the most critical ones in speciation analysis as it interferes species equilibria often irreversible [2]. Therefore, sampling concepts must preserve the original species information. Such techniques are often not available for the respective problem and must be adapted to the actual situation concerning element species of interest and matrix. *Table 2* summarizes some aspects important for sampling in speciation anaylsis.

Table 2. Summary for means of quality control in sampling

• sampling a representative aliquot;	• take urine samples from the middle stream;
• keep sampling time short;	• no use of stainless steel instruments for biological samples;
• the volume/surface ratio should be as possibly high to minimize wall effects at container surface;	• higher sample volume for soil samples is recommended to preserve representativity and species equilibria inside.
• contamination must be avoided;	

SAMPLE STORAGE AND SAMPLE PREPARATION

Sample typically cannot be analysed on site. Therefore a storage gets necessary, which as possibly is short. For replicate measurements from one sample and in general for most applications a longer storage time is gets necessary. For this reason, storage time must be kept to a minimum preferably at 4 °C or for prolonged storage freeze-drying or shock freezing at -80 °C may be useful. Clean room conditions and pre-cleaned vials etc. must be used throughout [3]. Pooled biological samples have some advantages when planning method development, as individual variation can be eliminated. Solid biological samples recently were in investigation for enzymatic extraction, employing protease (different types) or lysozyme. But also (hot) water extractions were compared to diluted HCl-leaching. When using yeast as a sample extraction efficiencies were quiet high, e.g. reaching 80-100% of total Se content of the sample after a protease treatment [4]. HCl leaching was successfully used for mimicking gastric juice digestion [5], but was found less suitable for speciation in a bacterial sample [11]. In some cases derivatisation provides advantages, too. As an example crown ether treatment of urine samples is mentioned [6]. High concentrated interfering cations from urine were removed and subsequent separation on cation exchange LC was performed successfully.

COMBINED TECHNIQUES OF SEPARATION AND DETECTION METHODS: HYPHENATED SYSTEMS

The most usual way for element speciation are combined and hyphenated systems. Species are first separated and subsequently the elements of the species get selectively detected. For increased quality control also molecule selective detection is hyphenated to separation devices. *Figure 1* gives an overview about typical combinations of separation and detection systems. Generally, the advantage of combined systems of separation modules and element selective detection modules is the extended variability and a broad applicability. Disadvantages are seen in increased complexity and thus in a risk of increased defectifity of the systems.

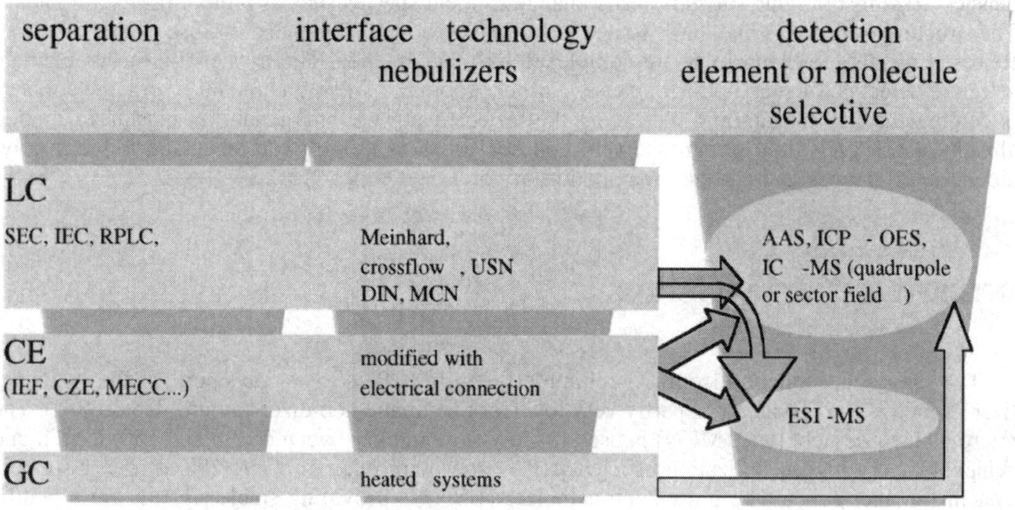

Fig. 1. Overview about speciation approaches: combined (-hyphenated techniques)

SEPARATION TECHNIQUES

LC - Liquid Chromatography

One of the most important advantages of liquid chromatography is the extended variability of separation mechanisms with the use of different mobile and stationary phases, providing nearly all necessary solutions for separation of element species. Therefore, a problem - related speciation analysis is possible, meeting the requirements for species stability and sufficient separation. Many stationary phases, however, or buffers and organic modifiers can denaturate native species. Chelating eluents or ion exchangers may cause recomplexation of free or labile bound metal species [2, 7]. Buffers can stabilize bio-molecules but may also alter species equilibria. There can be complexing tendencies or an input of metal contaminations [8]. For example size-exclusion chromatography (SEC) separates species depending on molecular size [9], where samples of unknown molecular mass get characterized in a mass-calibrated chromatographic system. SEC is a gentle method of chromatographic separation and normally does not result in a loss of element species or on-column alterations. Disadvantages are seen in the limited peak capacity and the fact that the stationary phase is not totally uncharged. Electrostatic effects, adsorption, hydrophobic interactions and species specific affinities or H-bridging were observed. Ion exchange chromatography provides high separation efficiency and wide applicability in speciation. IEC provides several possibilities to achieve separation even in complicated samples by optimizing at least three variables (pH, ionic strength, nature of the ion exchanger) [9]. Problems are known from the pore size of the resin particles possibly excluding analytes from separation or temperature related changes in selectivity. Often loosely bound metal ions are lost or replaced by other metals originating from the buffer [2]. Thus, IEC is predestined for the separation of kovalently bound element species of different valence states [10, 11, 12]. When using reversed phase chromatography (RPLC) a wide analyte spectrum may be analyzed, using the effective separation with high resolution of species and the flexibility caused by the multiple mobile phases including the possibility of adding also "ion pairing" reagents for the analysis of ionic/charged molecules [9]. In practice, however, the stationary phase exhibit ion exchange properties or undesired adsorption effects especially for basic analytes, which can be adsorbed tightly [9]. Typically, polar eluents show high complexing tendencies [9] whilst organic solvents easily change element species such as protein-metal complexes. The structure of proteins may be unfolded and complex bound elements are subsequently released. Released metals are likely to get re-complexed by other ligands. The hyphenation to an element selective detector, such as ICP-MS, causes further problems. The high amount of organic solvents is cooling the ICP-plasma and increasing the reflected power. This results in plasma extinction already at relatively low organic solvent concentrations. The high carbon intake induces polyatomic interferences and carbon precipitation on torch and cones [13].

INTERFACING LC TO ICP-MS

The sample introduction from LC column effluent to ICP-MS is always performed by a nebulizer. Nebulization efficiency is partly very low (for pneumatic nebulizers around 1-5%, [13]). The micro concentric nebulizer (MCN) is used for low flow rates between ca. 30 and 150 μL/min and achieves a very high nebulization efficiency. Together with microbore LC, this is a very suitable system for interfacing. The ultra sonic nebulizer (USN) is providing stable signals, but quantifications can be altered due to the salt content of the eluent. High salt concentrations are a more pronounced problem with this interface, as a crusting and finally clogging of tubings, transfer lines or sampler cones is appearing very rapidly. Several commercial USN have a built in desolvation system which removes this problem. Unfortunately, a couple of element species can be removed, too. LC needs buffers, often with an increased salt concentration, leading to clogging of the nebulizer and the cones. In RPLC organic modifiers must be employed, such as methanol or acetonitrile. They, too, are changing drastically the ionisation characteristics of the plasma even at low concentrations. Therefore, different sensitivity is observed for species eluting at different modifier concentrations. A slight increase in e.g. MeOH (up to around 10%) destabilizes the plasma and at higher concentrations the plasma is extincted (as long as no precautions are performed, see below) [14, 15]. The carbon input into the plasma produces a carbon layer on the torch. Therefore, the reflected power to the oscillating generator may get too high or flush overs may happen. Both effects are shutting down the plasma, too.

Solutions to these problems are a major concern in technical directed papers on speciation. New developments in column technology have been carried out, providing new columns with high separation efficiency even at low (buffer) salt concentrations (e.g. [16]). Even RPLC is employed now more in hyphenated systems, as the limitations caused by organic modifiers can be overcome. This is done via post-column dilution of the eluent [17, 18]. For iodine speciation this was accompanied with signal increase when taking HNO_3 as post-column reagent [10]. The most common method is to cool the transfer line and/or the spray chamber below 10 °C for lowering vapour pressure of organic modifiers. Methanol concentration up to 60% - 80% in the eluent were tolerated at flow rates of ca. 1.5 mL/min [19]. Especially, when using an USN, desolvating systems are likely to be employed in addition, e.g. membrane desolvation. This allows a methanol concentration up to 100% [20, 21], but some species may be removed, too. As a conclusion it is generally desirable to keep the total liquid intake into the interface and into the ICP-plasma as low as possible. Therefore, combinations of micro-LC with "low-flow-rate interfaces", such as MCN are a most promising way in LC-ICP-MS.

CE - CAPILLARY ELECTROPHORESIS

One of the most powerful separation devices is capillary electrophoresis. It provides most efficient separation of species by applying high voltage (20-30 kV) along the open-tube-column with low inner diameter. Analysis time is comparatively short and separation efficiency is very high (200000 - 700000 theoretical plates). Several different separation principles are available, distinguishing and separating the analytes according to different physico-chemical properties providing completely different characterization and identification mechanisms for element species. These modes are CZE, MECC, IEF, ITP and CEC. The variation in characterization is of significant importance, as species identification is rarely done by one single method but needs multidimensional strategies. The use of capillary electrophoresis in speciation is complex. It is employed either as a primary separation mechanism or as a secondary separation technique after e.g. HPLC, used in a second dimension for identification. In both cases it may be combined with non-selective (direct or indirect) UV-detection or hyphenated to ICP-MS and ESI-MS. The various combinations of CE techniques in different stages of multi-dimensional strategies and together with detectors, which are supplying information about isotopes and elements (ICP-MS) or molecules and structural compounds (ESI-MS/MS) makes capillary electrophoresis immense valuable in speciation analysis.

Typically, CE detection is performed by UV-based techniques (direct or indirect). The small diameter of separation capillaries are causing decreased optical detection sensitivity. Thus, interfacing technology is recommended. The interfacing to ICP-MS nowadays provide sub µg/L detection limits and the capability for multi-element monitoring [11]. However, at present an efficient interface to connect the capillary and the nebulizer of any plasma detector is still a challenge. In the past several interface designs have been published, providing acceptable operation and low detection limits. A good overview is given in reference [22]. Further species information is gained when coupling CE also to electrospray mass spectrometry for identification of bio-molecules. In contrast to the CE-ICP-MS coupling here direct species detection is provided. The MS/MS mode additionally provides structural information about element species. Problems and limitations of CE in speciation analysis are given by the small sampling volume (representativity, sensitivity) or by buffer components which may alter species stability. The high voltage can alter the integrity of elemental species. When using hyphenated techniques an additional suction flow can be forced on the "open tube" capillary. Technical solutions are working with reduced nebulizer capillary diameter or longer CE capillaries (both based on the law of Hagen-Poiseuille). There are also problems known for the detector, predominantly interfered mass signals due to polyatomic or isobaric interferences [23] in ICP-MS detection. A big problem when using ESI-MS detection is that only volatile buffers are possible. No free choice for seperation electrolytes is not given. Several severe problems in speciation are caused by the ionization process itself, which are ruled out in the ESI-MS section.

ELEMENT SELECTIVE DETECTION

Nowadays, inductively coupled plasma-atomic emission spectroscopy (ICP-AES) and ICP mass spectrometry (ICP-MS) are the most used detectors in element speciation. They provide multi-element capability and the high sensitivity [24]. On-line hyphenations are easily set up. The excitation - (ICP-AES) or ionisation - (ICP-MS) source is an inductively coupled plasma. Background compensation should be applied in any case. Sample introduction is performed via a nebulizer and spray chamber. ICP-AES system provide low detection limits down to the 10-100 µg/L level in hyphenated systems. Here superiority is given by ICP-MS, where detection limits down to the 20-100 ng/L level in hyphenated systems is reached [11, 16, 25] 20]. Isotopic and elemental information of species is gained [2]. There are quadrupole systems (qICP-MS) and highly resolving sector field ICP-MS (sf-ICP-MS) available. The latter is reported to improve detection sensitivity by a factor 10 - 100 [2] or even up to 1000 when equipped with a guard electrode [26]. Furthermore, the quadrupole mass filter provides a resolution of only 300 u. Therefore, "polyatomic interferences" are likely, especially in the mass range 40 - 80. The highly resolving sf-ICP-MS can distinguish between the interference and the element isotope, as the mass resolution is 7500 - 10000 u.

ESI-MS DETECTION

Electrospray ionization is a ionization process which may preserve the whole species intact under optimal circumstances. The element species itself may be monitored. ESI is suitable for extremely low flow rates. The success of this detection method is based on the capability to produce multi-charged ions from high molecular element species such as metallo-proteins and thus making the analysis of these compounds available, up to MW = 150000 - 200000. The possibility to couple this detector to LC or CE systems makes it additionally immense valuable [27]. Structural changes (mostly) do not appear as long as covalent bondings are present. In special cases (e.g. with selenium) stable element-organic molecules e.g. from bio-medical matrices can be analyzed. This was shown in literature for Se-species such as seleno-cystamine, seleno-cystine and seleno-methionine. Metallothioneins (Cd-MT, Zn-MT, Cu-MT) and their isoform, too, were examined widely. When applying collision induced dissociation (CID) together with a MS/MS system further structural information can be gained. Undesired ion-solvent clusters may be destroyed or the parent ions are fragmented into (molecule specific) daughter ions, selected by a second quadrupole [28]. No other detection technique is able to provide such detailed information about molecular weight and even structure of analyzed compounds. Unfortunately, several problems were faced, which were related to ion-solvent clusters, which split up one species into multiple signals, worsening detection limits and increasing spectral complexity. Electrolytic processes were observed resulting in generating new species or a transformation of species.

Examples

Finally, as an example some important elements in bio-medical speciation investigations are shortly summarized. These element species are selenium species and investigations on metallothioneins bound to different heavy metals. MT is characterized by showing isoforms and binding to several more heavy metals, such as Hg, Zn or Cu.

CADMIUM SPECIATION AND INVESTIGATIONS ON METALLOTHIONEINS AND OTHER METALLOPROTEINS

SEC has been used frequently to separate and characterize biological macromolecules and small organic molecules using hydrophilic column packings and aqueous or organic mobile phases. Mostly investigations of metallothioneins-1 and -2 bound to different metals such as Cu, Hg, Zn, Cd were performed. It turned out that SEC provides no sufficient separation, therefore anion exchange was subsequently employed. Reference [29] used a DEAE-5 PW SAX column with a 2-200 mM Tris-HCl, pH 7.4, gradient elution. For a quantitative elution a higher buffer concentration was necessary, causing problems with ICP-MS. After insertion of a hydride generator increased sensitivity was achieved for Cd in parallel to matrix/salt separation. Cd-Mt1 and Cd-Mt2 were found in rabbit liver standards, whereas Cu was associated with only Mt1. In urine three decomposition products of Mt1 were seen. Metallothioneins were investigated using a multidimensional analysis concept - the characterization/identification was performed with RPLC-ICP-MS in combination with RPLC-ESI-MS [30]. Finally, Chassaigne and Lobinski combined RPLC-ICP-MS with ICP-MS and ion-spray-MS for orthogonal characterization of Cd-metallothioneins [31]. Several Cd-containing sub-isoforms of Mt2 and unknown Cd- species were seen within 40 min. The Mt2 isoforms were further identified by ion-spray MS.

Selenium speciation

Selenium, both a toxic and an essential element, plays an important role in environmental analysis as well as in health studies. Excess Se intake can cause toxic reactions in living organisms [32]. However, selenium is also an essential trace-element for humans. A sufficient selenium supplementation can protect against several heart diseases and is discussed for prevention against cancer. The detoxification effects of Se are proved, and described widely for heavy metals [33, 34, 35]. Thyroid metabolism may be impaired, because many de-iodinases are Se-proteins [36].

Crews et al. [37] investigated cooked cod using SAX, and monitored the ^{82}Se isotope for Se-methionine, Se-cystine, Se(IV) and Se(VI) determination. They found 82% of total Se was an unknown species and 12% was Se (IV). Lafuente et al. [38] developed a Se-speciation method using ion-pairing RPLC combined with AAS, ICP-OES and ICP-MS, reaching the best LODs with the latter, at 0.1 µg/L. In native urine they found only one, unknown Se-species. Gammelgaard et al. [6], also investigated urine samples for Se-species. The extract was analyzed using a Dionex Ionpac CS 5 column at a flow-rate of 0.5 mL/min, interfaced to an ELAN 6000 ICP-MS. The total Se in different urine samples was found to be 47-81 µg/L, associated with SeM and minute amounts of TMSe. Birringer et al. [39] investigated Se-yeast and Se-garlic, using LC-ICP-MS. They found SeM and γ-glutamyl-Se-methylselenocysteine as the predominant species in both matrices. Gössler et al. [40] provided a cation-exchange ICP-MS method for Se-speciation in ten supplements. They found the total Se content to be consistent with the specifications, but the chemical forms were quite different. Se(VI) was the only Se-species in four supplements. Recently, Michalke et al. [16] provided an anion-exchange-ICP-MS method to analyse six Se standard compounds within 8 min, or several Se-species from a Se-containing bacterial sample within 12 min, at LOD around 0.1 µg Se/L. They found only minute amounts of SEM but higher concentrations of Se(IV) and SeC. The most prominent species were not identified, however. Later they used this method to examine children's sera for Se-species. For excluding the presence of methylated Se-species a RPLC-ICP-MS method was employed in addition [41].

CONCLUSION

Element speciation has become a field of increasing importance in analytical chemistry and gains relevance in bio-medical health care. The knowledge of the chemical species of the elements supports understanding of chemical and biochemical reactions and thus leading to more information about toxicity or essentiality. Problem related speciation analysis promises to become a key for an optimized supply of trace metals in health care. The analytical challenge of element speciation must lead to "true" speciation of the elements in a sample. Multiple investigations on preservation of species information during sampling, sample storage and preparation as well as separation and detection technologies have been carried out resulting in valuable method developments and quality control strategies. Typically, combined or hyphenated techniques are employed, consisting of suitable separation techniques and sensitive detection methods. Detection in speciation now typically uses element selective systems like ICP-MS or molecule sensitive detectors like ESI-MS. Orthogonal speciation concepts providing mostly high quality species information and species information of unknown compounds will be the basis for future speciation work. "Hot topics" in speciation are still Se-speciation and invetsigations on metallothioneins and it seems that both topics will remain in the focus of speciation interest in future.

REFERENCES

1. D. M. Templeton, F. Ariese, R. Cornelis, L.-G. Danielsson, H. Muntau, H. P. van Leeuwen, R. Lobinski, Pure Applied Chem. (2000) 72/8: 1453-1470.
2. Dunemann, L., Begerow, J. Kopplungstechniken zur Elementspeziesanalytik, VCH, 1995, Weinheim, New York, Basel, Cambridge, Tokyo.
3. Schramel, P., Hasse, S., Mikrochim. Acta, 1994, 116: 205-209.
4. Cassiot C, Szpunar J, Lobinski R, Potin-Gautier M (1999) Journal Anal Atom Spectrom 14: 645-650.
5. H.M. Crews, P.A. Clarke, D.J. Lewis, L.M. Owen, P.R. Srutt, A. Izquierdo, JAAS 11 (1996) 1177-1182.
6. Gammelgaard, B., Jons, O., Bendahl, L. JAAS, 2001, 16, 339-344.
7. G. Weber, Fres. J. Anal. Chem., 346 (1993) 639-642.
8. J. Arnaud, D. Andre, M.C. Bouillet, D. Kia, A. Favier, J. Trace Elem. Electrolytes Health Dis., 6 (1992) 81-90.
9. Mikes, O. High performance liquid chromatography of biopolymers and biooligomers, Elsevier, 1988, Amsterdam, Oxford, New York, Tokyo.
10. N. Gilon, M. Potin.- Gautier, *Astruc, M., J. Chrom.* A, 1996, 750: 327-334.
11. B. Michalke, H. Witte, P. Schramel, Biol. Trace Elem. Res., 78/1-3 (2000), 67-79.
12. E.H. Larsen, G. Pritzl, S.H. Hansen, J. Anal. At. Spectrom., 8 (1993) 1075-1082.
13. G.E. Batley, In: G.E. Batley, Ed. Trace Element Speciation: Analytical Methods and Problems, CRC Press, 1989, Boca Raton.
14. C.M. Andrle, N. Jakubowski, J.A.C. Brokaert, Spectrochim. Acta B 52/B2 (1997) 189-200.
15. H. Ding, L.K. Olson, J.A. Caruso, Spectrochim. Acta B 51B (1996) 1801-1813.
16. B. Michalke, H. Witte, P. Schramel, J. Anal. Atom. Spectrom. 16, (2001) 593-597.
17. H. Chasseigne, J. Szpunar, Analusis 26/6 (1998) M48-M51.
18. H. Chasseigne, R. Lobinski, Analytica Chim. Acta 359 (1998) 227-235.
19. H. Ding, K. Olson, J.A. Caruso, Spectrochim. Acta, Part B, 51 (1996) 1801-1807.
20. S. Lustig, B. Michalke, W. Beck, P. Schramel, Fres. J. Anal. Chem., 360 (1998) 18-25.
21. N. Jakubowski, C. Thomas, D. Stüwer, I. Dettlaff, J. Schram, J. Anal. At. Spectrom. 11 (1997) 1023-1027.
22. B. Michalke, in Rita Cornelis (Ed) Handbook of Speciation, John Wiley and Sons, 2002. in press.
23. ICP-MS Interferenz Tabelle, 1995, Finnigan MAT, Bremen.
24. Hill, S.J., Bloxham, M.J., Worsfold, P.J. J., Anal. At. Spectrom., 1993, 8: 499-515.
25. S.J. Hill, M.J. Bloxham, P.J. J. Worsfold,, Anal. At. Spectrom., 8 (1993) 499-515.
26. A. Prange, D. Schaumlöffel, J. Anal. At. Spectrom., 14 (1999) 1329-1332.
27. E. D. Lee, W. Mück, J. D. Henion, T. R. Covey: On-line capillary zone elctrophoresis in spray tandem mass spectrometry for the determination of dynorphons. J. Chromatogr., 458 (1988) 313-321.
28. M. M. Shahin: Mass spectrometric studies od corona discharges in air at atmosperic pressures. J. Chem. Phys., 46, 1966, 2600-2605.
29. H.G. Infante, M.L. Fernandez Sanchez, A. Sanz-Medel, JAAS 14 (1999) 1343-1348.
30. K. Polec, S. Mounicou, H. Chassseigne, R. Lobinski, Cell. Molecular Biol., 46/2 (2000) 221-225.
31. H. Chassaigne, R. Lobinski, Fresenius' J. Anal. Chem. 361 (1998) 267-273.
32. J. Neve, P. Therond, in P. Chaputs, Editor, Oligoelements en Medecine et Biologie, (1991), Lavoisier Paris, 425-453.
33. M.P. Rayman, Lancet, 356 (2000) 233.
34. L.C. Clark, G.F. Combs, B.W. Turnball, E.H. Slate, D.K. Chalker, J. Crow, L.S. Davis, R.A. Glower, G.F. Graham, E.G. Gross, A. Krongrad, J.L. Lesher, H.K. Park, B.B. Sanders, C.L. Smith, J. R. Taylor, JAMA, (1996) 276, 1957.
35. P.D. Whanger, J. Trace Elem. Electrolytes Health Dis., 6 (1992) 209-221.
36. J. R. Arthur, F. Nicol, G.J. Beckett, Biol. Trace Elem. Res., 33 (1992) 37-42.
37. H.M. Crews, P.A. Clarke, D.J. Lewis, L.M. Owen, P.R. Srutt, A. Izquierdo, J. Anal. Atom. Spectrom., 11 (1996) 1177-1182.
38. J.M.G. Lafuente, M.L.F. Sanchez, A. Sanz-Medel, J. Anal. Atom. Spectrom., 11 (1996) 1163-1169.
39. C.M. Birringer, E. Block, M. Kotrebai, J.F. Tyson, P.C. Uden, D.J. Lisk, J. Agr. Food Chem. 48/6 (2000) 2062-2070.
40. J. Zheng, W. Gössler, W. Kosmus, Trace Elem. Electrolytes 15/2 (1998) 70-75.
41. B. Michalke, Proceedings Ionen Fachtagung Trier 2002, in press.

Metal Ions in Biology and Medicine: vol. 7. Eds L. Khassanova, Ph. Collery, I. Maymard, Z. Khassanova, J.-C. Étienne.
John Libbey Eurotext, Paris © 2002 pp. 11-14.

Sod-like activity and lipophilicity of copper chelates with aminoacids and peptides

Gianella Facchin[1], María H. Torre[1], Inés Viera[1], Eduardo Kremer[1], Enrique J. Baran[2]

[1]*Química Inorgánica, DEC, Facultad de Química, UDELAR, Gral. Flores 2124, CC1157, Montevideo, Uruguay.*
[2]*CEQUINOR, Facultad de Ciencias Exactas, UNLP, Calle 47, 115, CC 962, La Plata, Argentina.*

ABSTRACT

Several copper complexes with dipeptides were prepared and characterized. Their superoxide dismutase (SOD)-like activity and lipophilicity were determined and compared with those of previously studied complexes of α-L-amino acids.

INTRODUCTION

Many copper(II) complexes of amino acids or small peptides have diverse pharmacological activities: antiinflammatory, anticancer, anticarcinogenic, anticonvulsant or antiulcerous [1]. In general, it can be contended that copper complexes of low molecular weight can be useful to facilitate transport of the metal to the active site of different enzymes. This allows activation or reactivation of copper-dependent enzymes, like superoxide dismutase (SOD). This enzyme catalyses the dismutation of the toxic superoxide radical. It has a central role in defending life forms that use oxygen, from oxidative damage. It has also been shown that many pharmacologically active copper complexes disproportionate superoxide anion in a manner similar to that of SOD.

Body membranes are mainly lipoid and are more easily penetrated by lipophilic molecules than by hydrophilic ones. The partition coefficient between organic solvent and water is and index of the lipid solubility of a compound. There is often a relationship between this and the rate of transfer of a molecule across a biological membrane [2].

MATERIALS AND METHODS

Synthesis of the complexes

Copper complexes with the amino acids Gly, L-Ala, L-Val, and L-Ile, and with the peptides L-alanine-L-glycine, L-alanine-L-valine, L-alanine-L-isoleucine, L-alanine-L-leucine, L-alanine-L-phenilalanine, L-alanine-L-tyrosine, and L-alanine-L-threonine, were synthesized and characterized as reported previously [3, 4, 5]. The obtained stoichiometries were Cu(amino acid)$_2$ and Cu(L-dipeptide) respectively.

Determination of SOD-like activity

The superoxide dismutase (SOD)-like activity was investigated by the method of Beauchamp and Fridovich as improved by Imanari et al. [3]. This method is based on the inhibitory effect of

SOD over the reduction of nitrobluetetrazolium (NBT) by the superoxide anion generated by the system xanthine/xanthine oxidase. All the reagents used in these assays were purchased from SIGMA. For comparative purposes the activity of native superoxide dismutase from bovine erythrocytes has also been determined.

Lipophilicity test

Lipophilicity tests were performed determining the partition coefficient of the complexes in physiological solution/n-octanol [2]. The copper concentration was measured by atomic absorption spectroscopy using a Perkin-Elmer 5000 instrument, with a Photron lamp for copper analysis.

RESULTS AND DISCUSSION

The Cu(II) concentrations of Cu-amino acids and Cu-dipeptides, necessary to produce 50% inhibition of the NBT reduction are shown in *Table 1.*

Table 1. Cu(II) concentrations required to yield 50% inhibition of the NBT reduction (IC_{50})

Compound	IC_{50} (µM)	Compound	IC_{50} (µM)
$Cu(Gly)_2.H_2O$	34	Cu (alagly)(H_2O)	124
$Cu(Ala)_2$	32	Cu(alaval)	79
$Cu(Val)_2$	30	Cu(alaile)	111
$Cu(Ile)_2.H_2O$	33	Cu(alaleu).2H_2O	22
Cu-SOD	0.010	Cu(alaphe)	5.0
		Cu(alatyr)	8.4
		Cu(alathr).1/2 H_2O	33

The IC_{50} values for Cu-amino acid complexes are very similar. On the other hand, those corresponding to the copper dipeptide complexes show a wider variation of IC_{50}. Some of them are more active than Cu-amino acid complexes, while others are less active.

The Cu(alaphe) and Cu(alatyr) complexes are the most active ones and they have an IC_{50} value close to those clinically interesting, according to the correlation found by Roberts and Robinson [6].

Cu-amino acid complexes have different crystalline structures, with distorted octahedral or distorted square pyramidal copper environment, with ligands in *cis* or *trans* configuration. However in solution the coordination of copper in all the studied complexes is essentially the same. This fact is supported by the similarity of the electronic spectra *(table 2).*

The IC_{50} values obtained in aqueous solution are also in agreement with a similar Cu(II) configuration for the different complexes.

On the other hand the crystalline structures of Cu(Ala Val), Cu(Ala Phe), Cu(Ala Thr) and Cu(Ala Ile), are very similar. In these complexes, the Cu(II) essentially has the same elongated pyramidal coordination, being equatorial *cis* coordinated by a N_2O_2 arrangement of ligand atoms, and axially by a carbonyl oxygen atom. The electronic spectra of aqueous solutions of Cu-dipeptides are similar as shown in *table 2*. This would be indicative of a similar Cu(II) environment, in solution.

It is remarkable that, despite the structural similarity, the SOD like activities of these Cu-dipeptide complexes are markedly different, suggesting an overall effect of the involved ligands.

Table 2. Wavelength of the absorption maxima (λ_{max}) and absortivity (a_M) of the visible spectra of aqueous solutions of Cu-amino acid and Cu-dipeptide complexes.

Compound	λ_{max} (nm)/a_M ($M^{-1}cm^{-1}$)	Compound	λ_{max} (nm)/a_M ($M^{-1}cm^{-1}$)
$Cu(Gly)_2.H_2O$	632/44.5	Cu (alagly)(H_2O)	635/56
$Cu(Ala)_2$	618/54.6	Cu(alaval)	629/98
$Cu(Val)_2$	612/53.3	Cu(alaile)	620/55
$Cu(Ile)_2.H_2O$	618/-	Cu(alaleu).$2H_2O$	627/72
		Cu(alaphe)	625/60
		Cu(alatyr)	629/70
		Cu(alathr).1/2 H_2O	630/80

The results of the lipophilicity assays are shown in *table 3*. The K-values (K) are defined as the ratio of the Cu(II) concentration in the organic phase over its concentration in the aqueous phase.

Table 3. Partition coefficients (K) between physiologic solution and n-octanol

Compound	K
$Cu(Gly)_2.H_2O$	0.052 ± 0.015
$Cu(Ala)_2$	0.066 ± 0.0015
$Cu(Val)_2$	0.082 ± 0.010
$Cu(Ile)_2.H_2O$	0.150 ± 0.010

The general trend observed in the tabulated K-values for Cu-amino acid complexes suggests that lipophilicity increases with ligand complexity. This trend follows a well known behavior, according to which in a homologous series the partition coefficients increase by added CH_2 group.

On the other hand for all the complexes with dipeptides the K-values are 0, as no partition was detected. This is consistent with the hydrophilic groups exposed in the external place of the first coordination sphere and with the water solubility of all the complexes.

CONCLUSIONS

This study allowed to attain preliminary data of pharmacological interest.

All the investigated Cu(II) complexes present a significant SOD-like activity, among which the the activity of Cu(Ala Phe) and Cu(Ala Tyr) is specially remarkable. These last complexes appear as potentially useful for the pharmacological applications in human or veterinary medicine [7]. The partition coefficients between physiologic solution and n-octanol may be useful for understanding of the permeability of these complexes across biomembranes.

ACKNOWLEDGEMENTS

This work is a part of a joint research project of the inorganic chemistry groups of Montevideo and La Plata (Argentina), under the auspices of the two Universities. The Uruguayan group thanks also PEDECIBA and INIA (grant INIA LIA 008) for support.

REFERENCES

1. Baran E. J., Química Bioinorgánica, McGraw-Hill, Madrid, 1994.
2. The Pharmaceutical Codex, The Pharmaceutical Press, London, 1979.
3. Totaro R M, Apella M C, Torre M H, Friet E, Viera I, Kremer E, Baran E J. Evaluation of Superoxide Dismutate-Like Activity in Some Copper (II) Complexes of Aminoacids. *Acta Farm Bonaerense* 1993; 12 (2) 73-78.
4. Facchin G., Torre M. H., Kremer E., Piro O. E., Castellano E. E., Baran E. J. Structural And Spectroscopic Characterization Of Two New Cu(II) Dipeptide Complexes. *Z. Naturforsch.* 2000, 55b, 1157.
5. Schimiedgen, R., Huber, F. Preparation and Characterization of Copper (II) Derivatives of Glycyl-β-Alanine and β-Alaninylglycine. Crystal Structure of Cu(II) -β-AlaGly.H_2O. *Z. Naturforsch.* 1993, 48 (3), 277-81.
6. Roberts N., Robinson P., Copper Chelates of Antirheumatic and Anti-Inflammatory Agents: Their Superoxide Dismutase-Like Activity and Stability. *British Journal of Rheumatology* 1985; 24: 128-136.
7. Smart M E, Cymbaluk N F, Christensen D A. A review of copper status of cattle in Canada and recomendations for supplementation. *Can Vet J* 1992; 33, 163-170.

*Metal Ions in Biology and Medicine: vol. 7.*Eds L. Khassanova, Ph. Collery, I. Maymard, Z. Khassanova, J.-C. Étienne.
John Libbey Eurotext, Paris © 2002 pp. 15-19.

Arsenic species analysis in wool of sheep and human hair

Andrea Raab, Jorg Feldmann

Department of Chemistry, University of Aberdeen, Meston Walk, Aberdeen AB24 1UE, Scotland, UK

SUMMARY

Hair and wool are part of the body metabolism and during their growth are influenced by the presence of toxins and drugs. The fibre is formed in the root; during which time there is a possibility of the inclusion of, for example arsenic species. Since hair proteins contain a large amount of sulfhydryl groups, and trivalent arsenic species have a high affinity to these groups, there is a high probability that these highly reactive arsenic species are enriched in hair. Trivalent arsenic species are under suspicion that they are highly carcinogenic, but their concentration in most body fluids is too low to measure them. Therefore hair and wool might be suitable materials to increase our knowledge about the metabolic changes of ingested arsenic species. To test this hypothesis we extracted wool from sheep feeding on seaweed (contains large amounts of organic bound arsenic) and human hair samples from people drinking arsenic contaminated well water. The fibres were extracted with water by boiling for 6 hours. After filtration the extracts were injected onto a HPLC column connected with an inductively coupled plasma mass spectrometer. For the identification of the separated species we used spiked samples. We were able to identify As(III), As(V), monomethylarsonic acid (MMA(V)) and dimethylarsinic acid (DMA(V)) in extracts of both kinds of fibre. The wool extract contained also monomethylarsonous acid (MMA(III)) and dimethylarsinous acid (DMA(III)). In addition to these identified species the wool extract contained a further four unidentified species.

INTRODUCTION

Hair and wool are part of the body metabolism and during their growth are influenced by the presence of toxins and drugs. The fibre is formed in the root during which time there is a possibility of the inclusion of heavy metal species being transported in the blood. The no longer metabolic active part of the fibre (outside the epidermis) is influenced by exogenic contaminations but not by endogenic ones. A long controversy exists about the usefulness of hair analysis in the determination of trace element status. The analysis of hair or wool for trace elements has probably a limited usefulness in large epidemiological studies concerning the exposure of humans or animals toward toxic elements, because the sample material is easily accessible [1]. This is of course under the assumption that no exogenic contamination has occurred or that it can be removed from the sample without disturbing the endogenic content.

A survey of the literature shows that arsenic is known to accumulate in hair after accidental ingestion of arsenic compounds or long term exposure through food and water [2, 3]. The concentration of arsenic ranges from 0.08-0.25 mg/g for people not contaminated with arsenic to more than 9 mg/g for people ingesting arsenic contaminated drinking water regularly [3]. Little is known about the arsenic species in hair from people ingesting inorganic arsenic via contaminated water or food. It is known that an increased ingestion of arsenobetaine, the main arsenic species in seafood, does not result in increased arsenic levels in hair.

The metabolism of organic arsenicals such as arsenoribosides, is less known. There are some hints that arsenoribosides are also metabolized in the body, partly to dimethylated arsenic [4, 5]. The pathway and its intermediates are not yet known, but since trivalent arsenic species might be formed in this pathway, a detailed research directed on the metabolism of organic arsenicals (e.g. arsenoribosides) seems to be justified.

The sheep of North Ronaldsay (Orkney Archipelago) gave us the opportunity to study the metabolites and intermediates of arsenoribosides in measurable concentrations. This is due to the fact, that they are restricted to the beach and live entirely on seaweed, in particular *Laminaria digitata* which is known to contain high level of arsenic (89.7 ± 0.3 mg/g) with the majority being arsenoribosides (> 98%). The pattern of arsenic species excreted via urine by this sheep is comparable to that of urine from humans ingesting seaweed. It is therefore possible to assume that the metabolic pathway of arsenoribosides is similar between sheep and humans. It is also known that sheep ingesting arsenic containing food accumulate arsenic in their wool similar to humans.

The aim of our study was to determine which arsenic species were detectable in the hair of people exposed to inorganic arsenic and the wool of sheep exposed to arsenoribosides.

MATERIALS AND METHODS

During summer 2000 wool was collected at North Ronaldsay from different sheep and stored individually in clean, sealed plastic bags until analysis. The sheeps' intake of arsenic from seaweed was calculated to be 35 ± 6 mg per day [5]. The human hair samples were collected from 2 families in Rasikpur Murutia, Nadia West Bengal India. The drinking water there contains about 5.5 mg/L, which means with a daily consumption of 4 L the take up is about 22 mg per day. For the sheep wool external contamination is very unlikely, whereas the exogenous contamination of human hair through hair washing with the arsenic contaminated water cannot be excluded.

For the determination of the arsenic species, fat-free wool and human hair were extracted with water by boiling for 6 hours. After that the extract was separated from the fibre, filtered and stored at 4°C until analysis. The column (PRP X 100 Hamilton (150 * 4.6 mm) + precolumn) was fitted in a commercial HPLC system with a 20 ml sample loop. We used a 30 mM ammoniumphosphate buffer at pH 6.0. The flow rate of 1.0 ml/min allowed a direct connection of the column outlet to the ICP-MS. For quantification of the chromatograms As(III), As(V), DMA(V) and MMA(V) containing solutions were used. All chemicals used were of analytical grade or higher. Arsenic was measured by inductively coupled plasma mass spectrometer (ICP-MS) (Spectromass 2000, Spectro Analytical Instruments, Kleve Germany). For quality control human hair GBW09101 reference material (Shanghai Institute of Nuclear Research Academia Sinica, Shanghai China) and NIST SRM 2670 (elevated level) (NIST Gaithersburg USA) were used since no reference material for arsenic speciation exists. A more detailed description of the sample handling can be found in [6].

RESULTS

The sheep ingested about 35 mg As daily most of which was in the dimethylated form. Seaweed contains mostly dimethylated arsenoribosides (98%, inorganic arsenic 2%). Compared to that, the intake of humans came nearly exclusively from the use of contaminated drinking water. This water contained about twice as much arsenate as arsenite.

The different arsenic species used as standards were baseline separated and eluted at 62 s As(III), 177 s As(V), 101 s MMA(V), 76 s DMA(V). In addition we reduced MMA(V) and DMA(V) to their corresponding trivalent forms using the method of Reay and Asher [7]. The reduced forms MMA(III) and DMA(III) were eluting at 360 and 610 s respectively.

In both human hair and sheep wool we found As(III), As(V), MMA(V) and DMA(V), the trivalent forms of DMA and MMA could only be detected in sheep wool. We could confirm the identity of DMA(V) in sheep wool by using electrospray mass spectrometry using ammonium carbonate buffer pH 8 instead of the phosphate buffer. The mean ratios of the different species are shown in *table 1*. The main species in SRM human hair GBW09101 were As(III), followed by As(V), whereas the sheep wool contained mostly dimethylated arsenic in its tri- and pentavalent form. The sum of the latter is quite constant from extraction to extraction; the amount of tri- and pentavalent can vary.

Table 1. Mean ratio of the different As-species in sheep wool (n=5) and human hair GBW09101 (0.57 mg/g cert.) in %, without and with treatment with 0.2 M CuCl in 0.2 M HCl 4 h at 37°C.

	sheep wool		**human hair CRM**	
		+ CuCl		**+ Cu Cl**
As(III)	18	19	52	48
As(V)	14	18	21	21
MMA(V)	5	9	7	30
DMA(V)	31	54	20	1
MMA(III)	1			
DMA(III)	31			
sum	740	1560	365	446

In sheep wool we found a mean recovery rate of only 11 ± 4% calculated from the total amount. The recovery of arsenic for human hair reference material was with 64 ± 8% higher. Treatment of the wool extract with 0.2 M CuCl in 0.2 M HCl (following the procedure of Styblo [8]) for 4 hours at 37° C increased the recovery to 26 ± 3%. It also oxidized all of the trivalent MMA and DMA. The recovery of arsenic in human hair SRM extract after treatment with CuCl was 78 ± 8%. The not eluting arsenic species are as yet unknown.

We analysed wool from 8 sheep and hair from 10 people (male, 4 female, 18-60 years) affected by chronic arsenic intake.

The species distribution of human hair samples from Rasikpur is shown in *table 2*. The sheep showed no large inter-individual variation in the species distribution (for details see [6]). The total amount of arsenic in human hair samples studied was between 1.1 - 8.7 mg/g hair. Most of the extracted arsenic is either As(III) or As(V) for human samples (sum inorganic As 69 ± 14%), where it is DMA(V) for the sheep (mean 71 ± 18%) [6].

DISCUSSION

Hair and wool contain inorganic arsenic and also the mono- and dimethylated form. In the extract arsenic is detected in its tri- and pentavalent forms. Previous experiments have shown that it cannot be decided in which valency the arsenicals are bound in the fibre. The finding that trivalent methylated species were only detectable in sheep wool does not mean that they cannot also be found in human hair. The increased recovery after the CuCl treatment, during which protein-bound arsenicals are released, shows that some of the inorganic and methylated arsenic is bound in complexes which do not elute under the chosen chromatographic conditions.

Table 2. Species distribution in percentage in human hair samples from Rasikpur

	As(III)	DMA	MMA	As(V)
P 115	47.7	18.3	7.1	26.9
P 114	44.2	36.9		18.8
P 113	16.9	40.7		42.4
P 112	7.0	40.2	2.9	49.9
P 111	16.5	37.5	2.9	43.2
P 110	26.1	9.7	1.3	63.6
P 109	18.5	12		69.5
P 108	24.9	20.2	3.9	51
P 107	40.2	9.9	3.8	46
P 106	34.2	10.1	2.7	52.9

The amount of inorganic arsenic is higher than expected, both in sheep wool and human hair. In the urine the sheep excreted, as do humans, arsenic was found to be mostly in dimethylated form. Since the intake of inorganic arsenic for sheep is low (only about 2%), the enrichment of inorganic arsenic can be the result of either a specific transport, which favors inorganic arsenic, or the specific binding of inorganic arsenic to molecules in the cells. Hair follicles are not directly supplied by blood vessels so the nutrients need to diffuse to the growing hair, which might be an explanation for the enrichment of inorganic arsenic. From in vitro incubation experiments it can be concluded that inorganic arsenic has a tenfold higher affinity for hair and wool [6 and unpublished results].

There are two possible explanations for the extremely low concentration of methylated species in human hair SRM GBW09101, one is that there are changes in the arsenic metabolism as the intake of arsenic increases which leads to a higher amount of methylated species in hair, the second that the amount of arsenic in this material is artificially increased by adding inorganic arsenic to the material.

In contrast to sheep the intake of humans is nearly exclusively in the form of inorganic arsenic, so the concentration of inorganic arsenic in the blood is probably higher than that of the sheep. The amount of methylated arsenic in human hair might depend on the methylation capacity of the individual. As can be seen in *table 2* some of the samples have higher amounts of methylated species compared to the others. Since samples from only ten individuals are analysed a statistical approach is not very useful. The hair samples from women seem to have a higher concentration of methylated species than that of the males, this might be a result of the individual methylation capacity, which could be determined from the ratio of methylated and inorganic arsenic in urine. Our results are in agreement with that reported by Yamauchi et al [9]. They found that the majority of arsenic in hair from workers exposed to inorganic arsenic (dust) is in its inorganic form with traces of dimethylated arsenic. There is one report about arsenic species found in hair of people ingesting organic arsenic species from seafood, saying that inorganic arsenic and dimethylated arsenic species have been determined [10]. Yamato et al. found a mean ratio of 74% for inorganic arsenic and 26% for dimethylated As. Their amount of inorganic arsenic is comparable to the samples analysed by us.

CONCLUSIONS

Human hair and sheep wool accumulate inorganic arsenic and mono- and dimethylated arsenic. Sheep wool has a tendency to enrich inorganic arsenic compared to the organic forms. The same conclusion can also be drawn from in vitro incubation experiments with human hair and sheep wool, where the uptake of As(III) and As(V) was nearly the same and tenfold higher than that of DMA(V) and MMA(V). Hair and wool may give some additional information about the metabolic pathway, especially since the extracted pentavalent arsenicals are probably transported into the cells as trivalent or reduced in there. If there is a correlation between the intake of arsenicals and their species distribution in the fibre, hair might make it possible to estimate the individual methylation capacity or reflect the form of arsenic intake. From our data it can be assumed that the ratios are depending on the methylation capacity and maybe on the excretion efficiency of the kidneys. If there is proof that the amount of methylated species in hair depends on the methylation capacity, this would enable us to get information about the amount of external contamination of the sample by comparing the relative amounts of methylated and inorganic species.

Acknowledgment: The authors are grateful to Dr C. Natarajan University of Aberdeen for the provision of the human hair samples.

REFERENCES

1. Hambidge K.M. Hair Analyses - Worthless For Vitamins, Limited For Minerals. *J. Am. Clin. Nutr.*, 1982, 36, 943-949.
2. Hindmarsh J.T., McCurdy R.F. Clinical and environmental aspects of arsenic toxicity. *Crit. Rev. Clin. Lab. Sci.*, 1986, 23, 315-347.
3. Das D., Chatterjee A., Mandal B.K., Samanta G., Chakraborti D., Chanda B. Arsenic In ground-water in 6 districts of West-Bengal, India - the biggest arsenic calamity in the world. 2. arsenic concentration in drinking-water, hair, nails, urine, skin-scale and liver-tissue (biopsy) of the affected people. *Analyst*, 1995, 120, 917-924.
4. Ma M., Le X.C. Effect of arsenosugar ingestion on urinary arsenic speciation. *Clin. Chem.*, 1998, 44, 539-550.
5. Hansen H.R., Feldmann J. Arsenic metabolism in the seaweed eating sheep from North Ronaldsay, ICE-BAMO, 2001.
6. Raab A., Hansen H.R., Zhuang L., Feldmann J. Arsenic speciation analysis in wool and hair. *Talanta*, 2002, in press.
7. Reay P.F. and Asher C.J. Preparation and purification of 74As-labeld arsenate and arsenite for use in biological experiments. *Anal. Biochem.*, 1977, 78, 557-560.
8. Styblo M., Hughes M.F., Thomas D.J. Liberation and analysis of protein-bound arsenicals. *J. Chrom. B*, 1996, 677(1), 161-166.
9. Yamauchi H., Takahashi K., Mashiko M., Yamamura Y. Biological monitoring of arsenic exposure of gallium arsenide-exposed and inorganic arsenic-exposed workers by determination of inorganic arsenic and its metabolites in urine and hair. *Am. Ind. Hyg. Assoc. J.*, 1989, 50, 606-612.
10. Yamato N. Concentrations And chemical-species of arsenic in human-urine and hair. *Bull. Environ. Contam. Toxicol.*, 1988, 40, 633-640.

*Metal Ions in Biology and Medicine: vol. 7.*Eds L. Khassanova, Ph. Collery, I. Maymard, Z. Khassanova, J.-C. Étienne.
John Libbey Eurotext, Paris © 2002 p. 20.

Serum selenium and manganese analysis by ICP-DRC-MS

Olga Piraner, Kathleen L. Caldwell, Robert L. Jones

Centers for Disease Control and Prevention, 4770 Buford Hwy. NE, MS F18, Atlanta, GA 30341

Selenium (Se) and manganese (Mn) are both essential nutrients to humans and are also known to be toxic at certain levels. Therefore, it is important to monitor the selenium and manganese content in biological matrices such as urine and serum. Normal U.S. serum Se levels are 95-165 μg/L. Normal serum Mn levels are 7.7-12.1 μg/L. Currently, selenium and manganese in biological matrices are determined by a number of different techniques, such as graphite furnace atomic absorption spectrometry (GFAAS), flow injection hydride generation electrochemical atomic absorption spectrometry (HG-AAS), or conventional inductively coupled plasma mass spectrometry (ICP-MS), all of which have limitations. In conventional ICP-MS, isotopic interferences are a problematic limitation. However, Dynamic Reaction Cell (DRC) technology using methane as the reaction gas can eliminate isobaric background and allows for the analysis of Se and Mn in serum. Additional improvements are achieved by adding ethanol, Triton X-100, and internal standards such as gallium for Se and rhodium for Mn. With the recent introduction of a Dynamic Reaction Cell in the ICP-MS (ICP-DRC-MS), analysis of serum selenium and manganese can now be done using the most abundant selenium isotope (^{80}Se), which results in a lower LOD, and ^{55}Mn, which is monoisotopic.

Metal Ions in Biology and Medicine: vol. 7. Eds L. Khassanova, Ph. Collery, I. Maymard, Z. Khassanova, J.-C. Étienne.
John Libbey Eurotext, Paris © 2002 pp. 21-25.

Proton microprobe technique in element analysis of biological liquids

M.E. Buzoverya[1], V.T. Punin[1], S.N. Abramovich[1], V.V. Chulkov[1], I.D. Gorlachev[2], S.N. Lyisukhin[2], V.N. Shabalin[3], C.N. Shatokhina[3]

[1]Russian Federal Nuclear Center - All Russia Scientific Research Institute of Experimental Physics, Sarov, Russia; [2]Institute of Nuclear Physics of the National Nuclear Center of the Republic of Kazakhstan Alma-Aty, Kazakhstan; [3]Scientific-Research Institute of Gerontology of RF, Moscow, Russia.

ABSTRACT

Early diagnostics of various diseases is an actual problem of the present-day medicine. Russian scientists S.N. Shatokhina and V.N. Shabalin are the world-first to develop the pathogenetic method of the urolithiasis early diagnostics - Litos system. It is based on the urine structure characteristics analysis on its transition into solid state through dehydration. Solid fixed bio-liquid structure in this case is formed by salts and organic elements dissolved in it.

The Litos method allows nephrolithogenesis and its activity level to be found out and salts participating in the process to be determined. One of the basic stages in this technique is a comparative analysis of chemical elements content in glass urine samples peripheral and central zones. At the present moment the element analysis by the Litos system is carried out with the X-ray spectral microanalysis.

In the report there is presented a possibility of the proton microprobe technique use for determining main lithogenic elements in glass bio-liquids by the Litos system. There is shown a good convergence of the results of samples proton microprobing and X-ray spectral analysis.

Development and enhancement of laboratory methods for bio-liquids examination is put in the first place in solving the problems of pathological states diagnostics. Diagnostics at disease early stages is of particular interest.

Till recently in medical practice biological tissue examination was applied only to cellular and higher forms of its organization. Study of bio-liquid morphology is a new trend in the area of analysis of living systems spatio-temporal structures [1]. Methodically it is based on the bio-liquid transformation into solid phase with the use of some special techniques. The dehydration results in film (facies) formation having a structure, peculiarities of which are defined by the whole complex of qualitative and quantitative parameters of the substances, present in the bio-liquid *(fig. 1)*.

Structural information contains data not only on concentration of the elements composing such a complicated system, as a bio-liquid is, but on the character of their interaction, as well. From here follows the logics of a diagnostic process aimed at the search of those molecular injuries taking place in the body at pathological process early stages, at pre-clinical stages of the disease.

Bio-liquid morphology as a new line actively wins positions in various branches of the clinical medicine - therapy, surgery, obstetrics and gynecology, pediatrics, ophthalmology, urology etc.

The present work goal was to study possibilities of proton microprobe method application in urolithiasis early diagnostics.

According to the world statistics 10-15% of the population suffers from urolithiasis. Up to the present this disease has been diagnosed only through its complication - nephrolith and urolith detecting. Russian scientists are the world-first to develop the pathogenetic method of the urolithiasis early diagnostics - Litos test-system. The method is patented [2], adopted by Ministry of Health of RF - N° 97/17-14-2, and is already used in clinical practice.

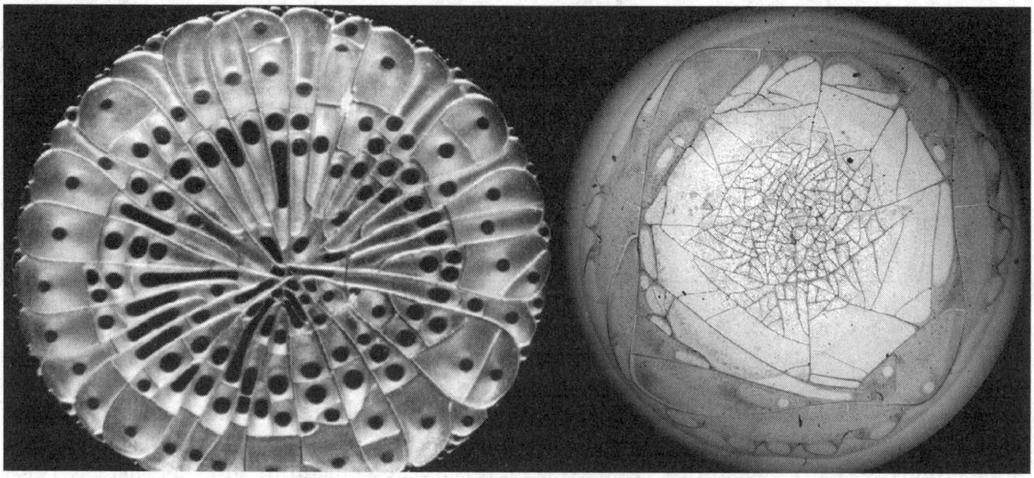

a) “Norm” b) Patient with cardio-vascular pathology

Fig. 1. Facies of blood serum

The method is based on the urine structure characteristics analysis on its transition into solid phase through dehydration. Solid fixed bio-liquid structure in this case is formed by salts and organic elements dissolved in it.

According to the Litos system the character of salts crystallization in the peripheral amorphous zone has distinguishing peculiarities in some cases and may be considered as a diagnostic feature. The peculiarities consist in the difference in peripheral zone saturation with salts crystals, indicating the different stages of the lithogenesis process *(fig. 2)*.

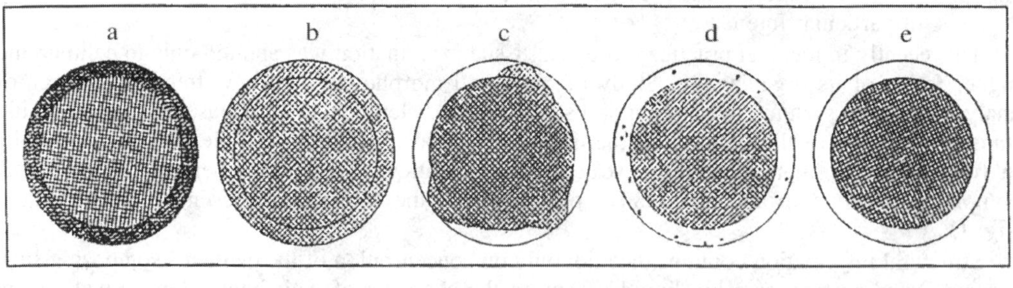

Fig. 2. Appearance of bio-liquid samples from patients suffering from urolithiasis with different urolithogenesis activity degree

a - sharply expressed; *b* - high; *c* - moderate; *d* - weak; *e* - with no urolithogenesis.

The important thing is to determine a kind of salts, participating in the lithogenesis. This can be done through a comparison of chemical elements concentrations (K,Na,Mg,P,S,Cl,Ca,Zn,Si) that are present in the peripheral and central zones of a bio-liquid dehydrated drop in patients with urolithiasis active stage. Today to perform the element analysis the Roentgen fluorescent microanalysis method with fluorescence excitation by electrons is used (RFE).

ADDENDUM

Par suite d'un incident technique, la liste des membres du Comité scientifique s'est trouvée erronée. Nous la republions donc dans son intégralité.

List and addresses of editors

Lylia Khassanova, Departement of Environmental Protection, Bashkir State University, Frunze Street, 32, 450074 Ufa, Bashkortostan, Russia.

Philippe Collery, Service de Cancérologie, Polyclinique Maymard, 20200 Bastia, France.

Ivan Maymard, Polyclinique Maymard, 20200 Bastia, France.

Zilara Khassanova, Department of Botany, Bashkir State Pedagogical University, October revolution Street, 3a, 450025 Ufa, Bashkortostan, Russia.

Jean-Claude Étienne, Centre Hospitalier Universitaire de Reims, 51092 Reims Cedex, France.

Chair persons

Khassanova L.A.

Collery Ph.

Honorary chair persons

Volgarev M.N.

Étienne J.-C.

Past chairman

Centeno J.

International Scientific Advisory Board

Alpoim M.C. (Portugal)
Anastassopoulou J. (Greece)
Andersen O. (Denmark)
Aposhian V.H. (USA)
Aro A. (Finland)
Arthur J. (UK)
Badawi A. (Egypt)
Banerjee A. (India)
Bao Sh. (China)
Beckett G. (UK)
Berthelot A. (France)
Blagoi Yu. (Ukraine)
Borella P. (Italy)
Bratter P. (Germany)
Burguera Jose L. (Venezuela)
Caroli S. (Italy)
Cebrian M. (Mexico)
Centeno J. (USA)
Chen C.-J. (Taiwan)
Combs G.F. (USA)
Conard B. (Canada)
Corbella J. (Spain)
Cornelis R. (Belgium)
Costa M. (USA)
Cser M.A. (Hungary)
Delinassios G. (Greece)
Deloncle R. (France)
Desoize B. (France)
De Wolff F.A. (Netherlands)
Dobrovolsky G.V. (Russia)
Domingo J. (Spain)
Durlach J. (France)

Ermakov V.V. (Russia)
Ershov Yu.A. (Russia)
Etcheverry S.B. (Argentina)
Evangelou A. (Greece)
Farzami B. (Iran)
Fields M. (USA)
Finkelman R.B. (USA)
Flores-Arce M.F. (Mexico)
Fomchenkov V.M. (Russia)
Fowler B. (USA)
Garfinkel D. (Israel)
Gibb H. (USA)
Gielen M. (Belgium)
Gogotov I.N. (Russia)
Golubkina N.A. (Russia)
Gromov B.V. (Russia)
Guillard O. (France)
Hackl E. (Ukraine)
Haroutiuniuan S. (Armenia)
Haywood S. (UK)
Hoel P.J. (Norway)
Hura C. (Romania)
Ilyaletdinov A.N. (Kazachstan)
Irgolic K. (Austria)
Ivanov A.Yu. (Russia)
Jimenez B. (Puerto Rico)
Kalfakakou V. (Greece)
Kamnev A.A. (Russia)
Kantola M. (Finland)
Karavaiko G.I. (Russia)
Kasprzak K.S. (USA)
Keppler B. (Austria)
Khassanova Z.M. (Russia)
Kisters K. (Germany)
Knekt P. (Finland)
Koudrine A.V. (Russia)
Kruse-Jarres J. (Germany)
Kumerova A. (Latvia)
Kvicala J. (Czechia)
Kylyvnyk K. (Ukraine)
Lando D. (Belarus)
Lebedev V.S. (Russia)
Limouris G. (Greece)
Littlefield N.A. (USA)
Llobet J. (Spain)
Longfellow D. (USA)
Lopes A.M. (Brazil)
Maximova I.A. (Russia)
Mazej D. (Slovenia)
Mazumder D.G.N. (India)
Meltzer H.M. (Norway)
Mendz G.L. (Australia)
Michalke B. (Germany)
Moore T. (New Zealand)
Mullick F.G. (USA)
Natochin Yu.V. (Russia)
Negretti de Bratter V. (Germany)
Nelson R. (USA)
Neve J. (Belgium)
Nielsen J.B. (Denmark)
Nowak G. (Poland)
Oller A.R. (USA)
Omu A.E. (Kuwait)
Pineau A. (France)
Pinevich A.V. (Russia)
Poirier L. (USA)
Prasad S.B. (India)
Rubikas J. (Lithuania)
Roussel A. M. (France)
Sadler P. (UK)
Sastre M. (Puerto Rico)
Selinus O. (Sweden)
Shukla S. (Italy)
Simonoff N. (France)
Skalny A.V. (Russia)
Solioz M. (Switzerland)
Stefanov V.E. (Russia)
Sundennan F.W. (USA)
Suzuki K.T. (Japan)
Tajmir-Riahi H.A. (Canada)
Tchernitchin T. (Chile)
Templeton D. (Canada)
Theophanides T. (Greece)
Thomassen Y. (Norway)
Thompson C. (USA)
Tseng C.-H. (Taiwan)
Tulub A.A. (Russia)
Vahter M. (Sweden)
Van der Voet G.B. (Netherlands)
Varfolomeav S.D. (Russia)
Vernet G. (France)
Vivoli G. (Italy)
Voschenko A.B. (Russia)
Waalkes M. (USA)
Weinstein P. (New Zealand)
Zahkan N.I. (Egypt)
Zaray G. (Hungary)
Zatta P. (Italy)
Zheng B. (China)

Organizing Committee

Bolshakov V.N. (Russia)
Maximova I.A. (Russia)
Stefanov A.V. (Russia)
Visulainen V.V. (Russia)

The present work main goal was to study the proton microprobe technique applicability to the bio-liquid analysis and to overcome methodical difficulties of biological object handling. The urine of practically healthy persons and of those suffering from urolithiasis was chosen to be the object of studies. The element analysis was carried out at the tandem electrostatic accelerator UKP-2-1 in the Institute of Nuclear Physics of the National Nuclear Center of the Republic of Kazakhstan, Alma-Aty.

As an example the data of analysis of samples from two patients suffering from urolithiasis are presented. In *table 1* there is presented element composition of the peripheral and central zones of the urine sample of the patient A, found with RFE technique. In *table 2* the results are presented of the same sample analysis with the Proton microprobe method.

The results of RFE and Proton microprobe for patient A show that mainly calcium prevails in the sample peripheral zone as compared to the central zone. Therefore, calcium oxalate is the main lithogenic salt. This conclusion has been later proved through the examination of the extracted nephrolith phase composition.

In *tables 3-4* there is presented an element content of the peripheral and central zones of the urine sample of the patient A. There one can see evident prevalence of two elements in the peripheral zone - phosphorus and calcium. Therefore, phosphorus and calcium are the main lithogenic elements, i.e. phosphate-calcium salts are formed. The urolith Roentgen-structural analysis has shown that the main phase is hydroxylapatite.

Table 1. Results of proton microprobing of the test-sample from patient A in urolithiasis active stage

activity	zone	Microelement content in relative units						
		Ca	**Cl**	**K**	**Mg**	**Na**	**P**	**S**
Urolithogenesis weak degree	periphery	453	3087	2197	831	4701	4276	2590
	center	210	9407	526	493	8905	3408	4382

Table 2. Results of RFE analysis of the test-sample from patient A in urolithiasis active stage

activity	zone	Microelement content in relative units								
		Ca	**Cl**	**K**	**Mg**	**Na**	**P**	**S**	**Zn**	**Si**
Urolithogenesis weak degree	periphery	2.2	30.0	34.0	-	6.0	6.5	20.0	-	0.9
	center	1.3	32.9	22.0	-	12.9	5.3	22.1	-	4.1

conclusion: **calcium oxalate**.

At present there is no unique conception of the lithogenesis origin in the body. In this connection the results of the topographic analysis of the main lithogenic elements in the system-structured drop of the bio-liquid under study, having specific characters for each person, is of great interest *(fig. 3)*. This will allow new data obtaining on the pathophysiologically significant elements distribution in the specific structural formations, revealing correlative relations between these elements. The experimental data obtained can be further used to study into different diseases pathogenesis and to reveal the conditions causing their origin.

So, at the present stage of investigations good agreement of the results of urine element composition determination with the Proton microprobe method and RFE-analysis is obtained.

CONCLUSIONS

1. Practical value of the information on the concretion composition obtained with the Litos system is that before the urolith formation in a patient with urolithiasis there can be revealed the type of the lithogenic salts and opportune prophylactic and therapeutic measures can be undertaken to prevent complications (nephrolith formation, inflammatory process development etc.).
2. There have been obtained coincident results for bio-liquid RFE and proton analysis.
3. There has been shown a possibility of detecting main lithogenic elements in the bio-liquid glass samples with the Proton microprobe method.
4. There have been outlined ways to adapt the method to obtaining quantitative information on the biological samples element composition.

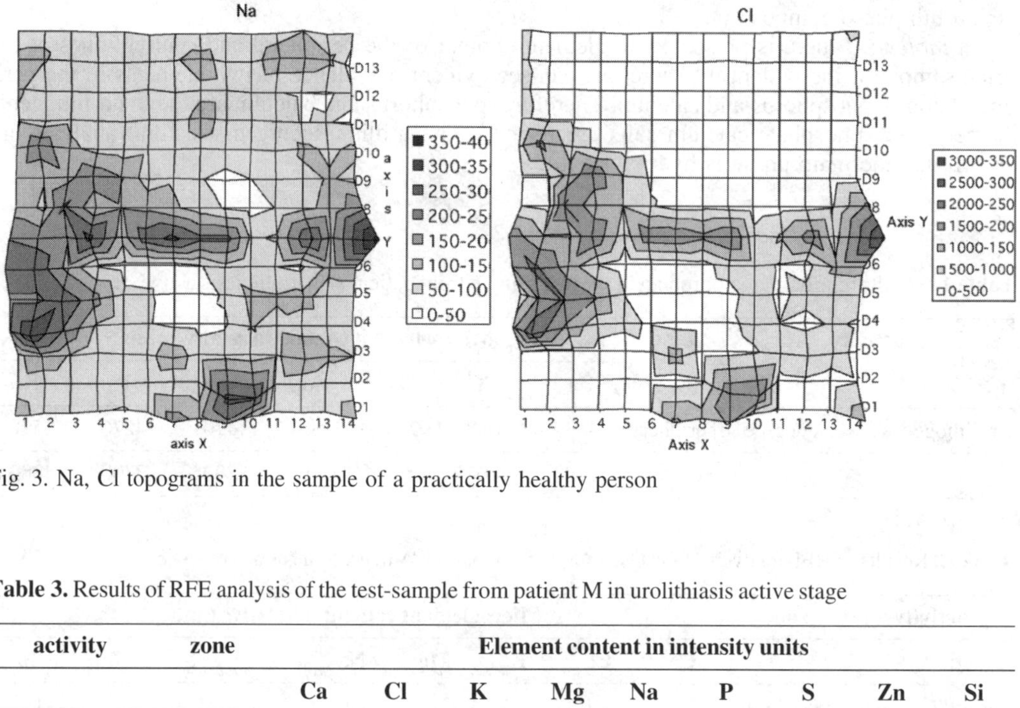

Fig. 3. Na, Cl topograms in the sample of a practically healthy person

Table 3. Results of RFE analysis of the test-sample from patient M in urolithiasis active stage

activity	zone	Element content in intensity units								
		Ca	Cl	K	Mg	Na	P	S	Zn	Si
Urolithogenesis in active stage of II degree	periphery	6.6	40.9	8.1	4.3	15.7	9.4	13.2	-	1.9
	center	1.0	45.8	21.4	-	8.7	4.1	11.2	-	1.5

Table 4. Results of proton microprobing of the test-sample from patient M in urolithiasis active stage

activity	zone	Microelement content in relative units						
		Ca	Cl	K	Mg	Na	P	S
Urolithogenesis in active stage of II degree	periphery	29	2430	516	287	2861	1732	1454
	center	7	4001	265	145	3803	891	1243

conclusion: **calcium-phosphate salts.**

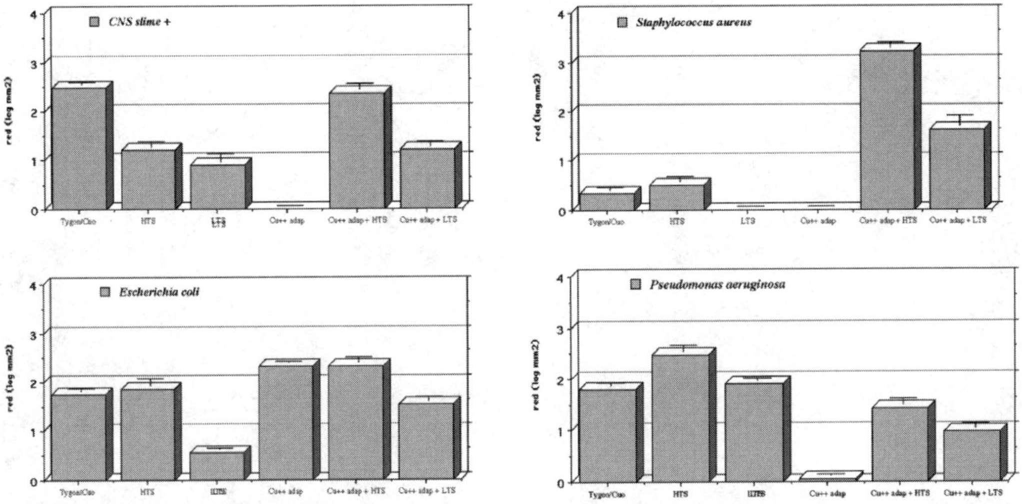

Fig. 1. Reduction, in log/mm^2 of bacteria number in biofilms
Column 1: Tygon® versus copper - other columns: assay versus control biofilm on copper
CNS: Coagulase Negative Staphylococcus HTS: High Toxic Shock LTS: Low Toxic Shock

Table 1. Sensitivity to copper ions (in log CFU/mm^2) of *E. coli* planctonic bacteria compared to bacteria included in a biofilm

	Planctonic bacteria		**Tygon® biofilm**		**Copper biofilm**	
Control	9,20	-	5,40	-	3,65	-
HTS	7,80	1,40	3,39	2,01	1,80	1,85
LTS	9,00	0,20	3,40	2,00	3,10	0,55
Cu^{++} adaptation	7,80	1,40	4,79	1,61	1,32	2,33

Table 2. Intrabacterial Cu^{++} content in *Staphylococcus aureus* and *Escherichia coli*

S. aureus			***E. coli***	
µg/cell	**Increase/control**		**µg/cell**	**Increase/control**
$2,68.10^{-10}$	-	Control	$8,23.10^{-11}$	-
$7,40.10^{-5}$	2.10^{5}	Cu^{++} adaptation	$1,22.10^{-1}$	10^{10}
$1,50.10^{-2}$	2.10^{8}	HTS	6,56	10^{11}
$6,90.10^{-1}$	2.10^{9}	Cu^{++} adapt + HTS	$7,05.10^{-2}$	10^{9}
$6,85.10^{-5}$	2.10^{5}	LTS	$1,40.10^{-1}$	10^{10}
$5,18.10^{-3}$	2.10^{7}	Cu^{++} adapt + LTS	$5,90.10^{-1}$	10^{10}

REFERENCES

1. V.N. Shabalin, S.N. Shatokhina. Human bio-liquids morphology, Moscow, Khrizostom, 2001, P. 304.
2. V.N. Shabalin, S.N. Shatokhina. A technique to determine urolothiasis activity degree, Patent 2061956, 1996.

II MOLECULAR AND BIOLOGICAL EFFECTS

*Metal Ions in Biology and Medicine: vol. 7.*Eds L. Khassanova, Ph. Collery, I. Maymard, Z. Khassanova, J.-C. Étienne.
John Libbey Eurotext, Paris © 2002 pp. 29-33.

The essential role of nickel affects physiological functions regulated by the cyclic-GMP signal transduction system*

Forrest H. Nielsen, Katsuhiko Yokoi and Eric O. Uthus

United States Department of Agriculture, Agricultural Research Service, Grand Forks Human Nutrition Research Center, Grand Forks, North Dakota, USA 58202-9034 (EOU & FHN), and Seitoku University, Faculty of Humanities, Department of Human Life and Culture, Matsudo 271-8555, Japan (KY)

ABSTRACT

Unequivocal acceptance of nickel as an essential nutrient awaits the definition of a specific biochemical function in higher animals. This acceptance may be forthcoming because findings have been accumulating suggesting that nickel has a role that involves cyclic nucleotide-gated (CNG) cation channels. Nickel potentiates cyclic guanosine monophosphate (cGMP)-gated cation channels in the rod outer segment of retinae, and desensitizes CNG channels of olfactory receptor cells [1]. A number of organs including some in the central nervous, urogentital, and reproductive systems contain CNG channels. If nickel is involved or needed for CNG channel function and/or for various guanylate cyclases upstream to CNG channels, nickel deprivation should affect blood pressure control, sperm physiology, and sodium metabolism. Thus, experiments were conducted with the objective of demonstrating that nickel deprivation affects spermatozoa and kidney in such a way to support the hypothesis that nickel has an essential role involving the cGMP signal transduction system. In an effort to enhance the effect or need for nickel, two stressors of systems in which CNG channels are important were used as treatment variables; these were N^G-nitro-L-arginine methyl ester (L-NAME), a nitric oxide (NO) synthase inhibitor that hinders blood pressure control, and dietary sodium chloride (NaCl) in an amount that acted as a stressor of sodium metabolism. In experiment 1, rats (15-16/group) were fed a nickel-deficient basal diet (27 ng Ni/g) supplemented with 0 and 1 ug Ni/g. After 5 weeks of feeding the experimental diets, L-NAME was added to the drinking water (0.5 g/L) provided to half of the rats in each group. Blood pressure was measured one week later. Epididymal sperm motility and density were measured three weeks later. In experiment 2, the treatments (8 rats/treatment) were supplements to the basal diet (27 ng Ni and 1 mg NaCl/g) of 0 and 1 ug Ni/g and 0 and 80 mg NaCl/g. After being exposed to the treatments for 9 weeks, blood pressures were measured. At 10 weeks, urine was collected to assess kidney function. At 16 weeks, sperm motility and production were determined. Nickel deprivation increased systolic blood pressure; decreased spermatozoa motility and density in the epididymides, epididymal transit time of spermatozoa, and testes sperm production rate; and induced kidney damage that resulted in hematuria and microalbuminuria. High NaCl exacerbated changes induced by nickel deprivation. The changes in blood pressure and sperm density caused by nickel deprivation were similar to those induced by L-NAME. Because L-NAME is a stressor of NO metabolism and excessive NaCl is a stressor of atrial natriuretic peptide (ANP) need, the changes suggest that nickel deficiency has detrimental consequences to the ANP-cGMP and NO-cGMP signal transduction systems, and support the hypothesis that nickel has an essential function involving CNG channels.

* This research was partially supported by a grant from the Nickel Producers Environmental Research Association (NiPERA), Durham, NC.

INTRODUCTION

Although nickel (Ni) has been characterized as being required for some enzymes in lower forms of life, an essential biochemical role for Ni in higher animals has not been found [2]. We have found that Ni deprivation increases urinary nitrate/nitrite excretion [3], an indicator of nitric oxide (NO) production. Nitric oxide activates guanylate cyclase, the enzyme that transforms guanosine triphosphate to cyclic guanosine monophosphate (cGMP). The opening of cyclic nucleotide-gated (CNG) cation channels regulated by cGMP is altered by Ni [1]. These findings prompted us to focus our attention on the possibility that Ni has a role in the cGMP signal transduction system. Cyclic GMP-gated channels are expressed in a number of tissues. In vascular smooth muscle, the intracellular portion of the atrial natriuetic peptide (ANP) receptor is guanylate cyclase. When ANP binds to the receptor, cGMP is produced which results in smooth muscle relaxation that alters blood pressure [4]. Ovarian follicular fluid and ANP attract human spermatozoa through cGMP production [5]. The cGMP opens a channel to increase intracellular calcium; this activates flagella movement of spermatozoa. Chronic inhibition of NO synthase induces hypertension, decreases natriuresis, and results in morphological and functional damage in the kidney [6]. Thus, if Ni is involved in or needed for cGMP-gated channel function and/or for guanylate cyclases upstream to the channels, Ni deprivation should have several consequences including altered blood pressure control, sperm physiology, and renal function. We conducted two experiments with the objective of showing that Ni deprivation has such consequences. In these experiments, two stressors of systems in which cGMP-gated channels are important were used as treatment variables; these were N^G-nitro-L-arginine methyl ester (L-NAME), a NO synthase inhibitor, and high dietary sodium chloride (NaCl), an effector of blood pressure. If Ni is involved in cGMP formation or function, we predicted that the response to Ni deprivation would be enhanced by these stressors.

MATERIALS AND METHODS

Two factorially arranged experiments using weanling male Sprague-Dawley rats were performed. In experiment 1, 32 rats were assigned to 4 weight-matched groups of 8 with 2 groups being fed the basal casein-sucrose-corn starch diet containing about 1 mg Na and 27 ng Ni per g and the other 2 groups fed the basal diet supplemented with 1 mg Ni/kg. At 5 weeks, L-NAME was added to the drinking water (0.5 g/L) of 1 each of the 2 dietary Ni treatment groups. In experiment 2, the treatment variables of 4 weight-matched groups of 8 were supplemental dietary NaCl at 0 or 80 g/kg and supplemental dietary Ni at 0 or 1 mg/kg. The rats in both experiments were housed individually in plastic cages in laminar air flow racks. The racks were located in a room maintained at 23° C and 47% relative humidity with a 12 hour light and 12 hour dark cycle. The rats had free access to food and deionized water.

In experiment 1, systolic blood pressure was measured by the tail-cuff method one week after the L-NAME treatment was initiated. In this experiment, 12 days after the L-NAME treatment started, rats began to die. Thus, the experiment was terminated 3 weeks after the L-NAME treatment began (8 weeks after experiment initiation). In experiment 2, systolic blood pressure was measured after 9 weeks on the treatments. One week later, the rats were placed in metabolic cages with free access to drinking water but not diet for 16 hours during which time urine was collected in plastic tubes kept on ice. The experiment was terminated at 16 weeks. In both experiments, the rats were anesthetized with ether, and blood was collected from the vena cava with a heparinized syringe and needle. The entire epididymis was used for the determination of sperm number and motility. Sperm counts were determined by a modified method of Reeves and Rossow [7]. In this method, the number of motile and immotile sperm were counted by using a hemacytometer. Sperm motility was expressed as a percentage of motile sperm in total sperm. In experiment 2, urinary hemoglobin and erythrocytes were qualitatively detected with an urine test strip; N-acetyl-D-betaglucosaminidase activity (NAG) was determined by using a commercially available kit; and albumin was measured by the bromcresol green binding method [8]. Data were analyzed by two-way ANOVA except where noted in the tables. A P value $\leq$ 0.05 was considered statistically significant.

RESULTS

In experiment 1, 4 rats in the Ni-deficient, L-NAME-supplemented group, and 2 rats in the Ni-adequate, L-NAME-supplemented group, died before the experiment was terminated. Thus, the number of rats available for analysis in these 2 groups were 4 and 6 respectively. In experiment 2, 2 rats in the Ni-deficient, NaCl-supplemented group, and 1 rat in the Ni-adequate, NaCl-supplemented group, died before the experiment was terminated. Thus, the number of rats available for analysis in these 2 groups were 6 and 7, respectively.

Table 1 shows the findings from experiment 1; final weight was not affected by the experimental treatments. Both Ni deprivation and L-NAME supplementation increased systolic blood pressure; thus the highest mean systolic blood pressure occurred in the Ni-deficient, L-NAME-supplemented group. Sperm motility was decreased by Ni deprivation. Sperm count and density were decreased both by Ni deprivation and L-NAME supplementation; thus the lowest mean values for these variables occurred in the Ni-deficient, L-NAME-supplemented group.

Table 1. Effects of dietary Ni and L-NAME on final weight, systolic blood pressure (SBP), and sperm motility, count and density in right epididymis (experiment 1)

Dietary				Sperm		
Ni[a]	L-NAME[a]	Weight	SBP	Motile	Count	Density
		g	mmHg	%	10^6	10^6/g
+	-	296±33[b]	126±11	6.8±2.7	153±15	425±45
+	+	289±46	173±31	-[c]	113±10	337±48
-	-	318±27	147±16	2.0±1.0	139±2	337±54
-	+	282±28	201±20	-[c]	78±17	245±26
Analysis of Variance - P Values						
Ni		0.58	0.003	0.0003[d]	0.003	0.003
L-NAME		0.14	0.0001	-	<0.0001	<0.0001
Ni × L-NAME		0.23	0.62	-	0.19	0.29

[a]Variables were supplemental dietary Ni at 0 or 1 mg/kg and 0 or 0.5 mg L-NAME/L drinking water (last 3 weeks of experiment). [b]Data are means ± SD. [c]Not measured. [d]Determined by t test.

Table 2 shows that in experiment 2, excessive NaCl significantly depressed final body weight. Both nickel deprivation and excessive NaCl significantly increased systolic blood pressure *(table 2)*; thus, the highest mean value occurred in the Ni-deficient group fed excessive NaCl. *Table 2* also shows that Ni deprivation significantly decreased sperm motility, count and density; excessive NaCl did not significantly affect these variables. However, the lowest mean values for these three variables occurred in the Ni-deficient group fed excessive NaCl.

The data in *table 3* show that abnormal renal function was found in experiment 2. Nickel deficiency induced hematuria and both Ni deficiency and excessive NaCl induced albuminuria. Excessive NaCl fed to Ni-deficient rats increased urinary NAG.

Table 2. Effects of Ni deprivation and excessive dietary sodium chloride (NaCl) on final weight, systolic blood pressure (SBP), and sperm motility, count and density (experiment 2)

Dietary				Sperm		
Ni[a]	NaCl[a]	Weight	SBP	Motile	Count	Density
		g	mmHg	%	10^6	10^6/g
+	-	425±45[b]	135±21	4.7±1.95	232±54	395±90
+	+	387±26	161±37	4.23±2.26	193±50	360±56
-	-	447±41	160±13	2.06±1.02	161±37	301±45
-	+	389±29	200±38	1.16±0.76	141±30	256±25
Analysis of Variance - P Values						
Ni		0.42	0.004	0.001	0.003	0.001
NaCl		0.002	0.004	0.26	0.13	0.11
Ni × NaCl		0.47	0.51	0.75	0.41	0.96

[a]Variables were supplemental dietary Ni at 0 or 1 mg/kg and NaCl at 0 or 80 g/kg. [b]Data are means ± SD.

Table 3. Effects of Ni deprivation and excessive sodium chloride (NaCl) on hematuria, urinary albumin, and urinary N-acetyl-D-β-glucosaminidase activity (NAG) (experiment 2)

Dietary		Urinary		
Ni[a]	NaCl[a]	Hematuria	Albumin	NAG
		% Positive	mg/hour	mU/hour
+	-	0	0.23±0.06	8.0±2.6
+	+	12.5	0.57±0.86	7.9±1.3
-	-	87.5	0.34±0.08	9.0±2.9
-	+	100	1.73±2.22	12.3±2.4
Analysis of Variance - P Values				
Ni		0.0001[c]	0.04	0.003
NaCl		0.09	0.05	0.06
Ni × NaCl		1.00	0.37	0.05

[a]Variables were supplemental Ni at 0 or 1 mg/kg and NaCl at 0 or 80 g/kg. [b]Data are means ± SD. [c]Determined by Fisher's exact probability.

DISCUSSION

Studies with several animal species indicate that spermatozoa motility involves cGMP. For example, bovine spermatozoa possess CNG channels that are ionized calcium entry points; these channels respond more sensitively to cGMP than to cyclic adenosine monophosphate [9]. Epididymal sperm density also can be related to cGMP action. In the freshwater turtle, the smooth muscle layer of the epididymis duct contains ANP receptors [10]; these receptors often enlist CNG channels to exert effects. The rete testis and ductuli efferentes testis isosmotically absorb more than 95% of the fluid leaving the testis in rats. The fluid absorption determines the epididymal sperm density [11]. Thus, the finding that Ni deprivation decreases both sperm motility and density is consistent with the hypothesis that Ni has a role that involves cGMP production or action.

The cGMP signal transduction system also has crucial roles in blood pressure control and kidney function. The disruption of the guanylate-cyclase-A receptor (ANP receptor) in mice results in a chronic elevation of blood pressure [4]. In the kidney, CNG channels occur in the inner medullary collecting ducts; the CNG channels are inhibited by cGMP that forms when ANP binds to guanylate cyclase-A receptors. Closure of CNG cation channels decreases sodium reabsorption and exerts natriuresis [12, 13]. In a test not reported here, we found that after an acute oral load of NaCl, natriuresis and the urinary cGMP/NO ratio were decreased by Ni deprivation, but urinary cGMP excretion was not affected. Those results plus those shown in *table 3* indicate that Ni deprivation decreases the natriuretic effect of cGMP.

In summary, Ni deprivation diminished sperm density and motility in the epididymis, increased systolic blood pressure, and induced hematuria and microalbuminuria. The changes in systolic blood pressure and sperm density and count induced by Ni deficiency were similar to those induced by the NO synthase inhibitor, L-NAME. Excessive NaCl apparently exacerbated the increased systolic blood pressure and renal function changes induced by Ni deprivation. The changes support the hypothesis that Ni deficiency has detrimental consequences to the ANP-cGMP and NO-cGMP signal transduction systems, and indicate Ni has an essential function related to CNG channels regulated by cGMP.

REFERENCES

1. Gordon SE & Zagotta WN. Localization of regions affecting an allosteric transition in cyclic nucleotide-activated channels. *Neuron*, 1995, 14: 857-864.
2. Nielsen FH. Ultratrace elements in nutrition: current knowledge and speculation. *J Trace Elem Exp Med*, 1998, 11: 251-274.
3. Nielsen FH, Yokoi K & Uthus EO. Marginal dietary pyridoxine and supplemental dietary homocystine and methionine affect the response of the rat to nickel deprivation. *Metal Ions in Biology and Medicine, vol 6*, Centeno JA, Collery Ph, Vernet G, Finkelman RB, Gibb H & Etienne JC, eds, John Libbey Eurotext, Paris, 2000, pp 525-527.
4. Lopez MJ, Wong SK, Kishimoto I, Dubois S, Mach V, Friesen J, Garbers DL & Beuve A. Salt-resistant hypertension in mice lacking the guanylyl cyclase-A receptor for atrial natriuretic peptide. *Nature*, 1995, 378: 65-68.
5. Anderson RA Jr, Feathergill KA, Rawlins RG, Mack SR & Zaneveld LJD. Atrial natriuretic peptide: a chemoattractant of human spermatozoa by a guanylate cyclase-dependent pathway. *Mol Reprod Dev*, 1995, 40: 371-378.
6. Navarro J, Sanchez A, Saiz J, Ruilope LM, Garcia-Estan J, Romero C, Moncada S & Lahera V. Hormonal, renal, and metabolic alterations during hypertension induced by chronic inhibition of NO in rats. *Am J Physiol*, 1994, 267: R1516-R1521.
7. Reeves PG & Rossow KL. Zinc deficiency affects the activity and protein concentration of angiotensin-converting enzyme in rat testes. *Proc Soc Exp Biol Med*, 1993, 203: 336-342.
8. Doumas BT, Watson WA & Biggs HG. Albumin standards and the measurement of serum albumin with bromocresol green. *Clin Chim Acta*, 1971, 31: 87-96.
9. Wiesner B, Weiner J, Middendorff R, Hagen V, Kaupp UB & Weyand I. Cyclic nucleotide-gated channels on the flagellum control Ca^{2+} entry into sperm. *J Cell Biol*, 1998, 142: 473-484.
10. Kim SZ, Kang SY, Lee SJ & Cho KW. Localization of receptors for natriuretic peptide and endothelin in the duct of the epididymis of the freshwater turtle. *Gen Comp Endocrinol*, 2000, 118: 26-38.
11. Hansen JC & Jones RC. In vivo microperfusion of the ductuli efferentes testis of the rat: flow dependence of fluid reabsorption. *Exp Physiol*, 1996, 81: 633-644.
12. McCoy DE, Guggino SE & Stanton BA. The renal cGMP-gated cation channel: its molecular structure and physiological role. *Kidney Int*, 1995, 48: 1125-1133.
13. Vandorpe DH, Ciampolillo F, Green RB & Stanton BA. Cyclic nucleotide-gated cation channels mediate sodium absorption by IMCD (mIMCD-K2) cells. *Am J Physiol*, 1997, 272: C901-C910.

The US Department of Agriculture, Agricultural Research Service, Northern Plains Area, is an equal opportunity/affirmative action employer and all agency services are available without discrimination.

*Metal Ions in Biology and Medicine: vol. 7.*Eds L. Khassanova, Ph. Collery, I. Maymard, Z. Khassanova, J.-C. Étienne.
John Libbey Eurotext, Paris © 2002 pp. 34-36.

Interaction of rhenium cluster compounds with human blood proteins

Shtemenko N.I., Gorelaya M.V, Alexandrova L.M.

Board of Biophysics and Biochemistry of Dniepropetrovsk State University, 49050 Dniepropetrovsk, 13 Naukoviy by-street, UKRAINE

ABSTRACT

Reaction between antigene (Ag - standart human blood serum) and antibody (Ab - antiserum against IgA, IgM, IgG) was studied in the presence of 1-dichlorotetra-μ-(i-butirato)dirhenium (III) in the range of concentration from 10^{-3} to 10^{-11} M. In all experiments with IgA and IgM no essential changes were discovered. The most shifts in characteristics of immunoprecepitation lines were noticed in experiments with IgG. We think that changes of antigenic properties of 1 may be explained by conformational shifts of proteins, which doesn't bring perturbation of complementarity of the antigen-antibody reactive sites.

INTRODUCTION

Investigation of biological activity of rhenium cluster compounds with organic ligands is very actual due to previously shown antitumor activity for some of them and low toxicity. Chemical properties of these substances: ability to realize practically all possible levels of oxidation from +7 to -1, presence of hydrophobic radicals and mobile chlorine ion in their structure and an unique quadrupol metal-metal bond proposed possible interaction with the range of biological molecules by different mechanisms. Recently we have shown cytostabilizing, antiradical, antihaemolytic and antioxidant properties of diclorotetra-(-(i-butirato)-dirhenium (III) - 1. But the very necessary step in involving a new substance to medical practice is investigation of its immunomodulative characteristics. The aim of this study was to study influence of 1 on the reaction antigene-antibody with human immunoglobulines IgA, IgG and IgM.

MATERIALS AND METHODS

Interaction between antigene-antibody was investigated by immunodiffusion method by Uchterlony. Standard human blood serum was taken as antigene (Ag). Monospecific serum against IgA, IgG, IdM served as antibodies (Ab). The optimal titre of serum for IgG, IgA and IgM was determined. Solutions of 1 were prepared in concentrations from 10^{-3} to 10^{-11} M.

The process was studied in two modifications: in conditions of preincubation of Ag in solutions of 1 (a) and in conditions of preincubation of Ab in solutions of 1 (b). Preincubation of Ag or Ab was accomplished during 1h under 37°C. Distance between starting point and the middle of a precipitation line (l) and width of it (d) were measured. Influence of 1 was appreciated as a shift of the middle of precipitation line in comparison with control meaning $dl = l_{exp} - l_{con}/ l_{contr}$, %. Positive value of dl means shifting of precipitation line to Ab, negative- to Ag. All experiments were repeated 6-12 times.

RESULTS

Our results showed, that in all experiments the precipitation line was certain and intensive that may characterize the influence on 1 as mild, not destructive on the reaction of highly specific reaction Ag - Ab. In such mild influence of 1 it was impossible to appreciate efficacy of 1 by common used procedures measuring the density of precipitate.

The most significant influence of 1 was observed for interactions between IgG-Ab. All values of shifts (dl) in both modifications of experiments were positive, that said about decreasing of mobility of Ab under influence of 1. Changing of shifts of precipitation line depended on concentration of 1. Thus, in the case of preincubation of Ag, the most intensive shifts were observed under 10^{-3} M and under 10^{-9} M concentrations of 1 *(fig. 1)*.

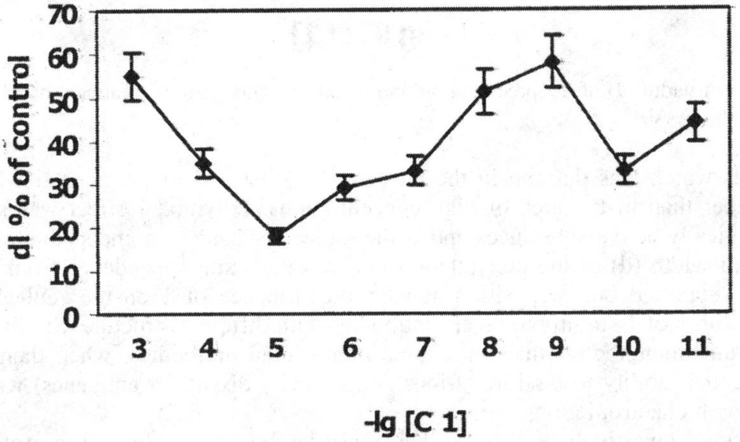

Fig. 1. Changing of precipitation line shifts (dl) in experiments with pretreatment of Ag with solutions of 1 for interaction Ig G-Ab

Thus we may say that the most influence of 1 on mobility of IgG-Ab took place under these concentrations of 1. The same, but in a little letter extent we observed for experiments with preincubation of Ab.

Increasing of the width (d) of precipitation lines may testify to some qualitative modifications in the reagents Ag - Ab under influence of 1. Changing of this parameter depended from concentration of 1 *(fig. 2)* at the same appropriateness as was shown above.

DISCUSSION

Influence of some platinum and palladium complexes on the interaction between IgG-Ab was studied and their destructive impact on the complex Ag-Ab was shown. Thus, preincubation of Ag in 10^{-3} M solution of $K_2[PtCl_4]$ led to decreasing of IgG-Ab interaction (to 11,7% in comparison with control). These facts led authors to conclusion about allergenic properties of platinum and palladium substances. Demonstration of mild influence of 1 on the IgG-Ab complex formation and very little influence of 1 on the IgA-Ab and IgM-Ab interactions may say about low toxicity of 1 and its gentle immunomodulative properties.

Dependence of mobility of Ab in all modifications of the experiment from concentration of 1 was very difficult to explain, especially in the diapason of low concentrations. Our previous works showed polymodal, i.e. dependent from concentration, effect of rhenium cluster substances with

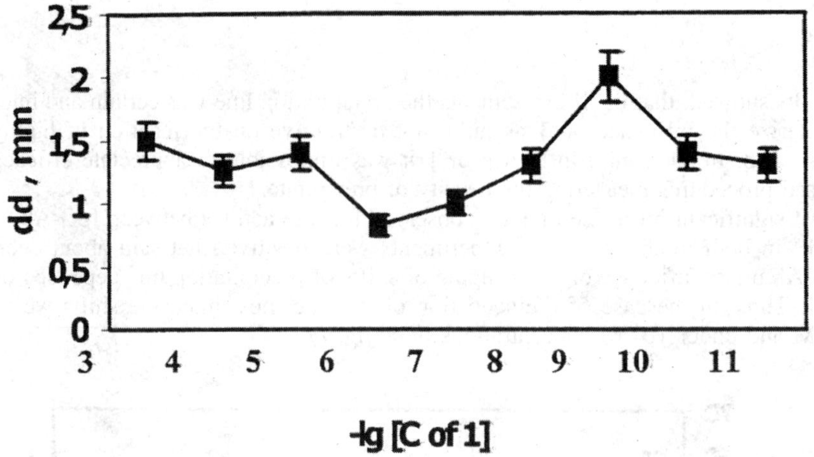

Fig. 2. Changing of width (d) of the precipitation line in experiments with pretreatment of Ab with solutions of 1 for interaction Ig-Ab

organic ligands which was definite in the area of 10^{-8} - 10^{-10} M of preincubation solutions and might be another than in the area of high concentrations. Polymodal effect was shown for the range of biologically active substances and is the molecular base of homeopathic effects.

Changing of width (d) of the precipitation line had the same dependence from concentration of 1 and that supported our preposition about mild influence of 1 on the studied process and demonstrated ability of 1 slightly to "feel" antibodies with different structure. Similar effects were shown for serum immunoglobulines and some monoclonal antibodies, when their polyreactive properties (i.e. their ability to bind to various antigenically dissimilar antigenes) was induced by the treatment with chaotropic ions.

Some structural investigations showed that transition metals coordinated to protein molecules via the imidazole rings of His residues thus modifying antigenic properties.

According to these facts and taking into account conformational plasticity of active centers of immunoglobulines, we consider that our data seems to be a result of changes in protein conformation called by coordination of 1 with His residues.

CONCLUSIONS

Mild immunomodulative properties of the rhenium cluster compound was shown on the IgG-Ab interactions. Changes of antigenic properties can be connected with conformational shifts of proteins, which doesn't bring perturbation of complementarity of the antigen-antibody reactive sites. We consider that Re atoms are coordinated to protein molecules via the imidazole ring of His.

REFERENCES

Eastland G.W., Yang G., Thompson T. Studies of rhenium carboxylates as antitumor agents. Part II. Antitumor studies of bis(propionato)diaquatetrabromodirhenium (III) in timor-bearing mice. *Methods Find. Exp. Clin. Pharmacol.*, 1983, **5**, N 7, 435-438.

Shtemenko A.V., Bovikyn B.A. Chemistry of binuclear Rhenium clusters. *Rhenium ans rhenium alloys. Pensilvania: TMS publications*, 1997, 189-197.

Shtemenko N.I., Piroshkova-Patalakh I.V., Shtemenko A.V., et al. Screening and testing strategy for biological activity of rhenium cluster compounds. *Metal ions in Biology and Medicine. Puerto Rico: John Libbey EUROTEXT*, 2000, **6**, 616-618.

Trynda L., Pruchnik F. Interaction of tetra-mu-acetato-dirhodium (II) with human serum albumin. *J. Inorg. Biochem.*, 1995, **58**, 69-77.

*Metal Ions in Biology and Medicine: vol. 7.*Eds L. Khassanova, Ph. Collery, I. Maymard, Z. Khassanova, J.-C. Étienne.
John Libbey Eurotext, Paris © 2002 pp. 37-40.

Spectroscopic study of iron(III) complexes with some indole derivatives

Alexei G. Shchelochkov[1]*, Alexander A. Kamnev[1]*, Petros A. Tarantilis[2] and Moschos G. Polissiou[2]

[1]*Laboratory of Biochemistry of Plant-Bacterial Symbioses, Institute of Biochemistry and Physiology of Plants and Microorganisms, Russian Academy of Sciences, 13 Prosp. Entuziastov, 410015 Saratov, Russia*
[2]*Laboratory of Chemistry, Department of Science, Agricultural University of Athens, Iera Odos 75, 11855 Athens, Greece*

ABSTRACT

Indole derivatives are known to play an important role in plant physiology and plant-bacterial interactions as a result of plant growth regulation effects (phytohormones of the auxin series) considering their biosynthesis by many soil microorganisms. The formation of their complexes with metal ions may essentially affect both their physiological and chemical properties, in particular, as a result of alterations in the electronic structure of the aromatic system. In the present work, the structure and coordination of iron(III) complexes with some indole derivatives were studied including indole-3-acetic acid (IAA; auxin), a product of IAA oxidation, and indole-3-butyric acid (IBA). For characterising the state of functional groups, Fourier transform infrared (FTIR) spectroscopy in the diffuse reflectance (DRIFT) mode was used. The results obtained show that iron(III) forms similar complexes with IAA and its oxidised derivative with a high degree of covalency. This is confirmed by the lower frequency of the stretching Fe-O mode (under 520 cm^{-1} in IAA complex), as well as by the presence of the carboxylic ν(C=O) band at 1710 cm^{-1} in both the IAA-related complexes (in free IAA, the ν(C=O) band is observed at 1703 cm^{-1}). The latter is absent in the IBA complex (as well as in many other carboxylates), which shows typical absorption of coordinated COO^- groups. Oxidation of IAA in air introduces a carbonylic group (1794 cm^{-1}) in the indole aromatic system.

Keywords: indole derivatives; indole-3-acetic acid oxidation products; indole-3-butyric acid; iron complexes; Fourier transform infrared (FTIR) spectroscopy

INTRODUCTION

Indole-3-acetic acid (IAA) and other indole derivatives are known to play an important role in plant physiology and plant-bacterial interactions as a result of their plant growth regulation effects (phytohormones of the auxin series) considering their biosynthesis by many soil microorganisms [1]. The formation of their complexes with metal ions may essentially affect both their physiological [2] and chemical properties [3], in particular, as a result of alterations in the electronic structure of the aromatic system [4].

In the present work, comparative investigations were performed on the structure and coordination of iron(III) in complexes with some indole derivatives including indole-3-acetic acid (IAA; auxin), a product of IAA oxidation, and indole-3-butyric acid (IBA). For characterising the state of functional groups, Fourier transform infrared (FTIR) spectroscopy in the diffuse reflectance mode was used.

MATERIALS AND METHODS

Iron(III) complexes with indole derivatives were obtained by mixing freshly prepared aqueous iron(III) nitrate solution with aqueous solutions of relevant organic acids partly neutralised with NaOH up to pH~6 (Fe-to-acid ratio 1:3). The resulting precipitates were filtered, washed well with bidistilled water and dried under vacuum (3 Torr) for 25 min. Oxidation of IAA was performed in dilute solution in the presence of $Fe(NO_3)_3$ excess (pH~3) overnight resulting in a coloured precipitate (according to our Mössbauer measurements [3], this reaction in air leads to the formation of an iron(III) complex which is soluble in ethanol and acetone). The reagents used and other preparation and experimental details were described elsewhere [3, 4].

For FTIR measurements in the diffuse reflectance infrared Fourier transform (DRIFT) mode, samples (dry powders) were placed in a Micro sampling cup (Spectra-Tech Inc., USA); spectra were recorded using a Spectra-Tech Diffuse Reflectance accessory against a KBr background on a Nicolet spectrometer (model Magna-IR 560 E.S.P.) with a total of 100 scans (resolution 4 cm^{-1}).

RESULTS AND DISCUSSION

FTIR spectra of pure IAA and its iron(III) complex were characterised in our previous work [3]. It has been found that the type of IAA coordination is similar to chelate where each IAA molecule binds iron(III) via one oxygen of the deprotonated carboxylic group (i.e., O=C-O-Fe) and the conjugated π-electronic system of the pyrrole moiety, which is in agreement with our UV-Vis spectroscopic data obtained in ethanolic solutions [4]. In particular, the very strong carboxylic ν(C=O) band in solid IAA, observed at 1703 cm^{-1}, upon coordination with iron(III) does not disappear (as is observed for many carboxylates with a symmetric COO^- group giving two bands, $\nu_{as}(COO^-)$ and $\nu_s(COO^-)$, at lower wavenumbers) but shifts to 1710 cm^{-1} [3]. This is typical for the carbonyl (which in pure IAA is likely to be involved in weak H-bonding and thus has a lower frequency) and shows that the $(O=C-O)-Fe^{III}$ bond in IAA complex is of rather high degree of covalency.

Quite similar features can be seen in the spectrum of IAA oxidation product *(fig. 1a)* showing ν(C=O) at 1710 cm^{-1}; note that the broad ν(NH) band has the same frequency (3403 cm^{-1}) as in the Fe^{III}-IAA complex [3]. The band at 1010 cm^{-1} in *fig. 1a* may then be ascribed to the ν(C-O) band of the coordinated carboxyl (1014 cm^{-1} in IAA complex [3]). Note that the typical aromatic ((CH) band at 3060 cm^{-1}, ν(CC) and in-plane δ(CC) at 1619 and 1456 cm^{-1}, and very strong narrow out-of-plane deformational γ(CH) band (749 cm^{-1}) in the benzene ring, as well as $\nu_{as}(CH_2)$ and $\nu_s(CH_2)$ bands of the side-chain methylene group (at 2926 and 2856 cm^{-1}, respectively) in *fig. 1a* correspond to the relevant bands in Fe^{III}-IAA complex (3056, 1619,1460, 746, 2925 and 2856 cm^{-1}, respectively [3]), which is indicative of similar structures of the benzene ring and the side chain in both the complexes of IAA and its oxidised product. However, the very characteristic strong band at 1794 cm^{-1} (see *fig. 1a*) indicates the presence of carbonyl in the latter, which is most probably at the C2 atom; its relatively high frequency might probably result from an increased ring strain of the five-membered heterocycle.

In the spectrum of Fe^{III}-IBA complex *(fig. 1b)*, the strong band at 1586 cm^{-1} is likely to represent the antisymmetric stretching mode $\nu_{as}(COO^-)$ of the coordinated carboxylate group. Its symmetric counterpart, $\nu_s(COO^-)$, which is usually weaker, may well be in the region of 1400-1300 cm^{-1}, thus overlapping with the strong bands of the aromatic in-plane ε(CC) (near 1450 cm^{-1}), methylene scissoring deformation $\delta(CH_2)$ (obviously appearing within some range of frequencies around 1460-1420 cm^{-1} owing to inequality of the three CH_2 groups in the IBA side chain) and some relatively strong combination bands around 1360-1320 cm^{-1} related to the indole moiety [5]. Also, the ν(C-O) band of the coordinated carboxyl (at 1010 cm^{-1} in *fig. 1a*) is virtually absent in the IBA complex, which confirms that the carboxylate group in the latter is more symmetric. Thus we propose that in the Fe^{III}-IBA complex the Fe-carboxylate bond is less covalent, which is in line with the higher ν(Fe-O) frequency (594 cm^{-1}) than in IAA complex (under 520 cm^{-1} [3]). Also, it is assumed that in IBA complex there is no coordination of iron with the aromatic π-system (in contrast to IAA complex [4]) owing to the unfavourably long carboxylic side chain. Instead, Fe^{III} is evidently coordinated, besides carboxylate, also by water and/or hydroxyl (like in Fe^{III}-anthranilate complex, see [3]), which accounts for an increased very broad absorption in the ν(OH) region (over ca. 3500-3200 cm^{-1}) with possible weaker contributions in the regions around 1700-1600 and 800-500 cm^{-1}; as well as for the strongly broadened ((NH) vibrations (around 3300 cm^{-1}; see *fig. 1b*) owing to the involvement of the pyrrole NH group in hydrogen bonding.

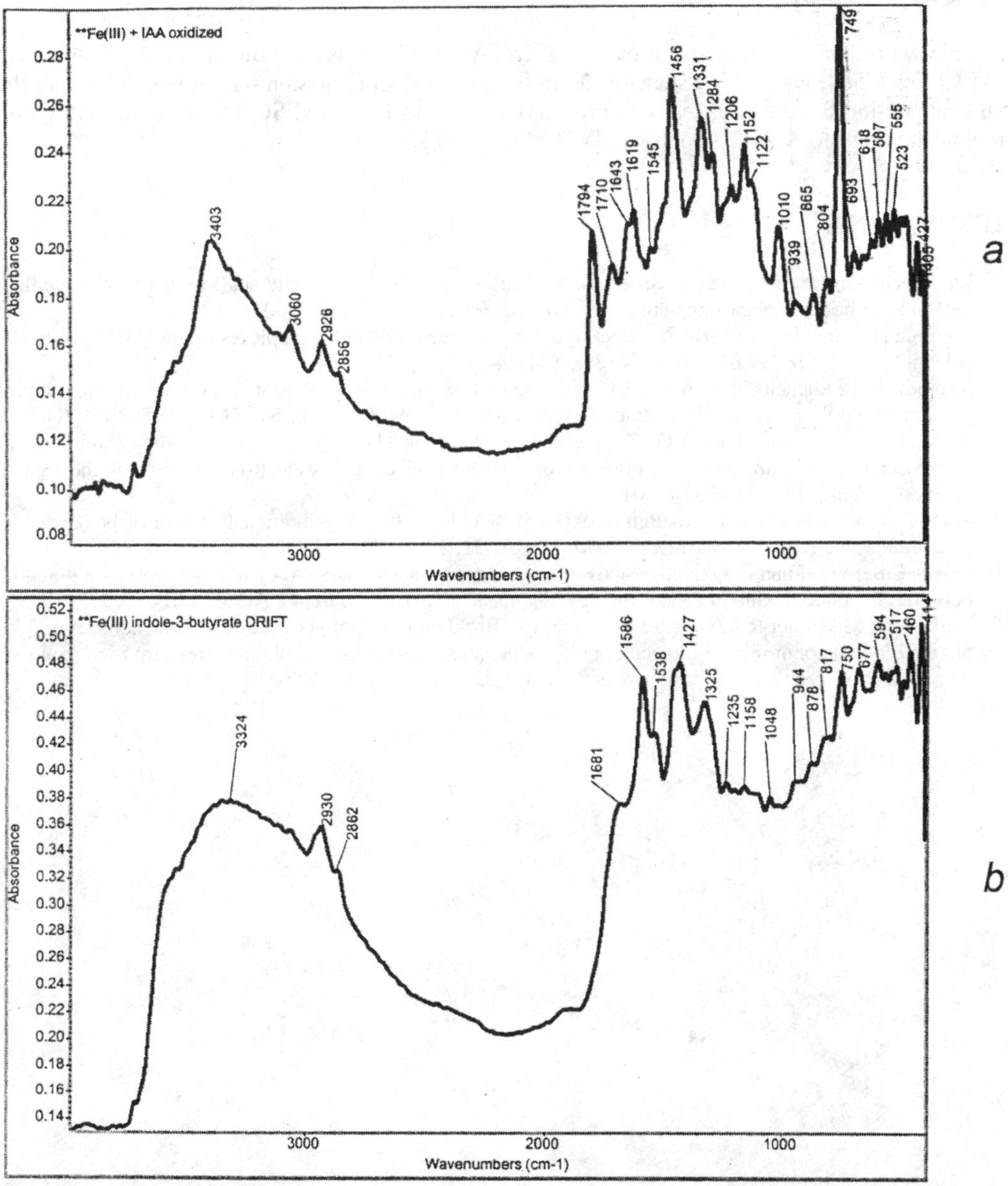

Fig. 1. FTIR spectra (in the DRIFT mode) of iron(III) complexes with *(a)* oxidation product of indole-3-acetic (IAA) and *(b)* indole-3-butyric acid (IBA)

In conclusion, the results obtained suggest that coordination of Fe^{III} with the aromatic π-system of the indole moiety is characteristic of IAA and its closely related derivatives with a relatively short (2C) side chain [3, 4], whereas already for the butyrate side chain (4C) the coordination type is essentially different, approaching that for simple carboxylates. The data on the coordination and oxidation processes involving IAA and related indole derivatives are of importance also considering their relevance to cytotoxicity [6] and iron-induced oxidative damage [7].

ACKNOWLEDGEMENTS

This work was supported in parts by INTAS (EC, Brussels, Belgium), NATO (Grant LST.CLG.977664) and the Russian Academy of Sciences' Commission (Grant No. 205 under the 6th Competition-Expertise of research projects), as well as by a UNESCO Short-term Fellowship in biotechnology to A.A.K. (Contract UVO-ROSTE 875.878.9).

REFERENCES

1. Lambrecht M., Okon Y., Vande Broek A., Vanderleyden J. Indole-3-acetic acid: a reciprocal signalling molecule in bacteria-plant interactions. *Trends Microbiol.*, **8**, 298-300 (2000).
2. Wenmian H., Jian F., Zhengzhi Z. Effect of auxinhormone lanthanide complexes on the growth of wheat coleoptile. *Biol. Trace Elem. Res.*, **64**, 27-35 (1998).
3. Kamnev A.A., Shchelochkov A.G., Perfiliev Yu.D., Tarantilis P.A., Polissiou M.G. Spectroscopic investigation of indole-3-acetic acid interaction with iron(III). *J. Mol. Struct.*, **563-564**, 565-572 (2001).
4. Kamnev A.A., Shchelochkov A.G., Tarantilis P.A., Polissiou M.G., Perfiliev Yu.D. Complexation of indole-3-acetic acid with iron(III): influence of coordination on the π-electronic system of the ligand. *Monatsh. Chem.*, **132**, 675-681 (2001).
5. Klots T.D., Collier W.B. Heteroatom derivatives of indene. Part 3. Vibrational spectra of benzoxazole, benzofuran, and indole. *Spectrochim. Acta*, Part A, **51**, 1291-1316 (1995).
6. Folkes L.K., Wardman P. Oxidative activation of indole-3-acetic acids to cytotoxic species - a potential new role for plant auxins in cancer therapy. *Biochem. Pharmacol.*, **61**, 129-136 (2001).
7. Karbownik M., Reiter R.J., Garcia J.J., Cabrera J., Burkhardt S., Osuna C., Lewinski A. Indole-3-propionic acid, a melatonin-related molecule, protects hepatic microsomal membranes from iron-induced oxidative damage: relevance to cancer reduction. *J. Cell. Biochem.*, **81**, 507-513 (2001).

Metal Ions in Biology and Medicine: vol. 7.Eds L. Khassanova, Ph. Collery, I. Maymard, Z. Khassanova, J.-C. Étienne.
John Libbey Eurotext, Paris © 2002 pp. 41-43.

Chromium salts - DNA interactions

Montrel M.M.[1], Shabarchina L.I.[1], Pleteneva T.V.[2], Ershov Yu.A.[2]

[1]*Institute of Theoretical and Experimental Biophysics RAS, Pushchino, Russia*
[2]*Sechenov's Moscow Medical Academy, Moscow, Russia*

INTRODUCTION

Polyvalent metal ions such as molybdenum, nickel, copper, chromium, cobalt, zinc are known as mutagenic and carcinogenic agents. Chromium compounds are the least studied in this respect despite of their wide application in a number of industrial processes.

We have studied the effect of three chromium salt influence on the structure of DNA double helix using IR-spectroscopy to detect DNA structural changes. $CrCl_3$, $Cr_2(SO_4)_3$, $K_2Cr_2O_7$ were used where chromium was tri- or hexavalent ion. IR-spectra of DNA films containing the above salts in an amount of 0.005 to 0.24 Cr ion per nucleotide were obtained for the samples placed in the atmosphere with relative humidity (r.h.) 0-93%. Chromium cations were found to interact with phosphate groups of DNA. Small amount of the salts in DNA films at $Cr/P<0.01$ prevents B-to-A transition of DNA induced by r.h. decreasing. The efficiency of the Cr salts to prevent the transition decreases in the row: $Cr_2(SO_4)_3 > K_2Cr_2O_7 \sim CrCl_3$. When $Cr_2(SO_4)_3$ or $K_2Cr_2O_7$ content in DNA exceeds 0.1 Cr/P, sugar-phosphate backbone of the double helix looses its regularity both with nucleic bases stacking, i.e. DNA denaturation occurs.

Metal ions are known to be obligatory structural components of biological macromolecules ensuring their normal functioning [1]. Such polyvalent metals as molybdenum, nickel, copper, chromium, cobalt, manganese and zinc acting on living organisms display mutagenic and carcinogenic activity [2]. To identify the mechanisms of such harmful effects of the metals, numerous investigations are provided in the field of metal ion influence on the structure and properties of nucleic acids [3-5].

The present work is focused in IR-spectroscopic study of DNA interaction with three different chromium salts, and the salts influence on the hydration and structural transitions of DNA.

Sturgeon sperm DNA with a molecular weight of 1.7×10^7 D was cleaned from heavy metal impurities by 2-day dialysis against 10 mM NaCl and 10 mM EDTA. $CrCl_3$, $Cr_2(SO_4)_3$ and $K_2Cr_2O_7$ chromium salts of chemically pure grade were used without further purification.

DNA films with different content of a chromium salt for IR experiment were prepared as follows: 250-300 µl of 5.4 mM DNA solution in 1 mM NaCl was placed on a fluoride support, then 2-100 µl of 2 mM solution of a chromium salt was added, mixed and then slowly dried at room temperature. The content of the chromium salt in the films varied from 0.005 to 0.24 Cr ions per nucleotide. IR spectra of the films were recorded with Specord M80 spectrophotometer. Water content of the films was evaluated from the relative intensity of OH and PO stretching vibration bands at 3400 and 1224 cm^{-1} [6].

RESULTS AND DISCUSSION

Infrared spectra of the DNA films containing different amount of $CrCl_3$, $Cr_2(SO_4)_3$ and $K_2Cr_2O_7$ are presented. DNA interaction with each of the salts changes its secondary structure due to Cr ion binding with phosphate groups or with nucleic bases of the polymer molecules and due to local pH decrease induced by the salts. The structural changes were reflected in the decrease of intensities of strong IR bands 1714 and 1088 cm^{-1} attributed to C=O stretching vibrations and symmetric vibrations of the PO_2^- groups respectively. In deuterated samples 1663 cm^{-1} band was appeared. IR-spectroscopic characteristics of the new DNA structure after complex formation with chromium ions differed from any of the known double helical structures. The spectral changes in the samples placed in the atmosphere of high relative humidity (r.h = 93%) pointed to disorder of base stacking and loose of regular helix structure of sugar phosphate backbone.

For DNA- $Cr_2(SO_4)_3$ samples a new intensive 1150 cm^{-1} band appeared in the IR spectrum. It may be attributed to the vibrations of sulfate groups since it is present in the spectra of films containing sodium sulfate and is absent in the spectra of the films with chromium chloride.

At r.h. = *75%* and chromium content Cr/P = 0.01 (for any of the salts indicated) 1188 cm^{-1} band attributed to A-form of DNA [7] disappeared together with a 14 cm^{-1} decrease of the frequency of anti-symmetric vibrations of phosphate groups. IR spectroscopic characteristics of DNA were the same as for its B-form. This means that all three chromium salts at low concentrations prevent the well-known DNA transition to the A form induced by humidity decrease.

Together with the common nature of the spectral manifestations of the interaction of the DNA molecules with the chromium ions forming part of all three salts, there are special features associated both with the nature of the anions and with the difference in the valences of chromium. These features consist primarily in the difference in the threshold values of the content of the salts in the films for which in the IR spectra denaturing changes in the structure of the DNA and prohibition of the transition to the A conformation begin to manifest themselves.

At higher chromium content when the ratio Cr/P exceeds 0.01 the signs of DNA denaturation can be detected in its IR-spectrum. These signs are 1088 and 1714 cm^{-1} band intensity decrease. The dependencies of the band intensities on a salt content permit one to compare the relative effectiveness of the three salts effect on DNA structure, $Cr_2(SO_4)_3$ prevents the B to A transition in DNA more effective than $K_2Cr_2O_7$ and the like relation is true for denaturation effect: lower content of the former salt (Cr/P = 0.01) than the latter one was enough to denature DNA.

More detailed information on the chromium salts influence on B-A transition in DNA was obtained from the dependencies of the absorption band parameters on relative humidity (r.h.). The usage of such dependencies is necessary because the presence of a low molecular weight salt in DNA film shifts the region of A-form existence along the r.h. scale and the value of this shift depends on the quantity of the salt.

We also observed the dependencies of the relative intensity of 1188 cm^{-1} band on r.h. for films containing chromium salts. The presence of this band in IR-spectrum as was shown earlier [7], indicates DNA is in A conformation. Since 1188 cm^{-1} band is placed on the slope of the wide band 1237 cm^{-1} belonging to the antisymmetric vibrations of the phosphate groups of the backbone, the ratio of the optical density in the maximum of the band to that one in the nearest minimum (1198 cm^{-1}) may be a quantitative characteristic of the A-form. Thus, if $D_{1188}/D_{1198} > 1.1$, the band is well marked and almost all the DNA molecules exist in A-form; if $D_{1188}/D_{1198} = 1$ and the band has been transformed to a shoulder, some of the molecules are in A-form; if $D_{1188}/D_{1198} < 0.8$ the band is absent and the molecules are not in A-form.

In the control samples DNA molecules are in A-form at r.h. from 58 to 75%. Addition of small amount of chromium sulfate (Cr/P = 0.01) leads to disappearance of 1188 cm^{-1} band and respectively A-form of DNA, in the whole range of r.h. from 0 to 93%. As may be seen from curves 4-9 the disappearance of the A-form is due not only to the chromium cations but also the sulfate anion. The A-form of DNA exists in the presence of $CrCl_3$ at Cr/P < 0.02 and almost the same amount of $K_2Cr_2O_7$ although the regions of its existence on the r.h. scale differ for these salts. Only at (Cr/P = 0.12) the $CrCl_3$ and $K_2Cr_2O_7$ salts absolutely prevent the double helix to adopt the A-form. Thus, the salts studied are ordered in the following row of effectiveness of B-A transition blocking: $Cr_2(SO_4)_3 > K_2Cr_2O_7 \sim CrCl_3$.

The results of the investigation of the hydration of DNA films containing chromium salts are presented in the *table*. Of the three salts studied only chromium chloride practically does not change the amount of hydrated water in the samples as compared with the control. DNA containing $Cr_2(SO_4)_3$ or $K_2Cr_2O_7$ is less hydrated over the entire r.h. range from 0 to 93%. Reduction of the water binding in the presence of $Cr_2(SO_4)_3$ and $K_2Cr_2O_7$ may be a consequence of the screening of some DNA hydration centers by these salt atoms and also of the close contact formation between neighboring macromolecules reducing their Van-der-Vaals surface accessible for the water molecules. In the r.h. interval from 0 to 40% when the secondary structure of DNA is still not formed the chromium salts either do not screen its hydration centers or screen ~ 1 center per nucleotide. This center is most probably placed near the phosphate group since the differences in the hydration of the samples with a different salt content are already present at r.h. = 2-10% where nucleic bases and ribose parts of DNA are not hydrated [8]. At r.h. > 60% the chromium salts strongly prevent the DNA hydration. This may be mainly due to aggregation effect of chromium cations on the polymeric molecules rather than the hydration centers screening.

Reduction of DNA hydration and change in the structure of the hydration shell in the presence of chromium salts may play an important role in biochemical reactions involving double stranded DNA by means of the change of the rates of the reactions.

Table 1. Hydration of DNA in the films (number of water molecules per nucleotide) in the presence of chromium salts for different values of relative humidity

Salt	Cr/P	2%	35%	58%	75%	80%	93%
Control	-	2,6	3,4	9,0	14,5	17,4	23,8
$CrCL_3$	0,005	2,5	3,5	8,4	12,3	14,2	22,6
	0,024	2,8	3,3	8,4	10,7	12,2	16,9
	0,12	1,9	2,5	5,0	13,7	15,8	20,3
$Cr_2(SO_4)_3$	0,05	2,5	3,8	7,8	10,7	13,5	17,3
	0,24	1,5	2,4	4,8	5,7	9,4	13,8
$K_2Cr_2O_7$	0,01	1,5	1,5	5,4	7,5	6,1	9,8
	0.07	1,6	1,6	5,7	9,1	8,7	18,2

REFERENCES

1. In: *Metal Ions in Biological Systems*, v. 8, Ed. H. Siegel, M. Dekker Inc., N.Y. & Basel (1979).
2. In: *Nucleic Acid-Metal Ion Interactions*, Ed. T. G. Spilo, 256 pp., Wiley, New York (1980).
3. H. Siegel, *Chem. Society Reviews*, 22, 255-267 (1993).
4. G.Albiser and *S. Yiemi}&t,J,Biomol. Struck Dynam.*, 2, 745 (1985).
5. H. Diebler, F. Secco and M. Venturim, *Biophys. Chem.* 2, 193 (1987).
6. M.A. Semenov, A.I. Gasan, T.V. Bolbukh, V.Ya. Maleev, *Biophysics*, 41, 1013 (1996).
7. B. I. Sukhorukov and M. M. Montrei, *Biophys. Chem.* 35, 47 (1990).
8. M.A. Semenov, B.I. Sukhorukov and V.Ya. Maleev, *Biophysics*, 26, 979 (1981).

*Metal Ions in Biology and Medicine: vol. 7.*Eds L. Khassanova, Ph. Collery, I. Maymard, Z. Khassanova, J.-C. Étienne.
John Libbey Eurotext, Paris © 2002 pp. 44-48.

Effect of cadmium and cadmium antagonists on corticomedullary enzymes of rat

Puri Vishamber Nath, Saha Soumya, Srivastava Mrigank

Division of Pharmacology, Central Drug Research Institute, Lucknow-226001, India.

ABSTRACT

Effect of Cadmium and the antagonist 24 on cardiovascular functions in relation to kidney cortico medullary enzymes were examined in Sprague-Dawley rats. Blood pressure,heart rate,serum, kidney cortical and medullary proteins, angiotensin converting enzyme(ACE), malondialdehyde (MDA), deoxynucleic acid (DNA), ribonucleic acid (RNA) and cadmium contents were determined in the rats. Cadmium acetate (1 mg/Kg i.v) produced hypertensive response which was associated with inhibition of serum ACE, while antagonist 24 (10 mg/Kg i.v) increased the ACE levels in rats. Hypertensive effect of this biometal has produced upregulation of ACE in cortical portion of kidney,while medullary levels of ACE were statistically significantly ($P<0.01$) reduced. Serum MDA levels were increased by cadmium treatment, while antagonist 24 attenuated increased activity of MDA to normal or below normal levels in cortical and medullary regions of kidney. Cadmium levels in control rats were 30±2 ng/ml,cadmium treatment increased cadmium levels to 310±14 ng/ml ($P<0.01$), while antagonist 24 reduced the cadmium levels to 270±70 ng/ml. Serum protein levels in serum,cortex and medulla were not significantly altered. Tissue DNA and RNA were altered but none of the values were statistically significant. These results indicate that cadmium ions significantly alter the cardiovascular and renal cortical, medullary enzymes ACE, MDA in rats. Cadmium antagonist 24 attenuate some of the cardiovascular and renal enzymes of rats.

INTRODUCTION

Cadmium (Cd) is a highly toxic environmental pollutant [1]. Hypertensive nature of this biometal has been worked out by several groups [2, 3, 4]. Extensive experimental and human data has accumulated indicating the association of cadmium ions with hypertension and nephrotoxicity [5, 6]. Chronic exposure to cadmium results in renal tubular, morphological changes, albuminaria, and increased urinary microglobulin and metallothionein [7]. Cadmium is known to produce reactive oxygen species (ROS) resulting in increased lipid peroxidation [8]. Reactive oxygen species and peroxynitrle produce cellular injury and necrosis via multiple mechanisms including peroxidation of membrane lipids, protein denaturation and DNA damage [9]. The effect of cadmium on corticomedullary regions of kidney in relation to angiotensin converting enzyme activity and oxidant enzymes has not been worked out as yet. Therefore, we investigated the effect of cadmium and its antagonist 24 (Dithiocarbamate) on corticomedullary enzymes in rats.

MATERIALS AND METHODS

Male Sprague-Dawley rats (210-260 g) bred in the CDRI animal house facility was used in this study. Rats were housed in polyethylene cages with free access to water and pellet diet in temperature controlled room (26±2°C). All experiments were carried out in accordance with guidelines of committee for purpose of control and supervision of experiments on animals (CPCSEA). Government of India, and were reviewed and approved by central drug research institute animal ethics committee. Three groups of rats, each group containing 6 rats were included in this study. Group I was treated with nornal saline (1 ml/Kg i.v) and formed as control. Group II was treated with cadmium (1 mg/kg i.v.) and group III was treated with antagonist 24 (10 mg/Kg i.v.). Rats were anesthetized with pentobarbitone (50 mg/Kg i.p), Jugular vein was cannulated by a polyethylene catheter (PE 60) for cadmium, antagonist and saline administration. Systemic blood pressure was recorded by placing a polyethylene catheter (PE60) in the common carotid artery. Arterial blood pressure was measured with a Stathum pressure transducer, connected to a Grass model 7 polygraph. Heart rate (HR) in anesthetized rat was recorded by increasing the paper speed. After rats had been in stable haemodynamic conditions for atleast 15 minutes, baseline values of blood pressure and heart rate were determined. Cadmium (1 mg/Kg i.v.) was administered and changes in BP and HR were recorded. Cardiovascular responses were observed for one-hour after which blood was collected from carotid cannula in sterlized glass tubes. Blood was centrifuged at 4000 r.p.m × 15 min. and serum was seperated. Kidneys were dissected out, perirenal fascia removed and from both the kidneys cortex and medulla were separated, half portion was homogenized in Tris - HCl (10%, pH 7.4) and another half in TCA (10%). The Tris buffer homogenate was centrifuged in 4000 r.p.m × 20 min., the supernatant was collected and the following biochemical estimations were done. Serum and angiotensin converting enzyme (ACE) was estimated by adding tripeptide Hippuryl-L-Histidyl-l- Leucine as described by Laberman [10]. Serum and tissue supernatant MDA was estimated as described by Ohkawa et al [11], using thiobarbituric acid reaction. Protein was estimated by modified method of Lowry [12] Proteins were estimated in the serum and tissue supernatant. DNA and RNA were extracted from the tissue (homogenized in TCA) by ether alcohol gradual centrifugation method. DNA was estimated by the method of Burton et al [13] using diphenylamine. RNA was estimated by colorimetric orcinol method described by Chomczynski et al [14]. Results were expressed as means ± SEM. Data was analyzed using Student's't' test. A value of P<0.05 was considered to be statistically significant.

RESULTS

Results of the present investigation indicate that cadmium produced hypertension in pentobarbitone anesthetized S-D rats. The BP increase in control group by the saline administration was 2.75±0.37 mmHg, where as, in cadmium treated rats the increase in BP was 37.5 ± 5.6 mmHg and the difference was statistically significant, (P<0.05). Antagonist 24 lowered the BP 6.4 ± 3.0 mmHg as compared to cadmium treated group (P< 0.05, *fig. 1*). Serum ACE activity in control group of rats was 10±1.9 U/ml, cadmium treatment reduced the ACE activity to 2.4 ± 0.15 U/ml, the difference was statistically significant (P<0.05). Antagonist 24 treatment produced increase in ACE levels 30.23 ± 15.87 U/ml *(table 2)*. Serum MDA levels were increased statistically significantly in cadmium treated group as compared to control group (P<0.05). However, antagonist lowered the serum MDA level significantly (P<0.05), as compared to cadmium treated group *(table 2)*.

Cadmium treatment decreased kidney cortical and medullary ACE levels 8.6 ± 3.6, 7.8 ± 1.2 U/mg, as compared to 10.3 ± 3, 12 ± 3 U/mg in control rats. Antagonist 24 reduced the medullary ACE 3.4 ± 1.1 U/mg and the effect was statistically significant (P<0.05). Kidney cortical ACE levels were increased 23.5±18.7 U/mg. Kidney cortex and medullary MDA levels were increased significantly (P<0.05) by cadmium treatment. However, antagonist 24 lowered MDA levels and difference was statistically significantly (P<0.05, *table 2*). Heart rate was decreased by cadmium administration, but by antagonist heart rate did not change significantly *(table 1)*.

Serum proteins were decreased by cadmium treatment, the effect was reversed by antagonist 24

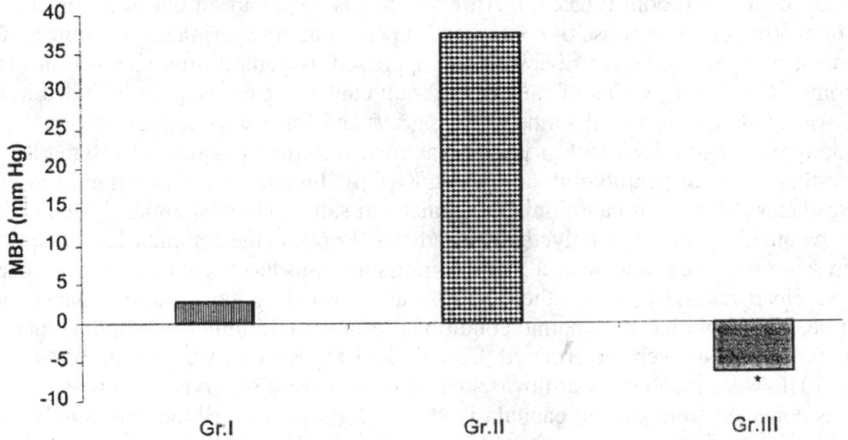

Fig. 1. Effect of cadmium & Antagonist 24 on BP of rats. Gr. I: Control, Gr. II: Treated with Cadmium (1 mg/Kg i.v.), Gr. III: Treated with Antagonist 24 (10 mg/Kg i.v.), Values presented are mean ± SEM, * $P<0.05$

Table 1. Effect of cadmium and antagonist 24 on Blood pressure (BP), Heart rate (HR) and serum, kidney cortex and medulla protein & RNA concentration of rats. Values represented are mean ± SEM

SN	Control	Cadmium (1mg/Kg i.v.)	Antagonist 24 (10mg/Kg i.v.)
BP (mm Hg) change	+2.75 ± 0.37	+37.5 ± 5.6*	-6.4 ± 3.0*
HR (B/min)	420 ± 7.57	336 ± 21.46	392 ± 24.52
Serum Protein (g%)	6.23 ± 0.2	5.32 ± 0.5	7.0 ± 0.4
K.Cortex Protein (g%)	2.0 ± 0.24	1.51 ± 0.19	1.9 ± 0.2
K.Medulla Protein (g%)	2.2 ± 0.28	1.82 ± 0.2	2.3 ± 0.2
K.Cortex RNA (µg/Mg)	257.07 ± 13.9	358.8 ± 39.18	336.42 ± 15.85
K.Medulla RNA (µg/Mg)	206.82 ± 12.32	262.51 ± 40.44	233.35 ± 16.29

* $P<0.05$ as compared to control.

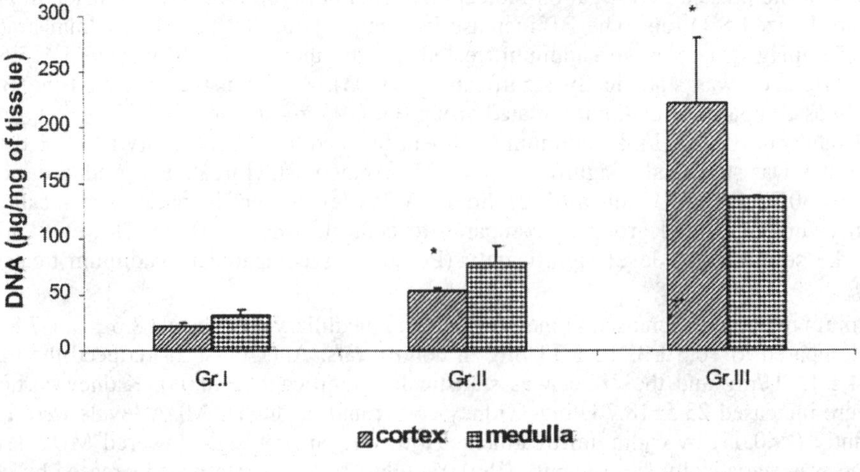

Fig. 2. Effect of Cadmium & antagonist 24 on DNA levels in Kidney of rats. Gr. I: Control, Gr. II: Treated with Cadmium (1 mg/Kg i.v.), Gr. III: Treated with Antagonist 24 (10 mg/Kg i.v.), Values presented are mean ± SEM, * $P<0.05$

Table 2. Effect of cadmium and antagonist 24 on ACE, MDA and cadmium concentration of serum, kidney cortex and medulla of rats. Values represented are mean ± SEM

SN.	Control	Cadmium (1mg/Kg i.v.)	Antagonist 24 (10mg/Kg i.v.)
Serum ACE(U/ml)	10.6 ± 1.9	2.4 ± 0.15*	30.23 ± 15.87*
K.Cortex ACE(U/ml)	10.3 ± 3	8.6 ± 3.6*	23.49 ± 18.73*
K.Medulla ACE(U/ml)	12.0 ± 3	7.8 ± 1.2*	3.38 ± 1.1*
Serum MDA (nM/ml)	0.14 ± 0.06	0.25 ± 0.02*	0.105 ± 0.03
K.Cortex MDA(nM/mg)	0.15 ± 0.01	0.25 ± 0.04*	0.08 ± 0.01
K.Medulla MDA (nM/mg)	0.14 ± 0.01	0.27 ± 0.03*	0:09 ± 0.01
Serum Cad(µg/ml)	0.03 ± 0.002	0.31 ± 0.14*	0.27 ± 0.07
K.Cortex Cad(µg/mg)	0.18 ± 0.03	0.32 ± 0.14*	0.17 ± 0.13
K.Medulla Cad(µg/mg)	0.20 ± 0.01	0.38 ± 0.16*	0.12 ± 0.07

* P<0.05 as compared to control.

(table 1). DNA levels were increased in cortex of antagonist 24 treated group statistically significantly (*fig. 2*, P<0.05) and RNA levels were increased by cadmium and antagonist 24 administration. Cadmium induced increase in serum, cortical and medullary concentration of cadmium (P<0.05), while antagonist 24 produced reduction in cadmium contents and changes were statistically significant as compared to cadmium treated groups. The maximum reduction in cadmium contents was observed in medullary region of rat kidney *(table 2)*.

DISCUSSION

Mechanism of hypertension and renal toxicity produced by cadmium still remains controversial. Design and development of cadmium antagonists appears to be a good approach to understand the mechanism of cardiovascular and nephrotoxicity produced by environmental pollutant like cadmium. In the present investigation an effort has been made to elucidate effect of cadmium and antagonist a dithiocarbamate compound 24 on some cardiovascular and renal enzyme system of rats. Kidney functions are well maintained due to several physiological and biochemical processes. Cortical and medullary portions of kidney have different osmotic gradient. Medullary portion is hyperosmotic due to impermeable ascending thin limb of loop of Henley. Interstitial fluid, ion and enzyme dynamics is different, due to which many toxicants like cadmium alter the various urinary enzymes ALP, LDH, GST, LDH etc. Bomhard and Rinke [15]. However, enough is not known how hyperosmolarity of medulla is maintained. Counter current multiplier hypothesis of Koko and Rector [16] is an attractive model and qualitatively accurate. Role of ACE and MDA and enzyme in the model has not been worked and our results have provided small window to look the contributions of enzyme on cadmium exposure to rats. Cadmium is known inhibitor of serum ACE, however, cortical and medullary ACE levels are reduced but antagonist 24 upregulated in cortical area, while medullary ACE levels were significantly decreased even when less cadmium was present in medulla. In contrast higher MDA levels in medulla indicated that free radicals might be responsible for some complex interactions between medulla and cortex of rats. These changes modulate the cardiovascular functions resulting hypertension on cadmium exposure of rats. Role of sodium ions and other solutes in relation to ACE and MDA remains to be worked out to design and develop better hypotensive agents which could reduce the cadmium burden and block free radical generation and some bioactive system like RAS, Prostaglandins, Endothelins. Neuropeptidases etc so as to reduce the cardiovascular morbidity and mortality caused by silent killer hypertension.

CONCLUSION

From this study it is thus concluded that cadmium produced hypertension and inhibition of ACE, while MDA levels were increased. However, ACE and MDA distribution in rat kidney was different on cadmium treatment. Cadmium antagonist 24 attenuated both the enzymes in cortex and medulla. Cadmium antagonist 24 cleared more cadmium from medulla suggesting that pharmacodynamics of cadmium is differentially regulated in medullary portions of rat kidney.

REFERENCES

1. Lall S. B., Das N., Rama R., Peshin S. S., Khattar S., Gulati K., Seth S.D. Cadmium induced nephrotoxicity in rats. Ind. J. Exp. Biol, 1997, 335, 151-154.
2. Schroeder H.A. and W.H. Venton. Hypertension induced in rats by small doses of cadmium. Am. J. Physiol. 1962, 202, 515-518.
3. Puri V.N. and Kapoor N.K. Cadmium induced hypertension and catecholamines in rats. In hypertensive mechanisms. The 4th International symposium on rats with spontaneous hypertension and related studies. Heidelberg, West Germany, 1981, PP 314-316.
4. Puri V.N. and Sur R.N. cardiovascular effect of cadmium on intravenous and intracerebroventricular administration in rats. Can. J Physiol. Pharmacol. 1983, 611, 1430-32.
5. Perry H.M. and Erlanger M.W. Circulating renin activity in the rat following doses known to induce hypertension. J. Lab. Clin. Med, 1973, 82, 399-404.
6. Kopp S.J., Glonek T., Perry H.M., Erlanger M., Perry E.F. Cardiovascular actions of cadmium at environmental exposure levels. Science, 1982, 217, 837-38.
7. Bakshi S.K., Chawla K.P., Khandekar R.N., Raghunath R. Cadmium and hypertension. JAPI, 1994, 42, 449-50.
8. Szuster-Cilsie lska A., Sachura A., Slot winska M., Kaminska T., Sneizko R, Paduch R, Abramezyk D. The inhibitory effect of zinc on cadmium induced cell apoptosis and reactive oxygen species (ROS) production in cell cultures. Toxicology. 2000, 14, 159-171.
9. Emannuela Mazzon. Effect of n-acetylcysteine on gentamicin mediated nephrotoxicity in rats. Eur. J. Pharm. 2001, 424, 75-83.
10. Laberman J. Evaluation of serum angiotensin converting enzyme (ACE) level in sarcoidosis. Am J. Med., 1974, 53, 365.
11. Ohkawa H., Ohishi N., Yagi K. Assay for lipid peroxidation in animal tissue by thiobarbituric acid reaction. Annals. Biochem. 1979, 95, 351-358.
12. Lowry O. H., Rosenbrough N.J., Farr A.L., Randall R.J. Protein measurement with Folin Phenol reagent. J. Biol. Chem. 1951, 193, 265.
13. Burton K. A study of the conditions and mechanisms of the DPA reaction for the colorimetric estimation of DNA. Biochem. J. 1956, 193, 265.
14. Chomczynski P., Saachi N. single step method of RNA isolation by acid guanidium and thiiocyanate phenol chloroform reaction. Am. Biochem. 1987, 162, 156-159.
15. Bomhard E.M., Maruhn D., Rinke M. Time course of chronic oral cadmium nephrotoxicity in Wistar rats: Excretion of urinary enzymes. Drugs and Chemical Toxicology. 1999, 22(4), 679-703.
16. Kokko J.P., Rector F.C jr. Countercurrent multiplication system without active transport in inner medulla. Kidney. Int. 1972, 2, 214-233.

*Metal Ions in Biology and Medicine: vol. 7.*Eds L. Khassanova, Ph. Collery, I. Maymard, Z. Khassanova, J.-C. Étienne.
John Libbey Eurotext, Paris © 2002 pp. 49-52.

Construction and cloning of pseudophytochelatin (PPC) gene for binding of heavy metals by PPC peptide

Chemeris D.A., Gimalov F.R., Nikonorov Yu.M., Sabirzhanov B.E., Chemeris A.V.

Institute of Biochemistry and Genetics of Ufa Science Centre of the RAS, 450054, Ufa, Russia

ABSTRACT

For constructing and cloning of pseudophytochelatin (PPC) gene with Met(GluCys)$_4$Gly amino acid sequences and TAG termination codon the corresponding oligonucleotides were synthesized which were selected in such a way that during their annealing and formation of double-stranded molecules cohesive ends suitable for directed cloning in phagemid vector pBluescript II KS(-) appeared. After stages of ligation and transformation of *E.coli* competent cells, search of recombinant colonies among transformants, production of single-stranded matrices, DNA sequencing was carried out by Sanger method, and it showed the presence of chemically synthesized PPC gene in a number of recombinant clones.

INTRODUCTION

The problem of plant resistance to heavy metals has been paid a tremendous attention for many years. A great number of reviews describing different aspects of plant life under the conditions of "metallic" stress may serve the evidence of acute interest to this problem, but their enumeration is hardly possible within the limits of this work. The main reasons of such interest lie in the fact that many heavy metals are hazardous ecological toxicants inhibiting on the one hand the plant growth, on the other hand producing harmful effect at human health when included into food chains. Metallothioneins of class III first revealed as cadystins in yeast *Shizosaccharomyces pombe* [1] and as phytochelatins in *Rauwolfia* cell culture [2] are likely to play a leading role in binding and rendering harmless of heavy metals in overwhelming majority of plants. Enzymatically synthesized cadystins or phytochelatins have a high cysteine content and the most typical among them is phytochelatin with the following amino acid sequence - (gamma-GluCys)$_n$Gly, where n is from 2 to 11. The existence of gamma-bond between glutamine acid and cysteine may be considered their peculiarity, because alpha-bonds are formed during the process of protein synthesis on ribosomes.

Nowadays at least 8 types of phytochelatin molecules, which differ mainly in variable amino acid at C-end, are known. Thus, in representatives of leguminous family there is beta-alanine instead of glycine [3], and there is serine in cereals [4]. There are variants of phytochelatin molecules without any terminal amino acid, which were designated desGly(gammaGluCys)$_n$ [5]. Recently phytochelatin with terminal amino acid Gln was revealed [6] in addition to an earlier known type with Glu [7]. It is necessary to point out here that together with the improvement of phytochelatin detection techniques [8, 9] an ever-growing number of their types is being revealed in nature. One and the same plant species is turned out to contain several types of phytochelatin molecules simultaneously, as, for example, it was shown for maize [10]. Proceeding from the above-mentioned facts we may come to the conclusion that information about the absence of phytochelatins in some plants based on the earlier studies may be called in question. Thus, besides plants and yeast phytochelatins have already been revealed in different groups of living organisms. Phytochelatin was detected in fungus Neurospora [11]. There is an information that phytochelatin induction was provoked by heavy metals in algae of sea phytoplankton [12]. The appearance of phytochelatin molecules was observed in fresh-water algae [13].

Notwithstanding the fact that phytochelatin structure was determined already in 1989 [14], almost a whole decade passed before the attempts of creation of synthetic genes of analogous compounds were fulfilled. Undoubtedly, gamma-bond between the residues of glutamine acid and cysteine serves as some obstacle to it in scientists' mind. The considerable similarity of the peptide complexes with cadmium, mercury and lead was revealed for both natural gamma-phytochelatins and artificial alpha-phytochelatins [15, 16]. It allows to hope for the possibility of using of artificial alpha-phytochelatins for binding of heavy metals, especially when the increase of resistance to cadmium of recombinant strain cells of *E.coli* carrying plasmids with synthetic phytochelatin genes has been lately demonstrated [17], and the increase of resistance directly depended on phytochelatin molecule length. The authors even constructed phytochelatin with 20 repeated motives GluCys, which doesn't exist in nature but imparts the greatest cadmium resistance to *E.coli* strain.

MATERIALS AND METHODS

Construction of a synthetic PPC gene. With the aim of the constructing of pseudophytochelatin gene coding peptide with the following amino acid sequence Met(GluCyc)$_4$Gly two complementary each other oligonucleotides phychA 5'-CTAGATGGAGTGCGAGTGCGAATGC-GAGTGCGGGTAG-3' and phychB 5'-GATCCTACCCGCACTCGCATTCGCACTCGCACTC-CAT-3' were synthesized. For formation of double-stranded structure these oligos in quantity of 20 ng each were mixed up in 50 μl of TM-buffer (10 mM Tris-HCl, pH8.0, 1 mM $MgCl_2$), were heated up to 70°C and then slowly cooled to +4°C.

Cloning of the synthetic PPC gene

For cloning of the artificial PPC gene phagemid vector pBluescript II KS(-) (Stratagene, USA) was digested by restriction endonucleases *Bam*HI and *Xba*I (both enzymes from MBI Fermentas, Lithuania) under the conditions recommended by manufacturer. Digestion of 1 μg of plasmid DNA was carried out in 30 μl volume overnight. Enzymes were removed by deproteinization with phenol-chloroform mixture. DNA was precipitated by 2 volumes of ethanol; the precipitate was pelleted in microcentrifuge at 14000 rpm and dissolved in 20 μl of 1x TE-buffer (10 mM Tris-HCl, pH8.0 and 1mM EDTA). Ligation of the vector prepared in such a way with synthetic double-stranded molecule of PPC gene was carried out in conventional ligase buffer in 10 μl volume with the help of T4 DNA ligase (MBI Fermentas, Lithuania) at +4°C overnight. Competent cells of *E.coli* XL1-Blue strain (Stratagene, USA) were transformed with ligase mixture and were plated on LB-agar with X-gal and IPTG. The grown white colonies were analysed using the alcaline lysis procedure for isolation of plasmid DNA and electrophoresis in 1% agarose gel.

Sequencing of the cloned PPC gene

For sequencing of cloned variants of PPC gene the selected colonies were cultivated in liquid LB-medium in the presence of helper phage M13K07 with the aim of production of single-stranded DNA matrices. Sequencing of single-stranded DNA was conducted by Sanger method using DNA polymerase Sequenase 2.0 (Amersham Pharmacia Biotech, Sweden). Sequencing gel-electorphoresis was carried out in Macrophor unit (LKB-Pharmacia, Sweden).

RESULTS AND DISCUSSION

The analysis of literature data showed that there were no serious obstacles to creation and functioning in plants of PPC genes and existence of pseudophytochelatins themselves with generalized formula Met(GluCys)$_n$Aaa (where n is number of repeated motives for GluCys and Aaa here is the corresponding amino acid from those typical of natural phytochelatins). Thus, at first we chose the variant of pseudophytochelatin with the following amino acid sequence - Met(Gly-Cys)$_4$Gly and its nucleotide sequence was deduced taking into account the frequency of occurrence of these or those codons in plants. For constructing and cloning of suggested PPC gene there were synthesized the corresponding complementary oligonucleotides 37 nucleotides long each. These oligonucleotide sequences were constructed in such a way that during annealing and formation of double-stranded molecule protruding GATC and CTAG nucleotides appeared on 5'-ends which were suitable for directed cloning in sites generated correspondingly by restriction endonucleases *Bam*HI and *Xba*I of phagemid vector pBluescript II KS(-). Constructed double-stranded DNA fragment besides 10 coding triplets contained also a TAG stop codon. After stages of ligation and transformation of *E.coli* competent cells, search of recombinant colonies among transformants, production of single-stranded matrices, DNA sequencing was carried out by Sanger method, and it showed the presence of chemically synthesized PPC genes in a number of recombinant clones. On the basis of this created PPC gene under the control of 35S promoter the constructions in binary Ti vectors with the aim of creation of tobacco transgenic plants will be prepared.

CONCLUSIONS

Nowadays transgenic plants with artificial phytochelatin genes haven't been created yet, however it is undoubtedly that such plants will appear in the nearest future. Though full coincidence of pseudophytochelatins to natural phytochelatins can't be achieved, we are faced with the task to ascertain the most optimal perspective variants of pseudophytochelatins. Thus, in order to reach maximal effect of binding, retention and transportation of heavy metals by transgenic plants we'll have to choice the most optimal pseudophytochelatin length, their C-end amino acid. It is highly probable that different plants may need such variations of PPC genes, which are suitable only to them. One cannot but take into account the existence of natural phytochelatins in plants and in order to obtain the best results it is necessary to try to achieve complementary action of natural phytochelatins and artificial pseudophytochelatins.

This work was supported by grants from the Russian Foundation for Basic Research (00-15-97810 and 02-04-97918) and INTAS (01-2170).

REFERENCES

1. Kondo N., Isobe M., Iami K., Goto T. Synthesis of metallothionein-like peptides cadystin A and B occurring in a fission yeast and their isomers. *Agric. Biol. Chem.*, 1985, 49, 71-83.
2. Grill E., Winnacker E.-L., Zenk M.H. Phytochelatins: the principal heavy-metal complexing peptides of higher plants. *Science*, 1985, 230, 674-676.
3. Grill E., Gekeler W., Winnacker E.-L., Zenk M.H. Homo-phytochelatins are heavy-metal-binding peptides of homo-glutathione containing *Fabales*. *FEBS Lett.*, 1986, 205, 47-50.
4. Klapheck S., Fliegner W., Zimmer I. Hydroxymethy-phytochelatins [(gamma-glutamylcysteine)$_n$-serine] are metal-induced peptides of the *Poaceae*. *Plant Physiol.*, 1994, 104, 1325-1332.
5. Mehra R.J., Tarbet E.B., Gray W.R., Winge D.R. Metal-specific synthesis of two metallothioneins and γ-glutamyl peptides in *Candida glabrata*. *Proc. Natl. Acad. Sci. USA*, 1988, 85, 8815-8819.
6. Kubota H., Sato K., Yamada T., Maitani T. Phytochelatin homologs induced in hairy roots of horseradish. *Phytochemistry*, 2000, 53, 239-245.

7. Meuwly P., Thibault P., Rauser W.E. Gamma-glutamylcysteinylglutamic acid - a new homolog of glutathione in maize seedlings exposed to cadmium. *FEBS Lett.*, 1993, 336, 472-476.
8. Vacchina V., Chassaigne H., Oven M., Zenk M.H., Lobinski R. Characterisation and determination of phytochelatins in plant extracts by electrospray tandem mass spectrometry. *Analyst*, 1999, 124, 1425-1430.
9. Yen T.Y., Villa J.A., DeWitt J.G. Analysis of phytochelatin-cadmium complexes from plant tissue culture using nano-electrospray ionization tandem mass spectrometry and capillary liquid chromatography/electrospray ionization tandem mass spectrometry. *J. Mass. Spectrom.*, 1999, 34, 930-941.
10. Chassaigne H., Vacchina V., Kutchan T.M., Zenk M.H. Identification of phytochelatin-related peptides in maize seedlings exposed to cadmium and obtained enzymatically in vitro. *Phytochemistry*, 2001, 56, 657-668.
11. Kneer R., Kutchan T.M., Hochberger A., Zenk M.H. *Saccharomyces cerevisiae* and *Neurospora crassa* contain heavy metal sequestering phytochelatins. *Arch. Microbiol.*, 1992, 157, 305-310.
12. Ahner B.A., Price N.M., Morel F.M.M. Phytochelatin production by marine phytoplankton at low free metal ion concentration, laboratory studies and field data from Massachusetts Bay. *Proc. Natl. Acad. Sci. USA*, 1994, 91, 8433-8436.
13. Pawlik-Skowronska B. Phytochelatin production in freshwater algae Stigeoclonium in response to heavy metals contained in mining water; effects of some environmental factors. Aquat Toxicol., 2001, 52, 241-249.
14. Grill E., Loffler S., Winnacker E.-L., Zenk M.H. Phytochelatins, the heavy-metal-binding peptides of plants, are synthesized from glutathione by a specific γ-glutamylcysteine dipeptidyl transpeptidase (phytochelatin synthase). *Proc. Natl. Acad. Sci. USA*, 1989, 86, 6838-6842.
15. Bae W., Mehra R.K. Metal-binding characteristics of a phytochelatin analog $(Glu\text{-}Cys)_2Gly$. *J. Inorg. Biochem.*, 1997, 68, 201-210.
16. Pickering I.J., Prince R.C., George G.N., Rauser W.E., Wickramasinghe W.A., Watson A.A., Dameron C., Dance I.G., Fairlie D.P., Salt D.E. X-ray absorption spectroscopy of cadmium phytochelatin and model systems. *Biochem. Biophys. Acta*, 1999, 1429, 351-364.
17. Bae W., Chen W., Mulchandani A., Mehra R.K. Enhanced bioaccumulation of heavy metals by bacterial cells displaying synthetic phytochelatins. *Biotechnol. Bioeng.*, 2000, 70, 18-524.

*Metal Ions in Biology and Medicine: vol. 7.*Eds L. Khassanova, Ph. Collery, I. Maymard, Z. Khassanova, J.-C. Étienne.
John Libbey Eurotext, Paris © 2002 pp. 53-58.

Characterization of lactoferrin from dog neutrophils

M. N. Berlov[1], G. M. Aleshina[2], P. A. Lodygin[1], E. S. Korableva[1], Yu. V. Andreeva[3], V. N. Kokryakov[1,2]

[1]St-Petersburg State University, 7/9 Universitetskaya Emb., 199034 Saint-Petersburg, Russia; [2]Institute of Experimental Medecine, 12 Ac. Pavlova Street, 111111 St-Petersburg, Russia; [3]Pavlov Institute of Physiology, 6 Makarova Emb., 199034 Saint-Petersburg, Russia

ABSTRACT

Background. Lactoferrin (Lf) is a glycoprotein of mammals localized in milk and other biological fluids and in specifiic granules of neutrophils. Lf is characterized by ability to form complexes with some metal ions, mainly with Fe^{3+}. According to the literature data some species including dog lack Lf in milk.

Aims. The aims of this work were to isolate Lf from dog neutrophils and to investigate its Fe^{3+}-binding properties in comparison with human Lf.

Methods. The procedure of Lf isolation included extraction of proteins from neutrophis by cetyltrimetylammonium bromide, ion-exchange chromatography on carboxymetylcellulose and gel filtration on Sephadex G-150. Immunological crossreactivity studies were performed using sandwich ELISA method in a system designed for human Lf. Iron-holding properties of human and dog Lfs were investigated by dissociation of Lf-Fe^{3+} complex at different pH values in presence of chelating agents (phosphate, citrate) at different concentrations.

Results. Lf from dog neutrophils was isolated and purified to homogeneity. Molecular mass of dog Lf was determined as 80-83 kDa. Dog Lf was revealed to be a glycoprotein as well as human Lf. Dog and human Lfs exhibit little degree of immunological crossreactivity. Iron-holding properties of dog and human Lfs are similar.

Conclusions. Dog neutrophils contain Lf. Dog Lf is similar to human Lf by some characteristics including Fe^{3+}-holding capacity.

INTRODUCTION

Lf is a member of the transferrin family of iron-binding proteins. Lf has many different functions in organism. Among them are regulation of iron transport from maternal milk (Iyer, Lonnerdal, 1993), anti-oxidative effect (Gutteridge et al., 1981), antimicrobial activity (Bullen et al., 1978; Ellison et al., 1988), anti-inflammatory activity (Elass-Rochard et al., 1998), immunomodulating effects (Brock et al., 2000). Such functions as iron transport regulation, anti-oxidative activity and in part antimicrobial activity directly depends on the ability of Lf to bind Fe^{3+}. The molecule of Lf includes two iron-binding sites. Iron binding occurs simultaneously with binding of bicarbonate anion.

It is interested that acording to the literature some species including dog do not contain Lf in milk (Masson, Heremans, 1971). We have tried to isolate Lf from dog neutrophils and to compare it with human Lf by iron-binding properties.

MATERIALS AND METHODS

Lactoferrin preparations. Dog Lf was purified from exudate neutrophils using method including extraction of cationic proteins with CTAB (cetyltrimethylammonium bromide) from whole neutrophils, ion-exchange chromatography on CM-cellulose, gel filtration on Sephadex G-150. Extraction was performed with 0.3% CTAB solution in 0.02 M sodium acetate buffer, pH 4.5, using a glass-Teflon homogenizer. The homogenate was centrifuged for 40-60 min at 25,000 g. The extraction procedure was repeated several times. For both chromatographic steps eluting buffer was 0.02 M sodium acetate buffer, pH 4.5, containing NaCl at concentration 0.2 M for ion-exchange chromatography or 1 M for gel filtration.

Lf from human milk was kindly provided by Dr E. Tikhonova (Institute of experimental medicine, Russia).

Electrophoresis. The purity of Lf preparations was chekced by electrophoresis in acid buffer (Panyjm, Chalkley, 1969). In the case of carbohydrate determination gel after electrophoresis was stained by Schiff's reagent. Molecular mass of Lf was estimated by disk-electrophoresis in presence of sodium dodecylsulphate (SDS) (Schagger, Von Jagow, 1987).

Immunological study. Immunological vicinity of dog and human Lf was estimated by sandwich ELISA method in a system designed for human Lf. Polyclonal rabbit antibodies for human Lf were purified by ammonium sulphate precipitation and by affinity chromatography using Lf immobilized on agarose. Horseradish peroxidase was conjugated with antibodies by periodate method.

Iron-holding properties. Free of iron Lf (apo-Lf) was prepared by dialysis of Lf against solution containing 0.2 M sodium citrate, 0.2 M sodium phosphate, 0.15M NaCl, pH 3.0-3.1.

Lf saturated with iron (holo-Lf) was prepared by dialysis of apo-Lf against mixture of 0.1 M sodium citrate and 0.05 M sodium bicarbonate, pH 8.4-8.5 followed by dialysis against the same solution containing $FeCl_3$ in 10-20X excess for complete saturation of Lf. The final dialysis was performed against 0.15 M NaCl and then the sample was centrifuged for 5 min at 10,000 g to eliminate opacity of solution causing by sedimentation of part of Lf near its isoelectric point (8.7).

The iron-holding capacity of Lf was studied by two methods. In the first case holo-Lf was diluted at concentration 3-5 mg/ml in 3 mM sodium citrate buffer (final concentration) with different pH values (3.0-5.0) containing 0.15 M NaCl. Samples were incubated overnight at 23°C.

The second method was modified method of Mazurier & Spik allowing to observe release of Fe^{3+} from two iron-binding sites distinctly (Mazurier, Spik, 1980). In this case holo-Lf was diluted at concentration 3-5 mg/ml in a mixture of 0.05 M sodium citrate, 0.2 M sodium phosphate, 0.15 M NaCl (final concentrations are indicated for all components) with different pH values (3.0-8.5). Phosphate and citrate ions posses chelating properties and stimulate release of Fe^{3+} from Lf. Samples were incubated overnight at 23°C.

Iron content in Lf was determined in both cases by measuring of absorbance at 465 nm. Absorbance value of holo-Lf in appropriate dilution was taken for 100% and the absorbance value of apo-Lf was taken for 0%.

RESULTS

Lf from dog neutrophils was isolated and purified to homogeneity by our method *(fig. 1)*. Molecular mass of both dog and human Lf was determined as 80-83 kDa (electrophoregram is shown on *fig. 2*). Dog lactoferrin was revealed to be a glycoprotein as well as human lactoferrin.

Dog Lf exerts little degree of immunological crossreactivity with human Lf. More than 1000X increase in concentration is required for dog lactoferrin to generate the same signal as human Lf.

In both variants of experiments we failed to demonstrate any differences in iron-holding capacity between dog and human Lfs. In first case no release of Fe^{3+} from Lf is observed up to pH 4.0 Between 4.0 and 3.0 pH values about 80% of Fe^{3+} is dissociate from complex. These results

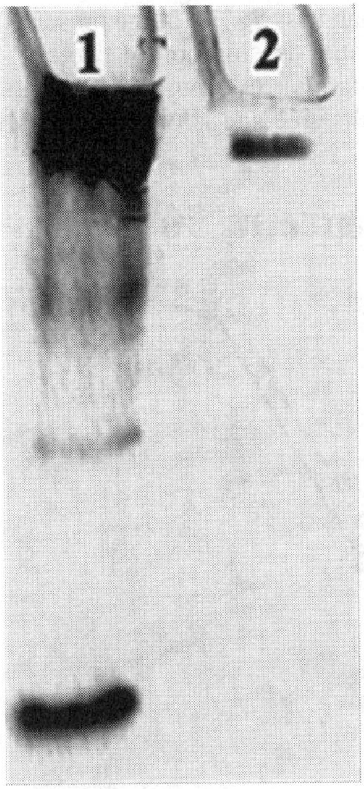

Fig. 1. Electrophoregram of purified dog Lf: 1) CTAB extract containing Lf; 2) Lf

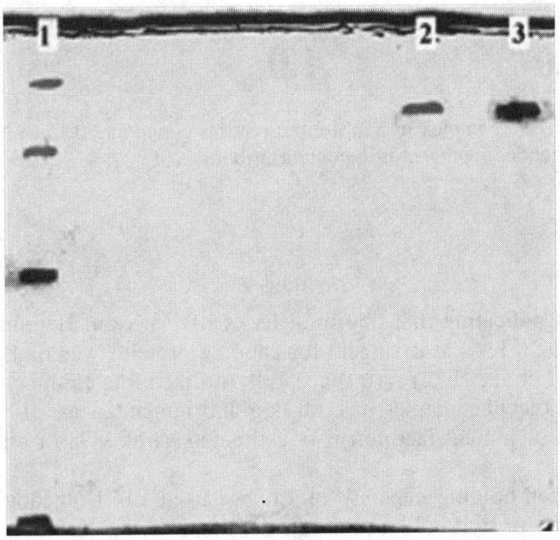

Fig. 2. Disk-electrophoresis of dog and human Lfs in the presence of SDS:
1) molecular mass standarts (from top to bottom): phosphorilase b (97 kDa), bovine serum albumine (66 kDa), ovalbumine (45 kDa), carbangidrase (30 kDa)
2) dog Lf
3) human Lf

are identical for both dog and human Lfs *(fig. 3)*. In the presence of high concentrations of chelating agents (citrate, phosphate) differential desaturation of two Fe^{3+}-binding sites is achieved. For both Lfs release of Fe^{3+} from the acid-labile site occurs at pH values more than 8.0 in our conditions. Dissociation of Fe^{3+} from the acid-stable site is observed at pH 4.0-5.0 *(fig. 4)*.

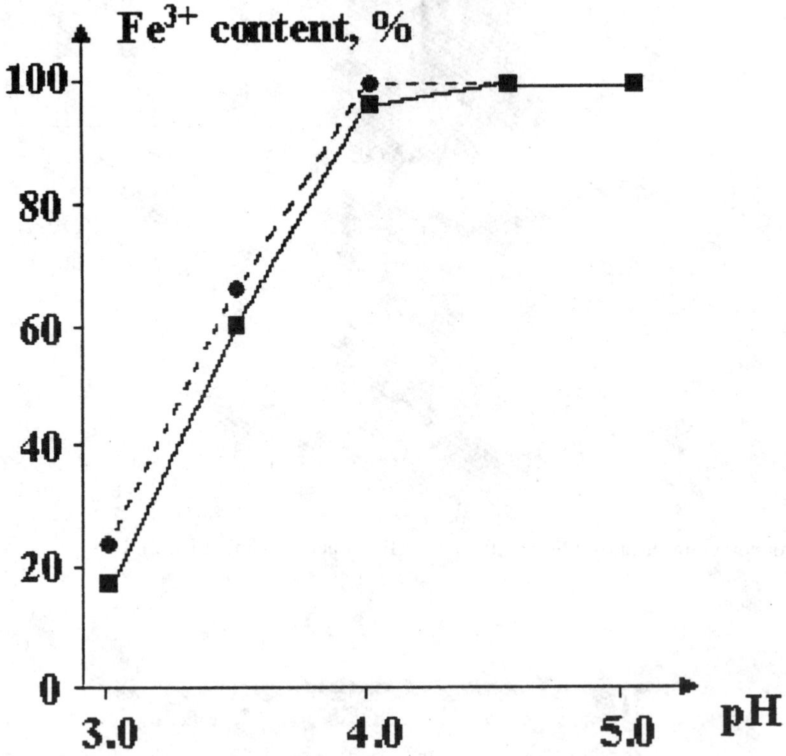

Fig. 3. Dissociation of Lf-Fe^{3+} complex in 3 mM citrate buffer containing 0.15 M NaCl. Balls represent iron content in dog Lf, and squaries represent iron content in human Lf

DISCUSSION

Despite of the data indicating that dog milk lacks Lf (Masson, Heremans, 1971) we isolated Lf from dog neutrophils. CTAB as extragent for cationic proteins was firstly used by Desser et al. for peroxidases (Desser et al., 1972). We succesfully utilized this compound for lactoferrin isolation. We determined molecular masses of both dog and human Lf as 80-83 kDa whereas in literature molecular mass of human lactoferrin is estimated as 80 kDa or slight lower (Rey et al., 1990).

In our studies of iron-holding capacity of Lfs we used Lfs from dog neutrophils and from human milk. The identity of iron-holdig properties of neutrophil and milk human Lfs was demonstrated earlier (Moguilevsky et al., 1985), thus our experiments were correct.

In context of possible clinical application of Lf as antibacterial or anti-oxidative agent comparative study of iron-binding properties of Lf from different species is important. It is known that some differences between Lfs from different species can exist: for example, bovine Lf releases iron more readily than human Lf (Baker et al., 2000). The other interesting fact is that Lf is extremly more effective in iron binding than serum transferrin despite of the identical structure of iron-binding sites. The absence of Lf in milk of dog promotes us to assume that dog Lf could exhibit some unusual properties. However we failed to demonstrate difference between dog and human Lfs in iron-holding capacity.

CONCLUSIONS

Whereas a little degree of immunological vicinity between dog and human Lfs is revealed, two proteins are similar by a number of physical and chemical paramaters including iron-holding capacity.

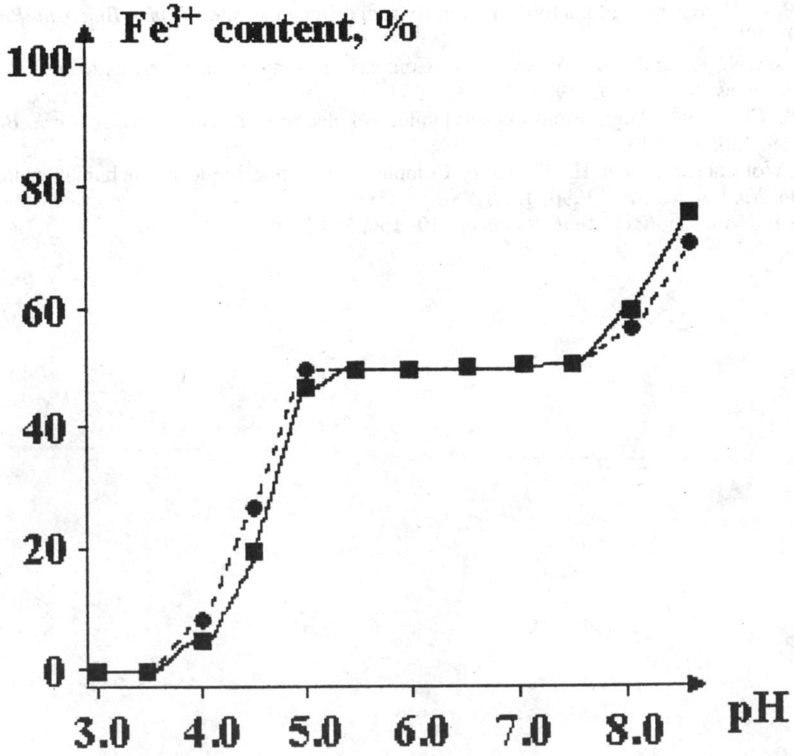

Fig. 4. Dissociation of Lf-Fe^{3+} complex in 0.05 M citrate - 0.2 M phosphate buffer containing 0.15 M NaCl. Balls represent iron content in dog Lf, and squaries represent iron content in human Lf

REFERENCES

1. Baker H. M., Anderson B. F., Kidd R. D., Shewry S. C., Baker E. N. Lactoferrin three-dimensional structure: a framework for interpreting function. *Lactoferrin: Structure, Fuction and Applications*, 2000, 3-15.
2. Brock J. H., Guillen C., Thompson C. Anti-inflammatory and immunoregulatory properties of lactoferrin. *Lactoferrin: Structure, Fuction and Applications*, 2000, 119-127.
3. Bullen J. J., Rogers H. J., Griffiths E. Role of iron in bacterial infection. *Curr. Top. Microbiol. Immunol.*, 1987, 80, 1-35.

4. Desser R. K., Himmelhoch S. R., Evans W. H., Januska M., Mage M., Shelton E. Guinea pig heterophil and eosinophil peroxidase. *Arch. Biochem. Biophys.*, 1972, 148, 452-465.
5. Elass-Rochard E., Legrand D., Salmon V., Roseanu A., Trif M., Tobias P. S., Mazurier J., Spik G. Lactoferrin inhibits the endotoxin interaction with CD14 by competition with the lipopolysaccharide-binding protein // Infect. Immun., 1998, vol. 66, p. 486-491.
6. Ellison R. T., Giehl T. J., LaForce F. M. Damage to the outer membrane of enteric Gram-negative bacteria by lactoferrin and transferrin. *Infect. Immun.*, 1988, 56, 2774-2781.
7. Gutteridge J. M. C., Paterson S. K., Segal A. W., Halliwell B. Inhibition of lipid peroxidation by the iron-binding protein lactoferrin. *Biochem. J.*, 1981, 199, 259-261.
8. Iyer S., Lonnerdal B. Lactoferrin, lactoferrin receptors and iron metabolism. *Eur. J. Clin. Nutr.*, 1993, 47, 232-241.
9. Mazurier J., Spik G. Comparative study of the iron-binding properties of human transferrins. I. Complete and sequential iron saturation and desaturation of the lactotransferrin. *Biochim. Biophys. Acta*, 1980, 629, 399-408.
10. Masson P. L., Heremans J. F. Lactoferrin in milk from different species. *Comp. Biochem. Physiol.*, 1971, 39B, 119-129.
11. Moguilevsky N., Retegui L. A., Masson P. L. Comparison of human lactoferrins from milk and neutrophilic leucocytes, *Biochem. J.*, 1985, 229, 353-359.
12. Panyim S., Chalkley R. High resolution acrylamide gel electrophoresis of histones. *Arch. Biochem. Biophys.*, 1969, 130, 337-346.
13. Rey M., Woloshuk S., Boer H., Pieper F. Complete nucleotide sequence of human mammary gland lactoferrin. *Nucl. Acids Res.*, 1990, 18, 5288.
14. Schagger H., Von Jagow G. *Anal. Biochem.*, 19, 166, 368-379.

*Metal Ions in Biology and Medicine: vol. 7.*Eds L. Khassanova, Ph. Collery, I. Maymard, Z. Khassanova, J.-C. Étienne.
John Libbey Eurotext, Paris © 2002 pp. 59-63.

Lipid oxidation and behavior are correlated in depleted uranium exposed mice

Wayne Briner & Daniel Davis

Department of Psychology University of Nebraska at Kearney, Kearney, NE, 68849 USA

ABSTRACT

DU exposure via drinking water produces behavioral changes in mice. DU exposure also produces increased lipid oxidation in the brains of mice. Lipid oxidation is correlated with behavior in DU exposed mice.

Key Words: depleted uranium, oxidation, lipid, brain, behavior.

INTRODUCTION

The status of depleted uranium (DU) as a toxic substance is under debate. It has been suggested that DU is approximately as toxic as lead [Doucet, 1994]. Work in some laboratories has shown indications DU accumulation in the CNS [Pellmar *et al.*, 1999a] and indications of neurotoxicity [Pellmar *et al.*, 1999b]. Previous work in this laboratory has shown that mice exposed to uranium during development differed from control animals on a variety of neurodevelopmental measures [Briner & Byrd, 2000]. One unresolved question is the mechanism by which uranium may exert its toxic effects. While there are probably other mechanisms, lipid peroxidation is common to many metals [Schaich, 1992]. Altered lipid oxidation may lead to a variety of changes in cell functioning, eventually leading to changes in behavior.

We have undertaken this study to determine if exposure to DU leads to behavioral changes in adult mice, and if this corresponds to lipid oxidation.

MATERIALS AND METHODS

Male adult Swiss-Webster mice were reared under standard laboratory conditions. The mice were exposed to DU acetate in drinking water at 4 dosage levels (0 (control, 19, 37, and 75 mg/L) for 2 weeks. After two weeks of exposure the animals were tested with a standard behavior assessment battery [O'Donoghue, 1996], which included an assessment of open-field behavior. After behavioral assessment the animals were killed and the brains removed. Lipid oxidation of brain tissue was measured using the thiobarbituric acid (TBA) assay (Ohkawa *et al*, 1976). Protein content of the sample will be determined using the commassie blue method and lipid oxidation expressed as a ratio to total protein content.

RESULTS

Mice exposed to DU did not demonstrate any decline in health that could be detected using the standardized toxicology battery for animals (data not presented). However, some behavioral

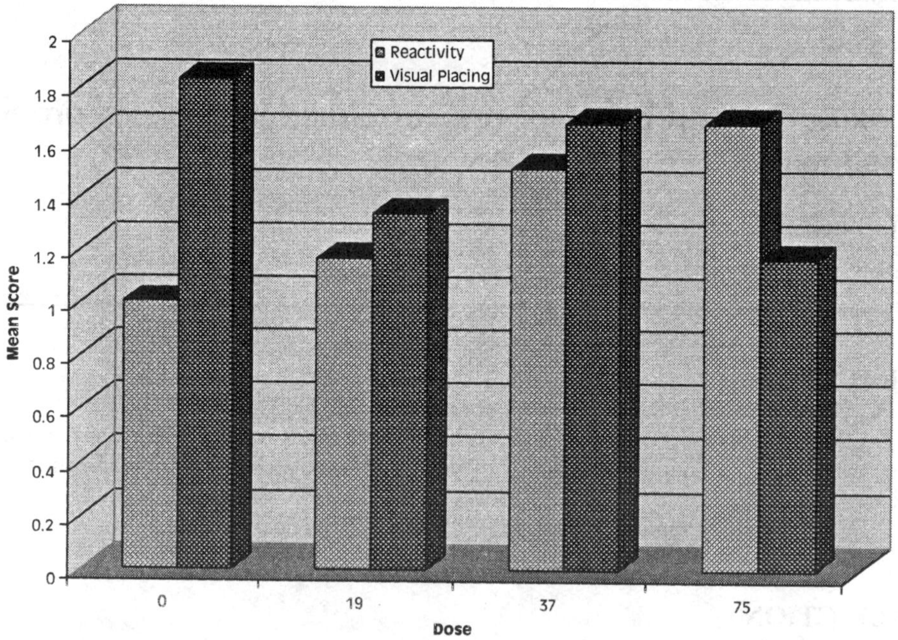

Fig. 1. Rearing

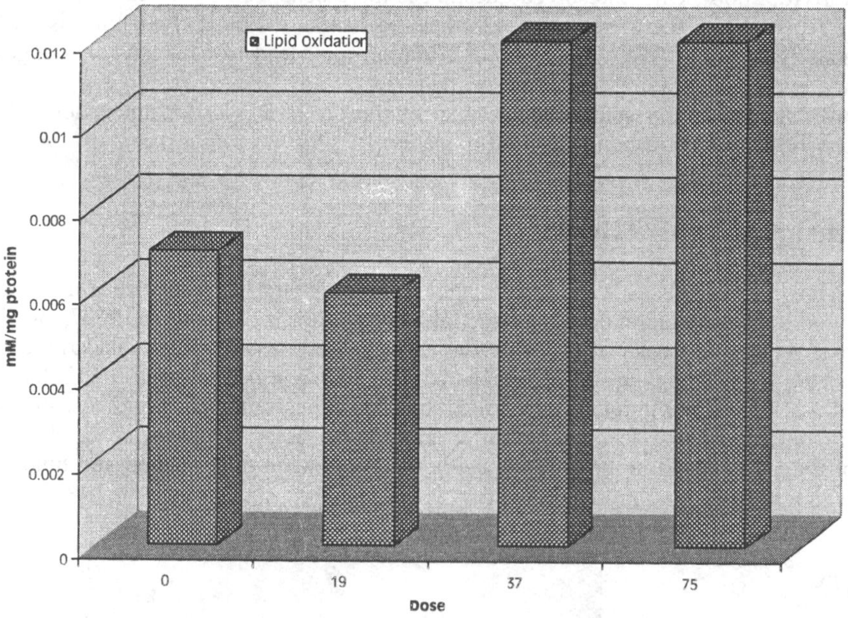

Fig. 2. Reactivity and Visual Placing

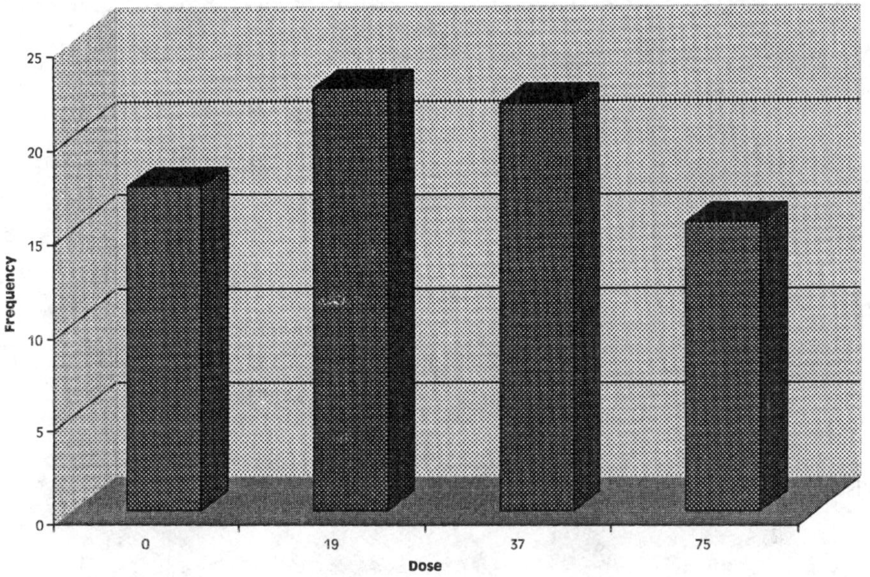

Fig. 3. Lipid Oxidation

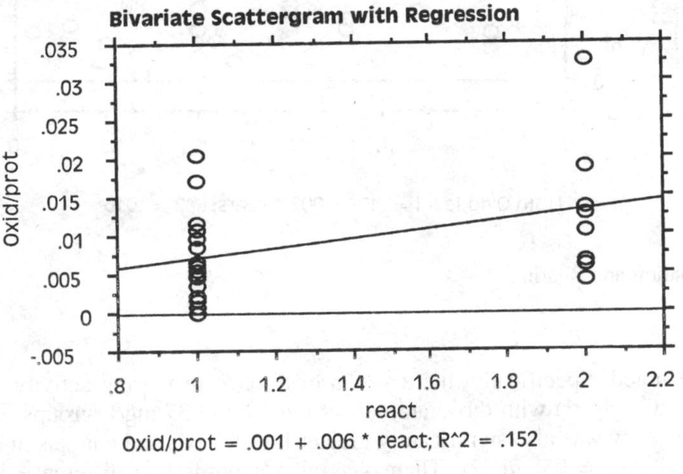

Fig. 4. Reactivity and Lipid Oxidation

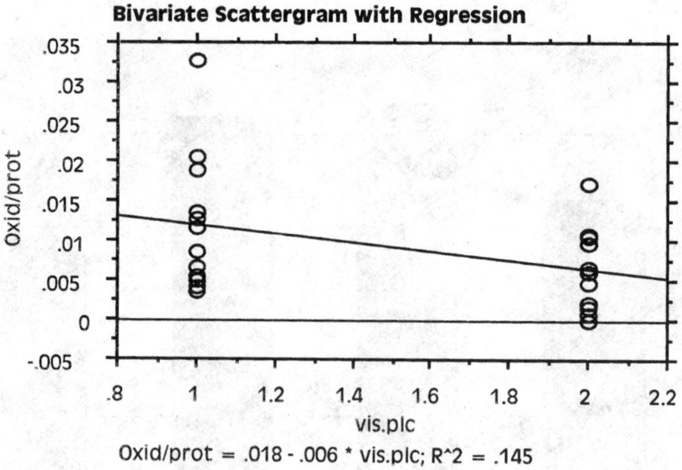

Fig. 5. Visual Placing and Lipid Oxidation

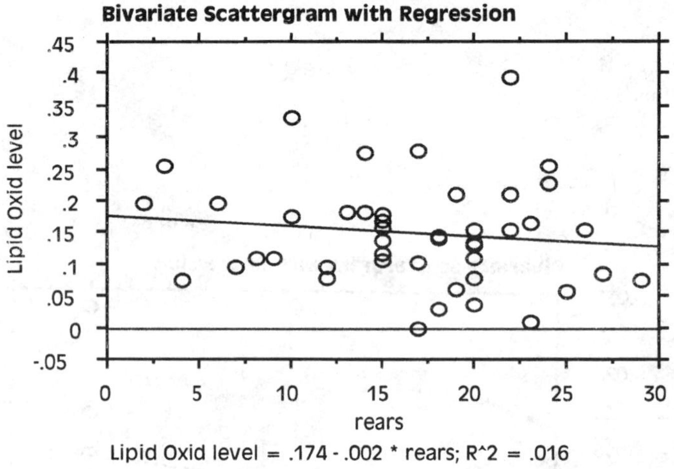

Fig. 6. Lipid Oxidation and Rearing

differences were noted. Specifically, there was a difference in rearing activity in the open field ($F(3,20)=3.33$, $p=.04$; *fig. 1*) with those animals at the 19 and 37 mg/L groups being more active than control. Reactivity was also significantly greater for DU exposed animals in a dose dependent fashion ($F(3,20)=3.03$, $p=.05$; *fig. 2*). There was also a borderline difference in visual placing behavior for the DU exposed animals ($X^2(6)=10.64$, $p=.10$; *fig. 2*).

Importantly, there was a borderline trend for lipid oxidation levels to be higher in the brains of DU animals, specifically at the 37 and 75mg/L dose ($r(22)=.26$, $p=.08$; *fig. 3*). Lipid oxidation is significantly correlated with the reactivity of the animal ($r(22)=.39$, $p<.05$, one tailed test; *fig. 4*) and the visual placing behavior ($r(22)=-.38$, $p<.05$ one-tailed; *fig. 5*). Rearing behavior showed a non-significant trend relationship with lipid oxidation levels *(fig. 6)*.

DISCUSSION

These data suggest that DU is absorbed through the GI system and that this type of exposure can produce behavioral changes in experimental animals. These behavioral effects are similar to those seen with exposure to other neurotoxic metals such as lead, mercury or cadmium. Exposure to neurotoxic metals typically produces changes in the animals' ability to modulate their response to environmental stimuli. This would be reflected in open-field activity, reactivity levels and visual placing responses.

The mechanism by which DU might produce neurotoxic changes is unclear. However, the general good health of the animals and the increase in lipid oxidation levels argues that the effects are mediated directly by exposure to DU, and not secondary to a systemic effect such as renal failure. The direct neurotoxic effect of DU on the CNS is further supported by the relationship between lipid oxidation levels and behavior.

CONCLUSIONS

These findings support the contention that DU has behavioral effects mediated by direct effects on the CNS. However, this is preliminary data, our laboratory will continue to increase sample sizes. Lipid oxidation is an indirect measure of the effects of DU and direct measurement of DU concentrations should be done to clearly draw a relationship between DU and its effects.

REFERENCES

1. Doucet, I. Desert Storm Syndrome: Sick soldiers and dead children. *Medicine and War*, 1994, 10, 183-194.
2. Briner W & Byrd K. Effect of depleted uranium on development of the mouse. In Centeno JA, Collery PH, Vernet G, Finkelman RB, Gibb H, Etienne JC (Eds). *Metal Ions in Biology and Medicine*, Paris, 2000, 6th, 459-461.
3. Pellmar TC, Fuciarelli AF, Ejnik JW, Hamilton M, Hogan J, Strocko S, Emond C, Mottaz HM, Landaur MR. Distribution of uranium in rats implanted with depleted uranium pellets. *Toxicological Sciences*, 1999a, 49, 29-39.
4. Pellmar TC, Keyser DO, Emery C, Hogan JB. Electrophysiological changes in hippocampal slices isolated from rats embedded with depleted uranium fragments. *Neurotoxicology*, 1999b, 20(5): 785-792.
5. O'Donoghue, JL. Clinical neurologic indicies of toxicity in animals. *Environ. Health Perspect.*, 1996, 104(suppl. 2), 323-330.
6. Ohkawa, H., Ohishi, N. & Tagi, K. Assay for lipid peroxides in animal tissues by thiobarbituric acid reaction. *Anal. Biochem.*, 1979, 95, 351-358.
7. Schaich, K. M. Metal and lipid oxidation. Contemporary issues. *Lipids*, 1992, 27, 209-218.

Metal Ions in Biology and Medicine: vol. 7. Eds L. Khassanova, Ph. Collery, I. Maymard, Z. Khassanova, J.-C. Étienne.
John Libbey Eurotext, Paris © 2002 pp. 64-67.

Metal complex composition in pharmacological modification of postradiation effects in hemopoiesis system

G.S. Grigorieva[1], N.E. Uzlenkova[2], N.F. Konakchovich[1], L.A. Frenkel[2]

[1]Institute of Pharmacology and Toxicology Academy Medical Sciences, 14 E. Potier Street, 03047 Kyiv, Ukraine; [2]Institute of Medical Radiology Academy Medical Sciences, 82 Pushkinskaya Street, 61034 Kharkiv, Ukraine

The unique role of metals in formation of hemopoiesis system is well known. But the significance of metal substances in positive tolerance promotion of oncology radiotherapy remains scantily explored. Nevertheless, we can expect the special success of the metal complexes as prospective drugs for pharmacological modification of post-radiation negative effects. It is connected with indispensable microelements functions in formation and protection of the body enzyme's pool and also with the estimated positive pharmacodynamics and low toxicity of metal complexes as compared with ion metal substances. The polymetal complex compositions are capable of optimal influence on the important homeostasis links. In the first instance such composition can reduce the risk of postradiation complications of hemopoiesis system.

The propose of present investigation is the corroboration of this concept at the analysis of hemopoietic competence of the original microelement composition (??). ?? contains d-biometals in a state of coordination compounds with amine carbonic acid (HL) and other essential metals - such as salts of oxygen acids. In the MC composition the microelement ratio is (mg/1 g): Fe: Zn: Mn: Cu: Co: Cr: Mo: Se: V: HL = 12.8: 15.2: 3.6: 3.3: 0.32: 0.30: 0.55: 0.22: 0.04: 366.0.

We proposed the route of CM creation by the principle new method of "summary synthesis". This method assumes the next reaction scheme:

Rapid contact

Phase 1: $\{(Fe^{3+}, Zn^{2+}, Mn^{2+}, Cu^{2+}, Co^{2+}, Cr^{3+})\ (\text{anti-ion})_n\} + nHL =$

$= \{(Fe^{3+}, Zn^{2+}, Mn^{2+}, Cu^{2+}, Co^{2+}, Cr^{3+})L_n + nH(\text{anti-ion})$

↑ → **MC**

Phase 2: Mo, V, Se
(oxygen acid salts)

Physico-chemical analysis excludes any interaction in MC between the individual ML_n complexes. At the same time synergism or summation of ingredient toxicity aren't stipulated by the MC creation: LD_{50} of MC is 2820 mg/kg.

The CM properties to the pharmacological modification of the hemopoiesis postradiation effects were investigated on rats at total X-ray irradiation in dose 6.2 Gy. In the radiobiological model selected the character of the MC protective action on pathogenic development of molecular-structuric, metabolic and morphologic changes of the blood system and the connective tissue was estimated.

It was established that the total body irradiation provokes appropriated alterations of the hemopoiesis system. At the first time, for irradiated animals the specific developments of leukopenia (from 3-th to 30-th days), anemia and trombocytopenia (from 7th to 14th days) are registered *(Pictures 1 and 2).*

The MC introduction (25 mg/kg/day; intra-abdominal route) stipulates the positive modification of the side postradiation blood effects *(Picture 1)*. The essential inhibition of dynamic of the leucopenia takes place. Simultaneously the anemia and trombocytopenia are absent in any periods of testing

(Picture 2). From the 7th day after irradiation the stimulation of marrow proliferative processes is observed and at 30th day - practically the total regeneration of the bone marrow cell pool.

The X-irradiation also significantly changes the biophysical data of the state of the structure and functions of the erythrocyte cell membranes: the indices of membrane steadiness and electrostability are diminished appreciably.

So the clamp current and tension are increased by 5-15%, the erythrocyte hemolysis time is decline, but the rate of hemolysis is accelerated. The ratio of membrane conductance before and after erythrocyte clamp is changes abnormally.

The MC introduction ensures the steadiness and electrostability of the erythrocyte membrane of irradiated animals integrally *(table)*.

Table. The MC effect on pharmacological modification of erythrocyte biophysic indices at total X-ray irradiation (6,2 Gy) ($x \pm S_x$)

Experiment mode	*Indices*				
	Clamp current, I (mk?)	**Cytoplasm resistance,** R (?m)	**Clamp tension,** U(mV)	**Relative conductance,** cond. unit	**Hemolysis time,** %
Biocontrol	363,4±6,0	2,84±0,25	0,69±0,02	4,4±0,4	100
7th day:					
- irradiation	336,1±3,6*	2,27±0,62	0,67±0,4	6,0±0,4*	50*
- irradiation + MC	358,8±4,3**	2,48±0,54	0,69±0,02	3,9±0,4	90**
14th day:					
- irradiation	343±4,8*	2,05±0,22*	0,65±0,02	5,4±0,5	77*
- irradiation + MC	353±8,7	2,56±0,44	0,68±0,01	3,5±0,4**	100**
30th day:					
- irradiation	340,6±7,6*	2,27±0,27	0,63±0,02	4,6±0,6	
- irradiation + MC	371,5±6,1**	2,70±0,27**	0,69±0,01*	3,5±0,5**	

* - is reliable to control, $p < 0,05$.
** - is reliable to MC, $p < 0,05$.

The action of X-ray irradiation also is accompanied by the growth of concentration of connective tissue metabolites in the blood. The content of total glykosamineglykans (GAG) increased doubly at the time from 3th to 30th days after irradiation, the rate of sulfuric GAG - increased from 170% to 280%. In early periods after irradiation the blood concentration of sialic acids as the components of the body “acute reaction” increased by 98%.

After the MC introduction the decrease of all connective tissue metabolite content is established in blood already at 3-th day after irradiation.

The hemocompetence MC effects at the X-radiation lesion are confirmed by morphological and histological investigations.

Analysis of data obtained gives evidence the mechanism of the modify blood MC action may be concerns with the stimulation of the body total unspecific protective reaction to lesion X-ray factors.

Thus, the results of experiments realized substantiate the next conclusions:

1. The original microelement composition synthesized as potential preparation possesses essential hemostimulative effect, inhibits the development of leucopenia of radioactive ethiology and assures the re-generation of the marrow cell pool at 7th day after irradiation.

2. The MC normalizes the biophysic characteristics and electrosta-bility of the erythrocyte membranes in irradiated animals.

3. The MC possesses the positive influence at metabolic processes: the content of main connective tissue metabolites in blood after X-irradiation is decreased.

Results obtained are not the self-depentent significance for positive pharmacocorrection of X-ray therapy, but also confirm the reality of metal base drugs future for medical practice.

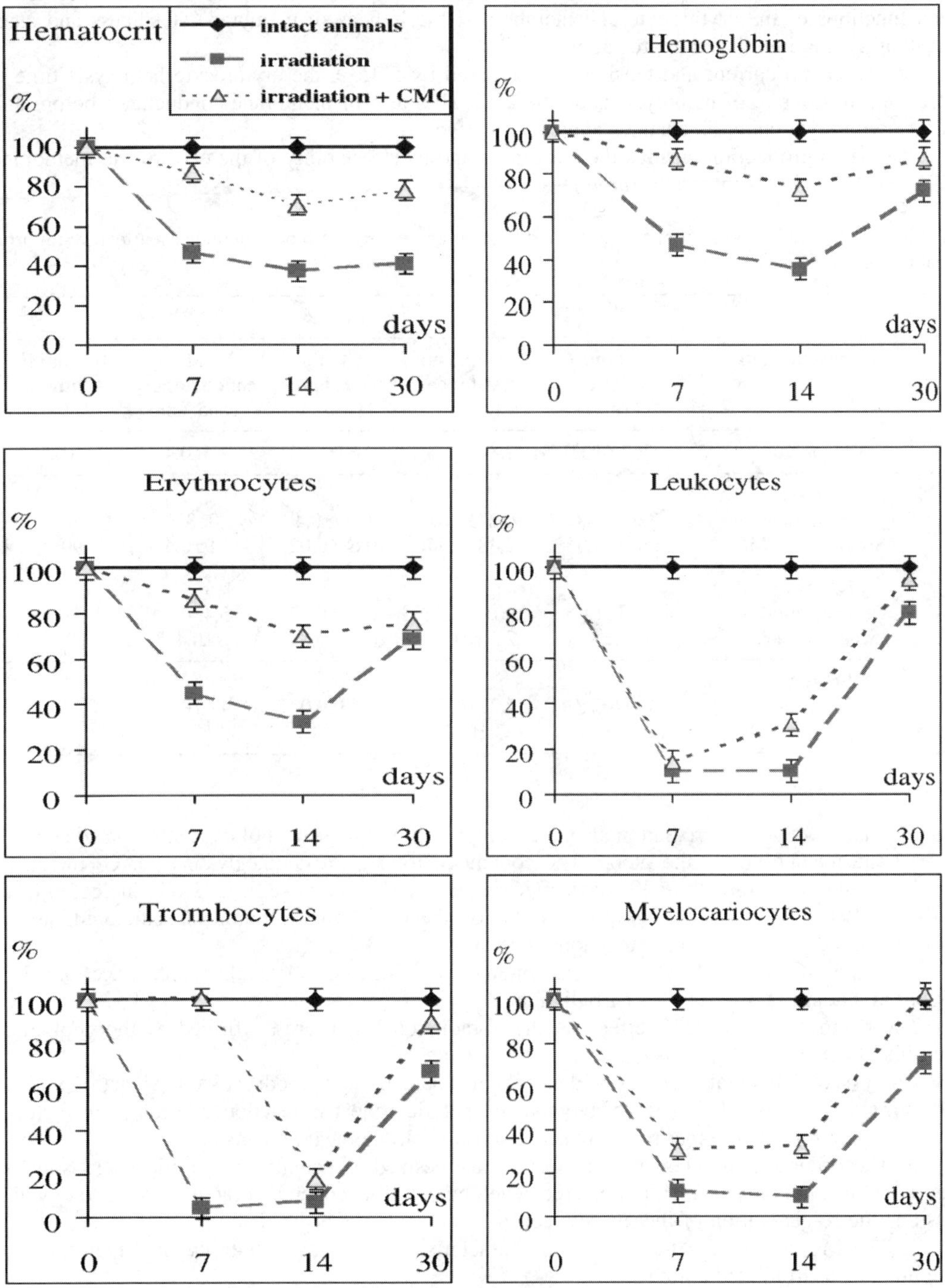

Picture 1. The MC influence on the dynamic of blood indices at total X-ray irradiation (6,2 Gy)

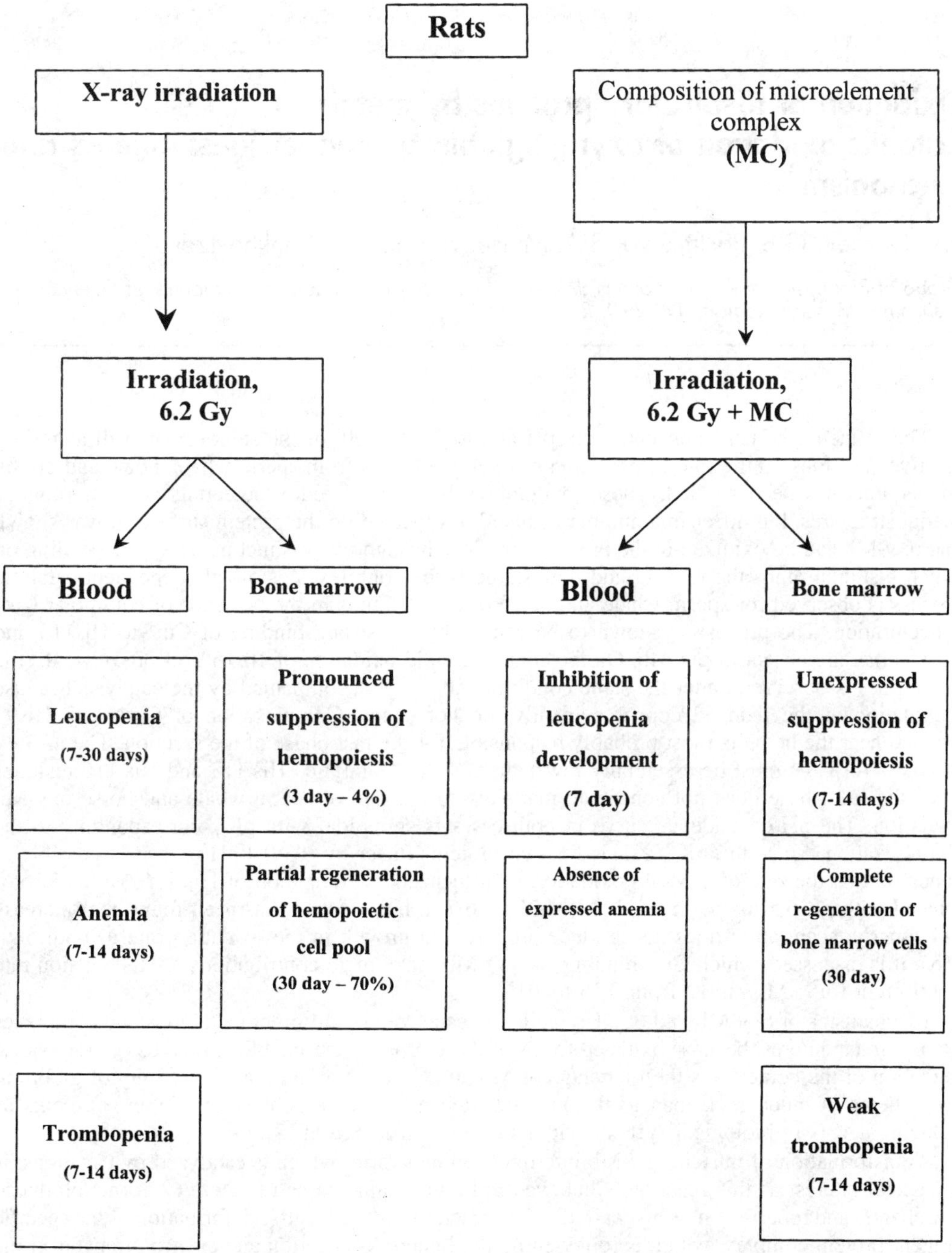

Picture 2. General scheme of MC influence on the alteration of hemopoiesis system at X-ray irradiation

*Metal Ions in Biology and Medicine: vol. 7.*Eds L. Khassanova, Ph. Collery, I. Maymard, Z. Khassanova, J.-C. Étienne.
John Libbey Eurotext, Paris © 2002 pp. 68-72.

Oxidation of respiratory proteins by metals. Catalitic oxidation of oxymyoglobin by copper ions: kinetics and mechanism

E.V. Goraev, G.B. Postnikova, S.A. Moiseeva and E.A. Shekhovtzova

Group of Biophysics of Redox Proteins, Institute of Cell Biophysics, Russian Academy of Sciences, Pushchino, Moscow Region, 142290, Russia.

The influence of Cu^{2+} concentration, pH and ionic strength of the solution, as well as redox-inactive Zn^{2+} ions on the rate of oxidation of oxymyoglobins from sperm whale, horse and pig by copper ions has been studied. These myoglobins have equal redox potentials and homologous spatial structures, but differ in a number of histidines located on the protein surface. It was shown that oxy-Mb can be oxidized in the presence of Cu^{2+} through two distinct pathways, depending on which histidine binds the reagent and how stable is the complex. A slow pH-dependent catalytic process is observed for sperm whale and horse oxymyoglobins in the presence of equimolar Cu^{2+} concentration. The process is shown to be caused by the strong binding of Cu^{2+} to His113 and His116 that are absent in pig Mb. On the contrary, rapid oxidation of 10-15% of pig oxy-Mb (the fast phase) is observed under the same conditions. It is not accompanied by the catalysis because reoxidation of the reduced copper evidently did not occur. Complexation of Cu^{2+} with His97 situated near the heme is most probably responsible for the fast phase of the reaction. The affinity of His97 for Cu^{2+} must be essentially lower than of the “catalytic” His 113 and His 116 residues since the fast phase does not contribute markedly to the rate of sperm whale and horse oxy-Mb oxidation. The pH-dependence curve in both cases is sigmoidal with pK_{eff} corresponding to the His ionization. Since the pK_{eff} values for two proteins differ by 0.3-0.4 pH unit, it is possible to conclude that the rate of oxy-Mb oxidation is influenced by ionization of His 116, which is protonated with pK 6.5 in sperm whale Mb and pK 6.7 in horse Mb. It has been found that increase in copper concentration does not produce proportional growth in the oxidation rate of both oxy-Mbs. It is discussed, which Cu^{2+} binding sites of Mb make main contributions to His reaction rate at different Cu^{2+}: Mb ratios, from 0.25 to 10.

Mechanisms of oxy-Mb and oxy-Hb oxidation catalyzed by different metal ions and complexes invoke greate interest because oxidated forms of these proteins are unable to bind oxygen. Besides, oxidation of the heme Fe is the first stage of Mb and Hb denaturation, as the affinity of globin to ferric heme is much less, than to ferrous one. Recently, the possible role of these proteins as reducing agents in biological systems had also been considered [1, 2].

Transformation of muscle oxy-Mb into oxidized met-form, which is catalyzed by Cu^{2+} ions, is of special interest as this reaction is believed to be the main saurce of active O_2 forms produced at ischemia and reperfusion of myocard [3]. The reaction proceeds through formation of the specific reagent-protein complex, which is followed by the ligand dissociation and electron transfer in the complex. Reduced copper Cu^{1+} is reoxidized then to Cu^{2+} by oxygen with participation protons of the medium, providing the closure of the catalytic cycle. Catalysis is not observed under anaerobic conditions.

The detailed physical mechanism of oxy-Mb oxidation by copper compounds is not clear now. One of the reasons is that sperm whale Mb containing maximal number of histidines was the sole object of its examination. Sperm whale Mb can bind up to six Cu^{2+} ions with different affinities.

From the equilibrium dialysis and NMR data, three binding sites have apparently the highest affinity for Cu^{2+}, those at His113, His116 and His48 (binding constants correspond to 10^5 - 10^6 M^{-1}).

In the present work, kinetics of oxidation of three oxy-Mbs, from sperm whale, horse and pig, by Cu^{2+} has been studied. These myoglobins have identical redox potentials and homologous spatial structures, but differ by amount of His residues located on their surface. In comparison to sperm whale Mb, horse Mb lacks His12 that is changed to Gln, while in pig Mb, three histidines, His12, His113, and His116 are replaced by Gln. The influence of pH, ionic strength, and Cu^{2+} and Zn^{2+} concentration on the rate of oxidation of all three oxymyoglobins is investigated.

MATERIALS AND METHODS

Sperm whale myoglobin (fraction IV) is isolated from skeletal muscles and purified, as described earlier [4]. Oxy-Mb has been obtained under aerobic conditions by reduction of met-Mb with sodium dithionite and separating the last by gel filtration through a Sephadex G-25 column. The MbO_2 concentration was determined spectrophotometrically, using extinction coefficients (M^{-1}??$^{-1}$) equal to 13600 at 543 nm and 14200 at 581 nm.

A complex of MbO_2 with Zn^{2+} is obtained by mixing of the protein solution with concentrated $ZnCl_2$ solution [5]. All experiments with zinc complexes of Mb (the $[Zn^{2+}]/[MbO_2]$ ratios are 2, 5, 10, and 20) were carried out in a Tris-malate buffer (1:1), in which, unlike in phosphate buffer, unsoluble salts of zinc are not formed.

The rate of oxymyoglobin oxidation has been studied spectrophotometrically on changes in absorption at 581 or 543 nm, using a biradial Specord UV-VIS spectrophotometer (Germany) with thermostated cuvette holder (at 20°C). For necessary $[Cu^{2+}]/[MbO_2]$ ratios, 10 µl of a concentrated $CuCl_2$ solution were added in a reaction mixture by the HAMILTON microsyringe.

The kinetic curves were monitored in a time interval, in which the reaction amplitude varies by 15-20%, covering usually its initial linear part. The reaction rate was characterized by an initial rate (v_0).

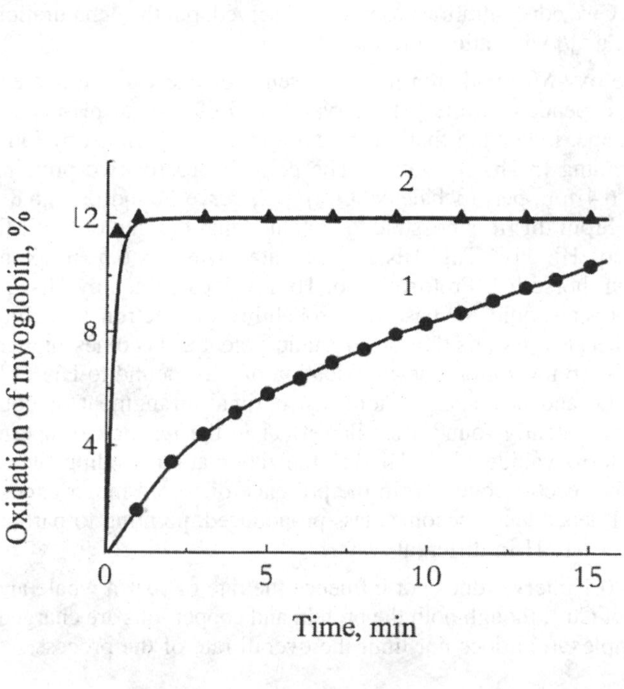

Fig. 1. Kinetics of oxidation of sperm whale oxy-Mb (curve 1) and pig oxy-Mb (curve 2) in the presence of one equivalent of $CuCl_2$. 0.01 M Tris-malate buffer (1:1), pH 7,5, 20°C. Initial MbO_2 concentration is 9.10^{-5} M.

RESULTS AND DISCUSSION

The character of the sperm whale and horse oxymyoglobin oxidation by copper, on one hand, and the pig oxy-Mb, on the other, is strongly different. The first two proteins are completely oxidized to met-Mb at equimolar Cu^{2+} concentration, a slow process is beeng observed in both cases (*fig. 1*, curve 1). At the same time, a fast, less than 1 min, oxidation of 10-15% from all amount of pig oxy-Mb is observed (fast phase), but its further transformation to met-Mb does not occur (*fig. 1*, curve 2). That is, in sperm whale and horse MbO_2, no fast phase of the protein oxidation is practically observed, while in pig MbO_2, a slow phase of the reaction is absent.

The results obtained suggest that oxy-Mb oxidation in the presence of Cu^{2+} can proceed by various pathways, depending on which histidine is complexed with the reagent and what is stability of the complex. The slow process is obviously explained by strong binding of Cu^{2+} to His113 and His116 identical for sperm whale and horse myoglobins, but absent in pig Mb. This process should prevail at small concentrations of Cu^{2+} when these very binding sites are saturated in the Mb structure (catalysis). The catalytic process is slow because it is most probably limited by the rate of Cu^{1+} reoxidation and dissociation of bound Cu^{2+} from myoglobin.

On the contrary, the fast oxidation of 10-15% of pig oxy-Mb (the fast phase) is not accompanied by the catalysis since reoxidation of the reduced copper in this case does not occur. Out of four histidines, His48, His81, His97, and His119, common for myoglobins studied, the Cu^{2+} binding to His97 closest to the heme (0.62 nm from it) is most probably responsible for the fast phase. The affinity of His97 to Cu^{2+} must be much less than His48, His81, His119, and the "catalytic" His 113 and His116 residues as only ~10-15% of pig protein is oxidized at equimolar Cu^{2+} concentration and, besides, no appreciable contribution of the fast phase to kinetics of the sperm whale and horse oxy-Mb is observed, though His97 is present.

Histidine 97 in Mb and Hb plays an important role in stabilization of the heme position. Therefore, its complexation with copper should result in a changed local conformation and stability of the heme cavity, thus increasing availability of intrinsic histidines, first of all of distal His64, to copper and resulting in turn in the protein denaturation. In our experiments, no denaturation of sperm whale and horse oxy-Mbs at the Cu^{2+} concentrations used was observed, but the denaturation of pig oxy-Mb did take place at the $[Cu^{2+}]/[Mb]$ ratio more than 1.

The rate of sperm whale and horse oxy-Mb oxidation in the presence of one Cu^{2+} equivalent at low ionic strength ($I = 0.01$) is pH dependent in the pH interval 5-8 (*fig. 2*). Both pH-dependencies have pronounced sigmoidal shape, indicating that the reaction rate is influenced by ionization of a group with pK_{eff} corresponding to His ionization. The pK_{eff} values for two proteins differ by 0.3-0.4 pH unit, being 6.3 - 6.4 for sperm whale MbO_2 (*fig. 2*, curve 1) and 6.7 - 6.8 - for MbO_2 from horse (*fig. 2*, curve 2). From this it is possible to conclude that the rate of oxy-Mb oxidation is influenced by ionization of His 116. This His is protonated with pK 6.5 in sperm whale Mb and with pK 6.7 in Mb from horse [6]. Protonation of His116 located nearby His113 which has the greatest affinity for copper should increase the probability of electron tunneling from the heme to the given Mb site because its positive electrostatic potential becomes higher. Besides, it is very probable that the His116 ionization rises reoxidation of Cu^{1+} bound to His 113, which proceeds with participation of O_2 and protons, providing an optimal arrangment of both reagents just like in enzymes. We have recently found a similar effect in the reaction of sperm whale oxy-Mb oxidation catalyzed by ferrocyanide where His 119 ionization at the binding site of anion $[Fe(CN)_6]^{4-}$ strongly increased the reaction rate [7]. In the presence of 5-fold molar excess of Cu^{2+}, the sigmoidal shape of the pH-dependence becomes less pronounced, pointing to participation of some other process dependent on pH in different way.

Increasing ionic strength in the 0 - 0.1 interval does not influence the rate of sperm whale and horse MbO_2 oxidation in the presence of Cu^{2+}, though both the protein and copper ions are charged. This implies that the Cu^{2+}- MbO_2 complexation does not limit the overall rate of the process.

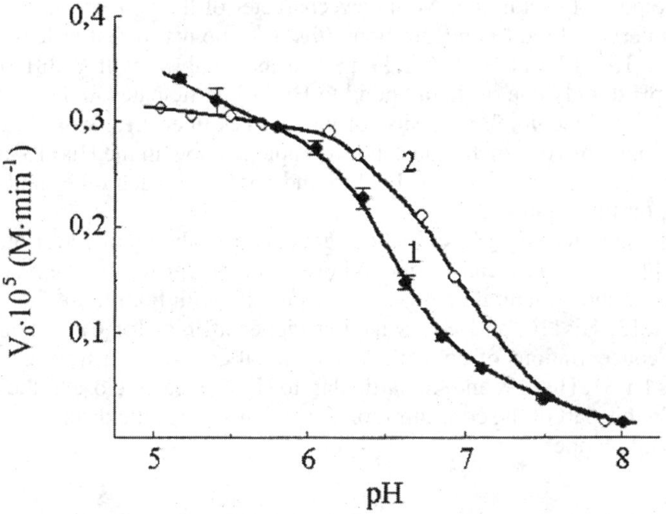

Fig. 2. The pH dependence of the rate of sperm whale MbO_2 (curve 1) and horse MbO_2 (curve 2) oxidation in the presence of one equivalent $CuCl_2$. 0.01 M Tris-malate buffer (1:1), 20ºC. Initial MbO_2 concentration is $2.25.10^{-5}$ M.

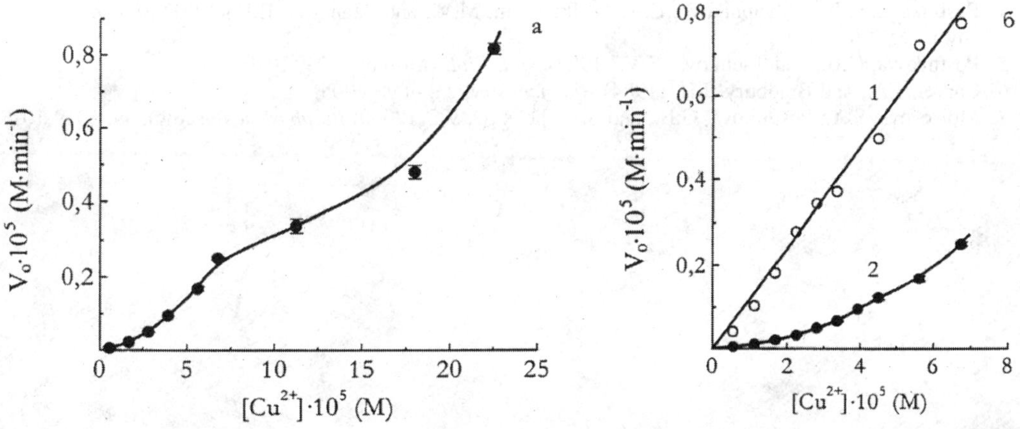

Fig. 3. a) The dependence of the sperm whale MbO_2 oxidation rate on Cu^{2+} concentration 0.01 M Tris-malate buffer (1:1), pH 7.5, 20ºC. MbO_2 concentration is $2.25.10^{-5}$ M.

b) The Cu^{2+} concentration dependence for sperm whale MbO_2 in the initial part of the curve, for $[Cu^{2+}]/[Mb]$ ratios from 0 to 3, at pH 6 (Curve 1) and at pH 7.5 (Curve 2). 0.01 M Tris-malate buffer (1:1), 20ºC.

The dependencie of the MbO_2 oxidation rate on $[Cu^{2+}]$ at pH 6 and 7.5 are complex and show no additivity in the increasing the rate with Cu^{2+} concentration *(fig. 3, a)*. The addition of redox inactive Zn^{2+} ions in various $[Zn^{2+}]/[MbO_2]$ ratios to the reaction mixture at different Cu^{2+} concentrations has no appreciable effect on the MbO_2 oxidation rate. At the first part of the concentration curve, up to $[Cu^{2+}]/[MbO_2]$ ratio equal to 3:1, the oxidation rate linearly grows with Cu^{2+} concentration at pH 6, while at pH 7.5, the dependence noticeably deviates from a linear one *(fig. 3, b)*.

Well-justified assumptions can be made about what Cu^{2+} binding sites give the main contributions to the reaction rate at variing the $[Cu^{2+}]$ concentrations. On the first part of the concentration curve, from 0 to 5 of the $[Cu^{2+}]/[Mb]$ ratios, His113, His116, and His48 have the greatest affinity for copper and are possibly saturated. Nonlinear character of the concentration dependence curve at pH 7.5 in this interval of Cu^{2+} concentrations *(fig. 3, b)* points to that at least two copper complexes, evidently at His113 and His116, take part in the catalysis with a different efficiency. At the same time, at pH 6, only one of them bound to His113 participates in the reaction. Protonation of His116 at pH < 7 and Coulomb repulsion of the charges interferes with binding of the second Cu^{2+} to His116, so that only one metal ion can be complexed now in the His113 and His116 region. It is in agreement with close positions of His113 and His116 to each other and with influence of His116 ionization on the catalysis.

The sharp increase in the oxy-Mb oxidation rate occurs in the second part of the concentration curve at the $[Cu^{2+}]/[Mb]$ ratio >8 can be due to both an involvement into the reaction of additional Cu^{2+} binding centers, and structural changes in the protein, with both these factors acting interdependently. As His113, His116, and His48 having higher affinity for Cu^{2+} should be completely saturated at large concentrations of copper, the additional contribution from Cu^{2+} bound to "non-catalytic" residues His81, His119, and, in particular, to His97 located close to the heme is probable. It is possible that in this part of the concentration dependence, even the distal His64 can be involved in complexation with copper.

REFERENCES

1. Augustin, M.A., and Yandell, J.K. (1979) *Inorg. Chim. Acta*, 37, 11-18.
2. Hegetschweiler, K., Saltman, P., Dalvit, C., and Wright, P.E. (1987) *Biochim. Biophys. Acta*, 912, 384-397.
3. Gunther M.R., Sampath V., and Caughey W.S. (1999) *Free Radic. Biol. Med.*, 26, 1388-1395.
4. Postnikova, G.B., Shlyapnikova, E.A., Volkenstein, M.V., and Atanasov, B.P. (1981) *Mol. Biol.* (Moscow), 15, 526-537.
5. Postnikova, G.B., and Tselikova, S.V. (1987) *Mol. Biol.* (Moscow), 21, 1040-1049.
6. Carver, J.A., and Bradbury, J.H. (1984) *Biochemistry*, 23, 4890-4905.
7. Moiseyeva, S.A., Postnikova, G.B., and Sivozhelezov, V.S. (2000) *Biophysics*(Moscow), 45, 1019-1028.

*Metal Ions in Biology and Medicine: vol. 7.*Eds L. Khassanova, Ph. Collery, I. Maymard, Z. Khassanova, J.-C. Étienne.
John Libbey Eurotext, Paris © 2002 pp. 73-74.

The role of metal ions in the chemistry of life

A.T. Bikkulova, G.M. Ishmuratova

Dumskaya ul., 3
Saint Petersburg, Russia 191011

"Life is a form of existence of albumen.
And the main property of this form
is metabolism with the environment"
F. Engels

Biosphere is the part of the Earth in which life exists. The famous Russian scientists V.I. Vernadsky, A.P. Vinogradov and others made a great contribution in the development of that science. V.I. Vernadsky said that life was a mutual, persistent flow of chemical elements between organisms and environment.

Nowadays more than 80 elements of the Periodical system by D.I. Mendeleev have been found out to play role in the organisms of people and animals.

Most of these elements are in the structure of ferments, vitamins, hormones and participate in biochemical and physiological processes, regulate colloid state of albumen in cells, osmotic pressure, catalyze processes of oxidation-reduction, etc.

So the study of the role of metal ions in the chemistry of life, medicine, biology, etc. is very actual.

The work "Bioelementology" (256 p.) is the result of the study and systematisation of the information of the role of metal ions in the chemistry of life which was conducted during many years.

Bioelementology explains the role of these ions, the role of their simple and complex compounds in the life of organisms, the influence of their abundance and lack in the biosphere on the health. It also studies the concentration of these elements in food-stuffs and how to correct your diet to fulfil their lack in the organism. Bioelementology also studies toxic influence on the health by the abundance of organic and inorganic compounds including metal ions, which can take place as a result of development in technology and industry. "Bioelementology" can be successfully used as a text-book for many specialists - chemists, biologists, physicians, technologists and others. It has already been confirmed by the fact that the book is being used by students and teachers. And among the decisions of a number of international Congresses there are some about the improvement of the education of specialists: chemists, biologists, technologists, physicians and so on.

Special position among inorganic toxic components have heavy metals. It's been found out that their toxic factor is more than those of other poisons and is about 150 balls, while for the carbon monoxide (CO) it is 10 balls and for some poisons - 30 balls, etc. The pollution of the biosphere with microcontents of toxic and highly toxic metals (Cu, Mn, Zn, Fe, Cr^{+6}, V, Hg, Cd, etc.) influence the health of population dramatically, that's why even very low concentration (10^{-4} - 10^{-9}%) in the biosphere is dangerous.

Decisions of above-mentioned congresses said that the task of the protection of the biosphere and health of organisms is not someone's else, but our common task.

The Academician V.I. Vernadsky pointed out that "... the protection of the biosphere for the present and future generations is in the hands of the humanity".

However, while the abundance of metal ions is toxic, some microcontent of them is considerably necessary for processes of life. They've have been called bioelements - metals of life. Such metals are divided into two groups according to their quantitative content in the organisms:

1) Microelements which content is about or less than hundredth parts of a percent. They are Fe^{+2} (in the hemoglobin); Cu^{+2}, Zn^{+2}, etc. (in the ferments), Co^{+2} (vitamin B_{12}) and others.

2) Macroelements which content is more than mentioned above. They are Ca^{+2} (2,5%), Mg^{+2} (0,07%), K^{+} (0,11%), Na^{+} (0,10%). Some people also mention in this group Fe^{+2}, Fe^{+3} (0,05%).

It should be mentioned that in the organism there are some microelements which concentration is less than 10^{-4}: Zi - $5*10^{-5}$; Be - $5*10^{-5}$; Ga - $1*10^{-5}$; Cs - $5*10^{-5}$; W $1*10^{-6}$; Au - $1*10^{-8}$; Hg - $5*10^{-6}$, etc. but their biological role hasn't been studied much. It is impossible to reproduce the whole content of a book "Bioelementology" in this article.

Here are some examples of biological role of some macroelements. Cations of Na^{+} and K^{+} are very important in the work of our heart and nerve system. The correlation between their concentration is very sufficient ($[Na^{+}]: [K^{+}] = 2:1$). Some vegetable products, for example rice, have such a correlation. Cations of K^{+} participate in the albumen and carbohydrate exchange, ferment processes, etc. The lack of cations of K^{+} provides a delay in the growth of an organism, the abundance of cations of K^{+} increases the weight.

More than 90% extracellular cations of metals in the organism are cations of Na^{+}. And K^{+} is typical intracellular element. Our blood contains 175 mg % of Na^{+}, 191 mg % of K^{+} and 5 mg % of Ca^{+}. In the process of metabolism the abundance of K^{+} or Na^{+} cations are removed by our organism with the help of our kidneys in a form of phosphoric, chamois and organic acids (urinary, milk acid). Sodium-vapor salt of urinary acid is low-soluble so it is often postponed in gristles and causes a disease - gout.

All these examples illustrate that the knowledge of the role of bioactive metal cations in a human organism and in biosphere has a great importance.

*Metal Ions in Biology and Medicine: vol. 7.*Eds L. Khassanova, Ph. Collery, I. Maymard, Z. Khassanova, J.-C. Étienne.
John Libbey Eurotext, Paris © 2002 pp. 75-81.

The formation of metals complexes and interaction with model cellular membranes

Kylyvnyk K.E.

Ukrainian State University of Chemical Technology, 8 Gagarin Ave, Dnipropetrovsk, Ukraine, 49005

ABSTRACT

The determination of correlation between the physical and chemical properties of coordination compounds of metals and their physiological activity remains the most important task for purposeful search of new medical products on the basis of metals complexes. Therefore the investigation of the interaction of metals complexes with model and cellular membranes with simultaneous analysis of structure is important for determination of interrelation between physiological activity and molecular structure of substances.

The influence of metals complexes structure on electrochemical properties of model membranes (bilayer lipid membranes BLM and monolayers) was investigated. The composition and share of each complex particle had been calculated with the help of step complexation constants, which were determined by the pH-metry, potentiometry and spectrometry of the concrete physiological conditions or used values from literary. For BLM - the influence of complexes on the electroconductivity, the membrane potential, the elastic properties were determined and for monolayers - the change of surface tension at complexes adding. In all cases the particles influencing on properties of membranes maximal were determined with the help of distributive diagrams.

INTRODUCTION

The medical properties of metals are known since the treatises of Avicenna and Hippocrates.

In the 60th years of the twentieth century the attempts to systematize a role of metals for living organisms and the assumptions concerning the role of complex coordination compounds in biosystems have appeared. Until recently the complexation has been used basically for a withdrawal of toxic metals excess from an organism. For this aim the appropriate non-toxic ligand (detoxicant) has been selected that chelate bonding with toxic metal in a complex and then it was removed from an organism.

The discovery of antitumor activity at platinum complexes, anti-inflammatory - at complexes of cuprum and gold, antianemic - at complexes of iron and zinc [1] has strengthened the interest of scientists to metals complexes, as to possible medical products [2]. It have accelerated the synthesis of new classes of coordination compounds of metals with various ligands including biological and to regular search of medicines on theirs basis [3].

The finding of correlation between physical and chemical properties of the obtained substances and their physiological activity remains the most important task for purposeful search of new medical products. The big variety of metal complexes and various forms of their existence in real physiological conditions make difficult the solving of this task.

In many cases the membrane-active forms and the mechanisms of functioning of carriers of metals ions are not determined. The adsorptive ability and the influence of complexes on phase transitions in membranes have been investigated insufficiently.

The plasmatic membrane is the first barrier on the way of substance passage into a living cell. The influence of the complexes on physical and chemical properties of the plasmatic membrane of a cell predetermines the possible further biological activity. It is proven fact [4] biological activity of metals complexes depends on the mechanism of their interaction with cellular membranes.

The lipids among which prevail phospholipids form the structural basis of membranes. Therefore the model lipid membranes (BLM and monolayers on interface "liquid-air") are convenient for investigation of interaction mechanisms of heavy metal ions with biological cellular membranes.

MATERIALS AND METHODS

For formation of model membranes we used the individual lipids (Ukraine) and the general fraction of phospholipids from bull brain.

Bilayer lipid membranes (BLM) were formed on an aperture with diameter of 1,1 mm in the teflon partition dividing two water solutions of electrolyte by Mueller-Rudin method [7].

The next methods were used:

- The method of current-voltage curve for the determination of specific conductivity and capacity of a membrane when modifying agent was added;
- The method of second harmonic (current minimum) at the investigation of adsorption of charge particles on the membrane surface;
- The method of third harmonic at studying change of the elastic properties of a membranes and other;
- The Langmuir monolayer technique (measurement of surface pressure and surface potential) was used for study of the influencing of metal complexes on surface properties of monolayers from membrane lipids.

The complex compound with the investigated structure undergoes changes at adding in the physiological solution. The changes degree depends on the electrolytic dissociation, hydrolysis, partial replacement of ligands etc. If the others ligands are present in the solution the mixed complexes will have formed and under certain conditions the full replacement of ligands is possible (depending on effective stability constants). Basically the ligands and the structure of complex determine which properties of membranes will change (electroconductivity, elasticity, membrane potential).

For determination of complex particle interacting with a membrane it is necessary to know the forms of complex ions existence at the given physiological conditions and the share of each form. The ion shares were found by the step constants of complexes equilibrium (Kn) taking into account the dissociation constants of protonate ligand forms [8]. The pH-metry, potentiometry and spectrometry were used for determination of Kn in physiological solution or used values from literary [9].

The mathematical approach [10] and values of step constants have allowed to determine the share of each ion depending on pH, the concentration of a complex and ligand for real conditions. For instance on the *fig. 1* the distribution diagram for Ni - ethylenediamine (En) complex is presented.

The share of all ions in solution containing Ni and different concentrations of En at pH=8 was calculated taking into account the formation both hydro- and En- complexes.

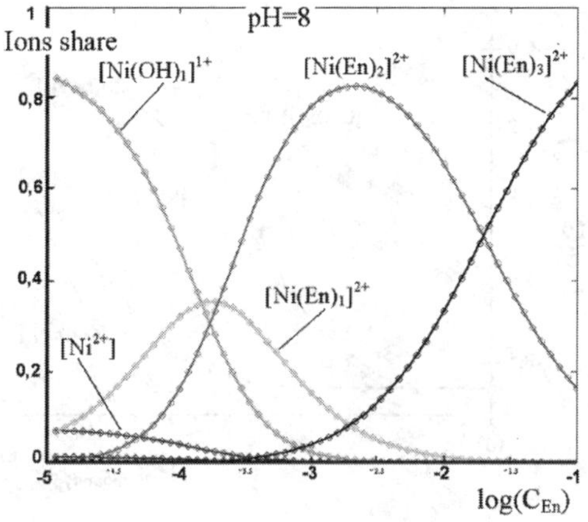

Fig. 1. The influence of En concentration on a share of complex ions in a solution. pH=8

RESULTS AND DISCUSSION

Aqua-complexes of metals are adsorbed on a surface of model membranes due to interaction with phosphate, carbonate, sulphide and hydrosulphide groups and change membrane potential, are reduce elasticity of BLM unsignificant (about 20%) and change a boundary potential difference of monolayers. In neutral and especial in alkaline solutions the influence on BLM is considerably reduced due to formation of hydroxide precipitates or hydroxo-complexes.

The influence on properties of model membranes increase at presence of ligands, forming a complex and varying hydrophilic-lipophilic balance of ions.

The investigation of ethylenediamine (En) and dipyridyl (bpy) complexes of Cu, Ni and Zn, cysteine complex Cr^{3+} and quinolinolato complex Ga^{3+} has shown, that in most cases the maximal interaction with model membranes is observed in the prevalence field of the coordinated saturated complexes with organic ligands.

1. The change of BLM conductivity under influence ethylenediamine, dipyridyl complexes of copper, nickel and zinc

The investigation of BLM conductivity change by method of volt-ampere curves (VAC) have shown, that complexes differently increase the electroconductivity depending on pH and concentration of ligand or metal ions.

However the analysis of VAC with taking into account the share of each ion (distributive diagrams) has shown the change of BLM conductivity depends *(fig. 2)* on concentration saturated ethylenediamine, dipyridyl complexes $[Me(bpy)_3]^{2+}$ and $[Me(En)_3]^{2+}$.

In the saturated complex ion the metal completely is surrounded by ligands and the conductivity a share of these ions in a solution determines change. So in the case of dipyridyl complexes with copper and zinc, the interaction with BLM does not depend by nature of central atom. And the difference in interaction of ethylenediamine and dipyridyl complexes is concerned with significant hydrophobicity of dipyridyl in comparison with ethylenediamine.

2. The influence of the cysteine complex of Cr^{3+} on elastic properties of BLM

The cysteine (H_2L) complex does not change BLM conductivity. But the complex influences on the membrane potential and elastic properties of BLM as evidenced by the current minimum method of second harmonic [11]. The dependences analysis of the amplitude of the second har-

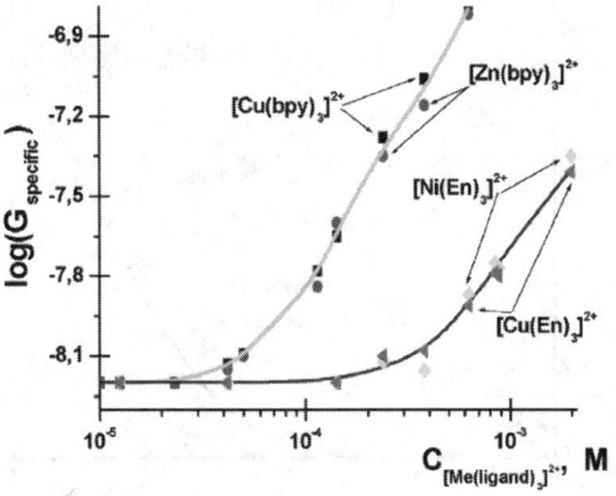

Fig. 2. The dependence of BLM conductivity on the saturated complexes concentration

monic current after addition of the complex $[Cr(H_2L)_6]^{3+}$ has shown, that the complex is quickly adsorbed on a membrane, changing the surface potential and then there is slower process resulting in decreasing of electrostrictive factor α. It testifies to weakening of elastic properties of lipid membrane. At the same time the membrane potential depends on the concentration of all kinds of positively charged ions ($[Cr(H_2L)_6]^{3+}$, $[Cr(H_2L)_n]^{3+}$, Cr^{3+}, $[Cr(OH)_1]^{2+}$...), and the electrostrictive factor depends only from saturated cysteine complex.

The maximal influence on elastic properties of BLM was observed in conditions where saturated cysteine complex ions of Cr^{3+} were prevailed in solution.

3. Influence of tris 8-quinolinolato complex of Ga^{3+} on surface tension of phosphatidylcholine monolayer

The interaction of quinolinolato complex of gallium with phospholipid has been investigated on the change of a surface tension of monolayer by Langmuir monolayer technique. The addition of tris 8-quinolinato gallium (III) (KP46) in the subphase (pH>3) on which surface the phosphatidylcholine (PC) monolayer was formed is resulted in smooth reduction of the monolayer area, which had identical surface tension *(fig. 3)*. It testifies about the reduction of surface activity of monolayer and the lipids leave a surface.

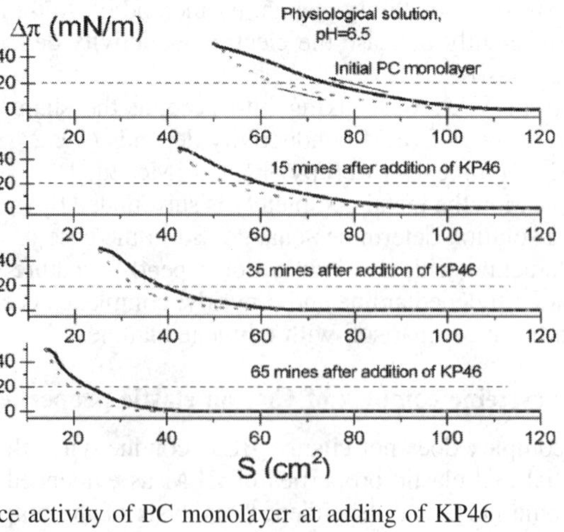

Fig. 3. The change of surface activity of PC monolayer at adding of KP46

If subphase pH was less than 2 the adding Ga^{3+} ions din't change the monolayer activity and the adding KP46 increased the area of the monolayer. The embedding of the complex in the monolayer and the increasing of the monolayer area was depending on concentration of the saturated complex. The pH increasing resulted to decomposition of complex and to formation of gallium hydroxide precipitate. The occurrence of new liquid-solid interface promoted redistribution of lipids from the monolayer and to decreasing of the monolayer area.

In all previous cases, the presence of organic ligands completely surrounding the central atom prevents hydrolysis and changes hydrophilic-lipophilic balance. The complex interaction with model membranes of cells is increasing.

However in some cases the inorganic ligands can influence on the properties of lipid membranes:

a) In case of the tetrameric complex formation containing double connected hydroxide groups ($[Zr_4(OH)_{12}(H_2O)_{12}]^{4+}$ and similar) the change of membrane conductivity is observed at presence of pH difference in solutions washing BLM.

b) The change of BLM conductivity in solutions containing Hg^{2+} depends on concentration of ions $HgHal^+$ (where Hal^- = Cl^-, Br^-, I^-). The permeability coefficients for the given particles have allowed making the assumption about permeability dependence on change of bound polarity in the row $HgCl^+$ - HgI^+.

4) Hydroxide ions as ligands and influence on properties of lipid membranes

The behaviour of Zr^{4+} ions in aqueous solution is variety and characterised by hydrolysis and polymerisation. The literature concerning study of this process is extensive. The majority of the researchers have found out that in acid solution the major complex is the tetrameric $[Zr_4(OH)_8(H_2O)_{16}]^{8+}$ containing double connected hydroxide groups. These complex ions are changing at pH increasing. It is connected with the reduction of polymerization at Zr-O-Zr bonds formation and the decreasing of ion charge at the replacement H_2O ligands on OH^-.

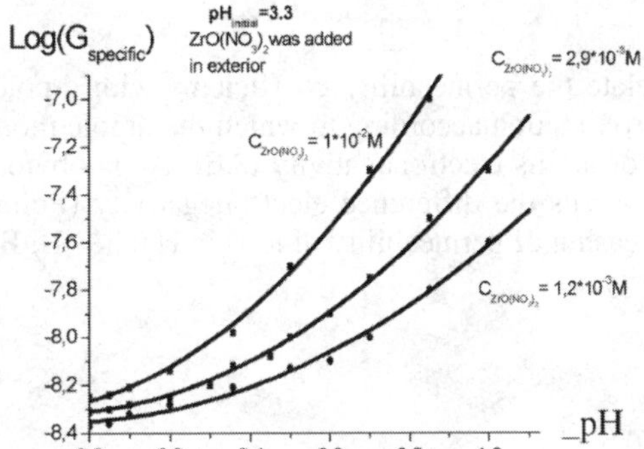

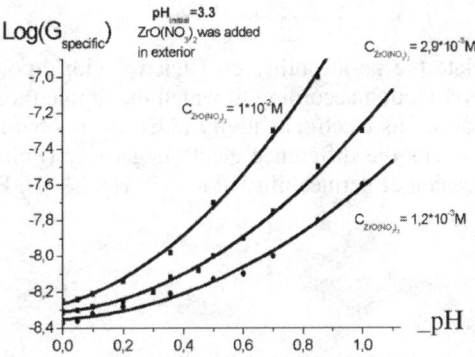

Fig. 4. Change of specific BLM conductivity on ΔpH and adding different quantities of $ZrO(NO_3)_2$

At adding $ZrO(NO_3)_2$ in neutral and alkaline solutions, which washing a membrane, the influence on electroconductivity and membrane potential is not detected. However at addition in acid solutions (pH<4) the change of the membrane conductivity is observed. This change depends on zirconium concentration and the conductivity are increasing greatly if the difference of solutions pH was in both side of membrane *(fig. 4)*. The increasing of ΔpH results in growth of LM conductivity more than on the order.

The experiments with various zirconil concentrations, different pH and the account of the present ions have allowed determining that the charged particles accumulate in a membrane. If there is a difference of solution pH in the both side of a membrane these particles increased membranes conductivity and acted as H^+ carriers. In this case the polymeric hydroxo-complexes are building into lipid membrane and the chain mechanism of protons transfer is formed *(fig. 5)*.

5) Halogenide ions, as ligands and influence on physicochemical properties of membranes

If the solution not containing halogenide ions, then Hg^{2+} ions are only adsorbing on BLM surface and changing the membrane potential. In the presence of halogenide ions (Hal^-) in a solution $HgHal^+$ appear which penetrate through a membrane, changing conductivity more than on 2 order. The calculation of $HgHal^+$ ions concentration with step constants of complexation has shown a direct connection with change of BLM conductivity and has allowed to determine the permeability coefficients: $?_{HgI+}$=7.9*10^{-6}; P_{HgBr+}=4.2*10^{-9}; P_{HgCl+}=1.5*10^{-10} cm/s.

With increasing of Hal^- ions concentration the neutral $HgHal_2$ molecules not changing conductivity are forming and then complex ions $[HgHal_3]^-$ and $[HgHal_4]^{2-}$ not passing through BLM. In result the influence on BLM conductivity is decreasing. When the $HgHal^+$ ions were removed from solution the membrane conductivity revert to the initial state.

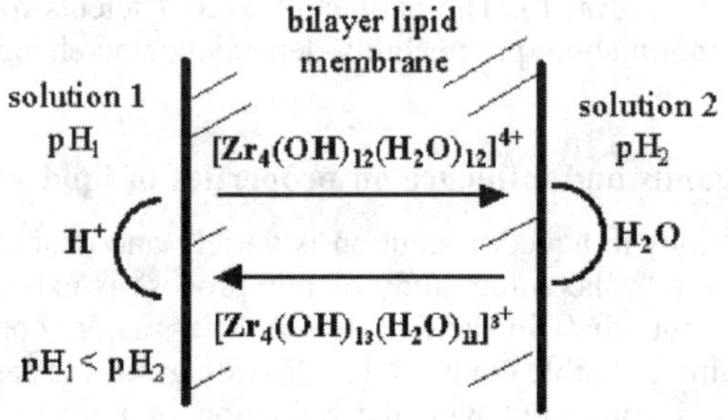

Fig. 5. The possible mechanism of H^+ carry through BLM

We have tried to correlate the permeability coefficients with dipole moment of $HgHal^+$. For this we have taken the approximation according to which the dipole moment for diatomic covalent bond equal the difference of atoms electronegativity (ΔE). As is obviously from *fig. 6* the points on dependence $lg(P_{HgHal+})$ versus the difference electronegativity (Poling) are on one line and it enables to assume that increasing of permeability of ions in $HgCl^+$ - $HgBr^+$ - HgI^+ row is connected with change of bond polarity.

CONCLUSIONS

Complexation, especially in the case when it is followed by saturation in bio-ligand, increases significantly the interaction with plasmatic membranes of cells and results in more various changes of their physical and chemical properties.

Probably, in actual biological systems a complex transport form "metal ion-lipophylic ligand-hydrophilic substrate" is formed that is of the principal significance during a transport through cell membranes.

REFERENCES

1. Poldinbger W., 1982, Editions "Roche", Basel, p. 114.
2. Rosenberg B., VanCamp L., Trosko J.E., 1969, Platinum compounds: a new class of potent antitumor agents. *Nature* 222, 385-386.

3. Bakhtian Ray, Ochiai Ei-Ichiro. Pharmacological application of inorganic complexes. *General Pharmacology* 32 (1999) 525-540.
4. Kriss E.E., Volchenkova I. et al. Coordination substances of metals in medicine. Kyiv: 1986.
5. Tien H.Ti, Ottova A.L. The lipid bilayer concept and its experimental realization: from soap bubbles, kitchens sink, to bilayer lipid membranes. *J. of Membrane Science* 4886 (2001) 1-35.
6. Langmuir I., Monomolecular layers of fatty acids at air-water interface, *J. Am. Chem. Soc.* 39 (1917) 1848.
7. Mueller P., Rudin D.O., Tien H.T., Wescott W.C. Reconstitution of cell membrane structure in vitro and its transformation into an excitable system, *J. Phys. Chem.*67 (1963) 534.
8. Rossotti F, Rossotti X, Determination of stability constant and other equilibrium constants in solution. M. Mir, 1965, 564 p.
9. Sillen L.G., Martel A.E. Stability constants of metal ion complexes. London. Chem. Soc. Burlington House, 1964, p. 754.
10. Butler I.N. Ionic equilibrium (A mathematical approach) Reading, Massachusetts 1964, p. 448.
11. Bovykin B.A., Kylyvnyk K.E., Zegzhda A.D. Influence of chromium (III) ions and the chromium-cysteine complex on bilayer lipid membranes. *Biophysics.* Vol. 44, N. 3, 1999, pp. 454-457.

*Metal Ions in Biology and Medicine: vol. 7.*Eds L. Khassanova, Ph. Collery, I. Maymard, Z. Khassanova, J.-C. Étienne.
John Libbey Eurotext, Paris © 2002 p. 82.

Interaction of metal ions and nanoparticles of silver with different biological active flavonoids

M.V. Kochetova[2], A.A. Revina[1], O.G. Tataurova[2], T.K. Lutsik[1], E.B. Haylova[1], O.G. Larionov[2]

[1]*A.N. Frumkin Institute of Electrochemistry of Russian Academy of Sciences, Moscow, Russia*
[2]*Institute of Physical Chemistry of Russian Academy of Sciences, Moscow, Russia*

ABSTRACT

It is currently topical to gain insight into the influence of metal ions on mechanism of major biological processes with participation of natural antioxidants-flavonoids (Fl).

In our work we used Fl, such as quercetin (Qu), rutin (Ru) and morin (Mo). It is well-known antioxidant activity of these substances is dependent on the presence of metal ions. So, it's to be interested to investigate of the structural effect of studied flavonoids on the stabilization of different complexes Fl and metal ions in solutions.

It's very important to note, that biological systems environment, consisting of membranes, cells etc. have very high level of organisation, in which metal ions could be present as polymetallic centers or clasters. That is why the reverse micellar solutions were choosen by us as a closed model for investigation of mechanisms of intramolecular interactions.

Our task is to investigate complexes of flavonoids with silver ions in molecular solutions and in micellar ones. Earlier we found the formation of silver nanoparticles in micellar system of Qu/ AOT/ iso-octane after adding of water solution of silver ions (Ag^+). These nanoparticles are very stable and they have intensive absorption in visible range. We obtained the similar nanoparticles in reverse micellas containing Ag^+ by radiation chemical method in which solvated electrons are as reducing agent.

In order to investigate the interaction of ions and nanoparticles of silver with flavonoids in dependence of their structures we used the methods of UV-VIS spectroscopy, RP-HPLC and TLC.

In all cases we observed the formation of complexes of silver ions with flavonoids in water-alcoholic solutions ($Ag^+ + Fl \rightarrow [Ag^+ ... Fl]$) and the formation of silver clasters ($Ag^+ Fl \rightarrow Ag_n^{m+}$) in reverse micellar solutions. Also we registered of complexes of silver clasters with flavonoids ($Ag_n^{m+} + Fl \leftrightarrow [Ag_n^{m+} ... Fl]$). The latter fact may be confirmed by the decreasing of intensity of initial optical adsorption band of Ag_n^{m+} ($\lambda \sim 420$ nm) and the appearance of new band with $\lambda \sim 295$ nm.

In the near future our investigations will be concerned with a study of the antioxidant activity of complexes of flavonoids with silver ions and nanoparticles.

ACKNOWLEDGEMENTS

This work was partially supported by RFFR N° 01-03-32783.

Metal Ions in Biology and Medicine: vol. 7. Eds L. Khassanova, Ph. Collery, I. Maymard, Z. Khassanova, J.-C. Étienne.
John Libbey Eurotext, Paris © 2002 pp. 83-88.

Lead-induced changes of physical state of human erythrocyte membrane? lipids *in vitro*

O.B. Oleksiuk, V.S. Finin*, E. I. Slobozhanina

*Institute of Photobiology of the National Academy of Sciences of Belarus, ul. Academicheskaya 27, Minsk, 220072, Belarus, sei@biobel.bas-net.by; *Belarussion State University, pr. F. Skorini 4, Minsk, 220050, Belarus*

ABSTRACT

We investigated effect of lead *in vitro* on physical state of human erythrocyte membrane lipids by lipophilic fluorescent probe 1,6-diphenyl-1,3,5-hexatriene (DPH) and spin labels 16-doxyl-stearat (16DS), iminoxyl-palmitinic acid (IP). It was shown that the effect of 2-10 mM lead acetate resulted to the increase of both DPH fluorescence intensity and DPH fluorescence polarization. It was supposed that microviscosity of lipid bilayer of erythrocyte membranes changes. From EPR - spectra parameters of 16DS and IP incorporated in erythrocyte membranes the effect of 2-10 mM lead acetate caused the increase of heterogeneity of lipid bilayer at the surface and deep inner membrane layers.

INTRODUCTION

Now lead is considered as one of the most dangerous poisons [1, 2]. Getting in an organism lead accumulates mainly in erythrocytes and then moves from a blood to other organs [3, 4]. In 1990 it was shown by Sugawara et al that lead added in blood plasma quickly incorporates in suspended erythrocytes at 37°C. The rate of lead incorporation in cells was 35 mcg (0,17 mM) 10 cells at hour. It was comparable to the value of the rate for isolated erythrocyte membranes. There are reasons to suppose that the incorporation of lead in erythrocyte is energetically dependent passive transport and Pb^{2+}-cytoplasmic labile component participates at the distribution of these ions [6-8]. In 1980 Ong and Lee shown that about 14% of total intracellular lead binds to a erythrocyte membrane [9]. Later it was demonstrated that at the incubation of erythrocytes in a blood plasma with 10 mg/dl of lead at 37°C (for 24 hours) 98% intracellular lead is in cytoplasma and only 2% is in a membrane fraction [5].

Studying Pb^{2+} - bound proteins in vivo it was found three polypeptides with Mm 240, 45 and < 10 ?D? accordingly [10]. The investigation of blood of the workers contacted with lead at the industry (the concentration of lead in blood on the average made 270 (from 97 up to 950 mcg/l)) allowed to establish that polypeptide with Mm 240 kDa contained 35-81% of lead bound protein, which had strong affinity to dehydrotase of δ-aminolevulinic acid (ALAD), 45 kDa polypeptide bound 12-26% of lead in blood and < 10 kDa polypeptide less - 1%, whereas the free lead was 2-45% [11]. It was shown by Xie Y. et al [12] that at the addition of 1.25 mcM Pb^{2+} to erythrocyte suspension in vitro the amount of Pb^{2+} in a ALAD fraction increased and there was Pb^{2+} - bound fraction at a level of the smaller size of molecules than hemoglobin fraction. The similar fraction occurred in a case of the chronic intoxication by lead of the workers and strongly depended on the time of a contact with lead at the industry. One of the clinical symptoms of the lead intoxication is anemia, which can develop not only because of infringement of hemoglobin synthesis, but also due to damages of erythrocyte membranes by ions of lead. Using isolated human erythrocyte membranes we have tried to find out, whether lead influences on a physical state of erythrocyte membrane lipid component *in vitro*.

MATERIALS AND METHODS

The experiments were carried out to erythrocytes of the donors (blood is obtained from Institute of Hematology and Transfusion of Ministry of Health Protection of Belarus). The erythrocytes were isolated from a blood by centrifugation and triply washed in an isoosmotic solution NaCl (5000 g, 5 min). The white membranes (ghosts) of erythrocytes were isolated by a method of Dodge and et al. [13]. In this case use of carefully washed white ghosts has a few advantages except simplicity of system there is an easy accessibility of ions to an internal surface of a membrane and easier control of experimental fluorescent parameters. Concentration of protein in ghosts of erythrocytes was determined by a micromethod of Lowry [14].

The incubation of erythrocyte membranes with 1-10 mcM of lead acetate was carried out in 10 mM NaP - buffer pH 7,4 containing 0,155 M NaCl at 37°C for 1 hour (C protein = 0,5 mg/ml) and then membrane samples were sedimentated by centrifugation (18 000-20 000 g, 15 min), supernatants were removed and received membranes resuspended in the same solution to initial protein concentration.

Change of physical state of membrane lipids was determined using fluorescent lipophilic probe 1,6-diphenyl-1,3,5-hexatriene (DPH) and spin labels 16-doxyl stearat (16DS) and iminoxyl-palmitinic acid (IP).

Hydrophobic fluorescent probe DPH was dissolved in tetrahydrofuran at the final concentration of 2 mM Before use a solution of DPH was diluted in 2000 times by 10 mM tris-HCl buffer(pH=7,4) and then treating of erythrocyte membranes were carried out as at the work [15]. DPH fluorescence intensity and DPH fluorescence polarization were determined at Ireg=428 nm and Iex=356 nm [16].

For registration of EPR-spectra of 16DS and IP of the erythrocyte membranes previously treated with lead acetate, 10 mcl (concentration 10 mcM) 16DS or IP in ethanol were added to 200 mcl of ghosts (C protein = 0,5 mg/ml). After 30 minute incubation at the room temperature free label was removed by centrifugation (18000-20000 g, 15 mines., 40C). Sediment was resuspended in 10 mM the NaP-buffer pH 7,4 + 0.155 NaCl to equal protein concentration at all samples [17].

The fluorescent and spectrophotometrical measurements were realized by the luminescent spectrophotometer LSF 222 (SOLAR, Belarus) and by the spectrophotometer (Specord M40, Germany). The registration of EPR-spectra was carried out at the radiospectrometer of 3 centimetric range designed on base of spectrometer RE 1306.

The following chemicals: DPH (Sigma, USA), lead acetate, NaH_2PO_4, Na_2HPO_4, HCl, TRIS and NaCl (CP (Reachim, CIS)) were used for our investigations. Spin labels 16DS and IP were synthesized at the BSU.

The received results were handled statistically by Student's criterion.

RESULTS AND DISCUSSION

The investigation of erythrocyte ghosts isolated from cells treated to 1-4 mcM lead acetate for 1 hour at 37°C shown that values of DPH fluorescence intensity and the DPH fluorescence polarization have not considerable differences from the control. However, after incubation of isolated erythrocyte membranes to 2-10 mcM lead acetate at the same conditions the DPH fluorescent parameters differed from control. The dependences of DPH fluorescence intensity and the DPH fluorescence polarization on concentration of lead acetate for erythrocyte membranes are presented at the *fig. 1*. You can see from the *fig. 1* that the DPH fluorescence intensity and the fluorescence polarization depend on concentration of lead acetate. Intensity increased by 20% but polarization estimated less considerably - by 10%. It corresponds to change of microviscosity of membrane lipids for human erythrocytes treated with lead in vitro.

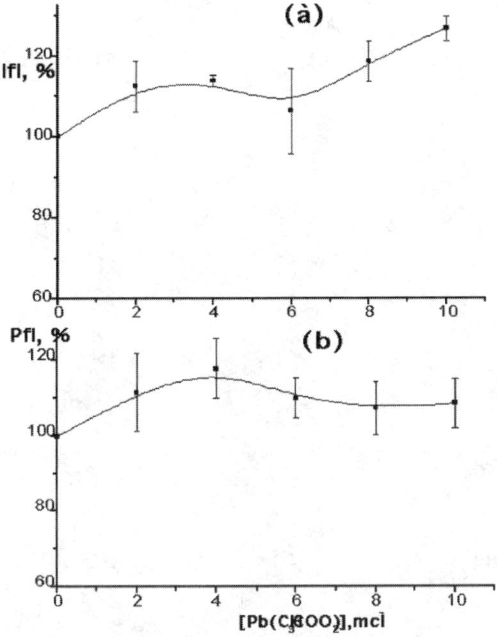

Fig. 1. DPH fluorescence intensity (a) and DPH fluorescence polarization (b) for isolated erythrocyte membranes treated to lead acetate for 1 hour 37°C. The average values are presented for 4 donors

Further we studied parameters of mobility for spin labels 16DS and IP in erythrocyte membranes treated to 2-10 mcM lead acetate. The EPR-spectrum of 16 П demonstrates heterogeneity of a radical microenvironment and is presented by a superposition at least of two lines of absorption as follows from the form of a highly field line of the spectrum *(fig. 2a)*. The change of 16DS mobility in membranes was estimated on parameter δ describing the contribution of more mobile molecules of 16DS at its total pool ($d = h^1_{-1}/h_0$). Parameter d increased by 10% for ghosts with rising of toxic concentration to maximum *(fig. 3)*. It points to increase of number of lipid regions at a less viscous state and testifies about increase of heterogeneity lipid bilayer in a deep inner layer of the erythrocyte membrane.

The character of the IP movement in the erythrocyte membranes was close to isotropic rotation (diffusion). Its EPR-spectrum is submitted at the *fig. 2b*. The IP mobility was estimated as a time of correlation [17]:

$t = 2{,}7.10^{-10}.H_0.(h_0/h_{-1}-1)^{1/2}$,

where - time of correlation at sec, H_0 - width of central component at gausses,

h_0, h_{-1} - amplitude central and highly field lines.

Our experiments have shown that the time of correlation depended on concentration of lead acetate and had a tend to increase with rising of toxic concentration *(fig. 4)*. It is known that the IP EPR-parameters reflect a physics-chemical state of lipids in a surface layer of a membrane [17]. The increase of time of correlation indicates to increase of heterogeneity and reduction of viscosity of lipid bilayer in a surface layer of a membrane.

For example using of spin labels 5DS and 16DS it has been shown that lead reduced of fluidity of lipids at different regions of erythrocyte membranes of the workers contacting to lead in absence of clinical anemia symptoms with concentration of lead in blood 66,25±14,93 mcg/100 ml [18]. With the help of DPH Cook L.R. et al found out that microviscosity of workers' erythrocyte membranes changed [19]. Besides at the effects of lead in vivo and in vitro on erythrocytes and their membranes redistribution of amount of phospholipids and cholesterol, change of fatty acids concentration [20-24] takes place. It is possible to assume that the change of a physical state of membrane lipids is a part of membrane lead effect.

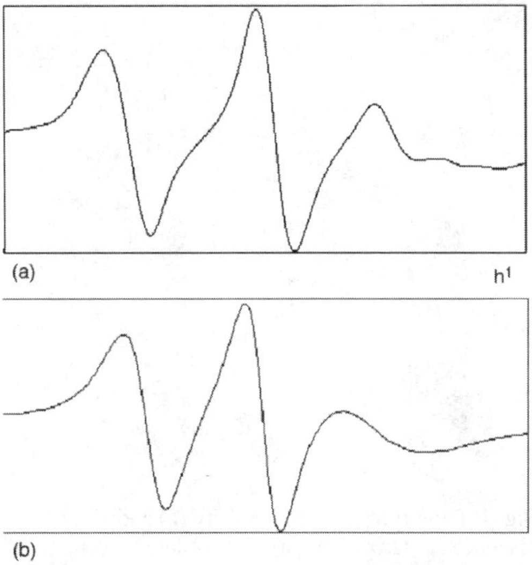

Fig. 2.

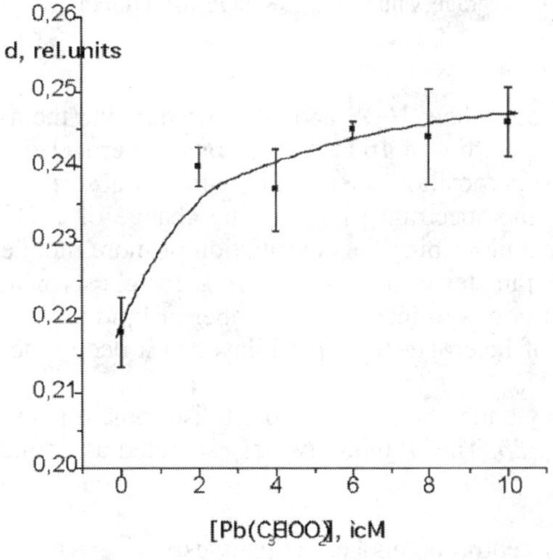

Fig. 3.

Earlier it have been shown by us that coefficient eximerization of a lipophilic probe pyrene (K_{ex}) incorporated in erythrocyte membrane isolated from erythrocytes treated by 6-10 mcM lead acetate increased by 20% on comparison with value K_{ex} for ghosts isolated from intact erythrocytes [25]. The results of this work confirm that the change of a physical state of lipids can have different directions to change for surface and deep inner layers of the erythrocyte membrane. As Pb^{2+} mainly influence on the protein membrane component and so such modification of a physical state by lead can be caused not only change of a ratio phospholipids/cholesterol but also infringements of protein-lipids interactions. It can result to decrease or increase of amount of more liquid regions in a membrane or by association them to large klasters.

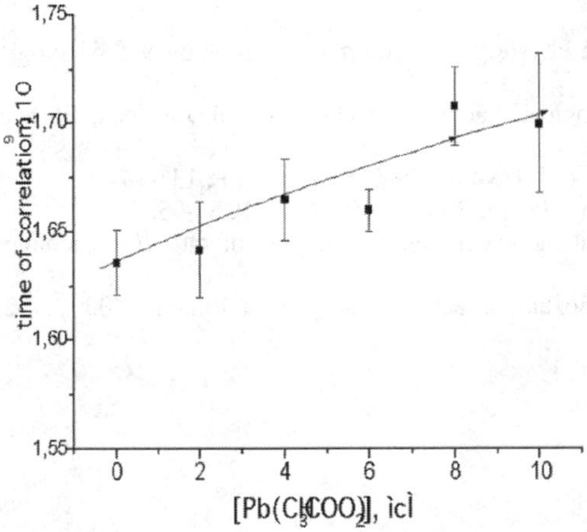

Fig. 4.

CONCLUSIONS

• It was shown that the effect of 2-10 mM lead acetate resulted to the increase of both DPH fluorescence intensity and DPH fluorescence polarization. This was supposed that microviscosity of lipid bilayer of erythrocyte membranes changes.

• From parameters of EPR - spectra 16DS and IP incorporated in erythrocyte membranes the effect of 2-10 mM lead acetate caused the increase of heterogeneity of lipid bilayer at the surface and deep inner membrane layers.

REFERENCES

1. Inorganic lead. Geneva: WHO, 1995. 300 p.
2. Toxicological Profile for lead. Atlanta: U.S. Dep. of Health (Human Services, 1999. 587 p.
3. Moskalev Y.I. Mineral exchange.-M: Medicine, 1985. 288 p. (in Russian).
4. Avtsin A.P., Javoronkov A.A., Rish M.A., Strochkova L.S. Microelementoses of the human: an etiology, classification, organopathology. - M.: Nauka, 1991, p. 496 (in Russian).
5. Sugawara E., Nakamura K., Fukumura A., Seky Y. // Kitasato Arch. Exp. Med. 1990. V. 63, N° 4. P. 5-23.
6. Simons T.J.B. // J. Physiol. (London). 1986. V. 378. P. 287-312.
7. Simons T.J.B. // J. Physiol. (London). 1988. V. 405. P. 105-113.
8. Simons T.J.B. // Pflugers Arch. 1993. V. 423, ? 3-4. P. 307-313.
9. Ong G.N., Lee W.R. // Br. J. Ind. Med. 1980. V. 37. P. 78-84.
10. Bergdahl I.A., Schutz A., Grubb A. // J. of analyt. atomic spectr. 1996. V. 11, N° 9. P. 735-738.
11. Bergdahl I.A., Sheleva M., Schutz A., Artamonova V.G., Skertving S. // Toxicol. Sei. 1998. V. 46, N° 2. P. 247-253.
12. Xie Y., Chiba M., Shinohara A., Watanabe H., Ihaba Y. // Ind. Health. 1998. V. 36, N° 3. P. 234-239.
13. Dodge J.T., Mitchell C., Hanahan D.J. // Arch. Biochem. Biophys. 1963. V. 100, N° 2. P. 119-130.
14. Markwell M.A.K., Haas S.M., Tolbert N.E. // Analyt. Biochem. 1978. V. 87. P. 206-210.
15. Kozlova N.M., Chernitski E.A., Slobozhanina E.I. // Experim. oncology. 1987. V. 9, N° 1. P. 50-52 (in Russian).
16. Chernitski E.A., Slobozhanina E.I. Spectral luminescence analys in medcine. Mn.: Nauka I texnika, 1989. 141p. (in Russian).
17. Lihtenshtein G.I. Method of spin labels at the molecular biology. M.: Nauka, 1974. 255 p. (in Russian).

18. Valentino M., Fiorini R.M., Curatola G., Governa M. // Int. Arch. of Occup. & Envir. Healh.- 1982. V. 51, N° 2. P. 105-112.
19. Cook L.R., Stohs S. J., Angle C.R., Hickman T. I., Maxell R. C. // Br. J. of Industrial Medicine.- 1987. V. 44, N° 12. P. 841-844.
20. Donaldson W.E., Knowles S.O. // Comp. Biochem. & Physilogy- C: Comp. Pharmacology & Toxicology. 1993. V. 104, N° 3. P. 377-379.
21. Iaverbaum P.M. // 3-d All-Union conference: biologically active substances of silicon, lead, tin and germanium. Abstracts. 1980. P. 99-100 (in Russian).
22. Amoruso M.A., Witz G., Goldstein B.D. // Toxicol. Ind. Health. 1987. V. 3, N° 1. P. 135-144.
23. Osterode W., Ulberth E. // J. of Toxicology & Envir. Health. 2000. V. 59, N° 2. P. 87-95.
24. Jehan Z. S., Motlag D.B. Metal induced changes in the erythrocyte membrane of rats. // Toxicology Letters. 1995. Vol. 78. P. 127-133.
25. Lukianenko L.M., Oeksiuk O.B., Kostin D.G., Slobozhanina E.I // Vesti N?S of Belarus. 2000. N° 2. P. 57-60 (in Russian).

Metal Ions in Biology and Medicine: vol. 7. Eds L. Khassanova, Ph. Collery, I. Maymard, Z. Khassanova, J.-C. Étienne.
John Libbey Eurotext, Paris © 2002 pp. 89-93.

Mechanisms of biological effects of metals mediated by interactions with nucleotide cofactors of proteins

V.E. Stefanov, A.A. Tulub

Department of Biochemistry, St. Petersburg State University, Universitetskaya nab.7/9, St. Petersburg 199034, Russia

ABSTRACT

To elucidate the mechanisms of biological effects of metals mediated by their interaction with nucleotide cofactors, quantum chemistry analysis of model systems, involving magnesium ion, was carried out. We undertook quantum chemistry calculations of six-coordinated Mg(2+) complexes with water, glutamic acid and ATP/GTP. For us, effects of magnesium ions are of special interest since they initiate biochemical processes leading to the process of physiologically meaningful self-aggregation of tubulin in microtubules. Acting on tubulin within the complex with GTP, Mg2+ coordinates two terminal phosphate groups and initiates transfer of a proton to a singly bound water molecule of the second and subsequent coordination shells. Realization of singlet (S) and triplet (T) states of the considered model complexes was investigated. It was shown that in the triplet state the magnesium complex concentrates its spin density on a coordinated water molecule (inner or outer coordination shell). Within the molecule a redox reaction occurs giving rise to hydrogen that is pushed away from the complex at a speed of ~ 125 m/s. In water solution, energy of the triplet state is higher than that of the singlet state. In the mixed environment composed of water, amino acids and ATP/GTP, energy of the magnesium complex in the triplet state is lower than that in the singlet state by 1.5 - 2.0 kcal/mol. A little difference in T and S states allows the Mg^{2+}-ATP/GTP complex to switch easily between two reaction mechanisms. Thus, magnesium complex may act via different mechanisms in the singlet and triplet states. The singlet mechanism suggests the proton transfer on the outer-shell water molecule and then on amino acid residues. This relatively slow transfer process obeys the "push-pull" mechanism and leads to ATP/GTP hydrolysis. Ejection of a hydrogen atom from the magnesium complex according to the triplet mechanism (under-barrier tunneling) is a fast process, which lifts most limitations on the velocity and distance range of the transferred energy in biological macromolecules and supramolecular structures.

Hydrogen atoms are essentially involved in energy transfer, initiation of biochemical reactions and membrane transport are largely dependent on formation and migration of hydrogen atoms [1-3]. Often these processes are connected with the interaction of proteins with inorganic cofactors [4-5]. Magnesium ion is one of the most common cofactors. In water and mixed water-protein solutions it forms stable six-coordinated Mg(2+) complexes [6, 7]. So far the role of Mg(2+) in biological processes has been considered from the view-point of its capacity to dissociate the coordinated water into proton and hydroxyl-ion [4, 6-9]. Water dissociation induced by Mg(2+) is possible, if the dissociation process is of singlet character [6, 8]. However, no hydrogen atom is yielded in this case. It was shown [10] that elimination of hydrogen occurs only if magnesium cation is singly charged - Mg(1+). The reaction may proceed in the gaseous phase and on the solid surface. However, it is hardly conceived in cellular solutions, because Mg(1+) is immediately oxidized there yielding cation Mg(2+) [3, 5, 11].

In the present work a triplet mechanism accounting for the elimination of a hydrogen atom from the water-Mg(2+) complex and mixed water-protein-Mg(2+) complex, containing ATP/GTP fragment and glutamic acid, is studied. The latter complex occurs in the structure of tubulin [4] and in a number of ATP/GTP-dependent proteins [12-13]. Ejection of a hydrogen atom from the magnesium-containing complex must be of paramount importance for biology. It explains the capacity of Mg(2+) to dissociate one of the coordinated water molecules in ATP/GTP-dependent proteins [4, 12, 13]. The released hydrogen atom can migrate at a long distance. The break-down of the magnesium complex results in the formation of a spatially separated pair of radicals, which, in turn, generate propagation of a spin wave connected with the translocation of the hydrogen atom. Propagation of the spin waves is a major factor responsible for polymerization of biological molecules [14, 15]. In our work we analyzed complexes $Mg^{2+}[H_2O]_6$, $Mg^{2+}[H_2O]_6[H_2O]$ (the seventh molecule of water imitates outer-shell water built into the protein matrix), $Mg^{2+}[H_2O]_4$GluATP and $Mg^{2+}[H_2O]_4[H_2O]$GluATP (the fifth molecule of water is outer-shell water). For the model the difference between ATP and GTP is not essential. Therefore, to simplify the notation we shall use only symbol ATP.

OBJECT OF INVESTIGATION AND CALCULATION METHODS

We used non-empirical quantum chemistry calculations: RHF (singlet states)/UHF(triplet and doublet states), 6-311++G**(p,d) basis set, Moller-Plesset perturbation theory technique MP4(FULL) and population analysis according to Lowdin [16, 17]. The initial optimization of geometry was performed by means of the density functional theory method (DFT) at the B3LYP level [18-21] with the use of GAUSSIAN-98 program. The original geometry of the complex $Mg^{2+}[H_2O]_6$ is assumed as in [7, 9, 22]. The original geometry the six-coordinated Mg(2+) complex with ATP and Glu the complex was taken from [4]. For the sake of simplicity we took into account γ-, β-terminal groups of ATP/GTP and mimicked the remaining part with a methyl group. The subsequent calculation of the singlet and triplet states for the optimized "frozen" frame of $Mg^{2+}[H_2O]_4$GluATP was performed by means of RHF/UHF method. Localization of the outer-shell water molecule in complexes $Mg^{2+}[H_2O]_6[H_2O]$ and $Mg^{2+}[H_2O]_4[H_2O]$GluATP is assumed according to the data [4, 12, 13]. To characterize the equilibrium configurations we used DFT:B3LYP method. On this basis, excited states were considered. To obtain charge and vibration characteristics of molecules we used RHF/UHF method. Thereafter the numbering of complexes will be as in the *table 1*.

RESULTS AND DISCUSSION

Let us consider six-coordinated Mg(2+) complexes with water in singlet (S), triplet (T) and doublet (D) states *(table 1)*. In all the complexes the value Q of charge on the magnesium atom is close to +2. The charge on the oxygen atom of the water molecule is close to -1, whereas on the atoms of hydrogen its value is about +0.5, with exception of complex 2. Within the accuracy of the calculation method the data obtained for complex 3 are close to those reported earlier [23]. The data for triplet and doublet states have been obtained for the first time.

Let us analyze processes that take place on excitement of complex **3** (transition from **3** to **1**). Emergence of the triplet state (complex **1**) modifies significantly properties of the six-coordinated complex of magnesium, these changes affecting only one fragment of the complex. The spin density concentrates on only one water molecule, but that can be any molecule in the complex. The charge on the oxygen atom rises from -1 to 0, while the charges on the hydrogen atoms decrease accordingly from +0.5 to 0 *(table 1)*. Hence, within one water molecule a redox reaction consisting of two half-reactions (1a) and (1b) proceeds:

$$O^{-1} - 1e \rightarrow O^0 \quad (1a) \quad 2H^{+0.5} + 1e \rightarrow H^0 \; (1b)$$

The molecule, formed within complex 1, immediately breaks down according to (2):

$$Mg^{2+}[H_2O]_5[H_2O] \rightarrow Mg^{2+}[H_2O]_5[O^{\delta-}H^{\delta+}] + H \; (2)$$

Table 1. Total energy E (a.u.) and efficient charges on atoms Q in 6- and 7-coordinated complexes of Mg(2+) in the singlet (S), triplet (T) and doublet (D) state

Complex	-E	Q[Mg]	$Q[O]_{H2O}$	$Q[H]_{H2O}$	$Q[O]_{OH}$	$Q[H]_{OH}$
1. $(Mg^{2+}[H_2O]_5[H_2O])_T$	458.789	1.716	{-1.007} -0.029	{0.519} 0.079		
2. $(Mg^{2+}[H_2O]_5[O^{\delta-}H^{\delta+}])_D + H_D$	458.821	1.939	{-1.005}	{0.520}	-0.209	0.295 0.000 (H_D)
3. $(Mg^{2+}[H_2O]_6)_S$	458.924	1.801	-1.008	0.520		
4. $(Mg^{2+}[H_2O]_6[H_2O])_T$	529.323	1.908	{-1.021} -0.074	{0.518} 0.032		
5. $(Mg^{2+}[H_2O]_6[O^{\delta-}H^{\delta+}])_D + H_D$	529.351	1.958	{-1.019)	{0.519}	-0.194	0.262 0.000 (H_D)
6. $(Mg^{2+}[H_2O]_6[H_2O])_S$	529.383	1.904	{-1.016} -1.123	{0.514} 0.508		
7. $(Mg^{2+}[H_2O]_2[H_2O]GluATP)_,$	1397.583	1.958	{-1.010} -0.070	{0.520} 0.032		
8. $(Mg^{2+}[H_2O]_2[O^{\delta-}H^{\delta+}]GluATP)_D + H_D$	1397.615	1.978	{-1.019}	{0.518}	-0.198	0.262 0.000 (H_D)
9. $(Mg^{2+}[H_2O]_3GluATP)_S$	1397.579	1.937	{-1.112}	{0.516}		
10. $(Mg^{2+}[H_2O]_3[H_2O]GluATP)_T$	1460.517	1.962	{-1.012} -0.072	{0.512} 0.034		
11. $(Mg^{2+}[H_2O]_3[O^{\delta-}H^{\delta+}]GluATP)_D + H_D$	1460.547	1.983	{-1.021}	{0.517}	-0.201	0.260 0.000 (H_D)
12. $(Mg^{2+}[H_2O]_3[H_2O]GluATP)_S$	1460.514	1.957	{-1.114} -1.120	{0.516} 0.508		

Braces for $Q[O]_{H2O}$ and $Q[H]_{H2O}$ correspond to the average values over five **(1, 2)**, six **(3-6)**, two **(7, 8)** and three **(9-12)** water molecules.

Here ↓- and ↓+ denote small charges emerging as a result of O-H-bond polarization caused by magnesium cation. These charges contribute to fixation of the OH fragment in the complex.

The dynamics of ejection of a hydrogen atom from complex **1** can be monitored with a computer in the course of geometry optimization of complex **1**. At each step of the optimization procedure the hydrogen atom is moved off from the complex more and more, the non-compensated spin density being retained both at the oxygen atom (s = +1/2) and hydrogen atom moving away from the complex (s = +1/2). Within the accuracy of the calculation of total energy ($\Delta = 10^{-8}$ kcal/mol), the hydrogen atom can move away from complex **1** at a distance of 4.34 nm. The observed effect is connected with the accumulated energy $\varepsilon = 0.811$ eV [ε = **Δ(2-1)**], acquired by the hydrogen atom during its ejection from complex **1**. We assume that this energy is converted into kinetic energy, so that the velocity of the hydrogen atom leaving the complex reaches 124.738 m/s. Even in a real solution the velocity can be rather big, which makes a long-distance transport of hydrogen atoms in the medium quite feasible.

Taking into account the difference in the energy of state **1** and **3** *(table 1)*, one can presume that transition from the state **3** to the triplet state **1** proceeds in the presence of an external irradiating source of high energy (photolysis, radiolysis). However, formation of **1** can also take place in relatively "mild" conditions. Thus, bonding of a seventh outer-shell water molecule to complex **3** (complex **4**) draws together the singlet and triplet levels to 37.578 kcal/mol *(table 1)*. The outer-shell water molecule is linked to the water molecule of the first coordination shell by means of a hydrogen bond. The following parameters characterize the optimized geometry of the complex: <O-H...O = 174.121^0, r[H...O] = 2.138, r[O-H] = 1.023 (dots denote the hydrogen bond). It should be noted that in this case the spin density concentrates on outer-shell water molecule. The properties of this molecule coincide with those of the water molecule from the first coordination shell, were described earlier *(table 1)*. The energy of the hydrogen atom detachment from the outer-shell water molecule is equal to 0.754 eV, which corresponds to the initial velocity 120.275 m/s. The value of charge on the magnesium atom in complexes **7-12**remains close to +2. The average distance Mg-OH_2 is 2.120. The distance Mg-O(P^γ) in the terminal γ-phosphate group is 2.098, while that in the β-phosphate group, Mg-O(P^β), is 2.142. The distance Mg-O(Glu) is equal to 2.124. The lengths of phosphodiesteric bonds are r[P^γ-O] = 1.598; r[P^β-O] = 1.624. They practically do not differ from those reported in literature [24]. Addition of another outer-shell water molecule to complex **7** does not affect the geometrical parameters of the complex.

For complexes with ATP and Glu the energy of the triplet state is lower than that of the singlet state: ΔE (7-9) = 1.54 kcal/mol; ΔE (10-12) = 1.92 kcal/mol *(table 1)*. In the triplet state the spin density concentrates on the inner-shell (complex 7) or outer-shell (complex 10) water molecule. Atoms of hydrogen and oxygen of the water molecule are not charged *(table 1)*, so that one of the hydrogen atoms is ejected from the complex. The energy of the hydrogen atom detachment is equal to 0.878 eV and 0.822 eV, and the initial velocity of the hydrogen atom is 129.788 m/s and 125.581 m/s for complexes 7 and 10, respectively. For complexes 1 and 4 these values are 0.811 eV, 124.738 m/s and 0.754 eV, 120.275 m/s, respectively.

The obtained data testify to a unique capacity of the cation Mg(2+) to affect bound water molecules manifested upon formation of the triplet excited complex. The emergence of atomic or molecular hydrogen in solution was observed earlier in the course of photolysis or radiolysis of water [25, 26] and in reactions of photosynthesis [37]. In all the above cases an external source of irradiation was necessary to generate atomic hydrogen. Magnesium cations can exert their effect in living cells in the dark demanding no external source of irradiation. The required energy appears spontaneously as a result of the formation of the solvate shell by the magnesium cation, water molecules being among six ligands included in the complex. Formation of a mixed complex containing water molecules, amino acid residues and ATP/GTP results in a lower energy of the triplet state as compared to the singlet state. The values of total energy of the two states being close to each other, transition between them is easy.

The magnesium complex initiates the reaction through different mechanisms in the singlet and triplet states. The singlet mechanism analyzed earlier in [4], suggests the proton transfer on the outer-shell water molecule and then on a number of amino acid residues surrounding ATP/GTP. The transfer process obeys the "push-pull" mechanism [4, 28]. It is relatively slow because of numerous energy barriers. It results in the breakdown of the phosphodiesteric bond O-P^β, which eventually leads to ATP/GTP hydrolysis. On the contrary, ejection of a hydrogen atom from the magnesium complex according to the proposed triplet mechanism [reaction (2)], is a fast process. Displacement of atomic hydrogen is connected with the under-barrier tunneling [29-31]. This removes most limitations on the velocity and distance range of the transferred energy (proton and electron are transferred as one particle, proton-coupled electron transfer [1, 2, 31]). Generally, singlet and triplet branches of reactions are realized in two different channels, implying the detachment of different hydrogen atoms (or cations) from the same water molecule within the magnesium complex. This aspect will be analyzed in subsequent publications.

The work was financially supported by the Russian Foundation for Basic Research (RFFI grant, project code 99-04-49836 and 00-03-32946). The authors are particularly grateful to Prof. Milligan (University of California) and assistance from Nasa Ecology for providing access to software (Gaussian 98) and Cray Station resources in the Computing Center of California Technology Institute.

REFERENCES

1. *Hammes-Schiffer S.*, Proton-coupled electron transfer. In: Electron transfer in Chemistry: Principles, Theories, Methods, and Techniques. Balzani V. Ed. Wiley-VCH: Weinheim, Germany, 2001. 457 p.
2. *Decornez H., Hammes-Schiffer S.* J. Phys. Chem. 2000, 104 A. P. 9370-9384.
3. *Fersht A.* Structure and mechanisms in protein science: a guide to enzyme catalysis and protein folding. N.Y.: Freeman. 1999. 614 p.
4. *Stefanov V.E., Tulub A.A., V.K. Pavlenko, Kutin A.A.* Anal. Quimica Int., 1998, 94. P. 250-257.
5. *Glusker J.P.* Adv. Protein Chem., 1991, 42. P. 1-73.
6. *Bock C.W., Katz A., Markham G.D., Glusker J.P.* J. Am. Chem. Soc. 1999, 121, P. 7360-7372.
7. *Bock C.W., Kaufman A., Glusker J.P.* Inorg. Chem. 1994, 33, P. 419-427.
8. *Beyer M., Williams E.R., Bondybey V.E.* J. Am. Chem. Soc. 1999, 121, P. 1565-1573.
9. *Trachtman M., Markham G.D., Glusker J.P., George P., Bock C.W.* Inorg. Chem. 1998, 37, P. 4421-4431.
10. *Sanekata M., Misaizu F., Fuke K., Iwata S., Hashimoto K.* J. Am. Chem. Soc. 1995, 117, P. 747-754.
11. *Cotton F.A., Wilkinson G.* Advanced Inorganic Chemistry. A Comprehensive Text, 4th ed., John Wiley and Sons: New York, 1980. 320 p.
12. *Kjelgaard M., Nyborg J., Clark B.F.C.* FASEB J. 1996, 10, P. 1347-1368.
13. *Smith C.A., Rayment I.* Biochem. 1996, 35, P. 5404-5417.
14. *Gurevich A.G., Melkov G.A.* Magnetic oscillations and waves. (Physmatlit: Moscow, in Russian) 464 p. (1994).
15. *Korolev G.V., Marchenko A.P.* Uspekhi Khimii (in Russian), 2000, 69. P. 447-475.
16. *He Y., Cremer D.* J. Phys. Chem. 2000, 104 A., P. 7679-7688.
17. *Lowdin P.-O.* Adv. Quant. Chem. 1970, P. 185-199.
18. *Lee C.*, Yang W., Parr R.G. Density functional theory of atoms and molecules. Oxford: Univ. Press. 1989. 345 p.
19. *Becke A.D.* J. Phys. Chem. 1993, 98, P. 1372-1377.
20. *Stephens P.J., Devlin F.J., Chabalowski C.F., Frisch M.J.* J. Phys. Chem. 1994, 98, P. 11623-11627.
21. *Siegban P.E.M., Blomberg M.R.A.* Theor. Chem. Acc. 1997, 97, P. 289-300.
22. *Markham G.D., Glusker J.P., Bock C.L., Trachtman M., Bock C.W.* J. Phys. Chem. 1996, 100, P. 3488-3497.
23. *Katz A.K., Glusker J.P., Markham G.D., Bock C.W.* J. Phys. Chem., 1998, 102 B, P. 6342-6350.
24. *Singer W.* Springer-Verlag: N.Y., 1984, 543 p.
25. *Slanger T.G., Black G.* J. Chem. Phys. 1982, 77, P. 2432-2440.
26. *Allen A.O.* The radiation chemistry of water and aqueous solutions. Princeton: D. Van Nostrand Comp. 1961. 456 p.
27. *Okamura M.Y., Feher G.* Annu. Rev. Biochem. 1992, 61, P. 861-878.
28. *Tributsch H., Pohlmann L.* J. Theor. Biol. 1996, 178, P. 17-28.
29. *LaVerne J.A., Pimblott S.M.* J. Phys. Chem. 2000, 104 A., P. 9820-9822.
30. *Goldansky V.I., Trakhtenberg L.I.* Tunneling phenomena in chemical physics (Nauka: Moscow, in Russian), 1984, 302 p.
31. *Soudackov A., Hammes-Schiffer S.* J. Chem. Phys. 2000, 113, P. 2385-2396.

*Metal Ions in Biology and Medicine: vol. 7.*Eds L. Khassanova, Ph. Collery, I. Maymard, Z. Khassanova, J.-C. Étienne.
John Libbey Eurotext, Paris © 2002 pp. 94-99.

Proteolysis of ceruloplasmin and copper transfer to lactoferrin

Pulina M.O.[1], Zakharova E.T.[1], Bass M.G.[1], Sokolov A.V.[1, 2], Solovyov K.V.[1, 3], Lodygin P. A.[1], Kokryakov V.N.[1, 2], Shavlovski M.M.[1], Vasilyev V.B.[1]

[1]Institute for Experimental Medicine, Saint-Petersburg, RUSSIA; [2]Saint-Petersburg State University; [3]Saint-Petersburg Technical University;
Address for correspondence: Vadim B. Vasilyev, Department of Molecular Genetics, Institute for Experimental Medicine, 12 Pavlov St., Saint Petersburg 197376 Russia.

ABBREVIATIONS

CP, ceruloplasmin (ferro-O_2-oxidoreductase; EC 1.16.3.1); LF, lactoferrin; PAG, polyacrylamide gel; PAGE, polyacrylamide gel electrophoresis; PMNG - polymorphonuclear granulocytes.

ABSTRACT

Acute phase reactants ceruloplasmin (CP), a copper protein, and lactoferrin (LF), an iron protein, form a complex both *in vitro* and *in vivo* [Zakharova et al., 2000]. Limited proteolysis facilitates release of copper ions from CP, which may have its deleterious effect in foci of inflammation. We were interested to study whether LF associated with CP affects its proteolysis and loss of copper. Neutrophil proteases elastase and cathepsin G efficiently cleave CP. Along with CP these two proteases are found in increased concentrations in the foci of inflammation where the share of proteolyzed CP may also increase. Incubation of apo-LF/CP complex with trypsin resulted in limited proteolysis of CP, while LF remained intact. Copper ions released from CP were incorporated in increasing amounts by apo-LF, which was evidenced by concomitant decrease of CP absorption at 610 nm and increase of the band at 435 nm corresponding to Cu-LF. Apo-LF did not incorporate copper ions if CP was intact. The observed mechanism might be of importance for protection against pro-oxidative transition metals in the foci of inflammation.

INTRODUCTION

Ceruloplasmin (CP, ferro-O_2-oxidoreductase, EC 1.16.3.1) is a copper-containing protein of mammalian plasma. The human protein is highly succeptible to limited proteolysis [1, 2], which is accompanied by loss of copper ions. CP is found in the foci of inflammation, where its limited proteolysis by the proteases of neutrophils is not improbable. Along with proteases the neutrophils secrete lactoferrin (LF) that is able to interact selectively with CP [3]. Since it was shown [4] that LF is able to bind copper ions, we tested the hypothesis that these ions released from CP can get incorporated into LF. We also studied the proteolytic pattern of CP obtained when the protein is cleaved by some neutrophil proteases.

MATERIALS AND METHODS

All solutions were prepared using deionized water and treated, when necessary, with Chelex-

100 to eliminate traces of heavy metals. Chemicals were purchased from Serva, Merck, Pharmacia and Sigma Chemical Co.

Human CP was purified as described in (1). Apo-LF was purified from breast milk using protocol of Zakharova *et al.* [3]. Apo-LF was saturated with copper according to Pulina *et al.* [5]. Proteins were usually dissolved in 10 mM phosphate buffer, pH = 7.4, containing 0.15 M NaCl (hereafter Buffer A) Protein concentrations were determined spectrophotometrically at 280 nm using the following coefficients: 1.12 ml mg^{-1} cm^{-1} for human apo-LF [6] and 1.61 ml mg^{-1} cm^{-1} for human CP [7]. Incorporation of copper into LF was followed by growth of absorption at 435 nm [4]. Copper release from CP was monitored by decrease of absorption at 610 nm.

Copper transfer from CP to LF was achieved as follows: a 50 mg ml^{-1} CP solution in buffer A was flown through a Chelex-100 column to eliminate the protein-unbound and loosely bound copper and then diluted to 15 mg ml^{-1} with Chelex-100-treated Buffer A. Apo-LF was added to CP at concentration providing the 2:1 molar ratio. The sample was split in four aliquots that were dialyzed overnight at 4 °C against the original Buffer A or Buffer A containing 0.001, 0.01 and 0.1 M $NaHCO_3$. The pH of the sodium bicarbonate-containing buffer solutions was 7.4, 7.8 and 8.5, respectively. Copper incorporation into LF was checked spectroscopically (see above) using Beckman DU-50 spectrophotometer.

Neutrophil extract was obtained from polymorphonuclear granulocytes (PMNG) as described by Roy *et al.* [8].

Cathepsin G and elastase were purified from neutrophilic leukocytes according to Desser *et al.* [9].

Limited proteolysis of CP was achieved by adding the extract from PMNG, cathepsin G, elastase and trypsin as 1:15, 1:25, 1:20 and 1:50, respectively.

Electrophoresis with SDS was carried out in PAG as described [10].

RESULTS

Since it is known that LF is able to tightly bind copper ions acquiring an absorption maximum at 435 nm [4], we tested the possibility of direct transfer of this metal from CP to LF within the complex. With this aim we prepared the complex using holo-CP and apo-LF and followed the occurrence and increase of the mentioned above absorption band to monitor the binding of copper to LF. The experiments were carried out at different concentrations of bicarbonate, since it is known that this anion facilitates the incorporation of iron and other metal ions into apo-LF.

In *fig. 1, A* the difference absorption spectra of the holo-CP/apo-LF complex after 6h incubation (spectrum 1) and of the copper-saturated LF (spectrum 2) *versus* apo-LF are shown. The analysis of spectrum 1 in comparison with spectrum 2 allows to assign the 435 nm feature (the insert represents a better resolution of the spectrum in this region) to the Cu-LF derivative within the CP/LF complex. These results are supported also by the data presented in *fig. 1, B*: spectrum 1 was obtained when the absorption of CP was recorded *versus* that of apo-LF; spectrum 2 represents the absorption of the apo-LF/CP complex *versus* CP. In this latter case the contribution of the type 1 copper of CP absorbing at 610 nm is completely subtracted and the absorption band at 435 nm (magnified in the insert) represents the only feature in the visible region of the LF/CP complex. It is worth noting that absorption band at 435 nm is observed when the complex is incubated against 0.1 M $NaHCO_3$, while the lower concentrations of sodium bicarbonate had no such effect.

To check the stability of CP under the experimental conditions, samples of CP were analyzed by SDS-PAGE before and after overnight incubation at different pH values (see Materials and Methods). The share of proteolytic fragments into which CP is typically split, increased after the incubation at pH=8.5. In contrast, an overnight dialysis of CP at other pH values did not cause noticeable changes in the SDS-PAGE pattern of the protein.

To test whether copper incorporation into LF could be facilitated by the loss of integrity of CP molecule accompanied by copper release, the CP/LF complex was subjected to limited tryptic proteolysis in Buffer A containing 0.01 M $NaHCO_3$, pH = 7.8. On account that LF had been found to be rather resistant to the tryptic digestion [11], this approach simulated the mentioned above spontaneous proteolysis of CP.

Absorption spectra of the CP/LF complex were registered 20, 60 and 100 min. after the beginning of the tryptic digestion on samples. In course of the proteolytic degradation, the absorbance at 435 nm, caused by copper incorporation into LF, increased while that at 610 nm attributed to the type I copper-binding centers decreased *(fig. 1, C)*.

The patterns of proteolytic cleavage of CP by the extract of PMNG, cathepsin G and elastase were compared to that resulting from trypsin treatment *(fig. 2)*. It is seen that regardless of protease applied, CP molecule is split into typical polypeptides, such as 19-kDa and 50-kDa fragments. With a rate somewhat slower PMNG extract produced similar fragments, suggesting that cathepsin G and elastase are the main reactants.

DISCUSSION

Both CP and LF are acute-phase reactants and are known to protect tissues from the damage caused by free oxygen radicals in the foci of inflammation where they are usually found [12, 13].

Considering the interaction of the two proteins, the possibility that LF can accept copper ions directly from CP was studied, especially on account that apo-LF readily binds copper ions from solution within a broad (5.0-10.0) range of pH [4]. Indeed, when the complex of CP with apo-LF was incubated in the presence of 0.1 M $NaHCO_3$ (pH = 8.5) apo-LF bound copper as was evidenced by an increasing intensity of absorption at 435 nm typical of Cu-LF. At pH = 7.4 and 7.8 no increment of absorption was evidenced. The spectra registered in the course of limited tryptic proteolysis demonstrated a more pronounced growth of absorption at 435 nm concomitantly with the loss of absorption at 610 nm.

It seems likely that in our study apo-LF scavenged the copper ions released into the buffer solution from CP that was partly degraded during the long incubation at pH = 8.5 or in the course of limited proteolysis.

It is known that partially proteolyzed CP interacts with LF no less efficiently than the intact protein [5]. Proteolysis of CP by cathepsin G and elastase yields at least large fragments that are similar, though not identical, to those resulting from trypsinolysis (see *fig. 2*). Indeed, trypsin cleaves CP at Arg-481 and Lys-887, producing, respectively 50-kDa and 19-kDa fragments. It is very likely that cathepsin G is able to cut CP at Val-483 and Val-888, while elastase might attack the protein at Phe-496 and Phe-889 [14]. The resulting fragments will differ from their tryptic homologues by a few N-terminal amino acids. Considering the primary structure of CP [14] and its conformation [15], the destruction of the three-nuclear copper cluster followed by the release of Cu^{2+} ions seems very likely upon CP cleavage by these proteases.

Our data allow suggesting that if CP within the CP/LF complex is proteolyzed and loses its copper, the possible deleterious effect of this reactive transition metal *in vivo* may be minimized once it is scavenged by LF. Such a mechanism might be realized in the foci of inflammation where CP is likely to get proteolyzed by cathepsin G and/or elastase from neutrophils. Suggestion that copper ions are bound by apo-LF *in vivo* is supported by finding minor amounts of Cu-LF in various body fluids [16]. Formation of such species *in vivo* is facilitated by the presence of the major part of LF in organism in its apo-form [17].

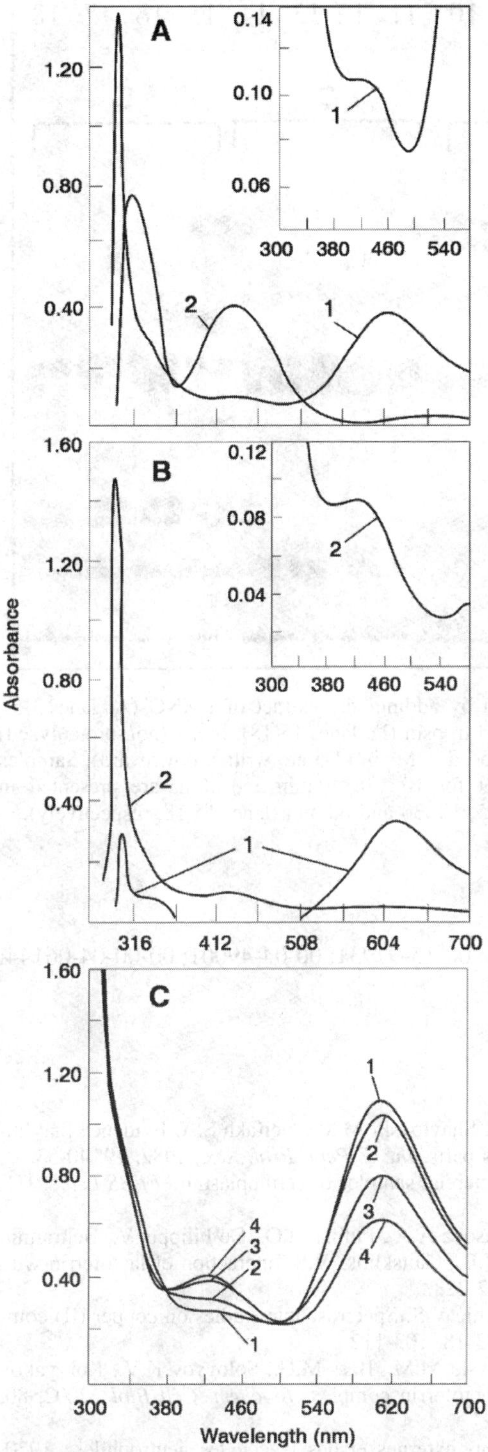

Fig. 1. Difference absorption spectra of:
A - apo-LF + CP *versus* apo-LF (spectrum 1) and copper-saturated LF *versus* apo-LF (spectrum 2) in 0.1 M $NaHCO_3$, pH=8.5; absorption band around 435 nm is magnified in the insert;
B - CP *versus* apo-LF (spectrum 1) and apo-LF + CP *versus* CP (spectrum 2); absorption band around 435 nm is magnified in the insert.
Wavelength numbers plotted along the abscissa in B are the same for A.
C - absorption spectra of the mixture CP + LF (2:1) in 0.1 M $NaHCO_3$, pH=8.5, registered *versus* Buffer A at different times of tryptic hydrolysis: 0 min (1); 20 min (2); 60 min (3); and 100 min (4).

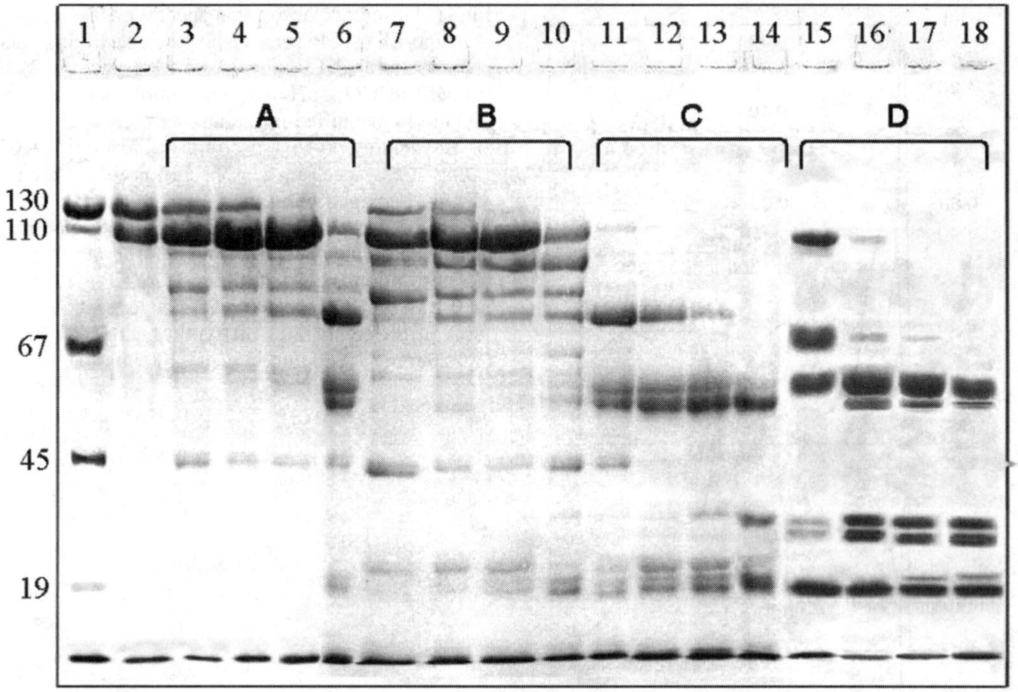

Fig. 2. Electrophoresis in SDS-PAG of CP proteolysed by adding: the extract of PMNG (A, lanes 3-6), cathepsin G (B, lanes 7-10), elastase (C, lanes 11-14) and trypsin (D, lanes 15-18). Intact (non-proteolyzed) CP was run on lane 2. Lane 1 - protein markers (the respective Mr in kDa are written downward). Samples of CP after treatment with respective proteolytic agent for 10, 30, 40 min and 12 hr are presented in groups A-C. In group D samples of CP were treated for 5, 10, 30 and 60 min (lanes 15-18, respectively).

AKNOWLEDGEMENTS

This work was partly supported by RFBR grants: 00-15-97931; 00-04-49001; 00-00-04-06144; 04-06146, and 02-04-49698.

REFERENCES

1. Prozorovski V.N., Rashkovetski L.G., Vasiliev V.B., Shavlovski M.M., Neifakh S.A. Evidence that human ceruloplasmin molecule consists of homologous parts. *Int. J. Pept. Prot. Res.*, 1982, 19, 40-53.
2. Ryden L. Evidence for proteolytic fragments in commercial samples of ceruloplasmin. *FEBS Lett.* 1971, 18, 321-325.
3. Zakharova E.T., Shavlovski M.M., Bass M.G., Gridasova A.A., Pulina M.O., De Filippis V., Beltramini M., Di Muro P., Salvato B., Fontana A., Vasilyev V.B., Gaitskhoki V.S. Interaction of lactoferrin with ceruloplasmin. *Arch. Bioch. Biophys.*, 2000, 374, 222-228.
4. Ainscough E.W., Brodie A.M., McLachlan S.J., Ritchie V.S. Spectroscopic studies on copper (II) complexes of human lactoferrin. *J. Inorg. Biochem.*, 1983 18, 103-112.
5. Pulina M.O., Zakharova E.T., Sokolov A.V., Shavlovski M.M., Bass M.G., Solovyov K.V., Kokryakov V.N., Vasilyev V.B. Studies of the ceruloplasmin-lactoferrin complex. *Biochem. Cell Biol.*, 2002, 80, 289-293.
6. Masson P.L. La Lactoferrine. Proteine des secretions externes et des leucocytes neutrophiles., 1970, Editions Arscia S.A., Bruxelles.

7. De Filippis V., Vassiliev V.B., Beltramini M., Fontana A., Salvato B., Gaitskhoki V.S. Evidence for the molten globule state of human apo-ceruloplasmin. *Biochim. Biophys. Acta*, 1996, 1297, 119-123.
8. Roy S.C., Singh V.K., More T. Electrophoretic detection of myeloperoxidase, protease, lactoferrin and lysozyme in buffalo polymorphonuclear granular acid extracts. *Vet. Res. Commun.*, 1997, 21, 325-334.
9. Desser R.K., Himmelhoch S.R., Evans W.H., Januska M., Mage M., Shelton E. Guinea pig heterophil and eosinophil peroxidase. *Arch. Biochem. Biophys.*, 1997, 148, 452-465.
10. Laemmli U.K. Cleavage of structural proteins during the assembly of the head of bacteriophage T4. *Nature*, 1970, 227, 680-685.
11. Brines R.D., Brock J.H. The effect of trypsin and chymotrypsin on the in vitro antimicrobial and iron-binding properties of lactoferrin in human milk and bovine colostrum. *Biochim. Biophys. Acta*, 1983, 759, 229-235.
12. Klebanoff S.J. Bactericidal effect of Fe2+, ceruloplasmin, and phosphate. *Arch. Biochem. Biophys.*, 1992, 295, 302-308.
13. Britigan B.E. Uptake of human lactoferrin by mononuclear phagocytes inhibits their ability to form hydroxyl radical and protects them from membrane autoperoxidation. *J. Immunol.*, 1991, 147, 4271-4277.
14. Takahashi N., Ortel T.L., Putnam F. Single-chain structure of human ceruloplasmin: the complete amino acid sequence of the whole molecule. *Proc. Nat. Acad. Sci. USA*, 1984, 81(2), 390-394.
15. Zaitseva I., Zaitsev V., Card G., Moshkov K., Bax B., Ralph A., Lindley P. The X-ray structure of human serum ceruloplasmin at 3.1 A: nature of the copper centers. *J. Biol. Inorg. Chem.*, 1996, 1, 15-23.
16. Wang C.-S., Chan W.Y., Kloer H.U. Comparative studies on the chemical and immunochemical properties of human milk, human pancreatic juice and bovine milk lactoferrin. *Comp. Biochem. Physiol.*, 1984, 78B, 575-580.
17. Goldsmith S.J., Eitenmiller R.R., Barnhart H.M., Toledo R.T., Rao V.N. Unsaturated iron-binding capacity of human milk. *J. Fd. Sci.*, 1982, 47, 1298-1304.

*Metal Ions in Biology and Medicine: vol. 7.*Eds L. Khassanova, Ph. Collery, I. Maymard, Z. Khassanova, J.-C. Étienne.
John Libbey Eurotext, Paris © 2002 pp. 100-103.

Effect of metal cation substitution with organic cations on the parameters of UV-melting curves of DNA

Kazaryan R.L., Petrov A.I., Sukhorukov B.I.

Institute of Theoretical and Experimental Biophysics RAS. 142290, Pushchino, Moscow region, Russia.

ABSTRACT

It has been well documented now that DNA-lipid interactions play an important role in many cellular processes. Therefore, the elucidation of physicochemical regularities of this interaction is an actual, but difficult task due to extremely low solubility of natural lipids. We have studied the influence of cationic amphiphiles, as easily soluble structural analogs of lipids, on the structure and stability of DNA double helix. High molecular weight surgeon sperm DNA was used in this study. The UV-melting curves of DNA with dodecyl-, tetradecil-, hexadecyltrimethylammonium bromide in 10^{-2} M NaCl were measured on the spectrophotometer VSU-2P. The shift of the melting curves of DNA with increase of concentration of cationic amphiphiles to the region of higher temperatures was demonstrated. As in the case of Na^+ cations this shift points to the stabilization of DNA double helix as a result of screening of the repulsion between negatively charged phosphate groups, but the stabilizing effect is about 3 order higher. In contrast to Na^+, the cationic amphiphiles result to the increase of transition width (ΔT) and to the decrease of hyperchromicity. Such behavior of melting curves is the consequence of hydrophobic fragment in amphiphile molecules providing for their specific binding to DNA in form of clusters. It provides for specific, sequence-dependent, binding of cationic amphiphiles in form of clusters. Such non-uniform distribution of amphiphile molecules along the DNA double helix results to strong heterogeneity stability manifested itself in the increase of ΔT. Besides, the formation DNA-cationic amphiphile complex results to the light scattering of solution, which is responsible for the observed decrease of hyperchromicity at UV-melting of DNA.

INTRODUCTION

It has been well documented now that metal cations play an important role in structural organization, thermal stability and functioning of DNA [1-3], in particular, much is known about the effect of replacement of one metal cation by another on the parameters of melting curve of DNA [4, 5]. At the same time there are only single studies on the effect on these parameters of the replacement of metal cations with organic ones. For example, it was shown [6, 7] that symmetric tetramethylammonium (TMA) and tetraethylammonium (TEA) cations bind specifically with DNA in small groove, showing higher affinity to AT than to GC base pairs though tetrabutylammonium (TBA) with longer hydrocarbon chains does not possess this property due to steric factors [4, 5]. TMA binding to DNA increases the melting temperature (T_m), up to the concentration of 3 M. Binding of TEA results to initial increase in T_m of DNA (up to 0.8 M) and then to its decrease. At TEA concentration 2.4 M and higher the GC-enriched sequences melt at lower temperature than AT-enriched was. Alkylammonium ions (for example, TBA) exert small effect on the stability of the double helix [8].

The effect of long-chained asymmetric N-trimethylsubstituted aliphatic amines with chain length C>8 has not been practically studied. The interest to such studies is determined by antiviral and antibacterial activity of this substances [9], their usage as the components of liposomal formulatione for transmembrane transfer of DNA [10], and they are structural analogs of natural lipids that necessitates the studies of their interaction with DNA for elucidation of physicochemical basis of complex formation DNA-lipid [11].

The aim of the present work was to study the thermal stability of DNA at the replacement of Na^+ with long-chain amphiphiles with length of hydrocarbone chain from 10 to 16, proceeding from UV-spectra and melting curves, as well as comparative analysis of the effect of Na^+, TMA^+ and long-chain cationic amphiphiles on the parameters of melting curves of DNA.

MATERIALS AND METHODS

DNA samples were extracted from milts of sturgeons, had a molecular weight of $1.7.10^7$, and contained less than 1% admixtures of proteins and RNAs. A DNA solution with the concentration of 1.5-2.0 mg/ml was dialized at 4°C against 0.1 M NaCl, 10 mM EDTA for a day and then against 10^{-2} M NaCl, 10^{-3} M Tris-HCl buffer (pH7.2) for three days. The determination of the nativity was carried out by UV spectra and melting curves. The hyperchromic effect at the wavelength 260 nm was 38-40%. The melting point in 0.1 M NaCl and phosphate buffer was 85°C. The DNA concentration in the solution was determined from UV spectra and it was 5.10^{-5} M (phosphate).

Dodecyltrimethylammonium bromide (DTAB) was from Aldrich, tetradecyltrimethylammonium bromide (TTAB) was from Serva hexadecyltrimethylammonium bromide was from Sigma and were used as purchased. The cationic amphiphiles effect on DNA was investigated in a 10^{-2} M NaCl solution containing 10^{-3} M Tris-HCl buffer (pH7.2). The UV spectra were recorded with a Specord UV/VIS spectrofotometer (Germany) and VSU2-P spectrofotometer (Carl Zeiss, Jena, Germany). Spectrofotometric melting curves were measured using a VSU2-P spectrofotometer equipped with a thermostat that controlled the sample temperature with an accuracy of 0.5°C. Melting curves were measured at incubation of the sample for 10 minutes at each fixed temperature. The data below were obtained for fresh solutions of complexes.

RESULTS AND DISCUSSION

UV-spectra of DNA complexes with long-chain N-trimethylsubstituted aliphatic amines

Typical changes of UV-spectra with increase of concentration of long-chain N-trimethylsubstituted derivatives of aliphatic amines have been shown. The view of concentrational dependence of optical density, A.

Sharp increase of optical density of DNA solution is observed when the concentration of cationic amphiphile reaches a fixed value. This is true for the peak of DNA absorption at λ_{max}=260 nm (*fig. 2*, curve a) and for the distant long-wave region $\lambda \geq 320$ nm, and there is no shift in λ_{max}. The reason for this sharp increase of A, starting from the fixed concentration of cationic amphiphile in complex formation of DNA with cationic amphiphile is the result of a cooperative replacement of Na^+ ions with organic cations, and for hexadecyltrimethylammonium bromide (HTAB), it corresponds to the concentration $7.0.10^{-6}$ M at room temperature and ionic strength 10^{-2} M. The values of concentration of cationic amphiphile corresponding to sharp increase of A are lowered with decreasing the length of aliphatic tail. For the same physicochemical conditions, these values are $2.0.10^{-3}$ M, $7.0.10^{-4}$ M and $8.0.10^{-5}$ M for decyltrimethylammonium bromide (DeTAB), dodecyltrimethylammonium bromide (DTAB) and tetradecyltrimethylammonium bromide (TTAB), respectively. Not only the results presented but the direct measurements [12, 13] evidence that the reason of the sharp increase of A is light scattering of the solution. It should be noted, that there are no the changes in absorption coefficient connected with disturbances of the secondary structure of DNA upon binding of cationic amphiphiles, which is supported by CD-data [14, 15]. After sharp increase of A a range of cationic amphiphile concentration is observed where optical density practically does not change, but further it is decreased. Optical density ceases to increase and reaches a plateau, since together with increase of A due to light scattering the decrease of A as the result of replacement of polar hydrate shell of DNA with hydrophobic one takes place [16, 17]. The decrease of A observed at high concentrations of cationic amphiphile is the result of aggregation and precipitation of complexes. In the case of HTAB it corresponds to the concentrations higher $8.0.10^{-5}$ M. The values of concentrations at which the lowering of A is observed, increase with shortening the length of hydrocarbon tail.

Now, let as consider how is the character of changes of spectral properties of solution of DNA-cationic amphiphile complexes from UV-data connected with the data on changes of parameters of pyrene probe, which is sensitive to the state of cationic amphiphile component of the complex.

At the concentrations called the concentrations of complex formation -C_d, the parameters of fluorescence of pyrene probe begin to change sharply evidences the formation of hydrophobic clusters [18]. The concentration of cationic amphiphile at which the sharp increase of A obtained from UV-spectra and the changes of the parameters of pyrene fluorescence coincide. At the concentration of cationic amphiphile corresponding to the concentration of complex formation the stoichiometry near 1:1, more exactly, 0.8 molecules of cationic amphiphile on 1 phosphate group was determined by using the ion-selective electrode [19]. The molecules of cationic amphiphile arrange as monolyer along the double helix forming the structure like "lamp-chimney brush" with some part of inverse molecules of cationic amphiphile in the complex, i.e. those contacted with water molecules by their cationic head, which results to additional stabilization of the complex in water surrounding [20].

Melting curves of DNA complex with long-chain N-trimethylsubstituted aliphatic amines

At concentrations of cationic amphiphiles $C<C_d$ practically there are no changes in parameters of melting curves of DNA. UV-melting curves of DNA without cationic amphiphiles and in the presence of different concentrations of TTAB can be observed. For the solution of DNA at ionic strength 10^{-2} M NaCl in the absence of cationic amphiphiles the following melting parameters were obtained: T_m=68-69°C, hypochromic effect, $\Delta H \cong 40\%$, and transition width, ΔT=11-15°C. The increase in T_m and ΔT, and lowering of ΔH takes place at addition of cationic amphiphiles at concentrations $C>C_d$. Let as consider the character of changes of these three parameters.

1. The increase of T_m of DNA at addition of long-chain N-trimethylsubstituted amines into solution evidences the stabilization of the double helix upon replacement of Na^+ ions with cationic amphiphiles. The same effect is observed at replacement of Na^+ ions with symmetric short-chain tetraalkylammonium cations in a wide range of their concentrations [21]. These substances according to Marmur interact with DNA at small groove, preferably on AT-rich sites, thus stabilizing the double helix. The reason of DNA stabilization with long-chain tetramethylsubstituted amines seems to be the hydrophobic contact between their hydrocarbon tails, though some contribution of the interaction of hydrophobic fragments on the small groove, analogous to short-chain tetraalkylammonium cations, occurs as well. The stabilizing affect of HTAB is essentially higher in comparison with the estimations from known equation [8, 22], to the results of the work [21], and to our data, for increase of T_m at the same value about 1000 times lower concentration of HTAB is necessary in comparition with Na^+ and 2000 time lower in comparition with TMA.

The increase of T_m, and hence, the stabilization of the double helix is the only common property characteristic of metal cations, symmetric alkylammonium cations and long-chain N-trimethylsubstituted amines.

2. The transition width, ΔT, is increased on addition of trimethylsubstituted amines at $C>C_d$, an asymmetry of melting profiles appeared, and at relatively high concentrations of N-trimethyl-

substituted aminer in solution (before opalescence and precipitation) an S-like melting curve transforms to double-step form, i.e. it is possible to choose a region of pre-melting. In the case of Na^+, TEA and TMA cations in solution the narrowing of melting curves and an increase in their symmetry observed. The reason seems to be an increase in inhomogeneity of stability of different sites of DNA upon the replacement of Na^+ cations on trimethylsubstituted amines [23].

3. Hyperchromic effect, ΔH, decreases on adding to the solution of long-chain N-trimethyl-substituted amines on the contrary to the effect of metal cations and short-chain symmetric tetraalkylammonium cations. It can be connected with: i) the appearance of light-scattering in the solution, ii) the formation of non-melting regions upon 95-100°C, iii) the replacement of water surrounding of DNA on hydrophobic one, resulting to the change in extinction coefficient of DNA, and iv) the precipitation of DNA associates.

The same changes are observed for the melting curves of DNA in the presence DTAB and HTAB.

Thus, cationic amphiphiles bound with DNA through electrostatic interaction provide due to brightly expressed hydrophobic contact between hydrocarbon chains the above mentioned character of the stabilization of DNA double helix, and the distinction from the character of the double helix stabilization by alkali metal cations and by symmetric short-chain alkylammonium cations.

ACKNOWLEDGEMENTS

K.R.L. acnowledges financial support of Robert Havemann Scolarship "Natural Scientists Iinitiative 'Responsibility for Peace'".

REFERENCES

1. Spiro T.G. (ed.), Nucleic Acid-Metal Interactions, Willey, New York, 1980.
2. Blagoi Yu. P., Galkin V.L., Gladchenko G.O., Metalocomplexes of nucleic acids in solutions. (rus), Naukova dumka, 1991.
3. Saenger W., Principles of nucleic acid structure. Springer, New York, 1984.
4. Zimmer Ch., Venner H., Naturwissenschaften. 1962, v. 61, N 8, p. 86.
5. Walwick E.R., Main R.K., Biochem. Biophis. Acta. 1962, v. 61, N 8, p. 876.
6. Daune M., In Structure and dynamics of Biopolymers (ed. Nicolini C.), 1987, Martinus Nijhoff Pablishers. P. 1-30.
7. Sapiro J.T., Stannard B.S., Felsenfeld G., Biochemistry, 1969, v. 8, N 8, p. 3233-3241.
8. Melchior W.B. Jr, Von Hippel P.H., Proc. Nat. Acad. Sci. USA, 1973, v. 70, N 2, p. 298-302.
9. Fredel D.L., Cationic Surfactants. Analitical and biological evalution (ed. Cross J, Singer E.J.), New York, Marcel Dekker, 1994, p. 31-60.
10. Lasic D.D., Liposomes in Gene Delivery. 1997, New York, CRC Press.
11. Shabarchina L.I., Sukhorukov B.I., Kuvichkin., Biofizika (rus), 1979, v. 24, p. 990-997.
12. Gorelov A.V., Kudryashov E.D., Jacquier J.-C., McLoughlin D.M., Dawson K.A., Phisica A, 1998, v. 249, p. 216-225.
13. Bathaie S.Z., Moosavi-Movahedi A.A., Saboury A.A., Nucleic Acids Res. 1999, v. 27, N 4, p. 1001-1005.
14. Pattarkine M.V., Ganesh K.N., Biochem. Biopys. Res. Comm. 1999, v. 263, N 1, p. 41-46.
15. Spink Ch.H., Chaires J.B., J. Am. Chem. Soc. 1997, v. 119, p. 10920-10928.
16. Bolton H.C., Weiss J.J., Nature, 1962, v. 195, p. 666-668.
17. Bakhshiev N.G., Spectroscopy of intermolecular interactions. Nauka L. 1972.
18. Sukhorukov B.I., Petrov A.I., Kazaryan R.L., Kuvichkin V.V., Biophisics, 2000, v. 45, N 2, p. 239-247.
19. Mel'nikov S.M., Sergeev V.G., Yoshikawa K., J. Am. Chem. Soc. 1995, v. 117, p. 9951-9956.
20. Sukhorukov B.I., Petrov A.I., Kazaryan R.L., Zurnal Fiz. Chim. 2002 (rus, in press).
21. Marmur J., Doty P., J. Mol. Biol., 1962, v. 5, N 1, p. 109-118.
22. Owen R.J., Hill L.R., Lapage S.P., Biopolymers 1969, v. 7, p. 503-516.
23. Lando D. Yu., Krot V.I., Frank-Kamenetskii M.D., Mol. Biol. (rus) 1975, v. 5, p. 856.

*Metal Ions in Biology and Medicine: vol. 7.*Eds L. Khassanova, Ph. Collery, I. Maymard, Z. Khassanova, J.-C. Étienne.
John Libbey Eurotext, Paris © 2002 pp. 104-107.

Surface-enhanced Fourier transform infrared spectroscopy of protein A conjugated with colloidal gold

Alexander A. Kamnev[1]*, Lev A. Dykman[1], Petros A. Tarantilis[2] and Moschos G. Polissiou[2]

[1]*Institute of Biochemistry and Physiology of Plants and Microorganisms, Russian Academy of Sciences, 13 Prosp. Entuziastov, 410015 Saratov, Russia*
[2]*Laboratory of Chemistry, Department of Science, Agricultural University of Athens, Iera Odos 75, 11855 Athens, Greece*

ABSTRACT

For staphylococcal protein A conjugated with colloidal gold (CG; monodisperse particles, mean diameter 30 nm), which is widely used in immunoassay, Fourier transform infrared (FTIR) spectroscopic studies were carried out using dried films prepared from CG/protein A sols on the surface of a standard ZnSe IR crystal window. FTIR spectra obtained for pure protein A and its CG bioconjugate have shown essential differences observed in some regions of characteristic vibration modes of protein A (involving, e.g. amide vibrations and some other bands) induced by the surface of CG particles. Comparative estimation of relative spectral intensities for pure protein A and its CG bioconjugate using FTIR spectroscopy in the attenuated total reflectance (ATR) mode has shown that CG induces the effect of surface-enhanced IR absorption (SEIRA) with moderate enhancement factors. The latter correspond to a relatively weak non-covalent (electrostatic and hydrophobic) binding of biopolymer molecules to the metal surface established upon bioconjugation in aqueous solution, which is nevertheless sufficient for stabilisation of the resulting colloidal sol. The FTIR spectroscopic data obtained provide evidence that protein A molecules are attached directly to the surface of CG particles. This condition is of primary importance for the synthesis of haptens with CG for subsequent immunization of animals, so that FTIR spectroscopy may be used as a rapid and sensitive technique for controlling the quality of such bioconjugates.

Keywords: protein A; colloidal gold; bioconjugates; Fourier transform infrared (FTIR) spectroscopy; surface-enhanced infrared absorption (SEIRA).

INTRODUCTION

Infrared (IR) spectroscopy, especially considering the greatly improved characteristics of modern Fourier transform (FT) spectrometers, has found its well-recognised place among a variety of instrumental techniques that give valuable structural information on the molecular level for complicated biological objects [1]. It is known that the effect of a thin layer of some metals (mostly Au or Ag) can result in surface-induced enhancement of FTIR absorption intensities of bands for functional groups adjacent to the surface (for a recent review see, e.g. [2]). This effect of surface-enhanced IR absorption (SEIRA) appears when the IR radiation induces an oscillating dipole in metal particles, aggregates or thin layers which, in its turn, sets up an electromagnetic field normal to the metal surface. The so-called surface selection rule, which is valid for thin layers close to

* Corresponding author. Tel./Fax: +7-(8452)-947303; E-mail: micbio@ibppm.saratov.su.

the metal surface, predicts that only those molecular vibrations which have transition dipole moment (TDM) components perpendicular to the metal surface will be enhanced, whereas those vibrations whose dipole moment change is parallel will not be observed [2]. This accounts for selective enhancement of various molecular vibration modes depending on the orientation of the functional groups involved relative to the surface plane. However, for non-flat or microscopically rough metal surfaces the surface selection rule has to be applied and interpreted with caution, in particular, owing to the possibility of non-uniform adsorption geometry and/or coexistence of different types of adsorbate species or different surface sites [3].

In SEIRA, the observed enhancement factors are usually of the orders of several units up to several hundreds. Though electromagnetic interactions play predominant roles in the SEIRA effect, chemical interactions of the adsorbed molecules with the surface (i.e., chemisorption or the formation of covalent bonds) may result in some additional enhancement [2]. The SEIRA effect may also include specific shifts of relevant bands and thus be sensitive to direct metal-molecule contact interactions or changes in molecular conformations occurring upon adsorption onto the metal surface.

In the present work, we studied FTIR spectroscopic properties of dry films prepared from sols of colloidal gold (CG) conjugated with staphylococcal protein A which are widely used in immunoassay [4], as compared to FTIR spectra of pure protein A. The results have shown that some characteristic bands of the protein are essentially affected by the CG surface. The spectroscopic data obtained provide evidence that the biomolecules are attached directly to the CG surface, which may be used as a rapid and sensitive method for controlling the quality of CG bioconjugates.

MATERIALS AND METHODS

Preparation of colloidal gold and its bioconjugate with protein A

Aqueous CG sol with monodisperse spherical particles (mean diameter 30 nm) was prepared from tetrachloroauric(III) acid (Aldrich, USA) using sodium citrate (Fluka, Switzerland) as a reducing agent at 100°C according to Frens [5]. The size of gold particles was controlled spectrophotometrically [6]. Conjugation was performed by mixing the gold sol (at pH 6.5) with staphylococcal protein A (Sigma, USA) taken in amounts exceeding by 20% the minimal amount necessary for stabilising the gold sol (estimated by disappearance of changes in the optical properties of the stabilised sol upon addition of 1% NaCl), with further separation of the bioconjugate by centrifugation (5 min, 4°C; 12000g; centrifuge Sigma 3K18, USA) and re-dissolving in bidistilled water. The concentration of protein A in the final sol (A_{520} = 1.0) was 6 μg ml^{-1}.

Sample preparation for FTIR measurements and acquisition of spectra

For preparing dry films used in FTIR measurements, 5 ml of CG/protein A conjugate was centrifuged as stated above, the supernatant was removed, the centrifugate was placed as a small drop at the surface of a ZnSe crystal support using ca. 10 μg of protein A for FTIR absorption (ZnSe IR crystal window, 13 mm diameter, 2 mm thick; Aldrich, USA) or as a narrow straight lane with ca. 20 μg of protein A for FTIR in the attenuated total reflectance (ATR) mode (standard ZnSe 45° flat plate Contact Sampler; 12 reflections; Spectra-Tech, USA) and dried in a thermostatted desiccator for 1 h at 45°C. For preparing dry films of pure protein A, 100 μl of its aqueous solution (1 mg ml^{-1}) were placed on a ZnSe support and dried as stated above. Spectroscopic measurements were performed using a Nicolet FTIR spectrometer (USA), model Magna-IR 750 (DTGS detector, Nichrome source; beamsplitter: KBr), with a total of 100 scans (resolution up to 4 cm^{-1}). Treatment of the resulting spectra was performed using the standard software OMNIC 3.1 (Nicolet, USA).

RESULTS AND DISCUSSION

The FTIR spectrum of pure protein A *(fig. 1a)* is quite typical for proteins and represents the vibrational regions for certain functional groups of protein macromolecules. In particular, the following bands corresponding to specific vibration modes [1] should be noted:

• a very broad absorption envelope (about 3500-2700 cm^{-1}) corresponding to stretching ((O-H) modes of hydrogen-bonded carboxylic, alcoholic and phenolic OH-groups in amino acid residues;

• broad stretching ν(N-H) amide-related bands centred at 3297 and 3071 cm^{-1};

• antisymmetric (ν_{as}) and symmetric (ν_s) stretching vibrations of $-CH_2-$ and $-CH_3$ groups of amino acid residues (*ca.* 3000-2800 cm^{-1});

• strong typical amide I and amide II bands resulting mainly from stretching ν(C=O) and in-plane bending δ(N-H) vibrations coupled to ν(C-N) in peptides, respectively, at 1656 and 1544 cm^{-1});

• CH_2 bending vibrations (1444 cm^{-1});

• stretching $C-NH_2$ of side chain primary amines (near 1400 cm^{-1}), as well as various weaker bands related to C-N, C-O, C-C-N, C-C-O, *etc.* stretching and bending vibrations of the protein backbone and amino acid residues (under 1350 cm^{-1}).

The FTIR spectrum of CG/protein A bioconjugate (dry film; *fig. 1b*) markedly differs from that of pure protein A. The major clearly visible spectroscopic differences induced by the CG surface involve the regions of amide-related vibrations. In particular, the ν(N-H) bands at 3297 and 3071 cm^{-1} (see *fig. 1a*) virtually disappeared, and instead of the strong well-resolved amide I and amide II bands (1656 and 1544 cm^{-1}; see *fig. 1a*), an asymmetric intermediate strong band appeared with a maximum at 1565 cm^{-1} and an essential non-resolved component at ca. 1595 cm^{-1}; a weaker shoulder at 1660 cm^{-1} is also present obviously related to the amide I component. In addition, the peak at 1408 cm^{-1} (see *fig. 1b*), which may probably be associated with the $\nu(C-NH_2)$ mode (observed at 1399 cm^{-1} in *fig. 1a*), essentially increased in intensity. Other changes include some redistribution of intensities of methylene $\nu_{as}(CH_2)$ and $\nu_s(CH_2)$ bands (2922 and 2851 cm^{-1}, respectively, in *fig. 1b*) as compared to the corresponding $\nu_{as}(CH_3)$ and $\nu_s(CH_3)$ bands of methyl groups (cf. 2960 and 2876 cm^{-1}, respectively, in *fig. 1a*), which may be caused by different orientations of the corresponding TDM components relative to the metal surface. However, it is difficult to compare TDM orientations for different vibrations relative to the CG surface plane, as the position and conformation of the protein molecules conjugated to CG particles are virtually unknown.

The FTIR spectroscopic changes observed in going from pure protein A to its CG bioconjugate are evidently caused by the effect of the surface of CG particles. Note that it was difficult to compare absorptivity values in FTIR absorption spectra of dry films of pure protein A and its bioconjugate in the standard transmission mode, which is necessary for assessing the enhancement factors, without standardizing the film thickness. Nevertheless, our estimation using FTIR-ATR spectroscopy (Kamnev *et al.*, in preparation) has shown that for CG/protein A bioconjugates the SEIRA effect is observed with enhancement factors of the order of 10 to 15. The latter relatively low values correspond to a relatively weak non-covalent (electrostatic and hydrophobic) binding of biopolymer molecules to the metal surface established upon bioconjugation in aqueous solution. This weak binding mode, which allows the nativity of biomacromolecules to be conserved, is nevertheless sufficient for stabilisation of the resulting colloidal sol for a long time preventing it from aggregation and subsequent flocculation, which is essential for various applications of CG bioconjugates [4].

In conclusion, it has been shown for the first time using dry films that certain characteristic IR absorption bands of protein A (e.g., amide I, amide II and some other vibration modes) are essentially affected by the CG surface. The FTIR spectroscopic data obtained also confirm that the protein molecules are attached directly to the CG surface. This condition is of primary importance for the synthesis of haptens with CG for subsequent immunization of animals [7], so that FTIR spectroscopy may be used as a rapid and sensitive technique for controlling the quality of such bioconjugates.

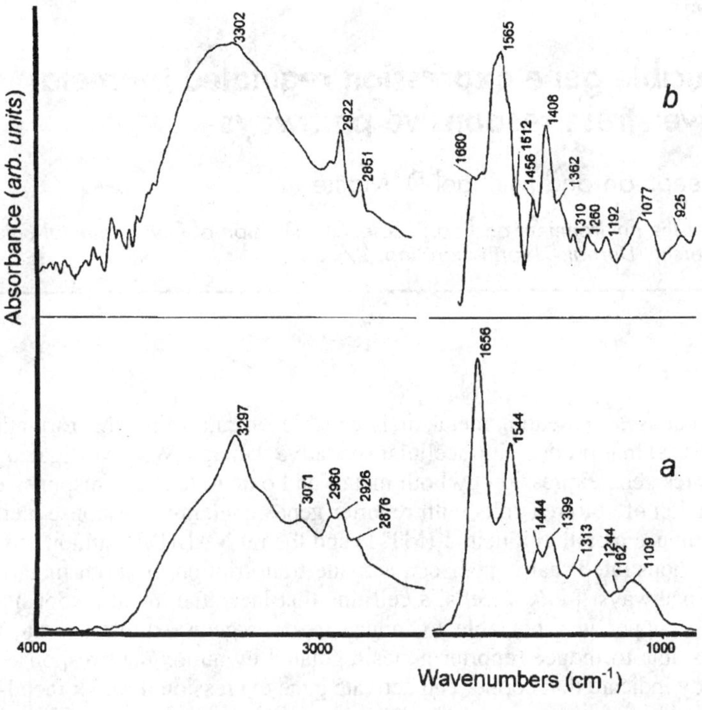

Fig. 1. Fourier transform infrared spectra of dry films of (a) pure protein A and (b) its bioconjugate with colloidal gold

ACKNOWLEDGEMENTS

This work was supported in part by INTAS (EC, Brussels, Belgium), NATO (Grant LST.CLG.977664), the Russian Foundation for Basic Research (Grant No. 01-04-48736) and the Russian Academy of Sciences' Commission (Grant No. 205 under the 6th Competition-Expertise of research projects), as well as by UNESCO Short-term Fellowships in biotechnology under Contracts UVO-ROSTE 875.878.9 (to A.A.K., 2000) and UVO-ROSTE 875.687.1 (to L.A.D., 2001).

REFERENCES

1. Mantsch H.H., Chapman D. (Eds.). *Infrared Spectroscopy of Biomolecules*, Wiley-Liss, New York (1996).
2. Osawa M. Surface-enhanced infrared absorption. In: Kawata S. (Ed.), *Near-Field Optics and Surface Plasmon Polaritons. Topics in Applied Physics*, Vol. 81, pp. 163-187. Springer, Berlin (2001).
3. Merklin G.T., Griffiths P.R. Effect of microscopic surface roughness in surface-enhanced infrared absorption spectrometry. *J. Phys. Chem. B* **101**, 5810-5813 (1997).
4. Dykman L.A., Bogatyrev V.A. Colloidal gold in solid-phase assays. A review. *Biochemistry* (Moscow) **62**, 350-356 (1997).
5. Frens G. Controlled nucleation for the particle size in monodisperse gold suspensions. *Nature Phys. Sci.* **241**, 20-22 (1973).
6. Khlebtsov N.G., Bogatyrev V.A., Dykman L.A., Melnikov A.G. Spectral extinction of colloidal gold and its biospecific conjugates. *J. Colloid Interface Sci.* **180**, 436-445 (1996).
7. Dykman L.A., Matora L.Yu., Bogatyrev V.A. Use of colloidal gold to obtain antibiotin antibodies. *J. Microbiol. Methods* **24**, 247-248 (1996).

Metal Ions in Biology and Medicine: vol. 7. Eds L. Khassanova, Ph. Collery, I. Maymard, Z. Khassanova, J.-C. Étienne.
John Libbey Eurotext, Paris © 2002 pp. 108-112.

Copper-inducible gene expression regulated by metal and oxidative stress responsive pathways

Jonathan H. Freedman and Michael D. Mattie

Nicholas School of the Environment and Earth Sciences, Division of Environmental Sciences and Policy, Duke University, Durham, North Carolina, 27708, USA

ABSTRACT

Although copper is an essential metal, it is capable of catalyzing the formation of reactive oxygen species (ROS) that produce intracellular oxidative damage. We investigated the hypothesis that copper regulates gene expression by both metal- and oxidative stress-responsive mechanisms. Transient transfection of cultured cells with reporter genes containing response elements from the promoters of the mouse metallothionein-1 (MT-1) and the rat NAD(P)H:quinone oxidoreductase 1 genes (NQO1) demonstrated that copper can activate transcription via both metal- and oxidative stress-responsive pathways. In *dko7* cells, a cell line that lacks the metal-responsive transcription factor-1 (MTF-1), copper was not able to induce reporter gene expression via metal response elements, but was able to induce reporter genes regulated by antioxidant response elements. The results of this study indicate that copper can activate gene expression through metal- and oxidative stress-responsive pathways.

INTRODUCTION

Transition metals such as copper are essential to biological processes. While it is critical for cells to acquire and maintain homeostatic levels of essential metals, copper overload induces pathologic effects that are consistent with intracellular oxidative damage. As a redox-active metal, copper is able to catalyze the formation of hydroxyl radicals via a Haber-Weiss or Fenton-like reactions (Walling 1975). At normal physiological concentrations, copper regulates the activity of several genes through a metal-responsive mechanism; but at higher concentrations, it causes alterations in gene expression through an oxidative stress-responsive mechanism.

Metal-induced stress responses commonly involve altered expression of many genes, which are responsible for maintaining metal homeostasis and protecting cellular components from damage (Zhu and Thiele 1996). One mechanism employed by cells to protect against excess metals is via the expression of metallothionein (MT). Elevated concentrations of many transition metals have been shown to elicit rapid induction of MT mRNAs and proteins in a wide variety of organisms and cultured cells (Hamer 1986, Kagi and Schaffer 1988, Schmidt and Hamer 1986). Additional functions ascribed to MTs include cellular detoxification and scavenging of free radical. A wide variety of stressors such as heat-shock, alkylating agents, oxidative stress, and UV irradiation induce MT transcription, suggesting that MT may be a general stress-response protein (Durnam and Palmiter 1984, Sato and Bremner 1993).

Studies of MT transcriptional regulation by metals have focused primarily on metal response elements (MREs), which are upstream regulatory elements found in the promoter regions of most metal-inducible MTs (Stuart et al. 1985). An additional *cis* regulatory element, the antioxidant response element (ARE), is involved in regulating oxidative stress-induced MT transcription (Dalton et al. 1994). AREs have been identified in the promoters of several ROS detoxification enzymes including glutathione-S-transferase and NAD(P)H:quinone oxidoreductase (Jaiswal 1994). ARE-mediated induction of detoxification enzymes is thought to be a critical mechanism involved in protecting the cell from electrophilic and ROS challenges. There is a paucity of data addressing the ability of redox cycling transition metals to activate transcription through the ARE.

As an essential metal, copper mediates several important biological processes through multiple signal transduction pathways. However, the cellular and molecular responses underlying copper-regulated gene expression and toxicity are poorly understood. The present study demonstrates that the treatment of cultured cells with copper results in the transcriptional activation of MT-I through both metal and antioxidant responsive pathways. This indicates that copper is capable of activating transcription through both metal- and oxidative stress-mediated mechanisms.

MATERIALS AND METHODS

Cell Culture and Reporter Gene Constructs

COS-7 cells (monkey kidney fibroblast) and MTF-1 knockout cells *(dko7)* (Radtke et al. 1994) were grown in Dulbecco's Modified Eagle's Medium supplemented with 10% fetal bovine serum, 2 mM L-glutamine, 100 units/ml penicillin, 100 μg/ml streptomycin, and 100 μM nonessential amino acids. Cells were maintained in a humidified incubator at 37°C with 5% CO_2, in 100 mm polystyrene tissue culture dishes. Chloramphenicol acetyltransferase (CAT) reporter genes containing regions of the mouse MT-1 promoter (-42/+63:CAT, -153/+62:CAT, $MREd_5$:CAT, and ARE_4:CAT) were provided by Dr. Glen Andrews (University of Kansas Medical Center). The region between -42 to +63, relative to the transcription start site in the MT-I gene, contains the TATA box and is defined as the minimal promoter. The proximal 153 bp of the MT-I promoter contains four functional MREs, as well as an ARE. $MREd_5$:CAT consists of five tandem copies of MRE_d fused to the MT-I minimal promoter, while ARE_4:CAT consists of four tandem copies of the ARE fused to the minimal promoter (Dalton et al. 1994). CAT reporter genes containing the ARE from the rat NQO1 gene (pARE and M1) were provided by Drs. Truyen Nguyen and Cecil Pickett (Schering Plough Research Institute). pARE contains the rat NQO1 ARE fused to the mouse glutathione-S-transferase minimal promoter. The M1 plasmid is similar to the pARE plasmid but it contains a single base mutation in the ARE sequence that abolishes responsiveness to oxidative stress (Favreau and Pickett, 1991).

Transient Transfection and Reporter Gene Assays

Cells were plated at a density of 8.0×10^4 cells/well in 24 well culture plates and allowed to grow for 18-24 hours before transfection. Cells were washed with Opti-MEM and then transfected using Lipofectin (COS-7) or LipofectAMINE *(dko7)* according to the manufacturer's instructions. Cells were transfected with 650 ng/well of one of the CAT-reporter constructs and 160 ng/well of pSV-βgal, as an internal control for transfection efficiency. Following transfection, cells were allowed to recover overnight. Treatment with copper sulfate was performed the following day as described in below. CAT protein concentration and β-galactosidase activity were determined using commercially available kits according to manufacturer's instructions. CAT protein levels were normalized to β-galactosidase activity.

Statistical Analysis

All statistical analysis was performed using StatView software. The results are presented as the mean ± SEM. The significance of mean differences was detected by analysis of variance (ANOVA) followed by Fisher's Protected Least Squares Difference *post hoc* test for individual comparisons. The criterion for statistical significance was set at $p < 0.05$.

RESULTS

Reporter gene assays utilizing CAT-fusion genes under the control of response elements, or portions of the promoter region from mouse MT-I gene were performed to investigate mechanisms by which copper activates transcription. COS-7 cells transfected with the four MT-I fusion genes were exposed to 0-600 μM $CuSO_4$ for 24 hours. The reporter gene activity was not induced at any concentration tested using the -42/+63-CAT, while significant increases were observed at 400 μM and above with the other MT reporter genes *(fig. 1A)*. To further investigate transcriptional activation by copper through an oxidative stress responsive pathway, COS-7 cells were transfected with the pARE and M1 reporter gene. The expression of pARE was increased significantly in cells exposed to 400 μM $CuSO_4$ for 24 hours *(fig. 1B)*. Copper did not induce gene expression in cells transfected with the non-functional ARE.

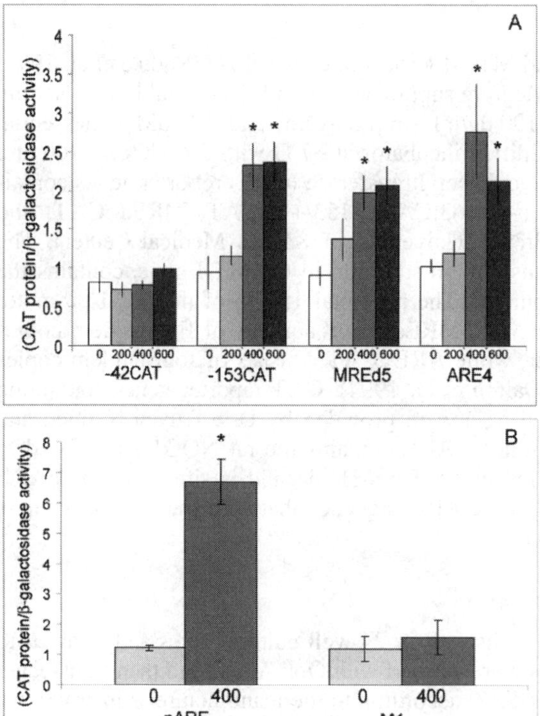

Fig. 1. Effect copper exposure on gene expression. **A:** COS-7 cells were transfected with MT-based reporter genes and exposed to 0-600 μM $CuSO_4$ for 24-h. **B:** Cells were transfected with NQO1-based reporter genes and exposed to 0 or 400 μM $CuSO_4$. Data are expressed as mean ± SEM. *Significantly different from cells not treated with copper, by ANOVA, $p < 0.05$, n = 3 observations

To further investigate copper-induced MT-1 transcription we used *dko7* cells, which do not express MTF-1. MTF-1 is the transcription factor that binds to MREs and it is required for basal and metal-inducible MT expression. Cells were transfected with the 153/+62:CAT, $MREd_5$:CAT, and ARE_4:CAT plasmids and exposed to 600 μM copper for 24 hours *(fig. 2)*. There was no significant induction of 153/+62:CAT and $MREd_5$:CAT reporter gene activity. However, there was a significant induction of ARE_4:CAT expression (5.3-fold). The *dko7* cells were also transfected with the pARE plasmid. In agreement with the earlier results, copper exposure induced expression of reporter gene activity using the pARE plasmid (4.6-fold).

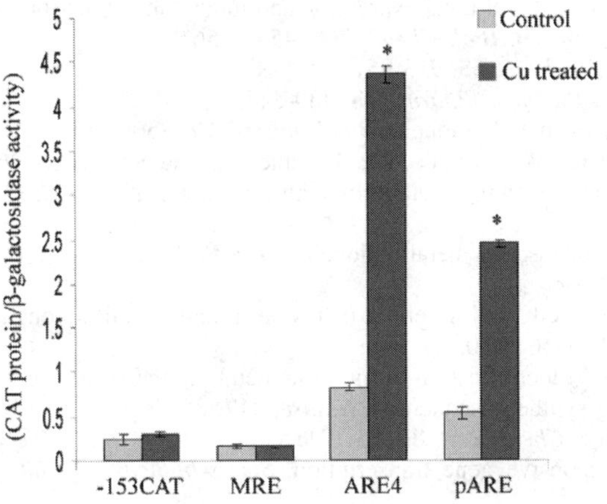

Fig. 2. Effect copper exposure on gene expression in a non-metal responsive cell line. *dko7* cells were transfected with MT-based *(-153CAT, MRE, ARE4)* and NQO1-based *(pARE)* reporter genes, and then exposed to 600 μM $CuSO_4$ for 24 h *(Cu treated)*. Data are expressed as mean ± SEM. *Significantly different from cells not treated with copper *(control)* by ANOVA, $p < 0.05$, n = 3 observations.

DISCUSSION

Previous studies have established that transition metals and oxidative stress induce MT-I expression (Dalton et al. 1994, Dalton et al. 1996, Bauman et al. 1991). Studies of MT-I transcriptional regulation have primarily focused upon zinc and cadmium activation of MT, MTF-1, and activation through MREs. Less attention has been paid to mechanisms involving induction via the ARE, or cooperative induction between both types of response elements. Reporter gene assays previously established that zinc, cadmium, H_2O_2, and tBHQ induce transcription through MREs, while the ARE sequence mediates induction by H_2O_2 and cadmium, but not zinc and tBHQ (Dalton et al. 1994, Dalton et al. 1996).

Several transition metals such as nickel, chromium, iron and copper are capable of free radical production and oxidative damage. Activation of the $MREd_5$:CAT and ARE_4:CAT reporter genes by copper provided initial evidence that copper may be acting as a dual activator of transcription. Induction of pARE, which contains only an antioxidant response element from the NQO1 gene, further supported oxidative stress-induced transcriptional activation by copper. The results of this study are consistent with previous models of MT transcriptional regulation and provide evidence that copper is capable of activating MT-I transcription via a metal- responsive mechanism that requires the presence of MTF-1. In addition, copper is capable of activating transcription through oxidative stress responsive mechanism. Further studies will be essential to understanding mechanistic differences underlying metal- and oxidative stress-mediated transcriptional activation.

REFERENCES

1. Bauman, J., Liu, J., Liu, Y.P. and Klaasen, C.D. (1991) Increase in metallothionein produced by chemicals that induce oxidative stress. *Toxicol. Appl. Pharm.*, **110**, 347-354.
2. Dalton, T., Palmiter, R.D. and Andrews, G.K. (1994) Transcriptional induction of the mouse metallothionein-I gene in hydrogen peroxide-treated Hepa cells involves a composite major late transcription factor/antioxidant response element and metal response promoter elements. *Nuc. Acids Res.*, **22**, 5016-5023.
3. Dalton, T.P., Li, Q., Bittel, D., Liang, L. and Andrews, G.K. (1996) Oxidative stress activates metal-responsive transcription factor-1 binding activity. Occupancy *in vivo* of metal response elements in the metallothionein-I gene promoter. *J. Biol. Chem.*, **271**, 26233-26241.
4. Durnam, D.M. and Palmiter, R.D. (1984) Induction of metallothionein-I mRNA in cultured cells by heavy metals and iodoacetate: evidence for gratuitous inducers. *Mol. Cell Biol.*, **4**, 484-491.
5. Favreau, L. and Pickett, C.B. (1991) Transcriptional regulation of the rat NAD(P)H:quinone reductase

gene. Identification of regulatory elements controlling basal level expression and inducible expression of planar aromatic compounds and phenolic antioxidants. *J. Biol. Chem.*, **266**, 4556-4561.
6. Hamer, D.H. (1986) Metallothionein. *Ann. Rev. Biochem.*, **55**, 913-951.
7. Jaiswal, A. (1994) Antioxidant response element. *Biochem. Pharm.*, **48**, 439-444.
8. Kagi, J.H. and Schaffer, A. (1988) Biochemistry of metallothionein. *Biochemistry*, **27**, 8509-8515.
9. Radtke, F., Heuchel, R, Georgiev, O., Hergersberg, M., Gariglio, M., Dembic, Z., and Schaffner, W. (1993) Cloned transcription factor MTF-1 activates the mouse metallothionein I promoter. *EMBO J.*, **12**, 1355-1362.
10. Sato, M., and Bremner, I. (1993) Oxygen free radicals and metallothionein. *Free Rad. Biol. Med.*, **14**, 325-337.
11. Schmidt, C.J., and Hamer, D. H. (1986) Cell specificity and an effect of ras on human metallothionein gene expression. *Proc. Natl. Acad. Sci., USA*, **83**, 3346-3350.
12. Stuart, G.W., Searle, P.F. and Palmiter, R.D. (1985) Identification of multiple metal regulatory elements in mouse metallothionein-I promoter by assaying synthetic sequences. *Nature*, **317**, 828-831.
13. Walling, C. (1975) Fenton's reagent revisited. *Acc. Chem. Res.*, **8**, 125-131.
14. Zhu, J. and Thiele, D.J. (1996) Toxic metal-responsive gene transcription. *Stress-Inducible Cellular Response*. Verlag Basel, Switzerland, pp. 306-320.

*Metal Ions in Biology and Medicine: vol. 7.*Eds L. Khassanova, Ph. Collery, I. Maymard, Z. Khassanova, J.-C. Étienne.
John Libbey Eurotext, Paris © 2002 pp. 113-115.

DNA as an enzyme and effect of metal ions in activation and inhibition, a route in mechanism of carcinogenesis

Bijan Farzami, Ramezan Sadeghi and Sayedeh Zahra Bathaie

Department of Clinical Biochemistry. School of Medicine Tehran University of Medical Sciences. P.O. Box 14155-5399 Tehran, Iran and Department of Clinical Biochemistry Tarbiat Moddares, University Tehran, Iran

ABSTRACT

In this report, it was found that in an electron transfer reaction in which an electron is transferred to an imidazole containing compound the rate is enhanced in presence of ***DNA*** and a specific metal ion. In this present study several metal ions were used to investigate the effect of metal ions in such reactions. The rate enhancement was associated with the turn over constant kp = 7 x 10^{-5} M/hr Mb. The use of Ni(II) and Pb(II) induced inhibition in the rate of electron transfer reaction in presence of DNA, although metal ions such as Mg(II), Cd(II), Zn(II), and Fe(II) caused activation of DNA. The rates of the reactions showed strong dependency on electronegativity of metal ions namely the increase in activity of DNA reaction due to each metal ion correlated with decreasing electronegativity of each metal ion. Correlation was also observed between the activity of DNA with metal conductance.

INTRODUCTION

The fluorescence activation by oxidation of diacetyl dichlorofluorescin (LDACF) to a nonfluorescent dichlorofluorescin (LDCF) and to fluorescent dichlorofluoroscein (DCF) has been used to detect ultramicroquantities of hydroperoxides i.e. lipid hydroperoxides as well as H2O2 detection in cellular processes [1, 2] and in enzyme system such as glucose oxidase [3, 4]. A method using LDCF was also employed to detect and evaluate the oxidative burst in neutrophiles [5]. Topographic distribution of oxidative stress temporal alterations in microcirculatory units of isolated perfused liver could also be studied by such method. Based on the nature of electron transfer process associated with the conversion of LDCF to DCF, we employed DNA and some imidazole containing compound to observe if such reaction components could interfere with the rate of the reaction. The assumption that was made for such a selection was based on the reports concerning the electrical potentials that was observed in double helical structure of DNA [6]. Other reports indicated that some aberrations in DNA structure is caused by obstruction of electrical flow through DNA [7]. The present report is an extension of a study in which a catalytic activity was observed in DNA [8].

MATERIALS AND METHODS

Materials: *2*', 7' - dichlorofluorocin diacetate (LDADCF) was a kind donation by Prof. P. Collery. imidazole derivatives were from Sigma Chemical Company. Hydrogen peroxide, Hematin and Metal salts were from Merck. (Germany). DNA was solvent extracted and was purified from calf Thymus.

Methods: A fluorometric method was used to study the rate of an oxidation process in which dichloroflourocin by losing an electron in the presence of peroxides and hematin is transformed to fluorescent dichlorofluoresceine. This reaction was enhanced several time in presence of ***DNA*** purified from calf thymus. The kinetics of catalysis was investigated further using several metal ions such as ***Pb(II)***, ***Ni(II)***, sodium hydroxide.

Sample preparation: 100 ml. of Tris-HCl buffer pH,7.5 was mixed with 14 ml. of hematin solution and boiled for 15 minutes. This solution was cooled on ice while purging with nitrogen. To prepare the medium for the reaction, a 2.7 ml. volume of this solution was mixed with 200 μl of activated dichlorofluorescin (4×10^{-5} M.) and 100 μl. Hydrogen peroxide (9×10^{-6} M). Ultimately DNA, imidazole derivatives and metal ions were added. Then the sample mixture was incubated under a nitrogen atmosphere in sealed vials at 50°C. for 45 or 10 minutes, depending on the speed of product formation. The reaction was cooled to room temperature and the relative fluorescence was determined using a spectrofluorophotometer (Shimadzu) with a 4 ml fluorescence cells. The excitation and emission wavelengths were at 500 and 520 nm, respectively, both with 2 nm band-width [1-4]. The optimal value of pH was determined in buffered solutions ranging in pH 7.0-9.0 with all the components of the reaction present. The assays were performed at each 0.2 unit pH intervals.

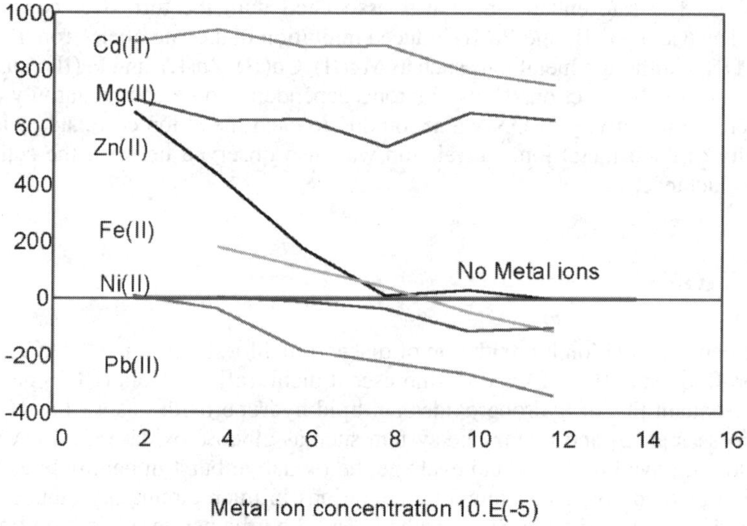

Fig. 1. Activity of DNA as a rate of increase in fluorescence vs. metal ions concentration. Zero line represents the reaction with no metal ions present

RESULTS AND DISCUSSION

In our experiments, the simultaneous use of DNA and imidazole derivatives in oxidation reaction of LDCF to DCF produced acceleration in the rate of the reaction that was absent without them. The metal ions produced effects that could portray the role of metal ions as the rate enhancer or inhibitors. DNA and N_trans-cynnamoyl imidazole as cofactors in presence of a metal ion such as Cd(II) accelerated the rate of the reaction from 0.55 μMol/min.ml to 5.7 μMol/min.ml or almost 10 times. These phenomena could be related to the catalytic potentials of DNA. The studies were carried out with some other metal ions such as Ni(II), Pb(II), Fe(II), Mg(II), Zn(II). Metal ions such as Pb(II) and Ni(II) produced inhibition in the rate of electron transfer reaction. Other metals such as Cd(II) and Mg(II) caused acceleration of the rate. Zn(II) and Fe(II) activated the reaction in lower concentration of metal ions but this effect was reversed in higher concentration of metal ions *(fig. 1)*. This effect could be attributed to the additional ligands that a metal ion could participate, in higher concentrations. The trend of metal activation or inhibition could be related to the structure of metal ions. Metal ions such as Cd(II), Mg(II) and Zn(II) that have filled inner orbitals could bind to phosphate groups in DNA and act as a conductive ligand between the center of reaction to convey the transfer of electrons to DNA. Based on the previous reports, DNA in double helical structure resembled a chord with electrical potentials [6]. A direct dependency was observed between the value of conductance of these metals and their induction of activity. Conversely the metal ions Pb(II) and Ni(II) produced inhibition. This effect, based on the previous assumption, could be assigned to their unfilled inner orbitals. It is noteworthy to mention that the trends in conductance with the induction of activities were held with the inhibitory metals *(fig. 2)*. A reverse relationship was observed between the electronegativity of metal ions and their activity *(fig. 3)*. The use of imidazole derivatives was revealing in that the conjugated system in N-trans cinnamoyl imidazole produced significant activation in DNA catalytic reaction in presence of a metal ion such as Cd(II).

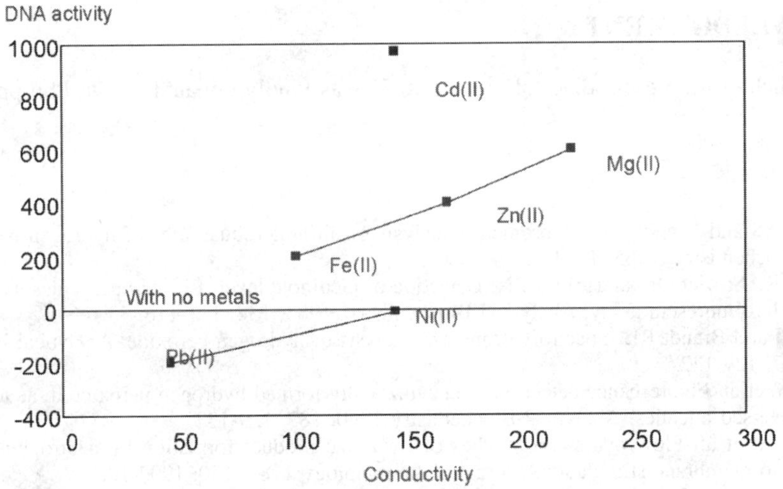

Fig. 2. DNA activity as a function of conductivity of metal ions

CONCLUSIONS

On the basis of the results obtained a mechanism is proposed in which the metal ion ***Mg(II)*** and ***Zn(II)*** or **Cd(II)** are acting as cofactors mediating the transfer of electrons through DNA chain. Metals such as ***Ni(II)*** and ***Pb(II)*** could act as electron pool to cause inhibition in such electron transfer reaction. The role of imidazole derivatives are to accelerate the rate of the reaction through mediating the transfer of electrons from the center of the reaction via the metal ion to the conductive chord of DNA. Therefore a process of catalysis can be evidenced. The fact that the metal ions which inhibit the reaction are those with carcinogenic potentials, points to the fact that restriction in the transport of electron through DNA may be related to the carcinogenic effect of these metals. Other reports indicated that this phenomena is observed when two or more guanine residues are placed adjacent to each other [9-10]. Considering the prime importance of such mechanism, a more detailed study of DNA catalysis, activation and inhibition is called for to clarify further both the nature of such phenomena and also its impact on the biological function of DNA.

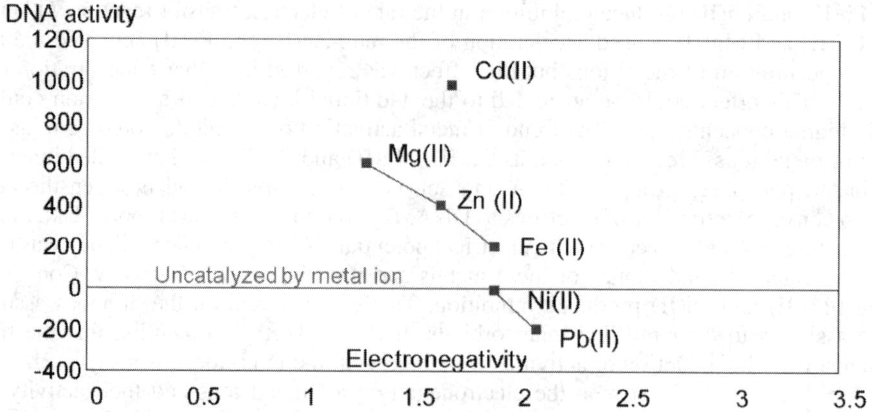

Fig. 3. DNA activity as a function of electronegativity of metal ions

ACKNOWLEDGEMENT

2', 7'- dichlorofluorescin diacetate (LADACF} was kindly donated by Dr. Philippe Collery.

REFERENCES

1. Kestone AS and Brandt. The fluorometric analysis of ultramicroquantities of hydrogen peroxide, Analytical Biochemistry, 1965, 11, 1-5.
2. Cathcart R., Shwiers E., and Ames BN: Detection of picomole levels of hydroperoxides using a fluorescent dichlorofluorescin assay. Analytical Biochemistry, 1983, 134, 111-116.
3. Black MJ and Brandt RB. Spectrofluorometric analysis of hydrogen peroxide, Analytical Biochemistry, 1990, 187, 129-132.
4. Frrer A.S. et al. Fluorescence detection of enzymztically formed hydrogen peroxide in aqueous solution and in reversed micelles. Analytical Biochemistry 1990, 187, 129-132.
5. Bass, D. A. et al. Flow cytometric studies of oxidative product formation by neutrophiles: A graded response to membrane stimulation. Journal of Immunology 1983, 130, 1907-10.
6. Arkin M.R. et al. Rates of DNA electron transfer between metallointercalators, Science 1996, V 273, p. 475.
7. Caghlan A. Electric DNA 1999 New Scientist p. 52.
8. Farzami, B. Shamsaie A. Farsam H. and Bathaie Z. The DNA catalysis of a carnosine based reaction and the inhibitory effect of Ni(II). Edits: Centeno J.A., Collery P. et al. Metal ions in Biology and Medicine, John Libbey Eurotext, Paris. 2000, vol. 6, p. 789.
9. Thomas J. Damage control, New Scientist, 1998, p. 36.
10. Hall D.B. et al. Oxidation DNA Damage through long range electron transfer, Nature V. 382, p. 731 (1996).

III CELL EFFECTS

Metal Ions in Biology and Medicine: vol. 7. Eds L. Khassanova, Ph. Collery, I. Maymard, Z. Khassanova, J.-C. Étienne.
John Libbey Eurotext, Paris © 2002 pp. 119-122.

Interaction of metal ions with cell membranes and molecular models

Mario Suwalsky[1,*], Fernando Villena[2], Beryl Norris[2], Hernan Cardenas[2], Paolo Zatta[3]

[1]Faculty of Chemical Sciences, [2]Faculty of Biological Sciences, University of Concepcion, Casilla 160-C, Concepcion, Chile, [3]CNR Center on Metalloproteins, Department of Biology, University of Padova, Italy.
**Corresponding author*

ABSTRACT

Although metals are normally used by cells, low levels of those that are nonessential, excess of essential and, particularly, toxic metals can produce serious harm. The cell membrane, as a diffusion barrier, protects the cell interior. Therefore, its structure and function are susceptible to be altered as a consequence of interactions with metal ions. In order to understand the molecular mechanisms involved metal ions such as Cu(II), Hg(II) and Al(III) were used to interact with human erythrocytes, isolated toad skin and molecular models of cell membranes. The latter, composed of multilayers of phospholipids commonly found in the outer and inner monolayers of cell membranes, were studied by X-ray diffraction. Examination by optical and scanning electron microscopy determined the type of shape change induced in erythrocytes by the metal ions. Accordingly to the bilayer couple hypothesis this information indicated in which monolayer the different metals were located. Electrophysiological measurements performed on toad skin revealed variations in the potential difference and short-circuit current after application of metal ion solutions, effects that interpreted the type and extent of the perturbation induced to the active transport of ions. The results obtained by these methods lead to the conclusion that the three metal ions altered the molecular structure of lipid bilayers, thereby modifying biophysical properties of cell membranes. However, the extent of these alterations was a function of the nature and concentration of each metal.

Key words: Metal ions; membrane; phospholipid bilayer; copper; mercury; aluminum.

INTRODUCTION

The cell membrane is a diffusion barrier, which protects the cell interior. Therefore, its structure and functions are susceptible to alteration as a consequence of interactions with heavy metals. Copper, mercury and aluminum are very common contaminants present in continental and sea waters. In aquatic environments copper exists in particulate, colloidal and soluble states, predominantly as metal and Cu^{2+} ions. Toxic effects of the latter upon membrane structure and functions have been reported, such as changes in the permeability and direct hemolysis of erythrocytes, decrease of neuronal membrane fluidity, changes in ionic channels and pumps, and destruction of the cell membrane of lymphocytes and monocytes [1]. In natural waters mercury can exist in three oxidation states: Hg^0, Hg^+, and Hg^{2+}; the distribution of these forms depends upon the pH, redox potential and availability of ligands which can form stable complexes. Several neurotoxic effects of mercury have been described at the cellular level: it modulates chloride channels, increases the release of neurotransmitters, and reduces the Na^+-K^+-ATPase activity [2]. Aluminum is the most abundant metal and the third most common element. However, despite its abundance, no useful biological function for it has been discovered. On the contrary, it is recognized as a toxic metal. In fact, compelling evidence has shown that abnormally high Al(III) levels are linked to pathologies such as dialysis dementia, iron-adequate microcytic anemia, osteomalacia and possible Alzheimer's disease [3]. The current work is a structural study undertaken in an attempt to understand the molecular mechanism of the interaction of Cu(II), Hg(II) and Al(III) with cell membranes. Because of the chemical and structural complexity of biological membranes, molecular models of cell membranes were used. The models consisted of multilayers of dimyristoylphosphatidylcholine (DMPC) and dimyristoylphosphatidylethanolamine (DMPE) which represent phospholipid classes located in the inner and outer monolayers of the human erythrocyte membrane, respectively [4]. Various concentrations of $CuCl_2$, $HgCl_2$ and $AlCl_3$ were incubated with DMPC and DMPE to examine their mechanism of action. The capacity of the metal ions to perturb their structures was determined by X-ray diffraction. In order to confirm whether the experimental results were reproducible in biological membranes, we explored the interaction of these salts with human erythrocytes, which were observed by scanning electron microscopy (SEM), and with isolated toad skin, in which electrophysiological parameters were measured.

MATERIALS AND METHODS

X-ray diffraction analysis of phospholipid multilayers

Synthetic DMPC and DMPE from Sigma and $CuCl_2$, $HgCl_2$ and $AlCl_3$ from Merck were used without further purification. About 1 mg of each phospholipid was introduced into special glass capillaries, which were then filled with distilled water and a concentration range of aqueous solutions of each salt. The specimens thus prepared were X-ray diffracted in flat plate cameras at 17 ± 2°C using Ni-filtered CuK radiation from a Philips generator.

The relative reflection intensities on films were measured by peak-integration using a Bio-Rad microdensitometer and image software.

SEM studies of human erythrocytes

Samples of blood mixed with a concentration range of $CuCl_2$, $HgCl_2$ and $AlCl_3$ were incubated at 37°C for 1 h, fixed overnight at 5°C with 2.5% glutaraldehyde, air dried at 37°C and gold coated. Resulting specimens were examined in an Etec Autoscan SEM.

Electrophysiological measurements on isolated toad skin

Pieces of abdominal skin were mounted in Ussing chambers. The short-circuit current (Isc) was monitored with non-polarizable Ag/AgCl electrodes connected to a voltage clamp circuit set to keep the potential difference (PD) across the skin at 0 mV. The PD was measured with calomel-agar electrodes. Each salt solution was added to the solution bathing the outer (mucosal) or the inner (serosal) surface of the skin in a range of concentrations.

RESULTS

Results of X-ray diffraction indicated that $CuCl_2$ in a concentration as low as 10 M induced a structural perturbation to the polar region of DMPC, which was completely disrupted by 1 mM $CuCl_2$. However, higher concentrations produced a reordering of DMPC bilayers *(fig. 1A)*. These effects can be respectively explained in terms of a first limited and then extended interactions between Cu^{2+} ions and DMPC phosphate groups. On the other hand, Cu^{2+} concentrations as high as 100 mM did not affect DMPE *(fig. 1B)*. The influence of Cu^{2+} ions on the physical properties of red cell membranes was studied by fluorescence spectroscopy on isolated resealed human erythrocytes. An ordering effect was observed at the polar group-water interface region, a result that agreed with that obtained by X-ray diffraction. It was also found that increasing concentrations of Cu^{2+} ions applied to either the mucosal or serosal surface of the isolated toad skin elicited a dose-dependent decrease of the Isc and PD, results that imply an inhibition of Na^+ transport across the epithelial cell membranes.

Exposure of DMPC to 10 M $HgCl_2$ induced a marked perturbation of the lipid structure, particularly of its polar head groups. However, 100 M $HgCl_2$ produced the same kind of molecular reordering as that observed with $CuCl_2$. On the other hand, increasing concentrations of $HgCl_2$ progressively perturbed both the polar and acyl chain regions of DMPE with equal intensity in a concentration-dependent manner. Human erythrocytes incubated with $HgCl_2$ showed changes from their normal discoid shape to both echinocytes and stomatocytes *(fig. 2)*. These effects mean that Hg^{2+} ions insert in both the outer and inner monolayers of the erythrocyte membrane, in agreement with the results of the X-ray experiments with DMPC and DMPE.

Experimental results indicate that $AlCl_3$ in the concentration range of 10-100 M induced the following structural and functional effects: i) perturbation of DMPC, and to a lesser extent of DMPE bilayers, ii) changes in the normal discoid shape of human erythrocytes to echinocytes due to the accumulation of Al(III) ions in the outer moiety of the red cell membrane, and iii) decrease in the Isc and in the PD of the isolated toad skin, effects that are in accordance with a time-dependent modulation of ion transport in response to changes in the molecular structure of the lipid bilayer.

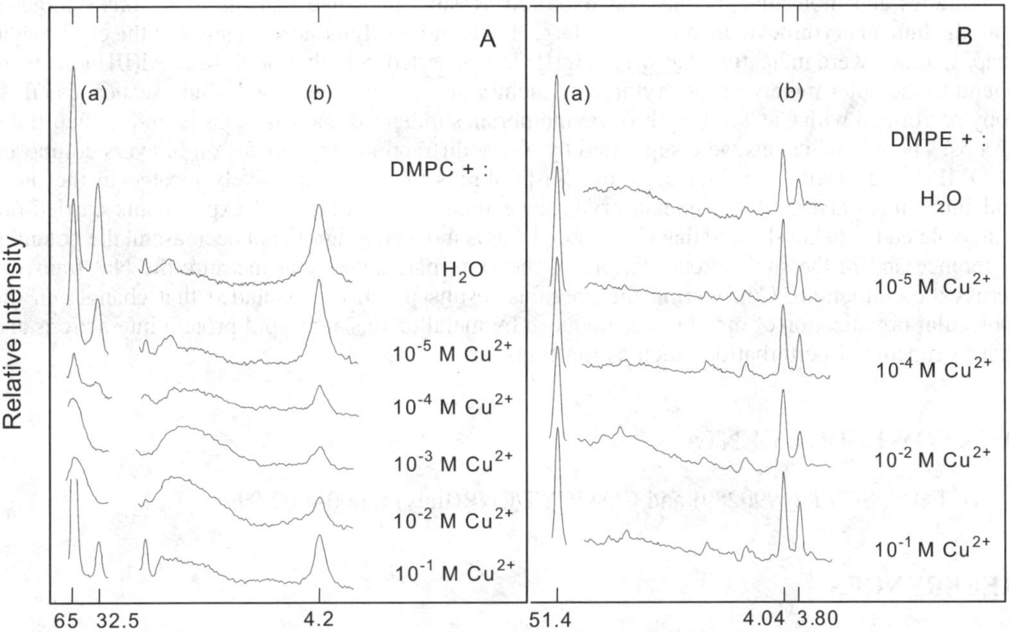

Fig. 1. Microdensitograms from X-ray diagramas of (A) dimiristoylphosphatidylcholine (DMPC) and (B) dimiristoylphosphatidylethanolamine (DMPE) with water and aqueous solutions of $CuCl_2$; (a) low-angle and (b) high-angle reflections.

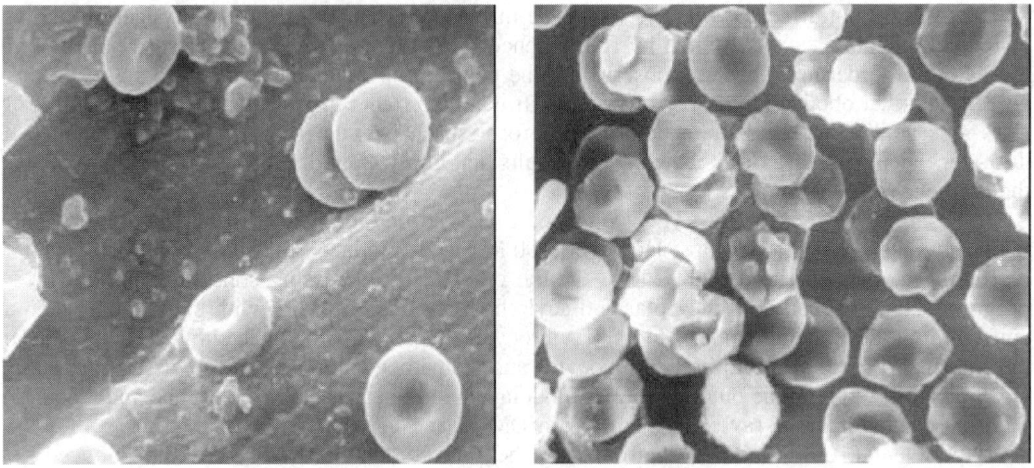

Fig. 2. Effect of Hg (II) on morphology of human erythrocytes. Scanning electron microscopy images of (A) untreated erythrocytes (2500x), and (B) erythrocytes incubated with 0.1 mM $HgCl_2$ (2000x).

DISCUSSION

In order to understand the molecular mechanisms of the toxicity of Cu(II), Hg(II) and Al(III) ions with cell membranes the interaction of their corresponding chloride salts with biological membranes and molecular models was explored. Results indicated that the three ions interacted with the human erythrocyte membrane. In fact, $HgCl_2$ and $AlCl_3$ induced changes in the erythrocyte shapes, which were indicative that while Hg(II) ions inserted in both monolayers, Al(III) ions were bound to the outer moiety of the erythrocyte membrane. On the other hand, fluorescence spectroscopy performed with $CuCl_2$ on erythrocyte membranes indicated that Cu^{2+} ions bound to their polar groups. These conclusions were supported by X-ray diffraction experiments on bilayers composed of DMPC and DMPE, representative of phospholipids that are respectively located in the outer and inner monolayers of the human erythrocyte membrane. Additional experiments carried-out with isolated toad skin showed that Cu^{2+} and Al^{3+} ions induced a significant decrease in the potential difference and in the short-circuit current of the skin, parameters that measure the Na^+ transport across the epithelium. Considering the previous results it can be concluded that changes in the molecular organization of membranes produced by metallic ions alter lipid-protein interactions and induce functional perturbations such as ion transport.

ACKNOWLEDGEMENTS

To FONDECYT (1990289) and CONICYT/CNR(Italy) (2000-5-02-099).

REFERENCES

1. Kiss, T., Osipenko O.N. Toxic effects of heavy metals on ionic channels. *Pharmacol. Rev.*, 1994, 46, 245-267.
2. Von Burg, R. Toxicology update. *J. Appl. Toxicol.*, 1995, 15, 483-493.
3. Yokel, R.A. The toxicology of aluminum in the brain. *Neurotoxicol.*, 2000, 21, 813-828.
4. Suwalsky, M. Phospholipid bilayers. *Polymeric Materials Encyclopedia*, 1996, 7, 5073-5078.

*Metal Ions in Biology and Medicine: vol. 7.*Eds L. Khassanova, Ph. Collery, I. Maymard, Z. Khassanova, J.-C. Étienne.
John Libbey Eurotext, Paris © 2002 pp. 123-126.

Zinc modulates the increased lipid peroxidation and apoptosis of rat Sertoli cell caused by induced hypomagnesaemia

Alexander Omu[1], Hussein Dashti[2], Abdullatif Al-Bader[3], Ayo Oriowo[4], Chacko Mathew[5], Tunde Fatinnikun[1]

Departments of obstetrics gynaecology[1], Surgery[2], Pathology[3], Pharmacology and toxicology[4], Electron microscopy unit[5], Faculty of medicine, Kuwait University, P O Box 24923, Safat, 13110 Kuwait

ABSTRACT

Background: Hypomagnesaemia has been shown to be associated with increased lipid peroxidation and apoptosis of cells. Zinc an essential component of superoxide dismutase, which is an important enzymatic antioxidant. The latter has an important function of scavenging free radicals and prevent lipid peroxidation.

Objective of study: To investigate the role of Zinc in modulating the lipid peroxidation and apoptosis by hypomagnesaemia of Rat Sertoli Cell.

Methodology: Three groups of Sprawley-Drew (SD) adult rat were used: (a) Five SD rats on normal diet as control (b) Five SD rats with induced hypomagnesaemia (c) Five SD rats with induced hypomagnesaemia and supplementation with Zinc. Blood was withdrawn from all the rats weekly and at the end of four weeks the rats were sacrificed and the testes harvested. In the sera, concentrations of Zinc, Mg, Cu, Se and Cd were estimated by Atomic Absorption Spectrophotometer (ASS), TNF, malonialdehyde and superoxide dismutase and caeruloplasmin were estimated by colorometric techniques.

After routine preparation of the testes, Electron Microscopy was used to study the pattern of apoptosis of the Sertoli cells with regards to blebbing of the cellular membranes and nuclear condensation of the chromatin.

Results: Hypomagnesaemia was associated with reduction of the testicular volume, and serum concentration of Zinc, Copper, Selenium and Magnesium, but increased level of Cadmium; decrease levels of superoxide dismutase and caeruloplasmin but increased in levels of malonylaldehyde and TNF. All the changes are corrected with Zinc supplementation. There was increased blebbing of the membrane and condensation of the nuclear chromatin of the Sertoli cells of the rat testes. These apoptotic changes were reduced with Zinc supplementation.

Conclusion: Hypomagnesaemia is associated with increase in serum malonylaldehyde and TNF and decrease in the enzymatic antioxidants, showing evidence of lipid peroxidation. These changes which can give rise to increased apoptosis of the Sertoli cells of the rat testis, are corrected by Zinc supplementation. These findings have important clinical implications in disease-induced hypomagnesiaemia.

INTRODUCTION

Oxygen is essential for all aerobic organisms, yet partially reduced forms of oxygen such as superoxide radical, hydrogen peroxide and hydroxyl ion are generated during oxygen metabolism [1]. These reactive oxygen species (ROS) induce oxidation stress for cells through cell membrane lipid peroxidation. Cells possess enzymatic systems that scavenge reactive oxygen species to prevent internal cellular damage [2]. Zinc and Copper are essential components of superoxide dismutase an important enzymatic antioxidant and Selenium containing Glutathione peroxidase, Caeruloplasmin, retinol and α-tocopherol [1] are important scavengers of free radicals and prevent lipid peroxidation.

The testicular spermatogenic epithelium, contain an extensive interactions between Sertoli cells and germ cells for the promotion of normal spermatogenesis. Each Sertoli extends from the basement membrane to the lumen of the seminiferous tubules. The nucleus contains finely dispersed ovoid or triangular chromatin with one or more deep infolding. It consists of a nucleolus flanked by a pair of karyosomes. The cytoplasm contains smooth and granular endoplasmic reticulum, spherical or elongated mitochondria, a Golgi apparatus, a number of microtubes, lysosomes and lipid droplets. A Sertoli-Sertoli junction complex consists of a tight junction composed of up to 50 or more lines of fusion of opposed membranes [3].

In the Sertoli cells, TNFα acts as a proinflammatory cytokine that induces IL-6 production and intracellular adhesion molecule-1 and vascular cell adhesion molecule-1 expression [4]. TNFα has been suggested to play a role in the local control of spermatogenesis, because in the Sertoli cells it regulates the production of lactate and induces apoptosis in a variety of transformed cell lines. TNFR-1 is responsible for transduction of the TNFα induced death signal, mediated by its cytoplasmic death domain and is usually associated with the activation of the caspase pathway [5]. Apoptosis (programmed cell death) is the physiological form of cell death that together with mitosis controls the number of cells in a given tissue [6]. Various factors regulate apoptosis among them, the potent Fas-Fas ligand (FasL) system, which belongs to the Tumour necrosis factor receptors (TNFR)-TNF superfamily. Fas and TNFR-1 share a 65 residue homology region in the cytoplasmic domains referred to as the "death domain" essential for transduction of the death signal [7]. The apoptotic threshold of a cell depends on the ratios of stimulating and suppressive factors of apoptosis. A requirement for the process of apoptosis is the trimerization of the Fas (CD 95 or APO-1) and TNFα molecules together form the DISC, the death inducing signaling complex that eliminate unwanted, injured, or virus-infected cells [8].

Objective of study: To investigate the role of Zinc in modulating the lipid peroxidation and apoptosis by hypomagnesaemia of Rat Sertoli Cell.

MATERIALS AND METHODS

Methodology: Three groups of Sprawley-Drew (SD) adult rat were used:

Six SD rats on normal diet as control (b) Six SD rats with induced hypomagnesaemia with frusemide(c) Six SD rats with induced hypomagnesaemia and supplementation with Zinc. Blood was withdrawn from all the rats weekly and at the end of four weeks the rats were sacrificed and the testes harvested. ELISA determined Atomic Absorption Spectrophotometer (ASS), and TNF-α estimated the serum concentrations of Zn, Mg, Cu and Se. Total antioxidant and Superoxide dismutase activities were determined with Randox kit (Randox Labs Ardmore, UK) and caeruloplasmin levels were measured by nephelometry utilizing monoclonal antiserum and control (The Array 360 protein systems, Beckman, Gagny, France). Retinol and α-tocopherol serum concentrations were determined simultaneously by high-performance liquid chromatography (HPLC). Serum Malonialdehyde was determined by a colorometric method.

After routine preparation of the testes, Electron Microscopy was used to study the pattern of apoptosis of the Sertoli cells with regards to blebbing of the cellular membranes and nuclear condensation of the chromatin.

Table 1. Effects of hypomagnesiaemia on the trace elements and the antioxidant activity in the rat serum

	HM n=6	**ZS** n=6	**Control** n=6
1. Trace elements (mg/L)			
Zinc	143±21	178±25	180±27*
Copper	141±20	94±18	96±23*
Selenium	4.4±1.8	7.8±2.4	7.5±2.5*
Magnesium	8.6±4.1	14.9±5.4	14.4±5.3*
2. Superoxide dismutase(mmol/L)	0.6±0.3	1.6±0.5	2.1±0.7**
3. Ceruloplasmin (mmol/dl)	27±3	22±5	25±7
4. TNFα (pg/ml)	34±6	12±5	14±5*
5. Retinol (μg/L)	0.4±0.3	0.6±0.4	0.7±0.4
6. α-tocopherol (μg/L)	1.5±0.8	6.2±1.5	5.9±1.7¶
7. Total antioxidant activity mmol/L	2.1±0.9	4.6±1.3	4.2±1.4*
8. Malonialdehyde nmol/ml	5.2±2.3	1.7±0.8	1.6±0.9¶

HM: Hypomagnesiaemia; ZS: Zinc Supplementation
* HM vs Control ($p < 0.05$); HM vs ZS ($p < 0.05$); Control vs ZS ($p > 0.05$).
** HM vs Control ($p < 0.01$); HM vs ZS ($p < 0.01$); Control vs ZS ($p < 0.5$).
¶ HM vs Control ($p < 0.01$); HM vs ZS ($p < 0.01$); Contro vs ZS ($P > 0.05$).

RESULTS

Hypomagnesaemia was associated with reduction of the testicular volume, and serum concentration of Zinc, Copper, Selenium and Magnesium ($p<0.05$), decrease levels of superoxide dismutase ($p <0.01$) and caeruloplasmin but increased in levels of malonialdehyde($p<0.01$) and TNF ($p<0.05$). As in *table 1*, all the above changes were corrected with Zinc supplementation. In induced hypomagnesaemia, there was increased blebbing of the cell membrane. The nucleus contained more sparsely dispersed chromatin, faint and with more deep infolding. The nucleolus and karyosomes are almost totally disappeared. The cytoplasmic and granular endoplasmic reticulum, mitochondria and Golgi apparatus are fewer. Similarly, the microtubes, lysosomes, lipid droplets and Sertoli-Sertoli junction complexes are also fewer. These apoptotic changes were reduced with Zinc supplementation.

DISCUSSION

This study has demonstrated that hypomagnesaemia is associated with increased apoptosis of the Sertoli cell of the rat testis. This involved nuclear and cytoplasmic changes in the cell. In addition, the serum levels of malonialdehyde and TNF are increased, while the natural, enzymatic and total antioxidant activities are decreased. It is tempting to speculate that hypomagnesaemia caused apoptosis of the Sertoli cell of the rat testis through two related pathways, namely through lipid peroxidation and the effect of TNF.

In the Sertoli cell, TNF has been shown to regulate the production of lactate and interleukin-6 and plays a role in the local control of spermatogenesis [4]. It is also known to induce apoptosis in the transformed cell lines through the TNF induced cell death signal mediated by the TNFR-1.

The death domain of TNF is associated with the activation of the caspase pathway [5]. These findings have important clinical implications in disease-induced hypomagnesaemia.

The changes that give rise to increased apoptosis of the Sertoli cells of the rat testis are prevented by Zinc supplementation. We have previously demonstrated that Zinc supplementation is associated with improved sperm parameters like sperm count, motility and morphology [9]. Zinc probably prevents apoptosis through its antioxidant activity, as a component of superoxide dismutase and its membrane stabilizing effect.

CONCLUSION

Hypomagnesaemia is associated with apoptosis of the Sertoli cell of the rat testis. These changes can be prevented by Zinc supplementation.

REFERENCES

1. Halliwell B. Current status review: free radicals, reactive oxygen species and human disease: a critical evaluation with special reference to atherosclerosis. Br. J. Exp Pathol 1989; 70: 737-741.
2. Kaul M, Siveski-Il Iskovic N, Thomas TP, Hill M, Khaper N, Singal PK. Probucol improves antioxidant activity and modulates development of diabetic cardiomyopathy. Nutrition 1995; 11: 551-554.
3. Ueno H, Mori H. Morphometrical analysis of Sertoli cell ultrastructure during the seminiferous epithelial cycle in rats. Biol Reprod 1990; 43:769-776.
4. Stephen JP, Syed V, Jegou B. Regulation of Sertoli cell IL-1 and IL-6 production in vitro. Mol Cell Endocrinol 1997; 134: 109-118.
5. Baker SJ, Reddy EP. Modulation of life and death by the TNF receptor superfamily 1998; 17: 3261-3270.
6. Malamisti-Puchner A, Sarandakou A, Tziotis J, Trikika P, Creatsas G. Evidence for a suppression of apoptosis in early postnatal life 2001; 80: 994-997.
7. Kerr JFR, Wylie AH, Cume AR. Apoptosis: a basic biological phenomenon with wide-ranging implications in tissue kinetics. Br. J. Cancer 1972; 26: 239-257.
8. Igney FH and Krammer PH. Signals and Diseases. Biochem 1999; 3: 24-27.
9. Omu AE, Al-Qattan F, Abdulhadi FM, Fatinikun MT, Fernandes S. Seminal immune response in infertile men with leukocytospermia Eur. J. Obstet Gynecol Reprod Biol 1999; 86: 195-202.

*Metal Ions in Biology and Medicine: vol. 7.*Eds L. Khassanova, Ph. Collery, I. Maymard, Z. Khassanova, J.-C. Étienne.
John Libbey Eurotext, Paris © 2002 pp. 127-131.

Effect of metal ions on cell cycle

A. Yamamoto, Y. Kohyama, and T. Hanawa

Biomaterials Center, National Institute for Materials Science, 1-2-1 Sengen, Tsukuba, Ibaraki, 305-0047, Japan

ABSTRACT

Metallic materials are currently used as medical devices such as bone-fixing plates and screws, artificial hip joints, heart pacemakers, cardiovascular stents, etc. When these devices are implanted into our body, they may corrode and wear, releasing metal ions and debris. Though these ions and debris can cause adverse reactions of surrounding tissues and organs, the mechanism of their toxicity has not been elucidated. To understand the toxicity of metal ions, cytotoxicity of metal ions has been evaluated using mammalian or human cells. In this study, the effect of metal ions on cell cycle was examined to approach the mechanism of metal ion cytotoxicity.

Cell cycle analysis was performed using a flow cytometer. First, 5.5×10^4 HeLa S3 cells were cultured in 4 mL of Eagle's MEM supplemented with 10% (v/v) FBS for 3 d before 40 mL portion of the metal salt solution in PBS(-) was added into the medium. After another incubation of 1 d, cells were treated with 0.05% (v/v) trypsin, 0.2% (v/v) triton-X100, 0.5% (w/v) RNase, and 0.005% (w/v) PI. Comparing to the control (without metal ions), Co^{2+} at 0.2 mM increased the percentage of the cells in gap 2-mitosis, suggesting that Co^{2+} inhibited cellular function relating to mitosis. Mn^{2+} at 0.4 mM increased the percentage of the cells in gap 0-gap 1, suggesting that the cell cycle stopped within gap 0-gap 1. Hg^{2+} at 0.1 mM increased the percentage of the cell in DNA synthesis. V^{3+} at 0.5 mM increased the percentages of the cells both in DNA synthesis and in gap 2-mitosis, which is the most popular case among 42 metal ions tested. Metal ions with relatively high cytotoxicity tend to increase the percentages of the cells gap 2-mitosis. The data obtained will contribute to the elucidation of the mechanism of the cytotoxicity induced by metal ions.

INTRODUCTION

In our daily lives, we have many occasions to touch the metallic materials used in tableware, kitchenware, accessories, stationeries, and medical devices such as surgical knives, injector needles, heart pacemakers, bone-fixing plates and screws, etc. These devices are used in various surroundings and may corrode and wear releasing metal ions and debris. When we take released metal ions and debris over the ranges which our bodies could accept, they may cause adverse reaction of our bodies such as inflammation, allergy, cancer, and the necrosis on organs.

Though the toxicity of metal ions and compounds has been examined for years by animal tests, cytotoxicity tests come to be performed widely to estimate the risk of metallic compounds to human beings in short period of time with low cost and high reproducibility using human cells. One of the most basic cytotoxicity tests is to evaluate the effect on cell proliferation and is studied for metal ions and compounds by several groups of researchers using several kinds of cells [1-5]. However, the mechanism of metal ion cytotoxicity has not been elucidated.

Proliferation is the very basic and common function for most kind of cells and proceeds along the cell cycle, consisting of four phases; gap 1, DNA synthesis, gap 2, and mitosis. When cellular proliferation is inhibited by metallic compounds, there should be any change on the cell cycle. In this study, the effect of metal ion on cell cycle was examined by cell cycle analysis using flow cytometer, to investigate the mechanism of metal ion toxicity on cell proliferation.

MATERIALS AND METHODS

Metal salts were used as the sources of metal ions because metal cations and counteranions are generally supposed to be dissociated from metal salts in an aqueous solution like a cell culture medium. Metal salts used are as follows; $AgNO_3$ (>99.9%), $Al(NO_3)_3 \cdot 9H_2O$ (>99.9%), $BaCl_2 \cdot 2H_2O$ (>99.9%), $BeSO_4 \cdot 4H_2O$ (>99.9%), $Bi(NO_3)_3 \cdot 5H_2O$ (>99.9%), $CdCl_2 \cdot 2.5H_2O$ (>99.9%), $CoCl_2 \cdot 6H_2O$ (>99.95%), $Cr(NO_3)_3 \cdot 9H_2O$ (>99.9%), CsCl (>99.9%), CuCl (>99.9%), $CuCl_2 \cdot 2H_2O$ (>99.9%), $FeSO_4 \cdot 7H_2O$ (>99.5%), $FeCl_3 \cdot 6H_2O$ (>99.9%), $GaCl_3$ (>99.9%), $HfCl_4$ (>99.5%), Hg_2Cl_2 (>99.9%), $HgCl_2$ (>99.9%), $InCl_3 \cdot 4H_2O$ (>99.9%), IrC14 (>99.5%), $LiCl \cdot H_2O$ (>99.9%), $MgCl_2 \cdot 6H_2O$ (>99.9%), $MnCl_2 \cdot 4H_2O$ (>99.9%), $MoCl_5$ (>99.9%), $NbCl_5$ (>95%), $NiCl_2 \cdot 6H_2O$ (>99.95%), $Pb(NO_3)_2$ (>99.9%), $PdCl_2$ (>99.9%), RbCl (>99.0%), $RhCl_3 \cdot 3H_2O$ (>99.5%), $RuCl_3$ (>90%), $SbCl_3$ (>99.9%), $SnCl_2 \cdot 2H_2O$ (>99.9%), $SrCl_2 \cdot 6H_2O$ (>99.9%), $TaCl_5$ (>90%), $TiCl_4$ (>99.0%), $TlNO_3$ (>99.9%), $Tl(NO_3)_3 \cdot 3H_2O$ (>99.5%), NH_4VO_3 (>99.0%), WCl_6 (>99.9%), $YCl_3 \cdot 6H_2O$ (>99.99%), $ZnCl_2$ (>99.9%), $ZrCl_4$ (>95%), and $K_2Cr_2O_7$ (>99.5%). Figures in parentheses indicate the purity of each metal salt. Hydrated water molecules are omitted in the following process. In advance of cell cycle analysis, each of the salts was dissolved in and diluted with sterile Dulbecco's phosphate-buffered saline without calcium and magnesium salts [PBS(-)] properly to prepare a metal salt solution, which would provide a desired final concentration of the salt in the medium.

Human uterine cervix epitheloid carcinoma HeLa S3 was used for cell cycle analysis. First, 5.5×10^4 HeLa S3 cells were seeded in 4 mL of Eagle's MEM supplemented with 10% (v/v) FBS in 60-mm tissue culture polystyrene dish and were cultured at 37 ºC in a CO_2-incubator. After the first incubation for 3 d, a 40 μL portion of the metal salt solution in PBS(-) was added to the dish. At least, 5 concentrations were tested for each metal salt. The same amount of PBS(-) was added to the control dish. Cells were cultured again in the CO_2-incubator for 1 d (second incubation). Then, the cells were treated with 0.05% (v/v) trypsin to be removed from the dish, 0.2% (v/v) triton-X100 to dissolve the cellular membranes for isolating nuclei, 0.5% (w/v) RNase to degrade the RNA, and 0.005% (w/v) PI to die the DNA in the nuclei. Then, the amount of DNA in each nucleus was measured by a flow cytometer (FACSCalibur HG, BD Japan, Tokyo). The count of the nuclei was plotted against the relative amount of DNA in each nucleus, and the percentage of the cells in each cell cycle phase was determined to compare to those of the control dish.

RESULTS AND DISCUSSION

The result of cell cycle analysis for the control dish was shown in *fig. 1* (a). Since the DNA amount inside a cell differs with cell cycle phases, the cells in G_0-G_1, S, G_2-M phases appeared on different values of the x axis. The percentages of the cells in G_0-G_1 phases, S phase, and G_2-M phases were 65%, 25%, and 10% on average, respectively. When the $CoCl_2$ was added to the dish to be 0.2 mM, the percentage of the cells in G_2-M phases were increased to 32% as shown in *fig. 1* (b), indicating that the cell cycle arrested or delayed at G_2-M phases. This fact suggests that Co^{2+} inhibited cellular function relating to mitosis. The increase of the G_2-M fraction is enlarged with the increase of $CoCl_2$ concentration (0.01-0.2 mM). This type of the effect on cell cycle was named as "G_2-M type" and was observed in the cases of $CdCl_2$, $RhCl_3$, and $BeSO_4$.

When the $MnCl_2$ was added to be 0.4 mM, the percentage of the cells in G_0-G_1 phases was increased to 74% as shown in *fig. 1* (c), indicating that the cell cycle arrested or delayed at G_0-G_1 phases. The increase of G_0-G_1 fraction is also enlarged with the increase of $MnCl_2$ concentration

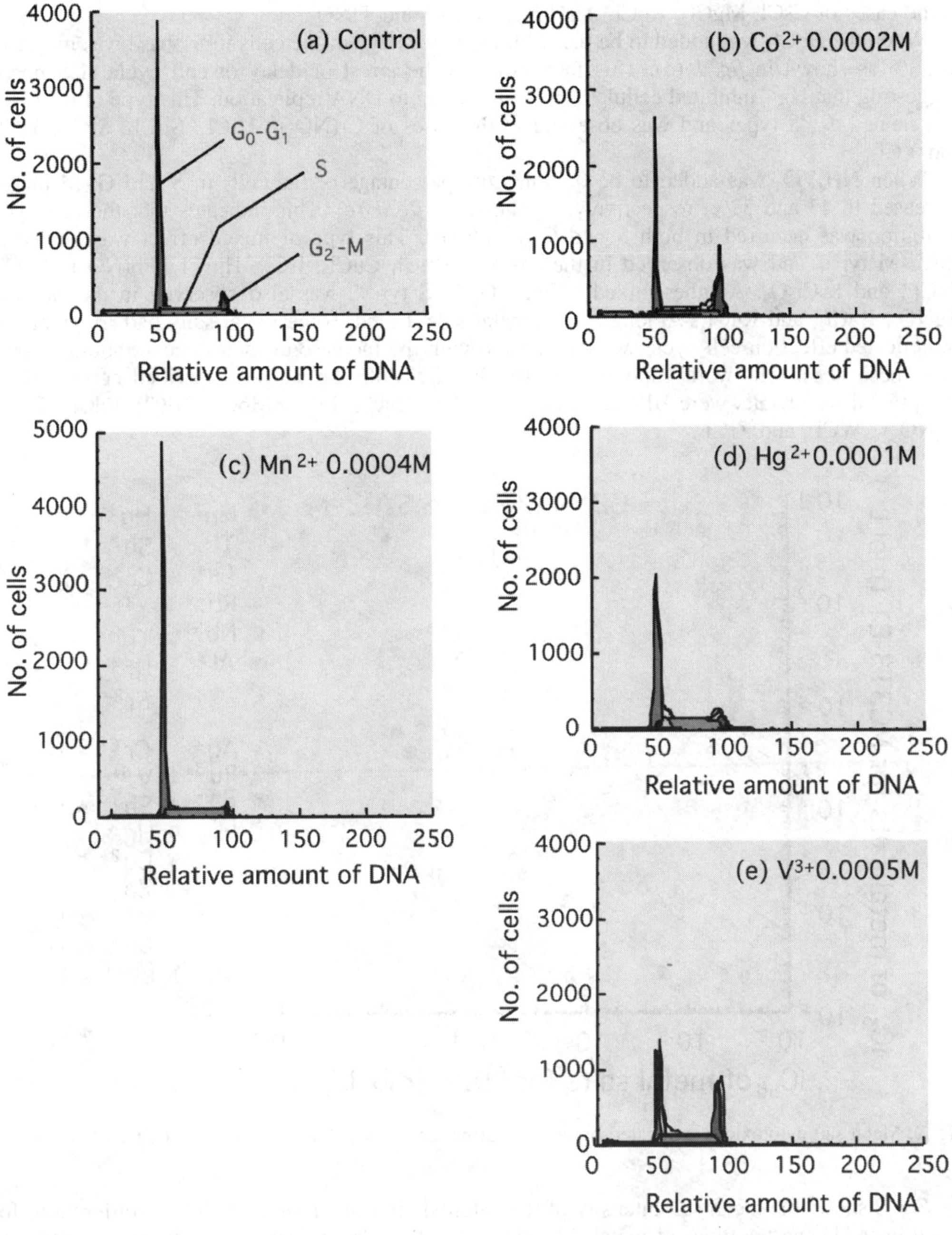

Fig. 1. The results of cell cycle analysis of HeLa S3 with and without metal salts

(0.05-0.4 mM). This type of the effect on cell cycle was named as "G_0-G_1 type" and was observed in the cases of CsCl, $MgCl_2$, RbCl, $Al(NO_3)_3$, $NbCl_5$, and $TlNO_3$.

When the $HgCl_2$ was added to be 0.1 mM, the percentage of the cells in S phase was increased to 33% as shown in *fig. 1* (d). This fact indicates the arrest or delay of cell cycle at S phase, suggesting that Hg^{2+} inhibited cellular function relating to DNA replication. This type of the effect was named as "S type" and was observed in the cases of $Cr(NO_3)_3$, $HfCl_4$, $SnCl_2$, $SrCl_2$, $TiCl_4$, and YCl_3.

When NH_4VO_3 was added to be 0.5 mM, the percentage of the cells in S and G_2-M phases increased to 43 and 32%, respectively, as shown in *fig. 1* (e). This indicates that the cell cycle inhibition was occurred in both S and G_2-M phases. This type of mixed effect was named as "S/G_2-M type" and was observed in the cases of CuCl, $CuCl_2$, $IrCl_4$, Hg_2Cl_2, $Pb(NO_3)_2$, $SbCl_3$, $ZnCl_2$, and $K_2Cr_2O_7$. Another mixed effect, "G_0-G_1/S type", was also observed in the cases of $AgNO_3$, $BaCl_2$, and $RuCl_3$. Among the 42 metal salts tested, some metal salts had irregular and complicated effect on cell cycle, which changed with the increase of their concentrations. Therefore, these metal salts were impossible to be classified into the 5 types described before. These exceptional metal salts were $Bi(NO_3)_3$, $FeSO_4$, $FeCl_3$, $GaCl_3$, $InCl_3$, $MoCl_5$, $NiCl_2$, $PdCl_2$, $TaCl_5$, $Tl(NO_3)_3$, WCl_6, and $ZrCl_4$.

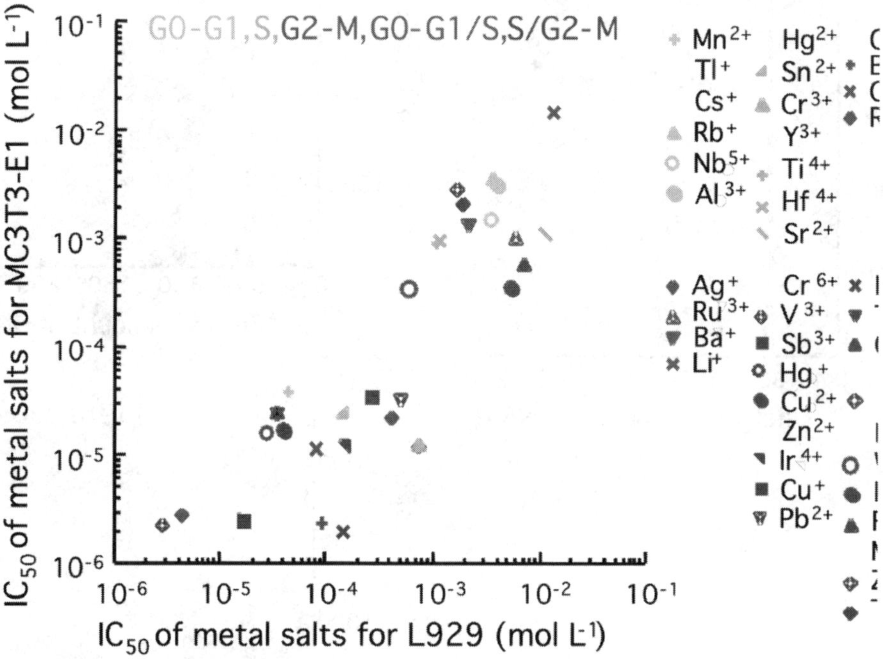

Fig. 2. Metal salt cytotoxicity evaluated by colony formation assay [1] and the types of cell cycle inhibition

Fig. 2 shows the metal salt intensity of the cytotoxicity (inhibition of cellular proliferation) for 2 cell lines [1] and the types of metal salt effect on cell cycle, indicating that the metal salts with relatively high cytotoxicity tend to be cause the "G_2-M" or "S/G_2-M type" effect on cell cycle. This relation suggests that the metal salts with high cytotoxicity may cause too serious damages to proceed over mitosis. According to the results of the *umu* test, a mutagenicity assay, CuCl, $CuCl_2$, $IrCl_4$, $K_2Cr_2O_7$, $RhCl_3$, and VCl_3 were mutagenic metal salts [6], which were categorized into "G_2-M" or "S/G_2-M type" in this study. The *umu* test evaluates the ability to cause DNA damages inducing SOS response [7]. Since the DNA in the nucleus is re-coiled on its replication,

S phase is expected to be the most sensitive phase for the DNA damages. However, the mutagenic metal salts causes G_2-M arrest as well as S arrest. This finding may suggest that there is a final checkpoint during G_2-M phases to stop seriously damaged cells proceeding into mitosis and producing malignant cells.

CONCLUSIONS

In this study, the cell cycle analysis was performed using a flow cytometer to examine the effect of metal salt cytotoxicity on cell cycle. As a result, the metal salt effect on cell cycle was classified into 5 major types; the increase of the percentage of the cells in G_0-G_1 phases, S phase, G_2-M phases, both of G_0-G_1/S phases, and both of S/G_2-M phases. "S/G_2-M type" was the most popular one among the metal salts tested, however some metal salts changed their types of effect with the increase of their concentrations. Metal salts with relatively high cytotoxicity tended to cause "G_2-M" or "S/G_2-M type" effect, suggesting that the serious damages caused by metal salts enforce the cells impossible to proceed over mitosis. Obtained data is expected to contribute to the elucidation of the mechanism of metal salt cytotoxicity.

REFERENCES

1. Yamamoto A., Honma R., Sumita M. Cytotoxicity evaluation of 43 metal salts using murine fibroblasts and osteoblastic cells. *J. Biomed. Mater. Res.*, 1998, 39, 331-340.
2. Yamamoto A., Honma R., Tanaka A., Sumita M. Generic tendency of metal salt cytotoxicity for six cell lines. *J. Biomed. Mater. Res.*, 1999, 47, 396-403.
3. Wataha J.C., Hanks C.T., Sun Z. Effect of cell line on in vitro metal ion cytotoxicity. *Dent. Mater.*, 1994, 10, 156-161.
4. Schedle A., Samorapoompichit P., Rausch-Fan X.H., Franz A., Fureder W., Sperr W.R., Sperr W., Ellinger A., Slavicek R., Boltz-Nitulescu G., Valent P. Response of L-929 fibroblasts, human gingival fibroblasts, and human tissue mast cells to various metal cations. *J. Dent. Res.*, 1995, 74, 1513-1520.
5. Hashimoto Y., Takeda S., Kayama K., Nakamura M. Cytotoxicity test of metal ion using serum-free cell culture. *The abstract of the symposium'92 of the Japanese society for biomaterials*, 1992, 79.
6. Yamamoto A., Kohyama Y., Hanawa T. Mutagenicity evaluation of forty-one metal salts by the *umu* test. *J. Biomed. Mater. Res.*, 2002, 59, 176-183.
7. Shinagawa H, Nakata A, Oda Y, Nakamura S, Kato T. *umu* test for the detection of environmental mutagens. *Toxicology forum*, 1985, 8, 580-586.

Metal Ions in Biology and Medicine: vol. 7. Eds L. Khassanova, Ph. Collery, I. Maymard, Z. Khassanova, J.-C. Étienne.
John Libbey Eurotext, Paris © 2002 pp. 132-137.

Biomarkers of sensitivity and effect associated with cadmium and mercury toxicity in human liver carcinoma ($HepG_2$) cells

Paul B. Tchounwou[1], Ali B. Ishaque[1], Dwayne Sutton[1], Nanuli Ninashvili[1] and Elaine Shen[2]

[1]*Molecular Toxicology Laboratory, NIH-Center for Environmental Health, School of Science and Technology, Jackson State University, Jackson, MS 39217, USA;* [2]*Xenometrix Research Laboratory, Xenometrix, Inc., Boulder, CO, USA*

ABSTRACT

In the past, demonstrating the biologic significance of chemical exposure has relied significantly on assessing and identifying the pathology induced by xenobiotic compounds. However, biomarkers of sensitivity have in recent years, increasingly being used in toxicology for detecting and predicting the harmful effects of chemical agents on the health and well being of humans and other life forms. This research was therefore designed to assess the cellular and molecular responses of human liver carcinoma cells following exposure to cadmium, and mercury; and to subsequently identify the potential biomarkers of sensitivity and effect associated with exposure to these metal compounds. Cytotoxicity was evaluated using the MTT-assay for cell viability, while the gene profile (CAT-Tox) assay was performed to measure the transcriptional activation of stress genes in thirteen different recombinant cell lines generated from $HepG_2$ cells. Cyctotoxicity experiments yielded LD_{50} values of 3.5±0.6, and 6.1±0.8 ug/mL upon 48 hrs of exposure to mercury, and cadmium, respectively; indicating that mercury was more toxic than cadmium to $HepG_2$ cells. A dose-response relationship was recorded with respect to both cytotoxicity and gene expression. Overall, nine (GSTYa, HMTIIA, *c-fos*, HSP70, CRE, p53RE, GADD153, GADD45, and GRP78) out of the thirteen recombinant cell lines tested for mercury showed inductions to statistically significant levels ($p<0.05$). Following exposure to cadmium, four of these stress genes (HMTIIA, HSP70, CRE, and XRE) were significantly induced and seven were marginally induced. No significant inductions ($p>0.05$) were observed for CYP1A1, and RARE. As expected, these results indicate that metallothioneins and heat shock proteins appear to be excellent candidates for biomarkers for detecting metal-induced proteotoxic effects at the molecular and cellular levels.

INTRODUCTION

Cadmium and mercury are non-essential elements that are released to the environment from natural sources as a result of natural phenomena such as erosion of mineral deposits, and volcanoes, but releases from human activities such as metal smelting, coal production, chemical production and use, and waste disposal can lead to substantial contamination of the environment. Cadmium is mainly used in the battery, alloy manufacturing and the electroplating industries. Roughly, 50,000 workers are occupationally exposed to cadmium in the United States [1]. Cadmium, apart from occupational concerns has a long biological half life. It is estimated that approximately 2700 to 6000 tons of elemental mercury are released naturally into the atmosphere by degassing from the earth's crust and oceans. Human activities, primarily the combustion of fossil fuels and industrial production, account for the release of 2000-3000 of mercury into the atmosphere [2]. Mercury enters hazardous-waste streams from electrical apparatus manufacturing, electrolytic production of chlorine and caustic soda, pharmaceuticals, paints, plastics, and mercury batteries. It is also used in the catalytic production of vinyl chloride and urethane, and occurs in 40% of metal-containing pesticides [2]. Exposure to cadmium and mercury occur primarily via the oral route (ingestion of contaminated foods), inhalation (mainly in occupational exposure), and dermal contact [1, 2].

Cadmium and mercury are known to produce a high degree of toxicity to humans and animals. The primary targets for cadmium and mercury toxicity are the nervous system, and the kidney. Both compounds have also been reported to cause gastrointestinal, hepatic, cardiovasular, hematologic, respiratory, developmental, genotoxic and carcinogenic effects [1, 2]. The first reported widespread outbreak of neurologic disorders associated with methylmercury-contaminated fish occurred in the Minamata area of Japan. The neurologic syndrome was characterized by a long list of symptoms including prickling, tingling sensation in the extremities (paresthesia), impaired peripheral vision, hearing, taste, and smell, slurred speech, unsteadiness of gait and limbs, muscle weakness, irritability, memory loss, depression, and sleeping difficulties [3]. The International Agency for Research on Cancer classifies cadmium as a probable human carcinogen. Elevated levels of respiratory, prostate and other cancers have been reported in cadmium-exposed workers; with the strongest evidence for lung cancer [4]. Although no epidemiological studies have linked mercury to human cancers, several investigations with experimental animals have reported increases in the incidence of forstomach squamous cell papillomas, thyroid follicular cell carcinomas, and renal tubule tumors in rats exposed to mercuric chloride [5].

Cadmium and mercury are both systemic toxicants showing significant toxicity in experimental animals, however, the biochemical and molecular mechanisms by which these heavy metals exert their toxic actions in humans remain to be elucidated. Therefore, the specific aims of this study were 1) to assess the transcriptional activation of stress genes in human liver carcinoma cells exposed to cadmium, and mercury, and 2) to subsequently identify the potential biomarkers of sensitivity and effect associated with exposure to these heavy metals.

EXPERIMENTAL DESIGN AND METHODS

The mammalian gene profile (CAT-Tox) assay was performed for measuring differential gene expression in the human hepatoma cell line, $HepG_2$. The test protocol described by Todd and coworkers was followed [6]. This test involves the use of different recombinant human liver cell lines generated by creating stable transfectants of different mammalian promoter-chloramphenicol acetyltransperase (CAT) gene fusions. Briefly, thirteen recombinant cell lines and the parental $HepG_2$ cell line were plated, one row each, over two 96-well microplates. The cell lines were dosed at 6 cadmium or mercury concentrations (0, 0.625, 1.25, 2.5, 5, and 10 ug/mL) in triplicates, and incubated at 37 °C, 5% CO_2, for 48 hours.

After the incubation period, the total protein was measured by the Bradford method at 600 nm using a microplate reader. A standard sandwich ELISA was performed and in the final step horse radish peroxidase catalyzed a color change reaction that was measured at 405 nm. The parental $HepG_2$ cell line was dosed in the same manner as the recombinant cell lines, and was used to perform a MTT-based cellular viability assay at 550 nm. The transcriptional fold inductions for each recombinant cell line at each cadmium/mercury concentration were calculated using the Xenometrix CAT-Tox software based on the optical density readings at 600 nm at 405 nm. The software also converted the 550 nm readings to cell viability percentages. The *Student* t-test was applied to determine whether differences in gene inductions between controls and chemically treated cells were significant ($p<0.05$).

RESULTS

Cyctotoxicity Assay

Figure 1 presents the experimental data obtained from this assay. The results yielded LD_{50} values of 3.5±0.6, and 6.1±0.8 ug/mL upon 48 hours of exposure to mercury, and cadmium, respectively; indicating that mercury was more toxic than cadmium to HepG2 cells. As shown in this figure, there was a strong dose-response relationship with respect to metal toxicity to human hepatocytes.

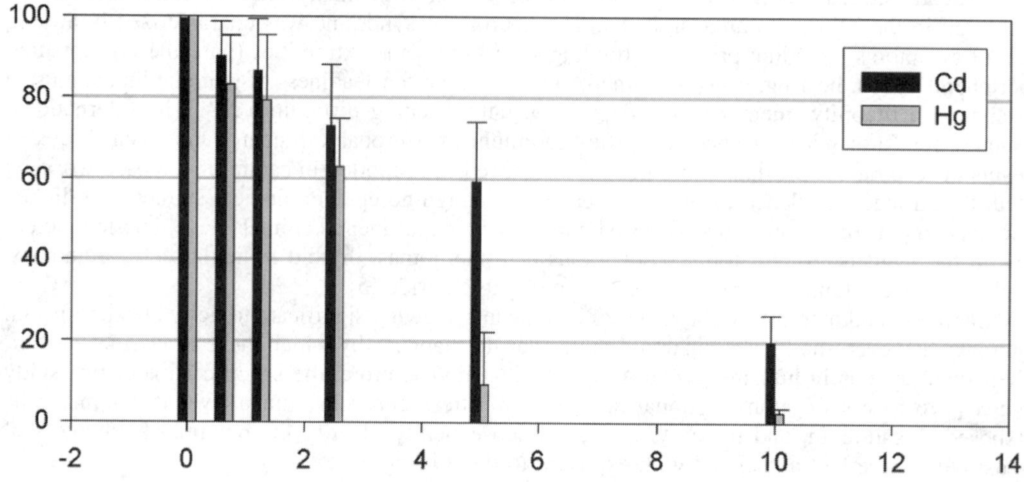

Fig. 1. Cytotoxicity of cadmium and mercury to HepG2 cells

Metallothionein Induction

Figure 2 presents the fold inductions of the metallothionein gene (HMTIIA) following exposure to cadmium and mercury. The data presented in this figure show a strong dose response relationship. Fold inductions of HMTIIA were 1.0±0.0, 54.3±20.0, 70.2±10.0, 80.8±11.0, and 91.3±26.0 in 0, 0.625, 1.25, 2.5, and 5.0 ug/mL cadmium, respectively. A biphasic response was observed with respect to mercury; showing a gradual increase in HMTIIA induction within the 0-2.5 ug/mL dose range, followed by a sharp decrease at 5 ug/mL due to the high rate of mortality (about 90%) of cells at this dose level *(fig. 1)*.

Heat Shock Protein Induction

Figure 3 presents the fold inductions of the 70 kDa heat shock protein, following exposure to cadmium and mercury. Data presented in this figure show a biphasic response for both cadmium and mercury. Cadmium exposure shows a gradual increase in fold induction of HSP70 within the dose range of 0-1.25 ug/mL, followed by a gradual decrease within the dose range of 1.5-5.0 ug/mL. A similar trend is also observed for mercury, showing a strong positive correlation within the dose range of 0-2.5 ug/mL, followed by a decrease in fold induction at 5 ug/mL, probably due to cell death at higher doses of mercury.

Induction of Other Gene Products

Following exposure to cadmium, four (HMTIIA, HSP70, CRE, and XRE) of the thirteen stress genes evaluated were significantly ($p<0.05$) induced, and seven (GSTYa, *c-fos*, NF-kB, p53RE,

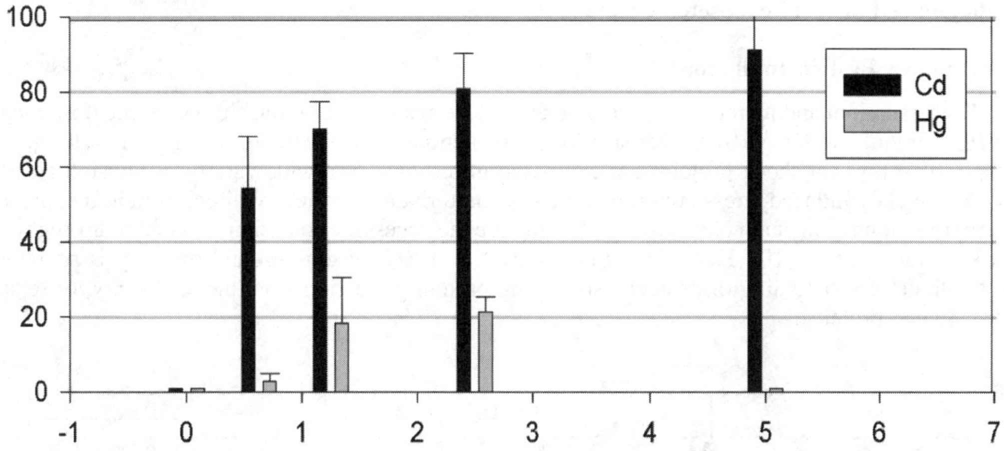

Fig. 2. Fold induction of HMTTA in HepG2 cells exposed to cadmium and mercury

GADD153, GADD45, and GRP78) were marginally induced. Test with mercury show that nine out of the thirteen recombinant cell lines tested showed inductions to statistically significant levels ($p<0.05$). At 2.5 ug/mL mercury, the average fold inductions were 5.2±0.9, 21.4±3.9, 7.0±6.2, 6.8±1.1, 2.7±1.0, 4.5±2.0, 7.5±6.0, 2.2±0.7, and 2.5±0.3, for GSTYa, HMTIIA, *c-fos*, HSP70, CRE, p53RE, GADD153, GADD45, and GRP78, respectively. No significant inductions ($p>0.05$) were observed for CYP1A1, and RARE.

DISCUSSION

Cytotoxicity Assay

Data obtained from this study clearly show that cadmium (Cd) and mercury (Hg) are highly toxic to human liver carcinoma cells; with mercury showing a higher cytotoxicity compared to cadmium. Previous studies in our laboratory have demonstrated that other metals such as arsenic (As), and lead (Pb) are relatively less toxic to $HepG_2$ cells [7]. The general order of increasing toxicity is Pb < As < Cd < Hg. It has been suggested that hepatocyte injury following metal exposure may be due to binding in the inner membrane and accumulation in the mitochondria, leading to the collapse of the mitochondrial membrane, followed by plasma membrane depolarization and cell death [8].

Metallothionein Induction

Induction levels of 1.0±0.0, 54.3±20.0, 70.2±10.0, 80.8±11.0, and 91.3±26.0 were recorded for HMTIIA at 0, 0.625, 1.25, 2.5, and 5.0 ug/mL cadmium, respectively; indicating a strong dose-response relationship with respect to metallothionein induction. A similar assertion could be made for mercury within the dose range of 0-2.5 ug/mL. It has been reported that the synthesis of metallothioneins (MTs) can be induced by a wide variety of metals. This metal-inducible system is of particular importance in metal detoxification since it provides an efficient feedback mechanism for controlling the concentrations of a high-affinity biological ligand which, in turn, controls the speciation of selected metals within the cells [9]. Therefore, the MT induction observed in this study represents a cellular defense mechanism against cadmium- and mercury-induced toxicity.

Hence, MTs may play a central role in protecting cells from potential damage as a result of cadmium and mercury exposure.

Heat Shock Protein Induction

Both cadmium and mercury showed a positive dose-response relationship within the dose range of 0-2.5 ug/mL, followed by a decrease in the induction level of HSP70 at 5 ug/mL. The genes responsible for heat shock proteins are expressed in response to a wide variety of physiologically and chemically induced stress conditions. Protein perturbations either by direct protein damage or disruption of nascent chain elongation or folding are the necessary conditions needed to up-regulate HSP70 transcription [10]. Hence, the induction of HSP70 by cadmium and mercury is probably to facilitate proper folding of proteins, signifying protein protection from metal toxicity, or repair of damaged proteins.

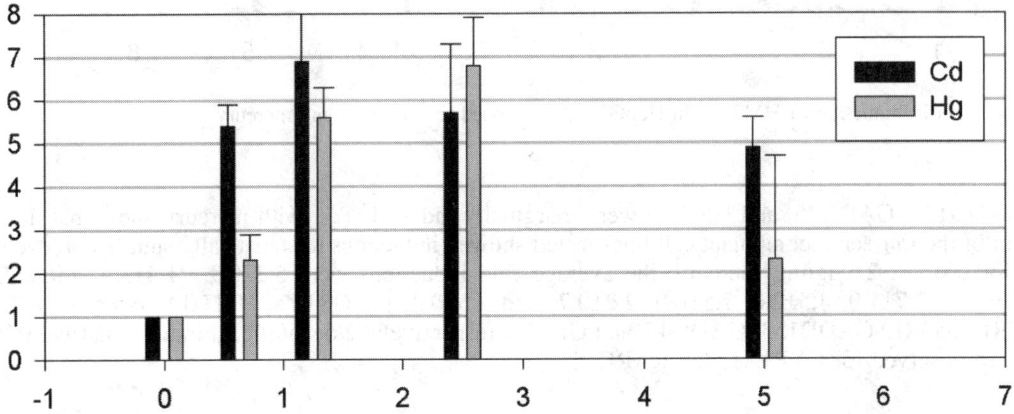

Fig. 3. Fold induction of HSP70 in HepG2 cells exposed to cadmium and mercury

Induction of Other Gene Products

Following exposure to cadmium, two other genes including cyclic AMP response element (CRE), and xenobiotic response element (XRE) were significantly ($p<0.05$) induced. The activation of CRE by cadmium makes this metal, an endocrine receptor agonist [11]. CAT expression by XRE construct suggests that cadmium chloride is indirectly acting as a ligand to Ah receptor, perhaps through breakdown or release of other cell products as a consequence of cadmium toxicity [12]. Also, cadmium was shown to marginally induce GSTYa, *c-fos*, NF-kB, p53RE, GADD153, GADD45, and GRP78. Test with mercury show that nine (GSTYa, HMTIIA, *c-fos*, HSP70, CRE, p53RE, GADD153, GADD45, and GRP78) out of the thirteen recombinant cell lines tested showed inductions to statistically significant levels ($p<0.05$). No significant inductions ($p>0.05$) were observed for CYP1A1, and RARE.

CONCLUSIONS

This research clearly demonstrates that cadmium and mercury have a high degree of cytotoxicity to HepG2 cells. Both metals also show a potential to cause protein damage (HTMIIA, HSP70 and GRP78), cell proliferation (*c-fos*, NF-kB), metabolic perturbation (CRE), growth arrest and DNA damage (GADD153, GADD45), and apoptosis (p53RE), and to undergo Phase II metabolism in the liver (GSTYa and XRE). As expected, these results indicate that metallothioneins and heat shock proteins appear to be excellent candidates for biomarkers for detecting metal-induced proteotoxic effects at the molecular and cellular levels.

ACKNOWLEDGMENTS

Research supported by NIH-RCMI Grant No. 1G12RR13459, and DOE-HBCU/MI-ETC Grant No. DE-FC04-90AL66158. Thanks to Dr. Abdul Mohamed, Dean of the JSU-School of Science and Technology, for his support and advise in this research.

REFERENCES

1. ATSDR. Toxicological Profile for Cadmium: TP-92/06. Agency for Toxic Substances and Disease Registry. Centers for Disease Control. Atlanta, Georgia, 1993.
2. ATSDR. Toxicological Profile for Mercury: TP-93/10. Agency for Toxic Substances and Disease Registry. Centers for Disease Control. Atlanta, Georgia, 1994.
3. Tsubaki T., Takahashi H. Recent Advances in Minamata Disease Studies. Kodansha, Ltd. Tokyo, Japan, 1986.
4. IARC. Monographs on the Evaluation of Carcinogenic Risks to Humans: Beryllium, Cad- mium, Mercury, and Exposure in the Glass Industry. International agency for Research on Cancer. Lyons, France, 1993, 119-238.
5. NTP Toxicology and Carcinogenesis Studies of Methyl Mercury in F344/N Rats and B6C3F1 Mice. NIH Publication No. 91-3139. National Toxicology Program. Research Triangle Park. North Carolina, 1993.
6. Todd M.D., Lee M.J., Williams J.L., Nalenzny J.M., Gee P. The CAT-Tox assay: A sensitive and specific measure of stress induced transcription in transformed human liver cells. *Fundamentals Appl Toxicol*, 1995, 28, 118-128.
7. Tchounwou P.B., Wilson B.A., Ishaque A.B. Scheneider J. Atrazine potentiation of arsenic-trioxide-induced cytotoxicity and gene expression in human liver carcinoma cells. *Mol Cell Biochem*, 2001, 222, 21-28.
8. Chang L.W., Suzuki T. *Toxicology of Metals*. CRC Press. Boca Raton. FL, 1996, 885-899.
9. Sanders B.M., Goering P.L., Jenkins K. The role of general and metal-specific cellular responses in protection and repair of metal-induced damage: stress proteins and metallothioneins. In: Chang LW ed. *Toxicology of Metals*. Boca Raton. Lewis Publishers, 1996, 165-187.
10. Mosser D.D., Theodorakis N.G., Morimoto R.I. Coordinate changes in heat shock element binding activity, and HSP70 gene transcription rates in human cells. *Mol cell Biol*, 1998, 8, 4736-4744.
11. Tchounwou P.B., Ishaque A.B. Scheneider J. Cytotoxicity and transcriptional activation of stress genes in human liver carcinoma cells exposed to cadmium chloride. *Mol Cell Biochem*, 2001, 222, 21-28.
12. Ma Q., Witlock J.P. The aromatic hydrocarbon receptor modulates the hepa 1c1c7 cell cycle and differentiated state independently of dioxin. *Mol Cell Biol*, 1996, 16, 2144-2150.

Metal Ions in Biology and Medicine: vol. 7. Eds L. Khassanova, Ph. Collery, I. Maymard, Z. Khassanova, J.-C. Étienne.
John Libbey Eurotext, Paris © 2002 pp. 138-143.

In vitro study of metal toxicity to renal epithelial cells

B. Gouget[1], L. Figard[1], F. Carrot[1], S. Lemaout[2], R. Gobain[2], H. Khodja[1]

[1] *Laboratoire Pierre Süe, CEA/CNRS UMR9956, CEA Saclay, 91191 Gif sur Yvette, France;*
[2] *Service de Biologie Cellulaire, CEA Saclay, 91191 Gif sur Yvette, France*

ABSTRACT

Heavy metals are important environmental pollutants. Although some of them have proved to be essential for cell growth, cadmium (Cd) or lead (Pb) do not present known biological functions and are extremely toxic even at very low concentration.

Preliminary results of the comparison of Cd and Pb toxicity to renal cells investigated on Madin-Darby canine kidney cells (the MDCK cell line) as a model of the distal tubule and collecting duct are reported. Cells are grown on two-compartment filters and are exposed to different concentrations of metal (1-50µM) for various times (1-24h). Although many studies dealt with Cd or Pb toxicity on MDCK cells, no simultaneous determination of concentrations and distributions of biological elements and toxic metals within renal epithelial cultured cells are reported to date.

INTRODUCTION

Nephron, the functional unit of kidney, is composed of different segments, each consisting of distinct cell types. These epithelial cells are responsible for renal work: regulation of the excretion and homeostasis of water and solutes. The glomerulus acts as filtering unit, the proximal and distal tubules as the regulatory or reabsorptive units for blood and urine. Their transport capacities make them frequent targets for xenobiotic toxics.

Continuous renal epithelial cell cultures have emerged as important tools to study *in vitro* nephrotoxicity. Epithelial transport and cellular uptake of toxic metals may be explored by culturing cells on semi-permeable supports, thereby separating the apical from the basolateral compartment by the cell monolayer and assuring maintenance of epithelial polarity and barrier frontier (tight junctions), characteristics of kidney cells *in vivo* [1].

Cadmium (Cd) is a highly toxic metal ion that causes severe damage to many organs including kidney, liver testis, lung and placenta. Because of its industrial use, Cd has become ubiquitous in the biosphere and enters the food chain, thus contaminating animals. Following oral exposure, Cd is absorbed through the gastrointestinal tract and transported via the blood circulation to various tissues. Kidneys are one of the major sites of Cd accumulation and many studies have investigated Cd uptake, Cd transepithelial transport and Cd toxic effects on kidneys [2]. Lead (Pb) poisoning results from environmental and occupational exposure to the metal and its compounds. Its biological effects, depending on the route, duration and intensity of exposure, include disruption of the heme synthesis system, impairment of the central and peripheral nervous systems, and reproductive and renal damage. Taken up through ingestion, Pb is absorbed from the small intestine and quickly enters erythrocytes. In addition, kidneys play an important role in regulating blood Pb levels, by accumulating more Pb than other soft tissues [3].

For this report, the Madin-Darbin canine kidney cell line (MDCK cells) has been used. Elsewhere well characterized [4], MDCK are assumed to be of distal tubular or cortical collecting duct origin and present a good alternative to *in vivo* studies [5]. Many studies investigated Cd and Pb toxicity on MDCK cells. But for the first time, metal uptake and changes in biological elements contents of renal cells were quantitatively monitored after cadmium or lead poisoning, using proton induced X-ray emission (PIXE) coupled with Rutherford backscattering spectroscopy (RBS), a very sensitive and multi-elemental technique. This technique is applied in nuclear microprobe analysis (NMPA), where a proton beam focused at the micron scale is rapidly scanned over the specimen, allowing construction of two-dimensional maps of trace and major elements in the sample. In this manner, the distribution as well as quantification of elements within cells can be correlated with the morphology of the sample [6].

MATERIALS AND METHODS

Cell cultures

MDCK cells, passages 50-70, were continuously grown at 37°C in an humidified atmosphere regulated at 5% CO_2. The culture medium was Dulbecco's Modified Eagle's Medium DMEM supplemented with 10% (v/v) inactivated fetal calf serum. At confluence, cells were subcultured with trypsine/EDTA 0.05% to dilution 10. For metal exposure, cells were grown on semi-permeable polyester membranes (Costar Transwell Clear) with 0.4 µm pore size and 1 cm^2 growth area. An initial equilibration period of membranes was processed before seeding by adding media to the plate well and to the membrane followed by incubation at 37°C for 1 h. Cd and Pb were administered as fresh dilutions of $CdCl_2$ and $(CH_3CO_2)_2Pb.3H_2O$ stock solutions in H_2O. They were directly added to the serum free culture medium after having rinsed cells twice with serum free and metal free culture medium. Cells were exposed to 1-50 µM ion metal for 1-24 h. For uptake determination, the toxic was administered to both the apical and the basolateral compartments. At the end of the incubation period, the toxic solution was aspired off, and the cells were washed twice with serum free and metal free culture medium. Determination of lactate dehydrogenase activity (using the LDH kit from Sigma) was monitored on the apical culture medium after centrifugation to remove all cells and detritus, since LDH release corresponds to the loss of integrity of the cell membrane [7]. Results were compared to cell viability assessed by the Trypan blue exclusion test.

Nuclear microprobe analysis

For analysis of cell monolayers using NMPA, two types of targets were prepared. For the first type, a cell suspension was seeded onto a very thin sterile Pioloform film (purchased from Agar) precoated with an attachment factor. Pioloform was preferred to Formvar film since it presents a higher mechanical and thermal stability as well as a lower mass thickness. Collagen was used as the recommended attachment factor for epithelial cells. In the course of 48 h of incubation, an epithelium at confluence was developed on the polymer film. Following the principle already described in [8], targets were carefully rinsed in PBS and very quickly in water, then frozen in isopentane chilled at -165°C by liquid nitrogen. After cryofixation, lyophilisation was achieved in 3 h from -10°C to 20°C under a vacuum of 0.37 mbar. Targets were stored under dry atmosphere until analysis. For the second type, cells were grown on two-compartment filters. When cells reached confluence (about 3 days), the epithelium was carefully rinsed in PBS and very quickly in water. Steps of cryofixation and lyophilisation were proceeded as for the first type targets.

Epithelia of MDCK cells (type 1 and type 2 targets) were analyzed with a 3 MeV proton beam of a few micrometers size and a beam current of the order of 1 nA using the scanning mode of the nuclear microprobe facility from the Pierre Süe Laboratory [9]. High speed scanning (1 ms/point) was processed in order to reduce the local heating of the sample. Large areas (200x200 µm2) were scanned: simultaneous analysis of several thousands of cells allowed determination of mean concentrations representative of the cell line. Elemental concentrations were deduced from the PIXE technique, normalized to the total mass of the sample derived from the protons which were elastically backscattered from C, N and O (RBS technique).

RESULTS AND DISCUSSION

Tests of viability, where cells from MDCK cell line were cultivated in presence of 1 to 50 µM $CdCl_2$ for 1 to 24 h were carried out. LDH determinations allowed to correlate results with those obtained with the Trypan blue exclusion method. *Figure 1* shows the variations of the percentage of mortality of MDCK cells for various concentrations of Cd: 50% of mortality is attained within 4 h after incubation with 50 µM $CdCl_2$, compared to 13 h with 20 µM and 22 h with 10 µM.

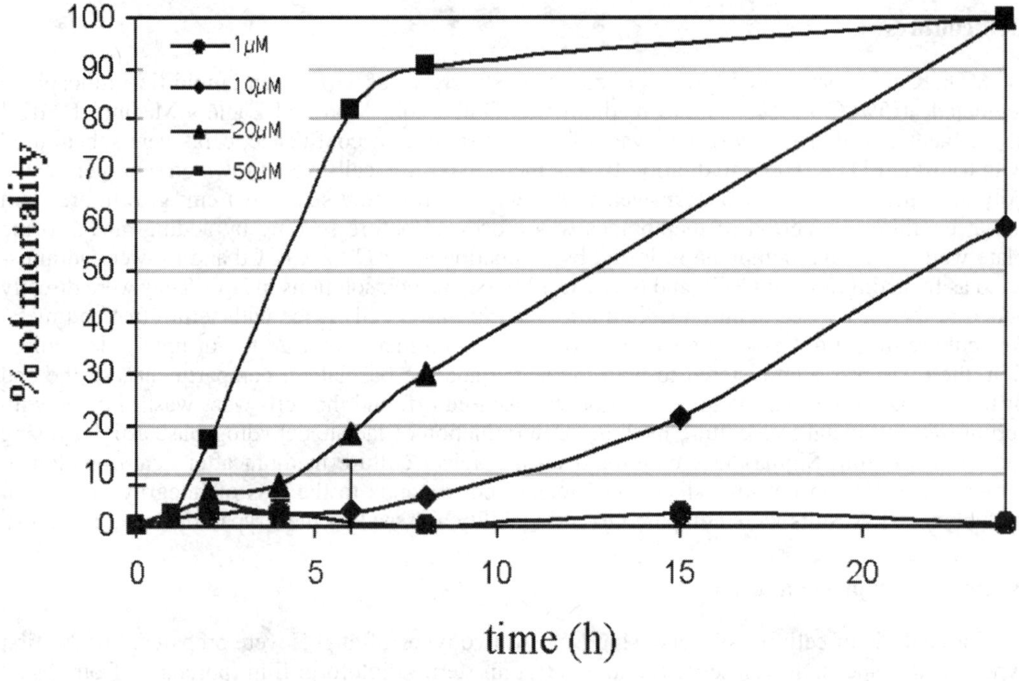

Fig. 1. Variations of the percentage of mortality of MDCK cells for various concentrations of Cd as a function of time (means and standard deviations of 3 separate experiments)

Nuclear microprobe analysis of monolayers grown on pioloform (type 1 targets) led to the determination of a mean mass thickness of 150 ± 13 µg/cm^2. This thickness was very reproducible between one measure to another: at confluence, epithelial cells form a monolayer with constant thickness. This thickness homogeneity, which is not found with other types of cultured cells, makes determination of elemental concentrations within cells easier.

Elemental distributions of K, Ca, Mn, Fe, Cu, Zn or Cd did not reveal special localization of biological elements or toxic metal within MDCK cells treated or not with $CdCl_2$. *Figure 2* presents the optical photograph and images of repartitions of Cl, K, Ca and metals within a few MDCK cells.

No Cd accumulation was measured in renal epithelial cells cultured on pioloform, whatever the Cd concentration in the culture medium. On the opposite, a Cd concentration of 797 ± 139 ppm (µg of Cd per g dry weight) has been measured in MDCK cells cultured on membranes (type 2 targets) after incubation with Cd 50 µM for 6 h. The difference between those results could come

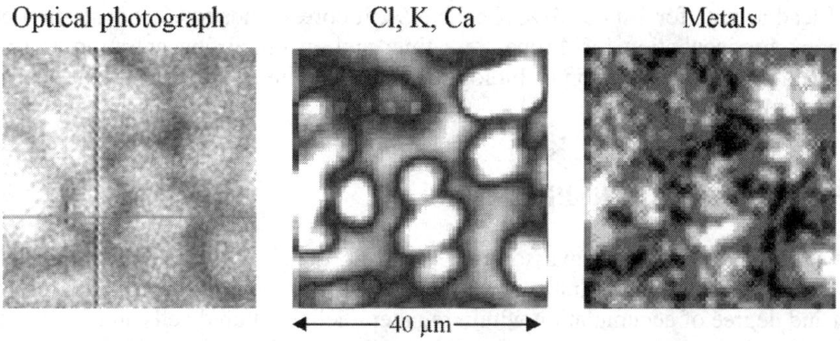

Fig. 2. Optical photograph and Cl, K, Ca and metals repartitions obtained by nuclear microprobe analysis of MDCK cells cultured with Cd 50 µM for 6 h (scanning of 40×40 μm^2)

from the lack of transepithelial transport through cells cultured on pioloform (non-permeable support) compared to cells cultured on semi-permeable membranes. Besides, it has been shown that cells accumulate much more Cd from the basolateral compartment than form the apical side [10]. Culture of cells on pioloform does not allow basolateral Cd exposure, explaining the difference of Cd accumulation on type 1 and type 2 targets.

With this latest specific culture technique (culture on membranes), results are in agreement with those already published: MDCK cells have been shown to accumulate high concentrations of Cd. Up to 12 nmoles Cd^{2+} per mg of protein have been measured after exposition to 20 µM Cd^{2+} for 8 h [11]. It has been demonstrated a lack of energy requirement for Cd uptake [12]: this could explain the high concentration of Cd accumulated in renal cells.

Highly significant differences in zinc concentrations in MDCK cells treated or not with Cd 50 µM for 6 h could be established. When adding Cd to the culture medium, intracellular Zn concentrations are decreased by a factor 3: they are decreased from 116 ± 4 to 41 ± 2 ppm in type 2 targets (membranes). No significant difference between biological (Cl, K, Ca) or other trace element concentrations (Mn, Fe, Cu) has been noticed between Cd-treated cells and controls. A common hypothesis in the literature is that Cd uptake involves competition with essential elements such as Ca or Zn. Whereas many studies assessed that Cd exposure can affect Ca^{2+} uptake by renal cells with decreased Ca^{2+} reabsorption [13], no variation in Ca contents within MDCK cells could be noticed in the conditions of our experiments. The physiological response that accompany Cd^{2+} uptake within renal epithelial cells involves Cd^{2+} displacement of Ca^{2+} from E-cadherin, a complex transmembrane protein [14]. But nuclear microprobe analysis only allows the determination of total intracellular calcium, and a variation of a part of membrane Ca or free Ca^{2+} cannot be measured using this technique.

Cd^{2+} has been described to interfere with zinc-containing proteins and enzymes, interacting with Zn^{2+} binding sites: this may be the case with the protein kinase C molecule, a zinc-finger protein [15]. Moreover, Cd-metallothionein has shown to be much less toxic than Cd^{2+} to renal epithelial cells *in vitro* [16] and loading of cadmium ions has shown to induce an increase of the Cd/Zn ratio in metallothionein [17]. Substitution of Cd to Zn on metallothionein might be a way for the cell to protect against Cd toxicity.

The same experiments carried out on type 2 targets incubated with Pb 30 µM for 4 h were carried out. Elemental distributions of K, Ca, Mn, Fe, Cu or Zn did not reveal particular localization of biological or trace elements within MDCK cells treated or not with Pb. Two-dimension maps of Pb distribution within the epithelium reveal localization spots with high concentrations of Pb. Although it has been reported that Pb^{2+} is more avidly taken up by cultured kidney cells than Cd^{2+} [18], the conditions of experiments described in this report led to mean concentrations (selecting areas without Pb-localization) of Pb^{2+} in cells of 255 ± 18 ppm (dry weight), that is 3 times lower than Cd^{2+} values. For comparison, an uptake of 4 ng Pb/10^6 cells has been measured after exposition to 2 μM of lead acetate for 1 h on MDCK cells [19]. It corresponds to a concentration of approximately 4 ppm, for incubation for 4 times less time and an initial concentration 15 times lower. Finally, no significant modification of biological or trace element (including zinc) concentrations could be noticed.

CONCLUSIONS AND PERSPECTIVES

Preliminary results of the comparison of Cd and Pb toxicity to MDCK renal cells have been presented in this paper. For the first time, it has been simultaneously determined the toxic metal repartition and degree of accumulation within renal epithelial cultured cells and its effects on other biological or trace elements concentrations and distributions. It has been demonstrated that Cd is highly and homogeneously incorporated within MDCK cells. In the conditions tested here, Pb is 3 times less accumulated, but many localizations could be observed. Another very interesting point is the significant decrease of intracellular Zn contents after Cd exposition. Hypotheses formulated in this report still need to be confirmed and will be the subject of future experiments.

REFERENCES

1. Handler J.S., Green N., Steele R.E. Cultures as epithelial models: porous-bottom culture dishes for studying transport and differentiation. *Meth. Enzymol.*, 1989,171, 736-744.
2. Friberg L., Elinder C.G., Kjellström I., Nordberg G. Cadmium and Health: A toxicological and epidemiological appraisal. Vols. 1-2, *CRC Press, Boca Raton, FL*, 1986.
3. Manton W.I., Cook J.D. High accuracy (stable isotope dilution) measurements of lead in serum and cerebrospinal fluid. *Br. J. Ind. Med.*, 1984, 41, 313-319.
4. Gaush C.R., Hard W.L., Smith T.F. Characterization of an established line of canine kidney cells (MDCK). *Proc. Soc. Exp. Biol. Med.*, 1966, 122, 931-935.
5. Gstraunthaler G., Steinmassi D., Pfaller W. Renal cell cultures: a tool for studying tubular function and nephrotoxicity. *Toxicol. Lett.*, 1990, 53, 1-7.
6. Llabador Y., Moretto P. (Eds.) Nuclear Microprobes in the Life Sciences. *World Scientific, Singapore*, 1998.
7. Decker T., Lohmann-Matthes M.L. A quick and simple method for the quantitation of lactate dehydrogenase release in measurements of cellular cytotoxicity and tumor necrosis factor (TNF) activity. *J. Immunol. Meth.*, 1988, 15, 61-69.
8. Gouget B., Sergeant C., Bénard J., Llabador Y., Simonoff M. N-*myc* oncogene amplification is correlated to trace metal concentrations in neuroblastoma cultured cells. *Nucl. Instrum. Methods Phys. Res.*, 2000, 170, 432-442.
9. Khodja H., Berthoumieux E., Daudin L., Gallien J.P. The Pierre Süe Laboratory nuclear microprobe as a multi-disciplinary analysis tool. *Nucl. Instrum. Methods Phys. Res.*, 2001, 181, 83-86.
10. Prozialeck W.C., Lamar P.C. Comparison of the cytotoxic effects of cadmium (Cd^{2+}) in high and low resistance strains of MDCK cells that express different levels of E-Cadherin. *Toxicol. In Vitro*, 1998, 12, 633-647.
11. Prozialeck W.C., Lamar P.C. Surface binding and uptake of cadmium (CD2+) by LLC-PK1 cells on permeable membrane supports. *Arch. Toxicol.*, 1993, 67, 113-119.
12. Endo T., Shaikh Z.A. Cadmium uptake by primary cultures of rat renal cortical epithelial cells: influence of cell density and other metal ions. *Toxicol. Appl. Pharmacol.*, 1993, 121, 203-209.
13. Wu X., Jin T., Wang Z., Ye T., Kong Q., Nordberg G. Urinary calcium as a biomarker of renal dysfunction in a general population exposed to cadmium. *J. Occup. Environ. Med.*, 2001, 43, 898-904.
14. Prozialeck W.C. Evidence that E-cadherin may be a target for cadmium toxicity in epithelial cells. *Toxicol. Appl. Pharmacol.*, 2000, 164, 231-249.
15. Berg J.M. More metal-binding fingers. *Nature*, 1986, 319, 264-265.
16. Blumenthal S., Lewand D., Krezoski S.K., Petering D.H. Comparative effects of Cd^{2+} and Cd-metallothionein on cultured kidney tubule cells. *Toxicol. Appl. Pharmacol.*, 1996, 136, 220-228.

17. Suzuki K.T. Studies of cadmium uptake and metabolism by the kidney. *Environ. Health Perspect.*, 1984, 54, 21-30.
18. Cherian M.G. Rat kidney epithelial cell culture for metal toxicity studies. *In vitro Cell Dev. Biol.*, 1985, 9, 505-508.
19. Dannon D.I., Olivi L., Bressler J. The role of anion exchange in the uptake of Pb by human erythrocytes and Madin-Darby canine kidney cells. *Toxicology*, 2000, 147, 101-107.

*Metal Ions in Biology and Medicine: vol. 7.*Eds L. Khassanova, Ph. Collery, I. Maymard, Z. Khassanova, J.-C. Étienne.
John Libbey Eurotext, Paris © 2002 pp. 144-148.

Ca^{2+} ions play key role in GS-induced activation of human platelets *in vitro*

Elene V. Hackl*, Sergey V. Gatash

Kharkov National University, Radiophysical Department, Chair of Molecular and Applied Biophysics, Svobody sq. 4, Kharkov 61077, Ukraine;
**Present address: Institute of Cell and Molecular Biology, University of Edinburgh, King's Buildings, Mayfield Road, Edinburgh EH9 3JR, UK;*
e-mail: ehackl@srv0.bio.ed.ac.uk

In our recent work [1] we have shown that the gramicidin S (GS) interaction with human platelets causes (depending on the GS concentration) swelling or activation of platelets. We have made a supposition that the mechanism of the GS-induced platelet activation involves Ca^{2+} ions. In the present work we want to demonstrate new results supporting this supposition and to propose a mechanism of the GS effect on platelets *in vitro*.

Gramicidin S is a cyclopeptide antibiotic; its molecule consists of two identical pentapeptides (Val → L-Orn$^+$ → Leu → D-Phe → Pro)$_2$ connected by the "head-tail" type and has two positively charged free aminogroups of L-ornitin residua [2, 3]. In a wide variety of environments, the GS molecule exists as a very stable amphiphilic antiparallel beta-sheet structure with a polar and a non-polar surface. This rather rigid structure is stabilised by four intramolecular hydrogen bonds, involving the four amide protons and the four carbonyl groups of the Val and Leu residues [4]. The spatial structure of the GS molecule is the important condition of the GS interaction with a membrane. The GS molecule has no specific protein receptor; it interacts directly with membrane phospholipids [2, 3, 5]. At the initial stage of interaction the GS molecule is fixed on a cell membrane interacting electrostaticly by means of NH_3^+-groups of L-ornitin with negatively charged residua of phosphate acid of lipid molecules. Then the lipid-GS complex is incorporated into a membrane changing the cyclopeptide conformation. The hydrophobic part of the GS molecule formed by Val and Leu residua is of considerable importance due to its capacity to hydrophobic interaction with a membrane. The lipid-GS complexes contact both between themselves and with membrane proteins due to hydrophobic interaction that lead to phase separation of phospholipids and to formation of a "liquid" zone of neutral lipids, which is very weak barrier for ions and molecules [2-5]. Thus GS at lower concentrations increases the permeability of model and biological membranes and at higher GS concentrations it causes membrane destabilization [2-5].

MATERIALS AND METHOD

In the work the venous blood of healthy donors of both sexes taken not later than for 24 hours before the experiment was used. The platelet-rich plasma (PRP) was prepared by centrifuging the whole blood for 10 min at 146 g (1000 rpm) at room temperature; the platelet poor plasma (PPP) was prepared by centrifuging the blood for 10 min at 1100 g (3000 rpm). The PRP samples were diluted with PPP to the platelet concentration about $2.5*10^5$ mm^{-3} (the initial optical density of the samples (D_0) was within 0.8-0.7 relative units).

The standard medical preparation of gramicidin S ("Pharmachim") was used in the present work. The initial 2-% ethanol solution of GS was diluted 30-50 times with 0.15 M NaCl (pH 7.4).

In the work the following chemical reagents were also used: ADP, adrenaline and thrombin ("Reanal", Hungary), $CaCl_2$ (reagent grade), papaverin ("Pharmachim"). To induce platelet swelling in hypotonic solutions we added the deionized water and PPP to PRP keeping constant the final platelet concentration in samples.

The platelet swelling and aggregation were studied by optical method according to Born [6, 7]. We used a photoelectrocolorimeter with a green light filter (λ = 540 nm). 1 ml of platelet rich plasma was placed into a specially designed thermostatic cuvette 3 mm thick and the change of the sample optical density with time was recorded.

RESULTS AND DISCUSSION

As we have shown recently [1] the character of the gramicidin S effect on platelets depends on the GS concentration. In small concentration ((65 µg/ml) GS induces the decrease of the sample optical density (OD) *(fig. 1, a)*. When rising the GS concentration the initial decrease of the sample OD is replaced by OD increase in 20-60 seconds after GS addition *(fig.1, b)*. Larger GS concentrations induce only increase of the optical density of a PRP sample *(fig. 1, c)*.

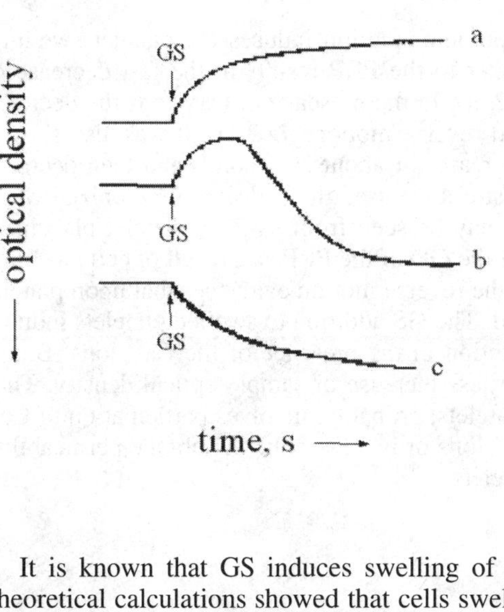

Fig. 1. Change of the PRP optical density upon GS addition. GS concentrations, µg/ml: 65 (a), 75 (b), 110 (c); temperature 22°C

It is known that GS induces swelling of different cells (for example, mitochondria [2, 4]). Theoretical calculations showed that cells swelling leads to the optical density decrease [8, 9]. As GS interacts only with membrane phospholipids, we can suppose that its effect on cells is non-specific. Thus the OD decrease observed *(fig. 1, a)* may be attributed to the platelet swelling induced by GS.

The main property of native platelets is the ability to form aggregates under the action of the aggregation reagents. There are many ways to block or to decrease significantly the platelet ability to aggregate, some of them are: *i)* incubation of platelets at high temperatures (45-50°C); *ii)* pre-incubation of platelets with papaverin (papaverin inhibiting phosphodiesterase prevents from cAMP → 5'AMP transition therefore the concentration of cAMP (the inhibitor of the platelet aggregation) in cells increases [10]); *iii)* pre-incubation of platelets with low concentrations of ADP to induce the platelet refractoriness. In all these three cases we have observed no platelet aggregation when ADP or thrombin were added to the platelet rich plasma samples (not illustrated). At the same time the GS addition to PRP samples leaded to the decrease of the sample OD, *i.e.* to platelet swelling. At that the swelling degree was approximately equal for native platelets and for treated platelets, which didn't respond to the aggregation reagent addition. So we can conclude that GS-induced platelet swelling is the non-specific process and does not depend on the platelet aggregation mechanism. Probably, swelling occurs due to increase of the non-specific membrane permeability when the GS molecules are incorporated into a membrane.

It was also shown theoretically that the change of the platelet shape from discoid to spherical (at a constant cell volume) should result in increase of the OD of a PRP sample [8, 9]. Such change of the platelet shape takes place upon platelet activation (stimulation), which precedes the platelet aggregation. It is known that the typical shape of resting platelets is discoid, upon activation they undergo the structure-morphological changes - first of all, the shape change to a globular form with pseudopodia (up to 5 μm long) [10, 11]. The OD increase (light transmission decrease) was observed experimentally during the first stage of platelet aggregation induced by ADP or adrenaline [10, 11]. In [1] we have proposed that at certain concentrations GS can induce the platelet activation. At that the GS concentration required to activate platelets is higher than that required to induce swelling *(fig. 1)*. Thus when rising the GS concentration platelet swelling is replaced by platelet activation.

After treatment with GS platelets do not lose their ability to aggregate under the aggregation inductor (ADP, adrenaline, thrombin, $CaCl_2$, H_2O_2, etc.) action. Thus the GS binding to platelet membranes (in the GS concentration interval studied in the work) does lead neither to platelet refractoriness nor to lose of membrane integrity nor to a significant inhibition of the membrane-associated receptors.

It is known that the platelet placement in a hypotonic solution induces the platelet swelling. Authors of [10] showed that addition of bidistil water to the PRP results in the fast decrease of the sample optical density due to cell swelling *(fig.2, a)*. In the presence of Ca^{2+} ions the decrease of the OD is replaced by the OD increase (so-called reverse motion, *fig.2, b)*. It was also shown, that in a hypotonic solution the platelet volumes increase for about 20 seconds and then decrease gradually. In the present work to induce hypotonic stress we used the addition of deionized water (to avoid any Ca^{2+} contamination) to the PRP. As may be seen from *fig.2, c*, platelet placement into the hypotonic solution induces rapid decrease of the OD of the PRP as a result of cell swelling. This result agrees with data of [10]. The absence of the reverse motion evidences that upon platelet swelling the intracellular Ca^{2+} have not been released. The GS addition to swelled platelets induces increase of the OD that is similar to the reverse motion in the presence of the Ca^{2+} ions. But in the presence of EDTA the GS addition cause much less increase of sample optical density. Thus we can suggest that the GS-induced activation of platelets, probably, involves participation of Ca^{2+} ions. The GS addition to a PRP sample releases Ca^{2+} ions or increases the membrane permeability for these ions; Ca^{2+} ions, in their turn, activate platelets.

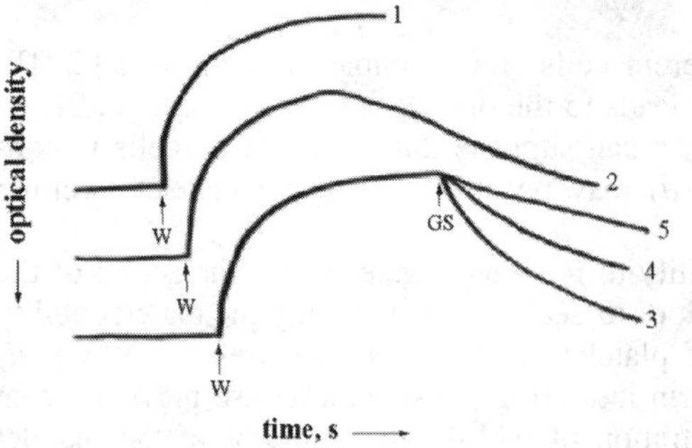

Fig. 2. Change of the PRP optical density with time for:
1 - platelet swelling in distil water; 2 - platelet swelling in a buffer containing Ca^{2+} ions; 3 - platelet swelling in deonising water followed by GS addition (3-5). GS was added to the PRP sample either alone (3) or after EDTA (4, 5). GS concentration - 50 μg/ml; EDTA concentration - 2 (4), 5 (5) mM; temperature 22^{0}C. (data 1, 2 from [10])

In the presence of EDTA degree of the GS-induced platelet activation is significantly decreased and at certain EDTA concentration the GS addition to a PRP sample does not cause the increase of sample OD (*i.e.* doesn't cause the platelet activation). At that EDTA does not prevent from the decrease of sample OD when GS adding *(fig. 3)*. Thus the presence of EDTA effects the GS-induced activation but does not effect the platelet swelling. This result support our suggestion that the GS-induced platelet swelling resulted in decrease of a sample OD is a non-specific process, while GS-induces platelet activation, which is characterised by the platelet shape change and resulted in increase of a sample OD, is a specifical process and involved Ca^{2+} ions as mediators.

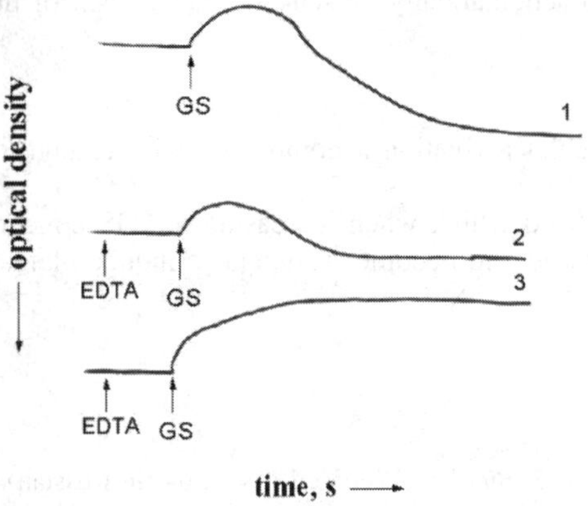

Fig. 3. Change of the PRP optical density with time upon the GS addition. GS was added to the PRP sample either alone (1) or after EDTA (2, 3). GS concentration - 110 μg/ml; EDTA concentration - 5 (2), 10 (3) mM; temperature 22⁰C

Figure 4 shows the time dependencies of the rat PRP optical density upon the GS addition. As it may be seen from the *figure*, in the absence of Ca^{2+} ions the GS addition induces only the increase of a sample OD attributed to the first step of platelet activation *(fig. 4, 4)*. Another situation was observed when PRP was pre-incubated with Ca^{2+} ions during 2-4 min before GS addition *(fig. 4, 1-3)*: GS addition caused the increase of the OD attributed to the platelet activation. Then this OD increase is replaced by significant decrease of the optical density as a result of platelet aggregation. In the latter case appearance of the platelet aggregates in a cuvette was observed even visually. It should be noted that Ca^{2+} ions themselves didn't cause any change of the samples OD (*i.e.* didn't cause platelet activation or aggregation). So we can say that GS induces the platelet activation at the expense of increase of the concentration of free Ca^{2+} ions but this concentration is not enough to induce platelet aggregation. When adding extra calcium ions from outside the total Ca^{2+} ion concentration was increased and became high enough to induce the platelet aggregation.

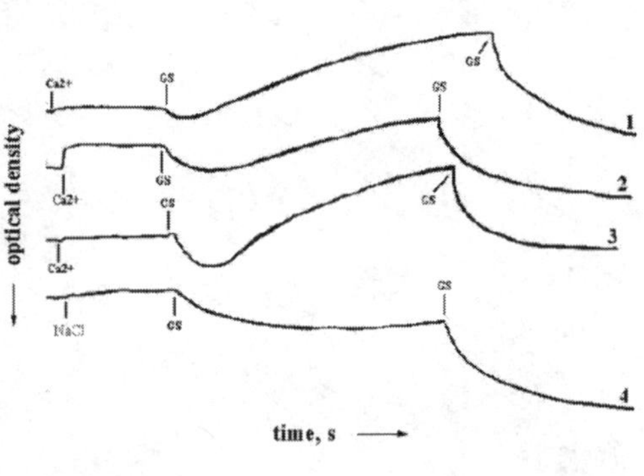

Fig. 4. Change of the rat PRP optical density with time upon the GS addition. Before GS addition the PRP samples were pre-incubated with Ca^{2+} ions (curves 1-3). In the curve 4 the PRP sample was pre-incubated with sodium chloride. GS concentration - 110 μg/ml; temperature 22⁰C

We can summarise that the GS addition to PRP induces two processes: platelet swelling resulted in decrease of sample optical density and platelet activation, which is characterised by change of platelet cell shape from discoid to spherical at a constant volume. Platelet activation results in increase of the PRP optical density. The fist process is a non-specific while the second one is a specific. With increase of the GS concentration platelet swelling is replaced by their activation. The higher GS concentration the stronger the GS-induced platelet activation is expressed. During platelet activation the change of platelet cell shape occurs as well as the increase of the concentration of free Ca^{2+} ions. But the platelet activation does not lead to platelet aggregation. Probably, the later requires a higher Ca^{2+} ion concentration.

Thus on the basis of the results obtained we can schematically present the mechanism of the GS effect on platelets *in vitro:*

1 platelet swelling

GS + PRP

2 increase of free Ca^{2+} ion concentration → platelet activation accompanied with change of cell shape from discoid to spherical.

At low GS concentrations only the way 1 is realised while, when increasing the GS concentration, the free Ca^{2+} ions concentration is also increased and become sufficient to induce platelet activation or even aggregation.

REFERENCES

1. E.V. Hackl, S.V. Gatash, V.P. Berest, *Visn. KhDU N 422, Biophys. Visnyk.*, 1998, 2, 64 (in Russian).
2. P.N. Korolev, V.G. Bulgakova, A.N. Polin *et al., Biolog. Nauki*, 1988, 7, 31 (in Russian).
3. E.J. Prenner, R.N. Lewis, R.N. McElhaney, *Biochim. Biophys. Acta*, 1999, 1462, 201.
4. R.N. Lewis, E.J. Prenner, L.H.Kondejewski et.al., *Biochemistry*, 1999, 38, 15193.
5. K. Takashi, K. Masakazu, T. Morikawa *et al., J. Pharmacobio. Dyn.*, 1989, 12, 124.
6. G.V.R. Born, *Nature*, 1962, 206, 927.
7. G.V.R. Born, *J. Physiol.*, 1970, 209, 487.
8. P. Latimer, G.V.R. Born, F. Michal, *Arch.Biochem.Biophys.*, 1977, 180, 151.
9. P. Latimer and F. Wamble, *Appl. Opt.*, 1982, 21, 2447.
10. A.B. Samal', S.N. Cherenkevich, N.Ph. Khmara, Platelet aggregation: methods of study and mechanisms, Minsk, *Universitetskoe*, 1990, p. 103 (in Russian).
11. A.B. Samal', S.N. Cherenkevich, N.F. Khmara, *Biofizika*, 1989, 34, 267 (in Russian).

Metal Ions in Biology and Medicine: vol. 7.Eds L. Khassanova, Ph. Collery, I. Maymard, Z. Khassanova, J.-C. Étienne.
John Libbey Eurotext, Paris © 2002 pp. 149-153.

In vitro evaluation of biocompatibility of dental metal materials on osteoblast cells in culture

M.C. Cortizo[1,2], M. Fernández L. de Mele[2], A.M. Cortizo[1]

[1]*Cátedra de Bioquímica Patológica, Facultad de Ciencias Exactas, Universidad Nacional de La Plata, 47 y 115 (1900) La Plata, Argentina;* [2]*INIFTA, Facultad de Ciencias Exactas, Universidad Nacional de La Plata, CC16 Suc 4, (1900) La Plata, Argentina*

ABSTRACT

The aim of this work was to evaluate the biocompatibility of the metal components of dental materials on the growth of osteoblast UMR106 cells in culture. Cells were exposed to Ag, Cu, Au, Pt, Pd and Ni/Ti alloy samples for different incubation periods. Number of surviving cells, mitotic index, cell morphology, and the expression of alkaline phosphatase activity as a marker of osteoblastic phenotype were investigated as parameters of biocompatibility. The metal ion content of the culture media was analyzed using atomic absorption spectrophotometry. Cu and Ag were the more cytotoxic elements, while Au, Pd, Pt and the Ni/Ti alloy were biocompatible in the osteoblastic culture. Cu and Ag induced strong morphological changes in the UMR106, after 48 hours, only a few cell survived (<5%). These results were correlated with the release of metal ions in the media. In conclusion, our study showed that the UMR106 cell culture system appears suitable for evaluating the biocompatibility of metallic dental materials.

INTRODUCTION

In clinical practice, different metal biomaterials are used in prothesis and orthodontic applications that place them into contact with oral tissues for long periods. Consequently, the alloys with lowest biological interactions with the oral epithelium, connective tissue, dental tissue and bone must be chosen to avoid biocompatibility problems (Wataha, 2000). Sometimes it is difficult to weigh up the toxic effect of each metal component through experimental assays made with the alloys (Mele and Cortizo, 2000). Thus, it is important to investigate the biocompatibility of each of the alloy components (Schedle *et al.*, 1998). Recent investigations have focus on the effects of different metal ions released from dental metal alloys on cells in culture (Wataha *et al.*, 1995, Messer *et al.*, 1999, Shaini *et al.*, 2000). It was also showed a correlation between the toxic effect of metal by cell proliferation assays and the levels of ions released in the medium (Grill *et al.*, 1997; Wataha *et al.*, 1995, Wataha and Hanks, 1996, Schedle *et al.*, 1995).

Although it would be clinically important to evaluate the behavior of dental metal alloys in vivo, many factors could influence experimental observations (Hwang et al., 2001, Grill *et al.*, 1997) which sometimes show great scatter of data. To overcome these problems, the biocompatibility has been estimated in vitro using fibroblastic cells, epithelial cells or osteoblast-like cells (Craig *et al.*, 1990, Shaini *et al.*, 2000, Roehlecke et al., 2001). Primary culture systems do not always exhibit reproducible results, in contrast, cloned cell line are much easier to be maintained in culture (Ahmad *et al.*, 1999).

The aim of the present study was to evaluate the biocompatibility of the components of dental metal materials on the growth and differentiation of UMR106 osteoblast cells in culture.

MATERIALS AND METHODS

Metal samples: pure Ag, Cu, Au, Pt, Pd, and Ni/Ti alloy wires (area = 0.157 cm^2) were tested after their sterilization in autoclave.

Cell culture: UMR106 cells were cultured in Dulbecco's Modifed Eagles Medium (DMEM) with 10% fetal bovine serum (FBS), as we have previously described (Cortizo and Etcheverry, 1995). Cells were subcultured using tripsin-EDTA and plated into 6 well/plate or 10 mm dishes. When cells reached 70% confluence, media was replaced by 0.5% FBS-DMEM and incubated for 24 or 48 h in the absence or the presence of different dental metal materials.

Metal ions release into culture media: 0.942 cm^2 metal samples were immersed in sterile DMEM for different periods between 4 and 76 h to assay the metal ion release through atomic adsorption spectrophotometry.

Cell survival assay: The viability of osteoblast-like cells was determined by counting the number of cells pre-stained with Giemsa as it was previously described (Cortizo *et al.*, 2001). A mitotic index (MI) was defined as the number of mitotic figures per field over the total number of cells per field.

Morphological assay: Morphological changes induced by different metal alloys were observed by light microscopy after staining with Giemsa (Cortizo *et al.*, 2001).

Osteoblastic differentiation: The activity of alkaline phosphatase (ALP) was evaluated as a specific marker of osteoblastic differentiation as we have previously described (Cortizo and Etcheverry, 1995). The protein content was assayed by the method of Bradford (Bradford, 1976).

Statistical analysis: At least two experiments were performed by duplicate for each experimental condition. Results are expressed as mean (SEM. Statistical differences were analyzed using Student's t test.

RESULTS AND DISCUSSION

The biocompatibility of different dental metal materials was evaluated throughout the cell survival, morphological changes and the alkaline phosphatase specific activity of UMR106 osteosarcoma line.

Cu and Ag were the more cytotoxic elements, inducing a significant decrease in the cell survival after 24 hours and almost not cells were found at the end of the culture of 48 hours *(table 1)*. Conversely, Au, Pd, Pt, and Ni/Ti behaved like compatible elements, without inducing changes in the cell growth. These results agree with previous observations of fibroblasts in which alloys of noble elements were compatible with cell viability, while Cu and Ag were the more toxic elements (Craig *et al.*, 1990, Wataha, 2000).

Recently, there have been controversies over possible biocompatibility problems of palladium (Wataha and Hanks, 1996). In sufficiently high concentrations of palladium ions toxic and allergic effects were observed. However, there is a low palladium dissolution in simulated biological media (Mele and Duffó, 2001). Present results confirm that the risk of biocompatibility problems is low.

Table 1. Effect of dental metal materials on UMR106 cell survival [% basal]

Incubation	Control	Cu	Ag	Au	Pd	Pt	Ni/Ti
24 hours	100 ± 10	59 ± 6**	70 ± 6*				
48 hours	100 ± 11	0.4 ± 0.2#	1.4 ± 0.7#	96 ± 11	100 ± 8	100 ± 12	110 ± 6

Differences vs Control are: * $p<0.05$; ** $p<0.01$; # $p<0.001$.

We also evaluated the effect of the metal ions on cell morphology and the percentage of mitotic figures. After 24 hours of culture, control UMR106 cells (media alone) showed a polygonal morphology with well stained nuclei and cytoplasms with processes connecting neighboring cells. Several mitotic figures were also observed, with a MI = 0.022 ± 5.10^{-3}. The presence of Cu and Ag induced a reduction in the number of surviving cells, as well as morphological alterations after 24 hours. The culture exhibited a heterogeneous distribution of cells, showing no osteoblasts in the surroundings of the metal. Cells in these areas showed a strong reduction in the cytoplasms, with lost of processes and a pronounced condensation of the nuclei. A longer incubation of the monolayer with Cu and Ag showed almost no surviving cells. Mitotic figures were no observed after 24 or 48 hour of incubation with both metal materials, indicating an arrest of the proliferative process under these culture conditions. On the other hand, Au, Pd, Pt and Ni/Ti support the growth of UMR106 cells without any morphological alterations.

The cell differentiation of UMR106 cells by the bone marker ALP was also assessed. Under basal conditions, the levels of ALP were 198 ± 22 and 414 ± 32 nmol / min.mg protein after 24 and 48 h of culture in 0.5% FBS-DMEM. As can be seen in *figure 1*, after 24 h of culture with Cu and Ag, cells still maintained their osteoblastic phenotype, with no significant alterations in the ALP activity. However, a longer incubation period (48 h) with these metals, a strong decrease in ALP of the few cells surviving was detected. Conversely, the noble elements as well as the Ni/Ti alloy were biocompatible with the osteoblastic differentiation. The ALP has been previously demonstrated to be a good parameter to estimate the biocompatibility of implant materials in osteoblasts and be very sensitive marker of cellular toxicity, reflecting the metabolism of osteoblasts (Itakura *et al.*, 1988).

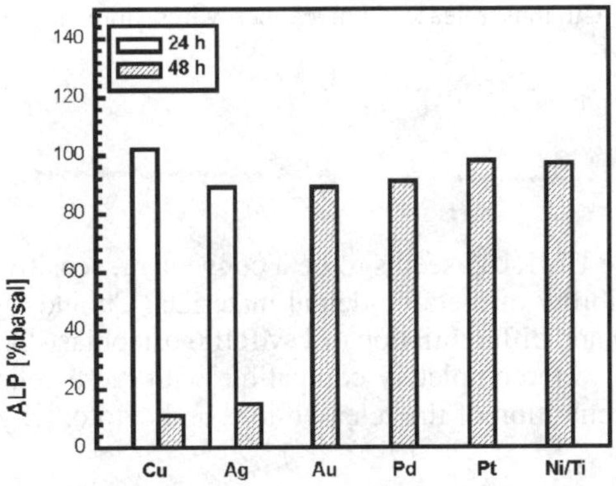

Fig. 1. Effect of dental metal materials on the ALP acitivity of UMR106 cells

Metal ions released from Au, Pt, Pd and Ni/Ti alloy were bellow the detectable limits. The level of Ag ions was also low but it became detectable when a large Ag sheet (area = 6 cm^2) was used. In this case 6 μg of Ag was detected after 48 h of immersion. Cu ions content for 0.94 cm^2 wires, after different exposure periods is shown in *fig. 2*. It can be observed that the dissolution rate decreases with time. Previous results show that Cu and Ag are the most likely agents of succinic dehydrogenase (SDH) depression, an assay used to evaluate metal cytotoxicity. Concentration of metal ions necessary to cause 50% of SDH depression are 4 μg/mL in the case of copper and 0.7 μg/mL in the case of silver (Wataha, 1995). Present results showed that concentration of copper ions are higher than 4 μg/mL. Notwithstanding that the concentration of silver ions is close to 0.1 μg/mL after 48 hours, present results show that it is able to produce changes in the number of surviving cells, mitotic index, cell morphology, and the expression of alkaline phosphatase activity.

It is worthy mention that the toxic action of copper is not limited to the copper ions release but is also able to generate oxygen species and hydroxyl radicals which are related to complex biocompatibility problems (Landolph, 1999). Release has been implicated as a primary cause of cytotoxicity with high-copper dental amalgams, and binary and ternary alloys of copper. However, there is not a lineal correlation between the decrease in the ALP activity, which was significantly lower after 48 h than after 24 h, and the concentration of copper ions, which rate of dissolution decreases with time. An ion concentration threshold may be needed to cause toxic effects.

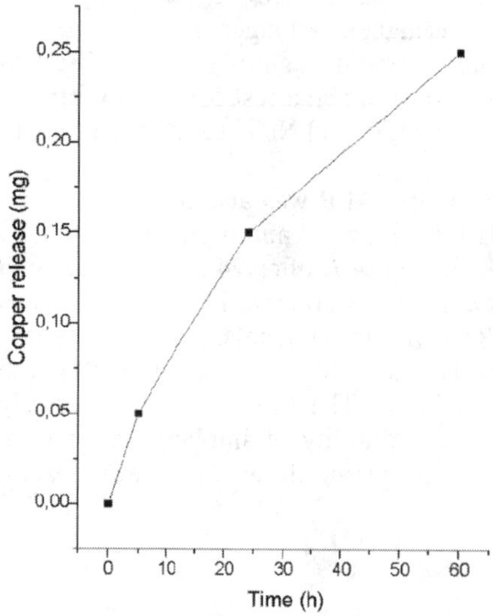

Fig. 2. Cu ions release from copper wires immersed in DMEM

CONCLUSIONS

Present results showed that the osteosarcoma line UMR106 seems to be a convenient, sensitive and reproducible system to evaluate the biocompatibility of metallic dental materials. Cu and Ag are the more cytotoxic elements affecting the growth and differentiation of UMR106 osteoblast-like cells, while the noble elements and the Ni/Ti alloy were completely compatible with these cells. There was not a linear correlation between the concentration of the released ions and cytotoxicity.

ACKNOWLEDGEMENTS

We thank Mr Raul Perez (CIDEPINT) for performing the metal levels in the media. This work was supported by grants from Universidad Nacional de La Plata, CONICET (PIP 4377/96), and the Agencia Nacional de Promoción Científica y Tecnológica, (PICT 6782). AMC and MFL de M are members of the Carrera del Investigador, CICPBA and CONICET, respectively.

REFERENCES

Ahmad M., McCarthy M.B., Gronowicz G. A in vitro model for mineralization of human osteoblast-like cells on implant materials. *Biomaterials*, 1999, 20, 211-220.

Bradford M. Rapid and sensitive method for quantitation of microgram quantities of protein utilizing the principle of protein-dye binding. *Anal Biochem*, 1976, 72, 248-254.

Cortizo A.M., Etcheverry S.B. Vanadium derivatives act as growth factor-mimetic compounds upon differentiation and proliferation of osteoblast-like UMR106 cells. *Mol Cell Biochem*, 1995, 145, 97-102.

Cortizo M.S., Allesandini J.L., Etcheverry S.B., Cortizo A.M. A vanadium/aspirin complex controlled release using a poly(β-propiolactone) film. Effects on osteosarcoma cells. *J Biomater Sci Polymer Edn*, 2001, 12, 945-960.
Craig R.G., Hanks C.T. Cytotoxicity of experimental casting alloys evaluated by cell culture tests. *J Dent Res*, 1990, 69, 1539-1542.
Grill V., Sandrucci M.A., Basa M., Di Lenarda R., Dorigo E., Martelli A.M., Bareggi R., Narducci P. The influence of dental metal alloys on cell proliferation adn fibronectin arrangement in human fibroblast cultures. *Arch Oral Biol*, 1997, 42, 641-647.
Huang C., Shin J, Cha J. Metal release from simulated fixed orthodontic appliances, *Am J Orthod Dentofacial Orthop*, 2001, 120, 383-391.
Itakura Y., Kosugi A., Sudo H., Yamamoto S. Development of a new system for evaluating the biocompatibility of implant materials using an osteogenic cell line (MC3T3E1). *Biomed Mater Res*, 1988, 22, 613-622.
Landolph J. Role of free radicals in metal-induced carcinogenenesis. In: Metal ions in biological systems (A. Sigel and H. Sigel eds), Marcel Dekker, Inc., New York, vol 36, pp. 445-483.
Mele M.F.L. de, Cortizo M.C. Biodeterioration of dental materials. Influence of bacterial adherence. *Biofouling*, 2000, 14: 305-316.
Mele M.F.L. de, Duffó G. Tarnish and corrosion of silver-based casting alloys in synthetic salivas of different compositions, J Appl. Electrochem, 2002, 32 (in press).
Messer M., Bishop S., Lucas L. Effects of metallic ion toxicity on human gingival fibroblast morphology. *Biomaterials*, 20, 1647-1657.
Roehlecke C., Witt M., Kasper M., Schulze E., Wolf C., Hofer A., Funk R.W. Synergistic effect of titanium alloy and collagen type I on cell adhesion, proliferation and differentiation of osteoblast-like cells. *Cells Tissues Organs*, 2001, 168, 178-187.
Schedle A., Samorapoompichit P., Rausch-Fan X.H., Franz A., Fureder W., Sperr W.R., Sperr W., Ellinger A., Slavicek R., Boltz-Nitulescu G., Valent P. Response of L.929 fibroblasts, human gingival fibroblasts, and human tissue mast cells to various metal cations. *J Dental Res*, 1995, 74, 1513-1520.
Schedle A., Samorapoompichit P., Fureder W., Rausch-Fan X., Franz A., Sperr W., Slavicek R, Simak S, Klepetko W., Ellinger A., Ghannadan M., Baghestanian M., Valent P. Metal ion-induced toxic histamine release from human basophils and mast cells. *J Biomed Mater Res*, 1998, 39, 560-567.
Shaini F., Shelton R., Marquis P., Shortall A. In vitro evaluation of the effect of freshly mixed amalgam and gallium-based alloy on the viability of primariy periosteal and osteoblast cell curtures. *Biomaterials*, 2000, 21, 113-119.
Wataha J.C., Malcolm C.T., Hanks C.T. Correlation between cytotoxicity and the elements released by dental casting alloys. *Int J Prosthodont*, 1995, 8, 9-14.
Wataha J.C., Hanks C.T. Biological effects of palladium and risk of using palladium in dental casting alloys. *J Oral Rehabil*, 1996, 23, 309-320.
Wataha J.C. Biocompatibility of dental casting alloys; Review. *J Prosthet Dent*, 2000, 83, 223-234.

Metal Ions in Biology and Medicine: vol. 7. Eds L. Khassanova, Ph. Collery, I. Maymard, Z. Khassanova, J.-C. Étienne.
John Libbey Eurotext, Paris © 2002 pp. 154-158.

Vanadium inhibits HaCaT cell proliferation but does not cause apoptosis

Evangelou, A[1]., Kolettas, E.[1,4], Tenopoulou, M.[2], Galaris, D.[2], Gonos, E. S.[3], Manos, G.[4]

[1]Laboratory of Physiology; [2]Laboratory of Biological Chemistry, University of Ioannina Medical School, 45100 Ioannina, Greece; [3]Laboratory of Cellular and Molecular Ageing, Institute of Biological Research and Biotechnology, NHRF, 11635 Athens, Greece and [4]School of Agricultural Technology, Technological Institute of Epirus, Greece.

ABSTRACT

Vanadium is a transition metal widely distributed in the enviroment. Vanadium compounds exert a variety of biological responses including antiproliferative responses through the activation of signalling pathways and ROS. As the human skin keratinocyte is a cell type most exposed to enviromental insults, a spontaneously immortalised human keratinocyte cell line (HaCaT) bearing mutant, transcriptionally inactive p53, was used to investigate the antiproliferative effects of vanadyl (IV) sulphate on their growth behaviour. Treatment of HaCaT cells with 25 to 200 μM $VOSO_4$ for 24 h induced inhibition of cell proliferation in a dose-dependent manner, reaching a saturation level at 100 μM vanadate, as determined by colony formation assays. Growth inhibition was linked to suppression of serum-induced expression of *c-fos* proto-oncogene, an early response gene involved in skin homeostasis and induction of the Cdk inhibitor $p15^{INKB}$. Further, there was a slight induction of *clusterin/apolipoproteinJ*, a gene expressed by surviving cells under stressful conditions, following treatment with vanadate. Constitutive expression of *clusterin/apolipoproteinJ* in HaCaT cells failed to rescue them from vanadium-induced growth inhibition or to alter the pattern of expression of c-fos mRNA. We are currently investigating the effects of vanadium compounds and the role of clusterin/apolipoproteinJ in vanadium- and H_2O_2-induced antiproliferative responses.

INTRODUCTION

Vanadium compounds exert preventive effects against chemical carcinogenesis on animals, by modifying, mainly, various xenobiotic enzymes, inhibiting, thus, carcinogen-derived active metabolites. Studies on various cell lines reveal that vanadium exerts its antitumor effects through inhibition of cellular tyrosine phosphatases and/or activation of tyrosine phosphorylases, thereby affecting signal transduction pathways leading to cell antiproliferative responses. Furthermore, vanadium compounds exert cytotoxic effects through the generation of reactive oxygen species (ROS), generated by Fenton-like reactions and/or during the intracellular reduction of V(V) to V(IV), mainly by NADPH, which contribute to the induction of apoptosis. Vanadium may also exert inhibitory effects on cancer cell metastastatic potential through modulation of cellular adhesive molecules, and reverse antineoplastic drug resistance [1-3].

The cellular antiproliferative responses are inhibition of cell proliferation and apoptosis. The former is almost certainly due to down-regulation of selected-positive acting cell cycle regulatory genes, including the *c-fos* proto-oncogene and cyclins, components of cyclin-dependent protein kinases (Cdks) and due to the induction of selected growth inhibitors, most notably the Cdk inhibitors $p15^{INKB}$, $p16^{INKA}$ and p21. Induction of apoptosis is characterised by DNA fragmentation and several genes have been shown to interfere with the apoptotic pathway induced by several different stimuli. Among these genes are the *bcl-2* family [4, 5] and *clusterin/apolipoproteinJ* [6-8].

Vanadium compounds have been found to induce apoptosis in cultured mammalian epithelial cells through p53 transactivation and hydrogen peroxide [9-11]. As epidermal cells are most exposed to enviromental insults, we used a spontaneously immortalised human keratinocyte line to investigate the effects of vanadyl sulphate on cell proliferation and apoptosis.

MATERIALS AND METHODS

Cell Culture and DNA transfections

HaCaT and Hela cells were cultured in DMEM supplemented with FCS, antibiotics and glutamine at 37°C, 5%CO_2. HaCaT cells were transfected with a *neo*-containing vector (pCDNA3.1) or an expression vector containing, in addition to *neo* gene, the entire human *clusterin/apoJ* cDNA (pCDNA3.1/*ApoJ*) with the DMSO-Polybrene method. Transfected cells were selected in G418 and expanded into mass culture.

Treatment of HaCaT cells with Vanadyl (IV) sulphate

For RNA or protein analysis, subconfluent HaCaT cells were treated with different doses of $VOSO_4$ (25, 50, 100, 200 μM) for 24 hours in complete growth medium. For colony formation assays, 200 HaCaT or Hela cells were plated per 60 mm dish. Following attachment for 36 hrs, the cells were treated with vanadate for 24 hrs. The medium was replaced with fresh complete growth medium and the cells were incubated for 14 days with regular media changes. Colonies were fixed and stained with crystal violet. Colony formation was expressed as %age of the number of cells plated.

RNA isolation and analysis

RNA was isolated from untreated and vanadate-treated HaCaT cells with the LiCl-Urea method and analysed by Northern blot hybridisation using ^{32}P-labelled cDNA-specific probes to a *c-fos* or $p15^{INKB}$.

Western blot analysis

Cells lysates were extracted from untreated and vanadate-treated HaCaT cells, according to standard protocols and analysed by SDS-PAGE followed by western blot using a monoclonal anti-clusterin-specific antibody (Quidel, USA).

RESULTS

Constitutive expression of clusterin/ApoJ fails to rescue epithelial cells from vanadate-induced inhibition of cell proliferation

HaCaT and Hela cells were transfected with pCDNA3.1 *(neo)* and the same expression vector carrying the entire human clusterin/ApoJ cDNA. Following selection in G418, the transfected cells were expanded into mass culture and analysed for the expression of clusterin/ApoJ under subconfluent and confluent conditions in the presence or absense of different doses of $VOSO_4$ *(fig. 1)*.

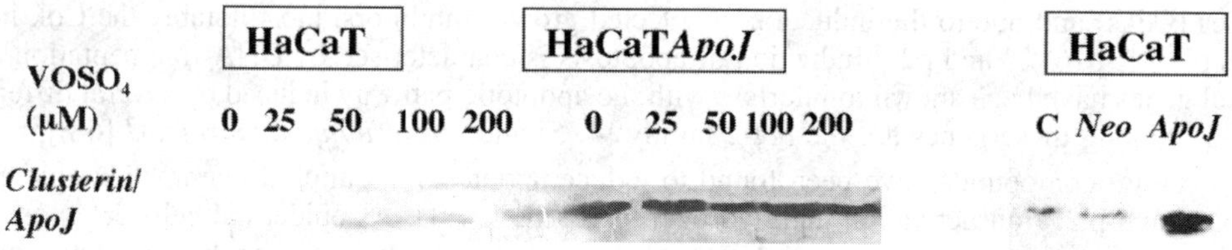

Fig. 1. Expression of clusterin/ApoJ in HaCaT cells

Cell lysates from untreated and vanadate-treated with the indicated doses for 24 hrs, subconfluent HaCaT and HaCaTApoJ cells (Panel A) and from confluent, untreated HaCaT, HaCaT*neo* and HaCaT*ApoJ* cells (Panel B) were analysed by western blot using an anti-human clusterin antibody.

Whereas untreated subconfluent HaCaT cells expressed very low to undetectable amounts of clusterin/ApoJ, vanadate used at 50 µM increased the synthesis of clusterin/ApoJ, the levels of which did not appear to change with higher vanadate concentrations (Panel A). In contrast, transfected HaCaT cells expressed markedly increased amounts of clusterin/ApoJ and addition of vanadate did not appear to further stimulate clusterin/apoJ synthesis (Panel A). Analysis of untreated, confluent cultures of HaCaT, HaCaT*neo* and HaCaT*ApoJ* cells showed that the clusterin/ApoJ-transfected HaCaT cells expressed significantly higher amounts of the protein than their control counterparts (Panel B).

Untransfected and *clusterin/ApoJ*-transfected HaCaT and Hela cells were treated with 25-200 µM $VOSO_4$ for 24 hrs and colonies were allowed to develop for 14 days and then fixed, stained and counted *(table 1)*.

Table 1. Colony Formation of HaCaT and Hela cells following vanadate treatment

Cell line	Plating efficiency (%) following $VOSO_4$-treatment (µM)				
	0	25	50	100	200
HaCaT	55.2	33.8	15.8	6.5	0
HaCaT*neo*	53.8	31.5	14.6	5.2	0
HaCaT*ApoJ*	39.0	24.0	10.1	4.4	0
Hela	47.0	28.0	13.5	0.0	0
Hela*neo*	46.0	27.5	11.7	0.0	0
Hela*ApoJ*	40.5	21.0	5.5	0.0	0

$VOSO_4$ inhibited both HaCaT and Hela cell proliferation in a dose-dependent manner. Overexpression of clusterin/ApoJ in both HaCaT and Hela cells failed to rescue them from vanadate-induced growth inhibition. Preliminary experiments showed that constitutive expression of clusterin/ApoJ inhibited the spontaneous apoptosis but it was not clear if vanadate increased apoptosis in HaCaT cells and whether clusterin/ApoJ blocked this response of HaCaT cells to vanadate. We are currently repeating these experiments with isolated clones of cells.

Vanadate downregulates c-fos proto-oncogene and induces $p15^{INKB}$ expression

Subconfluent HaCaT cells were serum-stimulated in the presence or absence of (0-200 µM) $VOSO_4$ for 24 hrs and total RNA extracted from untreated and vanadate-treated HaCaT cells was analysed by Northern blot hybridisation for the expression of *c-fos* proto-oncogene and $p15^{INKB}$ *(fig. 2)*, two of the genes implicated in cell cycle progression.

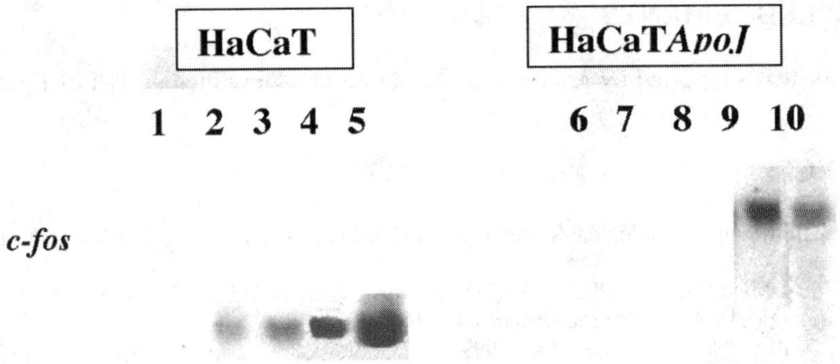

Fig. 2. Expression of *c-fos* proto-oncogene and the Cdk inhibitor $p15^{INKB}$ in untreated and vanadate-treated HaCaT cells. Lanes 1, 6: Control, Lanes 2, 7: 25 µM $VOSO_4$, Lanes 3, 8: 50 µM $VOSO_4$, Lanes 4, 9: 100 µM $VOSO_4$, Lanes 5, 10: 200 µM $VOSO_4$

Vanadate markedly inhibited the serum-induced expression of c-fos mRNA and induced the expression of the Cdk inhibitor $p15^{INKB}$, albeit in the presence of serum, in a dose-dependent manner. Constitutive expression of clusterin/ApoJ did not affect, at least, the vanadate-induced downregulation of *c-fos* gene expression.

DISCUSSION AND CONCLUSIONS

The results show that vanadate inhibits both HaCaT and Hela cell proliferation in a dose-dependent manner by affecting the expression of genes that regulate cell cycle progression.

The downregulation of the serum-induced expression of *c-fos* proto-oncogene is important for several reasons: (a) The *c-fos* proto-oncogene positively regulates cell cycle progression, (b) *c-fos* is implicated in skin homeostasis, and (c) in the absence of *c-fos* expression, benign cancer cells fail to upregulate angiogenic and proteolytic factors, and this may prevent their malignant conversion [12-14].

The induction of the Cdk inhibitor $p15^{INKB}$, a negative regulator of cell cycle progression, by vanadate in a dose-dependent manner suggests that vanadate interfers with the function of cyclins and this may be a mechanism contributing to the ceasation of cellular growth. As neoplastic transformation is accompanied by loss of regulation of cell cycle checkpoints in conjunction with aberrant expression of Cdks and/or cyclins and the loss of or mutation of Cdk inhibitors, one strategy is to inhibit malignant conversion and proliferation is inhibit Cdk activity or to enhance the function of Cdk inhibitors. Novel anticancer agents are now being developed to target these cell cycle control molecules [15-18]. The novel finding of the induction of $p15^{INKB}$ by vanadate and the low toxicity of vanadium compounds suggests that the vanadium-based drugs may be potential therapeutic agents for cancer treatment.

Although a preliminary finding, the failure of vanadate to induce apoptosis in HaCaT cells may be due to the lack of wild-type p53 expression by these cells, as p53 transactivation appears to be required for vanadium-induced apoptosis of epidermal cells [9, 10]. We are currently investigating the cytotoxicity and the role of novel vanadate complexes and clusterin/ApoJ in HaCaT and Hela cells and in different tumour cell lines using both sense and antisense *clusterin/ApoJ* transfectants.

ACKNOWLEDGEMENTS

This work was supported by grants from the Research Committee of TEI of Epirus, Greece.

REFERENCES

1. Rehder D. In: Metals in Biological Systems. Sigel H & Sigel A (eds), Marcel Dekker NY, 1995, Vol. 31, pp. 1-43.
2. Morinville A, Mayasinger D, Shaver A. (1998). *Trends Physiol Sci* **19**, 452-460.
3. Neachy BR. (1984). *Annu Rev Pharmacol* **24**, 501-524.
4. Gross A, et al. (1999). *Genes Dev* **13**, 18999-1999.
5. Korsmeyer SJ. (1999). *Cancer Res* **59**, 1639s-1700s.
6. Viard I, et al., (1999). *J Invest Dermatol 112*, 290-296.
7. Sensibar JA, et al. (1995). *Cancer Res* **55**, 2431-2437.
8. Miyake H, et al. (2000). *Cancer Res* **60**, 2547-2554.
9. Ye J, et al. (1996). *Mol Cell Biochem* **202**, 9-17.
10. Huang C, et al. (2000). *J Biol Chem* **275**, 32516-32522.
11. Ghosh P, et al. (2000). *Clin Cancer Res* **6**, 1536-1545.
12. Basst-Seguin N, et al. (1990). *J Invest Dermatol* **94**, 418-422.
13. Saez E, et al. (1995). *Cell* **82**, 721-732.
14. Yuspa SH. (1998). *J Dermatol Sci* **17**, 1-7.
15. Shapiro GI, Harper JW. (1999). *J Clin Invest* **104**, 1645-1653.
16. McDonald ER, El-Deiry WS. (2000). *Int J Oncol* **16**, 871-886.
17. Buolamwini JK. (2000). *Curr Pharm Des* **6**, 379-392.
18. Mani S, et al. (2000). *Expert Opin Investig Drugs* **9**, 1849-1870.

Metal Ions in Biology and Medicine: vol. 7. Eds L. Khassanova, Ph. Collery, I. Maymard, Z. Khassanova, J.-C. Étienne.
John Libbey Eurotext, Paris © 2002 pp. 159-162.

Synthesis, characterization and bioactivity of polyoxometalates on osteoblasts in culture

Irma L. Botto*[1,2], Daniel A. Barrio[3], María G. Eguzquiza[1], Carmen I. Cabello[1], Ana M. Cortizo[3], Susana B. Etcheverry[2,3]

[1]*CINDECA,* [2]*CEQUINOR,* [3]*Bioquímica Patológica, Facultad de Ciencias Exactas, Universidad Nacional de La Plata. 47 y 115 (1900) La Plata, Argentina*
**Author for correspondence*

ABSTRACT

Heteropoly- and isopolyoxometalates of V, Mo and W, represent a class of polyanionic compounds with a variety of important biological activities such as the inhibition of specific enzymes, antiviral and antitumoral effects. The K salt of the $[(PW_9O_{34})_2Co_4(H_2O)_2]^{10-}$ anion (I) has been obtained in aqueous solution from the precursor, $Na_8HPW_9O_{34}.24H_2O$ (II), and $CoCl_2$ in excess of KCl. Phases (I) and (II) were characterized by different techniques such as X ray diffraction (XRD), Vibrational spectrocopy (FTIR-Raman) and diffuse reflectance spectroscopy (DRS), scanning electron microscopy (SEM-EDAX) and thermal studies. Phase (I) is structurally related to the condensation of two fragments of (II) by a Co tetranuclear cluster in a sandwich type configuration. The biological activity on cell proliferation and morphological transformations, were investigated on rat osteosarcoma (UMR106) and mouse calvaria (MC3T3E1) -osteoblast-like cells in culture. Compound (II) did not show any effect on tumoral UMR106 proliferation while it induced the proliferation of MC3T3E1 cells in a bell like shape. On the contrary, phase (I) caused a strong inhibition on the UMR106 proliferation in a dose response manner. The latter observation correlated with the marked morphological changes caused by this compound on tumoral cells. In conclusion, compound (II) promoted the proliferation of non- transformed osteoblasts without any alteration of their morphology. This compound did not affect tumoral osteoblasts. On the contrary, phase (I) strongly inhibited tumoral osteoblast proliferation, inducing important morphological alterations. Hence, this compound is interesting from a pharmaceutical point of view for its potential use as an antitumoral drug.

INTRODUCTION

The polyoxometalates (POMs) are oligomeric aggregates of metal cations (usually d^o species) bridged by oxide anions that form by self assembly processes. Heteropolioxometalates (HPOMs) contain one or more p-, d- and f-block heteroatoms in addition to the other atoms. Structural and electronic properties of these type of compounds are easier to modify and this characteristic turn them attractive for applications in several fields of sciences. In fact, the polarity, redox potentials, surface charge distribution, shape, stability in physiological conditions, etc are attributes of POMs and HPOMs that render them attractive for their applications as pharmaceutical drugs [1]. The most significant therapeutical properties of these compounds are their antiviral and antitumoral activities, which have dominated the medicinal chemistry of them up to date [2]. They show antiviral activity both in vivo and in vitro against several RNA and DNA virus [3]. On the other hand, despite the number of POMs synthesized and characterized, there is limited information on their antitumoral activity.

From a mechanistic point of view, it is likely that many of the biological effects of POMs and HPOMs are related to the interactions with different enzymes [4]. In this context, it has been proposed that these compounds can affect different isoenzymes of phosphotyrosine phosphatases.

The aim of this work was to synthesize and characterize the $K_{10}[(PW_9O_{34})_2Co_4(H_2O)_2].20\ H_2O$ (I) and, $Na_8HPW_9O_{34}.24H_2O$ (II) phases in order to analyze their biological effect on osteoblast proliferation in two osteoblast-like cell lines in culture: the non-transformed MC3T3E1 derived from mouse calvaria and the tumoral UMR106 from a rat osteosarcoma.

MATERIALS AND METHODS

$K_{10}\ [(PW_9O_{34})_2Co_4(H_2O)_2].20\ H_2O^-$ (I) and its precursor $Na_8HPW_9O_{34}.24H_2O$ (II) were synthesized by precipitation, in aqueous solution. (I) was obtained from (II), Co(II) acetate and excess of KCl as starting materials whereas (II) is obtained from sodium tungstate, phosphoric and acetic acids. Both phases were characterized by XRD analysis, FTIR, Raman and Diffuse reflectance (DRS) spectroscopies and SEM and EDAX microscopy. Stability was proved by thermal and chemical studies.

MC3T3E1 and UMR106 osteoblast-like cells were grown in DMEM supplemented with 10% (v/v) of fetal bovine serum and antibiotics in a humidified atmosphere of 95% air /5% CO2, as previously described [5]. To test the effect of (I) and (II) on cell proliferation, the osteoblast-like cells were incubated overnight with different concentrations of the compounds in serum-free DMEM. The cellular proliferation was determined through the crystal violet assay [6].

The morphological studies were performed by light microscopy with Giemsa staining [5].

RESULTS AND DISCUSSION

$[(PW_9O_{34})_2Co_4(H_2O)_2]^{10-}$ (I), is derived of the condensation of two $[PW_9O_{34}]^{9-}$ units, (from the $[PW_{12}O_{40}]^{3-}$ Keggin structure) by a tetranuclear cluster of octahedral Co(II) polyhedra [7], with a sandwhich structure shown in *figure 1*.

The complex atomic arrangement has been analyzed by means of vibrational spectroscopy from the presence of different type of more simple structural fragments. FTIR spectrum shows the following typical bonds: PO_4 tetrahedral species of a truncated Keggin structure (1038 cm^{-1}), W-O_t (stretching modes of terminal groups) at 944 and 975 cm^{-1}, two types of W-O-W bridged stretchings of the internal bonds (881, 829 and 775 cm^{-1}) as well as Co-O-W bridge bonds (below 600 cm^{-1}).

DRS spectra of both phases (I and II) show typical $O^{=}\rightarrow W$ charge-transfer bands between 200 and 350 nm, corresponding to W-O octahedral groups. Co(II) d^7 electronic d-d transitions for octahedral polyhedra are clearly observed as a broad band centered at 560 nm, in agreement with that observed in other Co(II) oxidic environments. Thermal stability studies, carried out by DTA-TG analysis, show that the primary structure of the heteropolyanion remains unchanged up to 200°C. On the other hand, the stability in aqueous solution (in similar conditions to that observed in physiological environment) ensures that the original condensed polyhedral arrangement is retained upon dissolution. SEM and EDAX measurements also contribute to determine the crystal morphology and the W, P and Co contents.

FTIR spectra of phases I and II only show slight differences. Small structural distortions of the W-O and P-O units by the incorporation of Co(II) cluster can be observed. Hence, the metalic tetranuclaear cluster is occluded between two portions of the Keggin framework in a symmetric and stable structure.

HPOMs penetrate cell membranes and localize intracellularly which is a very important bio-

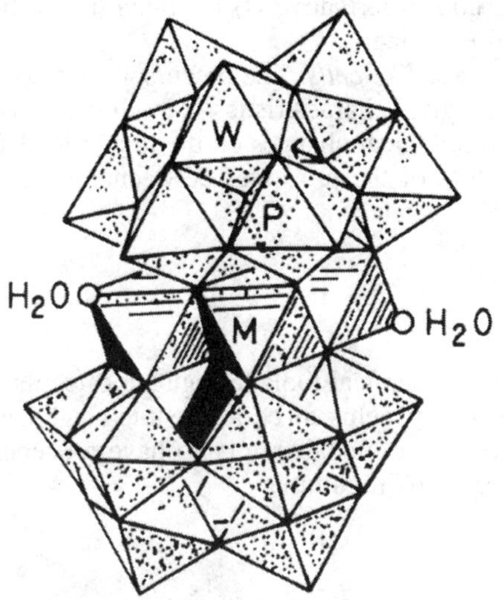

Fig. 1. (M = Co)

logical attribute. Besides, it is well known that the presence of bimetallic systems can increase the HPOMs reactivity by a synergic effect (which can also improve the anticancer and antiviral action mechanisms). For these reasons, it is interesting to know about structural availability of additional d^n-metals on the d^o ion (W) as well as the integral anionic stability. These parameters, known from structural and spectroscopic studies, are essential to analyze the toxicity of metal components and conclude about the drug efficacy.

BIOLOGICAL STUDIES

Effects of POMs on osteoblast proliferation. The precursor phase (II) caused an stimulatory effect in a bell-like shape on the proliferation of the non-transformed osteoblasts (MC3T3E1) without any effect on the tumoral cells, in the whole range of tested concentrations *(fig. 2)*.

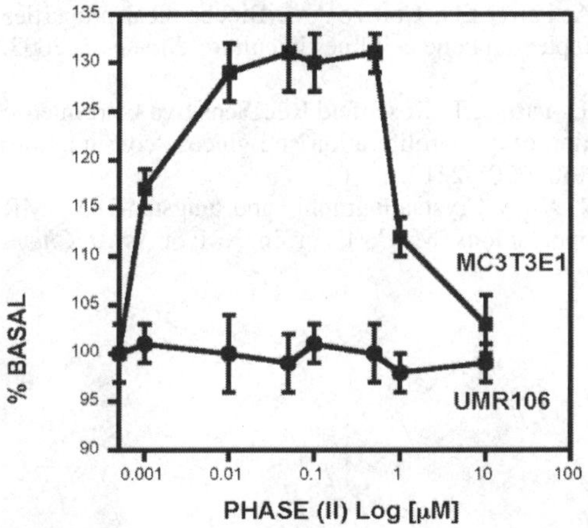

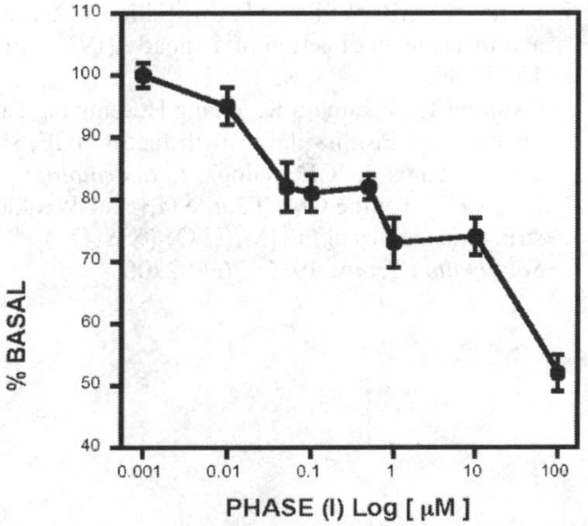

Fig. 2.

Fig. 3.

On the other hand, as can be seen from *figure 3*, the cobalt derivative (I) inhibited the proliferation of the tumoral cells (UMR106) in a dose response manner.

Effect of POMs on the morphological features of osteoblst-like cells. The precursor phase (II) did not produce any significant alterations in the morphological characteristics of both cell lines in culture. On the contrary, phase I caused important morphological changes on the tumoral cells. The nuclei and cytoplasm became round in shape, with loss of the processes between cells and with a marked decrease of the osteoblast interactions with the matrix.

CONCLUSIONS

The two phases described in this paper are interesting as chemical tools to regulate osteoblast proliferation. In particular, the sandwich-like derivative of Co seems to be a possible candidate for antitumoral treatment. Further studies will be carried out in our laboratory to have a deeper insight in the biological activity of these HPOMs in osteoblasts in culture.

ACKNOWLEDGEMENTS

ILB and SBE are members of the Carrera del Investigador, CONICET (Argentina). AMC and CIC are members of the Carrera del Investigador, CICPBA. DAB is a fellowship from CICPBA and MGE is a fellowship from ANPCyT, Argentina.

This work was supported by grants from UNLP and CONICET (PIP 1044/98).

REFERENCES

1. Rhule JT, Hill CL, Judd DA. Polyoxometalates in Medicine. *Chem Rev* 1998, 98: 327-357.
2. Baran EJ. La nueva Farnacoterapia Iniorgánica. XV. Comentarios sobre la Actividad Farmacologica de Algunos Polioxometalatos. *Acta Farm Bonaerense* 1992, 11: 155-160.
3. Yamamoto N., Schols D, De Clercq E., Debyser Z, Pauwels R., Balzarini J, Nakashima H, Baba M., Hosoya M, Snoeck R, Neyts J, Andrei G., Murrer B, Theobald D, Bossard G, Henson G, Abrams M, Picker D. Mechanism of anti-human immunodeficiency virus action of polyoxometalates, a class of broad-spectrum antiviral agent. *Mol Pathol* 1992, 42: 1109-1117.
4. Crans DC. Enzyme interactions with labile oxovanadates and other oxometalates. *Comm Inorg Chem* 1994, 16: 1-33.
5. Etcheverry SB, Williams PAM, Sálice VC, Barrio DA, Ferrer EG, Cortizo AM. Biochemical properties and mechanism of action of a vanadyl(IV)-Aspirin complex on bone cell lines in culture. *Biometals* 2002, 15: 37-49.
6. Okajima T, Nakamura K, Zhang H, Ling N, Tanabe T, Yasuda T, Rosenfeld RG. Sensitive colorimetric bioassay for insulin-like growth factor (IGF) stimulation of cell proliferation and glucose consumption: use in studios of IGF analogs. *Endocrinology* 1992, 130: 2201-2212.
7. Evans HT, Tourné CM, Tourné GF and Weakley TJR. X-ray Crystaallographic and tungsten-183 NMR structural studies of the $[M_4(H_2O)_2(XW_6O_{34})_2]^{10-}$ heteropolyanions (M=Co(II) or Zn, X=P or As). *J. Chem. Soc. Dalton Trans.* 1986: 2699-2706.

Metal Ions in Biology and Medicine: vol. 7. Eds L. Khassanova, Ph. Collery, I. Maymard, Z. Khassanova, J.-C. Étienne.
John Libbey Eurotext, Paris © 2002 pp. 163-167.

Chromium interference with mitochondrial bioenergetics

M.A.S. Fernandes[1], M.S. Santos[1], M.C. Alpoim[2], V.M.C. Madeira[2], and J.A.F. Vicente[3]

[1]Departamento de Zoologia, 3004-517 Coimbra, Portugal. [2]Departamento de Bioquímica, Apartado 3126, 3001-401 Coimbra, Portugal; [3]Departamento de Botânica, Faculdade de Ciências, Universidade de Coimbra, 3000-393 Coimbra, Portugal

ABSTRACT

The mechanism of Cr(VI)-induced toxicity in plants and animals was studied for mitochondrial bioenergetics in turnip root and rat liver mitochondria. Chromium(VI) interference with mitochondrial bionergetics of plant and animal mitochondria was shown by decrease of the succinate-dependent respiratory indexes, (RCR and ADP/O); alterations on oxygen-supported respiration by succinate on state 4, state 3 and uncoupled respiration with parallel $\Delta\Psi$ dissipation and decrease of the phosphorylation efficiency; inhibition of the NADH-ubiquinone oxidoreductase (complex I) and succinate-dehydrogenase (complex II); alterations on the mitochondrial ATPase activity (complex V); and membrane permeabilization to K^+. Although Cr(VI) interferes with mitochondrial bioenergetics of both plant and animal mitochondria, differences on the sensitivity to Cr(VI), $\Delta\Psi$ and phosphorylation efficiency, state 4 respiration and ATPase activities, and K^+ influx were detected between them. In conclusion, CrVI)-induced perturbations on mitochondrial bioenergetics may contribute to the basal mechanism underlying its toxic effects in plant and animal cells.

INTRODUCTION

Chromium(VI), an environmental contaminant, is widely recognized to exert toxic effects towards humans [1], animals [2] and plants [3, 4]. Although the mechanisms of Cr(VI)-induced toxicity are not clear, it is believed that oxidative stress plays an important role. Mitochondria provide most of the cellular energy (ATP) and yield many intermediate compounds involved in normal cellular metabolism. Therefore, perturbations of mitochondrial function may result in severe consequences for general metabolism and all the energy transducing processes that require ATP [5]. The strong decrease on the ATP levels detected in animal cells exposed to Cr(VI) has been closely correlated with the inhibition of mitochondrial respiration [6-8]. To clarify the mechanism of Cr(VI)-induced toxicity, the time and concentration-dependent effects of Cr(VI) on mitochondrial bioenergetics of mitochondria isolated from vegetal (turnip root) and animal (rat liver) sources were investigated.

MATERIALS AND METHODS

Turnip root (*Brassica napus*, L.) and rat liver mitochondria were prepared as previously described [9, 10]. The protein content was determinated by the biuret method [11]. Oxygen consumption was monitored polarographically at 25 °C with a Clark oxygen electrode, using succinate as respiratory substrate [9], and the respiratory control ratio (RCR) and ADP to oxygen ratio (ADP/O) calculated. Uncoupled respiration was initiated by the addition of 1 μM carbonyl cyanide *p*-

trifluoromethoxyphenylhydrazone (FCCP). The activities of the NADH-ubiquinone oxidoreductase (complex I) [12], the succinate dehydrogenase (complex II) [12], cytochrome *c* oxidase activity (complex IV) [12] and ATPase [12], as well as the mitochondrial transmembrane potential [12] and osmotic swelling were evaluated as previously described [13]. The results correspond to the mean ± S.E. of four independent experiments. The significance of the differences between the means was calculated by using the paired Student's *t*-test. Statistical significance was set at $p < 0.05$.

RESULTS

Dichromate significantly depresses RCR and ADP/O *(fig. 1)*. In turnip root mitochondria, dichromate does not affect state 4 respiration supported by succinate, but a strong stimulation occurs in rat liver mitochondria. State 3 and uncoupled respirations are inhibited in both turnip root and rat liver mitochondria *(fig. 2)*. However, rat liver mitochondrial respiration is more sensitive to dichromate than that of turnip root mitochondria *(figs. 1-5)*. Dichromate (up to 1 mM) significantly decreased the activity of rat liver mitochondria NADH-ubiquinone oxidoreductase (complex I), and succinate-dehydrogenase (complex II) of both tested mitochondria. However, rat liver complex II is more sensitive to Cr(VI) than turnip root mitochondria complex II. Both rat liver and turnip root cytochrome *c* oxidase (complex IV) were unaffected by Cr (VI). The ATPase activity was stimulated in rat liver, but inhibited in turnip root mitochondria *(fig. 3)*.

Dichromate depresses the membrane potential ($\Delta\Psi$) developed by succinate-supported respiration *(fig. 4)*, being the effect more pronounced in rat liver mitochondria, than in turnip root mitochondria. In contrast, the phosphorylation was slowed down by dichromate more extensively in turnip than in rat mitochondria *(fig. 4)*. Partial dissipation of turnip $\Delta\Psi$, by the action of the classical protonophore dinitrophenol to values lower than those induced by Cr(VI), had a comparatively smaller effect on the ADP phosphorylation rate *(fig. 4)*, suggesting that the Cr(VI)-induced decrease of phosphosphorylation rate in turnip root mitochondria is related to phenomena in additional to $\Delta\Psi$ dissipation, *i. e.* concerning the ATP-synthase complex.

Permeabilization to K_+ was probed in a 44% iso-osmolar medium of potassium succinate, being shown that dichromate induced mitochondrial membrane permeabilization to K^+ *(fig. 5)*. The negative antimycin effect is putatively not related with K^+ permeabilization, but a consequence of blocking the H^+ pumps required to drive K^+ inside resulting in accumulation of potassium succinate in the matrix. Permeabilization to H_+ was probed in a 44% iso-osmolar medium of potassium acetate, being shown that dichromate (up to 1 mM) does not increase the proton conductance of the inner mitochondrial membrane.

DISCUSSION

Cr(VI) interferes with mitochondrial bionergetics of plant and animal mitochondria by inducing perturbations in complex I and complex II of the electron transport system, in the phosphorylation system (complex V), and membrane permeabilization to K^+. However, differences in the sensitivity to Cr(VI), $\Delta\Psi$ and phosphorylation efficiency, state 4 respiration and ATPase activities, and K^+ influx were detected between plant and animal mitochondria.

Animal mitochondria respiration revealed more sensitive to Cr(VI) than plant mitochondria, as shown by the different Cr(VI) concentration ranges inducing similar alterations on mitochondrial O_2 consumption *(figs. 1 and 2)*, enzymatic activities *(fig. 3)*, $\Delta\psi$ and phosphorylation efficiency developed by succinate-supported respiration *(fig. 4)*. The higher sensitivities of rat mitochondria respiration to Cr(VI) as compared to turnip mitochondria may reflect the sensitivity of complex I and complex II of the mitochondrial respiratory system *(fig. 3)*.

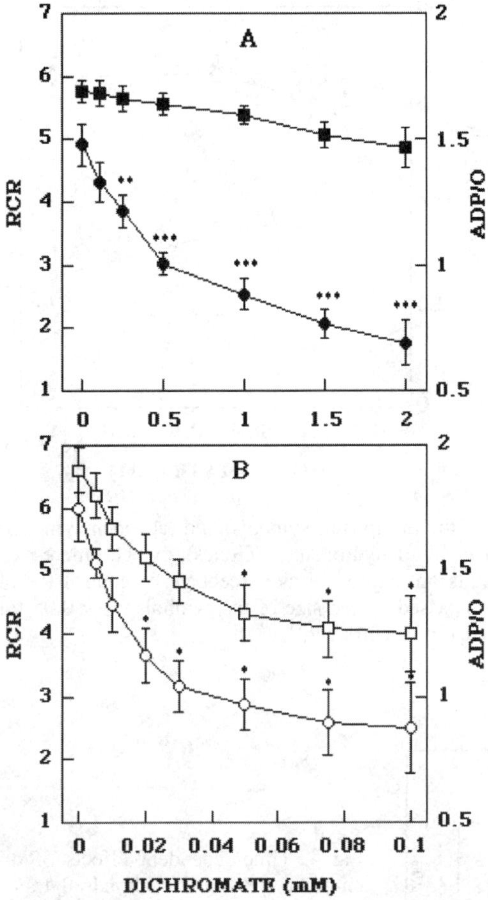

Fig. 1. Effects of dichromate on succinate-dependent respiratory indexes RCR (circles), and ADP/O (squares) of turnip root (**A**) and rat liver (**B**) mitochondria. $^{*}p < 0.05$; $^{**}p < 0.001$; $^{***}p < 0.0001$ for paired t-test when compared with the control (in the absence of dichromate)

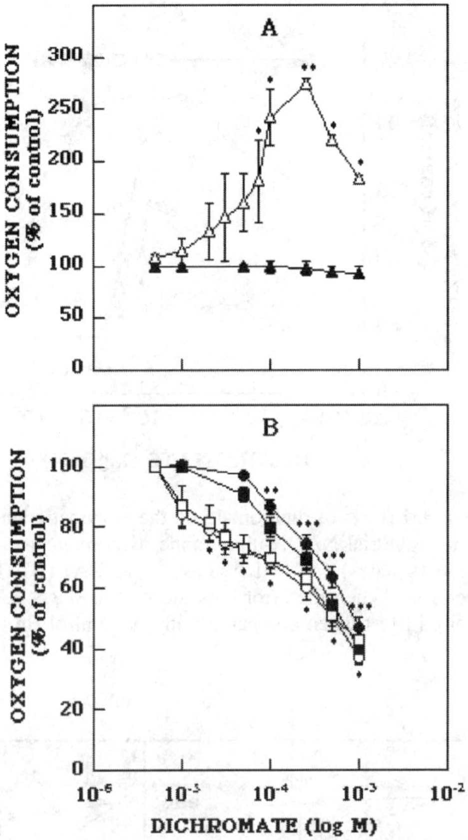

Fig. 2. Effect of dichromate on succinate-supported O_2 consumption in state 4 (**A**), and state 3 and uncoupled-respiration (**B**) of turnip root (full symbols) and rat liver (open symbols) mitochondria. Turnip root mitochondria: (H) state 4; (J) state 3; (B) uncoupled respiration. Rat liver mitochondria: (C) state 4; (E) state 3; (G) uncoupled respiration. When the error bars are not visible S.E. is encompassed by the size of the symbols. $^{*}p < 0.05$; $^{**}p < 0.001$; $^{***}p < 0.0001$ for paired t-test when compared with the control (in the absence of dichromate)

The highest sensitivity of rat mitochondria complex II to Cr(VI) supports the different data concerning Cr(VI)-induced dissipation of succinate-supported $\Delta\Psi$ in rat and turnip mitochondria *(fig. 4)*. However, it did not explain the accentuated decrease on the phosphorylation efficiency of turnip mitochondria *(fig. 4)*. This effect may be explained by the decrease of the ATP-synthase activity as reflected by a partial inhibition of the ATPase activity *(fig. 4)*.

Since mitochondrial membrane permeabilization to H^+ is insensitive to Cr(VI), the different Cr(VI) effects on state 4 respiration of rat and turnip may be explained by the different sensitivity of their ATPase to Cr(VI *(fig. 3)*. Thus, in rat mitochondria, the stimulation of the ATP hydrolysis promotes an increased ADP/ATP turnover, which dissipates the proton motive force. Therefore,

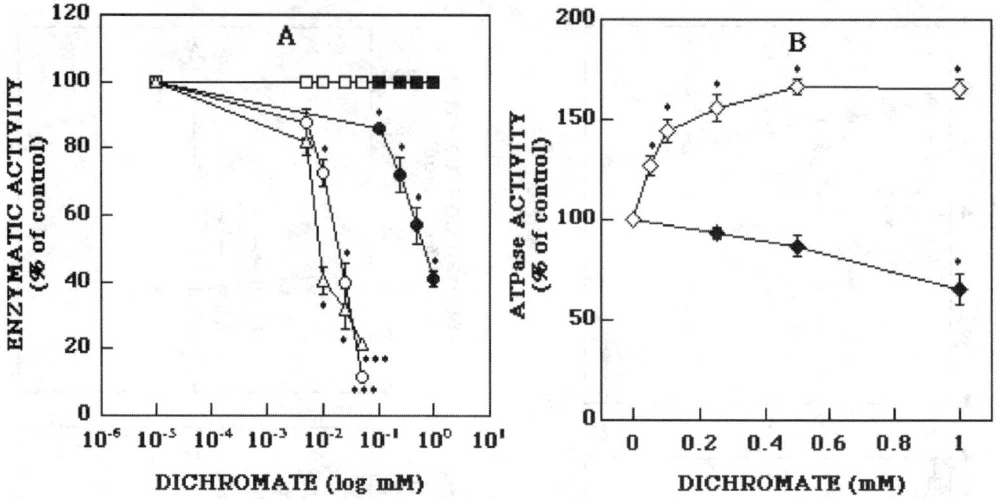

Fig. 3. Effects of dichromate on the enzymatic activities of the turnip (full symbols) and rat (open symbols) mitochondrial NADH-ubiquinone oxidoreductase (C), succinate-dehydrogenase (circles), cytochrome *c* oxidase (squares) (**A**) and ATPase (F and A) (**B**). The results are expressed as percentage of control (% of control). When the error bars are not visible S.E. is encompassed by the size of the symbols. *$p < 0.05$ for paired t-test when compared with the control (in the absence of dichromate)

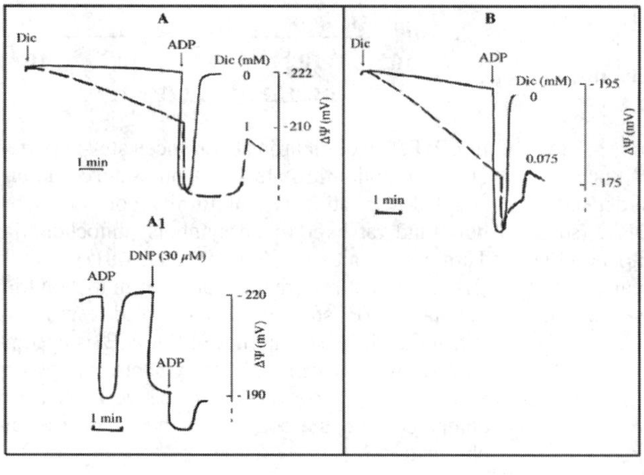

Fig 4. Time-dependent effects of dichromate on the mitochondrial transmembrane potential (Ψ) and phosphorylation efficiency of turnip (**A**) and rat (**B**) mitochondria energized with succinate. **A1**, turnip mitochondrial phosphorylation after partial Ψ dissipation by dinitrophenol (DNP) as a reference, for the evaluation of dichromate effects. The traces represent typical direct recordings from six independent experiments with different mitochondrial preparations

the H^+ pumps are stimulated and the state 4 respiration rate is potentiated. In turnip mitochondria, the absence of a Cr(VI) effect on the ATPase activity parallel the negligible stimulation of state 4 respiration *(fig. 2)*.

The difference sensitivity for the increased K^+ influx of rat and turnip mitochondria may be a consequence of the different sensitivity of plant and animal mitochondrial respiration to Cr(VI).

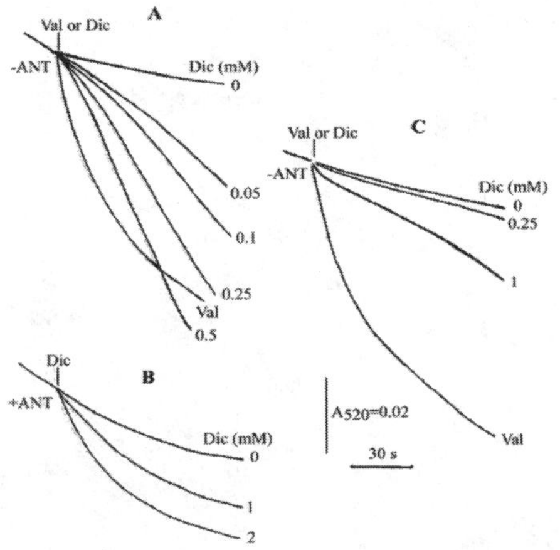

Fig. 5. Effects of dichromate on the mitochondrial respiration supported swelling of turnip in the absence (**A**) and presence (**B**) of antimycin A, and in rat in the absence of antimycin A (**C**), in potassium succinate medium. When required, 1 μM antimycin A was added (+ANT) to the reaction medium. - ANT, indicates the absence of antimycin A in the reaction medium. Dichromate (Dic) or 1 μM valinomycin (Val) were added after 50 s incubation. Depicted curves are representative of a at least three independent experiments with different mitochondrial preparations

CONCLUSION

In conclusion, although Cr(VI) interfers with mitochondrial bienergetics, important differences were detected for the Cr(VI) effects in plant and animal mitochondria. Cr(VI) interference with mitochondrial bionergetics may contribute to the basal mechanism underlying its toxic effects in plant and animal cells.

REFERENCES

1. Barceloux D.G. Chromium. *Clin. Toxicol.*, 1999, 37, 173-194.
2. Stohs S.J., Bagchi D., Hassoun E., Bagchi M. Oxidative mechanisms in the toxicity of chromium and cadmium ion. *J. Environ. Pathol. Toxicol. Oncol.*, 2001, 20, 77-88.
3. Hauschild M.Z. Putrescine (1,4-diaminobutane) as an indicator of pollution-induced stress in higher plants: barley and rape stressed with Cr(III) or Cr(VI). *Ecotoxicol. Environ. Safety*, 1993, 26, 228-247.
4. Appenroth K.J., Stöckel J., Srivastava A., Strasser R.J. Multiple effects of chromate on the photosynthetic apparatus of *Spirodela polyrhiza* as probed by OJIP chlorophyll *a* fluorescence measurements. *Environ. Pollut.*, 2001, 115, 49-64.
5. Palmeira C.M., Moreno A.J., Madeira V.M.C. Interactions of herbicides 2,4-D and dinoseb with liver mitochondrial bioenergetics. *Toxicol. Appl. Pharmacol.*, 1994, 127, 50-57.
6. Debetto P., Dal Toso R., Varotto R., Bianchi V., Luciani S. Effects of potassium dichromate on ATP content of mammalian cells cultured in vitro. *Chem. Biol. Interact.*, 1982, 41, 15-24.
7. Ryberg D., Alexander J. Inhibitory action of hexavalent chromium (Cr(VI)) on the mitochondrial respiration and a possible coupling to the reduction of Cr(VI). *Biochem. Pharmacol.*, 1984, 33, 2461-2466.
8. Messer R.L.W., Doeller J.E., Kraus D.W., Lucas L.C. An investigation of fibroblast mitochondria enzyme activity and respiration in response to metallic ions released from dental alloys. *J. Biomed. Mater Res.*, 2000, 50, 598-604.
9. Vicente J.A.F., Madeira V.M.C. The excellence of turnip mitochondrial fractions. *Biochem. Educ.*, 2000, 28, 104-106.
10. Gazotti P., Malmstron K., Crompton M.A. Laboratory manual on transport and bioenergetics. In *Membrane Biochemistry*, E. Carafoli, and G. Semenza, eds., 1979, pp. 62-69, Springer Verlag, New York.
11. Gornall A.G., Bardawill C.J., David M.M. Determination of serum proteins by means of the biuret reaction. *J. Biol. Chem.*, 1949, 177, 751-766.
12. Pereira C., Santos M.S., Oliveira C.R. Involvement of oxidative stress on the impairement of energy metabolism induced by Aβ peptides on PC12 cells: protection by antioxidants. *Neurobiology of Disease*, 1999, 6, 209-219.
13. Vicente J.A.F., Santos M.S., Vercesi A.E., Madeira V.M.C. Comparative effects of the herbicide dinitro-*o*-cresol on mitochondrial bioenergetics. *Pestic. Sci.*, 1998, 54, 43-51.

IV DISTRIBUTION

*Metal Ions in Biology and Medicine: vol. 7.*Eds L. Khassanova, Ph. Collery, I. Maymard, Z. Khassanova, J.-C. Étienne.
John Libbey Eurotext, Paris © 2002 pp. 171-174.

Comparative whole blood and gastric tissue distribution of selenium in patients with gastritis and *Helicobacter pylori* infection

José L. Burguera[1], Marcela Burguera[1], Carlos Rondón[1], Pablo Carrero[1], María L. Di Bernardo, Miguel A. Villasmil R.[2] and Luis M. Villasmil[2]

[1]*IVAIQUIM (Venezuelan Andean Institute for Chemical Research), Faculty of Sciences, [2]Faculty of Medicine. University of Los Andes, P.O. Box 542, Mérida 5101-A, Venezuela*

ABSTRACT

The gastric tissue and whole blood Se levels were examined in 149 subjects with different types of gastritis and with or without the presence of *Helicobacter Pylori* (HP) infection. The gastric and whole blood levels of Se in the subjects under study were 0.65 ± 0.52 µg g^{-1} and 136 ± 31 µg l^{-1}, respectively. HP was present in 100 subjects (48 males and 52 females) and it was absent in 49 patients (23 males and 26 females). The Se concentrations in gastric tissue was related with the kind of gastritis and the presence of HP infection and decreased in the different patients as follows: mild gastritis > erosive gastritis > chronic gastritis.

INTRODUCTION

The element selenium (Se) is an essential element. Se has been shown to be anti-tumorigenic in a variety of animal models [1] and several epidemiologic studies have linked Se status and human cancer [2]. Also, the Se concentration in the biopsies of patients with gastric ulcerations and with cancer has been respectively related to the severity of the inflammation process and to the cellular immune response; on the average, blood of cancer patients was reported to contain less Se than the blood of non-cancer patients [3]. Besides, the presence of HP infection might be related with different types of gastric damage, from peptic different gastritis types to gastric cancer [4].

The aim of the present study was to evaluate Se in whole blood (WB) and gastric tissue (GT) of 149 subjects aged 9-94 (mean 43 ± 17) years with different kind of gastritis with and without HP infection.

MATERIALS AND METHODS

The Se concentration was determined by electrothermal atomic absorption spectrometry (ETAAS) with iridium as a matrix modifier [3]. The WB and GT levels of Se were examined in 83 females aged 15-80 (mean 45 ± 17) years and 66 males aged 9-94 (mean 40 ± 18) years, 45 patients with mild gastritis aged 15-75 (48 ± 14), 76 patients with erosive gastritis aged 15-83 (62 ± 17) and 28 patients with chronic gastritis aged 28-94 (55 ± 23) years. The gastric tissue was collected and treated as previously described by our research team [3]. Blood samples were drawn by Vacutainer from the forearm veins into 30 ml heparinised polyethylene tubes (with a polyethylene stopper) and stored at about -4 °C. 15 U of heparin were added per ml of blood. The statistical evaluation of the results was carried out with SAS (Statistical Analysis System) software version 6.12 under UNIX using ANOVA with paired Student's t-test for the detection contrasts. The statistical significance was accepted for 95% confidence limit ($p \leq 0.05$).

RESULTS AND DISCUSSION

The Se concentrations in GT and WB samples from subjects with different types of gastritis with and without HP infection are shown in *tables 1 and 2*. Also, *figs. 1 and 2* respectively show the variation of Se in tissue and WB related to the kind of mucosal damage in males (M) and females (F) and with the presence of HP infection.

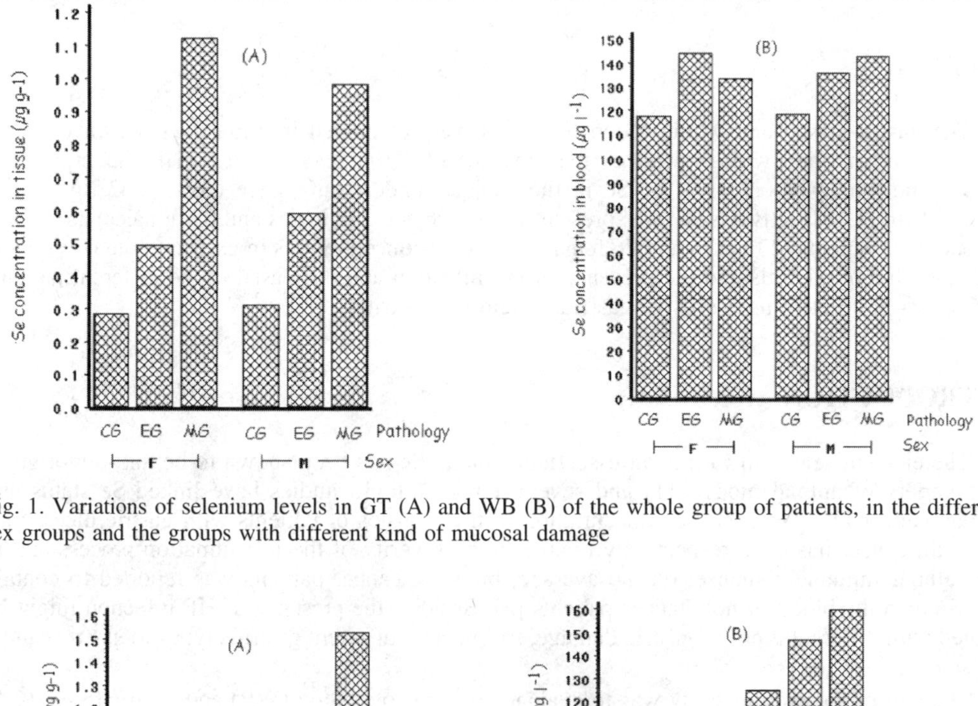

Fig. 1. Variations of selenium levels in GT (A) and WB (B) of the whole group of patients, in the different sex groups and the groups with different kind of mucosal damage

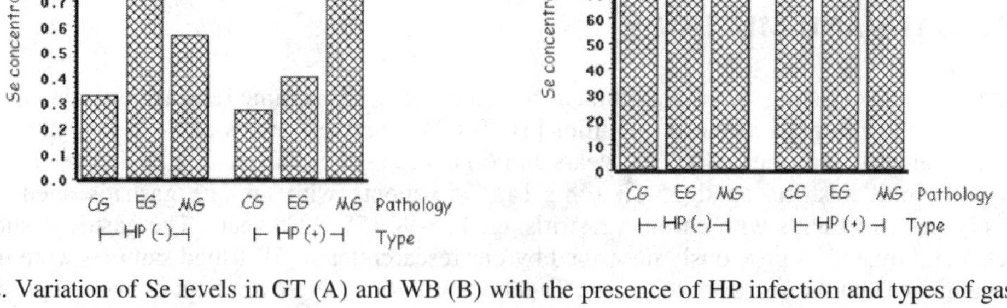

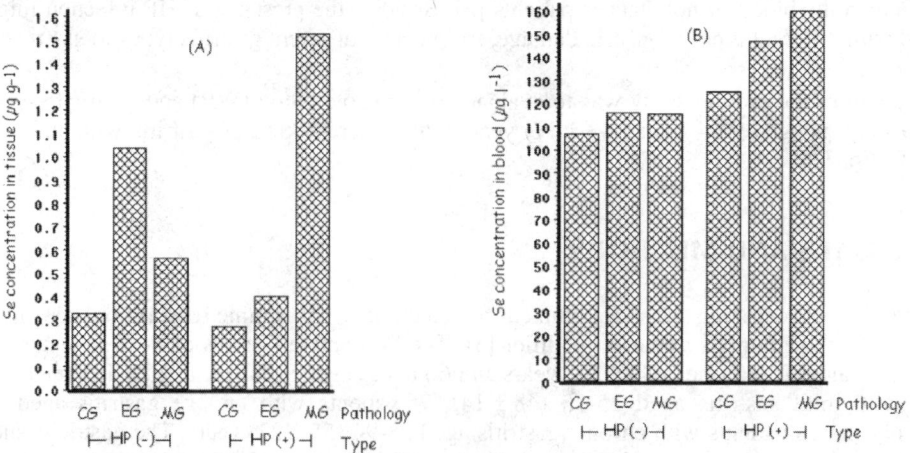

Fig. 2. Variation of Se levels in GT (A) and WB (B) with the presence of HP infection and types of gastritis

Table 1. Selenium concentration in gastric tissue and whole blood of patients

Subjects	Tissue (μg g^{-1})*	Whole blood (μg l^{-1})*	Sex	Age*	n
All	0.65±0.52	136±31	M+F	43±17	149
	0.69±0.47	135±34	M	40±18	66
	0.63±0.55	136±28	F	45±17	83
Mild gastritis	1.06±0.49	137±32	M+F	44±18	45
Chronic gastritis	0.30±0.04	118±24	M+F	40±16	28
Erosive gastritis	0.55±0.19	141±17	M+F	44±16	76

*Mean ± standard deviation. M = male and F = female.

Table 2. Effect of HP infection on Se levels of subjects with different kind of gastritis

Subjects	Tissue (μg g^{-1})*		Whole blood (μg l^{-1})*	
	HP(+)	HP(-)	HP(+)	HP(-)
Mild gastritis	1.53±0.49	0.53±026	160±44	115±28
Chronic gastritis	0.27±0.13	0.33±012	125±16	107±19
Erosive gastritis	0.40±0.19	1.00±0.47	147±17	122±26

*n of mild, chronic and erosive gastritis were for HP(+) =23, 17 and 60, and for HP(-) = 22, 11 and 16, respectively. HP(+) and HP(-) indicate the presence or absence of HP infection.

These results suggested that the following trends were statistically significant: (i) the Se levels in WB and gastric tissue were not related with the age ($p = 0.0003$), but they were related with sex ($p = 0.7690$); (ii) the kind of mucosal damage and the presence of HP did not have any influence on the Se levels in WB (in all cases $p < 0.0001$); (iii) females had lower values of Se in WB and GTs, regardless of the mucosal damage (in both cases $p < 0.0005$); (iv) the Se concentration in GT is related with the kind of mucosal damage and the presence of HP ($p = 0.54$) *(table 2 and fig. 2)*; (v) the Se concentration in GT decreased in the different HP(+) patients as follows: mild gastritis > erosive gastritis > chronic gastritis *(table 1 and fig. 1)*; in the last group the Se levels are lower in the patients with HP infection.

Our observations indicate that GT may become vulnerable to damage and ulceration with the presence of HP infection. The variation of the Se levels in subjects evaluated in this study was different with that previously found by our research team [3]. In the previous study, the selenium concentrations found in the gastric tissue of patients increased in the order: mild < chronic < erosive gastritis, whereas in this study the order in the whole group of patients *(table 1 and fig. 1)* was chronic < erosive < mild gastritis). However, in both cases, the high standard deviation in the groups with gastric ulcer may indicate that the cause of these conditions are heterogeneous, such as diet habits, drugs, and alcohol and toxic contaminants.

CONCLUSION

This study demonstrated that the kind of gastritis and the presence of HP affected the selenium levels in GT but not in whole blood. It seems reasonably to assume that blood modulates the mechanism in which selenium is involved. In this way, the immune system actively participates in surveillance where selenium is needed the most.

REFERENCES

1. El-Bayoumy K. The role of selenium in cancer prevention. In: De Vita V, Hellman S, Rosenberg S, eds., Practice of Oncology, 4th ed Philadelphia, Lippincott, 1991: 1-15.
2. Combs Jr GF, Clark LC, Turnbull BW. Evidence of cancer prevention by slenium in a randomized, placebo.controlled, clinical trial. In: Metal Ions in Biology and Medicine, Collery Ph, Brätter P, Negretti de Brätter V, Khassanova L, Etiene JC, eds., John Libbey Eurotext, Paris 1998, Vol. 5: 566-571.
3. Burguera JL, Villasmil LM, Burguera M, Carrero P, Rondón C, Matousek de Abel de la Cruz A, Brunetto MR, Gallignani M. Gastric tissue selenium levels in healthy persons, and non-cancer patients with different kind of mucosal damage. J Trace Elem Med Biol 1995; 9: 160-164.
4. Pakodi F, Abdel-Salam OM, Debreceni A, Mozsik G, Helicobacter Pylori. One bacterium and a broad spectrum of human disease. An Overview. J Physiol Paris 2000; 94: 139-152.

Metal Ions in Biology and Medicine: vol. 7.Eds L. Khassanova, Ph. Collery, I. Maymard, Z. Khassanova, J.-C. Étienne.
John Libbey Eurotext, Paris © 2002 pp. 175-180.

Blood lead (Pb) and selenium (Se) in chronic rhinitis and asthma bronchiale and in healthy Hungarian children and adults

Mária Ágnes Cser[1], Ilona Kovács[1], Edit Bocskai[1], Nóra Adányi[2], Ibolya Sziklai-László[3]

Bethesda Children's Hospital, 1146 Budapest, Bethesda St.3.[1], Central Food Research Institute, 1537 Budapest, P.O.B.393[2], Atomic Energy Research Institute, 1525 Budapest 114. P.O.B.49[3]

ABSTRACT

The trace element lead (Pb) is one of the most harmful pollutant in urban areas. Its adverse effects on health include upper and lower respiratory tract destruction since nearly 40% of Pb absorbs via lungs. Selenium (Se) is involved in immune protection actions in respiratory and alleregic diseases, and may protect against the toxic effects of Pb. The aim of the study was to analyse the possible relations between the effects of the two trace elements in healthy children (n=62), healthy adults (n=57) as well as in patients with chronic rhinitis (n=15 children, n=31 adults), acute bronchitis (n=34 children, n=24 adults) and in chronic asthma (n=13 children, n=15 adults). Both trace elements were measured with atomic absorption spectrophotometry by using element specific EDL lamps. Computing was performed by SPSS™. In the healthy population blood Pb concentrations increased with age. Under twenty years, it was 5-7 μg/100 ml, above that levels doubled in average. There was no correlation between blood Se and Pb levels, however nearly 30 per cent of healthy people showed toxic Pb levels (above 10 μg/100 ml). Children in the capital had higher Pb levels than others from an industrial city. The lowest (4.1 ± 0.5 μg/100 ml) Pb values were observed in rural area. Children with chronic rhinitis, acute bronchitis or chronic asthma had similar Pb levels than the healty age and sex matched controls. Different to these observations adults with bronchitis (15.2 ± 8 μg/100ml) or asthma (16.0 ± 7 μg/100 ml) had significantly higher Pb levels than the healthy group (8.9 ± 4 μg/100 ml, $p<0.001$). Se blood status of healthy Hungarians were lower, than in several European countries explaned by low soil Se content and low soil-plant-animal-human food chain. Se concentrations in humans increase with age in childhood, from 20 to 40 years Se levels are similar, thereafter a decline could be observed. Erythrocyte Se content and whole blood Pb concentrations showed a negativ correlation in adult asthmatics ($r=0.7692$, $y=-16x + 34$, $p<0.001$), but not in asthmatic children. This observation indicates that the Se concentration did not protect Pb absorption in the adult group of patients, who were exposed to Pb inhalation longer than the young patients. Results also indicate, that air pollution connected with Pb inhalation could be associated with developing chronic asthma bronchiale.

Key words: trace elements, selenium, plumbum, child, adult, rhinitis, bronchitis, asthma

INTRODUCTION

The number of allergic and asthmatic patients dramatically increased in the past years. World-wide, the prevalence of childhood asthma are increasing (Gergen) in the USA (Knorr) in Europe (WHO) and also in Hungary (Kovacs). Asthma is the most common chronic illness of childhood, affecting approximately 10% of children (NAEP). **Pb** is present in the atmosphere, results from the combustion of leaded gasoline. Lead containing dust affects the respiratory system, 40% of Pb will be absorbed via the lungs. Drinking water, the food chain could be contaminated by the global Pb pollution leading to accumulation in teeth, hair, bones, even in brain tissues. It is generally agreed, that blood Pb as a measure of absorbed Pb is the best indicator of current exposure (WHO 1977, 1980). **Se** is involved in immune defence mechanisms (Kiremidian), interferes with host defence and can influence the outcome of infections (Beck). Immunological alterations during Pb inhalation could induce respiratory tissue inflammation. The aim of the study was to investigate the relation of these elements in health and in acute inflammation like bronchitis and in chronic inflammations such as allergic rhinitis and bronchial asthma.

Patients

1-60 year old patients with acute bronchitis, chronic, allergic rhinitis or bronchial asthma as well as adults with industrial Pb exposition were investigated. Healthy controls were investigated during their annual check ups. Consent was given by all, local ethical commettee supported the observations.

Methods

Haematologic blood analyses were done by Cell-Dyn 3200 Abbott instrument (United Kingdom) with applied own PC software. **Selenium** was determined by AAS (Perkin Elmer 1100 B, HGA MHS-20 Lombeck). Internal pools were used, coefficients of variance were 3.2 and 3.4% within days, 4.1 and 3.7% between days. **Total blood Pb** was determined by flameless Perkin Elmer AAS, Zeeman 5000, HGA 500 with automatic sample-injector (AS-40) and graphit cuvett plattform (Ewers). Seronorm^{TM} was used as standard (level I, II., III. Nycomed AS Oslo, Norway). As reference material Standard CRM (Promochem GmbH,Wesel, Germany) served, given value: 126±4 ng/ml, measured values 126±2 ng/ml in 19 assays. Internal pool coefficient of variance was 3.9% within day and 4.42% between days.

RESULTS

In the healthy population blood Pb content increased with age *(table 1)*. Above the age of 16 years blood Pb concentrations were twice as high as under the age of 15 years. There was a significant correlation between the age and blood Pb concentration ($r=0.494$, $y=0.19age+5$, $n=164$ $p<0.001$). The distribution of blood Pb levels showed that nearly 30% of values were above the tolerable blood Pb concentration of 10 mcg/100 ml. Neither in healthy nor in ill patient were any association between blood Pb concentration and haematologic parameters. Se parameters in all three blood compartments are lower in healthy Hungarians compared to other healthy populations in several European countries (*table 2*, Cser 96).

Table 1. Blood lead (Pb μg/ 100ml) levels (M ± SD) in healthy Hungarians

Age	N	Pb
1-5	21	5.54 ± 2.49
5.1-10	27	6.87 ± 2.55
10.1-15	30	6.52 ± 2.51
15.1-20	16	17.45 ± 1.59
20.1-30	30	8.62 ± 4.12
30.1-40	50	7.73 ± 1.22
40.1-50	29	7.42 ± 1.31
50.1-60	22	6.53 ± 2.11

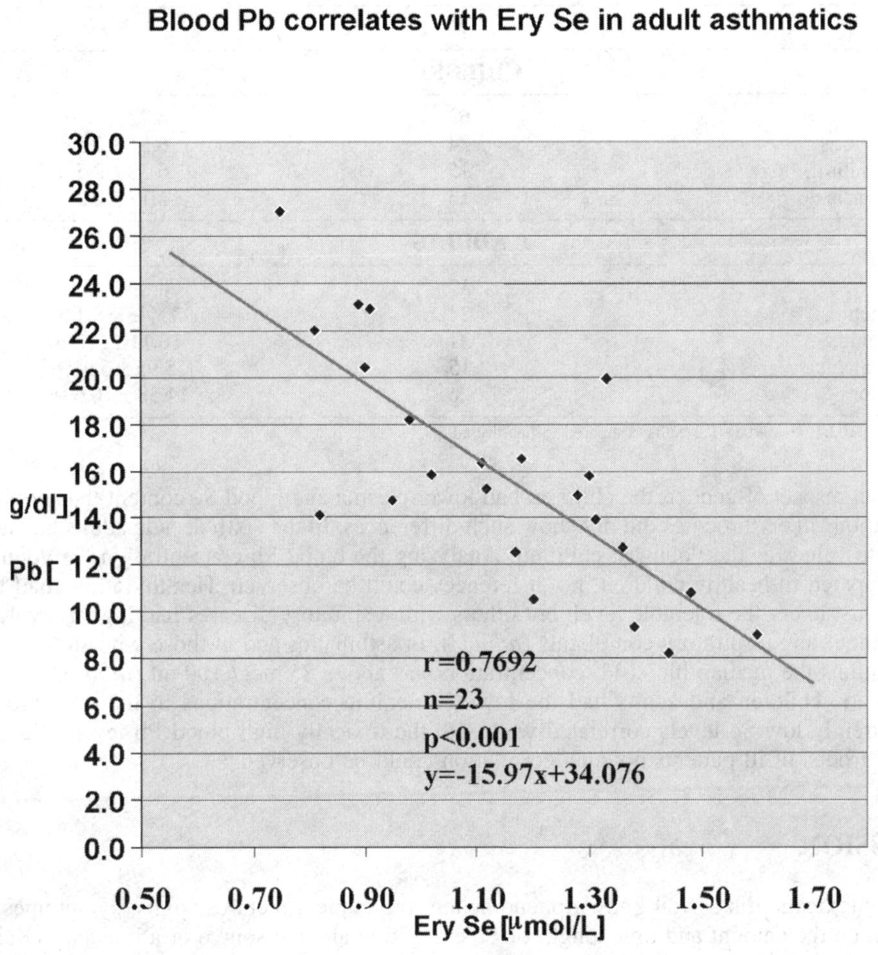

Fig. 1.

Table 2. Selenium (µmol/L) concentrations (M ± SD) in healthy Hungarians

Age	N	Plasma	Blood	Erythrocytes
1-5	42	0.55 ± 0.11	0.74 ± 0.13	1.06 ± 0.21
6-10	59	0.66 ± 0.10	0.83 ± 0.13	1.16 ± 0.19
11-15	60	0.65 ± 0.09	0.82 ± 0.11	1.12 ± 0.18
16-20	16	0.75 ± 0.09	0.92 ± 0.14	1.20 ± 0.25
21-30	30	0.78 ± 0.14	0.98 ± 0.15	1.26 ± 0.25
31-40	50	0.75 ± 0.13	0.96 ± 0.15	1.25 ± 0.24
41-50	29	0.72 ± 0.15	0.91 ± 0.16	1.18 ± 0.25
51-60	22	0.67 ± 0.19	0.83 ± 0.17	1.06 ± 0.23

Table 3. Blood PB (μg/ 100ml) levels (M ± SD) in respiratory diseases and Pb exposure

Groups	N	Pb
	CHILDREN	
Healthy	62	6.72 ± 2.76
Acute bronchitis	34	6.87 ± 2.55
Chronic rhinitis	15	6.52 ± 2.51
Asthma bronchiale	13	6.17 ± 2.23
	ADULTS	
Healthy	57	8.89 ± 4.20
Acute bronchitis	24	15.43 ± 8.14•
Chronic rhinitis	31	16.11 ± 3.1•••
Asthma bronchiale	15	15.95 ± 6.91•••
Pb exposition	26	14.58 ± 10.69••

• $p<0.05$, •• $p<0.01$, ••• $p<0.001$, results compared to healthy adults.

Without respect of gender, the children had lower plasma and blood Se content than adults. Se concentrations in erythrocytes did not show such differences. In the sixth decade adults had as low erythrocyte values as the youngest children. Analysing the blood Pb concentrations in young patients compared to healthy children, no differences could be observed. Healthy adults had blood Pb levels just under the tolerable level, but others with respiratory diseases had higher levels than adults without any respiratory complaints *(table 3)*. In asthmatics and in those who suffered from Pb expositions the median blood Pb concentration was above 25 mcg/ 100 ml *(table 3)*.

Asthmatic children and adults had the lowest selenium concentrations in their erythrocytes. These extremly low Se levels correlated well with the toxically high blood Pb levels *(fig. 1)*. In the other groups of ill patients no such correlation could be observed.

DISCUSSION

Lead intoxication is a well known phenomenon, the clinical pictures, signs, symptomes vary depending on the amount and time length of Pb exposition and the source of alimentary (Kákosy, Papp) or other environmental pollution (Alfani), active or passive nicotine abuse (Morisi). Gastrointestinal, respiratory, haemopoietic (Schwartz), bone, musle, kidney and nervous tissue can be affected (Lauwers). Disturbed skeletal growth (Abritti), hormonal dysfunctions (Gustafson), impared uptake of iodine (Siegel) could induce serious clinical dysfunctions. The leaded petrol still plays an important role in Hungary since the two tact cars did not disappeare. Exposure to the toxic metal lead starts in the humans as early as foetal life. During the second trimester Pb accumulated in the brain and kidney tissue of the foetuses (Lutz). Animal experiments proved that toxic Pb exposition markedly decreased DNA, RNA and protein content of cerebellar tissue (Nehru). Concomitant selenium administration decreased the lead induced histological alterations. Not only the neonatal, but the developing child/s brain is more sensitive to Pb exposition compared to the adults. The biological half life of Pb in the blood and connective tissue is 19-21 days but in the bone tissue it could be as long as twenty years (Bõse-O/Reilly). Therefore children run at higher risk since they store lead in the body at a faster rate than do adults (Duggan). Anaemia and neurologic complications are the best known long lasting effects of Pb exposure especially in children resulting lower IQ capacities, impared psychomotor and cognitive functions (Laraque, Needleman). The present study concentrated on the long lasting respiratory effects of Pb inhalation since dust, soil, paints are common inducers of chronic upper or lower restiratory inflammations. The fact, that in the capital 28%, in the industrual city 21% of healthy children had blood Pb concentrations well above the tolerable level suggest, that Pb exposure is very unfavorable (WHO 1977). Blood lead concentration in school children did not decrease in the past 6 years in Hungary (Cser unpublished fresh data). Blood lead concentration increased with age reflecting the legth of exposure similarly to other observations (Morisi). The time legth of exposure seemed to play an important role in developing chronic respiratory diseases. Selenium blood status has been found to be low even in healthy Hungarians based on the observation that plasma and erythrocyte glutathione peroxidase activity depended on plasma and erythrocyte Se content (Cser 1996). Such correlations could be observed only in populations living in Se defficient areas. Selenium is known to prevent intoxications with different heavy metal ions by decreasing the binding of these compaunds to tissue proteins (Goyer, He, Othman). The present study suggest, that the possible protective effect of selenium helped to postpone the development of respiratory inflammation children. On the other hand adults with chronic rhinitis and asthma had unsufficiently low Se content in their erythrocytes and they suffered from the highest high blood lead levels. These observations indicated that unsufficient selenium status could not prevent lead absorption which seemed to be connected with Pb inhalation induced respiratory diseases.

ACKNOWLEDGEMENTS

This work was supported by the grants from KFKI Condensed Matter Res. Centre (No. ICA1-CT-2000-70029) and partly supported by ETT No 054 / 99.

REFERENCES

Alfani A., Baldantoni D., Maisto G., Bartoli G., Virzo se Santo A. Time and site integrated biomonitoring of Pb, Cr,Fe,Cu,V and Cd in the urban area of Naples. J. Trace Elements Med. Biol. 1997, 11, 176-178.

Abbritti G., Muzi G., Cicioni C., Accatoli MP., Fiordi T., Morucci P. Effects of low doses of lead on children/s health. Ann. Ist. Super Sanita 1989, 25, 437-447.

Beck M.A. Selenium and host defence towards viruses. Proc. Nutr. Soc. 1999, 58, 707-711.

Boese-O/Reilly S., Kammerer S. Leitfaden Umwelt medizine. Gustav Fischer Verlag Luebeck 1997.

Cser Á., Sziklai-László I., Menzel H., Lombeck I. Selenium status and lipoproteins in healthy and diabetic children. J. Trace Elem. Electrolytes Health Dis. 1993, 7, 205-210.

Cser M.Á., Sziklai-László I., Menzel H., Lombeck I. Selenium and glutathione peroxidases in healthy Hungarian children and adults. Gyermekgyógyászat, 1966, 5, 384-394.

Ewers U., Brockhaus A., Dolgner R., Freier I., Turfeld M., Engelke R., Jermann E. Blutblei- und Blutcadmiumkonzentrationen bei 55-66 jährigen Frauen aus verschiedenen Gebieten Nordrhein-Westfalens - Entwicklungstrends 1982-1988. Zbl. Hyg. 1990, 189, 405-418.

Gergen P.J., Weiss K.B. Changing patterns of asthma hospitalization among children: 1979 to 1987. JAMA, 1990, 264, 1688-1692.

Goyer R.A. Toxic and essential metal interactions. Ann. Rev. Nutr. 1997, 17, 37-50.

Gustafson A., Hedner P., Schultz A., Skerfving S. Occupational lead exposure and pituitary function. Int. Arch. Occup. Environ. Health, 1989, 61, 277-281.

He B., Xu Z., Hao W., Wang S. Antagonistic action of organic selenium on lead poisoning. Wei Sheng. Yen. Chiu, 1998, 27, 229-232.

Kákosy T., Hudák A., Náray M. Lead intoxication epidemic caused by ingestion of contaminated ground paprika. Clinical Toxicology, 1996, 34, 507-511.

Kiremidjian-Schumacker I., Stoczky G. Selenium and immune responses. Environ. Res. 1987, 42, 277-303.

Kovács I., Cser M.Á., Stocker A. The detection of inflammative markers in atopic diseases (A gyulladás markereinek vizsgálata apópiás kórképekben) Medicina Thoracalis, 2000, 53, 16-20.

Knorr B., Matz J., Bernstein J.A., Nguyen H., Seidenberg B.C., Reiss T.F., Becker A. Montelukast for chronic asthma in 6-14-year-old children. JAMA, 1998, 279, 1181-1186.

Laraque D., McCormic M., Norman M., Taylor A., Weller SC., Karp J. Blood lead, calcium status and behavior in praeschool children. Am. J. Dis. Child. 1990, 144, 186-189.

Lauwers M.C., Hauspie R.C., Susanne C., Verheyden J. Comparison of biometric data of children with high and low leveles of lead in the blood. American J. Physical Antropol. 1986, 69, 107-116.

Lombeck I., Kasparek K., Hrbisch H.B., Feinendegen I.E., Bremer H.J. The selenium state of healthy children I. Serum selenium concentration at different ages; Activity of glutathione peroxidase of erythrocytes at differenet ages, Selenium content of food of infants. Eur. J. Pediatr. 1977, 125, 81-88.

Lutz E., Lind B., Herin P., Krakau I., Bui T.H., Vahter M. Concentrations of mercury, cadmiun and lead in brain and kidney of second trimester fetuses and infants. J. Trace Elements Med. Biol. 1996, 10, 61-67.

Morisi G., Patriarca M., Carrieri MP., Taggi F. Lead exposure: assessment of the risk for the general Italian population. Ann. Ist. Super Sanita, 1989, 25, 423-435.

Mushak P., Crocetti AF. Determination of numbers of lead-exposed American children as a function of lead source: integrated summary of a report to the U.S. Congress on childhood lead poisoning. Environ. Res. 1989, 50, 210-229.

National Asthma Education and Prevention Program. Expert panel report, Bethesda, Md, National Institute of Health USA 1995.

Needleman HL., Gatsonis CA. Low-level lead exposure and the IQ of children. JAMA 1990, 263, 673-678.

Nehru B., Dua R., Iyer A. Effect of selenium on lead-induced alterations in rat brain. Biol. Trace Elem. Res. 1997, 57, 251-258.

Othman AI., El Missiry M.A. Role of selenium antagonist in lead tooxicity in male rats. J. Biochem. Mol. Toxicol., 1998, 12, 345-349.

Papp Cs., Kákosy T., Soós G. Alimentáris eredetü ólommérgezés családi elõfordulása. (Familiar lead poisoning from alimentary source). Med. Universalis, 1995, XXVIII, 357-364.

Schwartz J., Landrigan PJ., Baker EL., Orenstein WA., von Lindern IH. Lead-induced anemia: dose-response relationship and evidence for a treshold. Am. J. Public Health,1990, 80, 165-1678.

Siegel M., Forsyth B., Siegel L., Cullen MR. The effect of lead on thyroid function in children. Environ. Res. 1989, 49, 190-196.

WHO 1977, Environmental Health Criteria 3. Lead, Geneva.

WHO 1980, Recommended Health-based Limits in Occupational Exposure to Heavy Metals. Tech. Rep. Ser. 647, Geneva.

Metal Ions in Biology and Medicine: vol. 7. Eds L. Khassanova, Ph. Collery, I. Maymard, Z. Khassanova, J.-C. Étienne.
John Libbey Eurotext, Paris © 2002 pp. 181-184.

On the distribution of some mineral levels in the diabetes rats (GK-RATS)

Hobara T, Takita M, Wakamoto U, Kunitsugu I, Kobayagawa S, Sugiyama S, Yamada T, Okuda M

Department of Public Health, Yamaguchi Medical School, 1-1 Minamikogushi Ube, 755-8505 Japan

ABSTRACT

To observe the distribution of mineral concentration in the several organs for diabetes, we analyzed the concentration of Ca, Mn, Mg, Zn, Cu, and Fe in the blood, urine and organs (cerebrum, cerebellum, lung, heart, liver, pancreas, spleen, kidney, testis, prostate, muscle and fat) in the control and GK(Gotho-Kawasaki)-rats, and compared with the data of these two groups.The concentration of Mg and Mn were not significant differences between the controls and the GK rats in the blood. However, in the urine, GK-rats of these minerals were significant higher than those of the controls. Moreover, in the pancreas, Zn levels of GK rats was higher than that of controls, and in the kidney, testis, and intracutaneous fat, this levels of GK rats were lower than those of controls. In the Ca levels, cerebrum, lung, kidney, testis and muscle of GK-rats were significantly higher than those of the controls. In the Mn levels, cerebrum, pancreas and prostate in GK-rats were higher than those of controls. Moreover, another many interesting differences were observed between these two groups. From these results, it is indicate that many mineral distribution was disturbed with a body in the diabetes. It is necessary to control the mineral distribution in a body by the nutrition or drug at the treatment procedure of diabetes patients.

INTRODUCTION

There are many reports that indicate a correlation between mineral distribution in the organs, and the overall nutritional condition and occurrence of certain diseases in human beings. It is well known that the diabetes disturbed the mineral balance in the body. Moreover, Mg and Zn deficiency were caused the arteriosclerosis or heart disease. The experimental diabetes rats which were made by dose of streptozotosin or alloxon, were changed with the distribution of mineral balance. This diabetes were caused by the disturbance of the pancreas for drug administration and were different from the nutritional diabetes. From these reasons, to observe the mineral distribution of the nutritional diabetes in the body, we examined the mineral balance of several organs in the GK rats.

MATERIAL AND METHODS

Five male Sprague-Dawley rats (about 200 g) were obtained from Charles River Japan (SPF) and five male Goto-Kawasaki rats were obtained from.

This rat was an autogenesis diabetic rat. These ten rats were used the experiments.

These two groups of rats were fed with the diet containing minerals about 1.3g of Ca, 0.25g of Mg, 10.9 mg of Fe, 0.78mg of Cu, 5.0mg of Zn and 0.92g of P per 100g, respectively. Rats were fed these diets at three weeks under conditions of controlled temperature (23 + 2 C), humidity

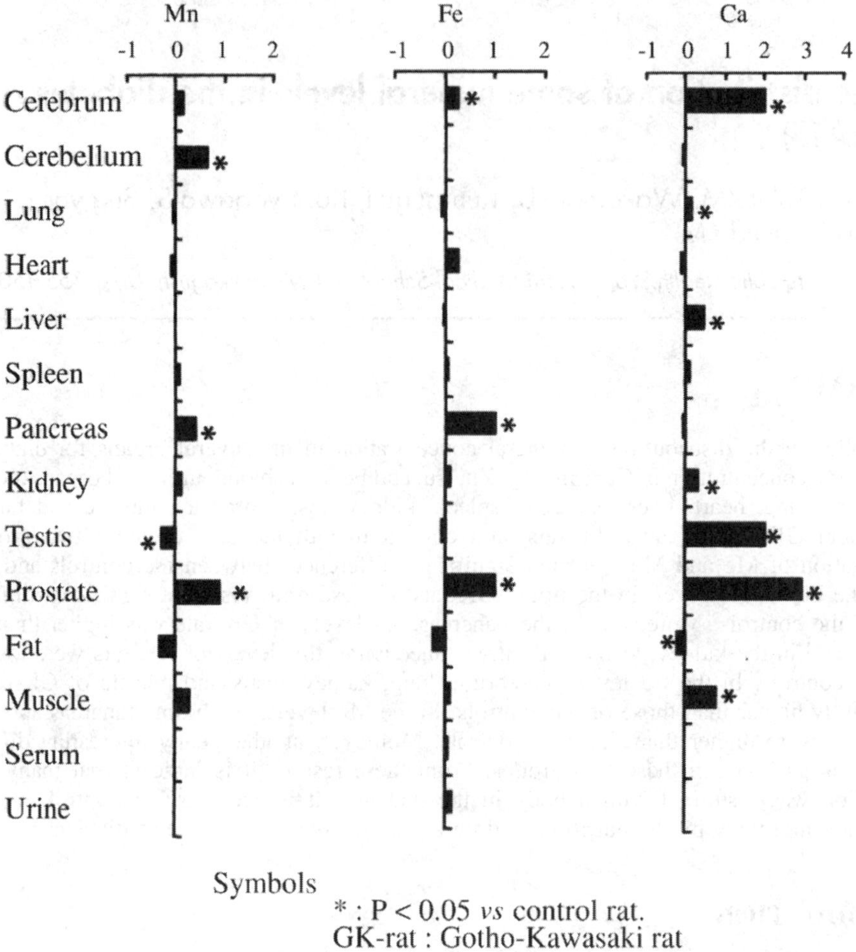

Fig. 1. Increase or decrease ratio of Calcium, Iron and Manganese in GK-rats for those of normal rats

(60 + 10%), lighting (light on; 7:00 A.M. to 7:00 P.M.) and ventilation (ventilated 13-15 time per hour).

After anesthesia by ether asphyxiation, blood was drawn from the vena cava superior of each rat. Twelve organs were then extracted: cerebrum, cerebellum, lung, heart, liver, spleen, pancreas, kidney, prostate, testis, muscle and fat. At the same time urine blood and serum were extracted. The mineral contents of blood, serum, urine and above organs were analyzed using the Zeeman background-corrected flameless atomic absorption method using Hitachi Z-8100. (Ca, Mg, Fe, Mn, Zn, Cu). P was analyzed by the Spectrophotometer (Hitachi 7250). Blood sugar and another data (RBC, WBC, Ht, AST, ALT, r-GTP, LDH,BUN, AL-P, etc.) were analyzed with Auto Analyzer (Hitachi 7250).

All data were presented as means + SD. The data were compared by ANOVA and Newman-Keul's test. P values < 0.05 were regarded as statistically significant.

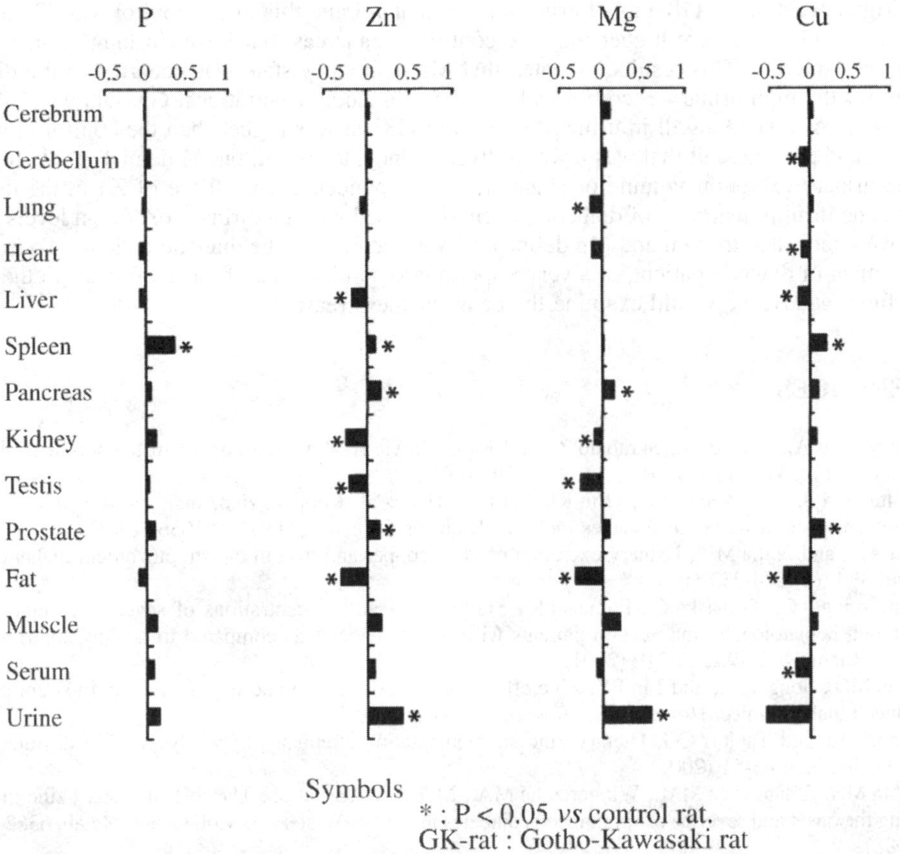

Fig. 2. Increase or decrease ratio of Copper, Magnesium, Zinc, and Phosphate in GK-rat for the normal rats

RESULTS AND DISCUSSION

In the blood sugar levels, the mean levels of GK-rats was 185 + 28 mg/dl, and that of control was 122 + 5 mg/dl. The blood sugar of GK rats was significant higher than that of control rats. Another data was no significant between these two groups.

Figure 1 shows the increased ratio of Ca, Fe and Mn in GK rat for that of control rats. In Ca levels, cerebrum, lung, liver, kidney, testis, prostate and muscle of GK rats were significant higher than those of the control. It is well known that many diabetes patients had osteoporosis with the complication. This data shows that the loss of Ca from the bone was deposited in the low Ca concentration organs in the body. In Fe levels, cerebrum, pancreas, prostate of GK rat were significant higher that those of the control. In Mn levels, cerebellum, pancreas, prostate of GK rat were significant higher than those of the control. The function of these minerals in diabetes was not clarified in the present time.

Figure 2 shows the GK vs Control rat ratio of Cu, Mg, Zn and P in the organs. In Cu levels, spleen and prostate of GK-rat were significant higher than those of the control. On the other side, GK-rats of Cu concentration was lower than the control group in the cerebellum, heart, liver, fat and serum, respectively. Car N et al. reported that serum Cu concentration of the diabetic patients were significant lower than those of the control. This reports indicated the same tendancies of our experiments. In P concentration, GK-rat was higher than the control in spleen.

In Mg concentration, GK-rats of urine was 639.8 ug/dl and this of the control was 435.6 ug/dl. Mg levels of GK-rats were higher than the control in pancreas, but lower in lung, kidney, testis and fat, respectively. This results indicated that Mg deficiency state was occurred in the diabetic patients for the high urinary excretion value of Mg. In Zn concentrations, GK rats was 2.4 ug/dl and the control was 1.9 ug/dl in urine. Zn levels of GK rat was higher than the control in spleen, pancreas, and prostate, but that was lower in liver, kidney, testis and fat. Minami T. et al., reported that the urinary excretion volume of sugar was correspondence with those of Zn in the diabetic patients and insulin treatment of diabetic patient decreased with the urinary excretion levels of Zn. These were indicated that Zn and Mg deficiency was occurred in the diabetic patients. To examine the treatment of diabetic patient, it is very important to administrate of these minerals in the body. In the future study, we would examine the detail of these reasons.

REFERENCES

1. Car N., Car A., Granic M., Skrabalo Z., and Momcillovic B. Zinc and copper in the serum of diabetic patients. Biol. Trace. Elem. Res. 32: 325-329 (1992).
2. Walter RM Jr., Uriu-Hare JY., Olin KL., Oster MH., et al. Cooper, zinc, manganese and magnesium status and complications of diabetes mellitus. Diabetes Care, 14 (11) 1050-1056 (1991).
3. Lau AL., and Failla ML., Urinary excretion of zinc, copper and iron in the streptozotocin-diabetic rat. J. Nutr. 114: (1) 224-233 (1984).
4. Ekmekcioglu C., Prohaska C., Pomazal K., Steffan I., et al. Concentrations of seven trace elements in different hematological matrices in patients with type 2 diabetes as compared to healthy controls. Biol. Trace. Elem. Res. 79: 205-219 (2001).
5. Chen MD., Song YM., and Lin PY. Zinc effects on hyperglycemia and hypoeptinemia in streptozotocin induced diabetic mice. Horn. Metab. Res. 32: 107-109 (2000).
6. Simon SF., and Taylorr CG. Dietary zinc supplementation attenuates hyperglycemia in db mice. Exp. Biol. Med 226: 43-51 (2001)
7. Tobia MH., Zdanowicz MM., Wingertzahn MA., McHerry AB., et al., The role of dietary zinc in modifying the onset and severity of spontaneous diabetes in the BB Wister rat. Mol. Genet. Metab. 63: 205-213 (1998).

*Metal Ions in Biology and Medicine: vol. 7.*Eds L. Khassanova, Ph. Collery, I. Maymard, Z. Khassanova, J.-C. Étienne.
John Libbey Eurotext, Paris © 2002 pp. 185-188.

Hepatic deposition of butyltin in humans

Jesper Bo Nielsen[1] and Jakob Strand[2]

[1]*Institute of Public Health, University of Southern Denmark, DK-5000 Odense C, Denmark.*
[2]*Department of Marine Ecology, National Environmental Research Institute, Risø, DK-4000 Roskilde, Denmark*

ABSTRACT

Health effects following exposure to butyltin compounds have focussed on effects on the immune system, but endocrine effects of butyltin have been documented in a variety of marine species. The information on human exposure to butyltin compounds and hepatic deposition is limited. Intake of marine food is the main source of butyltin exposure to humans and recent data indicate that butyltin concentrations in marine food may exceed 1000 ng/g. The present study includes 18 consecutively sampled human livers analysed for butyltin compounds. Dibutyltin (DBT) concentrations varied between 0.8 ng/g and 28.3 ng/g with a mean concentration of 9.0 ng/g. Significantly lower concentrations of mono butyltin (MBT) was observed ranging from 0.3 ng/g to 4.7 ng/g with a mean value of 1.6 ng/g. Age and DBT/MBT ratio were significantly associated. We conclude that younger men have more recent exposures or have a lower capacity to debutylate DBT than older men and would therefore potentially be more susceptible to butyltin toxicity. Given the inter-person variability observed in our limited group of men, we can not exclude that thresholds for either immunotoxicity or effects on the endocrine system may occur due to exposure to butyltin compounds alone or in combination with other environmental toxicants with similar target organs.

INTRODUCTION

Butyltin compounds occur widely in the marine food chains and in food products (Belfroid et al 2000), the exposure source being mainly decades use of tributyltin (TBT) as an antifouling agent for ships. Even after recent restrictions in the use of TBT for smaller vessels, deposition of TBT in sediments will remain an important exposure source for marine biota for many years to come due to its relative stability under anoxic sediment conditions (Sarradin et al. 1995).

Intake of marine food is the main source of butyltin exposure to humans. Previously, the average concentrations of butyltin (ΣBT) in marine food have ranged from 100-1500 ng/g with highest concentrations present in cultured fish and molluscs (Kannan et al. 1995). The most recent study from Denmark indicates that levels of total butyltin in have decreased. The highest concentrations were observed in the liver with dibutyltin (DBT) being the dominant butyltin species *(table 1)*.

Health effects following exposure to butyltin compounds has mostly focussed on effects on the immune system (Snoeij et al. 1987). Specifically, butyltin compounds inhibit the T cell system, and natural killer cells seem to constitute a main target (Whalen et al. 1999). Further, endocrine effects have been documented in a variety of marine species, e.g., imposex in snails, and inhibited egg hatching, malformations and reduced growth in crustaceans, molluscs and fish (Matthiessen and Gibbs 1998).

The available information on butyltin deposition in humans is limited to two studies on hepatic deposition in 4 Japanese (range: 59-96 ng/g; Takahashi *et al.* 1999) and 9 Polish people

Table 1. Butyltin concentrations (ng butyltin ions/g ww) in two fish species and porpoise. Results are from Danish seawaters and stated as the observed range of concentrations measured (data from Strand and Jacobsen 2000)

Species	Area	TBT	DBT	MBT	TPhT
Cod (liver)	Coastal	4.9-5.3	2.4-3.4	< 0.3	7.1-9.5
Flounder (liver)	Coastal	0.5-16.4	2.3-102	0.6-13	1.6-22
Flounder (muscle)	Coastal	0.1-0.7	0.2-1.2	0.3-0.5	< 0.2
Porpoise (liver)	Coastal	21-147	54-965	1.6-22	0.3-5.1
Cod (liver)	Harbour	30-47	41-42	2.0-2.1	9.2-10.9
Cod (muscle)	Harbour	3.8-13.5	1.9-13	< 0.3-5.0	0.2-0.3
Flounder (liver)	Harbour	8.6-58	249-1239	23-60	5.1-32
Flounder (muscle)	Harbour	7.4-50	3.8-22	1.2-5.1	0.9-2.5

(range: 2.4-11 ng/g; Kannan and Falandysz 1997), the Japanese participants having a considerable higher dietary intake of fish.

Knowledge on hepatic deposition of butyltin compounds in humans and on the molar ratio between butyltin species is limited, but of importance for future risk assessments. We therefore collected liver samples form 18 Danish men, and measured TBT, DBT, MBT, and triphenyltin (TPhT).

MATERIALS AND METHODS

Liver samples: Liver samples were collected consecutively at the Institute of Forensic Medicine, SDU, University of Southern Denmark from 18 deceased males aged 21-82 years, all unexpected deaths. The males were anonymous and information was available on age, liver and kidney pathology, whereas information on exposure to butyltin or dietary habits was not available.

Chemical analysis: Following digestion of the liver samples with hydrochloric acid and extraction, the organic tin compounds were detected with a modified VARIAN 3500 gas chromatograph with a pulsed flame photometric detector (GC-PFPD) (Walnut Creek, CA, USA). The method has been thoroughly described in Jacobsen *et al.* (1997).

RESULTS

Concentrations of TBT and TPhT were all below levels of determination. DBT concentrations varied between 0.8 ng/g and 28 ng/g with a mean concentration of 9.0 ng/g *(table 2)*. Significantly lower concentrations of MBT were observed ranging from 0.3 ng/g to 4.7 ng/g with a mean value of 1.6 ng/g *(table 2)*. The total butyltin burden was estimated by adding the concentrations of TBT, DBT, and MBT. ΣBT ranged from 1.1 ng/g to 33 ng/g with an average concentration of 10.7 ng/g *(table 2)*. MBT is the primary metabolite of DBT and the molar ratio between the two may be used to indicate metabolic capacity for the degradation process of DBT. The molar ratio between DBT and MBT varied between 1.1 and 12 *(table 2)*, and there was a significant and negative association between age of the donors and the DBT/MBT ratio ($p = 0.012$; $R^2 = 0.335$).

Table 2. Butyltin concentration (ng butyltin ions/g liver, wet weight) in 18 Danish males. Results are stated as means with the ranges in parenthesis

	TBT	DBT	MBT	ΣBT	TPhT	DBT/MBT (molar ratio)
Mean	< 0.3	9.0	1.6	10.7	< 3	5.6
Range		(0.8-28)	(0.3-4.7)	(1.1-33)		(1.1-12)

*: ΣBT = TBT + DBT + MBT.

DISCUSSION

The present study includes the up till now largest number of consecutively sampled human livers analysed for butyltin compounds. The previous studies have not allowed discussion of inter-person variability, relation between butyltin species, or associations between butyltin and age.

DBT appears to be the main butyltin species deposited in the human liver. The large inter-person variability with a more than 25-fold difference between the highest and lowest hepatic DBT concentration indicate extensive differences in exposure, as age alone does not explain this variation (data not shown). The absence of measurable concentrations of TBT irrespectively of DBT concentrations indicates that TBT not only have a very fast clearance from the blood (Kannan et al. 1999), but also that the hepatic debutylation of TBT occurs fast in humans. Evidence from experimental animals demonstrates that the toxicity of butyltin compounds decreases along with the debutylation process (Whalen et al. 1999). The ratio between DBT and MBT may therefore not only indicate the time since exposure (the higher the ratio, the more recent the exposure), but may also indicate the ability of the individual to debutylate DBT. Thus, an increased ratio between DBT and MBT may indicate an increased susceptibility and higher risk of toxicity due to a reduced capacity to debutylate, or a more recent exposure.

A tolerable daily intake for TBT based on exptrapolations from effects on the immune function in rats has been calculated to 0.25 µg TBT/kg b.w. (Penninks 1993). As the daily intake of seafood in Western Europe has been estimated to vary between 40 and 60 g, the average intake may approach one third of the tolerable daily intake. Given the inter-person variability observed in our limited group of men, we can not exclude that thresholds for either immunotoxicity or effects on the endocrine system may occur due to exposure to butyltin compound alone or in combination with other environmental toxicants with similar target organs.

REFERENCES

1. Belfroid, A.C., Purperhart, M., Ariese, F. Organotin levels in seafood. *Mar. Pol. Bull.* 2000; **40**: 226-232.

Jacobsen, J.A., Struer-Lauridsen, F., Pritzl, G. Organotin speciation in environmental samples by capillary gas chromatography and pulsed flame photometric detection (PFPD). *J. Appl. Organomet. Chem.* 1997; **11**: 737-741.

Kannan, K., Tanabe, S., Iwata, H., Tatsukawa, R. Butyltins in muscle and liver of fish collected from certain Asian and Oceanian countries. *Environ. Pollut.* 1995; **90**: 279-290.

Kannan, K., Falandysz, J. Butyltin residues in sediment, fish, fish-eating birds, harbor porpoise and human tissues from the Polish coast of the Baltic Sea. *Mar. Poll. Bull.* 1997; **34**: 203-207.

Kannan, K., Senthilkumar, K., Giesy, J.P. Occurrence of butyltin compounds in human blood. *Environ. Sci. Technol.* 1999; **33**: 1776-1779.

Matthiessen, P., Gibbs, P.E. Critical appraisal of the evidence for tributyltin - mediated endocrine disruption in mollusks. *Environ. Toxicol. Chem.* 1998; **17**: 37-43.

Penninks, A.H. The evaluation of data-derived safety factors for bis(tri-n-butyltin)oxide. *Food Additives Contam.* 1993; **10**: 351-361.

Sarradin, P.M., Lapaquellerie, Y., Astruc, A., Latouche, C., Astruc, M. Long term behaviour and degradation kinetics of tributyltin in a marina sediment. *Sci. Total Environ.* 1995; **170**: 59-70.
Snoeij, N.J., Penninks, A.H., Seinen, W. Biological activity of organotin compounds - an overview. *Environ. Res.* 1987; **44**: 335-353.
Strand, J., Jacobsen, J.A. (2000). Forekomst af organiske tinforbindelser i planter og dyr fra danske farvande: Akkumulering og fødekæderelationer, The National Environmental Research Institute, Denmark, Research Note no 135. (In Danish).
Takahashi, S., Mukai, H., Tanabe, S., Sakayama, K., Miyazaki, T., Masuno, H. Butyltin residues in livers of humans and wild terrestrial mammals and in plastic products. *Environ. Pollut.* 1999; **106**: 213-218.
Whalen, M.M., Loganathan, B.G., Kannan, K. Immunotoxicity of environmentally relevant concentrations of butyltins on human natural killer cells in vitro. *Environ. Res. A* 1999; **81**: 108-116.

Metal Ions in Biology and Medicine: vol. 7. Eds L. Khassanova, Ph. Collery, I. Maymard, Z. Khassanova, J.-C. Étienne.
John Libbey Eurotext, Paris © 2002 pp. 189-192.

Intramuscular injection of desferrioxamine lowers brain aluminum concentration in patients with Alzheimer disease

Theo P.A. Kruck[1,3], S.S. Krishnan[2], Donald R.C. McLachlan[1], and Maire E. Percy[1,3]

[1]*Dept. of Physiology, University of Toronto, Toronto ON, Canada M5S 1A8;* [2]*The Toronto Hospital, Toronto, Canada;* [3]*Neurogenetics Laboratory, Surrey Place Centre, 2 Surrey Place, Toronto ON, Canada M5S 2C2*

ABSTRACT

Although genetic risk factors are involved in Alzheimer disease (AD), there also is strong evidence that environmental aluminum (Al) is involved in pathogenesis. Based on the hypothesis that Al may act as a toxic factor in AD, a pilot clinical trial was initiated to determine if intramuscular injection of the trivalent metal chelator desferrioxamine (DFO) (500 mg, twice daily, 5 days per week) was effective in reducing brain Al levels and slowing down mental deterioration. Autopsy brains became available from 6 DFO-treated patients who died from causes unrelated to drug administration. By comparing Al levels in autopsy brains from DFO treated and minimally (or not) treated AD patients, it became possible to address the hypothesis that DFO treatment will lower brain levels of Al. Three brains were from AD patients who had received less than 7 DFO injections (minimally DFO treated), 3 were from AD patients who had received 47, 76 and 108 DFO injections (extensively DFO treated). Tissue Al levels were determined by electrothermal absorption spectroscopy. Samples from minimally treated patients had a mean value of 4.04 µg Al/g dry weight for 21 neocortical regions in each brain, whereas samples from the same 21 brain regions of extensively treated DFO treated patients had a mean value of 2.67 µg Al/g.. The difference between the mean Al levels of the two groups was statically significant ($P < 0.001$). There was no difference between the mean Al level for the minimally treated group and that for a group of 7 untreated AD patients. This study, though small, suggests that DFO treatment can reduce Al levels in the AD brain. Because the muscular injection of 125 mg of DFO twice daily for 5 days a week over 2 years was subsequently found in an independent phase II clinical trial to reduce the rate of deterioration in daily living skills in AD patients in excess of 50%, over the 2-year period of the study, the beneficial effect of DFO treatment may have resulted, at least in part, from reduction of the level of brain Al.

INTRODUCTION

Alzheimer disease (AD) is a slowly progressing, fatal, multifactorial disorder. Three genes have been identified (amyloid precursor protein, presenilin-1 and presenilin-2) that, when mutated, cause familial forms of AD - usually early onset. The late onset and most common forms of AD, which occur sporadically, likely are caused by various genetic susceptibility, metabolic and/or environmental factors. Aluminum (Al) has been implicated as a cofactor in generating the AD condition. In neocortical samples of the AD brain, there are increased amounts of Al with foci of unusually high concentrations [1, 2]. Aluminum has been localized in AD brain in neurofibrillary tangle bearing neurons [3, 4]. Desferrioxamine (DFO) a strong iron (III) and Al (III) chelating drug has been shown *in vitro* to remove Al from neocortical nuclei [5] - one of the sites where

Al may exert toxic effects by interfering with transcription [6]. Also, in autopsy brain from renal dialysis patients who had elevated Al in blood, AD-like changes in the processing of tau protein were found - truncated tau fractions and hyperphosphorylated tau associated with a decreased proportion of normal tau protein [7]. A positive correlation between Al levels and truncated tau was noted, and a role for Al in the development of AD-like pathology in patients subject to prolonged Al exposure was suggested. The evidence suggests that Al may be an important factor in the pathogenic process of AD. If Al is a significant cofactor, it must be present in elevated, toxic amounts at sites of pathological importance. An effective treatment to arrest the disease should remove Al from its sites of toxic action and removal of Al should precede beneficial clinical or pathological changes. That Al might be involved in the pathogenesis of AD has gained strong support from the results of a double blind, placebo controlled phase II clinical trial in which DFO was administered intramuscularly (IM) at a dose of 125 mg twice a day, five days a week, for 2 years. This clinical trial showed a 50% reduction in the average rate of decline of living skills [8]. No patients died in the latter study. However, in a study employing a higher dose of DFO, several patients died of causes unrelated to drug treatment and it became possible to address the hypothesis that DFO lowers the concentration of Al in the neocortex.

METHODS AND PATIENTS

Inclusion criteria for the higher dose DFO study included a history of progressive dementia, a diagnosis of probable AD, and Hachinski scores of 4 or less. In this double blind, placebo controlled trial, elderly patients were given DFO - 500 mg IM twice daily - five days per week. Six patients died of causes unrelated to drug treatment. Brains were collected at autopsy with special precautions to prevent any metal contamination. These were bisected in the sagittal plane; one half was fixed in Al free formalin and the other half was frozen at - 90°C for trace metal analysis. All six brains revealed microscopic evidence of neocortical neuron neurofibrillary degeneration, neuritic plaques with amyloid cores, hippocampal pyramidal cell granulovacuolar degeneration, and a Bielchowsky staining pattern compatible with the pathological diagnosis of AD. Under conditions excluding trace metal contamination, 21 specific cerebral grey matter regions were carefully dissected from the frozen brains excluding blood vessels and supporting tissue. The cortical samples were taken from each brain at pre-determined, stereoscopically fixed locations. After drying at 80°C, the samples (40-60 mg) were ashed in platinum crucibles at 650°C for 18 h. The aluminum content of samples dissolved in 5% ultra pure nitric acid was measured by electrothermal absorption spectroscopy.

RESULTS

Of the 6 individuals who died of causes unrelated to AD in the higher dose DFO clinical trial, 3 individuals (group C, age at death 76, 81 and 84 years) had received 0, 5 and 7 IM injections of 500 mg, respectively, and another 3 patients (group D, age at death 76, 81, and 83 years) had received 47, 76 and 108 IM injections of 500 mg, respectively, up to 48 hour prior to death. Al concentrations were measured in the 21 neo-cortical samples taken from each brain. The sample concentrations were rank ordered in descending order of magnitude for each brain and subjected to analysis by the nonparametric statistical Kolmogorov-Smirnov-test (testing for different sample group means). Population means for group C (minimally treated) and group D (extensively treated) were 4.04 μg Al/g dry weight and 2.67 μg Al/g dry weight, respectively, and the probability of equal sample means was $p<0.001$. These data were compared with sample Al brain concentrations, obtained from similar stereotactic regions for randomly collected AD and normal, unaffected autopsy brains, reported in an earlier study [1]. The Al sample means for the AD patient autopsy brains (group A), age 52-70 years and non-AD brains ranging in age from 18-54 (group B), were 3.8 μg Al/g and 1.96 μg Al/g respectively, which are significantly different with a probability of equal means, <0.001. To test whether there was a significant difference in Al distribution between the patients in this study (minimally treated, group C) and AD brains from slightly younger untreated AD brains (group A), sample data for these two groups were compared. Analysis showed that there was no significant difference p>0.95. The effectiveness of DFO in mediating Al removal from AD brains was examined further by comparing Al concentrations of extensively DFO treated AD brains from group D with the combined groups A (AD brains) and C (minimally DFO treated AD brains). Again there was a significant difference in mean Al concentrations between group A+C (mean =3.92) and group D (mean = 2.67 μg Al/g dry weight). The DFO treatment related reduction of the mean Al concentration from 4.04 μg AL/g to 2.67 μg Al/g amounts to 67% in neocortical Al concentration within 4.5-10.5 weeks. Furthermore, extensive treatment with DFO compared to minimal treatment with DFO not only was associated with a significant reduction in the mean Al levels, but also with a reduction in the number of loci with unusually high Al concentrations. The highest Al levels of the minimally DFO treated group C and the extensively DFO treated group D were 13.9 μg Al/g and 6.05 μg Al/g, respectively, with corresponding standard deviations of ± 2.27 and ± 1.06 μg Al/g.

DISCUSSION

Notwithstanding significant advances made in the understanding of pathological events associated with the AD condition, no long term effective treatment has yet evolved that significantly slows and arrests the disease. A previous clinical treatment trial was the first to show that it is possible to retard the course of AD by treating patients intramuscularly with the Fe and Al chelator, DFO [8]. However, during the duration of the 24 month clinical treatment none of the treated patients died, and consequently no brains became available for Al analysis. Hence, the Al studies on the current autopsy brains become important considerations in studies involving a link between Al exposure and AD and it treatment.

DFO treatment of elderly patients resulted in a significant reduction of neocortical Al concentrations, also evidenced by the different distribution of Al in the minimally treated and the more extensively treated patient brains. The extensive DFO treatment resulted in a general reduction of the mean neocortical Al content, a reduction in maximal Al levels, and a reduction in standard deviation, relative to less extensive or no treatment. These results demonstrate that Al can be removed from the brain by drug intervention. Unfortunately, the minimum toxic intracellular or extracellular Al concentration, which is pathogenic, is not known. Work in this laboratory (unpublished) indicates that in rabbits the LD_{50} lies between 5.5-6.0 μg Al /g dry weight. Results from the present study show that DFO treatment for 4.5-10.8 weeks reduced the mean neocortical Al concentration by 67%, towards the basal non-AD neocortical Al concentrations. The Al reduction in this study parallels the reduction in the mean rate of functional decline that was observed over the first 6 months of the DFO chelation treatment trial [8]. However, the continuing decline in function that was observed after 6 months of treatment in the published study [8] may mean that the Al levels remaining were still in the toxic range.

SUMMARY

We have shown that: (i) mean Al levels in AD neocortical tissue are significantly higher than in non -AD controls, (ii) treatment with the trivalent chelator, DFO, results in a significant reduction in both mean Al levels and the number of loci with unusually high Al concentrations; and, (iii) reduction of neocortical Al parallels the reduction or stabilization of the rate of functional decline observed in a published independent clinical trial of DFO in AD patients. Our studies provide evidence that Al may indeed be an active promoter of the AD pathogenic process.

REFERENCES

1. Crapper DR, Krishnan SS, Quittcat S. 1976. Aluminum, neurofibrillary degeneration and Alzheimer's disease. Brain 9: 67-79.
2. Xu N, Majidi V, Marksbery WR, Ehmann WD. 1992. Brain aluminum in Alzheimer's disease using an improved GFAAS method. Neurotoxicol 13: 735-744.
3. Good PF, Perl DP, Bierer LM, Schmeidler J. 1992. Selective accumulation of aluminum and iron in the neurofibrillary tangles of Alzheimer's disease: a laser microprobe (LAMMA) study. Ann Neurol 33: 121-127.
4. Lovel MA, Ehmann WD, Marksbery WR. 1993. Laser microprobe analysis of brain aluminum in Alzheimer's disease. Ann Neurol 33: 36-42.
5. Kruck TPA, McLachlan DRC, Bergeron C, Lukiw WJ. 1993. Aluminum in neocortical nuclei. Removal by shuttle chelation and relevance to Alzheimer's disease pharmacotherapy. Pergamon Press 87: 255-256.
6. Lukiw WJ, Kruck TPA, McLachlan DRC. 1987. Alterations in human linker-histone-DNA binding in the presence of aluminum salts in vitro and in Alzheimer's disease. Neurotoxicol 8: 255-256.
7. Harrington CR, Wischik CM, McArthur FK, Taylor GA, Edwardson JA, Candy JM. 1994. Alzheimer's disease-like changes in tau processing: association with aluminum accumulation in brains of renal dialysis patients. Lancet 343: 993-997.
8. Crapper-McLachlan DR, Dalton AJ, Kruck TPA, Bell MY, Smith W, Kalow W. Andrews DF. 1991. Intramuscular desferrioxamine in patients with Alzheimer's disease. Lancet 337: 1304-1308.

*Metal Ions in Biology and Medicine: vol. 7.*Eds L. Khassanova, Ph. Collery, I. Maymard, Z. Khassanova, J.-C. Étienne.
John Libbey Eurotext, Paris © 2002 pp. 193-196.

The copper levels in the plasma and dopamine beta-hydroxylase activity in a chronic mild stress model of depression in rats

M. Grabowska[1], M. Schlegel-Zawadzka[2], G. Nowak[3,4], M. Papp[4]

[1]Institute of Medical Biochemistry, Jagiellonian University, 7 Kopernika, 31-034 Cracow, [2]Department of Food Chemistry and Nutrition, and Department of Continuing Education and Postgraduate Studies and [3]Laboratory of Radioligand Research Jagiellonian University, 9 Medyczna, 30-688 Cracow, [4]Institute of Pharmacology, Polish Academy of Sciences, 12 Smetna, 31-343 Cracow; Poland

ABSTRACT

Dopamine β-hydroxylase (DBH) catalyzes the beta-hydroxylation of dopamine into norepinephrine. Its decreased activity has been reported in unipolar psychotic depression. DBH is known to use copper ions for catalytic activity. Our objective was to study the effect of chronic mild stress (CMS) in the rat model of depression on the content of copper in the plasma and the DBH activity. DBH was isolated from blood plasma. It's activity was measured by the Nagatsu and Udenfriend method, modified by Grabowska and Guminska, based on the oxidation of tyramine to octopamine. The octopamine was oxidized to p-hydroxybenzaldehyde. The absorbance of the second was measured spectrophotometrically at 330 nm. The copper content in the plasma was measured by AAS method. The copper content in the plasma in the chronic mild stress of the rat's depression model remains stable (0.87-1.11 mg/l), whereas the DBH activity is altered (22.84-73.14 umol/min/l). There exists a negative significant correlation between the content of copper in the plasma and the activity of DBH in the rats subjected to CMS ($r=-0.48$, $p<0.05$).

INTRODUCTION

Dopamine β-hydroxylase (DBH) is one of the enzymes participating in endogenous catecholamine formation [1, 2]. DBH catalyzes the biosynthesis of noradrenaline by stereospecyfic oxidation of dopamine in the β-position. It is localized in the synaptic vesicles of sympathetic nerve terminals and in the storage grains of the adrenal medulla chromaffin cells. The enzyme in chromaffin granules occurs in a soluble form and a form confined to the surrounding membrane [1, 3]. The secretion of neurotransmitter from these organelles via exocytosis is accompanied with DBH release [2]. The discovery of this enzyme activity in the plasma opened a possibility of circulating DBH activity measurements as an indicator in a clinical evaluation of the sympathetic nervous system stimulation in man [2, 4].

DBH requires copper ions for catalytic activity. The stechiometry of copper activation has been a matter of discussion, but most of the recent literature agrees on a model with two copper ion per active site [5]. Decreased DBH activity has been reported in unipolar psychotic depression [6].

Our objective was to study the effect of chronic mild stress (CMS) in the rat model of depression on the content of copper in the plasma and the DBH activity.

MATERIALS AND METHODS

DBH was isolated from blood plasma. DBH activity was measured by the Nagatsu and Udenfriend method, modified by Grabowska and Guminska [7] based on the oxidation of tyramine to octopamine. The octopamine was oxidized to p-hydroxybenzaldehyde. The absorbance of the second was measured spectrophotometrically at 330 nm. The results were expressed in umol/min/l. The copper concentration in the plasma was determined using flame atomic absorption spectrometry (AAS) (Perkin Elmer 5100 PC equipped with a 5100 ZL Zeeman Furnace Module). The measurements were taken after 2 days from the beginning of the experiment and followed in the intervals 1, 2 and 5 weeks of undergoing stress. All procedures were conducted according to NIH Animal Care and use Committee guidelines.

CMS procedure

Male Wistar rats (approx. 200g) were housed singly in plastic cages (40x25x15 cm) with food and water freely available, and maintained on a 12h light-dark cycle (lights on at 8 a.m.). One group of animals was subjected to a CMS procedure for a period of eight consecutive weeks. Each week of stress regime consisted of: two periods of food and water deprivation, two periods of 45^0 cage tilt, two periods of intermittent illumination (light on and off every 2h), two periods of soiled cage (200 ml water in sawdust bedding), two periods of paired housing, two periods of low intensity stroboscopic illumination (150 flashes/min), and two periods of no stress. All stressors were of a 12-14h duration and were applied continuously, day and night. The second group (sham) of animals was housed in a separate room and had no contact with the stressed animals [8].

Group differences were assessed using the unpaired *t*-test, one-way analysis of variance (ANOVA) followed by Bonferroni Multiple Comparisons Test (BCMT) or Fisher's Exact Test. Relationships between variables were assessed using Spearman's rank order correlation coefficient. Data were deemed significant when $p<0.05$.

RESULTS

DBH activities were expressed in plasma in control rats and after 2 days, 1, 2 and 5 weeks after chronic mild stress experiments. Results of the study are presented in the *table 1.*

Table 1. The copper content and DBH activity in the rat's plasma in a chronic mild stress model of depression

Duration of time	**Cu content in the plasma [mg/l]**		**DBH activity in the plasma [umol/min/l]**	
	Control	**CMS**	**Control**	**CMS**
2 days	0.84±0.13	0.87±0.90	58.75±13.74	39.70±8.08*
1 week	0.99±0.12	0.90±0.08	58.87±22.72	73.14±15.05
2 weeks	0.99±0.12	0.90±0.08	53.49±7.81	60.24±9.40
5 weeks	1.01±0.25	1.11±0.10	54.61±21.20	22.84±9.82*

* p<0.05 - significant difference vs. control

The copper content in the control plasma after 2 days of experiments expressed in mg/l was 0.84 ± 0.13 (0.72 -1.10). After 1 and 2 weeks copper contents were unchanged: 0.99 ± 0.12 (0.83 - 1.16), while after 5 weeks a small increase was observed - 1.00 ± 0.25 (0.72 - 1.43). The copper content in the plasma in the chronic mild stress of the rat's depression model remained stable.

DBH activity in control plasma, expressed in umol/min/l, after 2 days (58.75 ± 13.74) and 1 week of experiment (58.87 ± 22.72) was stable. After 2 and 5 weeks of experiment a slight decrease in DBH activity was observed (from 53.49 ± 7.81 to 54.61 ± 21.20). DBH activity after 2 days of chronic mild stress experiment was significantly diminished (to 39.70 ± 8.08). After 7 days the increase in DBH activity was observed (to 73.14 ± 15.05), as well as after 2 weeks of CMS experiment (to 60.24 ± 9.40). The most spectacular and statistically significant decrease in DBH activity was noticed after 5 weeks of experiments.

There exists a negative significant correlation between the content of copper in the plasma and the activity of DBH in the rats subjected to CMS ($r=-0.48$, $p<0.05$).

DISCUSSION

The pathophysiology (aetiology factors, pathogenesis) of depression is unknown. Animal models of depression are used to study the mechanism of depression. Interest in a copper role in depression is not widespread, although this element is an important component of enzymes that take part in the catecholaminergic pathways involved in the pathophysiology of depression [6, 9]. Serum DBH can be used not only as an indicatory enzyme for various diagnostic purposes, especially for the diagnosis of the sympathetic system stimulation, when it appears in the blood together with noradrenaline [2, 4], but also serum can be used as a source of target enzyme for more detailed studies of various signal molecules.

In our experiments the copper content in the plasma in the chronic mild stress of the rat's depression model was stable, whereas the DBH activity was altered. In the beginning of experiment DBH activity was lower then in control, together with time of experiment (1 and 2 weeks) the significant increase in DBH activity was observed, and after 5 weeks of experiment a statistically significant decrease in DBH activity was noticed. It could confirm, that after long-time lasted stress a specific adaptation of this enzyme appeared. DBH activity seems to be a very sensitive factor of depression, which can manifested earlier then changes in copper content in the plasma in the chronic mild stress of rat's depression model.

CONCLUSIONS

The copper content in the plasma in the chronic mild stress of the rat's depression model remains stable, whereas the DBH activity is altered. There exists a negative significant correlation between the content of copper in the plasma and the activity of DBH in the rats subjected to CMS ($r=-0.48$, $p<0.05$).

REFERENCES

1. Stewart L.C., Klinman J.P. Dopamine β-hydroxylase of adrenal chromaffin granules: structure and function. *Ann. Rev. Biochem.*, 1988, 57, 551-592.
2. Kopin I.J., Kaufman S., Viveros H., Jacobowitz D., Lake C.R., Ziegler M.G., Lovenberg W., Goodwin F.K. Dopamine β-hydroxylase. Basic and clinical studies. *Ann. Int. Med.*, 1976, 85, 211-223.
3. Markoglou N., Wainer I.W. Synthesis and characterization of immobilized dopamine beta-hydroxylase in membrane-bound and solubilized formats. *J. Biochim. Biophys. Methods.*, 2001, 48, 61-75.
4. Geffen L. Serum dopamine β-hydroxylase as an index of sympathetic function. *Life Sci.*, 1974, 14, 1593-1604.

5. Abudu N., Banjaw M.Y., Ljones T. Kinetic studies on the activation of dopamine beta-monooxygenase by copper and vanadium ions. *Eur. J. Biochem.*, 1998, 257, 622-629.
6. Landowski J., Lysiak-Szydlowska W. Low serum dopamine β-hydroxylase activity in patients with panic disorder. *Psych. Pol.*, 1998, 32, 781-788.
7. Grabowska M., Guminska M. Effect of buthobendin on dopamine β-hydroxylase *in vitro*. *Pol. J. Pharmacol.*, 1996, 48, 39-45.
8. Schlegel-Zawadzka M., Zieba A., Dudek D., Krosniak M., Szymaczek M., Nowak G. Serum trace elements in animal models and human depression. Part II. Copper. *Hum. Psychopharmacol. Clin. Exp.*, 1999, 14, 447-451.
9. Meyers B.S., Alexopoulos G.S., Kakuma T., Tirumalasetti F., Gabriele M., Alpert S., Bowden C., Meltzer H.Y. Decreased dopamine β-hydroxylase activity in unipolar geriatric delusional depression. *Biol. Psychiatry*, 1999, 45, 448-452.

Metal Ions in Biology and Medicine: vol. 7. Eds L. Khassanova, Ph. Collery, I. Maymard, Z. Khassanova, J.-C. Étienne.
John Libbey Eurotext, Paris © 2002 pp. 197-200.

Serum selenium concentrations in children with celiac disease

RM. Pérez-Beriain[1], JF. Escanero[2], A. García de Jalón[1], ML. Calvo Ruata[1], MD. Zapatero[1], T. Pérez-Beriain[3], M. Guerra[2]

[1]*Section of Nutrition and Metals. Biochemistry department. Miguel Servet Universitary Hospital. Pº Isabel la Católica 1-3. 50009 Zaragoza, Spain.*
[2]*Phisiology department of University. C/ Domingo Miral s/n. 50006 Zaragoza, Spain.*
[3]*Veterinary School. C/ Miguel Servet 177. 50013 Zaragoza, Spain.*
The mail will be remitted to:
Dr. Ángel García de Jalón Comet.
Jefe de Servicio de Bioquímica Clínica
Hospital Universitario Miguel Servet
Paseo Isabel la Católica 1-3
50009 ZARAGOZA (Spain)

ABSTRACT

Many studies relate a deficit of selenium in patients with celiac disease, compared with the healthy population. Dietary supplementation with antioxidant molecules may offer some benefit and deserves further investigation.

PATIENTS AND METHODS

Selenium levels were evaluated in the serum of 6 children with celiac disease and in a 28 age-matched healthy controls (2-10 years).

Serum selenium levels were measured by atomic absorption spectrophotometry (AAS) with a graphite furnace and Zeeman background corrector.

The statistical test used, was the Mann-Whitney U test for two independent samples.

RESULTS AND CONCLUSIONS

The concentrations of serum selenium were significantly ($p<0.05$) lower in patients (64.05 μg/l) compared with controls (75.74 μg/l).

Serum selenium in CD patients is decreased. In patients at free diet the deficit of selenium can be attributed to malabsorption, while in patients at gluten-free diet it may be due to the diet itself.

BACKGROUND

The diagnosis of celiac disease (CD), or gluten-sensitive enteropathy, is established by the histologic demonstration of small-intestinal villus atrophy with inflammation and hyperplastic crypts in a patient ingesting gluten, followed by clinical remission after withdrawal of gluten from the diet [1].

An important role for free radicals in a number of gastrointestinal disease has been demonstrated [2].

Selenium is the component of the glutation peroxidase one of the most important enzymes in the antioxidant protection of the organism [3].

Many studies relate a deficit of selenium in patients with celiac disease, compared with the healthy population [3, 4, 5].

The malabsorption of selenium was found to play also an important role in the decreased protecting capacity against oxidative stress effects in celiac diseases of childhood.

Dietary supplementation with antioxidant molecules may offer some benefit and deserves further investigation [6].

AIMS

The following study attemps to establish the relationship between celiac disease and serum selenium levels.

PATIENTS AND METHODS

Subjets

Selenium levels were evaluated in the serum of 6 patients with celiac disease confirmed by biopsy and in a 28 age-matched healthy controls (2-10 years).

Instrumentation

Serum selenium levels were measured by atomic absorption spectrophotometry (AAS) with a graphite furnace and Zeeman background corrector (Perkin Elmer 4110 ZL), using Pd(NO3)2 solution as matrix modifier.

Statistical analysis

The statistical calculations were carried out using SPSS statistics program. The statistical test used, was the Mann-Whitney U test for two independent samples for equality of means.

RESULTS

The distribution of the selenium levels in the two groups are defined to the following illustration *(fig. 1)*:

The results of the serum concentrations of selenium in the groups *(table 1)* were as follows:

The means comparison using the Mann-Whitney U test for two independent samples shows a statistically significant difference with a $p<0.05$ *(fig. 2)*.

CONCLUSIONS

Serum selenium in CD patients is decreased. In patients at free diet the deficit of selenium can be attributed to malabsorption, while in patients at gluten-free diet it may be due to the diet itself.

These findings seem to suggest that, besides the adoption of a strictly gluten-free diet, the possibility for an adjuvant therapy with selenium may be considered for celiac disease patients.

Table 1.

	N	Selenium (μg/l)	C.I. (95%)
Celiac children	6	64.05	56.00-72.10
Control children	28	75.74	71.67-79.82

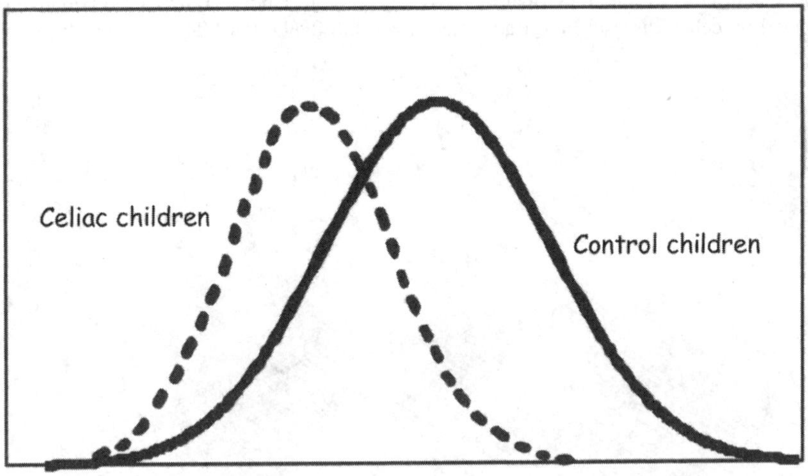

Fig. 1.

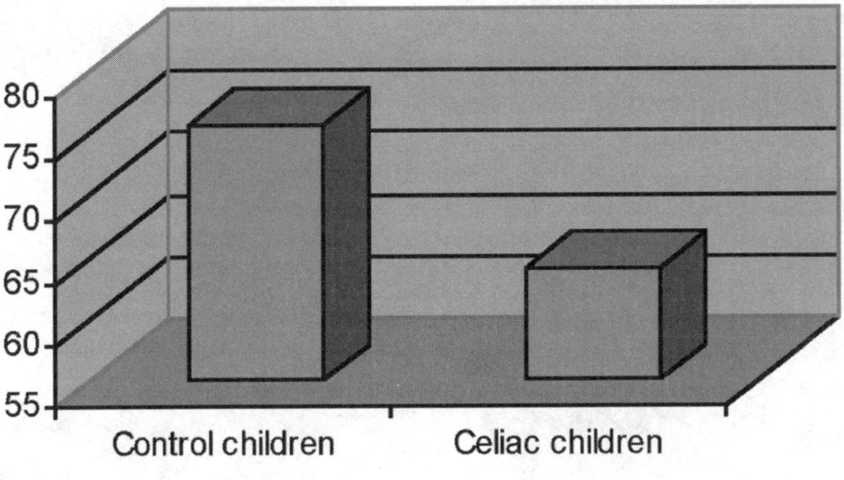

Fig. 2.

REFERENCES

1. Walker-Smith JA, Guandalini S, Schmitz J, Shmerling DH, Visako. Revised criteria for diagnosis of coeliac disease. Arch Dis Child. 1990; 65. P 909-11.
2. Rivabene R, Mancini E, De Vincenzi M. In vitro cytotoxic effect of wheat gliadin-derived peptides on

the Caco-2 intestinal cell line is associated with intracelular oxidative imbalance: implications for coeliac disease. Biochim Biophys Acta. 1999 Jan 6. 1453(1). P 152-60.
3. Boda M, Nemeth I. Selenium levels in erythrocytes of children with celiac disease. Orv Hetil. 1989 Sep 24. 130(39). P 2087-90.
4. Hinks LJ, Inwards KD, Lloyd B, Clayton BE. Body content of selenium in coeliac disease. Br Med J (Clin Res Ed). 1984 Jun 23. 288(6434). P 1862-3.
5. Cortigiani L, Nutini P, Caiulo VA, Ughi C, Ceccarelli M. Selenium in celiac disease. Minerva Pediatr. 1989 Nov. 41(11). P 539-42.
6. Odetti P, Valentini S, Aragno I, Garibaldi S, Pronzato MA, Rolandi E, Barreca T. Oxidative stress in subjets affected by celiac disease. Free Radic Res. 1998 Jul. 29(1). P 17-24.

Metal Ions in Biology and Medicine: vol. 7. Eds L. Khassanova, Ph. Collery, I. Maymard, Z. Khassanova, J.-C. Étienne.
John Libbey Eurotext, Paris © 2002 pp. 201-203.

Plasma copper and zinc content in patients with chronic renal insufficiency: influence of diet

Juan Llopis[1], Elena Planells[1], Cristina Sánchez[1], Manuel Larrubia[3], Pilar Aranda[1], Concepción Asensio[2], Pilar Galindo[2], Antonio Pérez de la Cruz[2], José Mataix[1]

[1]*Department of Physiology, School of Pharmacy, University of Granada, E-18071 Granada;* [2]*Hospital Universitario Virgen de las Nieves, Granada;* [3]*Sanaví, S.A., Lachar, E-18327 Granada, Spain*

ABSTRACT

Alterations of trace element metabolism in chronic renal failure (CRF) has been reported, but the mechanisms responsible for these changes are poorly understood. Changes in plasma concentration of copper and zinc in patients with CRF were studied before dialysis. The patients were given nutritional therapy consisting of a balanced diet adjusted to their needs, or prepared from manufactured (Sanavi®) low-protein dietary products (LPDP). Forty-two patients with CRF were included in the study on the basis of the following criteria: blood creatinine >3 mg/dL, creatinine clearance < 20 min, and stable clinical condition. The patients were divided into three groups of 14 patients each. In Group1 (G1) patients consumed their regular diet, those in G2 consumed a balanced diet adjusted to their needs, and those in G3 consumed a balanced diet consisting of LPDP. Copper and zinc concentrations were measured by atomic absorption spectrometry. At the beginning of the study (day zero), mean plasma concentrations were higher than 1.70 mg Cu/L and lower than 1.50 mg Zn/L, in three groups. After 6 months, plasma Cu concentration was lower in patients who consumed LPDP ($p<0.05$), but no significant change was seen in G1 or G2. Plasma Zn concentration showed a tendency to decrease in three groups, becoming significant in G2 ($p<0.05$).

INTRODUCTION

Actually, the benefits what mean to continue an hypoprotean diet on the chronic renal failure (CRF), are proved [1-5]. However, the dietary prescription implies some practical problems, such as the limitation on the consumption of many foods and the insufficient caloric and micronutritional (mineral and vitamin) contribution. Alteration of trace element metabolism in renal failure has been frequently reported, but mechanisms responsible for these changes are poorly understood and the contribution of trace element deficiency or toxicity to the symptoms of renal disease is uncertain [6].

Patients with CRF included in the experimental group were given nutritional therapy consisting of a balanced diet adjusted to their needs, prepared from manufactured low-protein dietary products (LPDP) with aspect and texture similar to the foods of usual consumption. Influence of diet on plasma magnesium content in patients with chronic renal failure (CRF) before dialysis, was evaluated.

In the present study, changes in plasma concentration of Cu and Zn in patients with CRF were studied before dialysis.

PATIENTS AND METHODS

Patients and diets

The experimental group patients were given nutritional therapy consisting of a balanced diet adjusted to their needs (300-350 mg Mg/day) prepared from manufactured (Sanavi®) low-protein dietary products (LPDP) with aspect and texture similar to the foods of usual consumption (milk substitute, flours, bread, confectionery, Italian pastas, biscuits, speed desserts, etc) which give variety and sufficient calories to the diet for making it more attractive and agreeable to the patients subject to protean restriction.

Forty-two patients with CRF were included in the study on the basis of the following criteria: blood creatinine >3 mg/dL, creatinine clearance <20 min, and a stable clinical condition.

The patients were divided into three groups: group 1 (G1) patients (n=14) consumed their regular diet, group 2 (G2, n=14) consumed a balanced diet adjusted to their needs and group 3 (G3, n=14) consumed a balanced diet consisting of LPDP.

The diets were designed on the basis if the results of a nutritional questionnaire that included 24 hour recall (recorded on three different days, one of them festive) and frequency of consumption. Diets were elaborated following CRF recommended intakes: Protein: 0.6-0.8 g/kg/day (7% Kcal); Calcium: 500-600 mg/day; Potassium: 1800-2600 mg/day; Sodium: 1-3 g/day; Phosphorus: 0.5-0.7 g/day. At twice-monthly follow-up visits, blood and urine samples were obtained.

Analytical techniques

The Cu and Zn concentrations were measured by atomic absorption spectrometry (AAS) (Perkin Elmer Aanalyst 300), in wet-mineralised samples (NO_3H/ClO_4H, 10:1).

Human Serum (Certified Reference Material seronorm™ 201405, Asker-Norway, UK) was used for quality control assays.

Statistical analyses

The data for control and Mg-deficient animals were compared with Student's *t* test. All analyses were done with the SPSS software package. Differences were considered significant at the 5% probability level.

RESULTS AND DISCUSION

At the beginning of the study (day zero), mean plasma concentrations were higher than 1.70 mg Cu/L and lower than 1.50 mg Zn/L, in three groups, with no significant differences between the groups for either ion. After 6 months, plasma Cu concentration was lower in patients who consumed LPDP but no significant change was seen in G1 or G2 *(table 1).*

Table 1. Zinc and copper concentrations in plasma

	mg Zn/L (Initial)	**mg Zn/L (6 months)**	**mg Cu/L (Initial)**	**mg Cu/L (6 months)**
GROUP 1 (G1) n=14	1.27±0.10	1.10±0.15	1.72±0.17	1.63±0.12
GROUP 2 (G2) n=14	1.39+0.11	1.04±0.10*	1.72±0.16	1.58±0.13
GROUP 3 (G3) n=14	1.25±0.13	1.09±0.15	1.93±0.16	1.39±0.11*

* Initial *vs*. 6 months, $p<0.05$

After 6 months of balanced nutritional therapy, plasma Zn concentration showed a tendency to decrease in three groups, becoming significant in G2, although, plasma Zn was into normal range in three groups [6, 7] *(table 1)*. Application of LPDP diet to patients with chronic renal failure (CRF) before dialysis, was evaluated positively on plasma copper content after 6 months of nutritional treatment. These results suggest that Cu and Zn deserve to be checked in CRF patients in a routine way.

REFERENCES

1. Kopple J. Nitrogen metabolism. In: Massry S., Sellers A., Eds. *Clinical aspects of uraemia and dialysis.* Springfield: Charles C. Tomas, 1976; 241-273.
2. Brener B., Meyer T., Hostetter T. Dietary protein intake and the progressive nature of pathogenesis of progressive glomerular sclerosis in ageing, renal ablation and intrinsic renal disease. *N Engl J Med* 1982; 307: 652-659.
3. May RC, Kelly RA, Mitch WE. Patophysiology of uraemia. In: Brenner BM, Rector FC, eds. *The kidney.* Philadelphia: WB Saunders Company, 1991; 1.997-2.108.
4. Harter HR. Review of significant findings from the National Cooperative Dialysis study and recommendations. *Kidney Int* 1983; 23(suppl. 13): 107-112.
5. Maschio G., Oldrizi L., Tessitore N., D'Angelo A., Valvo L., Lupo A., Loschiavo D., Fabris A., Gammaro L., Rugio C., Panzetta G. Effects of dietary and phosphorus restrictions on the progression of chronic renal failure. *Kidney Int* 1982; 22: 371-376.
6. Mitch WE, Klahr S. Handbook of Nutrition and the Kidney. 1998 Lippincott-Raven, Publishers. Philadelphia. New York.
7. Sauberlich E. Laboratory test for the assessment of nutritional status. 1999. CRC Press. Boca Raton.

*Metal Ions in Biology and Medicine: vol. 7.*Eds L. Khassanova, Ph. Collery, I. Maymard, Z. Khassanova, J.-C. Étienne.
John Libbey Eurotext, Paris © 2002 pp. 204-206.

Influence of zinc, copper and magnesium intake with diet on these elements level in the blood serum of young men doing military service in Polish army

Klos Anna[1], Klos Krzysztof, Bertrandt Jerzy[1], Stezycka Elzbieta[1], Schlegel-Zawadzka Malgorzata[2]

Department of Infectious Diseases and Allergology Central Clinical Hospital Military School of Medicine, 128 Szaserów St., 00-909 Warsaw, Poland,
[1]Military Institute of Hygiene and Epidemiology, 4 Kozielska St., 01-163 Warsaw, Poland,
[2]Department of Food Chemistry and Nutrition Collegium Medicum, Jagiellonian University, 9 Medyczna St., 30-688 Kraków, Poland

ABSTRACT

The aim of the work was estimation of zinc, copper and magnesium content in food ration that is the base for nutrition planning for most men serving in Polish Army. Level of zinc, copper and magnesium was estimated in blood serum of soldiers fed this diet as well. Content of m/a elopements in basic food ration was calculated using calculation software FOOD 2. Level of zinc, copper and magnesium in blood serum was estimated by atomic absorption spectrophotometry method. Obtained results were compared to obligatory in Poland norms. It was found that content of examined elements in food ration, including technological losses occurring during food processing, was for zinc 18.9 mg, copper 1.93 mg and for magnesium 590.4 mg. They met the recommended requirements in 118.1%, for zinc, from 77.0% to 96.3% for copper and 159.4% for magnesium. Average zinc content in blood serum of examined people was 10.95±1.7 µmol/l. This value is included in norm limits of 8-23 µmol/l. Copper level in blood serum was 13.26±2.87 µmol/l and reached lower norm's limit (13-22 µmol/l). Average magnesium content was 0.82±0.54 mmol/l. and was included in the norm's limits of 0.7-1.1 mmol/l. Low copper content found in daily food ration may be one of the reasons of its low level in blood serum of examined men.

Key words: food ration, level in blood serum

INTRODUCTION

Mineral elements play many different roles in human organism. They are contained in fundamental bones building materials and participate as enzymes activators [1]. Epidemiological researches show significant meaning of zinc, copper and magnesium in some diseases pathogenesis. There are many facts proving mineral elements role in young organism's functioning and development, immunity and susceptibility to diseases and reactions to stress and environment pollution [2]. Many authors [3, 4, 5, 6, 7] underline relation between unbalanced mineral elements supply and incidence of chronic diseases. Proper magnesium supply decreases risk of atherosclerosis development. Disturbed relations between zinc and copper increase risk of ischaemic heart disease occurrence. Increase of Zn:Cu quotient value in food rations favours and decrease of this ratio prevents against ischaemic heart disease [5].

Food is the main source of elements for human being. Constant elements level regulation in the blood confirms importance of keeping proper elements amount in the organism [8].

Young men begin their military duty at age of 19-21, they come from different social environments and nutrition habits in their homes are not always in accordance with proper nutrition rules. Therefore the fundamental task in their alimentation during military service is to ensure diet of standard value guaranteeing fulfilment of all nutrition requirements of young organism endangered to stress and huge physical exercises.

AIM OF WORK

The aim of the work was estimation of zinc, copper and magnesium content in food ration, on which nutrition of most men doing military service in Polish Army is planned. Level of these elements in blood serum of men fed this ration was estimated as well.

MATERIAL AND METHODS

Content of these elements in basic food ration was estimated based on calculation software FOOD 2. Zinc, copper and magnesium level in blood serum was indicated by atomic absorption spectrophotometry.

Obtained results were compared to obligatory in Poland norms.

RESULTS AND DISCUSSION

It was found that content of examined elements in analysed food ration amounted for zinc 21.01 mg, copper 2.14 mg and 656 mg magnesium. It is generally known that during food preparation technological losses occur that cause reduction of these elements amount in meals by 10%. Considering these technological losses analysed food ration contained 18.91 mg zinc, 1.92 mg copper and 590.4 mg magnesium. The examined elements met the requirements of the recommended norms in following amounts: zinc 118.1%, copper 77.0-96.3% and magnesium 159.4% *(table 1)*.

Table 1. Mean zinc, copper and magnesium content in food ration

	Zinc mg	**Copper mg**	**Magnesium mg**
Mean metals content in food ration	21.01	2.14	656.0
Mean metals content in food ration after technological loss subtraction	18.91	1.92	590.4

Analysis of elements content in blood serum is one of the fundamental methods to estimate human organism's saturation in these elements. However this method is not the best (invasiveness, short-term assessment, increase of some metals content after meals consumption, constant level of elements thanks to homeostasis etc.) but is still the most popular.

Mean zinc content in blood serum of examined amounted 10.95±1.7 μmol/l. These value was included in the norm's limits of 8-23 μmol/l. Copper level in blood serum was 13.26+2.87 μmol/l and was close to the lower norms' limit (13-22 μmol/l). Mean magnesium content was 0.82μ0.54 mmol/l and was included in the norm's limits of 0.7-1.1 mmol/l.

Researches on daily food rations given for consumption to young men in National Fire Department College showed that zinc met the requirements in 106%, copper - 101% and magnesium - 107%. Daily food rations planned for alimentation of teenagers going to group of agriculture schools in Poznan met the requirements for zinc in 91%, copper - 108% and magnesium in 90% [2].

Daily food rations recreated based on data about mean yearly food consumption obtained from researches on home budgets, contained 0,83-0,94 mg copper and 7,61-8,39 mg zinc [9].

Mean zinc, copper and magnesium content in daily food ration of students from Medical Academy amounted as follows: 13.9 mg, 1.57 mg and 385.2 mg. Magnesium content lower than recommended in Polish norms was found in 27.5%, zinc in 41.7% and copper in 85.8% daily food rations [10].

Wójciak *et al.* [1] investigating zinc, copper and magnesium content in blood serum of adults from Wielkopolska region showed that these elements level was similar to proper values.

CONCLUSIONS

1. It should be stated that low copper content in daily food ration might be one of reasons of its low content in blood serum of examined men.
2. Zinc and magnesium content in daily food ration and in blood serum of examined men met the requirements of the norms for young men.

REFERENCES

1. Wójciak R., Smigiel-Papinska D., Przyslawski J., Krejpcio Z., Gawecki J. Poziom wapnia, magnezu, cynku, miedzi i zelaza w surowicy krwi doroslych ludzi. Zyw. Czlow. Metab. 2000, 27, 66-69.
2. Olejnik D., Krejpcio Z., Smigiel-Papinska D., Wójciak R., Gawecki J., Wisniewska J. Zawartosc wybranych skladników mineralnych (Ca, Mg, Zn, Cu, Fe) w calodziennych racjach pokarmowych mlodziezy starszej. Roczniki PZH 1999, 50, 4, 361-367.
3. Mielcarz J., Majewski W., Petelski J., Uryszek W. Wapn, magnez, miedz i cynk w surowicy mezczyzn z miazdzyca tetnic udowych. Roczniki PZH 1995, 46, 59-64.
4. Kulikowska E., Moniuszko-Jakoniuk J., Minuk K. Rola cynku w procesach fizjologicznych i patologicznych organizmu. Pol. Tyg. Lek. 1991, 46, 470-477.
5. Szajkowski Z. Badania nad zawartoscia i wzajemnymi relacjami wybranych skladników mineralnych w calodziennych racjach pokarmowych wytypowanych populacji z Regionu Wielko[polski. Zyw. Czlow. Metab. 1996, 23, 66-72.
6. Anke M., Glei M., Muller R., Dorn W., Vorman J., Anke S. Macro, trace and ultratrace element intake of adults in Europe: Problems and dangers. J. Commodity Science, 2000, 39, 119-139.
7. Prasad A.P. Recognition of zinc-deficiency syndrome. Nutrition 2001, 17, 67-69.
8. Versieck J., Cornelis R. Normal levels of trace elements in human blood plasma of serum. Anal. Chim. Acta 1980, 116, 217-254.
9. Wojtasik A., Iwanow K., Rutkowska U., Kunachowicz H. Jakosc zdrowotna krajowych racji pokarmowych - badania analityczne i ocena teoretyczna. Zyw. Czlow. Metab. 2000, 27, 2, 115-129.
10. Trafalska E., Grzybowski A. Realizacja zalecen zywieniowych przez studentów Lódzkiej Akademii Medycznej. Zyw.Czlow. Metab. 2001, 28, supl. 418-423.

*Metal Ions in Biology and Medicine: vol. 7.*Eds L. Khassanova, Ph. Collery, I. Maymard, Z. Khassanova, J.-C. Étienne.
John Libbey Eurotext, Paris © 2002 pp. 207-211.

Effect of magnesium supplementation on magnesium, calcium, zinc, iron and copper balance in rats

Regina Oledzka, Dorota Skrajnowska

Department of Bromatology, Medical University of Warsaw, Warsaw, Poland

ABSTRACT

Background : Dietary imbalance of minerals, especially magnesium has been frequently described in the literature. The simplest solution is to correct a slight magnesium deficiency with an appropriate diet, but often used method is to supplement the deficiency by using pharmaceutical products containing magnesium.
Aim : The aim of the study was to investigate the balance of magnesium, calcium, zinc, iron and cooper in rats which diet had been supplemented with magnesium carbonate and after its discontinuation.
Materials and methods : Male Wistar rats, aged 4-5 weeks were given LSM laboratory chow and deionized water ad libitum. Magnesium carbonate suspended in 2% arabic gum was administered for 4 weeks to the experimental groups by gavage in doses of: 2,5 mg; 5,0 mg; 10,0 mg and 20,0 mg Mg/kg b. w./24h and for another 4 weeks the magnesium supplementation was discontinued. A control group was given 2% arabic gum. During the experiment the animals were four times placed in metabolic cages for 2, 4, 6 and 8 weeks for determination of minerals balance. The balance was calculated as the retention coefficient. The results were statistically evaluated.
Results : It was shown that after two weeks of magnesium supplementation in a dose of 2,5 mg Mg/kg b. w./24h there were no significant differences in the coefficient of apparent retention for all estimated minerals. The higher doses influenced the balance of magnesium and zinc which was increased and the balance of iron and cooper which was decreased. Supplementation with magnesium (for 4 weeks) caused the changes in apparent retention of iron (marked reduction) and zinc (increase). After 4 weeks of discontinuations of magnesium supplementation reduction of the coefficients of apparent retention for iron, cooper and calcium in all experimental groups was noted.
Conclusions : The results obtained in this study show that oral supplementation of magnesium in doses of 2,5 mg; 5,0 mg; 10,0 mg and 20,0 mg Mg/kg b. w./24h during 4 weeks causes the changes of the balance for iron, cooper and calcium in rats.

INTRODUCTION

In recent years, there has been a dramatic increase in the demand for preparations to add vitamins, micro- and macronutrients, “energy” supplements etc. to our everyday diet. A time-consuming life-style makes following a well-balanced diet difficult, people tend to consume food that does not provide sufficient calories while the use of artificial fertilizers deprives many foodstuffs of their nutritional value [1, 2]. The easiest solution would be to maintain appropriate nutritional status by adoption of a proper diet, but tempted by advertisements people are eager to reach for commercially available nutritional supplements. Some of these must be actually taken at certain stages in our lives, including folic acid, vitamin C, vitamin D, calcium or magnesium [3]. However, it remains to be answered whether the supply of high doses of single nutrients does not affect the homeostasis of the body and its normal physiological functions. For example, it is well known that high doses of vitamin C are related to enhanced formation of urinary calculi while excessive magnesium intake may "wash" calcium out of the body. On the other hand, the actual bioavailability of foodstuffs, especially of micro- and macronutrients is affected by a number of factors, including the degree of tissue saturation, the size of tissue stores (deficiency or excess of particular elements), the presence of competitive ions and of substances either inhibiting or facilitating their absorption, stress, and associated hormone secretion, consumption of coffee, tea or alcohol, or coexisting disease [4, 5, 6]. As a result, dietary intakes may be inadequate to meet the requirements of the body. In such circumstances, the need for dietary supplements cannot be questioned, but they should be used with caution, bearing in mind their potential adverse effects.

AIM OF THE STUDY

Hypomagnesemia is a common nutritional and metabolic disorder [7, 8]. According to the World Health Organization, the recommended dietary allowance is 5-6 mg Mg/kg/24h [7]. A long-term supply of one element is likely to affect the levels of other macro- and micronutrients. The aim of the present study was to determine, using an animal model, the effect of oral supplementation with magnesium given in doses of 2.5; 5.0; 10.0 and 20.0 mg Mg/kg/24h to rats fed a standard diet, on the retention of magnesium, calcium, zinc, iron and copper during the period of magnesium intake and after its discontinuation.

EXPERIMENTAL PART

Male Wistar rats, age 4-5 weeks, body weight 60-70 g were used in the study. They were fed the LSM laboratory chow and given demineralized water *ad libitum*. The animals were divided into five groups (4 study groups and a control group). For 4 weeks, the animals were given by an intragastric tube a suspension of magnesium carbonate in 2% arabic gum in the amounts equivalent to 2.5 mg Mg/kg/24h (group 1), 5.0 mg Mg/kg/24h (group 2), 10.0 mg Mg/kg/24h (group 3) and 20.0 mg Mg/kg/24h (group 4), while the recommended daily allowance was 5.0 mg Mg/kg/24h [9]. Over the following 4 weeks the experiment was continued, but the animals did not receive magnesium. Control animals were given 2% arabic gum. During the course of the experiment the animals were housed in metabolic cages (2 animals per cage), at 2, 4 and 6 weeks or at 2, 4, 6 and 8 weeks, i.e. three or four times. For 36 hours, urine and feces were quantitatively collected from each pair of rats and subjected to magnesium and iron balance studies. The balance was calculated as the apparent retention rate according to the following formula:

intake - (fecal excretion + urinary excretion) × 100%

intake

Simultaneously, laboratory chow (standard diet) intake and body weigh gain were monitored. The element contents were measured by the ASA atomic spectrometry method.

DISCUSSION

The apparent retention rate of magnesium was the highest compared to controls in study groups 2, 3 and 4 at week 2 of the experiment *(table 1)*. After 28 days of intake no significant differences were observed. The excess of magnesium following multiple dosing is probably removed mainly via the kidneys and the gastrointestinal tract, and as a result of hormonal regulation [10, 11]. After a 14-day interval in magnesium supplementation, i.e. at 6 weeks of the experiment another short-lasting increase in the apparent retention rate of magnesium was observed in the same study groups. At 8 weeks of the experiment, however, there were no differences found in the magnesium retention between the study groups and controls.

Magnesium doses used did not ultimately effect any permanent change in the apparent retention rate of magnesium, although they produced some disturbances in the retention of calcium. In our study, the effect of magnesium supplementation was investigated in animals fed a well-balanced diet with adequate content of magnesium and other nutrients (a standard laboratory chow - LSM granules). In these conditions, at 28 days of supplementation, a statistically significant decrease in the apparent retention rate of calcium was observed in the group receiving the highest dose of magnesium, i.e. 20.0 mg Mg/kg/h. Although at the other doses there were no changes in the calcium retention during the period of magnesium intake, at the end of the experiment, at 4 weeks following discontinuation of the magnesium supplementation, a significant decrease in the apparent retention rate of calcium was observed in all study groups *(table 2)*. That probably reflected a sustained competition for the intestinal absorption between magnesium and calcium. A long-term supply of dietary calcium is known to produce reduced absorption and retention in the rat [11, 12] as a result of formation of insoluble calcium - magnesium - phosphorus complexes in the gut. That effect, however, is related to the diet content of phosphorus. Our study demonstrated that also prolonged intake of magnesium supplements for no apparent medical indications disturbs long-term calcium metabolism.

Our study also revealed that magnesium supplemented in the dosage described considerably modified iron retention. In the initial period of magnesium intake (week 2) a significant decrease in the apparent retention rate of iron was noted only in the group receiving 20.0 mg Mg/kg/24h. On the other hand, in the fourth week of magnesium supplementation the apparent retention rate was found to fall for iron in all study groups, while in group 3 the apparent retention rate of iron became negative. Even after a 4-week interval in magnesium supplementation (week 8), negative apparent retention rates of iron were observed in all study groups. The disturbed iron homeostasis persisted for a long time after magnesium discontinuation, which reflected the body's inability to rapidly adapt to changed magnesium intake. In the experimental conditions described, the excretion of iron with the feces and urine was excessive while the retention of iron was markedly decreased. It is known that the homeostasis of iron is controlled predominantly by intestinal absorption, mainly in the duodenum. Although there are many identified factors which affect the absorption, magnesium has not been clearly confirmed to influence the intestinal absorption of iron. In a reverse experimental system, Kimura et al [13] investigated tissue retention of iron in the conditions of dietary magnesium deficiency. They found that magnesium deficit may lead to inefficient utilization of iron and its storage in the tissues, enhanced by administration of oral iron supplements. Our results confirm the interaction of magnesium and iron. Dietary magnesium supplements also produce disturbances in iron metabolism, which persist even after a 4-week interval in magnesium supplementation.

Our study reveals that magnesium supplemented at the doses of 5.0; 10.0 and 20.0 mg Mg/kg/24h exercises a beneficial effect on zinc retention *(table 1)*. The effect persisted even in the initial period (week 4) after magnesium supplementation was discontinued, which confirms strong synergy between the two ions. The effects of long-term magnesium intake, such as a positively increased zinc balance are observed long after discontinuation of the supplementation (up to 8 weeks in study group 4). Zinc is absorbed in the small intestine, the process being affected also by the presence of Fe, Cu, Mg, Co and P cations, and excreted by the kidneys and with feces [14, 15]. With zinc deficit, iron is known to play a compensatory role. In our study, after 4 weeks of continuous magnesium intake, irrespective of the dose used, unlike calcium and iron ions, zinc excretion with feces was significantly reduced, which may indicate the enhanced absorption of zinc. Many studies have confirmed a competitive action of zinc with respect to iron, copper and cadmium, achieved by binding to similar sites in the gut. Additionally, it is possible that zinc reduces iron binding to ferritin and in this way affects iron metabolism.

A 2-week supplementation with magnesium significantly decreased the apparent retention rate of copper compared to controls at a dose 20.0 mg mg/kg/24h *(table 2)*. However, at 28 days after discontinuation of supplementation in all doses used a significant decrease in the apparent retention rate of copper was observed.

Table 1. Apparent retention rates of magnesium and zinc (%)

Study groups	**Experimental Weeks**			
	Week 2	**Week 4**	**Week 6**	**Week 8**
	Apparent Retention Rates of Magnesium (%)			
Control	45.30 ± 7.88	53.04 ± 15.97	11.53 ± 2.76	52.23 ± 10.07
Study group 1	51.98 ± 11.41	56.87 ± 9.03	13.03 ± 3.69	40.50 ± 10.52
Study group 2	58.92 ± 10.99*	47.68 ± 12.66	33.43 ± 7.18*	37.60 ± 2.36
Study group 3	55.41 ± 8.82*	57.33 ± 8.41	35.57 ± 10.80*	40.04 ± 4.35
Study group 4	60.74 ± 9.66*	60.87 ± 12.33	52.08 ± 10.54*	47.50 ± 5.76
	Apparent Retention Rates of Zinc (%)			
Control	21.63 ± 8.52	22.12 ± 11.77	20.14 ± 5.75	25.13 ± 7.38
Study group 1	21.10 ± 6.13	37.21 ± 9.45*	28.60 ± 8.03	26.93 ± 3.42
Study group 2	35.12 ± 11.06*	18.60 ± 5.97	34.16 ± 10.83*	31.77 ± 10.26
Study group 3	38.80 ± 11.67*	56.12 ± 12.67*	36.74 ± 8.45*	40.87 ± 7.16
Study group 4	30.78 ± 7.78	39.17 ± 9.22*	45.22 ± 11.37*	43.50 ± 2.0*

* A statistically significant difference compared to control ($p \leq 0.05$)

Table 2. Apparent retention rates of calcium, iron and copper (%)

Study groups	**Experimental Weeks**		
	Week 2	**Week 4**	Week 8
	Apparent Retention Rates of Calcium (%)		
Control	74.33 ± 8.94	65.25 ± 10.13	43.37 ± 6.07
Study group 1	79.75 ± 5.22	67.65 ± 6.49	17.60 ± 6.14*
Study group 2	78.73 ± 5.57	60.13 ± 7.43	25.30 ± 4.85*
Study group 3	66.30 ± 13.52	55.62 ± 17.31	20.43 ± 5.71*
Study group 4	67. 61 ± 10.27	37.46 ± 10.01*	22.10 ± 4.32*
	Apparent Retention Rates of Iron (%)		
Control	49.71 ± 3.81	53.56 ± 8.56	43.93 ± 11.83
Study group 1	48.49 ± 7.79	41.95 ± 9.39*	-103.1 ± 17.21*
Study group 2	53.23 ± 11.98	40.60 ± 7.84*	-16. 73 ± 3.95*
Study group 3	46.23 ± 6.52	-22.64 ± 7.61*	-24.83 7.39*
Study group 4	33. 35 ± 9.30*	-37.04 ± 11.26*	-26.70 ± 1.22*
	Apparent Retention Rates of Copper (%)		
Control	36.20 ± 8.02	27.43 ± 5.64	56.43 ± 11.91
Study group 1	42.34 ± 6.29	36.33 ± 8.84	-14.66 ± 3.43*
Study group 2	40.12 ± 9.20	24.07 ± 6.92	7. 53 ± 1.36*
Study group 3	35.11 ± 8.26	30.62 ± 10.13	21.00 ± 3.74*
Study group 4	27. 74 ± 7. 53*	29.54 ± 8.37	-11.93 ± 4.20*

* A statistically significant difference compared to control ($p \leq 0.05$)

Summing up, the findings reveal that multiple-dose magnesium intake in the doses described produced several disturbances in the retention of the investigated elements, especially of iron, both during magnesium supplementation and after its discontinuation.

CONCLUSIONS

Multiple-dose magnesium intake in nearly all doses used lowers the apparent retention rate for iron. The changes were observed over almost entire study period, i.e. both during magnesium supplementation and after its discontinuation. Compared to its effect on iron, magnesium intake in the doses described slightly differently affects the apparent retention of calcium and copper. In all study groups, with the exception of the group with the highest magnesium intake, the reduced retention rates for calcium and copper became obvious only after magnesium supplementation had been discontinued. On the other hand, the apparent retention rate for zinc remained increased compared to controls during the entire period of magnesium intake and in the initial period after the discontinuation. Magnesium retention, on the other hand, was increased exclusively in the initial period of supplementation.

REFERENCES

1. Bednarek W., Lipinski W. Zapotrzebowanie na magnez jeczmienia jarego nawozonego zróznicowanymi dawkami nawozów mineralnych. Biul. Magnezol. 1994; 4, 12-14.
2. Harling B., Van Delft W. Changes in the mineral composition of food as a result of cooking in hard and soft water. Arch. Environ. Health 1998; 36, 33-35.
3. Meyer K. A., Kushi L. H., Jacobs D. R., Slavin J., Sellers T. A., Folsom A. R. Carbohydrates, dietary fiber, and incident type 2 diabetes i older women. Am. J. Clin. Nutr. 2000; 71, 4, 921-930.
4. Creedon A., Flynn A., Cashman K. The effect of moderately and severely restricted dietary magnesium intakes on bone composition and bone metabolism in the rat. Brit. J. Nutr., 1999; 82, 63-71.
5. Hunt J.R., Roughead Z.K. Adaptation of iron absorption in men consuming diets with high or low iron bioavailability. Amer. J. Clin. Nutr., 2000; 71, 94-102.
6. Princi T., Artero M., Malusa N., Uxa N., Livia V., Reina G. Serum and intracellular in intoxicated chronic alcoholic and control subjects. Drug and Alkohol Dependence, 1997; 46, 119-122.
7. Durlach J., Bac P., Durlach V., Rayssiguier Y., Bara M., Guiet-Bara A. Magnesium status and ageging: an update (Review). Magn. Resear. 1998; 11, 25-42.
8. Wojtasik A., Rupner J., Rutkowska U., Stolarczyk A., Iwanow K., Kunachowicz H., Socha J. Spozycie magnezu i stan odzywienia tym pierwiastkiem dzieci chorych na celiaklie - Badania wstepne. Biul. Magnezologiczny, 1999; 4, 2, 442-447.
9. Edited by Griffith and Farris. Lippincott Company, The rats in laboratory i nvestigatoion, 1942; 97.
10. Kuhn J., Jost V., Wieckhorst G. et al. Renal elimination of magnesium as a parameter of bioavilability of oral magnesium therapy. Methods. Find. Exp. Clin. Farmacol. 1992; 14, 269-272.
11. Miura T., Matsuzaki H., Suzuki K., Goto S. Long - term high intake of calcium reduces magnesium utiliazation in rats. Nutr. Research, 1999, Vol. 19, No. P, 1363-1369.
12. Smith L. G., Burns P. A., Schanier R. J. (1992). Calcium homeostasis in pregnant women receiving long - term magnesium sulphate therapy for preterm labor. Am. Jour. of Obster. And Gynecol., 1, 167, (45-51).
13. Kimura M., Yokoi K. Iron accumulation in tissues of magnesium - deficient rats with dietary iron overload. Biol. Trace Elem. Resear. 1996, 51, 177-197.
14. Solomons N. W., Ruz M. Zinc and iron interaction: concepts and perspectives in the developing world. Nutr. Research, 1997, Vol. 17, 1 177-185.
15. Salgueiro M., Zubillaga M., Lysionek A., Sarabia M., Caro R., DE Paoli T., Hager A., Weili R., Boccio J. Zinc as an essential micronutrient: A review. Nutr. Research, 2000, Vol. 20, 5, 737-755.

Metal Ions in Biology and Medicine: vol. 7. Eds L. Khassanova, Ph. Collery, I. Maymard, Z. Khassanova, J.-C. Étienne.
John Libbey Eurotext, Paris © 2002 pp. 212-216.

Helicobacter pylori related to whole blood and gastric tissues bismuth levels in patients with different types of gastritis

M. Burguera*[1], J. L. Burguera[1], M.Y. García[1], M. L. Di Bernardo[1], O. M. Alarcón[1], P. Carrero[1], C. Rondón[1], L. M. Villasmil[2], M. A. Villasmil[2]

[1]*Faculty of Sciences (Venezuelan, Andean Institute for Chemical Research),* [2]*Faculty of Medicine, Gastroenterology Unit of University Hospital, Los Andes University, P.O. Box 542, Mérida 5101-A, Venezuela*

ABSTRACT

This study involved 67 subjects (38 female and 29 male) with ages between 14 and 80 years (43.51±17.04) suffering different types of gastritis (mild (MG): n = 22; erosive (EG): n = 33; and chronic (CG): n = 12). Positive HP appeared in 44 of the subjects; 10 with MG, 27 with EG and 7 with CG. Bi global levels for the remaining subjects were 10.25±6.15 $\mu g\ l^{-1}$ and 0.34±0.18 $\mu g\ g^{-1}$ in blood and gastric tissue, respectively. Bi concentrations in both samples were significantly lower in CG when compared to the other types of gastritis ($p = 0.002$). In all cases there was a slight decrease of the Bi levels in gastric tissue with age ($r = -0.20$). There was no relationship between the Bi concentration in either matrix nor between the prevalence of HP and the sex. However, Bi concentration in both matrixes seem to be related to the severity of the inflammatory process, being lower where the tissue is damaged most (CG).

INTRODUCTION

Helicobacter Pylori (HP) infection has been postulated to participate in ethiopathogenesis of different types of gastritis which can progress to peptic ulcer disease and gastric cancer. The sequence of events by which the bacterium might cause gastric damage is still not wholly elucidated but its eradication seems to have a beneficial impact on human health [1]. Long-term bismuth (Bi) salts oral administration was very popular in the treatment of gastritis, given their antimicrobial effect in humans. However, recurrence of HP, clinical relapse and many toxic effects have been associated with the administration of solitary Bi therapies due to the fact that they act suppressing but not eliminating the organism [2, 3]. These drawbacks conduced to the substitution of Bi compounds by strong antibiotics during many years. The financial burden and the emergence of drug resistant strains needed further reconsideration. In recent studies Bi compounds have been used with conventional antibiotics, producing elimination of the organism, histological improvement and amelioration of symptoms for periods longer than one year [4, 5]. There are many papers in the specialized literature dealing with the administration of three or four drugs therapies but there is a lack of data on the determination of bismuth in biological samples from patients with not reported Bi-based treatments. This might be due to the analytical difficulties in determining minutes amounts of Bi in such complex matrixes.

Our research team recently developed a simple and sensitive method for the determination of Bi in biological samples [6] which was used in this work to evaluate the basic concentrations of Bi in whole blood and gastric tissues from patients suffering different types of gastritis without any known Bi-related treatment and to correlate it to the presence of HP.

MATERIALS AND METHODS

Urease test and atomic absorption spectrometry with electrothermal atomization (ETAAS) was used to determine HP and Bi, respectively in two small samples of the stomach lining (biopsies) obtained during endoscopy. Bi was also determined in whole blood sampled from the same patients suffering different types of gastritis. All diagnosis of gastritis were confirmed endoscopically. The statistical evaluation of the results was carried out with SAS (Statistical Analysis System) software version 6.12 under UNIX using ANOVA with paired Student's t-test for the detection of contrasts. The statistical significance was accepted for 95% confidence level ($p = 0.05$). Bi concentrations in the blood and tissue samples are reported as mean ± standard deviation (SD) expressed in $\mu g\ l^{-1}$ and $\mu g\ g^{-1}$, respectively. Presence and absence of HP is indicated respectively by (+) and (-) signs.

SAMPLING AND SAMPLES PROCESSING

During the first visit to the gastroenterological unit, the patient answered a clinical interview and was submitted to an upper gastrointestinal endoscopy. Two antral specimens of gastric tissue were taken, one for a rapid urease test (Ballard Medical Products) and the other for the analytical determination of Bi [6]. Also 5 ml of whole blood was sampled with stainless steel needles from the forearm vein in heparinized, metal-free Vacutiner® tubes. Several blood pools were prepared from samples submitted to our laboratory for the analysis of heavy metals and other trace elements, selected at random without knowledge of origin and pathology except that the individuals were not taking any bismuth-containing drugs and were used for optimization and comparison purposes.

The whole blood samples were diluted 1: 4 with 0.1% (v/v) Triton X-100 + 0.01 mol l^{-1} Suprapur® nitric acid just before analysis. Stomach biopsies were first weighed then dried for 24 hours by lyophylization and then wet digested using 100 µl of concentrated Suprapur® nitric acid. Details of the analytical methodology were given elsewhere [6].

RESULTS AND DISCUSSION

The results of this study are shown in *table 1*. The experiment involved 69 subjects (40 female and 29 male) with ages between 14 and 80 years suffering different types of gastritis (mild (MG): $n = 22$; erosive (EG): $n = 33$; and chronic (CG): $n = 14$). Positive HP appeared in 46 of the subjects; 10 with MG, 27 with EG and 9 with CG. Two of the patients with CG, both female (46 and 53 years of age) with positive HP had elevated levels of Bi in blood (83.9 and 62.85 $\mu g\ l^{-1}$) and in tissue (0.90 and 0.68 $\mu g\ g^{-1}$) when comparing with the rest of the subjects in the group. An examination of their clinical files revealed recent treatment with bismuth subcitrate, therefore these values were excluded from further calculations.

Although the results obtained are highly dispersed (coefficient of variation values above 50%), there was a slight, but not significant tendency towards decreasing blood and tissue Bi levels with age *(fig. 1, A and B)*.

The gastrointestinal damage grows in the sense CG > EG > MG and depends antagonically on age as shown in *table 1* and *Fig. 2*.

Bi concentrations in both samples were significantly lower in CG when compared for instance to MG ($p = 0.004$). Additionally, Bi concentrations in blood from subjects with HP (+) are lower than for those with HP (-) while the behavior for tissue is opposite *(Fig. 3)*. The fact that Bi accumulates in the gastric tissue of the patients with milder impairment (MG and EG) or that there is a shift from blood to tissue, suggests that this element might have a protecting role against gastrointestinal mucosal injury. Regardless the mechanism involved, Bi blood and tissue concentrations have not been mutually related for neither pathology. With respect to the sex, it is interesting to note that Bi levels were higher in female in all cases.

Table 1. Summary of Bi concentrations* in whole blood and gastric tissue related to type of gastritis, age and prevalence of HP

Gastritis Type	[Bi] in Blood ($\mu g\ l^{-1}$)	[Bi] in tissue ($\mu g\ g^{-1}$)	Age*, years	HP
All groups	10.25 ± 6.15	0.34 ± 0.18	43.51 ± 17.04	(+), (-)
	9.57 ± 5.89	0.35 ± 0.19	46.07 ± 14.15	(+)
	11.57 ± 6.55	0.30 ± 0.14	38.61 ± 11.01	(-)
MG n = 22	11.00 ± 7.09	0.31 ± 0.16	34.68 ± 14.69	(+), (-)
	13.36 ± 8.34	0.31 ± 0.16	41.80 ± 14.24	(+)
	9.04 ± 5.46	0.31 ± 0.16	28.75 ± 12.72	(-)
EG n = 33	9.99 ± 5.91	0.38 ± (0.17	45.94 ± 16.42	(+), (-)
	8.64 ± 4.89	0.40 ± 0.18	45.21 ± 14.66	(+)
	16.03 ± 6.79	0.27 ± 0.11	46.00 ± 16.90	(-)
CG n = 12	9.62 ± 5.24	0.26 ± 0.19	53.00 ± 16.53	(+), (-)
	7.70 ± 3.00	0.22 ± 0.22	52.63 ± 10.70	(+)
	12.30 ± 6.82	0.31 ± 0.14	55.25 ± 17.86	(-)

* Mean (Standard Deviation).

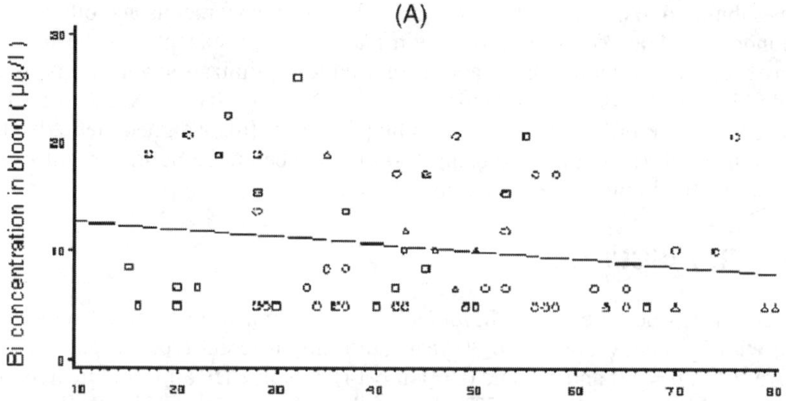

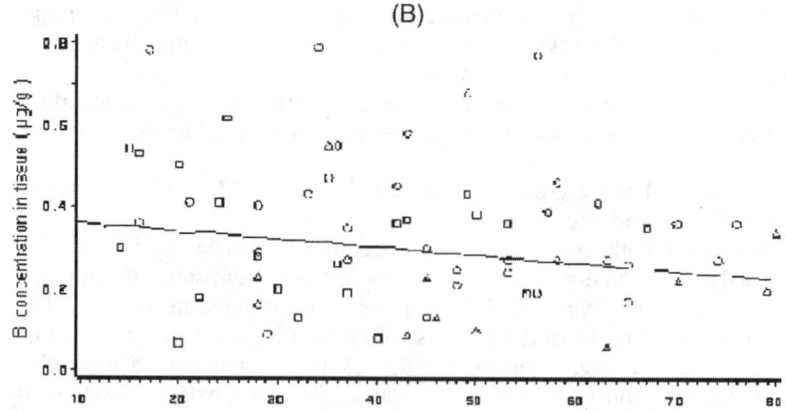

Figure 1. Age distribution of Bi in blood (A) and gastric tissue (B).

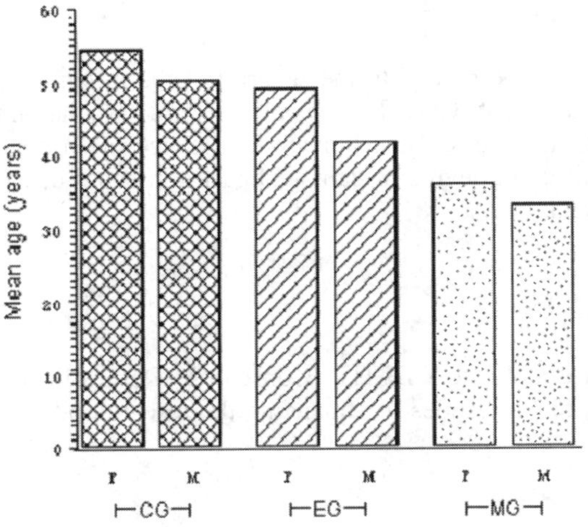

Figure 2. Age-pathology-sex relationship

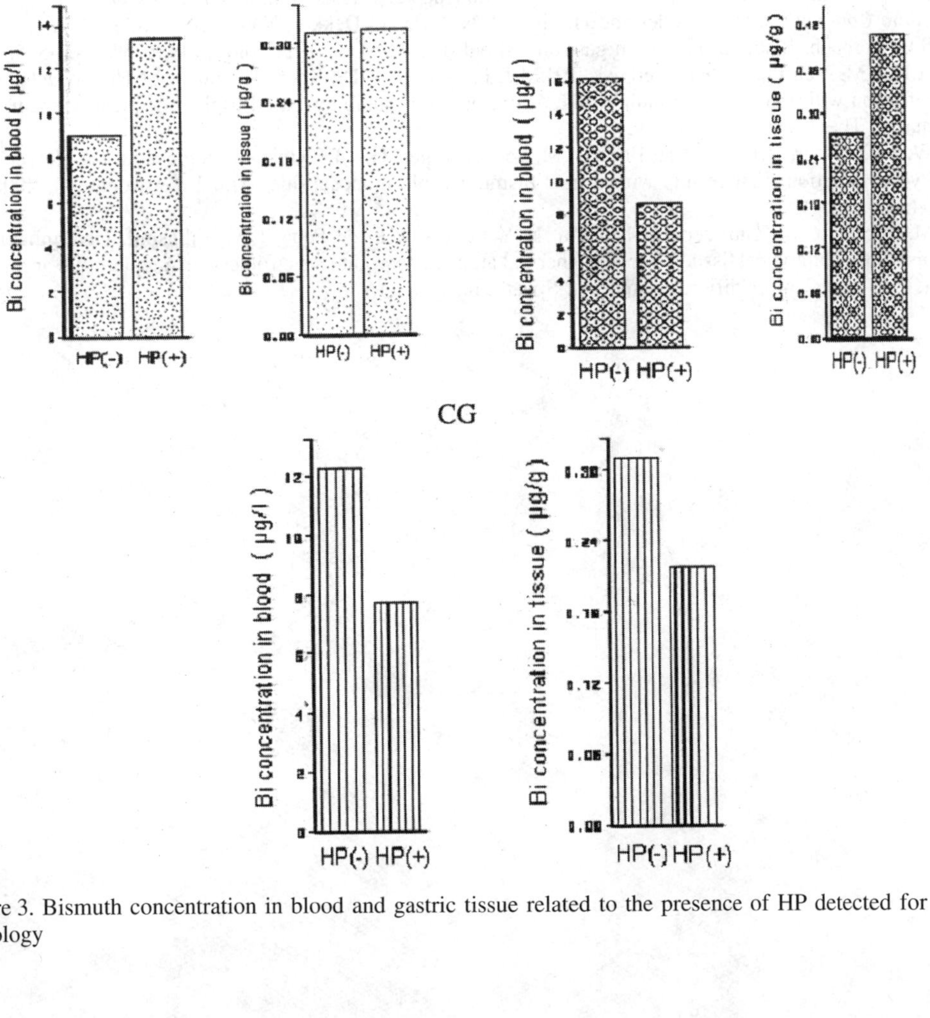

Figure 3. Bismuth concentration in blood and gastric tissue related to the presence of HP detected for each pathology

CONCLUSIONS

Bi concentration seems to be related to the severity of the inflammation process, being lower where the tissue is damaged most (CG). There is also certain association between the presence of HP and the evolution of the illness. The clinical significance of Bi concentrations in blood and gastric tissue related to different degree of damaged gastric mucosa deserves further investigation in a more numerous population.

ACKNOWLEDGMENTS

The research team is grateful to CDCHT (Consejo de Desarrollo Científico, Humanístico y Tecnológico), University of Los Andes for financing the research project C- 877-98-09-A.

REFERENCES

1. F. Pakodi, O.M. Abdel-Salam, A. Debreceni, G. Mozsik, Helicobacter Pylori. One bacterium and a broad spectrum of human disease. An overview. J. Physiol. Paris 2000, 94, 139-52.
2. D.W. Thomas, T.F. Hartley, P. Coyle, S. Sobecki, Chapter 11, Bismuth, in *Handbook on Toxicity of Inorganic Compounds*, H. G. Seiler and H. Sigel (Eds.), Marcel Dekker, New York, USA, 1987.
3. S.L. Gorbach, Bismuth therapy in gastrointestinal diseases. Gastroenterology, 1990, 99, 363-375.
4. A.H.A.M. Van Oijen, A.L. Verbeek, J.B.M.J. Jansen, W.A. De Boer, Treatment of Helicobacter Pylori infection with ranitidine bismuth citrate or proton pump inhibitor-based triple therapies. Aliment. Pharmacol. Ther. 2000, 14, 991-999.
5. W.A. De Boer, R.J.X.M. Van Etten, P.M. Schneeberger, G.N.J. Tytgat, A single drug for Helycobacter Pylori infection: first results with a new bismuth triple monocapsule. Am. J. Gastroenterol. 2000, 95, 641-645.
6. M. Burguera, J.L. Burguera, C. Rondón, M.Y. García, Y.P. de Peña, L.M. Villasmil, Determination of bismuth in biological tissues by electrothermal atomic absorption spectrometry using platinum and tartaric acid as chemical modifier. J. Anal. At. Spectrom., 2001, 16, 1-7.

*Metal Ions in Biology and Medicine: vol. 7.*Eds L. Khassanova, Ph. Collery, I. Maymard, Z. Khassanova, J.-C. Étienne.
John Libbey Eurotext, Paris © 2002 pp. 217-221.

Regional distribution of aluminium in the rat brain: influence of vitamin E

M. G. Abubakar, A. Taylor and G. A. Ferns

Centre for Clinical Sciences & Measurement, School of Biomedical & Life Sciences, University of Surrey, Guildford Surrey GU2 7XH UK, and Royal Surrey County Hospital, Egerton Rd, Guildford, Surrey, GU2 7XH, UK

ABSTRACT

We have investigated the effects of the antioxidant vitamin E on the regional accumulation of aluminium (Al) in the rat brain. Al was administered i.p. as aluminium lactate at a dose of 10 mg Al/Kg body weight, 5 times a week for 4 weeks. Male Wistar albino rats were either treated with aluminium alone, or together with a dietary supplement of 5, 15, and 20 mg/g, of vitamin E (n=5 per group). Further groups received either 20 mg/g vitamin E supplemented diet or normal chow alone (n=5 per group). Animals were killed by anaesthetic overdose, the brain removed and dissected into identifiable regional segments, namely cortex, cerebellum, medulla, hippocampus, striatum and hypothalamus, for measurement of Al by atomic absorption spectrometry.

Al content was increased in all the regions of the brain examined, in the Al-treated rats compared to the controls ($p<0.01$). The regional accumulation of Al, in decreasing order of magnitude was; cerebellum > cortex > hippocampus > medulla > striatum. Concomitant dietary vitamin E significantly reduced ($P<0.01$) the Al accumulation in the various regions of the brain. These results suggest that dietary vitamin E may influence the regional accumulation of Al in the brain. Hence if Al-induced neurotoxicity is mediated by pro-oxidant events, vitamin E may offer some protection.

Keywords: aluminium, vitamin E, rat brain, regional, distribution.

INTRODUCTION

It has been hypothesised that aluminium neurotoxicity is the cause of dialysis dementia (Salusky *et al.*, 1991). The presence of aluminium in dialysis fluids may contribute to this neuropathology. Brain tissue is susceptible to peroxidative damage because of its high lipid content, high oxygen tension, low cellular mitotic rate, low antioxidant capacity and high iron content (Evans, 1993). In this communication we report the influence of vitamin E, an antioxidant, on aluminium distribution in the various regions of rat brain.

METHODS

Experimental animal model

Male Wistar albino rats weighing 170-180g were randomized to one of the following experimental groups, with 5 animals per group:

(a) Rats were injected with an aqueous solution of aluminium lactate (Johnson Matthey GmbH, Zeppelinstraβe 7, Karlsruhe), at a dose of 10 mg aluminium per Kg body weight per day given as an i.p. injection five times per week for 4 weeks, and received a standard chow diet during this time.

(b) A control was injected with similar volume of normal saline over the same period and received a standard chow diet *ad libitum*.

(c) Rats were injected with aluminium at the same dose as for (a) but received a chow supplemented with 5, 15, and 20 mg/g, of (-tocopherol (Sigma Chemical Company Poole, Dorset).

(d) A further group received 20 mg/g vitamin E supplemented diet without receiving i.p. aluminium.

All animals were individually weighed and examined daily, with food and water intakes recorded. They were all housed in polypropylene cages to avoid extraneous trace element contamination. Animals were killed by anesthetic overdose using sodium pentabarbitone (100 mg/Kg). Blood was withdrawn via the dorsal vena cava into lithium heparin tubes. The brain was quickly excised, rinsed with cold 0.14 M NaCl and dissected into identifiable regional segments, namely cortex, cerebellum, medulla, hippocampus, striatum and hypothalamus for measurement of aluminium. Brain tissue was removed, weighed and stored in aluminium-free plastic containers at -80°C, until analysis.

ALUMINIUM ANALYSIS

The samples were digested with 90% nitric acid and 70% perchloric acid for measurement of aluminium as previously described (Taylor and Walker, 1992). Aluminium in plasma or in digested brain samples was determined by electro-thermal atomic absorption spectrometry (ETAAS).

RESULTS

Figure 1 shows the tissue levels of Al in different regions of the brain of rats from the control and Al-treated groups receiving normal chow. There was a statistically significant (P<0.01) accumulation of Al in the various regions of the brain of Al-treated rats compared to the control group. Al-accumulation was found to be greatest in the cerebellum followed by the cortex, hippocampus, medulla, and striatum.

Co-administration of dietary vitamin E was associated with a significant (P<0.01) reduction in Al concentration in all regions of the brain experimental *(fig. 2-7)*.

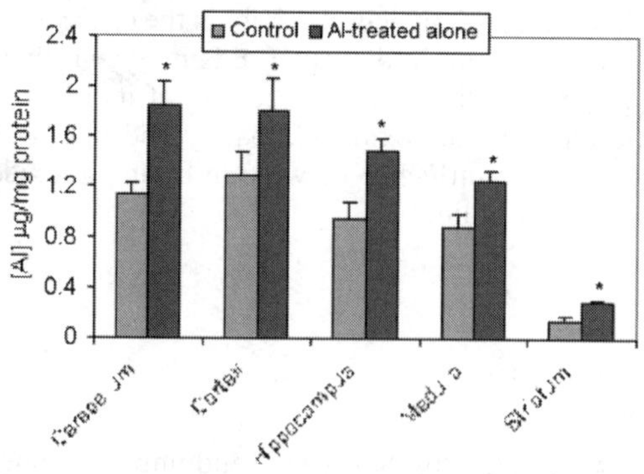

Figure 1. Aluminium distribution in various regions of the brain
* P < 0.01 compared to the control.

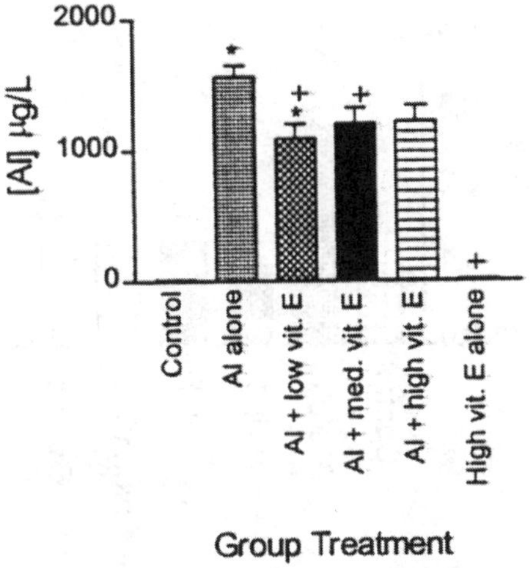

Figure 2. Plasma aluminium level in different group treatment. Values are means ± SEM for n = 5.
* $P < 0.01$ compared to the control.
+ $P < 0.001$ compared to Al alone.

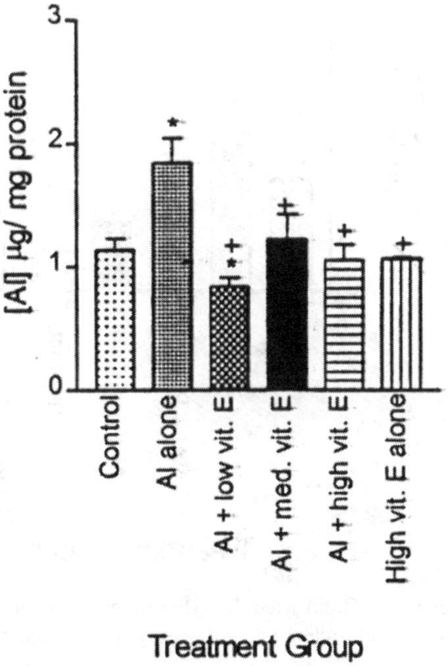

Figure 3. Brain Cerebellum Aluminium level in different group treatment. Values are means ± SEM for n = 4.
* $P < 0.01$ compared to the control.
+ $P < 0.001$ compared to Al alone.

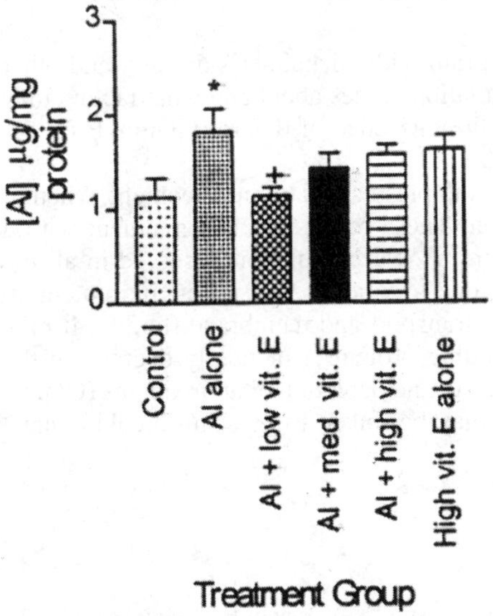

Figure 4. Brain Cortex Aluminium level in different group treatment. Values are means ± SEM for n = 4.
* $P < 0.01$ compared to the control.
+ $P < 0.001$ compared to Al alone.

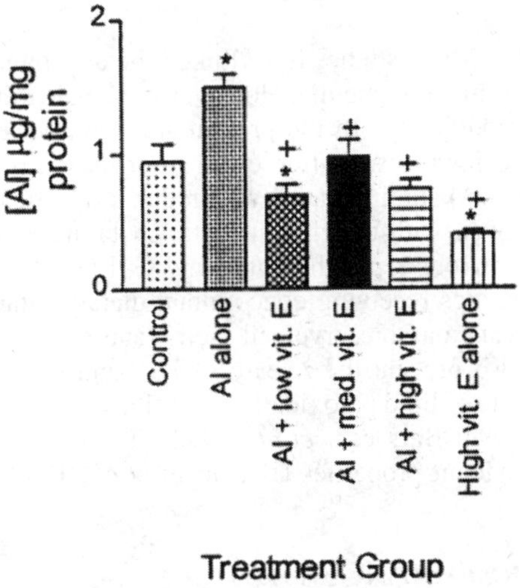

Figure 5. Brain Hipocampus Aluminium level in different group treatment. Values are means ± SEM for n = 4.
* $P < 0.01$ compared to the control.
+ $P < 0.001$ compared to Al alone.

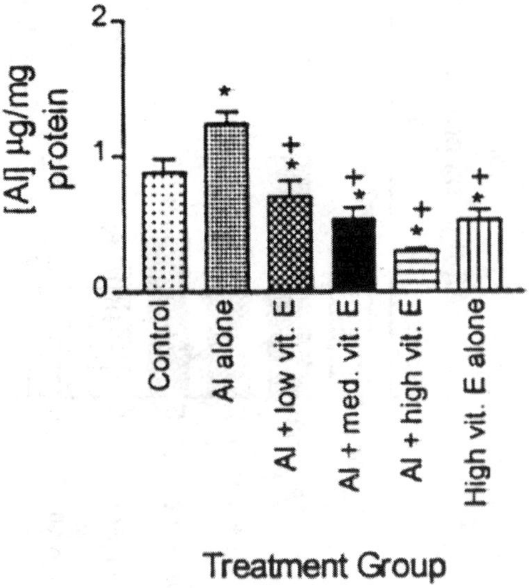

Figure 6. Brain Medulla Aluminium level in different group treatment. Values are means ± SEM for n = 4.
* P < 0.01 compared to the control.
\+ P < 0.001 compared to Al alone.

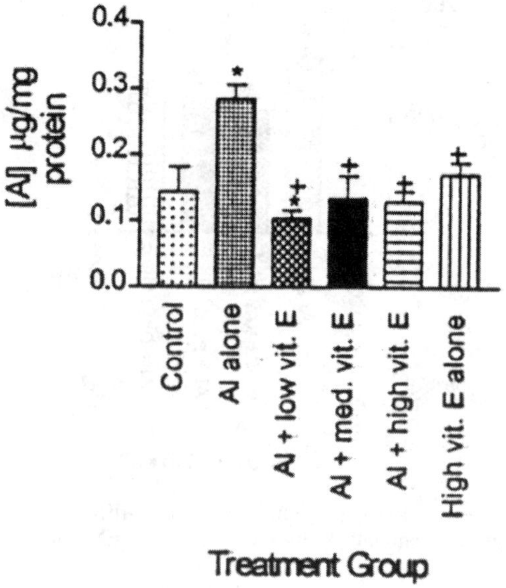

Figure 7. Brain Striatum Aluminium level in different group treatment. Values are means ± SEM for n = 4.
* P < 0.01 compared to the control.
\+ P < 0.001 compared to Al alone.

DISCUSSION

Many studies have linked the accumulation of Al in brain with Alzheimer's disease, and other forms of dementia. However it is unclear how Al accumulation comes about and what factors may modify this. In the present study we have investigated the influence of dietary vitamin E on the regional distribution of Al in of the rat brain.

Al administration resulted in a significant increase in both plasma and brain Al content, whether given with or without dietary vitamin E supplementation. These results are in agreement with a previous report by Julka and Gill (1996). The attenuation in Al accumulation observed in all the groups receiving concomitant dietary vitamin E may be due to vitamin E's scavenging potential. This may preserve cell membrane function including ion transport and membrane fluidity. It may also prevent the release of Fe^{2+} and Mg^{2+} from their binding proteins, potentially decreasing the rate of lipid peroxidation. Fe^{2+} ions causes membrane fragility and depletion of antioxidants (Oteiza, 1994; Barsacchi *et al.*, 1992). The effects of vitamin E may be linked to its chain breaking antioxidant properties (Barsacchi *et al.*, 1992).

REFERENCES

1. Barsacchi R., Pelosi G., Maffei S., Baroni M., Salvatore L., Ursini F., Verunelli F., Biagini A. (1992). Myocardial vitamin E is consumed during cardiopulmonary bypass: indirect evidence of free radical generation in human ischemic heart. *Int J Cardiol* 37(3): 339-343.
2. Evans P. (1993). Free radicals in brain metabolism and pathology. *Brt. Med. Bull.* 49: 577-587.
3. Julka D. and Gill K. D. (1996). Altered calcium homeostasis: a possible mechanism of aluminium-induced neurotoxicity. *Biochem. Biophys. Acta* 1315: 47-54.

4. Oteiza P.I. (1994). A mechanism for the stimulatory effect of aluminum on iron-induced lipid peroxidation. *Archives of Biochemistry and Biophysics* 308: (2) 374-379.
5. Salusky I.B., Foley J., Nelson P., Goodman W.G. (1991) Aluminum accumulation during treatment with aluminum hydroxide and dialysis in children and young adults with chronic renal disease. *N Engl J Med* 324(8): 527-531.
6. Taylor A., and Walker A.W. (1992). Measurement of aluminium in clinical samples. *Ann Clin Biochem* 29 (Pt 4): 377-389.

V METAL IONS AND MICROORGANISMS

Metal Ions in Biology and Medicine: vol. 7. Eds L. Khassanova, Ph. Collery, I. Maymard, Z. Khassanova, J.-C. Étienne.
John Libbey Eurotext, Paris © 2002 pp. 225-230.

Stimulation of uptake of some metals and radio-nuclides with help of phosphate mobilizing bacteria

Irina Shtangeeva[1], Antti Vuorinen[2], Sophie Ayrault[3], Tatiana Lissitskaia[4]

[1]St. Petersburg University, Universitetskaya nab., 7/9, St. Petersburg 199034 Russia; [2]Department of Geology, University of Helsinki, P.O.Box 11, FIN-00014, Finland; [3]Laboratoire Pierre Sue CEA-CNRS 91191Gif sur Yvette Cedex France; [4]St. Petersburg Institute of Technology, Moskovsky pr., 26, St. Petersburg Russia

ABSTRACT

The potential of phosphate mobilizing bacteria *Cellulomonas sp.*32 and *Mycobacterium sp.*12 for stimulation of uptake of metal ions was studied under greenhouse conditions. Short-term pot experiments were conducted to assess the direct and indirect effects of the bacteria on biogeochemical situation in rhizosphere and element concentrations in roots and leaves of wheat *Tricum vulgare (vill) Horst*, oats *Avena Sativa* L. and barley *Hordeum vulgare* L. Microbiological treatment of the plants may be potentially used to enhance uptake of different metals and radio-nuclides. Inoculation of the plants with *Mycobacterium* and *Cellulomonas* had also a positive effect on uptake of potassium by the wheat and oats, while K concentration in the barley was not affected after the microbiological treatment. More significant variations in elemental composition of the experimental plants were observed after inoculation of wheat seeds with *Mycobacterium*. It seems that *Mycobacterium* was more proficient in solubilization and absorbing of the metals than *Cellulomonas*. The microbiological treatment of seeds also caused variations in relationships between elements in the experimental plants. Both quantity and quality of the relationships were affected.

INTRODUCTION

Numerous studies describing uptake of metals by plants often completely ignore the impact of soil microflora on the process of the metal uptake. Meanwhile, soil micro-organisms may be effective mediators of mobilization and immobilization of the metals in rhizosphere. The role of bacteria in rhizosphere chemistry is particularly interesting. The root-colonizing bacteria excrete own exudates into surrounding soil. The organic compounds can catalize specific oxidation-reduction reactions that alter mobility of the metal ions [Wielinga et al., 1999]. The micro-organisms are able to facilitate movement of the metals that would otherwise be unavailable for the plants [Domelly and Fletcher, 1994]. In some cases the bacteria are perhaps more proficient than roots at solubilization and absorbing of certain metals [Jackson, 1993]. This is particularly significant since bioavailability and toxicity of the metals depend on their solubility. Although now a lot of data is available both on ion uptake by different species of plants and attraction of the metal ions to bacterial cell walls, the system "soil - bacteria - plant" as a network is not fully understood. Besides, the rhizosphere colonization is specific to certain plant species. The relationships between the host plant and its micro-organisms have a marked role in the uptake of different elements. Therefore, variations in diversity of rhizospheric micro-organisms may have a significant impact on the process of the metal uptake. It seems that inoculation of seeds with micro-organisms may be more effective for enhancement of metal bioavailability than microbiological treatment of soil.

Seed germination is an important stage of the plant growth. The conditions of the germination may have a crucial significance for further life of the plant. The main goal of the research was to study an effect of treatment of plant seeds with new strains of bacteria on biogeochemical processes in the rhizosphere soil and uptake of metal ions by the plants. To assess the potential of the bacteria for stimulation of metal uptake, naturally occurring bacteria have been isolated and two strains selected for their ability to solubilize metals from different substrates.

MATERIALS AND METHODS

Seeds of wheat *Tricum vulgare (vill) Horst*, oats *Avena Sativa* L. and barley *Hordeum vulgare* L. were used as an experimental material. One part of the seeds (control) was germinated for six days on a moist filter paper at room temperature. The other part of the seeds was germinated under the same conditions, but additionally the seeds were treated either by a culture of phosphate mobilizing bacteria *Cellulomonas sp.*32 SPBTI or by *Mycobacterium sp.*12 (concentration of the bacteria in the water solution was 10^5-10^6 cell/ml). Uniform germinated seedlings were transferred to pots filled with soil taken from two sites in St. Petersburg - a contaminated area in centre of the city and a clean zone in a park, 25 km from St. Petersburg. We conducted three experiments. Experiment 1. Seeds of barley and oats were treated by *Cellulomonas* and seedlings of the plants were grown in the contaminated and clean soils. Plants and soil (from surface of plant roots) were sampled three times - within 5, 10 and 19 days after beginning of the experiment. Experiment 2. Seeds of wheat were treated by *Mycobacterium.* Wheat seedlings have been grown in both soils and harvested four times - within 9, 13, 20 and 27 days after sowing. Experiment 3. Seeds of wheat were treated by *Mycobacterium.* Germinated seedlings were grown for seven days in pots filled with clean soil. One part of the pots was amended with thorium nitrate and the other part was kept as a control. The plants and soil were sampled three times - within 2, 4 and 7 days after sowing.

To remove dust and particles of soil from surface of the experimental plants, all the plants were rinsed carefully by water and then air-dried. Soil samples were also air-dried up to constant weight. Concentrations of 26 elements (Na, K, Ca, Sc, Cr, Fe, Co, Zn, As, Br, Rb, Ag, Sb, Cs, Ba, La, Sm, Eu, Tb, Yb, Lu, Hf, Ta, Au, Th and U) in soil, roots and leaves of the experimental plants were determined by instrumental neutron activation analysis. The finest (<0.2mm) soil fraction was used to conduct a series of sequential leaching procedures [Shtangeeva et al., 2001] to determine mobile fraction of elements. Concentrations of Al, Cr, Mn, Fe, Co, Ni, Cu, Zn and Pb in soil leachates were determined by ICP-AES. ICP-AES was also used to study content of exchangeable cations (Ca, K and Mg) in 0.01M CsCl soil leachates. A statistical treatment of the experimental data included calculation of mean concentrations of the elements, correlation analysis and analysis of variances to estimate statistically significant differences between groups of the samples (Statistica for Windows 5.5 Software package).

RESULTS AND DISCUSSION

Inoculation of the seeds with *Cellulomonas* and *Mycobacterium* resulted in certain changes in concentrations of different elements in the experimental plants and soil solutions. More significant variations were observed after treatment of wheat seeds by *Mycobacterium.* In this case concentrations of Rb, Cs and La in roots of treated with the bacterium plants decreased compared to those in non-treated plants and concentrations of Na and Ca in the leaves of treated with *Mycobacterium* wheat became less than those in non-treated plants. The differences were statistically significant ($P<0.01$). Concentration of Zn in leaves of barley treated by *Cellulomonas* was less ($P<0.01$) than that in non-treated plants. Concentrations of Th in roots and K in leaves of oats treated by *Cellu-*

lomonas were higher ($P<0.01$) than those in the non-treated plants. As is seen from *fig. 1*, in this case concentration of exchangeable K^+ in the soil leachates decreased ($P<0.001$). Thus, *Cellulomonas* was favourable for supply of the plants with potassium. Similar variations in K content were observed after treatment of wheat seeds with *Mycobacterium*. However, treatment of barley seeds did not change significantly concentrations of K neither in soil leachates nor in different parts of the barley.

The ability of micro-organisms to accumulate large amount of potassium is well-known [Leonard, 1983]. Growth rate of bacteria is also dependent upon K^+ [Hughes, 1974]. As a consequence of metabolic activity of the plants and micro-organisms, concentration of K^+ in the soil solution near root surface can decrease, thereby inducing desorption of K^+ held on the external surface of soil particles. Therefore, increase of K content in the plants and concurrent decrease of its concentration in the soil is quite appreciable. Concentrations of other exchangeable cations (Ca^{2+} and Mg^{2+}) in soil leachates also decreased after cultivation of the experimental plants. However, it seems that such a decrease was mainly caused by metabolic activity of plants themselves. It was reported that under some conditions ion uptake by plant roots can be stimulated by bacteria, possibly by providing chelating agents or plant regulators to promote active ion transport, but under other conditions they can be inhibitory, either by competing for the nutrients or producing phytotoxic compounds [Lynch, 1982].

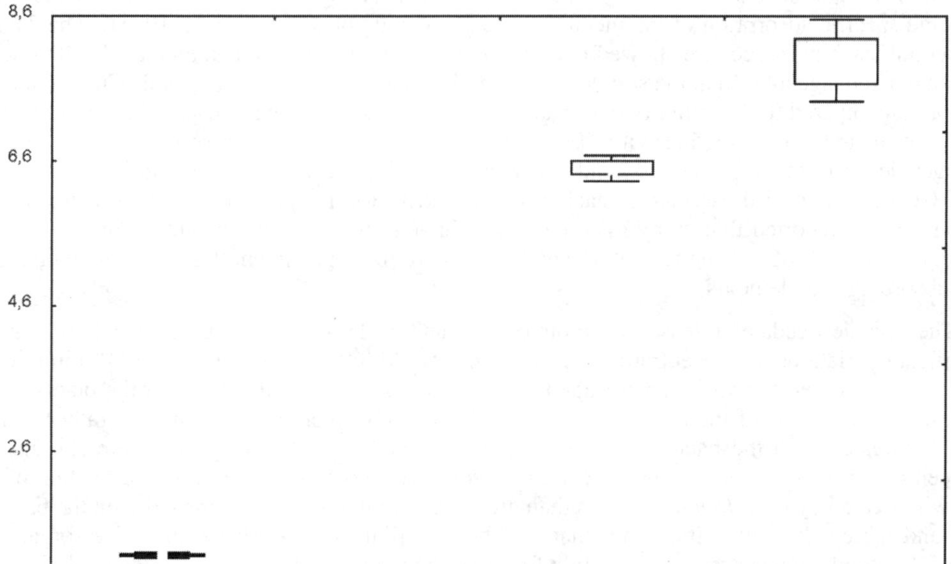

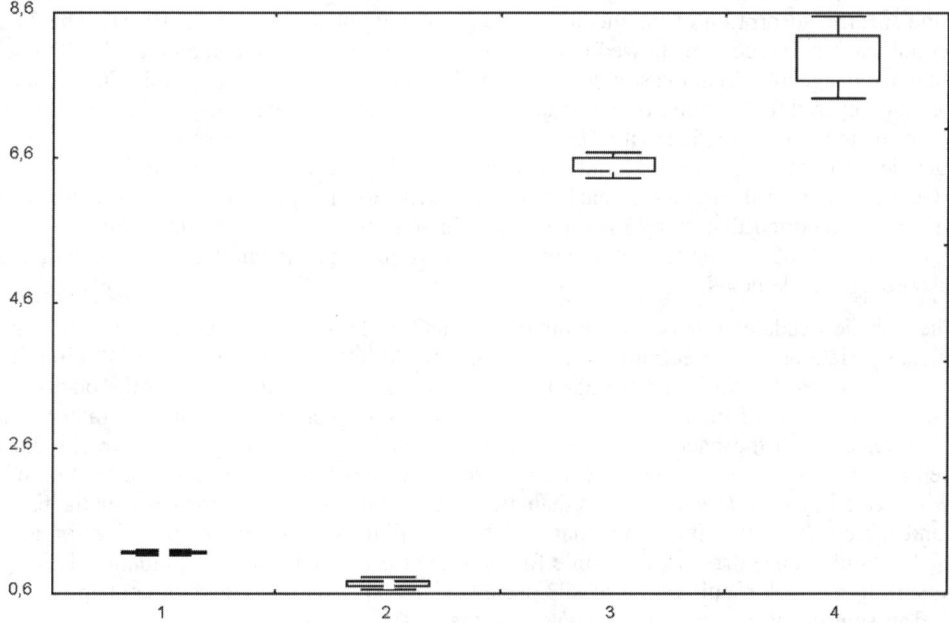

Fig. 1. Concentration of exchangeable K^+ in soil leachates after cultivation of contaminated soil with non-treated (1) and treated with *Cellulomonas* (2) oats seedlings and K content in leaves of non-treated (3) and treated with *Cellulomonas* (4) oats

Comparison of elemental composition of plants grown in contaminated and clean soils showed that concentrations of many elements (Cr, Fe, As, Sb and also Na and Br) in roots of plants grown in the contaminated soil increased significantly compared to those in roots of the plants grown in the clean soil. It is interesting that concentrations of so-called heavy metals in leaves of the plants remained at approximately the same level as in the leaves of the plants grown in clean soil. However, concentrations of Na and Br in the leaves were higher and K content was less ($P<0.005$) than those in leaves of the plants grown in clean soil. Therefore, treatment of the plants by bacteria resulting in the stimulation of K uptake by leaves is favourable for survival of the plants in negative environmental conditions.

Microbial treatment of seeds had also an effect on the relationships between elements in the roots and leaves of the experimental plants. Both quantity and quality of the relationships were affected. For example, total number of statistically significant correlations between different elements in leaves of treated with *Cellulomonas* plants was 35, and in leaves of non-treated plants the number of such correlations was 23. In this case most part of the correlations was different in groups of treated with the bacteria and non-treated plants, and only correlations between Cr and Eu, Sb and Br, As and Eu were statistically significant both in leaves of treated and non-treated plants. In roots total number of statistically significant correlations was nearly the same both in ordinary plants and in plants treated before sowing with *Cellulomonas* - 33 and 34, respectively. Again the biggest part of the relationships was unique to one or other group of the plants. Only statistically significant correlations between Fe and Sc, Ca and Co, Ca and Br, Co and Br were an inherent characteristic of both groups: roots of non-treated and treated with the bacteria plants.

The effectiveness of metal ion phytoextraction is highly dependant on availability of the ions for the plant uptake. Therefore, it was necessary to assess mobile fractions of the metals (since anthropogenic metal contamination of soil generally leads to accumulation of free metal ions and soluble metal complexes in the soil solution and gives a rise to exchangeable forms of the metal ions and specific adsorption of the metals on inorganic soil constituents [Salt, 1995]). The data of sequential leaching procedure showed that after cultivation of wheat in contaminated soil concentration of exchangeable Zn in the soil decreased, and concentration of exchangeable Cu, in contrast, increased compared to those in non-cultivated soil. The same effect was observed after cultivation of wheat treated before sowing with *Mycobacterium*. However, in this case concentration of exchangeable Pb in the soil was less, and concentrations of loosely bound Mn and Zn were higher ($P<0.01$) compared to those in soil leachates after cultivation of the soil with non-treated wheat. It was already reported that many microbes can solubilize metals from insoluble compounds and minerals [Gadd, 1986]. It seems that *Mycobacterium* are more proficient than roots at solubilizing and absorbing the elements.

The organic exudates released by roots of oats and barley into the contaminated soil caused significant variations in concentrations of many metals (Al, Cr, Fe, Co, Ni, Cu, Zn, Pb) in the soil leachates. However, the variations resulted mainly from physiological activity of the plants themselves: concentrations of the metals in the soil leachates after plant cultivation were rather similar both for treated with the bacteria and non-treated plants. The only exceptions were Zn and Pb. Concentrations of these exchangeable cations in soil leachates were higher after cultivation of the soil with treated by *Cellulomonas* plants than after growing in the soil non-treated with the bacteria oats and barley. It is also important that metabolic activity of the plants and micro-organisms affected not only concentrations of mobile forms of the metals, but also total amounts of elements in the soil. In particular, cultivation of all species of the experimental plants in contaminated soil resulted in significant decrease of As content in the soil.

Lastly, we conducted a short-term experiment to study the potential of the microbial treatment of plants for phytoremediation of soil contaminated with radio-nuclides. *Fig. 2* illustrates dynamics of thorium concentrations in soil artificially contaminated with Th and in different parts of wheat grown in the soil (after inoculation of the wheat seeds with *Mycobacterium*). It is known that owing to their high complexing ability all radio-nuclides can be bound tightly to solid surfaces [Mortvedt, 1994]. Therefore, one may expect that addition of Th to the experimental soil may result in quite rapid absorption of Th by soil organic matter and Th adsorption on the surface of soil particles. The observed variations in Th content in the soil might be caused solely by physiological activity of the experimental plants and micro-organisms. Although at first glance it seems strange that in roots the initial increase is followed by a decrease in Th concentration, this may point to the fact that part of the Th was transferred to leaves of the experimental plants.

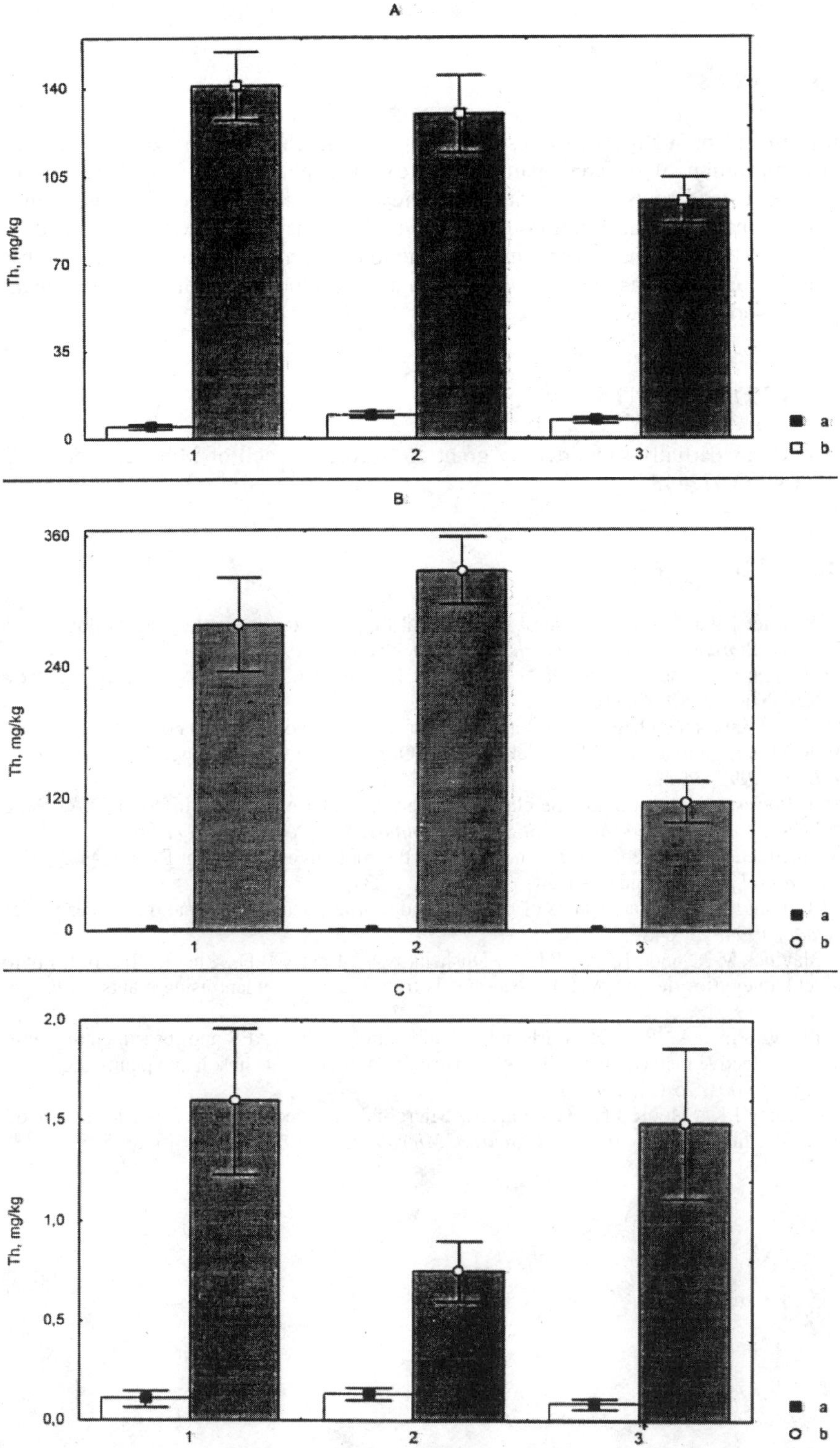

Fig. 2. Dynamics of Th concentration in soil (A), roots (B) and leaves (C) of wheat grown in clean (a) and artificially contaminated with Th (b) soils. 1, 2 and 3 - dates of sampling: two, four and seven days after sowing of the wheat in the soils

CONCLUSIONS

Treatment of plants with bacteria resulted in significant changes in rhizosphere chemistry and elemental composition of different parts of the experimental plants. Concentrations of certain elements in leaves and roots treated with the bacteria changed compared to those in non-treated plants. At the same time concentrations of several toxic elements (for example, As and Pb) in soil decreased. Both strains of the bacteria may be potentially used for removal of metals from contaminated soil, though it seems that *Mycobacterium* are more proficient than *Cellulomonas* at solubilizing and absorbing the elements.

ACKNOWLEDGEMENTS

This work was partially supported by grant of Nordic Council of Ministers and NATO C&L grant N EST.CLG.977448.

REFERENCES

Domelly P.K., Fletcher J.S. Potential use of mycorrhizal fungi as bioremediation agents. In: Anderson T.A., Coats J.R. (eds). *Bioremediation through rhizosphere technology*, 1994, 93-99.

Gadd G.M. Fungal response towards heavy metals. In: Herbert R.A., Codd G.A. *Microbes in extreme environments*, New-York, 1986, 83-110.

Hughes M.N. The inorganic chemistry of biological processes, London - New-York, 1974.

Jackson W.R. Humic, Fulvic, and Microbial Balance: Organic Soil Conditioning, *An Agricultural Text and Reference Book*. 1993, 34-42.

Leonard R.T. Potassium transport and the plasma membrane-ATPase in plants. In: Robb D.A., Pierpont W.S. *Metals and micronutrients: uptake and utilization by plants*, New-York, 1983, 71-86.

Lynch J.M. Interactions between bacteria and plants in the root environment. In: Rhodes M.E., Skinnen F.A. (eds). *Bacteria and plants*, London - New-York, 1982, 1-23.

Mortvedt J.J. Plant and soil relationships of uranium and thorium decay series radio-nuclides - a review. *J.of Environ. Qual.*, 1994, 23, 643-650.

Salt D.E., Blaylock M., Nanda Kumar P.B.A., Dushenkov V., Ensley B.D., Chet I., Raskin I. Phytoremediation: A novel strategy for the removal of toxic metals from the environment using plants. *Biotechnol.*, 1995, 13, 468-474.

Shtangeeva I., Vuorinen A., Rietz B., Carlson L. Combination of ICP-AES and instrumental neutron activation analysis as effective methods for studying of distribution of elements in soil and plants. *J.of Geostandards and Geoanalysis*, 2001, 25 (2), 1-9.

Wielinga B., Lucy J.K., Moore J.N., Gannon J.E. Microbial and geochemical characterization of fluvially deposited sulfidic mine tailing. *Appl. and Environ. Microbiology*, 1999, 65 (4), 1548-1585.

Metal Ions in Biology and Medicine: vol. 7. Eds L. Khassanova, Ph. Collery, I. Maymard, Z. Khassanova, J.-C. Étienne.
John Libbey Eurotext, Paris © 2002 pp. 231-236.

Spectroscopic aspects in studying the impact of heavy metals on rhizobacteria and their role in bacterial metabolism

Alexander A. Kamnev[1*], Lyudmila P. Antonyuk[1], Petros A. Tarantilis[2], Moschos G. Polissiou[2], Leonid A. Kulikov[3] and Yurii D. Perfiliev[3]

[1]*Laboratory of Biochemistry of Plant-Bacterial Symbioses, Institute of Biochemistry and Physiology of Plants and Microorganisms, Russian Academy of Sciences, 13 Prosp. Entuziastov, 410015 Saratov, Russia*
[2]*Laboratory of Chemistry, Department of Science, Agricultural University of Athens, Iera Odos 75, 11855 Athens, Greece*
[3]*Laboratory of Nuclear Chemistry Techniques, Department of Radiochemistry, Faculty of Chemistry, Moscow State University, 119899 Moscow, Russia*
* *Corresponding author. Tel./Fax: +7-(8452)-947303; E-mail: micbio@ibppm.saratov.su.*

ABSTRACT

Modern spectroscopic techniques are highly useful in studying diverse processes in bacterial cells related to their interactions with metal ions. Spectroscopic data for whole cells, supramolecular structures or isolated cellular constituents can reflect structural and compositional changes occurring in the course of metabolic responses induced by metal ions. This information on the molecular level is of importance for basic studies on mechanisms of bacterial tolerance to heavy metals and their impact on bacterial metabolism, as well as for applied multidisciplinary research in the fields related to biotechnology, bioremediation, agriculture, biogeochemistry, *etc*. In the present work, some examples are presented which illustrate the application of different spectroscopic techniques for monitoring metal-induced metabolic changes in soil bacteria, as well as for obtaining structural information for metal-activated bacterial enzyme. Thus, the effects of a range of heavy metals on the plant growth-promoting rhizobacterium *Azospirillum brasilense* are discussed as revealed using vibrational spectroscopy of whole cells and isolated cell walls. In particular, Fourier transform infrared (FTIR) spectroscopy provides information on the overall structure and state of functional groups, their involvement in metal binding and/or hydration. As its counterpart, FT-Raman spectroscopy gives complementary data on the state of less or non-polar functional groups. Emission (^{57}Co) Mössbauer spectroscopy (EMS) can be used to monitor binding of cobalt(II) by bacterial cells at its trace (physiological) concentrations and its further metabolic transformations. These studies were performed in the presence of radioactive $^{57}Co^{II}$ in suspensions of *A. brasilense* cells rapidly frozen prior to measurements. Also, it has been shown for the first time how EMS can be used to probe the active sites of $^{57}Co^{II}$-activated enzyme. This is demonstrated using an example of glutamine synthetase, a key enzyme of nitrogen metabolism, isolated from *A. brasilense*, activated by $^{57}Co^{II}$.

Keywords: Fourier transform infrared (FTIR) spectroscopy; FT-Raman spectroscopy; emission (^{57}Co) Mössbauer spectroscopy; bacterial metabolism; heavy metals; *Azospirillum brasilense*.

INTRODUCTION

The advantages of using modern spectroscopic techniques in studying diverse processes of interaction of bacterial cells with metal ions can hardly be overestimated. Spectroscopic data for whole cells, supramolecular structures or isolated cellular constituents can reflect structural and compositional changes occurring in the course of metabolic responses induced by metal ions. Such information on the molecular level is of primary importance for basic studies on mechanisms of bacterial tolerance to heavy metals and their impact on bacterial metabolism [1-7], as well as for applied multidisciplinary research in the fields related to biotechnology, bioremediation, agriculture, biogeochemistry, *etc.* [8]. It is essential that in many cases spectroscopic techniques are non-destructive or may be readily adapted to minimise sample preparation.

In the present work, some examples are presented which illustrate the application of different spectroscopic techniques for monitoring metal-induced metabolic changes in soil bacteria, as well as for obtaining structural information for metal-containing active sites of a bacterial enzyme. In particular, the effects are discussed of a range of heavy metals on the plant growth-promoting rhizobacterium *Azospirillum brasilense* revealed using vibrational spectroscopy of whole cells and isolated cell walls. Fourier transform infrared (FTIR) spectroscopy [3, 4, 6] provided information on the overall structure and state of functional groups, their involvement in metal binding and/or hydration, while the FT-Raman technique [5] gave complementary data on the state of less or non-polar functional groups. As an example of a specifically metal-sensitive technique, emission (^{57}Co) Mössbauer spectroscopy (EMS) [7] was used to monitor the binding and metabolic transformations of cobalt(II) traces by live cells of *A. brasilense*, as well as to probe the active sites of $^{57}Co^{II}$-activated glutamine synthetase (GS), a key enzyme of nitrogen metabolism, isolated from this bacterium.

MATERIALS AND METHODS

Preparation of bacterial cultures

A. brasilense strains Sp7 and Sp245 (the Collection of IBPPM RAS, Saratov, Russia) were cultivated in a standard mineral medium as described elsewhere [4-7] either without (control) or with $CoCl_2$, $CuSO_4$ or $ZnSO_4$ added up to 2.0×10^{-4} M.

Sample preparation for FTIR and FT-Raman measurements and acquisition of spectra

FTIR spectroscopic measurements were performed using a Perkin-Elmer spectrometer (Model 2000) with a total of up to 100 scans (resolution 4 cm^{-1}; KBr pellets). FT-Raman spectra of cell samples mixed with KBr (Merck) were obtained using a Nicolet spectrometer (model Magna IR 750) with a FT-Raman accessory (1064 nm excitation; sample power ca. 0.48-0.50 W; 250 scans; resolution 4 cm^{-1}). Other details were reported earlier [4-6].

Sample preparation for emission Mössbauer measurements and acquisition of spectra

For EMS studies of Co^{2+}-treated cells, strain Sp245 was grown up to the mid-exponential growth phase (ca. 2.4×10^{8} cells/ml). Aliquots of the cell suspension (1 ml) were placed into PTFE sample holders each containing 1 mCi of $^{57}CoCl_2$ (final $^{57}Co^{2+}$ concentration ca. 2 μM), and after 2 or 60 min the samples were rapidly frozen in liquid nitrogen. Prior to EMS measurements, the intact cell samples were freeze-dried to enhance the Mössbauer effect. For EMS measurements with GS (EC 6.3.1.2.), $^{57}CoCl_2$ (1 mCi) was added to 1 ml of aqueous solution of the cation-free enzyme (up to ca. 24 $^{57}Co^{2+}$ ions per enzyme molecule [7]). The solution was incubated for 1 h at ambient temperature and rapidly frozen in liquid nitrogen. EMS measurements and data treatment were performed as described earlier [7].

RESULTS AND DISCUSSION

Vibrational spectroscopy

A typical vibrational spectrum of whole bacterial cells represents the overall cellular composition. It is controlled mainly by two major factors: (*i*) the relative abundance of all the cellular components which contribute to the spectrum (equally valid for both IR and Raman spectra) and (*ii*) the physical regularities which determine peak intensities for different vibration modes. Factor (*ii*) is essentially different for IR and Raman spectroscopies. Thus, polar functional groups (including hydroxo groups and water) give strong IR absorption and weak or very weak Raman scattering, whereas less polar or non-polar moieties (e.g., C-H, C-C, C=C, *etc.*) give much stronger Raman bands and weak IR absorption. In view of that, these two counterparts of vibrational spectroscopy provide complementary information [2-6] which may be very essential for highly complicated biological systems, including bacterial cells.

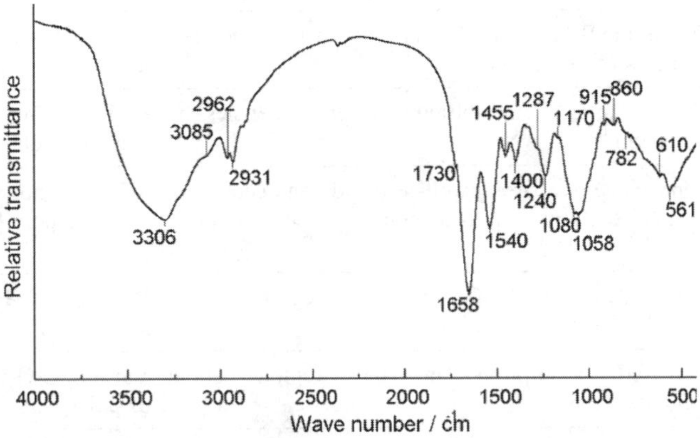

Fig. 1. Fourier transform infrared spectrum of vacuum-dried whole cells of *Azospirillum brasilense* Sp7 grown in a standard medium

The FTIR spectrum of *A.brasilense* Sp7 whole cells *(fig. 1)* is typical for bacterial cells and includes the following characteristic regions: very broad strong absorption typical for stretching O-H and N-H vibrations of carboxylic, alcoholic and phenolic hydroxyls and amides (over ca. 3500-2700 cm^{-1}); weaker but more definitely located stretching C-H bands mainly of methylene and methyl groups in lipid-containing moieties and amino acid residues (3000-2800 cm^{-1}); characteristic region of different stretching C=O vibrations of carboxyls, polyesters and polypeptides (1750-1600 cm^{-1}; note the typical amide I and amide II bands at 1658 and 1540 cm^{-1}, respectively, attributed mainly to peptidic stretching C=O and bending N-H vibrations in proteins); and the further so-called fingerprint region (commonly under 1800 cm^{-1}) including C-H bending (around 1450 cm^{-1}), phosphate (1240 cm^{-1}) and polysaccharide vibrations (*ca.* 1100-950 cm^{-1}), *etc.*

Previously we found [3-6] that *A. brasilense* grown in the presence of heavy metals accumulated the foreign cations from the medium. This seems to be a typical feature for bacteria which are capable of binding heavy metal cations; in Gram-negative bacteria (including azospirilla) this process is mediated primarily by capsular polysaccharide (PS) and lipopolysaccharide (LPS) constituents [4, 5]. However, it was found that heavy metals induced metabolic changes in bacteria which could be detected using vibrational spectroscopy.

Some relevant spectroscopic changes induced by heavy metals are summarised in *table 1*, together with possible band assignments and interpretation. It should be noted that, besides an increased overall hydration of metal-stressed bacterial cells related to bound water, including the hydration effects involving amide and phosphate vibrations (which may also be caused by metal binding), accumulation of polyester compounds is reflected by a number of relevant spectroscopic signs (see *table 1*). An enhancement in biosynthesis and accumulation of polyhydroxyalkanoates

(PHA) as an energy and/or carbon storage material is known to occur in some bacteria under nutrient imbalance or other stress conditions; these thermoplastic biodegradable polymers are commercially attractive (see, e.g. [6] and references therein).

Table 1. Characteristic changes in vibrational spectra of whole cells of Azospirillum brasilense induced by heavy metals ($Co2^+$, $Cu2^+$, Zn^{2+}) [4-6]

Spectroscopic changes induced by heavy metals	Possible assignment, interpretation and comments
FTIR spectra	
Enhanced very broad absorption in the region 3400-2700 cm^{-1}	νOH); enhanced hydration of bacterial cells (bound water; see also FT-Raman)
The band centred at about 3300 cm^{-1} shifted to lower frequencies (by 15-40 cm^{-1})	νNH); H-bonding (due to water) and possible metal binding of amide groups
The amide II band (1540 cm^{-1}) shifted to higher frequencies (by up to 8 cm^{-1})	δNH); H-bonding and/or possible metal binding of amide groups
The band at 1240 cm^{-1} shifted to lower frequencies (by ca. 10 cm^{-1})	$\nu_{as}(PO_2^-)$; increased hydration of phosphate moieties (may be induced by metal binding)
Appearance of a strong well-resolved band centred at ca. 1727 cm^{-1} (present as a shoulder in control cells)	νC=O) in polyesters; accumulation of poly-ester compounds (poly-β-hydroxybutyrate and probably phospholipids (PLs))
Increased absorption in the region about 1460-1440 cm^{-1}	Various δ(C-H) modes (accumulation of polyesters; see above)
Increased absorption in the region about 1150-1000 cm^{-1}	Various C-O-C and C-C-O vibrations (accumulation of polyesters; see above)
Increased absorption in the region about 750-700 cm^{-1}	Rocking CH_2 mode (accumulation of polyesters; see above)
FT-Raman spectra	
Increased very broad non-specific "humps" in the regions around 3400-2700, 1800-1200, under 600 cm^{-1}	Typical regions of water vibrations (weak and non-specific in Raman); enhanced overall hydration of bacterial cells
Decreased intensity of a weak shoulder at 3010-3000 cm^{-1} (induced by Co^{2+} and Cu^{2+} but not by Zn^{2+})	ν(=CH-) in unsaturated fatty acid (UFA) residues (may reflect a decrease in UFA content in bacterial membrane PLs)
Appearance of a new medium or weak band at 945-943 cm^{-1}	C-C-O vibrations (accumulation of polyesters; see above)

Among other heavy metal-induced effects, broadening of the amide I band, which is rather symmetric and narrow in control cells (with a maximum at 1658 cm^{-1}; see *fig. 1*), may reflect some some alterations in the secondary structure of cellular proteins to which the amide I band is known to be sensitive [6]. It is also noteworthy that Co^{2+} and Cu^{2+} (but not Zn^{2+}) induced some changes in FT-Raman spectra in the region of unsaturated hydrocarbons (see *table 1*), which may reflect a decreasing degree of unsaturation of fatty acid chains in membrane phospholipids, thus decreasing their susceptibility to well-known metal-induced lipid peroxidation as well as membrane permeability [5].

Emission Mössbauer spectroscopy

Mössbauer (nuclear g-resonance) spectroscopy is a unique method sensitive to the chemical state and coordination of a Mössbauer-active nuclide. In EMS, most widely used is the nuclear transformation of radioactive ^{57}Co via electron capture by its nucleus, to ^{57}Fe; thus the ^{57}Co-containing sample is used as a source. The sensitivity of EMS is several orders of magnitude higher than that of the absorption variant, which allows the state of cobalt species to be probed at its trace concentrations, which is essential for biological systems.

The EM spectrum of *A. brasilense* Sp245 cells frozen after 2 min of contact with 2×10^{-6} M $^{57}CoCl_2$ *(fig. 2)* shows two high-spin components corresponding to oxidation state +2 *(table 2)*, neither of which, however, corresponds to $[^{57}Co(H_2O)_6]^{2+}$ that gives $\Delta>3.2$ -mm/s (similar measurements in the cell-free supernatant showed part of $^{57}Co^{II}$ present as the hexaaquo complex). Thus, sorption of cobalt traces by the bacterial cells is evidently rapid and complete. Note that the yield of stabilised nucleogenic $^{57}Fe^{III}$ component as a consequence of after-effects (58%, see *table 2*, sample 1) is typical for very dilute rapidly frozen aqueous solutions of $^{57}CoCl_2$ [7]. Therefore, Co^{II} initially bound to bacterial cell surface biopolymers is likely to contain coordinated water.

Table 2. Mössbauer parameters[a] for $^{57}Co^{II}$-containing *Azospirillum brasilense* Sp245 cells and $^{57}Co^{II}$-activated glutamine synthetase (GS) from this bacterium [7] (T=80 K)

Sample and preparation procedure	Oxidation state[b]	δ,[c] mm/s	Δ,[d] mm/s	Γ,[e] mm/s	S_r,[f] %
1. Bacterial cells (frozen 2 min after incubation with 2 μM $^{57}CoCl_2$)	+2	1.24(3)	3.08(6)	0.70(10)	19(1)
	+2	1.14(3)	2.35(9)	0.83(13)	23(1)
	+3	0.35(5)	1.26(8)	1.43(12)	58(1)
2. Bacterial cells (frozen 60 min after incubation with 2 μM $^{57}CoCl_2$)	+2	1.22(4)	2.84(7)	0.88(10)	38(1)
	+2	1.00(5)	2.03(9)	0.52(25)	8(1)
	+3	0.26(5)	1.55(7)	1.36(16)	54(1)
3. GS (cation-free enzyme solution frozen 60 min after incubation with 2 μM $^{57}CoCl_2$)	+2	1.08(2)	3.08(8)	0.48(5)	18(1)
	+2	1.05(2)	2.39(6)	0.75(8)	60(1)
	+3	0.34(10)	1.12(20)	1.25(30)	22(1)

[a]Errors (in the last digits) are given in parentheses. [b]For the nucleogenic ^{57}Fe components stabilised after nuclear decay of the parent $^{57}Co^{II}$. [c]Isomer shift (relative to α-Fe; converted to the normal absorption convention). [d]Quadrupole splitting. [e]Full line width at half maximum. [f]Relative areas of spectral components (represent relative contents of the corresponding Fe forms assuming a common recoilless fraction for all forms in a sample).

The presence of two Co^{II} forms (with different δ and Δ; see *table 2*, sample 1) may be connected with the availability of different ligands at the cell surface. It is interesting to note that, after 1 h of incubation of live cells with $^{57}Co^{II}$, the Mössbauer parameters changed (cf. samples 1 and 2 in *table 2)*. The observed decrease in Δ for the daughter $^{57}Fe^{II}$ component with δ=1.22-1.24 mm/s and its increased content (38%) in sample 2, as well as the appearance of the $^{57}Fe^{II}$ component with δ=1.00 mm/s) and Δ=2.03 mm/s reflect further transformation of $^{57}Co^{II}$ following its primary sorption. This could arise from the formation of species with a lower coordination number (with O-donor ligands), as well as with more covalent coordination bonds (involving, e.g. N-donor atoms) for the latter component, which may also account for a slight decrease in the yield of the stabilised daughter $^{57}Fe^{III}$ component in sample 2. It should be noted that similar EMS measurements for dead cells or cell-free supernatant gave distinctly different parameters (not shown). Thus, it may be inferred that the observed changes reflect metabolic transformation of trace Co^{II} in *A. brasilense* Sp245 occurring already within an hour.

Similar EMS measurements performed for the first time for a bacterial enzyme, GS activated by $^{57}Co^{II}$, have shown the presence of two forms of cobalt(II) bound at the active sites (*table 2*, sample 3) with different coordination of Co^{II} and essentially different affinities (note the higher proportion of the Fe^{II} component with Δ=2.39 mm/s). This conclusion is in line with the literature data on similar GS isolated from other bacteria (see [7] and references therein). Note also that the yield of the stabilised nucleogenic $^{57}Fe^{III}$ component resulting from after-effects is much lower (22%) for sample 3 (see *table 2*). This may be ascribed to a more strong binding of parent $^{57}Co^{II}$ at the GS active sites by certain amino acid residues (Glu and His [7]) as ligands with high electron-donor properties.

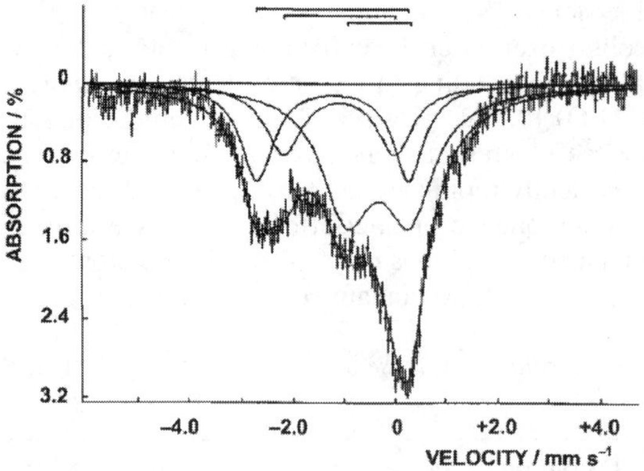

Fig. 2. Emission Mössbauer spectrum of *Azospirillum brasilense* Sp245 live cells frozen after 2-min incubation with trace $^{57}CoCl_2$ (dried; T = 80 K)

ACKNOWLEDGEMENTS

This work was supported in parts by INTAS (EC, Brussels, Belgium), NATO (Grant LST.CLG.977664) and the Russian Academy of Sciences' Commission (Grant No. 205 under the 6th Competition-Expertise of research projects), as well as by Short-term Fellowships to A.A.K. from UNESCO in biotechnology (Contract UVO-ROSTE 875.878.9; 2000) and from EMBO (ASTF 9826; 2001).

REFERENCES

1. Nelson W.H. (Ed.). *Modern Techniques for Rapid Microbiological Analysis,* VCH, New York (1991).
2. Naumann D., Keller S., Helm D., Schultz Ch., Schrader B. FT-IR spectroscopy and FT-Raman spectroscopy are powerful analytical tools for the non-invasive characterization of intact microbial cells. *J. Mol. Struct.*, **347**, 399-405 (1995).
3. Kamnev A.A., Risti M., Antonyuk L.P., Chernyshev A.V., Ignatov V.V. Fourier transform infrared spectroscopic study of intact cells of the nitrogen-fixing bacterium *Azospirillum brasilense*. *J. Mol. Struct.*, **408/409**, 201-205 (1997).
4. Kamnev A.A., Antonyuk L.P., Matora L.Yu., Serebrennikova O.B., Sumaroka M.V., Colina M., Renou-Gonnord M.-F., Ignatov V.V. Spectroscopic characterization of cell membranes and their constituents of the plant-associated soil bacterium *Azospirillum brasilense*. *J. Mol. Struct.*, **480-481**, 387-393 (1999).
5. Kamnev A.A., Tarantilis P.A., Antonyuk L.P., Bespalova L.A., Polissiou M.G., Colina M., Gardiner P.H.E., Ignatov V.V. Fourier transform Raman spectroscopic characterisation of cells of the plant-associated soil bacterium *Azospirillum brasilense* Sp7. *J. Mol. Struct.*, **563-564**, 199-207 (2001).
6. Kamnev A.A., Antonyuk L.P., Tugarova A.V., Tarantilis P.A., Polissiou M.G., Gardiner P.H.E. Fourier transform infrared spectroscopic characterisation of heavy metal-induced metabolic changes in the plant-associated soil bacterium *Azospirillum brasilense* Sp7. *J. Mol. Struct.*, in press (2002).
7. Kamnev A.A., Antonyuk L.P., Smirnova V.E., Serebrennikova O.B., Kulikov L.A., Perfiliev Yu.D. Trace cobalt speciation in bacteria and at enzymic active sites using emission Mössbauer spectroscopy. *Anal. Bioanal. Chem.*, **372**, 431-435 (2002).
8. Kamnev A.A., van der Lelie D. Chemical and biological parameters as tools to evaluate and improve heavy metal phytoremediation. *Biosci. Rep.*, **20**, 239-258 (2000).

*Metal Ions in Biology and Medicine: vol. 7.*Eds L. Khassanova, Ph. Collery, I. Maymard, Z. Khassanova, J.-C. Étienne.
John Libbey Eurotext, Paris © 2002 pp. 237-241.

Effects of heavy metals on the plant-associated bacterium *Azospirillum brasilense*: endophytic and non-endophytic strains

Anna V. Tugarova[1*], Alexander A. Kamnev[1*], Lyudmila P. Antonyuk[1], Petros A. Tarantilis[2] and Moschos G. Polissiou[2]

[1]Laboratory of Biochemistry of Plant-Bacterial Symbioses, Institute of Biochemistry and Physiology of Plants and Microorganisms, Russian Academy of Sciences, 13 Prosp. Entuziastov, 410015 Saratov, Russia
[2]Laboratory of Chemistry, Department of Science, Agricultural University of Athens, Iera Odos 75, 11855 Athens, Greece
** Corresponding author. Tel./Fax: +7-(8452)-947303; E-mail: micbio@ibppm.saratov.su*

ABSTRACT

The plant-associated soil bacterium *Azospirillum brasilense* attracts worldwide attention owing to its plant growth-promoting activities. Among hundreds of its strains known up to date, wild-type strain Sp245 is capable of colonising the root interior (i.e. facultative endophyte), whereas others are non-endophytes colonising the root surface only. Thus, the different ecological niches occupied by these strains suggest that their responses to environmental conditions might differ as well. In this study, we compared the metabolic responses of *A. brasilense* strains Sp245 and Sp7 to several heavy metal cations (Co, Cu, Zn) present in the medium in tolerable concentrations (up to 0.2 mM) taken up by the bacteria. Structural features of whole cells were studied using vibrational (Fourier transform infrared (FTIR) and FT-Raman) spectroscopic techniques. The results obtained show that all the heavy metals studied are significantly accumulated in the cells, albeit up to different levels. In strain Sp7, the heavy metals induced noticeable metabolic responses revealed in vibrational spectra of whole cells, consisting in an enhanced accumulation of polyester compounds, as well as some effects on the state of certain functional groups. In contrast, the response of the endophytic strain Sp245 to heavy metal uptake was found to be much less pronounced. These dissimilarities in their behaviour may be caused by different adaptation abilities of the strains to stress conditions owing to their different ecological status.
Keywords: heavy metal uptake; bacterial metabolism; *Azospirillum brasilense*; Fourier transform infrared (FTIR) spectroscopy; FT-Raman spectroscopy.

INTRODUCTION

The plant-associated nitrogen-fixing soil bacterium *Azospirillum brasilense* attracts worldwide attention owing to its plant growth-promoting activities [1, 2]. Even within this species, there are endophytic strains (e.g., wild-type strain Sp245 capable of colonising the root interior) and non-endophytes colonising the root surface only [2]. Thus, the different ecological niches occupied by these strains suggest that their metabolic responses to environmental conditions might differ as well.

Previous studies [3-6] have shown that azospirillum is tolerable to several conventionally toxic heavy metals (at concentrations under 1 mM), some of which (e.g., Co, Cu, Zn, *etc.*), however, in trace amounts are involved in diverse enzymatic activities and therefore are necessary for bacteria. The aim of this study was to compare the metabolic responses of endophytic and non-endophytic strains to several heavy metals using vibrational spectroscopic techniques as a probe for structural and compositional changes in whole bacterial cells.

MATERIALS AND METHODS

Azospirillum brasilense (wild-type strains Sp7 and Sp245; the Collection of IBPPM RAS, Saratov, Russia) were cultivated in a standard phosphate- and malate-containing medium with 0.5 g/l NH_4Cl (control) or in the same medium with $CoCl_2$, $CuSO_4$ or $ZnSO_4$ added up to 2.0×10^{-4} M, harvested by centrifugation, washed and dried as described elsewhere [4-6]. For FTIR in the transmission mode and for FT-Raman measurements, cell samples were mixed with KBr (Merck) or, for diffuse reflectance infrared Fourier transform (DRIFT) measurements, used as dry finely ground powder in a Micro sampling cup (Spectra-Tech Inc., USA). FTIR studies were performed using a Perkin-Elmer (Model 2000) or (for DRIFT) a Nicolet spectrometer (model Magna-IR 560 E.S.P.) with a total of up to 100 scans (resolution 4 cm^{-1}). FT-Raman spectra were obtained using a Nicolet spectrometer (model Magna IR 750) with a FT-Raman accessory (1064 nm excitation; sample power 0.48-0.50 W; 250 scans; resolution 4 cm^{-1}). Other details of spectra acquisition were reported earlier [4-6].

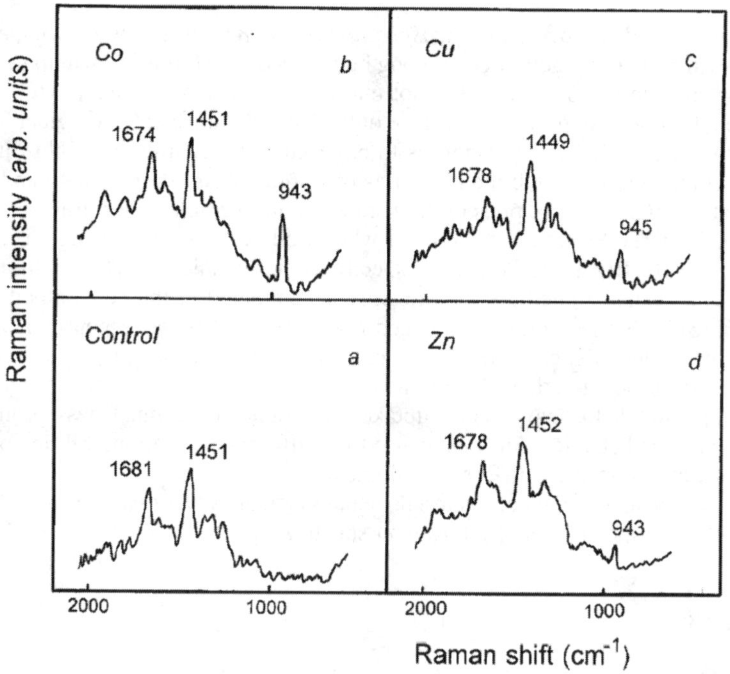

Fig. 1. FTIR spectra of *A. brasilense* Sp7 cells grown *(a)* in a standard medium (control) and in the presence of 0.2 mM Co^{2+} *(b)*, Cu^{2+} *(c)* or Zn^{2+} *(d)*

RESULTS AND DISCUSSION

Comparison of FT-Raman spectra of whole cells of the non-endophytic strain *A. brasilense* Sp7 grown in a standard medium (control) and in the presence of 0.2 mM Co^{2+}, Cu^{2+} or Zn^{2+} shows

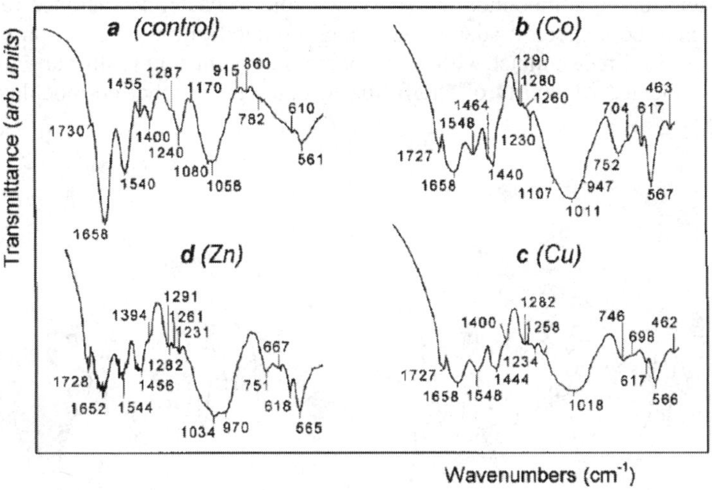

Fig. 2. FTIR spectra of *A. brasilense* Sp7 cells grown *(a)* in a standard medium (control) and in the presence of 0.2 mM Co^{2+} *(b)*, Cu^{2+} *(c)* or Zn^{2+} *(d)*

that, besides an enhanced overall hydration and (for Co and Cu only) some decrease in the unsaturation degree of fatty acid residues revealed in the region of stretching C-H vibrations [5], for metal-stressed cells a new band is observed at ca. 945 cm^{-1} in the region of C-C-O vibrations *(fig. 1)*. Also, in FTIR spectra of metal-stressed cells there appears a well-resolved ν(C=O) band at ca. 1727 cm^{-1} featuring polyester carbonylic groups, which is present as a relatively weak shoulder in control cells *(fig. 2)*. These metal-induced changes, together with an increased FTIR absorption in the regions of C-O-C and C-C-O vibrations (1150-1000 cm^{-1}), CH_2 bending (1460-1440 cm^{-1}) and rocking vibrations (ca. 750 cm^{-1}; cf. *fig. 2a-d*) provide evidence for the accumulation of polyester compounds in cells of strain Sp7 as a response to metal stress. Note that the ν(C=O) band with a maximum under 1730 cm^{-1} corresponds to poly-3-hydroxybutyrate (PHB), whereas other polyhydroxyalkanoates (PHAs) give bands at 1732-1740 cm^{-1} [6]. Thus, considering the asymmetry of the ν(C=O) band at ca. 1727 cm^{-1} (see *fig. 2b-d*), PHB seems to dominate, which has been documented for azospirilla under unfavourable conditions (see [5, 6] and references therein), but the presence of other PHAs is also possible. Some other effects observed in metal-stressed cells include an enhanced overall hydration noticeable in the ν(O-H) region (not shown), with corresponding shifts of the $\nu_{as}(PO_2^-)$ band of phosphates from 1240 cm^{-1} (see *fig. 2a*) to 1234-1230 cm^{-1} (see *fig. 2b-d*).

In order to compare the response to heavy metal stress of the endophytic strain Sp245, the latter was grown under identical conditions, and the harvested cells were studied using FTIR spectroscopy in the DRIFT mode. It can be seen from the spectra *(fig. 3)* that the presence of 0.2 mM Co^{2+}, Cu^{2+} or Zn^{2+} in the standard NH_4^+-complemented medium did not result in any essential alterations in cell composition. In particular, the ν(C=O) region around 1730 cm^{-1} is quite similar both for the control *(fig. 3a)* and metal-stressed cells *(fig. 3b-d)* represented by a relatively weak shoulder as compared to the typical amide I and amide II bands of cellular proteins at ca. 1650 and 1540 cm^{-1}, respectively. Some redistributions of band intensities can only be noted in the polysaccharide (PS) region (1150-950 cm^{-1}); however, these are not specific, as PSs in azospirilla are known to be highly variable.

Thus, the results obtained demonstrate that the response of the endophytic strain Sp245 to a moderate heavy metal stress is much less pronounced than that of the non-endophytic strain Sp7. These dissimilarities in their behaviour may be caused by different adaptation abilities of the strains to stress conditions owing to their different ecological status. In particular, an enhanced accumulation of polyester storage compounds, which play a role in bacterial tolerance to environmental stresses, in Sp7 may be a specific adaptation strategy related to the localisation of this strain on the rhizoplane, i.e. in direct contact with rhizosphere soil components. In particular, this corresponds to the documented capability of strain Sp7 to outcompete other co-inoculated strains [2].

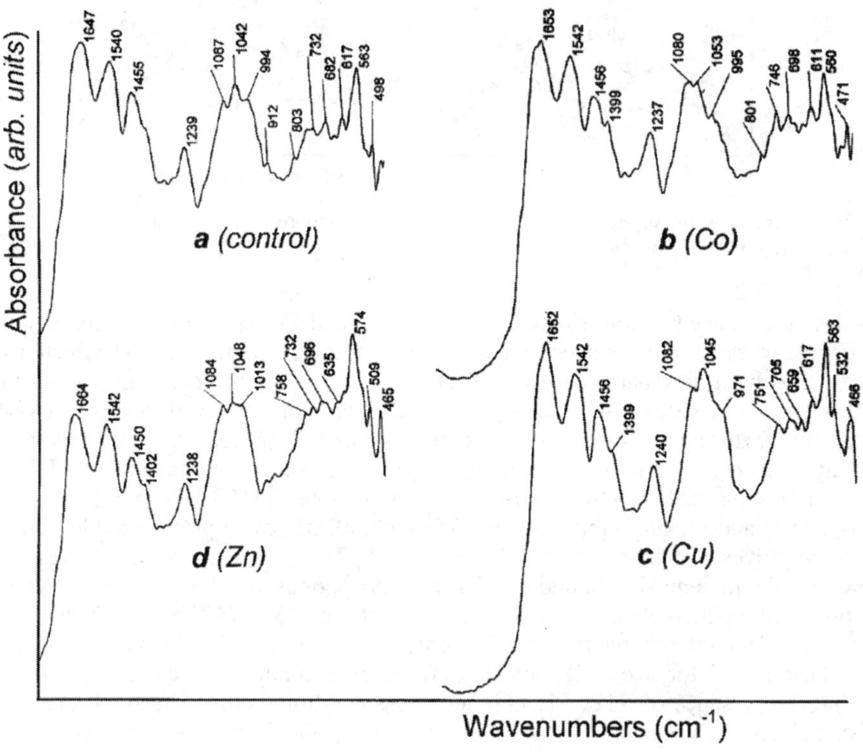

Fig. 3. FTIR spectra (in the DRIFT mode) of *A. brasilense* Sp245 cells grown *(a)* in a standard medium (control) and in the presence of 0.2 mM Co^{2+} *(b)*, Cu^{2+} *(c)* or Zn^{2+} *(d)*

ACKNOWLEDGEMENTS

This work was supported in parts by INTAS (EC, Brussels, Belgium), NATO (Grant LST.CLG.977664) and the Russian Academy of Sciences' Commission (Grant No. 205 under the 6th Competition-Expertise of research projects). A.A.K. appreciates support by Short-term Fellowships from UNESCO (in biotechnology; Contract UVO-ROSTE 875.878.9) in 2000 and from EMBO (ASTF 9826) in 2001.

REFERENCES

1. Steenhoudt O., Vanderleyden J. *Azospirillum*, a free-living nitrogen-fixing bacterium closely associated with grasses: genetic, biochemical and ecological aspects. *FEMS Microbiol. Rev.*, **24**, 487-506 (2000).
2. Kirchhof G., Schloter M., Aßmus B., Hartmann A. Molecular microbial ecology approaches applied to diazotrophs associated with non-legumes. *Soil Biol. Biochem.*, **29**, 853-862 (1997).
3. Kamnev A.A., Renou-Gonnord M.F., Antonyuk L.P., Colina M., Chernyshev A.V., Frolov I., Ignatov V.V. Spectroscopic characterization of the uptake of essential and xenobiotic metal cations in cells of the soil bacterium *Azospirillum brasilense*. *Biochem. Mol. Biol. Int.*, **41**, 123-130 (1997).
4. Kamnev A.A., Antonyuk L.P., Matora L.Yu., Serebrennikova O.B., Sumaroka M.V., Colina M., Renou-Gonnord M.-F., Ignatov V.V. Spectroscopic characterization of cell membranes and their constituents of the plant-associated soil bacterium *Azospirillum brasilense*. *J. Mol. Struct.*, **480-481**, 387-393 (1999).
5. Kamnev A.A., Tarantilis P.A., Antonyuk L.P., Bespalova L.A., Polissiou M.G., Colina M., Gardiner P.H.E., Ignatov V.V. Fourier transform Raman spectroscopic characterisation of cells of the plant-associated soil bacterium *Azospirillum brasilense* Sp7. *J. Mol. Struct.*, **563-564**, 199-207 (2001).
6. Kamnev A.A., Antonyuk L.P., Tugarova A.V., Tarantilis P.A., Polissiou M.G., Gardiner P.H.E. Fourier transform infrared spectroscopic characterisation of heavy metal-induced metabolic changes in the plant-associated soil bacterium *Azospirillum brasilense* Sp7. *J. Mol. Struct.*, in press (2002).

Metal Ions in Biology and Medicine: vol. 7. Eds L. Khassanova, Ph. Collery, I. Maymard, Z. Khassanova, J.-C. Étienne.
John Libbey Eurotext, Paris © 2002 pp. 242-246.

Effect of fumonisin B_1 ON NaA zeolite effectiveness to adsorb aflatoxin B_1

Kikot A.[4], Magnoli C.[1], Chiacchiera S.[2], Dalcero A.[1], Miazzo R.[3], Basaldella E.I.[4]

[1]Dpto. de Microbiología e Inmunología. [2]Dpto. de Química y Física, Fac. de Ciencias Exactas, Físico Químicas y Naturales. [3]Dpto de Producción Animal, Fac. de Agronomía y Veterinaria. Universidad Nacional de RíoCuarto,AP#3(5800), Río Cuarto,Córdoba. [4]Centro de Investigación y Desarrollo en Procesos Catalíticos (CINDECA)CONICET-CIC-UNLP, Calle 47 N°257, (1900) La Plata, Argentina

ABSTRACT

Synthetic zeolite NaA was evaluated for its ability to adsorb aflatoxin B_1 (AFB_1) from aqueous solutions in the presence of fumonisin B_1 (FB_1). Physicochemical studies *in vitro* demonstrated that fumonisin B_1 (FB_1) and AFB_1 are markedly adsorbed at pH = 2 and 39°C. It was observed that FB_1 did not modify the adsorption capacity of AFB_1 under equimolar concentrations of both toxins at pH 2 and 39°C. However, when concentrations of FB_1 were 5 to 30 times higher than those of AFB_1, a noticeable decrease of AFB_1 adsorption capacity was observed. Model fitting to adsorption isotherms allowed to calculate association constants under co-adsorption conditions and a parameter related to interaction between adsorbed molecules. Under these conditions, FB_1-FB_1 was an attractive interaction and was statistically more significant than FB_1-AFB_1 interaction that was of atractive nature. With regard to AFB_1, only the attractive interaction with itself resulted significant. Association constants for toxin mixtureses were 10 fold lower than those obtained when pure toxins were used. At pH values similar to the in vivo conditions, these results indicate that fumonisin levels higher than AFB_1 ones decreased site availability on the adsorbent. Even though FB_1 has a noticeable adsorption capacity, when pH reaches neutrality FB_1 is desorbed to the medium. At this pH, AFB_1 cannot be adsorbed. Thus, FB_1 presence should be taken into account since relative amounts of FB_1 might markedly decrease zeolite Na-A capacity to prevent aflatoxicosis.

Key words: *Zeolite naA. Aflatoxin.*Fumonisis.*Adsorbtion.*

INTRODUCTION

Micotoxins are fungal secondary metabolites which occur worldwide and contaminate wheat, corn, soy beans and sorghum, which are normally used for poultry rations [1]. Natural incidence studies carried out over a period of three years on poultry feeds samples from Córdoba, Argentina, showed the presence of AFB_1 and FB_1 all over the samples. The levels ranged from 10-197 and 136-4270 ppb, respectively [2].

Aflatoxins (AF) are produced by the *Aspergillus* section *flavi*: *A. flavus* var. *flavus, A. Flavus* subsp. *Parasiticus*, and *A. Nomius*. They cause a variety of effects in poultry, including the decrease in both the BW gain and feed utilization efficiency. In poultry, AFB_1 is associated whith liver damage, poor performance, and immunosuppression.

Fumonisins, toxic metabolites of certain *Fusarium* molds, are capable of causing leukoencephalomalasia in horses and pulmonary edema in swine. FB_1 is suspected of producing pulmonar edema and promoting hepatic and esophagic carcinogenesis causing much lost to the poultry industry.

The harmful effects caused by toxin contamination require practical and effective low-cost strategies to detoxify contaminated food and feed. One of the most encouraging approaches to solve the problem has been the addition of nutritionally inert sorbents to feedstufs in order to reduce the bioavailability of the toxins in the gastrointestinal tract of the birds [3].

Our previous *in vitro* and *in vivo* studies showed the efficacy of synthetic zeolites NaA to prevent aflatoxicosis [4]. As FB_1 is frecuently present in poultry feeds that also conteins AFB_1, the aim of our present work was to evaluate the capacity of NaA zeolite to sorb AFB_1 in the presence of high concentrations of FB_1.

Zeolites are cristalline aluminosilicates compounds that are classified according to common features of the framework structures. Particularly, the zeolite structure known as A type is a specific arrangement in which the unit cell contains 24 thetrahedra, 12 AlO_4, and 12 SiO_4. When fully hydrated, there are 27 water molecules, and there are also one monovalent cation for each Al present.

These cations are interchangeable. The energy of interaction with the adsorbate and the adsorption capacity of these zeolites strongly depend on type, size, position and number of accessible cations in the zeolite, as well as on the zeolite structure.

MATERIALS AND METHODS

Reagents

Chromatographic standards of AFB_1,FB_1, FB_2 and FB_3 (Sigma Chemical Co.,St. Louis, MO, USA), purity>99%, and deionized HPLC quality water were used in the preparation of working solutions. HPLC grade solvents from Sintorgan were utilized without further purification.

Commercial argentine kaolinite (48.50% w/w SiO_2; 36.94% Al_2O_3; 0.03% Fe_2O_3), water glass (26.10% w/w SiO_2; 8.24% Na_2O; 65,66% H_2O), sodium aluminate (36.50% w/w Al_2O_3; 29.60% Na_2O; 33.90% H_2O), sodium hidroxide (Carlo Erba, analytical grade) and deionized water, were the chemicals used in the synthesis of NaA zeolite.

Zeolite hydrothermal synthesis and characterization

Synthetic zeolite NaA was prepared and characterized as previously reported [5]. In that method, reactions mixturers with appropriate compositions were submited to 90ºC, in closed polypropylene vessels, at autogenous pressure. After several hours, solids were separated by centrifugation, washed with water and dried in a furnace.

These crystaline solids were characterized by X-ray diffraction (XRD) using a Philips PW 1732/10 diffractometer. FT-IR spectra of solids were obtained with a Bruker IFS 66 equipment. Particule size and morphology was estimated with a Philips 505 scanning electron microscope.

In vitro Studies

Instrumental. The HPLC systems were either a 2125 LB pump (Bromma, Sweden) connected to a MPF-44B Perkin Elmer spectrofluorimetric detector and a Perkin Elmer LCI/100integrator (Perkin Elmer, Norwalk CT, USA) or a Varian 5000 with a UV-visible fixed wavelenght 2550 detector. The UV-visible spectra were performed on a 8453 Hewlett Packard spectrophotometer.

Analytical method for fumonisins determination. Fifteen grams of finely ground culture were extracted with 50 ml of acetonitrile - water solution (1:1) and shaken for 30 min. The mixture was filtered through a Whatman Nº 4 filter paper. Ten ml of the filtered extract were applied to a Bond-Elut strong anion exchange (SAX) cartridge (Varian, Harbor City, CA) fitted to a Supelco solid-phase extraction (SPE) manifold (Supelco, Bellefonte, PA), previously conditioned by a successive passage of methanol (5 ml) and methanol/water solution (3:1, v/v) 5 ml. The flow rate was kept below 2 ml/min. The cartridge was then washed with methanol/water solution (3:1, v/v) 8 ml, followed by methanol (3 ml), and the fumonisins were eluted with 0.5% acetic acid in methanol (14 ml). The eluate was evaporated to dryness at 40 °C under a moderate stream of nitrogen, and stored dry at 4 °C until HPLC analysis.

The diluted extract was quantitatively determined by a modified high performance liquid chromatography method proposed by Shephard [6].

Analytical method for AFB_1 determination. The AFB_1 was determined by UV-visible spectrometry.

Adsorption assays. A stock of methanolic toxin solution was prepared. The working solutions were made from the stock by dilution with buffered water. Samples of sorbent (100 mg) were weighted into clean glass tubes (two replicates per samples), and a total of 2.0 ml of each working solution was added at timed intervals, so that the total incubation period was equal to one hour for each tube. Controls of the toxins without sorbent were performed, and blanks with buffered water and sorbent were also carried out. All tubes were shacked for 1 h at 39.5 ± 0.1 °C. After the incubation time, the solutions were centrifuged for 5 min at 16,060 g and of the supernatant were decanted carefully into a clean tube and kept refrigerated to perform toxin quantification. Percent sorption of toxin was estimated from the amount of unbound (free) toxin remaining in the supernatant after incubation.

RESULTS AND DISCUSSION

Effect of pH on the NaA zeolite FB_1-adsorption ability

As expected from the acid-base nature of FB_1, a strong pH dependence (not shown in the text) was found. The highest adsorption occurs from FB_1 solutions at pH 4. However, a little bit of the toxin was effectively adsorbed at pH 2, while at pH 6 there was not adsorption. This pH dependence is probably attributed to the changes in the charge of FB_1, which has both acid and basic groups in their structure. Changes in the charge density of the toxin molecules probably affect coulombic interactions with the anionic zeolite network and its counterions.

Aflatoxin and Fumonisins adsorption isotherms

Considering that AFB_1 is irreversibly adsorbed on NaA zeolite at pH 2 -similar to physiological conditions of muscular stomach of poultry- and that FB_1 has only a moderate reversible adsorption at pH 2, it is necessary to determine if FB_1 presence may alter the preventive effect of NaA zeolite on aflatoxicosis. Thus with that rational, *in vitro* co-contamination of AFB_1 and FB_1 studies were performed at pH 2.

Studies were carried out with concentrations of FB_1 between 5 to 30 times higher that those of AFB_1, because previous incidence studies showed that natural levels of FBs were higher than AFB_1 ones. *Figures 1 and 2* show obtained results.

As seen in *figure 2*, FB_1 presence considerably decreased maximum adsorption capacity of AFB_1, what evidenced a lower site availability that can be attributed to the blockage of sites due to FB_1 adsorption.

A mathematical model that considered common sites and interaction between adsorbed molecules interpreted the experimental results.

Model fitting to adsorption isotherms allowed to calculate association constants under co-adsorption conditions and a parameter related to interaction between adsorbed molecules. *Table 1* shows parameters estimated from the model AFB_1 only exhibits attractive interaction with itself. Besides, FB_1 shows an FB_1-FB_1 attractive interaction with a higher statistical significance than FB_1-AFB_1 interaction, that is also of attractive nature.

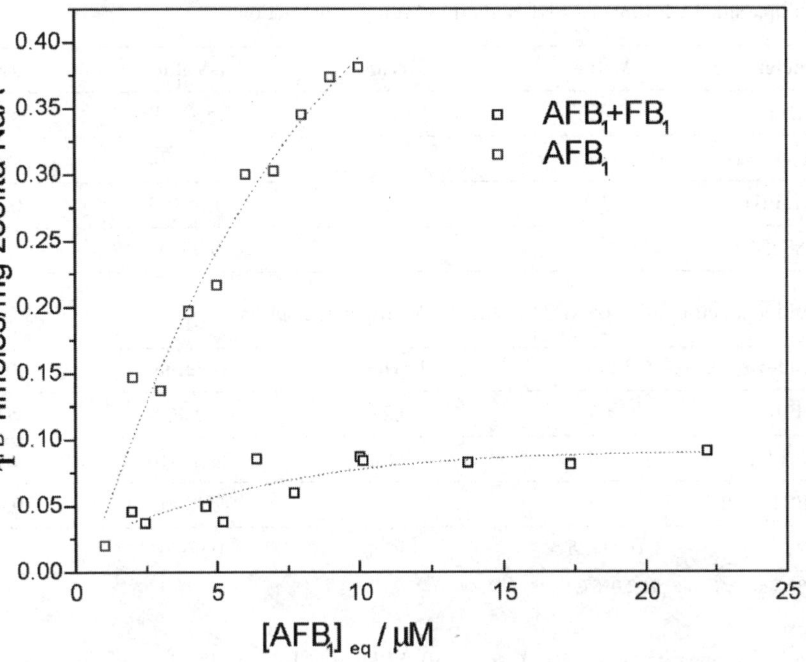

Fig. 1. Effect of FB_1 addition on adsorption isotherm of AFB_1 at pH=2 and 39ºC

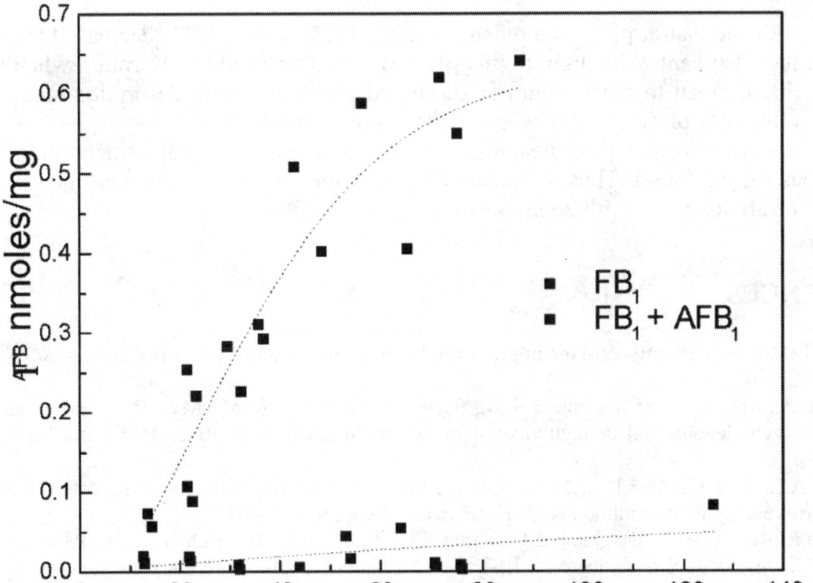

Fig. 2. Effect of AFB_1 addition on adsorption isotherm of FB_1at pH=2 and 39ºC

Table 1. Multiparametric fitting of AFB1 isotherm - Multiple interaction

Parameter	Value	Error	t-Value	Prob>|t|
- ln βAFB1	-7.2	0,3	-25.28533	<0,0001
2 aAFB1, AFB1	-14	2	-8.97374	<0,0001
2 aAFB1, FB1	-0.9	0.6	-1.55493	0,14593
$R^2 = 0.88254$		R	0.9394	

Table 2. Multiparametric fitting of AFB1 isotherm - Multiple interaction

Parameter	Value	Error	t-Value	Prob>|t|
- ln βFB_1	-6,7	0,2	-28,50674	<0,0001
-2 aFB1, FB1	-6.4	0.6	-11,17161	<0,0001
-2 aFB1, AFB1	-3,3	1	-2.33003	0,03807
R2	0.93125	R	0.9650	

Temperature 39,5 ± 0,1°C and pH 2,0 ± 0,1.

Association constants were 2059 and 750 M^{-1} for AFB_1 and FB_1, respectively. They were 10 times for AFB_1 and 5 times for FB_1 lower than association constants obtained with pure toxins.

CONCLUSIONS

Results indicate that at pH 2, fumonisins in higher levels than AFB_1 decreased the availability of sites on the adsorbent. Although, at this pH, FB_1 has a noticeable adsorption when pH reaches neutrality, FB_1 is freed to the medium as discussed previously with desorption studies. Besides, FB_1 desorption takes place at a pH where AFB_1 cannot be adsorbed.

These findings have practical meaning, because along gastrointestinal tract similar pHs to the assayed ones can be found. Thus, co-contamination with FB_1 should be taken into account when prevention of aflatoxicosis with zeolites is faced.

REFERENCES

1. Miller J.D., Fungy and mycotoxins in grain implications for stored product research. *J. Stored Prod. Res.* 1995, 31, 1-16.
2. Dalcero A., Magnoli C., Chiacchiera S.M., Palacio G., Reynoso M.M., Mycoflora and incidence of aflatoxin B_1, zearalenone and deoxynivalenol in puultry feeds in Argentine. *Mycopathologia.* 1997, 137, 179-184.
3. Ramos A.J., Fink-Gremmels J., Hernández E., Prevention of toxic effects of mycotoxins by means of nonnutritive adsorbent compounds. *J. Food Prot.* 1996, 59, 631-641.
4. Miazzo R, Rosa C.A.R., De Queiroz Carvalho E.C., Magnoli C., Chiacchiera S.M., Palacio G, Saenz M., Kikot A., Basaldella E.I., Dalcero A. Efficacy of synthetic zeolite to reduce the toxicity of aflatoxin in broiler chicks. *Poultry Science.* 2000, 79, 1-6.
5. Basaldella E.I, Kikot A, Tara J.C., Effect of aluminium concentration on crystal size and morphology in the synthesis of NaA zeolite. *Material.Letters..* 1997, 31, 83-86.
6. Shephard G.S., Sydenham E.W., Thiel P.G., *J. Liq. Chromatogr.* 1990, 13, 2077-2080.
7. Grant P., Phillips T., *J. Agric. Food Chem.*, 1998, 46, 599-605.

*Metal Ions in Biology and Medicine: vol. 7.*Eds L. Khassanova, Ph. Collery, I. Maymard, Z. Khassanova, J.-C. Étienne.
John Libbey Eurotext, Paris © 2002 pp. 247-251.

Biosorption of metal ions by microorganisms and their consortia with aqueous plants

Gogotov I.N., Zorin N.A., Tikhonov K.G.

Institute of Basic Biological Problems RAS, Pushchino, 142290, Russia

Phototrophic and chemotrophic microorganisms of different taxonomic groups and their consortia with water plants (aquatic fern *Azolla*, water hyacinth) are capable to accumulate metal ions of Ni, Pt, Ru, Cu, Cr, Pb, Zn, Si and Au. This ability enables to use them for purification of agricultural and industrial waste water from toxic heavy metals and to obtain rare trace metals. For this purpose it is possible to use growing cultures and immobilized cells, their biopolymers or enzymes (hydrogenase). Purple non-sulfur bacteria (*Rhodobacter* spp., *Rhodopseudomonas* spp.) are able to accumulate Cu, Zn, Ni and Hg., showing various resistance to these metals. Biopolymer of purple sulfur bacterium *Ectothiorhodospira shaposhnikovii* accumulated more than 99% Cu and Zn from diluted waste water of electrolysis. Green algae *Chlorella* spp. and water plants *Azolla*, duckweed and water hyacinth showed higher ability to a biosorption of metal ioins. Cyanobacteria *(Spirulina platensis)* and plants (amaranthus) may be recommend as producers of homeopathic drugs containing necessary metals (Se, Zn, Ni etc.) for microelemental diseases.

Many of metal ions play the important role in a metabolism of microorganisms, plants, animals and man [1, 2]. They are required for synthesis of pigments, electron carriers, enzymes, vitamins and other cell components. Elevated concentrations or deficit of a number of metal ions induce a toxic effect on organisms, causing various diseases or suppressing their growth. Chemotrophic and phototrophic microorganisms, as well as their consortia with plants and lichens, accumulate efficiently the metals from an environment in the amount higher their requirements for growth, that can be used for bioremediation of waste water and soils, and for the development of technologies of obtaining dissipated expensive metals and homeopathic drugs as well [1-4]. The amount of metals accumulated by them can be different. The absorption mechanisms also vary from simple physicochemical process of absorption and desorption, up to highly specific systems of a transport of metal ions in the cells. More than 16 metals are necessary for the growth of microorganisms and plants and their metabolism. However, biological functions of many metals and mechanism of their operation are not clear. The purpose of this presentation is the analysis of the literary and own data on efficiency of a biosorption and stability to ions of metals of the growing cultures, suspensions of native (immobilized) cells, their biopolymers and enzymes.

BIOSORPTION OF METALS BY PHOTOTROPHIC MICROORGANISMS

Purple bacteria

The purple not sulfur bacterium *Rhodobacter capsulatus* and different species of *Rhodopseudomonas* spp absorb ions of Cu, Ni and Cr. from waste water up to 80-90% [1]. Purple sulfur bacterium *Ectothiorhodospira shaposhnikovii*, making a biopolymer [6], is able to absorb more then 90% Cu, Ni and Zn at a contents 100 mg/l of metal ions in the medium. The metal-saturated biopolymer in it reached 25-30% of a dry mass. According to our data, in the presence of 214 g/l

Ni^{2+} and 0,00371 g/l Cu^{2+} their absorption was much higher by a biopolymer (60-70%), than by intact cells (30-40%) at pH 6-7.

Metals, absorbed by the cells of purple bacteria or their biopolymers, can be isolated from them, at low pH (~ 2). The reduction of ions metals up to a metal state is possible in the presence of hydrogenase isolated from purple sulfur bacterium *Thiocapsa rosopersicina*, if the intermediate low potential carrier of electrons (ferredoxin, cytochrom C_3 or methylviologene) will be added in a the reaction mixture. Hydrogenase or microorganisms, containing it, reduced ions of Ni^{2+}, Pb^{2+}, Pd^{2+} and Ru^{3+} in these conditions. However, reduction of Cd^{2+}, Co^{2+} and Cu^{2+} ions did not take place, though there are no thermodynamic limitations for this reaction. The rate of metal ions reduction, measured by consumption of H_2, was low (~ 100 nmol/hour) due to the toxic effect of metal ions on the hydrogenase. Investigation of stability of the hydrogenase from *T. roseopersicina* to metal ions has shown, that it considerably depends on temperature and time of an incubation [8, 9]. Inhibition of this enzyme in the presence of Cd^{2+} (K_i = 0,95) and Ni^{2+} (K_i = 3,8 mM) depends on pH too. In the presence of Ni^{2+} and Cd^{2+} ions inactivation of hydrogenase is reversible, whereas Cu^{2+} and Hg^{2+} ions induced an irreversible inhibition [1, 8, 9].

A great number of species of the family *Rhodospirillaceae* are investigated with respect of the stability to oxyanions of heavy metals [3]. Species of this family maintain redox-control due to reduction of tellurite (TeO_3^{2-}). For *Rh. sphaeroides* the mechanism of resistance to this group of tellurite involves membrane -bound FAD-dependent metal oxyanion reductase (MOR). As a result the intracellular precipitation of metal and significant hydrogen evolution under anaerobic photoheterotrophic conditions occur. The salts of TeO_3^{2-} and SeO_3^{2-} (~ 1 µg/ml), added to the cultural medium, kept it from the growth of Gram-negative bacteria.

Cyanobacteria

Many cyanobacteria have transport systems of metal ions from the environment in the cells. A transport system of Ni^{2+} with K_m = 17 nM, enabling to concentrate it approximately by 2700 times is known for *Anabaena cylindrica*. Absorption of Ni^{2+} depends on membrane potential and decreases in the dark or in the presence of metabolism inhibitors [1]. Cells of *Anacystis nidulans* absorb actively Cd^{2+}. This absorption is inhibited completely by Ca^{2+} and Zn^{2+}. Low concentrations $HgCl_2$ (6 µm) influence on the energy transfer inside phycobilisomes in *Spirulina platensis* and its increased concentrations (18 µm) suppress electron transport from water to methylviologene. The absorption of metals in the cells of *Cyanidium caldarum*, grown in the light, depends on E_h: (mg/g dry weight): Cu (72, E_h 40 mV) > Fe (68, E_h 400 mV) > Zn (3,56, E_h 280 mV) > Cr (1,95, E_h 40 mV) > Pb (0,43, E_h 40 mV).

Passive accumulation of Mn, Co, Zn, Ag, Sn, Sn, Cs, Hg, Np, Pu and Am was determined for *Synechococcus* sp. With concentration factors from 0 (Cs and Np) up to ~ 10^6 in the row: Pu ~ Hg ~ Sn > Am > Zn > Co > Cs ~ Np [1]. Biomass of *Phormidium valderianum* is efficient for biosorption of toxic metals. It absorbs 65-70% of Cd and Co from solution, containing 25 mM each metal ions. Immobilized into polyvinyl particles cyanobacteria absorbed Cd and Co actively. About 80% bound metal ions were desorbed with 0,1 N HCl.

Alga

The cells of *Chlorella* and *Scenedesmus* sp. bind most actively Au^{3+}, Ag^+ and Hg^{2+} at pH 2. Irrespectively, live or dead cells are used for biosorption, it takes place their marked affinity to ions of gold [1, 2]. The cells of *C. pyrenoidosa* immobilized on the column with silicagel absorb the gold in reversible way. More than 50 cycles of filling the column with tetrachloraurate solution at pH 1,5 and washing it off with HCl and acidic thiourea were replicated without loss of sorbent efficiency. The experiments performed with Au^+-tetrathionate Na and *Chlorella vulgaris* on the same column showed that both algae could be used both for efficient biosorption of gold and its removing. They may be used for extraction of gold from very diluted solutions, including ocean waters.

Relative value of biosorption to general absorption and absorbing capacity of metal ions varies greatly for different types of algae. For *Ankistrodesmus brauni* and *Chlorella vulgaris* the binding of Cd with cell walls satisfies 80% from general absorption. *Chlorella vulgaris* is the main absorbing environmental component for other metals, including uranium [1]. For living cells of *Chlorella vulgaris* the efficiency of binding metals occurs as follows: $UO_2^{2+} > Cu^{2+} > Zn^{2+} > Cd^{2+} > Ni^{2+} > Sr^{2+}$. For cell walls of *Vaucheria* sp. the order of accumulation was: $Cu^{2+} > Sr^{2+} > Zn^{2+} > Cd^{2+}$[1], and for marine algae the following order was obtained: $Hg^{2+} > Ag^{+} > Zn^{2+} > Cd^{2+}$ [2].

Au^{3+}, Au^{+} and other complexes of gold could be efficiently bound by algae biomass; concentration of gold in them could reach 10% of dry weight. Au^{3+} was shown to reduce up to Au^{+} with further elementary gold in colloid state.

Biosorption of metals can be efficient even in the presence of other cations or protons. Ca^{2+}, Mg^{2+}, Na^{+}, Mn^{2+}, Zn^{2+}, Co^{2+} and Ni^{2+}, but not K^{2+}, inhibit the absorption of Cd^{2+} by *Chlorella vulgaris*; Cd^{2+} hinders binding of Mn^{2+}. In various algae the absorption of two-valent ions of Cd, Cu, Zn and Mn is mainly decreased at low pH. However, it depends on concentration of element. *Chlorella vulgaris* was shown to be able to absorb Au^{3+}, Ag^{+} and Hg^{2+} at pH 2. Density of cells can also influence on efficiency of biosorption and it is reduced with increasing biomass concentration.

Plant and fungi *Pleurotus ostreatus*

Our data showed that aqueous fern *Azolla caroliniana*, representing symbiotic association with cyanobacterium *Nostoc azollae*, purple (*Rhodospirillum* sp.) and chemotrophic (*Artrobacter* sp.) bacteria, as well as other consortia of aqueous plants with microorganisms (*Eichornia crassipes, Salvinia herzogii*) absorbed actively of metal ions *(table 1, 2)*.

Table 1. Content of metals (mg/kg) in *Chlorella, Azolla, Eichornia* and *Pleurotus ostreatus*

Cultures	Cu		Zn		Pb		Cd	
Chlorella	sp.2083¹	254²	6771¹	827²	208¹	25²	**104¹**	13²
A. caroliniana	734	86	1405	163	70	8	**7,03**	0,82
*A. caroliniana**	1786	124	3075	214	79	5,52	**19,84**	1,38
E. crassipes, leaves	1216	83	1419	97	54	3,7	**13,5**	0,92
E. cassipes, stem	1024	102	1706	170	51	5,1	**3,4**	0,34
E. crassipes, leaves**	2410	151	3615	227	120	7,6	**12,05**	0,76
E. crassipes, stem**	1612	95	9677	569	323	19	**32,3**	1,9
Pl. ostreatus	n.d.	1,0	n.d.	13,7	n.d.	no	**n.d.**	no

• Growth at 1 - 20 ppm; ** Biomass in blooming; 1) Ashes; 2) Dry biomass.

The data obtained testify that Azolla absorbed actively metals at their low concentration in the medium (1-20 ppm). The absorption of metal ions by Azolla occurred in two stages: active and passive. The metals actively absorbed during the grown of Azolla in ponds or other water reservoirs. Passive absorption takes place when the effluent passes through biomass in filter. In these conditions heavy metals are concentrated in biomass of Azolla as much as 500-1000 times within 2-7 days of the growth and the content of metals in it reaches 1% of Cu, Cd, Zn, U, Ag and 0,3% of Cr^{6+} and Ti. About 40-60% of heavy metals were removed from the water.

Total content of metal in Azolla ashes reached 5%. Active absorption was carried out at pH 4-6 with waste water of electronic and electrolysis industry. At passive absorption the filters made from dried Azolla, bound 3,2% Ag, 4,2% Cd, 2,7% Ni, 2,1% Zn and 3,9% Cu, containing up to 3000 ppm metal. Total content of metal in Azolla ashes reached 20. Up to 99% Cd and Ni was removed from industrial waste water, containing 1383 ppm Cd and 2026 ppm Ni. Maximal amount bound by Azolla metal reached 2,8%. Up to 74% Cr^{6+} was removed by Azolla from waste water, containing it in concentration of 137 ppm. Passive processes of metal binding run at pH 3-9 and they are applied in the case of high concentrated metal wastes in electronic and metallurgic industry. *Chlorella* absorbed most actively Cu, Zn, Pb and Cd, and *A. caroliniana* -nickel *(table 1)*. The content of metals in plants ashes depended both on the type of plants, and its separate parts (laves, stem) and the stage of plant development.

Table 2. Efficiency of Ni^{2+} absorption by various sorbents

Sorbent	Maximal concentration, mg/l	pH	Absorption, mg/l
Absorbite	100	6,5-7,3	8,1
Azolla caroliniana	1000	6,5	43,4
Eichornia crassipes	2	5,0-6,6	11,6
Salvinia herzogii	2	5,0-6,6	14,4

Increasingly more attention is given in many countries to cultivation of amaranthus due to high concentration of protein, balanced by irreplaceable amino acids, oil, pectin, pigments, vitamins and other physiologically active compaunds, as well as high biological activity. In its seeds, as showed our researches with *Amaranthus cruentus* Don Pedro, besides with high concentration of protein (18-23%), oil (~ 10%), phosphorus and calcium there were metals, important for vital activity of the living organisms, such as: (mg/kg): B - 9; Na -24; Mg - 2800; Se - 0,1; Al - 24; K - 5700; Ca - 177; Mn - 33; Fe - 119; Cu - 7; Zn - 62; Ba - 7. This fact is important for use of amaranthus flour as an additive in various products of diet therapy.

INTRACELLULAR ACCUMULATION OF METALS AND THEIR TRANSPORT

Absorption of metals dependent on metabolism, is the more weak process than passive absorption. It is suppressed at low temperatures, absence of energy source, by inhibitors of metabolism and uncouples of an electron - transport system [1, 2]. Nature and composition of culture medium and physiological state of the cells have an influence on the rate of their absorption ability [10]. For the metals, required for the growth and metabolism of microorganisms, there are transport systems of various specificity for their accumulation from the medium. However, even unessential metals are also absorbed by such systems. So, in some bacteria and algae Cd enters through Mn - absorbing system. In cyanobacterium *Anabaena sphaeroides* Zn^{2+} and Co^{2+} suppress the active transport of K^+ and Na^+, disturbing permeability of the cell membrane. There is a dependence of resistance of this cyanobacterium on the growth phase. Its cells exhibit maximum sensitivity to Co at the initial phase of the growth, and minimum are in the end. It should be mentioned that Co ions are an order more toxic that Zn ions (0,05 mg/l). Green algae *Solenastrum* sp. is more resistant to toxic effect of Zn^{2+} (0,12 mg/l). At present the mechanisms of transport participating in accumulation of heavy metals by microorganisms are not yet understood. The amount of metal accumulated at energy - dependent transport may be much higher than at passive absorption, though there are exceptions. *Ch. vulgaris, Ch. pyrenoidosa, S. platensis* and *Cyanidium caldarum* bind Ca^{2+} and Mg^{2+} weakly, and this hinders binding of heavy metals ions, that is an important advantage of these algae above ion - exchange resins at their remove from hard and waste water [10].

Energy - dependent absorption of Cr by cyanobacterium *Anabaena doliolum* exhibits two-phase behaviour and dependence on concentration of chrome [4]. Unlike other metals which are mainly in cationic forms, Cr presumably exists in the form of oxyanion CrO_4^{2-} which is not blocked by anionic components of bacterial shell. There are only a few data concerning ion chrome transport in algae. The data available testify that green algae retain more chrome, as well as Al and Fe, as compared to brawn and red algae. Epiphytic algae, living on plants, exhibit high affinity to atmospheric pollutants and are capable of accumulating heavy metals from air. The content of Cr and Pb was higher in epiphytic alga *Pleurococcus* sp. in places near the highways.

Cr can enter the cells of yeast through nonspecific anion carriers, permease systems, transferring various anions, like sulfate and phosphate [4]. Some tolerant to chromate mutants of *Neurospora crassa* show strong regenerating properties at sulfate transport. The researches conducted showed that toxicity of CrO^{2-} is due to the specific antagonism to sulfate absorption, whereas the toxicity of Cr^{3+} is the result of antagonism to Fe transport.

As for transport and accumulation of metals and their forms by consortia of aqueous plants with microorganisms no reliable data available. We *(table 1, 2)* and other authors gave only the efficiency of biosorption of a number of metals by some organisms and their organs (roots, leaves, stalks) and showed the possibility of their application for purification industry and agricultural waste water [1-3, 6]. The data obtained are also important for obtaining medical drugs for treatment of microelemental diseases [7].

REFERENCES

1. Gogotov I.N., Zorin N.A., Zadvorny O.A. Accumulation of metals by phototrophic microorganisms and their extraction. In: Ecology and soils. Moscow: POLTEX. 1999. V.3. P.238 - 251 (in Russian).
2. Gadd G.M. Accumulation of metals by microorganisms and algae. In: Biotechnology (H.-J.Rehm ed.). VCH: Weinheim. 1988. V.6 b. P.401-433.
3. Moore M.D., Kaplan S. Members of the family Rhodospirillaceae reduce heavy-metal oxyanions to maintain redox poise during photosynthetic growth. ASM News. 1994. V.60. P.17-23.
4. Cervantes C., Campos-Garcia J., Devars S. et al. Interactions of chromium with microorganisms and plants. FEMS Microb. Rev. 2001. V.5. P.335-347.
5. Sag Y., Acikel U., Aksu Z., Kutsal T. A comparative study for the simultaneous biosorption of Cr (VI) and Fe (III) on C.vulgaris and R.arrhizus: application of the competitive absorption models. Process Biochem. 1998. V.33. P.273-281.
6. Schmeichen H., Wittig H., Martin S. Enfernung von Schwermetallen aus Abwassern mittels Sorption am Biopolymer des Microorganuismus Ectothiorhodospira shaposhnikovii. BioEngineering. 1992.?1. S.38-41.
7. Metal ions in biological systems. Concepts on metal ion toxicity(H. Sigel, A. Sigel eds.). Moscow: Mir. 1993. 366 P.
8. Serebryakova L.T., Zorin N.A., Karpilova I.F., Gogotov I.N. Temperature dependence of anaerobic oxidation of metals mediated by bacterial hydrogenases. Appl. Biochem. Microbiol. 1997. V.33. P.282-285.
9. Zadvorny O.A., Zorin N.A., Gogotov I.N. The effect of metal ions on hydrogenase of purple sulfur bacterium Thiocapsa roseopersicina. Biochemistry. 2000. V.65. P.1525-1529 (in Russian).
10. Ahlf W. Recovery of metals from acid waste water by Cyanidium caldarium. Appl. Microbiol. Biotechnol. 1988. V.238. P.512-513.

*Metal Ions in Biology and Medicine: vol. 7.*Eds L. Khassanova, Ph. Collery, I. Maymard, Z. Khassanova, J.-C. Étienne.
John Libbey Eurotext, Paris © 2002 pp. 252-256.

Physiologic studies in Cr(VI)-resistant and Cr(VI)-reducing *Ochrobactrum* spp. 5 bvl-1

Rita Branco[1], M.C. Alpoim[2], V.M. C. Madeira[2] and P.V. Morais[1]

[1]*Instituto do Ambiente e Vida, 3004- 517 Coimbra, Portugal.*
[2]*Departamento Bioquímica, Faculdade de Ciências e Tecnologia da Universidade de Coimbra, Apartado 3126, 3001-401 Coimbra. Portugal.*

ABSTRACT

The increasing interest in the bioremediation potential of bacteria lead us to characterized strain 5-bvl-1, previously isolated from a chromium-contaminated environment, for its ability to resist and to reduce Cr(VI) either in the presence or in the absence of other toxic metal contaminants. The strain was identified as *Ochrobactrum tritici* by comparative analysis of the 16S rDNA gene sequence and DNA-DNA hybridisation (>70% homology). The bacterium was resistant to a broad range of antibiotics and to several metal ions such as Cr(VI), Ni^{2+}, Co^{2+}, Cd^{2+} and Zn^{2+}. The strain was able to grow, under aerobic conditions, in as much as 10 mM Cr(VI). Cr(VI) increasing concentrations in the medium decreased the growth rate and the maximum growth yield. Strain 5 bvl-1 was also able to reduce Cr(VI). Although there is a measurable Cr(VI) reduction during the exponential growth phase of the culture, it only become more evident when the culture reached the stationary phase. High Cr(VI)-reduction yields and low Cr(VI)-uptake levels were achieved on cultures obtained by using a starting low cell density and a medium containing 1 mM Cr(VI). Our results suggest that Cr(VI) reduction is an energy demanding process and therefore depends on the metabolic state of the cells and the energy available. The results obtained also hold for the possibility that different Cr(VI)-reduction mechanisms are ought to be operating. The findings that the other strains of the genus in spite of not being able to grow in the presence of Cr(VI) were able to reduce Cr(VI), corroborate the idea that the Cr(VI)-reduction ability does not confer Cr(VI)-resistance.

INTRODUCTION

Hexavalent chromium (Cr(VI)) is a strong oxidant, and this property is likely related to its toxicity for most organisms. Micro-organisms are generally the first category to be exposed to environmental heavy metal contamination, therefore, they acquire a variety of mechanisms to detoxify metal ions. Bacterial resistance to Cr(VI) has been found in several bacterial species [1, 2] and recently we reported the isolation of several Cr(VI)-resistant bacteria belonging to the β-*Proteobacteria*, high G+C Gram-positive bacteria and genus *Acinetobacter* [3].

Serious concerns about the toxicity of Cr(VI) compounds necessitates recovery and reuse of chromium from industrial wastes or at least rendering it to a less toxic form. The use of bacteria in bioremediation of Cr(VI)-contaminated soils and ground-waters has achieved growing attention, because the biological reduction of Cr(VI) usually generates an insignificant quantity of chemical sludge. Recently, a Cr(VI)-resistant and Cr(VI)-reducing bacterium, was isolated from a chromium-contaminated wastewater treatment plant from a tannery industrial area, and since there is an interest in the bioremediation potential of bacteria, the isolated strain, was identified using quimiotaxonomic tools and DNA sequence as *Ochrobactrum*, and was fully characterized for its ability to resist and to reduce Cr(VI) either in the presence or in the absence of other toxic metal contaminants.

MATERIALS AND METHODS

Bacteria characterisation was done by FAME analysis [4] and API 50CH, G+C content [5], 16S rDNA sequencing [6] and DNA-DNA hybridization [7]. Bacteria was grown in buffered mineral medium [4], and growth was assayed for Cr(VI) concentrations ranging from 0-10 mM. Metal resistance induction was tested in presence of different Cr(VI) concentrations (1 to 3 mM), using as inocula cells grown, in the absence or in the presence of 2 mM Cr(VI).

The strain Cr(VI) reduction was assayed for Cr(VI) concentrations ranging from 0-10 mM and the Cr(VI) concentration was determined as reported [4].

Resistance and reducing experiments were also performed with a starting cell density adjusted either to 0.8 or 2.0 OD and inocula grown in the presence of Cr(VI) for strain 5-bvl-1 or in the absence of Cr(VI).

The resistance of strain 5 bvl-1 to $NiCl_2$ (Ni^{2+}), $CoCl_2$ (Co^{2+}), $CdCl_2$ (Cd^{2+}) and $ZnSO_4$ (Zn^{2+}) was tested by growing the strain in the presence of 1 mM of each metal salt. The resistance of the strain to the simultaneous presence of 1 mM Cr(VI) and a 0.5 mM concentration of each metal salt was also evaluated.

RESULTS

Comparative analysis of the 16S rDNA gene sequence identified strain 5bvl-1 as *O. tritici* and DNA-DNA hybridization confirmed the identification (>70% homology). The bacterium was resistant to a broad range of antibiotics. In contrast to other strains of the genus *Ochrobactrum* this strain was able to grow, under aerobic conditions, in as much as 10 mM Cr(VI) and the presence of 1 mM Cr(VI) in the medium did not affect the growth yield although a decrease in growth rate was observed. Cr(VI) concentrations higher than 1 mM reduced both the growth rate and the maximal OD achieved. Strain 5 bvl-1 was also able to resist to 1 mM Ni^{2+}, Co^{2+}, Cd^{2+} and Zn^{2+}. The presence of Zn^{2+} did not affect the growth rate. In contrast, either Ni^{2+} and/or Co^{2+} increased the lag phase and decreased the growth rate. Except Co^{2+}, that decreases the growth yield, none of the other metal salts affects the growth yield. When the strain was incubated simultaneously with 1 mM Cr(VI) and 0.5 mM of one of each of the other metals, the growth rate was similar to that observed in the presence of 1 mM Cr(VI), and just Co^{2+} and Cd^{2+} decrease the growth yield *(fig. 1)*.

Metal resistance induction was performed using as inoculum cells grown in the absence or in the presence of 2 mM Cr(VI) *(fig. 2)*. The growth rate and the growth yield were the same independently of using as inoculum cells grown in the presence or in the absence of Cr(VI). The decreased of the growth rate was only related with the increase of Cr(VI) in the growth medium.

The strain was also able to reduce Cr(VI). The Cr(VI)-reduction yield of the strain was dependent on Cr(VI) concentration, on the population growth curve and of the extent of the experiment. Increasing Cr(VI) concentrations decreased the percentage of Cr(VI) reduction, but when considering the total amount of Cr(VI) reduced, increasing concentrations of Cr(VI) in the medium, from 1 mM to 3 mM, increased the total amount of Cr(VI) reduced. For 1 mM Cr(VI) it was observed the total disappearance of Cr(VI) from the medium. Reducing or removing the exponential phase of the growth curve (increasing the starting cell density) also lead to a decrease in Cr(VI) reduction percentage *(fig. 3)*. As illustrated in *fig. 3* the strain Cr(VI)-reduction yields increased notoriously after 24 h incubation. For exposition times to Cr(VI) up to 24 h, growing cultures showed lower Cr(VI) reduction yields than cells suspensions, but the opposite was observed for exposition times longer than 24 h.

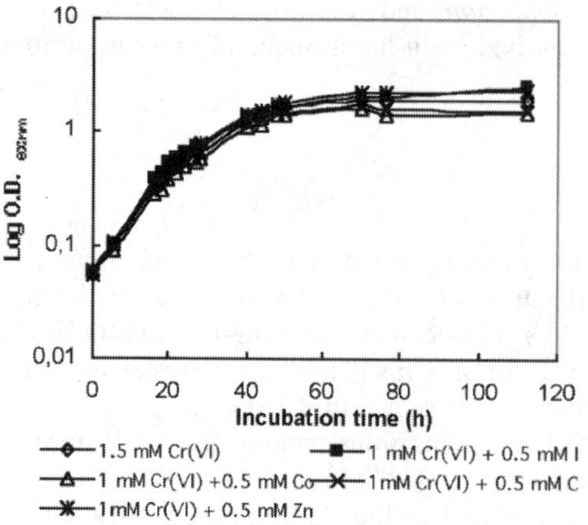

Fig. 1. The resistance of the strain to the simultaneous presence of 1 mM Cr(VI) and a 0.5 mM concentration of each metal salt. Growth was assayed by measuring turbidity with a spectrophotometer at 600 nm

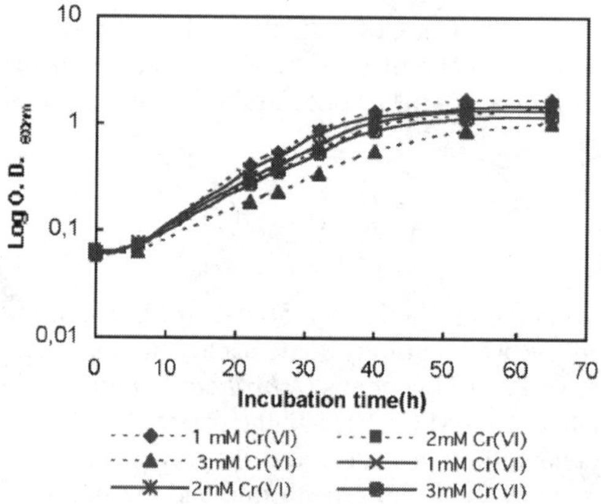

Fig. 2. The effect of the presence of Cr(VI) in the growth of strain 5-bvl-1. Cells used as inoculum were grown in the absence (dashed lines) or in the presence of 2 mM Cr(VI) (solid lines). Growth was assayed by measuring turbidity with a spectrophotometer at 600 nm

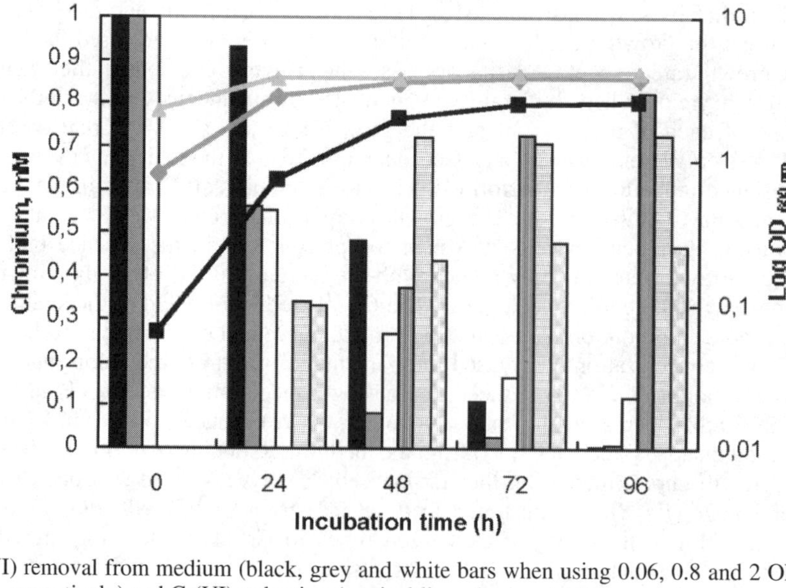

Fig. 3. Cr(VI) removal from medium (black, grey and white bars when using 0.06, 0.8 and 2 OD inoculation cell density, respectively) and Cr(VI) reduction (vertical lines, horizontal lines and dots when using 0.06, 0.8

and 2 OD inoculation cell density, respectively) by strain 5-bvl-1 inoculated in 1 mM Cr(VI) at different cell densities: 0.06 (■), 0.8 (□) and 2.0 (π). Growth was assayed by measuring turbidity with a spectrophotometer at 600 nm. Cr(VI)-reduction was assayed by the diphenylcarbazide and Cr(VI)-uptake by absorption spectroscopy as already reported [4]

The reduction efficiency of strain 5-bvl-1, expressed as the level of Cr(VI)-reduction per unit of OD, was also dependent on, the same factors that affect the Cr(VI) reduction. The Cr(VI)-reduction efficiency increased up to 3 mM Cr(VI), then remained almost constant for concentrations ranging between 3 to 7 mM and increased again for increasing concentrations of Cr(VI) from 7 to 10 mM. Maximum Cr(VI)-reduction efficiency was achieved for 3 mM Cr(VI).

The other strains of the genus *Ochrobactrum* were not Cr(VI)-resistant but were able to reduce Cr(VI).

DISCUSSION

The strain identified as *Ochrobactrum tritici* unlike *O. tritici* type strain is not susceptible to polymyxin B and is not able to assimilate gluconate. The similar pattern of resistance to antibiotics of strain 5 bvl-1 and the type strains of *O. antropi* and *O. intermedium* suggests that, apparently, the resistance to chromium does not seem to be linked to antibiotic resistance, in spite of the reports that heavy metal pollution may contribute to increase antibiotic resistance through indirect selection [8].

The findings that, in contrast with the other species of the genus *Ochrobactrum* and even with the type strain of the species *O. tritici*, strain 5-bvl-1 is able to grow in presence of Cr(VI) is a strong indication that Cr(VI)-resistance is an acquired characteristic. Furthermore, the observed dependence of the strain growth yield and growth rate on Cr(VI) concentration, probably indicates that, increasing the Cr(VI) concentration, will lower the energy available for growth. This suggests that Cr(VI)-reduction is an energy demanding process with increasing Cr(VI) concentrations requiring more energy for the reduction of Cr(VI), to the lower oxidation state Cr(III) in order to prevent Cr(VI)-induced cytotoxicity and thus ensuring cell survival.

Strain 5-bvl-1 was also resistant to 1 mM of Ni^{2+}, Co^{2+}, Cd^{2+}, and Zn^{2+}, and the results obtained indicate that Cr(VI) and the other metals resistance effects seems to be not cumulative. Our results agree with the findings reporting that Cr(VI) resistance is due to reduced accumulation, whereas that of Co^{2+}, Ni^{2+} and Zn^{2+} result from inducible energy-dependent cation efflux systems [9].

The fact that the other strains of the genus *Ochrobactrum*, in spite of not being resistant to Cr(VI), are able to reduce Cr(VI), as strain 5 bvl-1 supports the idea that the chromate reduction ability found for several bacteria does not confer resistance to chromate [10].

Independently of the growth conditions of the inocula (growth in the presence or absence of Cr(VI)), and the presence or absence of an exponential phase, the Cr(VI) disappearance from the medium increased notoriously after 24 h incubation time of the bacteria to Cr(VI). This increase results not only from the time-dependent increase on cell numbers but also from the increase on Cr(VI)-reduction efficiency of the strain, which, may reflect different mechanism operating and/or the cells physiological status because when a standard growth was performed Cr(VI)-reduction was just significant at late exponential phase. In fact, the observed differences on Cr(VI)-reduction yields observed in stationary phase cultures and cell suspensions of the strain with the same cell numbers is explained by differences on Cr(VI)-reduction efficiencies. In addition, the finding that, the Cr(VI)-reduction efficiency increases with increasing Cr(VI) concentrations up to 3 mM, remains constant for Cr(VI) concentrations ranging between 3 to 6 mM, and then increases again for Cr(VI) concentrations up to 10 mM, is another indication that probably more than one mechanism is responsible for this strain Cr(VI)-reducing ability.

CONCLUSIONS

In the presence of 1 mM Cr(VI) the highest growth rate and reduction yield were obtained. The results obtained with the other strains of the genus *Ochrobactrum* allowed us to state that the Cr(VI) reduction ability does not confer resistance to Cr(VI).

REFERENCES

1. McLean J., Beveridge T.J. Chromate reduction by a Pseudomonad isolated from a site contaminated with chromated copper arsenate. *Applied and Environmental Microbiology*, 2000, 67, 1076-1084.
2. Peitzsch N., Eberz, G., Nies, D.H. *Alcaligens eutrophus* as a bacterial chromate sensor. *Applied and Environmental Microbiology*, 1998, 64, 453-458.
3. Francisco R., Alpoim M.C., Morais, M.P. Diversity of chromium-resistant and reducing bacteria in a chromium-contaminated activated sludge. *Journal Applied Microbiology*. In Press.
4. Branco R., Alpoim M.C., Morais P.V. A new bacteria resistant to several metal ions and able to reduce hexavalent chromium. *Metal Ions in Biology and Medicine*, 2000, 6, 661-663.
5. Mesbach M., Premachandran, U., Whitman, W.B. Precise measument of the G+C content of deoxyribonucleic acid by high-performance liquid chromatography International Journal Systematic Bacteriology, 1989, 39, 159-167.
6. Morais P.V., Mesquita C., Andrade J.L., Costa M. Investigation of persistent colonization by Pseudomonas aeruginosa-like strains in a spring water bottling plant. 1997, 63, 851-856.
7. De Ley J., Cattoir H., Reynaerts A. *Eur. J. Biochem.*, 1970, 12, 133-142.
8. Leff L., McArthur J.V., Shimkets J. Spatial and temporal variability of antibiotic resistance in freshwater bacterial assemblages. *FEMS Microbiology Ecology*, 1993, 13, 135-144.
9. Nies D., Silver S. Metal ion uptake by a plasmid-free metal-sensitive *Alcaligenes eutrophus* strain. *Journal Bacteriology*, 1989, 171, 4073-4075.
10. Lovley D., Dissimilatory metal reduction *in Annual Review of Microbiology*, 1993, 47, 263-290.

*Metal Ions in Biology and Medicine: vol. 7.*Eds L. Khassanova, Ph. Collery, I. Maymard, Z. Khassanova, J.-C. Étienne.
John Libbey Eurotext, Paris © 2002 pp. 257-262.

Toxicity of copper ions in the presence of vitamin C for the plasma membrane of *Escherichia coli* K-12 cells

Ivanov A. Yu.[1], Khassanova L. A[2], Khassanova Z. M[3], Gavryushkin A. B.[4], Markeliya L. Yu[3], Fesenko E. E.[1]

[1]Institute of Cell Biophysics, Russian AS, 142290 Pushchino, Russia; [2]Department of Environmental Protection of Bashkir State University, 32, Frunze Street, 450072, Ufa, Russia; [3]Department of Botany of Bashkir State Pedagogical University, 3a, October Revolution Street, 450025, Ufa, Russia; [4]State Research Institute of Applied Microbiology, 142279 Obolenck, Russia

ABSTRACT

The influence of copper ions in the presence of vitamin C (ascorbate) on the plasma membrane (PM) of *Escherichia coli* K-12 cells was studied. By the method of electroorientational spectroscopy, disturbances in the barrier properties of PM were found to occur in the presence of a catalytic copper-ascorbate system. Addition of H_2O_2 to this system increased the degree of cell damage although the introduction of hydrogen peroxide to intact cells and cells preincubated with copper ions did not produce any change in membrane permeability. The inhibition of endogenic catalase by preincubation of the cells with sodium azide did not affect the degree of damage to PM. The metal chelating agent EDTA suppressed completely the cell damage, whereas the well-known scavenger of OH^* radicals dimethyl sulfoxide, did not show protective properties under those conditions. When copper ions in the binary catalytic system were replaced with a large organic cobalt complex, hydroxycobalamine, ROS showed no toxic effect on PM. It is suggested that the toxic effect of copper ions observed in the presence of ascorbate is provided by hydroxyl radicals generated by the catalytic system on the surface of PM of the cells, whereas the bulky hydroxycobalamine complex cannot penetrate the cell wall.

INTRODUCTION

Reactive oxygen species (ROS) play the key role in the development of oxidative damage to cells. Earlier there were reports on the ROS-induced toxicity of copper (II)-ascorbate complexes for bacteriophages [Samuni et al.,1983], cancer cells [Bram et al., 1980], hepatocytes [Walt et al., 2000], as well as of copper (II)-adrenalin [Aronovitch et al., 1991] and copper (II)-catechol complexes for PM of *E. coli* cells [Hoshino et al., 1999]. At the same time, there is little information on the molecular mechanisms of damage to prokaryotic cells caused by ROS generated by metal-reducer systems, in particular on the changes in the barrier properties of bacterial membranes. The change in the barrier properties of membranes on ROS attacks has been judged from the passive leakage or loss of intracellular K^+ and ATP [Aronovitch et al., 1991; Hoshino et al., 1999]. Recently we used the method of high-frequency electroorientational spectroscopy to assess the damaging effect of OH^* radicals generated in a hydrogen peroxide-Fe^{2+} system in the presence of reducers on the barrier properties of PM of *E. coli* [Ivanov et al., 2000, a, b]. The chief agent that directly accomplishes the oxidative degradation of biosubstrates is believed to be the hydroxyl radical OH^*. Some authors however assume that hydrogen peroxide generated in such binary catalytic systems may be involved in the toxic effect [Hoshino et al., 1999].

The present study was carried out to investigate by electroorientation spectoscopy of cells the influence of ROS generated in a binary catalytic copper-ascorbate pair on the barrier properties of PM of *E.coli* K-12. Evidence is presented that the main toxic agent in this system is the hydroxyl radical.

MATERIALS AND METHODS

A wild strain of the heterotrophic bacterium *E. coli* K-12 was used. Cells were grown for 4 to 5 h (the exponential phase of growth) in a shaken flask (150 osc / min) at 37°C in a liquid nutrient medium, M9, with addition of 0.1% of yeast extract. Then the cells were precipitated and washed with distilled water by centrifugation. Washed cells were stored in a dense suspension (10^{10} cell/ml) throughout the experiment (2-4 h) at 4° C. Bidistilled water of a specific electrodonductivity of 1.2 10^{-4} Sm/m was used. Prior to the experiment, a water-salt solution with minor additions of 1.10^{-5} n HCl and $1.6.10^{-5}$ M NaCl (pH 5.4 and specific electroconductivity $1.2.10^{-3}$ Sm/m) was prepared. The acidification of the original water solution to pH 5.4 was dictated by that at neutral pH values, copper ions are hydroxylated and become more toxic for PM [Ivanov et al., 1997] and, hence, may mask the direct effect of HO* radicals on cells. Besides, at pH 5.4 of the medium the disturbances of the barrier properties of PM of cells are better revealed. Prior to measurements, cellular suspensions (5.10^{7} cell/ml) were prepared. Copper ions were introduced to cell suspensions, and the cells were preincubated at room temperature for 15 min. Then ascorbate (Asc) was added to the suspensions, the cells were incubated for 30 min at 25°C, the specific electroconductivity of suspensions was adjusted to $1.6\text{-}1.8.10^{-3}$ Sm/m with 0.01 M sodium chloride solution, and EO measurements were performed. In experiments with hydrogen peroxide the latter was introduced simultaneously with ascorbate. In some of the experiments, the cells, before being incubated with copper, were treated with sodium azide, to inactivate the endogenous catalase. The antioxidants sodium ethylenediaminetetraacetate (EDTA) and dimethyl sulfoxide (DMSO) were added to the suspension before ascorbate. The catalytic system hydroxycobalamine-ascorbate was introduced as follows: the cells were preincubated with 20-25 μM hydroxycobalamine at room temperature for 15 min, then 200-250 μM ascorbate was added, and the suspension was incubated at 37°C for 30 min. Then the cells were sedimented and placed to a medium with a specific electroconductivity of 1.6 10^{-3} Sm/m and pH 5.4. The electroorientational spectra (EO-spectra) of cells were obtained by measuring the relative change in cell suspension optical density due to orientation variations of cells in a uniform alternating electric field of a fixed frequency within 0.5-10 MHz, field intensity 8000 V/m. The EO measurements were carried out on an experimental device developed at the State Research Center of Applied Microbiology (Obolensk, Russia). Detailed analysis and interpretation of EO measurements are presented in [Miroshnikov et al., 1986]. Damage to PM of cells was judged from the course of the high-frequency decline of their EO-spectra by calculating the ratio of the β (beta) value of the EO effect of cells at a field frequency of 5 MHz to that at 0.5 MHz, and then the per cent of intact cells was calculated according to the formula:

$\alpha_\beta = \beta - \beta_1 / \beta_0 - \beta_1.100\%$,

where β_0 and β_1 are the ratios of EO effects for intact and completely inactivated (70°C, 15 min, water bath) cells [Fomchenkov et al., 1986]. The following fresh prepared reagents were used: 0.01 M base solution of cupric chloride ($CuCl_2$) (sigma, USA), 0.1 M solution of ascorbic acid adjusted by KOH to pH 5.4; hydrogen peroxide (3%) (Reachim Russia); base 0.01 M solutions of dimethyl sulfoxide (DMSO) (Sigma, USA); sodium ethylendiaminetetraacetate (EDTA); sodium azide (NaN_3) (Reachim, Russia); hydroxycobalamine (vit B_{12b}) (Moscow P.A. Herzen Research Institute of Oncology, Russia).

RESULTS AND DISCUSSION

In preliminary experiments, we determined the concentrations at which copper ions were non-toxic for PM of cells studied but served as catalysts of ascorbate oxidation and ROS formation. Preincubation of cells with copper enabled the metal to settle down on the membrane surface. This provided the immediate vicinity of the target molecule in PM to short-lived ROS whose generation was triggered by addition of ascorbate. EO-spectra of *E. coli* K-12 cells: (1) intact, (2) incubated with copper ions, (3) preincubated with copper ions and then treated with 100 μM ascorbate are presented in *fig. 1*. It can be seen that addition of ascorbate to the cells preincubated with copper ions made the high-frequency decline in the spectrum of these cells to shift to lower frequences. Similar changes in the high-frequency spectrum of cells were also observed upon simultaneous introduction of copper ions and ascorbate to cell suspension but were less pronounced. Incubation of cells with only copper ions for a time interval equal to the overall time of treatment of cells in the system copper-ascorbate did not change the EO-spectrum. It should be noted that introduction of ascorbate alone to the cell suspension did not change as well the character of the EO-spectrum.

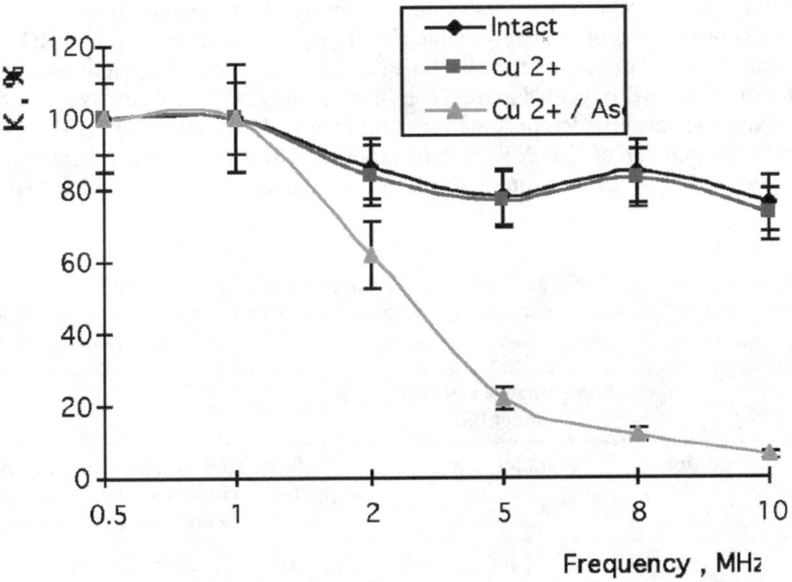

Fig. 1. EO-spectra of intact (1), incubated (45 min) with copper ions (2 μM) (2), preincubated (15 min) with copper ions (2 μM) and then incubated (30 min) with ascorbate (100 μM) (3) *Escherichia coli*K-12 cells $K = (\Delta D_1 / \Delta D_0) \times 100\%$, where ΔD_1 is the change in optical density of cell suspension on switching of the electric field; ΔD_0 is the same for the suspension of intact cells at a field frequency of 0.5 MHz

Previously changes in the high-frequency region of the EO-spectrum were reported for various kinds of bacterial cells with disturbed barrier properties of their PM due to the action of heavy metals [Ivanov et al., 1992; Khassanova, 1996]. Changes in PM permeability of cells are accompanied by exit of potassium and some low-molecular compounds to the extracellular medium (of low ion concentration), which causes a decrease in the electroconductivity of the cytoplasm and the cell as a whole. Theoretically [Miroshnikov et al., 1986], this phenomenon causes the above-mentioned changes in the high-frequency decline of the EO-spectrum.

Fig. 2 presents the data on the extent of damage to PM of *E.coli* K-12 cells treated with copper (1, 2 μM) or with catalytic pairs of various concentration ratios of components (1, 2 μM Cu^{2+} / 50, 100 μM ascorbate, respectively). Copper ions themselves at concentrations used did not produce any change in the barrier properties of PM of cells, whereas the percentage of cells with intact PM markedly decreased after addition of ascorbate to the cell suspension. This effect was probably related to attacking the cell membrane by ROS generated in the system Cu^{2+} - ascorbate. The value of the toxic effect depended on both of the components of the catalytic system but primarily on the dose of copper ions.

Simultaneous introduction of ascorbate and H_2O_2 to cells preincubated with copper ions increased the number of cells with damaged PM *(table)*. When intact cells or cells preincubated with copper ions were treated with much higher H_2O_2 concentrations but with no ascorbate, the barrier properties of PM were not disturbed.

Suspensions of cells preincubated with copper ions were treated with antioxidants: EDTA or DMSO prior to addition of ascorbate. EDTA inhibited completely damage to PM of the cells. With DMSO, no protective effect on PM was observed *(table)*.

In order to determine what effect the endogenous catalase, the enzyme of antioxidative protection of cells, exerts on the damage of PM by exogenous ROS, the cell suspension, prior to being treated with the catalytic pair, was incubated with sodium azide, inhibitor of catalase. The experiments did not reveal any relation between the activity of endogenic catalase and the degree of disturbance of the barrier properties of PM under the action of exogenous ROS *(table)*.

To assess the contributions of relatively stable (O_2^-, H_2O_2) and short-lived (OH^*) ROS generated in catalytic metal-reducer pairs to the toxicity for PM, the copper ions were replaced with a large organic cobalt complex, hydroxycobalamine. The high toxicity of the binary system hydroxycobalamine-ascorbate was recently demonstrated for tumor cells [Lemeshko, 2001]. The introduction of ascorbate to the suspension of *E. coli* K-12 cells preincubated with hydroxycobalamine triggered reactions with the formation of the same radicals but no damage to PM was observed therewith *(table)*.

Table Effect of H_2O_2, EDTA, DMSO, and NaN_3 on the number of *E. coli* K-12 cells with intact PM for intact suspensions and those treated with catalytic systems copper (II) - ascorbate and hydroxycobalamine - ascorbate

	Percentage of cells with intact PM	
Cell treatment conditions	**intact cells**	**cells treated with the catalytic system copper-ascorbate (ratios of components are: 2 mkM / 100 mkM or 1 mkM / 50 mkM*)**
control	100	18±5; 45±5*
50 mkM?$_2$?$_2$	-	33±5*
200 mkM?$_2$?$_2$	100	-
100 mkM EDTA	100	100
200 mkM DMSO	100	16±5
200 mkM NaN_3	100	43±5*
25 mkM Vit.B_{12b}/ 200 mkM?sc	100	-

Analysis of our results indicates that the catalytic system copper (II)-ascorbate changes the membrane permeability of *E. coli* K-12 cells, whereas each of its components individually at the same concentrations does not *(fig. 2)*. A similar picture was observed when *E. coli* cells were treated with a catechol-copper complex [Hoshino et al., 1999]. Obviously, the main contribution to the toxicity of such systems is made by ROS they generate. In our studies, the number of cells with damaged PM was primarily determined by copper (II) ion dose *(fig. 2)*, which confirms the significance of redox transformations of Cu^{2+} for membrane damage. The same is evidenced by the high degree of protection of PM against ROS by a chelating agent, EDTA *(table)*. The binding of copper ions by the chelator "switches off" the process of generation of membranotoxic ROS.

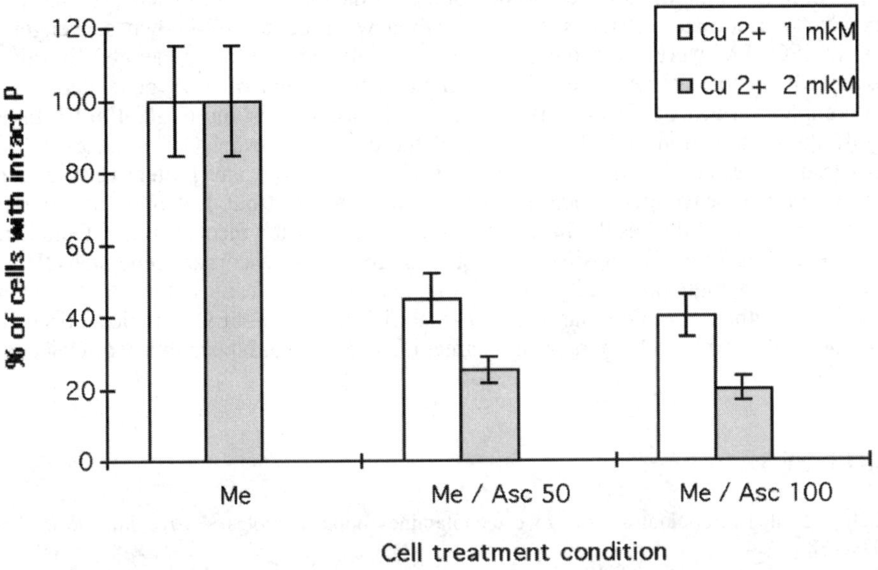

Fig. 2. The number of *E. coli* K-12 cells (%) with intact PM for cell suspensions (1) incubated with copper ions (1, 2 μM) and (2) preincubated with the same concentrations of copper ions and then incubated in the presence of two ascorbate doses (50 and 100 μM)

The authors of [Hoshino et al., 1999] believe that an important role in the disturbance of the barrier function of bacterial cell PM under the action of a catalytic metal-reducer pair is played by hygrogen peroxide formed locally on PM. An increased exit to the outer medium of intracellular ATP and K^+ pointing to a damage to PM of *E. coli* cells upon addition of H_2O_2 to the adrenaline-copper complex has been reported [Aronovitch et al., 1991]. The results of our studies have cast some doubt on the idea of a direct damage of PM by H_2O_2. Treatment of cells with increased concentrations of hydroperoxide (200 μM) in the absence of ascorbate did not cause changes in membrane permeability because the initiation of the recycling redox system that provides the generation of membranotoxic ROS becomes impossible without a reducer. The increase in the number of cells with damaged PM observed in the experiments with a simultaneous action of a copper-ascorbate complex and H_2O_2 on the cells is evidently due to that here the exogenous hydrogen peroxide acts as an additional substrate for generation of OH^* radicals.

Some authors [Hoshino et al., 1999; Medentsev et al., 2001] pointed out that there exists a correlation between the level of survival of cells exposed to H_2O_2 and the activity of endogenous catalase. In our case, the dissociation of hydrogen peroxide by endogenous catalase might affect the general level of OH^* radicals and hence the degree of disturbance of the barrier properties of PM. But differences in the character of influence of the copper-ascorbate complex on PM of intact and sodium azide-treated cells were not revealed. Most probably, this is related with the location of the ROS generator on the surface of or inside PM, which excludes the availability of endogenous catalase for ROS.

The location of the catalytic complex determines the possibility of interaction of oxygen radicals with membrane targets. The absence of membranotoxic effects in the case of treating the cells with the catalytic system hydroxycobalamine - ascorbate points to the significance of spatial vicinity of ROS and membrane target. The large-sized organic complex of cobalt is probably unable to overcome the outer membrane of the cell. The remoteness of target object, PM, makes rather difficult the interaction of short-lived ROS with PM components.

Thus, the results of our investigations confirm the idea of the leading role of the most reactive ROS, OH^* -radical, in the disturbance of membrane permeability of cells treated with the catalytic complex Cu^{2+} - ascorbate. At the same time, when we used a well-known scavenger of OH^* -radicals, DMSO, PM were not protected against the attacks of ROS generated by this system. The absence of the protective effect of DMSO upon the action of ROS generated by a Cu^{2+} - catechol complex on PM of *E.coli* cells was reported earlier [Hoshino et al., 1999]. In addition, in studying the toxic action of OH^* - radicals of the catalytic system Cu^{2+} - ascorbate on bacteriophages [Samuni et al., 1983], the authors did not find as well any protective effects of other known OH^* -radical scavengers, such as sucrose, mannitol, tert-butyl alcohol, and polyethylene glycol. They interpreted the results in terms of the "site-specific" mechanism of Fenton reaction. The cyclic redox-reaction of a transition metal in the presence of ascorbate generates OH^* -radicals which react, with a high probability, with biological macromolecules thus producing multiple lesions. It is the authors' opinion that this "site-specific" formation of OH^* radicals occurring near the target molecules explains both their high damaging capacity and the inability of OH^* scavengers to prevent the damage.

REFERENCES

Aronovitch J. et al. Bactericidal activity of catecholamine copper complexes. *Free Rad. Res. Com.* 1991. 12-13: 479-488.

Bram S. et al. Vitamin C preferential toxicity for malignant *melanoma* cells. *Nature*. 1980. 284: 629-631.

Fomchenkov V.M. et al. Electric characteristics of bacterial cells measured when the barrier function of cytoplasmic membrane is disordered. *Microbiologiya* (Moscow). 1986, 55: 754-759.

Hoshino N. et al. Damage to the cytoplasmic membrane of *E. coli* by catechin-copper (II) complex. *Free Rad. Biol. and Med.* 1999, 27: 1245-1250.

Ivanov A. Yu. et al. The influence of heavy metals ions on the electrophysical properties of *Anac. nidulans* and *E. coli. Microbiologiya* (Moscow). 1992, 61: 455-463.

Ivanov A. Yu. et al. Toxic effect of hydroxylated heavy metal ions on the plasma membrane of bacterial cells. *Microbiologiya* (Moscow). 1997, 66: 588-594. a.

Ivanov A. Yu. et al. Fe^{2+} induced hydroxyl radicals effects on the plasma membrane of *Escherichia coli* K-12 cells. *Metal Ions in Biology and Medicine*. John Libbey Eurotext (Paris). 2000, 6: 490-494. b.

Ivanov A. Yu. et al. Comparative efficiency of antioxidants upon the protection of bacterial plasma membrane from active forms of oxygen. *Biofizika* (Moscow). 2000, 45: 660-665.

Khassanova L. A. Electrophysical analysis and physiological-biochemical particularities of cell damages by heavy metal ions. Thesis (St. Petersburg). 1996.

Lemeshko V.V. Research of the mechanism cytotoxicity of operation hydroxycobalamine (vitamin?12b) in a combination with ascorbic an acid on tumor cells *in vitro*. Thesis (Puschino). 2001.

Miroshnikov I. A., Fomchenkov V.M., Ivanov A.Yu. Electrophysical analysis and cell separation, Nauka (Moscow). 1986, 198 p.

Medentsev A.G. et al. Adaptation of the phytopathogenic fungus *Fusarium decemcellulare* to oxidative stress. *Microbiologiya* (Moscow). 2001, 70: 34-38.

Samuni A. et al. On the cytotoxicity of vitamin C and metal ions. A site-specific Fenton mechanism. *Eur. J. Biochem.* 1983, 137: 119-124.

Walt N.T. et al. The role of free radicals in copper mediated toxicity in hepatocytes. *Metal Ions in Biology and Medicine*. John Libbey Eurotext (Paris). 2000. 6: 481-483.

Metal Ions in Biology and Medicine: vol. 7. Eds L. Khassanova, Ph. Collery, I. Maymard, Z. Khassanova, J.-C. Étienne.
John Libbey Eurotext, Paris © 2002 pp. 263-267.

Sensitivity of *Fusarium solani* plasma membrane to heavy metal ions and reactive oxygen species

Ivanov A.Y.[1], Khassanova L.A[2], Khassanova Z.M[3], Kuzin A.I.[4], Gavryushkin A.B.[5], Markeliya L.Y.[3], Azidbekyan P.P.[4]

[1]Institute of Cell Biophysics, Russian Academy of Sciences, 142290 Pushchino, Russia; [2]Department of Environmental Protection of Bashkir State University, 32, Frunze Street, 450072, Ufa, Russia; [3]Department of Botany of Bashkir State Pedagogical University, 3a, October Revolution Street, 450025, Ufa, Russia; [4]State Research Institute of Genetics, Ministry of Industry, Science and Technology, 1st Dorozhny Proezd, 1, 113545 Moscow, Russia; [5]State Research Institute of Applied Microbiology, Russian Academy of Sciences, 142279 Obolenck, Russia

ABSTRACT

The method of electroorientational spectroscopy was used to study disturbances in the barrier properties of the plasma membrane (PM) of fungus *Fusarium solani* cells under the action of heavy metal (HM) ions (Ag^+, Hg^{2+}, Cu^{2+}) and reactive oxygen species (ROS) (hydroxyl radicals) generated in copper- and hydroxycobalamine-ascorbate systems. The following order of these HM toxicity (at studied concentrations) for the *Fusarium solani* PM was shown: $Ag^+ > Hg^{2+} > Cu^{2+}$, which correlated with the affinity of these metals to the functional groups of membrane proteins. When the cells were attacked by ROS, the changes in their membrane permeability were observed only for the catalytic pair copper-ascorbate, probably, due to the difficulties for the large complex of hydroxycobalamine to penetrate through the cell wall, that is why formed in this binary system hydroxyl radicals were inaccessible to the PM.

INTRODUCTION

One of the most dangerous fungus of cereal crops (leading to yield decrease up to 30%) is *Fusarium solani* from *Fusarium* family. HM salts demonstrate a strong fungicidal effect. An increase of the toxic effect of low concentrations of transition HM was observed in media containing reducing agents. In this case HM ions act as catalysts of formation of reactive oxygen species (ROS), primarily hydroxyl OH radicals characterized by a high damaging capacity. Earlier a high toxicity of copper-ascorbate complexes was shown for DNA and proteins [Chiou,. 1983], poliomyelitis virus [Peloux et al., 1962] and cancer cells [Bram et al., 1980]. A cytotoxic effect observed upon injection of a hydroxycobalamine-ascorbate system to a cancer cell culture [Lemeshko, 2001]. The action of such catalytic pairs on pathogenic fungi has not been studied.

t is known that one of the first reaction of a plant to infection is activation of processes of hydrogen peroxide generation to form reactive oxygen radicals which suppress the pathogenesis [Harding et al., 1998]. The aim of the present work was to study, using the method of electroorientation spectroscopy, the permeability changes in the plasma membrane of fungus *Fusarium solani* cells treated by HM ions or attacked by ROS generated by the catalytic metal-reducer systems.

MATERIALS AND METHODS

Fusarium solani conidia were obtained from a pure 10-day fungus culture grown on the solid nutrient medium (bactoagar-20g, glucose-15g, potato broth up to 1liter). Agar culture was cut to 1×1cm squares and placed in the flask with 100 ml of sterile water. The flask was shaken during 30 min. Cells were washed by distilled water and stored during 3-4h at 4°C. Aqueous media with appropriate pH and electroconductivity were prepared on a distilled water by using 0.01 n HCl, 0.01 M NaCl and Tris solutions. The pH values, chosen for each metal, corresponded to the accumulation of the maximal amount of hydroxylated forms responsible for toxicity [Ivanov et al., 1997]. The medium pH used in the experiments with catalytic systems was 5.4 [Fomchenkov et al., 1986]. There are no hydroxylated HM forms in the medium at this pH. Cell suspensions with optical density 0.2-0.25 (cuvette 1cm, 540 nm) were incubated with HM ions during 15-30 min at room temperature. In the case of catalytic systems, cells were preincubated 10-15 min with copper or hydroxycobalamine ions, the reaction of ROS formation was started by introduction of ascorbate, after which the cells were incubated during 30 min. Prior to electroorientation (EO) measurements, the specific electroconductivities of cell suspensions were adjusted to 0.002 Sm/m with 0.01 M sodium chloride solution. The EO spectra of cells were obtained by measuring the relative changes in cell suspension optical density occurring due to varying orientation of cells in the uniform alternating electrical field with fixed frequency 0.5-10 MHz and intensity 80 V/cm. The EO spectra were registered on an experimental device developed at the State Research Center of Applied Microbiology (Obolensk, Russia). according to EO measurements [Miroshnikov et al., 1986; Khassanova et al., 1996]. The following reagents were used: Tris (oxymethyl) amino methane, NaCl, HCl, ascorbic acid ("Reachim", Russia), silver chloride, mercuric chloride (II), cupric chloride (II) (Sigma, USA), hydrocobalamine (vit B_{12b}) (Moscow Herzen Research Institute of Oncology, Russia). All measurements were made at room temperature.

RESULTS AND DICUSSION

Analysis of the EO spectra of *Fusarium solani* cells *(fig. 1)* points to a shift in the high-frequency decline of the spectrum of Ag^+ treated cells to lower frequencies.

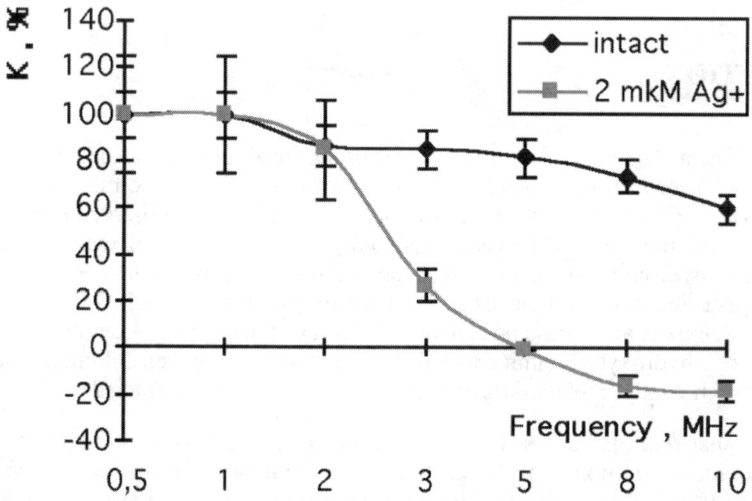

Fig. 1. EO spectra of intact and silver cation-treated (2 μM) *Fusarium solani* cells
$K = (\Delta D_1 / \Delta D_0) \times 100\%$, where ΔD_1 is the change in optical density of cell suspension on switching on the electric field; ΔD_0 is the same for the suspension of intact cells at a field frequency of 0.5 MHz

The character of changes in the EO spectrum is explained by the decrease in the efficient cell electroconductivity due to the leakage of cytoplasmic ions and low-molecular compounds, which suggests disturbances in the barrier properties of PM under the action of Ag^+. Earlier it was shown similar changes in the EO spectra of various bacteria after HM treatment [Ivanov et al., 1992; 1997; Khassanova et al., 1996]. The extent of PM damage of cells exposed to various reagents was assessed from the (value *(fig. 2)*. A damaging effect on PM similar to that of Ag^+ was also registered for Hg^{2+} after prolonged incubation of cells at Hg^{2+} concentrations exceeding the acting concentration of Ag^+. The β value of cells treated with Cu^{2+} in a wide range of concentrations (1-25 μM) remained at the level of control values, which suggested no damage to PM.

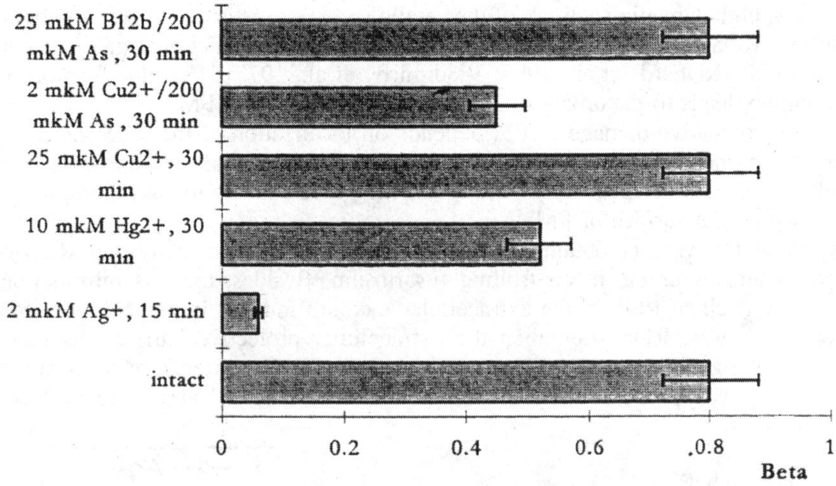

Fig. 2. Effect of various reagents on the β (beta) value of fungus *Fusarium* solani cells. The cells were treated with all the reagents, except for Cu^{2+} cations, in a medium of pH 5.4. In the case of Cu^{2+}, *a medium of pH 7.0 was used

The toxic effect of transition metals increased when they were included in a metal-reducer system catalyzing ROS formation [Peloux et al., 1962; Hoshino et al., 1999]. When a low concentration of copper (2 μM) was used in the catalytic pair copper-ascorbate, the (value of *Fusarium solani* cells dropped *(fig. 2, 3)*, which pointed to disturbances in the barrier properties of their PM. In this case the degree of the toxic action on PM was near to the changes produced by Ag^+, but after a more prolonged incubation. When however the cells were treated with the catalytic system hydroxycobalamine (B_{12b})-ascorbate, no disturbances in membrane permeability was revealed even with high (25 μM) hydroxycobalamine concentrations.

In earlier experiments, a rather wide scatter in the toxic HM concentrations for various groups of bacteria was found [Ivanov et al., 1992; Khassanova, 1996; Ivanov et al., 1997]. The toxicity threshold was largely dependent on the structure of the cell wall, which was clearly demonstrated for *Lactobacillus plantarium* 195 with a complete and incomplete cell wall [Ivanov et al., 2000]. A decrease in β value *(fig. 2, 3)* suggesting disturbances in the barrier properties of PM was observed for *Fusarium solani* conidia only with rather high HM concentrations, which can be explained by the powerful cell wall.

The results of the present study indicate that in cells treated with HM ions, the changes in membrane permeability were registered for silver and mercury salts but not for copper. These results are in agreement with the data on the action of HM ions on bacterial cells for which the toxicity series of HM is analogous, $Ag^+ > Hg^{2+} > Cu^{2+}$ [Ivanov et al., 1992; Khassanova et al., 1998]. Taking into account the high affinity of HM for the SH groups, we can explain these results by the availability of Ag^+ and Hg^{2+} but not Cu^{2+} to proteins reponsible for the permeability of PM and by inactivation of the proteins by these two cations. The difference in the damaging effect of Ag^+ and Hg^{2+} ions are probably due to differencies in ion radius and charge value. Despite the high toxicity of Ag^+ and Hg^{2+} for *Fusarium solani* cells, the use of salts of these HM for disinfection of cereals and treatment of infected plants is practically unsuitable because this is associated with considerable expenditures and possible side effects on crops.

The problem of control of pathogenic fungi can be solved with the help of catalytic ROS-generating pairs. The results presented in *fig. 2, 3* point to a high membranotropic effect of the system copper-ascorbate. This effect is comparable with the damaging action of silver and mercury salts, moreover, the copper concentration used is several times lower than the toxic concentration of mercury. The high damaging activity of the catalytic complexes is determined by the interaction of the resulting ROS, primarily hydroxyl OH radicals with the functional groups of membrane proteins and lipids [Richards et al., 1998; Vladimirov et al., 1972]. The modification of the protein-lipid complex leads to disturbances in the barrier properties of PM.

The value of oxidative damage to PM depends on localization of the ROS generator. This is attested by the absence of any fungicide effect in the case of a binary system using as a catalyst a large hydroxycobalamine (B_{12b}) complex *(fig. 2)* which was unable to pass through the cell wall and settle down on the surface of PM.

The experimental evidence obtained suggests that moderate-sized catalysts in ROS-generating binary systems are preferable in controlling fusariosum. Besides, there is information that the accumulation by itself of ROS in the extracellular medium induces in plants increased tolerance for unfavorable factors. ROS strengthen the extracellular protective barrier thus hindering the penetration of pathogenic microorganisms to a plant cell. The occurrence of ROS in the external medium induces the expression of genes triggering other, delayed and more perfect forms of protective reactions.

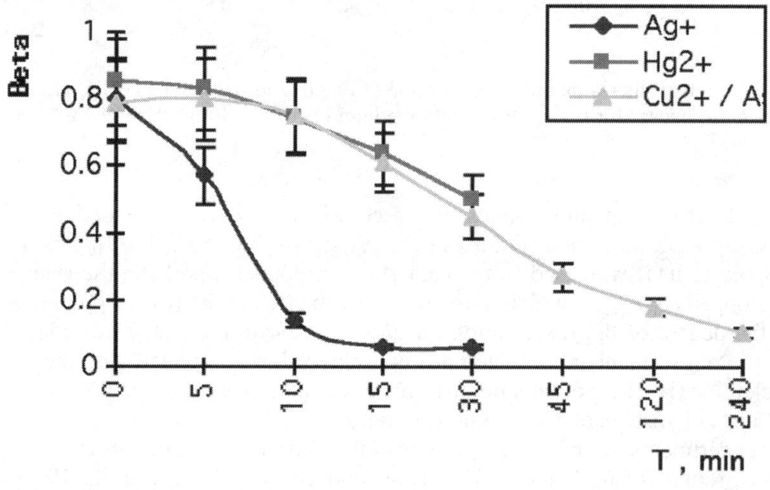

Fig. 3. Temporal changes in the β value of fungus *Fusarium solani* cells. The changes reflect the degree of damage to their PM in the systems: ; Ag^+ 2 μM; □, Hg^{2+}, 10 μM; Δ, Cu^{2+}, 2 μM / Asc 200 μM

REFERENCES

Aronovitch J., Godinger D., Czapski G. Bactericidal activity of catecholamine copper complexes. Free Rad. Res. Com. 1991. 12-13: 479-488.

Bram S., Froussard P., Guichard M., Jasmin C., Augery Y., Sinoussi-Barre F., Wray W. Vitamin C preferential toxicity for malignant melanoma cells. Nature. 1980. 284: 629-631.

Fomchenkov V..M., Mazanov A. L., Chugunov V. A., Chaykina L.K., Azhermachev A. K. Electric characteristics of bacterial cells measured when the barrier function of cytoplasmic membrane is disordered. Microbiologiya (Moscow). 1986, 55: 754-759.

Hoshino N., Kimura T., Yamaji A., Ando T. Damage to the cytoplasmic membrane of E. coli by catechin-copper (II) complex. Free Rad. Biol. and Med. 1999. 27: 1245-1250.

Ivanov A. Yu., Fomchenkov V.M., Khassanova L. A., Kuramshina Z. M., Sadikov M. M. The influence of heavy metals ions on the electrophysical properties of Anac. nidulans and E. coli. Microbiologiya (Moscow). 1992. 61: 455-463.

Ivanov A. Yu., Fomchenkov V.M., Khassanova L. A., Gavryushkin A.V. Toxic effect of hydroxylated heavy metal ions on the plasma membrane of bacterial cells. Microbiologiya (Moscow). 1997, 66: 588-594.

Ivanov A. Yu., Gavrjushkin A.V., Khassanova L. A., Collery Ph., Choisy C., Etienne J. C., Khassanova Z. M. Fe2+ induced hydroxyl radicals effects on the plasma membrane of Escherichia coli K-12 cells. Metal Ions in Biology and Medicine. John Libbey Eurotext (Paris). 2000. 6: 490-494. a.

Ivanov A. Yu., Novoselov V. I., Gavryushkin A.V., Fesenko E. E. Comparative efficiency of antioxidants upon the protection of bacterial plasma membrane from active forms of oxygen. Biofizika (Moscow). 2000. 45: 660-665. b.

Khassanova L. A. Electrophysical analysis and physiological-biochemical particularities of cell damages by heavy metal ions. Thesis (St. Petersburg). 1996.

Lemeshko V.V. Research of the mechanism cytotoxicity of operation hydroxycobalamine (vitamin?12b) in a combination with ascorbic an acid on tumor cells in vitro. Thesis (Puschino). 2001.

Miroshnikov I. A., Fomchenkov V.M., Ivanov A. Yu. Electrophysical analysis and cell separation, Nauka (Moscow). 1986, 198 p.

Medentsev A.G., Arinbasarova A. Yu., Akimenko V.K. Adaptation of the phytopathogenic fungus Fusarium decemcellulare to oxidative stress. Microbiologiya (Moscow). 2001. 70: 34-38.

Samuni A., Aronovitch J., Godinger D., Chevion M., Czapski G. On the cytotoxicity of vitamin C and metal ions. A site-specific Fenton mechanism. Eur.J. Biochem. 1983, 137: 119-124.

Walt N. T., Evans G. S., Tanner M. S. The role of free radicals in copper mediated toxicity in hepatocytes. Metal Ions in Biology and Medicine. John Libbey Eurotext (Paris). 2000. 6: 481-483.

Metal Ions in Biology and Medicine: vol. 7.Eds L. Khassanova, Ph. Collery, I. Maymard, Z. Khassanova, J.-C. Étienne.
John Libbey Eurotext, Paris © 2002 pp. 268-271.

Involvement of the bacterium *Azospirillum brasilense* in wheat tolerance to cadmium

Nina V. Bezverkhova[1], Vera I. Safronova[2], Lyudmila P. Antonyuk[1], Andrey A. Belimov[2]

[1]*Laboratory of Biochemistry of Plant-Bacterial Symbioses, Institute of Biochemistry and Physiology of Plants and Microorganisms, Russian Academy of Sciences, Saratov 410015, Russia*
[2]*Laboratory of Ecology of Symbiotic and Associative Rhizobacteria, All-Russia Research Institute for Agricultural Microbiology, Saint Petersburg 196608, Russia*

ABSTRACT

Cadmium is known to cause severe inhibition of plant growth. Some plant growth-promoting rhizobacteria were reported to be capable of decreasing partially the toxicity of Cd for barley. We studied the ability of the rhizobacteria *Azospirillum brasilense*, strains Sp7 and Sp245, to reduce Cd toxicity for wheat in hydroponic culture. Since one of the wheat lectins (wheat germ agglutinin, WGA) is involved in plant response to some kinds of environmental stress and is also essential for plant-bacterial interactions, the second aim of this work was to elucidate whether Cd stress could induce synthesis of WGA in wheat plants. The capacity of *A. brasilense* to decrease Cd toxicity for wheat was found to depend on the Cd concentration and on the wheat variety. Thus, when the $CdCl_2$ concentration in the plant growth medium was 2 mg/l, both strains diminished Cd toxicity for the wheat cultivar Saratovskaya 52; however, they gave no effect with the other three varieties used. At 8 mg/l $CdCl_2$, the inhibition of plant growth was significant (by 82-85%), and inoculation with bacteria did not decrease the toxicity of Cd. It was found that the presence of Cd in the plant growth medium caused an increase in the concentration of WGA in wheat roots. The enhanced level of WGA was retained during the whole period of plant cultivation with cadmium. The results obtained suggest that *A. brasilense* is capable of decreasing partially the toxicity of Cd for wheat, and WGA is probably involved in plant response to cadmium stress.

Key words: cadmium; wheat; *Azospirillum brasilense*; wheat germ agglutinin; rhizobacteria; heavy metal stress.

INTRODUCTION

Heavy metal pollution, including cadmium pollution, is a big and important environmental problem. In terms of environmental pollution, capacity for its accumulation in food and toxicity, cadmium is one of the most dangerous and harmful heavy metals [1]. Cd, as differentiated from Cu, Zn or Co, is not involved in the metabolic processes in living organisms. Entering into the cell, cadmium cations bind to SH-groups of proteins, that causes, in its turn, inactivation of enzymes, structural and regulatory proteins. One of the important factors of Cd and other heavy metals cytotoxicity is the disturbance of barrier properties of the membranes; it brings about efflux of potassium cations and a drop in the cytoplasmic conductivity [1-2].

The possibility of plants to resist harmful effects of heavy metals is linked to activation of a complex of adaptive and defense reactions. Plants are known to increase the synthesis of phytochelatins, metallothioneins and stress proteins under the influence of cadmium and some other heavy metals [1]. One of the proteins, which is probably involved in the defense of wheat plants from different kind of stress, is the lectin wheat germ agglutinin (WGA). It is well documented that WGA accumulates in wheat plants under various unfavourable conditions [3]. In addition, the establishment of effective symbioses of wheat and plant growth-promoting rhizobacteria *Azospirillum brasilense* is evidently mediated by WGA [4].

It is common knowledge that positive influence of plant growth-promoting rhizobacteria (PGPR) on plant organism is manifold. Bacteria supply their host plants with phytohormones, in particular indole-3-acetic acid (auxin), with combined nitrogen (owing to N_2 fixation), *etc.* (see [4] and references therein). A question of PGPR capacity to reduce heavy metal toxicity has come under study only in recent years. Thus, some PGPR were reported to be capable of decreasing partially the toxicity of cadmium for barley [5] and toxicity of nickel, zink and lead for some othe plants [6].

The aim of this study was to estimate whether the rhizobacteria *A. brasilense,* strains Sp7 and Sp245, could reduce cadmium toxicity for wheat in hydroponic culture. The second aim of this work was to elucidate whether cadmium stress could induce synthesis of WGA in wheat plants.

MATERIALS AND METHODS

A. brasilense, strains Sp7 and Sp245, from the Collection of IBPPM RAS (Saratov, Russia), were grown during 48 h at 37° C on solid synthetic medium containing (in g/l): K_2HPO_4 3.0, KH_2PO_4 2.0, NaCl 0.1, $MgSO_4 \cdot 7H_2O$ 0.2, $CaCl_2$ 0.02, $FeCl_3$ 0.02, $MnSO_4 \cdot 7H_2O$ 0.1, $Na_2MoO_4 \cdot 2H_2O$ 0.002, NH_4Cl 0.5, sodium malate 5.0, yeast extract 0.1 (pH 8.0). The cultures were rinsed off from plates with 0.9% NaCl, and the suspensions obtained were added to the plant growth medium to the final concentration of 10^5 cells/ml.

Seeds of *Triticum vulgare*, varieties Saratovskaya 29, Saratovskaya 38, Saratovskaya 42 and Saratovskaya 52 were superficially sterilised in 96% ethanol for 15 min and then grown on plates moistened with distilled water. For estimation of WGA content in roots of 4-day seedlings of the 4 aforementioned varieties, the wheat plants were grown on plates in the dark at 22°C. In all other cases the plants were grown 12 or 13 days in hydroponic culture on the medium containing (in mg/l): $CaCl_2$ 50, K_2HPO_4 15, $MgSO_4$ 150, KCl 40, FeEDTA 1, $Ca(NO_3)_2$ 10. $CdCl_2$, 2 mg/l or 8 mg/l, was added to some experimental variants. Some growth parameters, including wet weight and root length, were monitored to assess the cadmium toxicity and ability of azospirilla to decrease its harmful effect.

Content of WGA in wheat roots was estimated using the commonly accepted method based on the ability of the lectin to agglutinate rabbit erythrocytes, i.e. in the reaction of hemagglutination (RHA). RHA was performed in standard multiwell plates; each well contained 50 µl of 1% suspension of trypsin-treated rabbit erythrocytes, and 100 µl of extract were analysed in successive two-fold dilutions. To obtain WGA-containing extracts, the roots were frozen, crushed and then the lectin was extracted with 0.05 M HCl, supplemented with 5.5 mM EDTA and 5 mM ascorbic acid. The pH of the extract was adjusted to 7.0. WGA content was judged from the titer, i.e. the final dilution of the extract which gave RHA. A minimal titer in our experiments was 2 (the extract dilution 1:4) and maximal one was 11 (the extract dilution 1:2048).

RESULTS AND DISCUSSION

From 16 varieties of soft spring wheat that had been previously assessed for WGA content in their seeds, four varieties were selected, two with a low WGA level and two with high level of this protein in their seeds. Four-day seedlings of selected varieties were grown, and the content of WGA in the extracts of the seedling roots was estimated. The two varieties with a low level of

WGA were found to have a low level of the lectin in their roots as well (Saratovskaya 42 and Saratovskaya 38, the titers were 1:32? 1:64, respectively). The two other varieties, Saratovskaya 52 and Saratovskaya 29, had a high level of the lectin in their roots (the titers were 1:384? 1:256, respectively). Therefore, the content of WGA in the roots of Saratovskaya 42 was 8-fold less than that in Saratovskaya 29 and 12-fold less than that in Saratovskaya 52.

Wheat plants of selected varieties were grown in hydroponic culture in the presence of 2 mg/l and 8 mg/l $CdCl_2$. To estimate the toxicity of cadmium, growth parameters of plants were compared with those for plants grown in the absence of Cd. Each of the 4 varieties was also grown in the presence of *A. brasilense* Sp7 or Sp245 only and in the presence of each of the two strains together with Cd (2 mg/l and 8 mg/l) in order to assess whether the bacteria can step down the cadmium toxicity. We also cultivated the wheat plants with the bacteria only, because the plant growth-promoting effect of inoculation is known to be not always stable. Some of the results obtained (for Cd concentration of 2 mg/l) are shown in *table 1*.

Table 1. Effect of cadmium (2 mg/l) on wet weight of the roots of four wheat varieties grown in the presence and in the absence of *Azospirillum brasilense*

Factors	Wet weight of wheat roots, mg per plant			
	Saratovskaya 42	Saratovskaya 38	Saratovskaya 29	Saratovskaya 52
	Low level of WGA		High level of WGA	
-	118	126	102	111
Cd	71	79	77	68
A. brasilense Sp7	128	114*	117**	128***
A. brasilense Sp245	122	134	120**	119
Cd + *A. brasilense* Sp7	72	74	84	81*
Cd + *A. brasilense* Sp245	73	67*	80	80*

Notes. For values given in lines 3-6, statistically significant differences as compared to the corresponding controls are asterisked: * - statistically significant at $P<0.05$; ** - at $P<0.01$; *** - at $P<0.001$.

Cadmium was found to be toxic for wheat plants at both the concentrations used. At 8 mg/l $CdCl_2$, the inhibition of plant growth was significant (by 82-85%, data not shown) while at 2 mg/l $CdCl_2$ the toxicity of Cd was less expressed. For the varieties with a low WGA level, the presence of 2 mg/l $CdCl_2$ brought about an inhibition of the plant growth (estimated by wet weight of the wheat roots) by 40-42% *(table 1)*. In the case of the varieties with a high WGA level, the inhibition of growth was lower (32-34%). As is evident from the data given in *table 1*, plant growth-promoting effect of inoculation is exhibited only in the case of the wheat plants with a high WGA level. The varieties with a low lectin level either did not respond to inoculation, or their response manifested itself as inhibition of plant growth (Saratovskaya 38, inoculation with *A. brasilense* Sp7).

The capacity of *A. brasilense* to decrease Cd toxicity for wheat was found to depend on the Cd concentration and on the wheat variety. Thus, when the $CdCl_2$ concentration in the plant growth medium was 2 mg/l, both the strains tested diminished Cd toxicity for Saratovskaya 52, the variety with the maximal WGA level *(table 1)*. In the case of the other three varieties used and $CdCl_2$ concentration of 8 mg/l, no defense effect of *A. brasilense* was recorded *(table 1)*. Moreover, strain Sp7 that inhibited growth of Saratovskaya 38 in the absence of cadmium was found to increase the toxicity of this pollutant *(table 1)*. When the concentration of $CdCl_2$ was 8 mg/l and the inhibition of plant growth was significant (by 82-85%), inoculation with bacteria did not decrease the toxicity of Cd (data not shown).

It was found that the presence of Cd in the plant growth medium caused an increase in the concentration of WGA in wheat roots *(fig. 1)*. As can be seen from *fig. 1*, the content of WGA in roots was dependent on the plant age. Five-day wheat seedlings had the maximal level of the lectin, and as the age of plants increased, the WGA content reduced. When the plants grew in the presence of cadmium, the dependence of the WGA content on the age of plants was similar; however, they had a higher level of this protein in their roots. The enhanced level of WGA was retained during the whole period of plant cultivation with cadmium *(fig. 1)*.

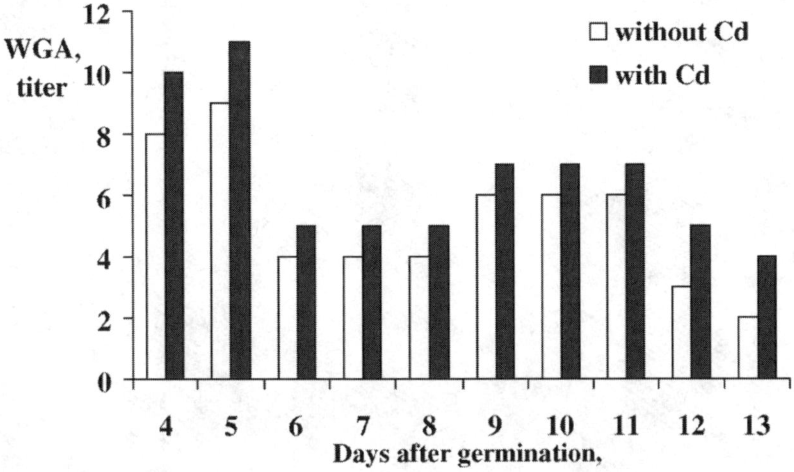

Fig. 1. WGA content in the roots of wheat, var. Saratovskaya 29, grown in the presence and in the absence of 8 mg/l Cd

The results obtained suggest that *A. brasilense* is capable of decreasing partially the toxicity of Cd for wheat. An important point is that capability of bacteria to diminish cadmium toxicity is connected with their growth-stimulating properties and with the content of WGA in the plant. Our findings infer that WGA is likely to be involved in plant response to cadmium stress.

ACKNOWLEDGEMENTS

This work was supported in parts by INTAS (EC, Brussels, Belgium), NATO (Grant LST.CLG.977664) and the Russian Academy of Sciences' Commission (Grant No. 205 under the 6th Competition-Expertise of research projects).

REFERENCES

1. Sanita di Toppi L, Gabrielli R. Response to cadmium in higher plants. *Environ Exp Bot* 1999, **41**, 105-130.
2. Ivanov AYu, Gavryushkin AV, Siunova TV, Khasanova LA, Khasanova ZM. Investigation of heavy metal resistance of some *Pseudomonas* strains. *Microbiology (Moscow)* 1999, 68: 313-320.
3. Shakirova FM, Bezrukova MV. Effect of 24-epibrassinolide and salinity on the levels of ABA and lectin. *Russ J Plant Physiol* 1998, **45**, 388-391.
4. Antonyuk LP, Ignatov VV. The role of wheat germ agglutinin in plant-bacteria interactions: a hypothesis and the evidence in its support. *Russ J Plant Physiol* 2001, **48**, 364-369.
5. Belimov AA, Dietz KJ. Effect of associative bacteria on element composition of barley seedlings grown in solution culture at toxic cadmium concentrations. *Microbiol Res* 2000, **155**, 113-121.
6. Burd GI, Dixon G, Glick BR. Plant growth-promoting bacteria that decrease heavy metal toxicity in plants. *Can J Microbiol* 2000, **46**: 237-245.

Days after germination

VI METAL IONS AND PLANTS

*Metal Ions in Biology and Medicine: vol. 7.*Eds L. Khassanova, Ph. Collery, I. Maymard, Z. Khassanova, J.-C. Étienne.
John Libbey Eurotext, Paris © 2002 pp. 275-278.

Technetium species induced in maize

C. Sergeant, M. Simonoff, T. Kijhniak

CNRS - UMR 5084 - Chimie Nucléaire Analytique et Bioenvironnementale, C.E.N., Le Haut Vigneau, B.P. 120, 33175 Draguignan Cedex, France

ABSTRACT

Plants have the ability to accumulate the long lived fission product ^{99}Tc. In this work, an attempt was made to separate and characterize technetium species induced by cereals plants grown on soil containing $Tc^{VII}O_4^-$ nutrient solution. Combination of data obtained with selective extraction (ammonium sulfate, acetone, diethyl ether and 8-hydroxyquinoline in chloroform) and use of Liquid Scintillation gave us insight into Tc species in maize.

INTRODUCTION

Technetium has only radioactive isotopes. Nuclide Tc^{99} with half live of 2.13×10^5 years, β^- emitter with $E\beta^-_{max}$ = 292 keV is a fission product of U^{235} and Pu^{239}, the most important from an environmental point of view. Tc present on soil results from anthropogenic activities (nuclear power generation, nuclear weapon testing program, nuclear medicine). Soil is directly implicated in the radionuclide exposition for man by irradiation and inhalation. Moreover, as a support for terrestrial cultures, it can play a key role in the transfer of radionuclides to plants, directly eaten by man or cattle. Two processes can be involved: the phytodisponibility offered by soil and the soil-plant transfer. The cumulated amount of Tc in soil now is about 500 $mBq.m^{-2}$ (Holm *et al.* 2000). Higher concentrations levels in soil and plants can be found in contaminated areas (Chernobyl exclusion zone 1.1 to 14.8 $Bq.kg^{-1}$ dry soil (Uchida *et al.* 2000, Uchida *et al.* 1999)).

Tc has 9 oxidation states from -I to VII, the most stable ones being: zero, VII, IV. In solution TcO_4^- is the most stable form in a large range of pH. This pertechnetate ion is reduced by Zn, HCl, H_2SO_4 diluted to form $TcO_2.2H_2O$. The oxide Tc_2O_7 is soluble in aqueous solution. H_2SO_4 precipitates Tc VII as Tc_2S_7 from acid pertechnetate solutions.

In anaerobic conditions, Tc has a strong tendency to give complexes with organic matter. In aerobic media, the most stable form, and the most mobile for Tc is $Tc^{VII}O_4^-$. Lixiviation experiments show that 96% of introduced Tc as TcO_4^- is present in the effluent with sandy soils (Sheppard and Hawkins 1991). Reduced forms of Tc (dioxide, sulphurs) have a very low solubility. In reducing conditions, Tc forms associated to soluble organic matter have been identified as the most frequent Tc form, not so phytodisponible than TcO^{4-} (Yanagisawa *et al.* 1997a and b).

Several factors influence the Tc adsorption on the soil constituents: oxydo reduction potential, organic matter, microflora (Denys 2001). In aerobic conditions, Tc absorption is very low (Kd near zero Lkg^{-1}) (Echevarria *et al.* 1997); in reductive one, Tc absorption is significant (Kd range in anaerobiosis from 18 to 68 $L.kg^{-1}$). Organic matter gives complexes, Tc complexes with humic acids are soluble and characterized by very slow kinetic exchange of ligands. Microflora can create reducing conditions, even in aerobic soils, and immobilize Tc as TcO_2 or sulphurs, or give Tc complexes with organic matter.

Moreover TcO_4^-, associated to organic matter have been identified as Tc vector from soil to plants Tc complexes. There is a competition between TcO_4^- and NO_3^- ions for transport in the plant

(Krijger *et al.* 2000, Echevarria *et al.* 1997, 1998, Lembrechts *et al.* 1988). On tomato plants, TcO_4^- and NO_3^- use the same transporter (Krijger *et al.* 2000). When Tc is fixed by roots it moves to atmospheric parts (Cataldo *et al.* 1983) mostly in leaves, without accumulation in stalk (Shinonaga *et al.* 1998). Accumulation in seeds is low. For flooded rice, less than 1% is contained in seeds. As to corn, 95% Tc is accumulated in leaves, less than 1% in seeds (Echevarria *et al.* 1997). Concerning ray grass, Tc accumulation in leaves can reach 78% of the input (Echevarria 1996). Tc would be mostly located in chlorophyllian organs where it would be reduced by photosynthetic processes (Myttenaere *et al.* 1986).

Tc speciation in plants is difficult and oxidation of species bound to Tc, transformed in TcO_4^-, is suspected. Recent studies have used ^{99m}Tc and ^{95}Tc (γ emitter) to examine oxidation processes during the sample preparation (Krijger *et al.* 1999). From experiment, Tc move to leaves by xylem as TcO_4^- where it can be associated to organic compounds (Krijger *et al.* 1999, Echevarria *et al.* 1997, Cataldo *et al.* 1989, Lembrechts and Desmet 1986, Muller and Meer-Bekk 1986). About ten Tc organic samples have been identified in spinach leaves, and more than 80 to 90% had a molecular mass less than 6000 $g.mol^{-1}$ (Lembrechts and Desmet 1986). In soya bean leaves most Tc bound to polypeptides had molecular weight between 4000 and 10000 Da (Vandecasteele *et al.* 1982). In leaf, Tc metabolism would be associated to sulphur; in vitro Tc would replace sulphur in cystein-type amino acid via cystein synthetase (Cataldo *et al.* 1989). Other studies confirm the affinity of Tc for containing sulphur groups, like cystein and glutathione.

We used several extraction techniques (Garten and Lamax 1989, Farago 1994, Harms *et al.* 1999) to separate the different kind of proteins resulting of the Tc incorporation in maize cultivated on enriched TcO_4^- soil.

MATERIAL AND METHODS

Cultivation of maize on enriched Tc^{99} soil

Maize seeds were germinated on earth enriched in Tc^{99}, as TcO_4^-. After 6 months, leaves were cut, dried at room temperature, then powdered after grinding and kept under nitrogen in seal box till extraction. Tc^{99} activity was 1 $kBq.g^{-1}$ (dry weight).

Tc^{99} Protein Extraction

Maize powder (0.5g) was homogenized in 25 mL of HEPES buffer (70 $mmol.L^{-1}$ N- (2-hydroxyethyl)-piperazine-N'-(2 ethanesulphonic acid) and 8 $mmol.L^{-1}$ NaCl, pH 7.0) using a Potter-Elvehjem homogenizer at 0 to 5°C.

Homogenate was divided in five aliquots of 5 mL, in order to apply several extraction techniques to differentiate between classes of Tc species (Harms *et al.* 1999). Tc radioactivity was measured by a Beckman Liquid Scintillator.

- Homogenate I was centrifuged (20 mn, 15000 g, 4°C); supernatant S1 contains free technetium proteins $NP_1 + NP_2 + HP_1 + HP_2 + T$. Pellet was resuspended in 1M NaOH; after centrifugation we obtain supernatant SP_1 and a pellet.
- Homogenate II was treated by ammonium sulphate (final concentration 2.1 $mol.L^{-1}$)) to precipitate Tc proteins. After centrifugation we obtain supernatant S2, containing fractions $NP_1 + NP_2 + T$, acetone final concentration 80% w/w is added. Supernatant SS2 after centrifugation contains $NP_2 + T$.
- Homogenate III is treated by acetone (final concentration 80% w/w) to precipitate hydrophilic protein, then centrifuged to separate supernatant $NP_1 + HP_1 + T$ from pellet.
- Homogenate IV is treated by diethyl ether to extract non polar Tc species.
- Homogenate V is treated by 4% w/w 8-hydroxyquinoline (H-8-Q) in chloroform to extract labile Tc^V species (Hwang *et al.* 1984).

RESULTS

The different equations obtained allowed as to evaluate several classes of Tc species in the homogenate of maize leaves. Results are given in next table.

Species	Distribution	Description
T	32%	Insoluble Tc species
NP_1	4.5%	Nonprotein Tc species soluble in 80% w/w acetone
NP_2	10.6%	Highly polar nonprotein Tc species
HP_1	2.3%	Tc bound to hydrophobic protein
HP_2	13.3%	Tc bound to hydrophilic protein
T	35.5%	TcO_4^-

DISCUSSION

After centrifugation of the homogenate I, 68% of the total Tc stayed in supernatant, the remaining Tc being in the pellet 32% (species T). This is in very good agreement with other results obtained for maize (Zee Mays L. 31%) and not too different to soya bean 39% (Vandecasteele *et al.* 1986).

Most of the Tc species in the pellet after centrifugation of the homogenate I (species T) are extracted into 1 $mol.L^{-1}$ NaOH, suggesting Tc is bound to hemicellulose (poorly defined polysaccharides of the cell wall excluding cellulose, very soluble in alkaline solution).

After treatment of homogenate II by ammonium sulphate supernatant contains $NP_1 + NP_2 + T$. If submitted to acetone extraction supernatant contains soluble Tc species (non protein Tc species NP_1 and T) and highly polar non protein Tc species NP_2, in the pellet. Tc species that precipitate in ammonium sulphate are insoluble Tc species and hydrophobic or hydrophilic proteins.

Chloroform addition to homogenate III extracts directly NP (non protein Tc species soluble in acetone), HP1 (Tc bound to hydrophobic protein) and free TcO_4^-. There was no observation of very hydrophobic Tc species since no radioactivity was extracted in diethyl ether. The assay to extract labile Tc^V species from homogenate V by H-8-Q in chloroform (Hwang *et al.* 1984) failed since all technetium stayed in aqueous phase.

CONCLUSION

Next step of this research will consist in a combination of selective extraction and chromatographic techniques (paper chromatography and Phosphorimager) in order to identify the Tc valences in these proteins.

ACKNOWLEDGMENTS. We are very grateful for ANDRA support of this study and Dr Echevarria, ENSAIA Nancy, for supplying maize leaves.

REFERENCES

Cataldo D.A., Wildung R.E., Garland T.R. Root absorption and transport behavior of Tc in soybean. *Plant Physiol.*, 1983, 73, 849-852.

Cataldo D.A., Garland T.R., Wildung R.E., Fellows R.J. Comparative metabolic behavior and interrelationship of Tc and S in soybean plants. *Health Physics*, 1989, 57 (2), 281-287.

Denys S. "Prédiction de la phytodisponibilité de deux radionucléides (^{63}Ni et ^{99}Tc) dans les sols", Thèse de doctorat, Nancy 2001.
Echevarria G. (1996). Contribution à la prévision des transferts sol-plante des radionucléides. Thèse de Doctorat de l'INPL, Vandœuvre-les-Nancy, 222 pp.
Echevarria G., Vong P.C., Leclerc-Cessac E., Morel J.L. Bioavailability of ^{99}Tc as affected by plant species and growth, application form, and soil application. *J. Environ. Qual*, 1997, 26, 947-956.
Echevarria G., Vong P.C., Morel J.L. Effect of NO^{3-} on the Fate of TcO^{4-} in the Soil-Plant System. *Journal of Environmental Radioactivity*, 1998, 38, 163-171.
Farago M.E. Phytochemistry: Methods for the study of inorganic species in plant tissues. In M.E. Farago (ed.) Plants and the chemical elements. VCH, Weinheim, Germany, 1994, 265-283.
Garten C.T. Jr. and Lomax R.D. Technetium-99 cycling in maple trees: Characterization of changes in chemical form. *Health Phys.*, 1989, 57, 299-307.
Harms A.V., Van Elteren J.T., Claessens H.A. Technetium speciation: non-size effects in size-exclusion chromatography. *J. Chromatogr.*, 1996, A 755, 219-225.
Harms A.V. and Van Elteren J.T. Search for plant-induced technetium species with size-exclusion and reversed-phase ion-pair chromatography. *J. Radioanal. Nucl. Chem.*, 1998, 228, 139-144.
Harms A.V., Krijger G.C., Van Elteren J.T., de Goeij J.J.M. Characterization of Technetium Species Induced in Spinach. *J. Environ. Qual.*, 1999, 28, 1188-1194.
Holm E., Gäfvert T., Lindahl P., Ross P. In situ sorption of technetium using activated carbon. *Applied Radiation and isotopes*, 2000, 53, 153-157.
Hwang L.L.-Y., Ronca N., Solomon N.A., Steigman J. Extraction of Tc(V) from radiopharmaceuticals. *Int. J. Appl. Radiat. Isot.*, 1984, 35, 825-830.
Krijger G.C., Vliet P.M.V., Wolterbeek H.T. Metal speciation in xylem exudate of Lycopersicon esculentum Mill-Tc. *Plant and Soil*, 1999a, 212, 165-173.
Krijger G.C., Harms A.V., Leen R., Verburg T.G., Wolterbeek H.T. Chemical forms of technetium in tomato plants: TcO^{4-}, Tc-cystein, Tc-glutathione and Tc-proteins. *Environmental and Experimental Botany*, 1999b, 42, 69-81.
Krijger G.C., Kollöffel C., Wolterbeek H.T. Effect of nitrate on uptake of pertechnetate by tomato plants. *J. Environ. Qual.*, 2000, 29, 866-870.
Lembrechts J.F. and Desmet G.M. Accumulation of Tc-bio-organic complexes in spinach plants in relation to growth. *In Technetium in the environment.* (G. Desmet et C. Myttenaere Eds), Elsevier Applied Science Publishers, Londres, 1986, 295-360.
Lembrechts J.F., Van Loon L.R., Van Ginkel J.H., Desmet G.M. Interpretation of soil-to-plant transfer on the basis of soil solution chemical composition. *In impact des accidents d'origine nucléaire sur l'environnement.* Actes du IVe Symposium International de Radioécologie de Cadarache, CEA, Cadarache, 14-18 Mars 1988, Tome 1, Chap. D, 170-178.
Müller H. and Meer-Bekk C.T. Technetium uptake by Sinapis alba. *Journal of Radioanalytical and Nuclear Chemistry*, 1986, 99 (1), 181-185.
Myttenaere C., Vandecasteele C.M., P. Roucoux P., Lietart E.A., Itschert A., Mousny A. Processus biologiques responsables de l'accumulation du ^{99}Tc par les végétaux. *In Technetium in the environment.* (G. Desmet and C. Myttenaere Eds). Elsevier Applied Science Publishers, Londres, 1986, 281-300.
Sheppard M.I. and Hawkins J.L. A linear sorption/dynamic water flow model applied to the results of a four-year soil core study. *Ecological Modeling*, 1991, 55, 175-201.
Shinonaga T., Ambre S., Yamaguchi I. Uptake rate of trace elements by soybean plants. *Journal of Radioanalytical and Nuclear Chemistry*, 1998, 236 (1-2), 133-137.
Uchida S., Tagami K., Wirth E., Rühm W. Concentration levels of technetium-99 in forest soils collected within the 30-km zone around the Chernobyl reactor. *Environmental Pollution* , 1999, 105, 75-77.
Uchida S., Tagami K., Rühm W, Steiner M., Wirth E. Separation of Tc-99 in soil and plant samples collected around the Chernobyl reactor using a Tc-selective chromatographic resin and determination of the nuclide by ICP-MS. *Applied Radiation and Iso*topes, 2000, 53, 69-73.
Vandecasteele C.M., Delmotte A., Roucoux P., Van Hove C. Technétium et organisme diazotrophes: toxicité, localisation, facteur de transfert. In Environmental Migration of Long-Lived Radionuclide. Actes de Symposium Knoxville, USA, 21-23 Juillet 1981, IAEA, Vienne, 1982, 275-286.
Yanagisawa K and Maramatsu Y. Transfer of technetium from soil to plants. *Radiochemistry*, 1997a, 39 (4), 316-320.
Yanagisawa, K., Maramatsu Y., Ban-Nai T. Behavior of technetium in paddy soils. *Journal of Radioanalytical and Nuclear Chemistry*, 1997b, 226 (1-2), 221-223.

*Metal Ions in Biology and Medicine: vol. 7.*Eds L. Khassanova, Ph. Collery, I. Maymard, Z. Khassanova, J.-C. Étienne.
John Libbey Eurotext, Paris © 2002 pp. 279-283.

Metal phytoextraction by cereals

Sorochan O.A., Shtemenko N.I.

Dniepropetrovsk National University, Board of Biophysics and Biochemistry13 Naukoviy by-str., Dniepropetrovsk 49050, Ukraine

SUMMARY

Recently we have shown that application of growth regulators led to sharp increase of root exudative activity of some cultural *(Zea mays L.)* and wild cereals during first stages of germination. In this work we present results obtained in experiments with *Lollium perenne L. Festuca rubra L*, grown on sterile sand and on soils contaminated with great quantities of Zn and Pb salts with application of previously described growth regulator FeSuc (unstechiometric mixture of Fe-succinate). Detailed analysis of amino acid content of root exudates of cereals showed that wild specie had more certain amino acids (cysteine, aspartic and glutamic acids and their amides, serine) in root exudates than cultural ones. These amino acids had more possibility for chelation due to existance of one more polar or ionogenic functional groop. Seeds of cereals were treated with FeSuc and planted on soils contaminated with salts of Zn and Pb. It was shown that during 15 days of germination quantity of metals in primary leaves increased approximately in twice and decreased in soil, especially in upper layers (on 7-10%). Amino acids in root exudates of plants may be chelate agents as an α-amino acid can act like a bidentate ligand, forming a five-membered heterocyclic ring with suitable metal cations and in a such way increasing mobility of metal ions. Thus, it was shown that stimulation of root exudative activity by pretreatment with a growth regulator may be succesful in cleaning of soils and basicly may be a good instrument for phytoremediation.

Key words: cereals, Zn, Pb, phytoexctraction, remediation.

INTRODUCTION

Due to ancient and actual industrial runoffs, an accumulation of inorganic pollutants is frequently observed. Depending on their mobility and biodisponibility, heavy metals contained in dredged sediment deposits can be transferred to the soil, the groundwater and the vegetation and in this way, they may enter the food chain. Therefore, it becomes urgent to solve the problems linked to these deposits, specially about their impact on the surrounding environment, including the flora, the fauna and consequently Man.

As a matter of fact, some plant species can adapt to metallic pollution by developing protection mechanisms such as: precipitation of heavy metals in rhisosphere, sequestration of the pollutants in the roots and immobilization in the vacuoles or in the cell walls [1]. In other respects, vegetation is also able to set up mechanisms altering the chemical and physical extent of the rhisosphere in order to enchance micronutrient bioavailability, such as: acidification, production of chelating compounds, changes in the ionic concentrations of such compounds using selective absorption [2]. These mechanisms have a certain impact on heavy metals. Among mentioned above we focuse our attention on amino acid exchange of phytoremediative plants, especially on amino acids as chelating agents in root exudates.

Our previous works [3, 4] showed that pretreatment of seeds of cultural and wild cereals with the growth stimulator - (unstechiometric mixture of Fe-succinate - FeSuc) before planting enchanced Zn phytoextraction, plant uptake and translocation of the metal ions from roots to green parts of tested plants. This study focuses on elucidating of common features of growing cereals on Zn- and Pb-rich soils and on biochemical mechanisms that may be involved in this process.

MATERIALS AND METHODS

Seeds of maize *(Zea mays i L.)*, Pioneer 3978 and wild cereals *(Festuca rubra L., Lollium perenne L.)* were germinated in a vessel with temper on previously cleaned sterile sand at room temperature during 15 days. Seedlings were irrigated daily with deionized water (a) or with 8% solution of FeSuc (b). Leachates were placed on the glass column with ion exchange resin to collect free amino acids. At the end of the experiment ion exchange column was washed by strong ammonium solution with following evaporation for amino acids analysis on automatic amino acids analyzer AAA-339 (Czech).

Seeds of *Festuca rubra L.* and *Lolium perenne L.* were treated with strong water solution of FeSuc before planting in soil (c) of black earth type and in the same soil, which was previously treated with solution of $ZnSO_4$ (650 mg $ZnSO_4$ per 1 kg of soil - d) and Pb (360 mg $PbNO_3$ per kg of soil - e). Control experiments (control) without FeSuc and salts of metals were done in the same conditions. After 15-days of growing control, and (c, d, e) plants were dried, free amino acid content, Zn and Pb in plants and in three layers of soil (upper - 1-2 cm (2), middle - 7-8 cm (3) and lower - 15 cm (4) were measured by ICP-AES metod on spectrophotometer AAS-30 (Germany).

RESULTS

During growing of seeds on clean sand in conditions, described above (a, b), total quantity of amino acids in root exudates increased in 1,4 - 2 times under influence of FeSuc, that depended from a specie. It is interesting to note, that quantitatively change of growing was followed by exudation of those amino acids, that had one more ionized or polar functional group *(table 1).*

Table 1. Content of some free amino acids in root exudates of germinating seeds in control (a) and experimental (b - with FeSuc) conditions in % to total sum, n = 4, P = 0,05

Amino acid	Name of a specie					
	***Zea mays* L.**		***Festuca rubra L.***		***Lolium perenne L.***	
	a	**b**	**a**	**b**	**a**	**b**
?sp+ Glu	20.58±0.9	24.11±0.7	15.78±0.78	17.96±0.89	17.35±0.9	25.94±0.79
S?r	27.39±1.4	28.27±1.41	7.89±0.39	12.16±0.10	1.83±0.09	8.26±0.41
Phe	16.22±0.9	18.74±0.93	8.89±0.44	5.61±0.28	6.81±0.34	11.56±0.56
Cys	0.53±0.1	1.71±0.08	9.61±0.47	13.52±0.67	12.49±0.6	13.31±0.06
Tre	3.03±0.2	3.77±0.18	2.41±0.12	16.07±0.8	14.57±0.7	13.52±0.67
Lys	11.65±0.9	14.94±0.74	11.91±0.59	8.47±0.42	7.51±0.37	10.10±0.5

Complex influence of contamination and growth stimulation led to decreasing of sizes of leaves and roots of experimental plants on 30% in average. Even during such a short period of growth they accumulated Zn and Pb twice more than in control conditions *(table 2)*.

Table 2. Quantity of Zn and Pb in plants *Festuca rubra L.* (1) and in upper (2), middle (3) and low (4) layer of soil in % to experiment without FeSuc

Experimental conditions	Zn				Pb			
	1	2	3	4	1	2	3	4
FeSuc (?)	121.46±10.63	95.74±9.56	95.83±9.62	85.72±8.51	141.85±13.34	67.39±7.24	89.79±8.63	82.78±8.38
FeSuc + Zn (d)	243.75±19.35	85.07±8.99	82.04±8.23	72.74±7.79	184.78±12.41	83.43±8.21	85.90±8.62	84.11±8.38
FeSuc + Pb (e)	147.88±9.48	83.43±8.21	85.90±8.62	84.11±8.38	171.76±10.72	95.54±9.59	91.88±8.83	69.49±7.35

Total quantity of free amino acids in primary leaves increased approximately in 6 - 7,5 times *(table 3)* under influence of Zn-contamination.

Table 3. Some amino acids of primary leaves of cereals grown in control (c) and experimental (d, e) conditions mg% (1) and in % to total sum (2)

Amino acid	Lolium perenne L.				Festu?? rubra L.			
	c		d		c		e	
	1	2	1	2	1	2	1	2
Asp	10.09±1.01	7.83	34.60±4.65	3.85	6.69±0.23	2.22	79.66±4.68	3.76
Tre	4.83±0.49	3.39	30.52±3.48	3.39	1.04±0.02	0.35	25.64±2.13	1.21
Tyr	10.46±1.55	7.99	138.82±10.5	15.73	2.81±0.05	0.93	14.34±1.96	0.67
Lys	2.42±0.31	1.88	48.15±5.06	5.35	3.35±0.01	1.11	21.9±2.34	1.03
Arg	5.67±0.62	4.39	80.50±8.23	8.95	0.55±0.01	0.18	4.7±0.38	0.22
Total sum	128.92		899.57		301.54		2113.72	

Especially significant was accumulation of free hydroxy amino acids Tyr, Tre, basic amino acids Lys and Arg and acidic amino acid Asp.

By determining of free amino acid content in primary leaves in experiments c, d and e we have found a common reply of plants to the influence of both heavy metals, that is demonstrated by data of accumulation of free aspartate *(fig. 1)*.

DISCUSSION

Ionized and polar amino acids have one more functional group that can potentially act as additional site of interaction with heavy metal ions. We guess that changing of free amino content in root exudates under influence of FeSuc to the side of exudation of amino acids with such functional groops may be very important fact that could explain altering of rhisosphere by exudative process in conditions of FeSuc application. As more polyfunctional substances are exudated, as

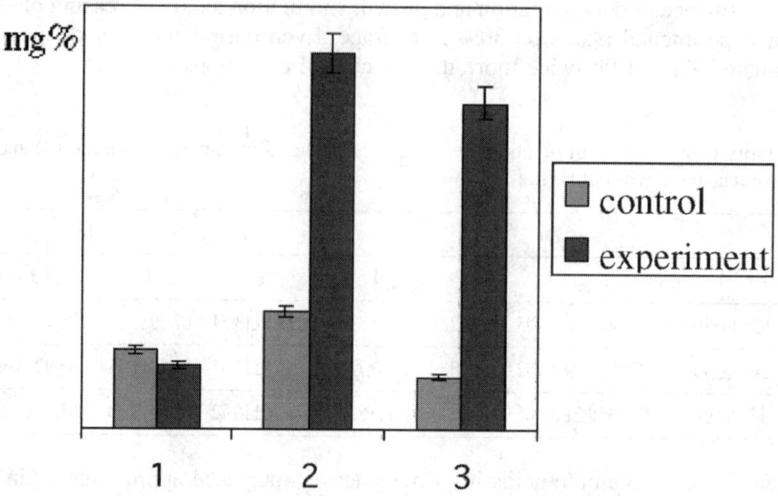

Fig. 1. Quantity of free aspartate in primary leaves of *Festuca rubra L.* in control (1) - c conditions of growth and in experiments with Zn- (2) - d and Pb- (3) - e contamination, mg%

more intensive is the process of modification of rhisosphere by enhancing mobility and bioavailability of metals.

Further experiments were accomplished in condition close to natural. It is impossible to appreciate real exudation in these experiments due to the complex influence of the soil (microorganisms, humic acids, etc.). But we consider that effect of growth regulator was the same as in sterile conditions and revealed in intensive exudative activity. Process of heavy metal absorption by plants from soil was more effective in the case of FeSuc application. It was demonstrated by measuring of metal quantity as in whole plants, as in different layers of soil. Absorption coefficients under influence of FeSuc changed from 1,33 to 9,05 and from 1,66 to 15.01 for Lollium perenne L. and Festuca rubra L. accordingly. Thus, stimulation of growth (the only condition of surviving in such quantities of toxic metals) led to a 8-10 times growth of the absorption process.

Influence of several factors on a plant cell led to increasing of free amino acids [4, 5], especially on the first stages of germination of plants. Accumulation of Asp was especially sharp and common for tissues of the studied wild cereals *(table 3, fig. 1)*. Aspartic acid is a multifunctional amino acid and its accumulation under influence both toxic metal and the growth regulator may be explained by 1) its transport function; in the process of adaptation it was necessary to support reparative processes for which transport of metabolites is necessary; 2) FeSuc contains residues of succinate - a metabolite of Krebs cycle and its introduction led to acceleration of oxaloacetate-aspartate transamination pathway; 3) as a multifunctional amino acid it could take part in different biosynthetic processes. Accumulation of other amino acids in free state was specific for wild cereals and requires a particular consideration.

CONCLUSIONS

Several calculations show that heavy metal extracting plants will need 100 years or more for some contaminated soils to yeild acceptable heavy metal concentrations. From the other hand, phytoremediation techniques does not cost much money, and may be accelerated by application of some growth regulators. We think that it is better to remediate slowly but certainly.

Wild cereals that are recommended for phytoremediation, have unusial accumulation of free amino acids. Primary leaves of Festuca rubra L. and Lollium perenne L. may be interesting models for investigations of amino acid exchange and other phytochemical experiments.

REFERENCES

1. Baker AJM, Brooks RR. Terrestrial higher plants which hyperaccumulate Metallic elements. A review of their distribution, ecology and phytochemistry. *Biorecovery*, 1989, 1, 81-126.
2. Salt DE, Blaylock M, Kumar NPBA, Dushenkov V, Ensley BD, Chet I and Raskin I. Phytoremediation: a novel strategy for the removal of toxic metals from the environment using plants. *Biotechnology*, 1995, 13, 468-474.
3. Shtemenko N, Pyshnaya O, Mazurenko E, Pakhomov A, Misiura A. Free amino acids in root exudates of cereals during the process of growth stimulation and phytoremediation. *Proceedings of the Seventh International FZK/TNO Conference on Contaminated Soil 18-22 September 2000, Lepzig, Germany*, 2000, 2, 1270-1271.
4. Shtemenko NI, Sorochan OO. Free amino acids of primary leaves of Lollium perenne L. during growing on Zn-rich soil. *ISEB 2001 Meeting Phytoremediation, Leipzig, Germany*, 2001, 126-127.
5. Shtemenko NI, Sorochan OO. Free amino acids of maise graine on the early phases of germination. *Physiology and biochemistry of cultural plants*, 2001, 33, 5, 441-446.

*Metal Ions in Biology and Medicine: vol. 7.*Eds L. Khassanova, Ph. Collery, I. Maymard, Z. Khassanova, J.-C. Étienne.
John Libbey Eurotext, Paris © 2002 pp. 284-288.

Reduced heavy metal levels in leaves of sludge-treated tobacco plants

D. Douma[1], A. Katsaraki[2], P. Vezyraki[3], G. Papadopoulos[1], A. Evangelou[3] and V. Kalfakakou[3]

[1]*Technological Educational Institute of Epirus, Plant Production Department Arta, Greece*
[2]*Statistics-Computer Department, University Regional Hospital of Ioannina, 45110, Greece*
[3]*Environmental Physiology Unit, Exp. Physiology Laboratory, Faculty of Medicine, University of Ioannina, Ioannina 45110,Greece*

ABSTRACT

Anthropogenic interventions, such as inorganic fertilization, produced soil aggravation by heavy metals whereas organic fertilization, by means of sewage sludge application, produced conflicting results in relation to heavy metal soil pollution.

Heavy metal bioaccumulation in plants is related to species physiology and plant part, environmental conditions and essential metal antagonism, substrate's metal burden and bioavailability. Tobacco leaves insulted by metals via substrate pollution construct a major health issue for smokers due to heavy metal toxic effects on humans. The present work aims to compare the effects of inorganic and organic (treated domestic sewage sludge) fertilization on two Greek oriental type tobacco varieties in relation to heavy metal bioassimilation. Two (2) tobacco varieties (I and II) were grown on soils of different pH (5.5 and 7.4), mixed with either anaerobically treated sewage sludge (pH: 7.1) at various proportions (0%, 10%, 20% and 40%) or inorganic fertilizer. The determination of Cd,Zn,Cu,Pb and Cr was conducted in tobacco leaves by means of atomic absorption spectrophotometry (AAS).Plants grown on acid soils presented significantly lower Cd levels in both varieties after sludge treatment in comparison to inorganic fertilizer application ($p<0.05$).Plants grown on alkaline soils presented significantly lower Cd, Cu and Pb levels, only in variety I, after high sludge proportion treatment (20% and 40%) in comparison to inorganic fertilizer application. Sludge proportion 40% resulted in statistically higher Zn levels in both varieties and soil pH regime. Variety I plants grown on sludge, presented significant correlation between essential and non essential metals.

In conclusion, metal bioavailability in tobacco leaves is depended on substrate's features and variety's genetic traits, hence fertilizing practices should consider various parameters in order to protect public health.

INTRODUCTION

Heavy metal presence in tobacco blends is significantly correlated to increased levels of certain toxic metals in the tissues and organs of tobacco smokers as well as to chronic health effects, especially in reference to carcinogenesis [1]. Elevated Cd, Pb and Cr concentrations in the blood, adipose tissue, prostate gland, kidneys, muscles and lungs, induce direct premutational injuries or indirect mechanisms of oxidative stress, that may lead to tumorigenesis [2, 3, 4, 5]. Heavy metal concentrations in cigarettes vary according to tobacco origin, variety, cultivation practices, crop management and the analytical methods applied for metal determination [2, 6, 7]. Cd and Pb levels present significant differences between various countries due to anthropogenic interventions such as fertilizers application, air pollution and discharges deposition [8, 9, 10]. Sludge application in the cultivation of various plant species in relation to soil and consequently plant aggravation by heavy metals, is an ongoing research with more or less dubious conclusions [11, 12, 13].

MATERIAL AND METHODS

40 seedlings of the Greek variety I (Tsebeli) and 40 seedlings of the Greek variety II (Mirodata) were transplanted in acid soil (pH1=5,5) and alkaline soil (pH2=7,4).

Plants were separated in 10 groups of substrates (4 plants / group):

Group 0% - pH1	= 0% sludge + 100% acid soil
Group 10% - pH1	= 10% sludge + 90% acid soil
Group 20% - pH1	= 20% sludge + 80% acid soil
Group 40% - pH1	= 40% sludge + 60% acid soil
Group IF - pH1	= inorganic fertilizer + 100% acid soil
Group 0% - pH2	= 0% sludge + 100% alkaline soil
Group 10% - pH2	= 10% sludge + 90% alkaline soil
Group 20% - pH2	= 20% sludge + 80% alkaline soil
Group 40% - pH2	= 40% sludge + 60% alkaline soil
Group IF - pH2	= inorganic fertilizer + 100% alkaline soil

IF group, was treated by N-P-K inorganic fertilizer [11-15-15] at 30g/pot.

Acid digestion of plant tissues was conducted at 180° C with HNO_3 and $HClO_4$ solution, followed by infiltration. The concentration of Cd, Zn, Pb, Cu and Cr, was determined by means of Atomic Absorption Spectophotometry (Perkin-Elmer mod.560).

RESULTS

In tobacco variety I, sludge application in proportions 10%, 20% and 40% and on acid soil (pH1=5,5), significantly decreased Cu and Cr levels and increased Pb level in comparison to control (0% sludge proportion) *(fig. 1, table 1)*. Sludge application on alkaline soil (pH2=7,4), decreased Cu and Cr as well as Cd and Pb. In tobacco variety II, sludge application in proportions 10%, 20% and 40% and on acid soil significantly decreased Cd, Cu and Zn levels in comparison to control plants (0%). The application of sludge on alkaline soil in variety II as well significantly decreased Cu and Cr concentrations and increased significantly Zn levels.

Sludge application, on acid soil, resulted in significantly less Cd burden in both varieties and higher Zn levels in variety I, in comparison to inorganic fertilizer application. Sludge application on alkaline soil, resulted in significantly lower Cd,Cu and Pb levels in variety I whereas in variety II significantly lower Cu levels and higher Zn levels occurred, in comparison to inorganic fertilizer application.

Sludge proportion 40% in all varieties and soil proportions, significantly reduced Cd and increased Zn levels compared to inorganic fertilizer.

The application of high sludge proportions (20%, 40%) on variety I produced significantly positive correlation between Zn and Cr, Pb, Cd, as well as between Cu and Pb, Cd *(table 2)*. Contrary wise, in variety II, correlation between metals was observed only in control (0%) plants.

DISCUSSION

Sludge application on acid or alkaline soil, decreased heavy metal contents in both varieties compared to control plant group (0%) *(table 1)*. Besides heavy metal sludge concentration, the bioavailability of soil, depended mainly on organic matter and pH, is considered an important factor of heavy metal uptake by plants [14]. Sludge proportion (40%) decreased Cd levels and increased Zn levels in both varieties since although Cd uptake by plants is depended on Cd concentration in soil, Zn abundance may decrease Cd uptake *(fig. 1)*. Nevertheless, it is reported that Zn addition in soil sometimes may increase Cd uptake if the latest exists in high levels in the soil [15]. Metal species and soil pH produce differences in metal uptake profile within the two tobacco varieties *(fig. 1, table 2)*. Hence, besides soil parameters, the plant species and variety, the metal species and the essential metal concentrations must be taken under consideration in sludge application as a fertilizer [16]. Sludge application in comparison to IF, produced significantly decreased Cd, Cu, Pb levels and significantly higher Zn levels in plants, while Cr concentrations were not affected. The above may be attributed to high levels of toxic elements met in sedimentary parent phosphate rock where from phosphate fertilizers originate [10].

Table 1. Differences between substrates (Mann-Witney U-test)

		VARIETY?					VARIETY?				
pH	**Substrates**	**Cd**	**Cu**	**Cr**	**Zn**	**Pb**	**Cd**	**Cu**	**Cr**	**Zn**	**Pb**
	0% - 10%	N.S.	P<0,05	P<0,05	P<0,05	P<0,05	P<0,05	P<0,05	N.S.	P<0,05	P<0,05
	0% - 20%	N.S.	P<0,05	P<0,05	N.S.	P<0,05	P<0,05	P<0,05	N.S.	P<0,05	N.S.
	0% - 40%	N.S.	P<0,05	P<0,05	N.S.	P<0,05	P<0,05	P<0,05	N.S.	P<0,05	N.S.
5,5	0% - IF	P<0,05	P<0,05	P<0,05	P<0,05	P<0,05	NS	P<0,05	NS	P≈0,05	N.S.
	IF - 10%	P<0,05	N.S.	N.S.	N.S.	N.S.	N.S.	N.S.	P≈0,05	N.S.	N.S.
	IF - 20%	P<0,05	N.S.	N.S.	P<0,05	N.S.	P<0,05	N.S.	N.S.	N.S.	N.S.
	IF - 40%	P<0,05	N.S.	N.S.	P<0,05	N.S.	P<0,05	N.S.	N.S.	N.S.	N.S.
	0% - 10%	P<0,05	P<0,05	P<0,05	N.S.	N.S.	N.S.	P<0,05	P<0,05	N.S.	N.S.
	0% - 20%	P<0,05	P<0,05	P<0,05	N.S.	P<0,05	N.S.	P<0,05	P<0,05	P<0,05	N.S
	0% - 40%	P<0,05	P<0,05	P<0,05	N.S.	P<0,05	P<0,05	P<0,05	P<0,05	P<0,05	N.S
7,4	0% - IF	N.S.	P<0,05	P<0,05	N.S.	N.S.	N.S.	P<0,05	P<0,05	P<0,05	N.S.
	IF - 10%	N.S.	P<0,05	P<0,05	N.S.	N.S.	N.S.	P<0,05	N.S.	P<0,05	N.S.
	IF - 20%	P<0,05	P<0,05	N.S.	N.S.	P<0,05	N.S.	P<0,05	N.S.	P<0,05	N.S
	IF - 40%	P<0,05	P<0,05	N.S.	P<0,05	P<0,05	P<0,05	P<0,05	N.S.	P<0,05	N.S

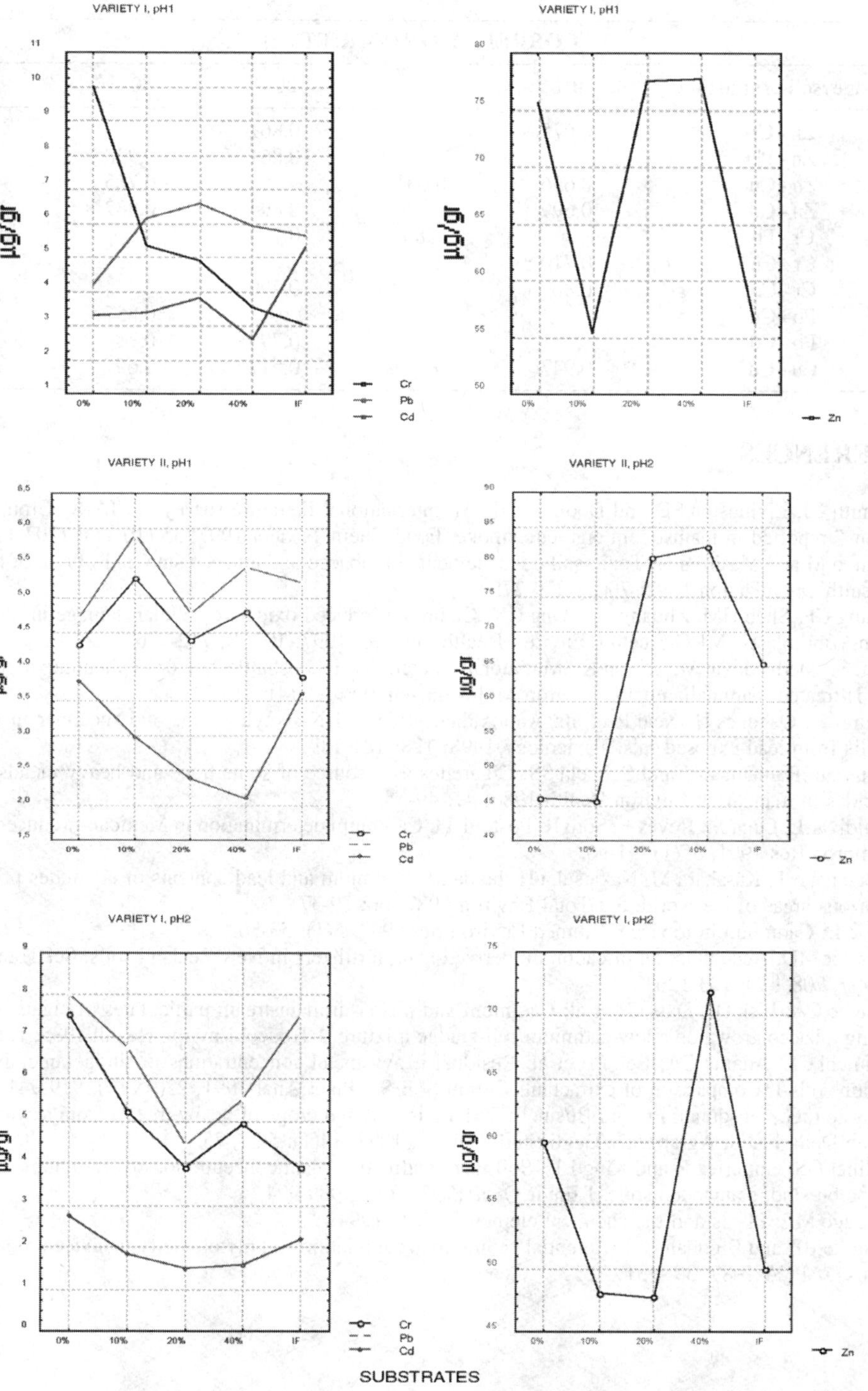

Fig. 1. Heavy metal concentrations in different substrates

Table 2. Heavy metal corellations (Spearman) in variety I at different substrates

	CORRELATION (VARIETY I)				
Sludge /soil proportions	**0%**	**10%**	**20%**	**40%**	**IF**
Zn - Cr	0,675		0,861		
Zn - Pb			0,861	0,930	
Zn - Cu	0,636	0,681		0,765	
Zn - Cd	0,699		0,894	0,887	0,672
Cr - Pb		586	0,764		0,669
Cr - Cu	0,710				
Cr - Cd	0,809		0,790		
Pb - Cu				0,857	
Pb - Cd			0,795	0,784	
Cu - Cd	0,947		0,517	0,657	

REFERENCES

1. Smith CJ, Livingston SD and Doolittle DJ: An international literature survey of "IARC Group carcinogens" reported in mainstream cigarette smoke. Food Chem Toxicol 1997; 35 (10-11): 1107-113069.
2. Chiba M and Masironi R: Toxic and trace elements in tobacco and tobacco smoke. Bulletin of the World Health Organization 1992; 70(2): 269-275.
3. Yang CF, Shen HM, Zhuang ZX, Ong CN: Cadmium-induced oxidative cellular damage in human fetal lung fibroblasts (MRC-5 cells). Environ Health Perspect 1997; 105(7): 712-716.
4. Liu S, Medvedovic M, Dixon K: Mutational specificity in a shuttle vector replicating in chromium (VI)-treated mammalian cells. Environ Mol Mutagen 1999; 33(4): 313-319.
5. Gurer H, Ozgunes H, Neal R et al.: Antioxidant effects of N-acetylcysteine and succimer in red blood cells from lead-exposed rats. Toxicology 1998; 128: 181-189.
6. Mussalo-Rauhamaa H and Salmela SS: Cigarettes as a source of some trace and heavy metals and pesticides in man. Arch Environ Health 1986; 41: 49-55.
7. Saldivar L, Luna M, Reyes E, Soto R, Fostoul TI: Cadmium determination in Mexican-produced tobacco. Environ Res 1991; 55(1): 91-96.
8. Watanabe T, Kasahara M, Nakatsuka H, Ikeda M: Cadmium and lead contents of cigarettes produced in various areas of the world. Sci Total Environ 1987; 66: 29-37.
9. Yue L: Cadmium in tobacco. Biomed Environ Sci 1992; 5(1): 53-56.
10. Taylor MD: Accumulation of cadmium derived from fertilizers in New Zealand soils. Sci Total Environ 1997; 208(1-2): 123-126.
11. Bache CA, Lisk DJ, Dass GJ et al: Cadmium and nickel in mainstream particulates of cigarettes containing tobacco grown on a low-cadmium soil-sludge mixture. J Toxicol Environ Health 1985; 16: 547-552.
12. Mulchi CL, Adamu CA, Bell PF et al: Residual heavy metal concentrations in sludge-amended coastal plain soils-I. Comparison of extractants.Comm Soil Sci Plant Anal 1991; 22(9&10): 919-941.
13. Gigliotti G, Giusquiani PL and Businelli D: Fate in soil and crops of heavy metals from sewage-sludge. Ann Della Fac di Agrar Univ Degli Stud di Perug 1993; 14(1): 563-575.
14. Miner GS, Gutierrez R and King LD: Soil factors affecting plant concentrations of cadmium, copper and zinc on sludge-amended soils. J Envir Qual 1997; 26(4): 989-994.
15. Farago ME: Plants and the chemical elements, VCH 1994.
16. Clarke BB and Brennan E: Differential cadmium accumulation and plytotoxicity in sixteen tobacco cultivars. JAPCA 1989; 39: 1319-1322

Metal Ions in Biology and Medicine: vol. 7. Eds L. Khassanova, Ph. Collery, I. Maymard, Z. Khassanova, J.-C. Étienne.
John Libbey Eurotext, Paris © 2002 pp. 289-295.

Effect of different growth media on uptake of elements by wheat

Irina Shtangeeva[1], Sophie Ayrault[2], Tatiana Lissitskaia[3]

[1]*St. Petersburg University, Universitetskaya nab., 7/9, St. Petersburg 199034 Russia;*
[2]*Laboratoire Pierre Sue CEA-CNRS 91191 Gif sur Yvette Cedex France;*
[3]*St. Petersburg University of Chemical Technology, Moskovsky pr., 26, St. Petersburg Russia*

ABSTRACT

Short-term tests have been performed to study the influence of different media (soil and water culture solutions) on uptake of trace and macro-elements by roots and leaves of wheat seedlings. Concentrations of most part of the elements (except potassium, bromine and rubidium) in roots were always higher than in leaves. In many cases relationships between elements in the experimental media differed significantly from that in the plants. The most important differences between plants grown in soil and water solutions were found for roots, while elemental composition of leaves kept at rather stable level. Nevertheless, in spite of very different concentrations of elements in the experimental media, plants were able to maintain their elemental composition at more or less determined level.

INTRODUCTION

During recent years significant progress has been made in understanding the processes of uptake and pathways of ions in plants, the kinetics of nutrients in roots and their translocation to leaves. At present, the ability of plant roots in binding ions from the surrounding environment is a widely recognized phenomena that has a number of implications [Vasconcelos and Leal, 2001]. Among others, the important role of plant roots is to selectively acquire ions from surrounding medium and prevent the penetration of particular elements to leaves [Jiang et al.; 2000 Shtangeeva, 1995]. In fact, the effectiveness of ion uptake by plants depends greatly upon the roots. Moreover, to date, most knowledge regarding the ionic mechanisms in higher plants originates from studies on plant roots [Shabala, 2000], though it seems that leaves accumulate solutes by other means than the roots [Karley et al., 2000]. Unfortunately, because of influence of numerous factors affecting the movement of elements through soil to roots, effect of ion concentrations in the environment on uptake and kinetics of the ions in plants have been often studied in solution culture [Mackay and Barber, 1984]. Meanwhile, there is no doubt that soil and water (including nutrient solutions) differ widely in bioavailability of nutrients and the ability to supply plants with micro- and macro-elements in quantities sufficient for optimal plant growth. The aim of the research was to study the effect of surrounding media on the uptake of macro- and micro-nutrients by roots and transfer of the elements from roots to leaves. Short-term tests have been performed to assess variability of element uptake by wheat seedlings grown in different water solutions and in soil. Multivariate statistical treatment of experimental data was used to estimate similarities and differences of element behaviour in plants grown in different media and classify the plants according to their ability to uptake the ions.

MATERIALS AND METHODS

Experiment was performed in 2000 in a naturally illuminated greenhouse. Seeds of *Tricum vulgare (vill) Horst* were germinated for six days on a moist filter paper at room temperature. Uniform germinated seedlings were divided into four equal parts and transferred to pots. The pots were filled with soil, doubly distilled water, water taken from a spring and nutrient solution of Hoagland. Spring was situated in a park, 25 km from St. Petersburg. Soil was taken near the spring from top (0 - 10 cm) soil horizon. First series of soil and water samples was taken before the beginning of the experiment. Plants were harvested three times - after 2, 4 and 7 days of cultivation. Water and soil from surface of roots were taken simultaneously with the experimental plants. After sampling water was placed in plastic bottles and kept in a fridge at 4°C. Soil was dried at room temperature up to constant weight. In order to remove dust and small particles of soil from surface of leaves and roots, all plants were washed carefully by water and also dried at room temperature up to constant weight.

Instrumental neutron activation analysis was used to determine concentrations of 26 elements (Na, K, Ca, Sc, Cr, Fe, Co, Zn, As, Br, Rb, Ag, Sb, Cs, Ba, La, Sm, Eu, Tb, Yb, Lu, Hf, Ta, Au, Th and U) in soil and different parts of plants. Concentrations of Na and K in water samples were determined by liquid ion chromatography. ICP-MS was used to determine concentrations of 24 elements (Li, Ti, V, Cr, Mn, Fe, Co, Ni, Cu, Zn, As, Se, Rb, Sr, Mo, Cd, Sb, Cs, Ba, La, Eu, Dy, Pb and U) in water samples. Before analysis a half of each water sample was filtered through ester cellulose 0.22 µm syringe filter. Another part of the sample was analysed in its natural state, without filtration. A statistical treatment of experimental data included calculation of mean concentrations of elements, correlation analysis and analysis of variances to estimate statistically significant differences between groups of the samples. In addition, cluster analysis and principal component analysis (PCA) were carried out. The data for PCA was normalised to unit concentration to avoid misclassifications caused by different order of magnitudes of variables (Statistica for Windows 5.5 Software package).

RESULTS AND DISCUSSION

Concentrations of most part of elements in doubly distilled water were less than in water taken from spring and, especially, compared to nutrient solution of Hoagland. There were no statistically significant differences between concentrations of all elements studied in filtered and non-filtered distilled water. Concentrations of Sr in non-filtered and filtered nutrient solutions were 84.2 µg kg^{-1} and 80.3 µg kg^{-1}, respectively. The difference was statistically significant ($P<0.05$). Mn content in filtered spring water (0.21 µg kg^{-1}) was also less compared to concentration of Mn in non-filtered water (0.38 µg kg^{-1}). It is surprising that after filtration concentration of Cu in spring water increased from 0.76 µg kg^{-1} (in non-filtered water) to 0.98 µg kg^{-1}. For both last cases (Mn and Cu in spring water) the differences were statistically significant ($P<0.01$).

Concentrations of elements in roots and leaves of experimental plants are shown in *table 1*. The most important differences between element concentrations were found in roots of plants grown in different media, while elemental composition of leaves kept at rather stable level. There were certain regularities in distribution of elements in different parts of the plants. Roots: concentrations of Ca, Br and Cs were quite similar in plants grown in all media; roots of plants grown in spring water were enriched with Na, Co and Rb (only Na content was higher in spring water than in distilled water and nutrient solution); concentrations of Sc, Cr, rare earth elements (REE) and Th in roots of plants grown in soil were higher than those in plants grown in water; high concentrations of K, Fe and Ba in roots were specific feature of plants that grew in nutrient solution; concentrations of K, Sc, Fe and Co in roots of plants growing in distilled water were much less than those in roots of other experimental plants; in spite of sufficiently high (in comparison with other water media) concentration of Zn in Hoagland's solution, content of Zn in roots of plants grown in the solution was less than in roots of the plants grown in soil and other liquid media. The highest contents of Zn were found in roots of plants grown in distilled water and water taken from a spring, though concentrations of Zn in these media were much less than in nutrient solution and soil; contents of Ag and Au in roots of all water-grown plants were approximately the same and exceeded concentrations of these elements in roots of plants grown in soil in several times. Leaves: for many elements there were no statistically significant differences between plants grown in different media. In addition to Ca, Br and Cs that already demonstrated sufficiently constant level of concentrations in roots, contents of Cr, Fe, Ag, Sb, REE, Au and Th were also quite similar in leaves of all experimental plants; concentrations of Na and Co in leaves of plants grown in spring water were higher than those in other plants; only concentrations of K and Ba in leaves of plants that grew in Hoagland's solution were higher than concentrations of these elements in leaves of other experimental plants; content of K decreased not only in roots, but also in leaves of plants grown in distilled water. It was reasonable to expect this effect since uptake of K^+ can be strongly reduced under negative environmental conditions [Kuiper et al., 1994].

Thus, in spite of very different concentrations of elements in the experimental media, plants maintained their elemental composition at more or less determined level. At the same time there were specific differences in uptake of elements by roots of plants growing in different conditions and rather similar behaviour of elements in leaves. The concentrations of most part of elements in roots were always higher than in leaves. Only concentrations of Br and Rb were quite similar in all parts of the plants and potassium content was always higher in leaves than in roots. It is known that roots possess special mechanism controlling ion uptake and transfer of elements to shoots [Briat and Lebrun, 1999]. In many cases roots can accumulate large amounts of different elements and serve as a barrier protecting upper parts of the plants from environmental stress.

However, in spite of even careful washing of plants after sampling the problem of possible contamination of the plants by dust and soil still exists. Therefore, it was important to verify whether the more high concentrations of Sc, Cr, REE and Th in roots of soil-grown plants really reflected the peculiarities of physiology of plants growing in soil or this was a result of contamination of the roots by small particles of soil. It would appear reasonable that in case of contamination of the roots by soil one might expect higher concentrations not only these (listed above) elements, but also others, in particular, Co and especially Fe. Besides, there should be quite similar correlation between elements in both soil and roots of plants growing in the soil. Meanwhile, there were no observed increased concentrations of Fe and Co in roots of plants grown in the soil (compared to all other experimental plants). Moreover, cluster analysis of soil samples and samples of roots of plants growing in the soil produced two different groups *(fig. 1)*. As a typical example one can demonstrate a high correlation between Na and K in soil, while in roots of plants grown in the soil Na and K were found in different groups. This last finding can be explained by quite distinct biological role of K and Na in plants. Potassium is a mineral nutrient required in large amount by plants [Schachtman and Weihong, 1999]. On the other hand, Na is an essential mineral element only for a few plants [Brownell, 1979]. In particular, this results in protection of leaves of young plants from Na accumulation by depletion of Na from the xylem stream. As is seen from *table 1*, Na content in leaves was usually 10 times less than in roots.

Our results also indicate that in most cases behaviour of elements in experimental media and in different parts of plants may be rather different. For example, there was statistically significant ($P<0.05$) negative correlation between Co and Na in the soil, positive ($P<0.01$) correlation between these elements in roots and there was no correlation between Co and Na in leaves. In fact, there was no similar correlation between any two elements in both any medium and all parts (roots and leaves) of the plants grown in the medium. It is also interesting that there were often quite different correlation between elements in different water media.

Statistical treatment of experimental data by method of principal component analysis showed that roots of the plants were clearly separated into different groups, mainly by the first PC *(fig. 2)*.

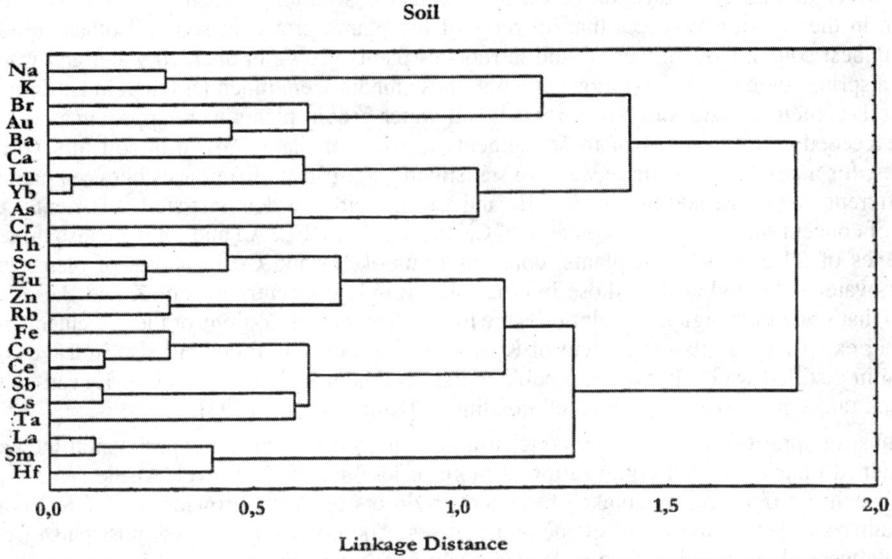

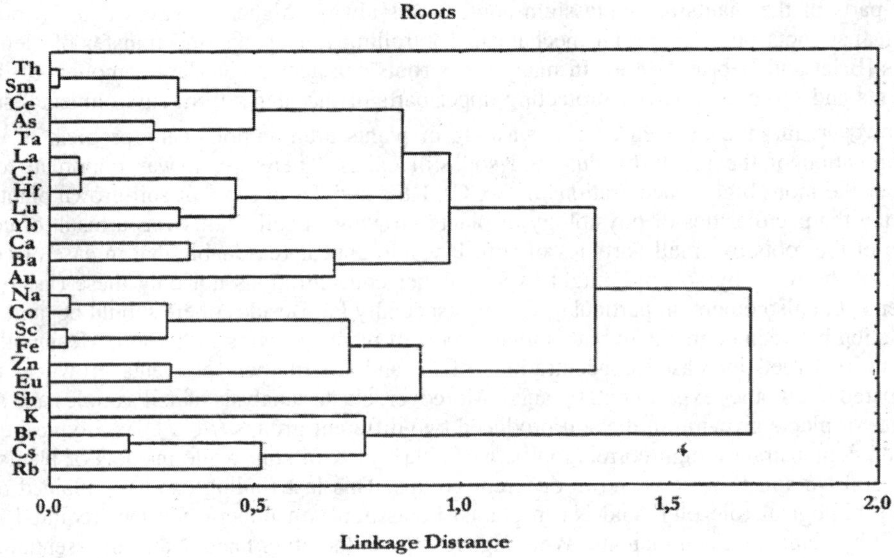

Fig. 1. Cluster analysis of soil samples and samples of roots of plants grown in the soil

Sc, Cr, Ag, La, Sm, Au and Th were responsible for the separation. Roots of soil-grown plants were found to be more close to roots of the plants grown in nutrient solution, while groups of plants grown in distilled water and spring water were more close to each other and separated from the two first groups. Another situation was found in leaves. Only leaves of water-grown plants formed own groups. Again groups of plants grown in spring and double distilled water were more close to each other than plants grown in nutrient solution. In this case only two elements (Br and Co) were highly correlated with the first PC. The difference in mechanisms of uptake of elements by roots and leaves (probably, resulting from rather different homeostatic processes that maintain the correct concentrations of elements in different compartments of the plants) is additional

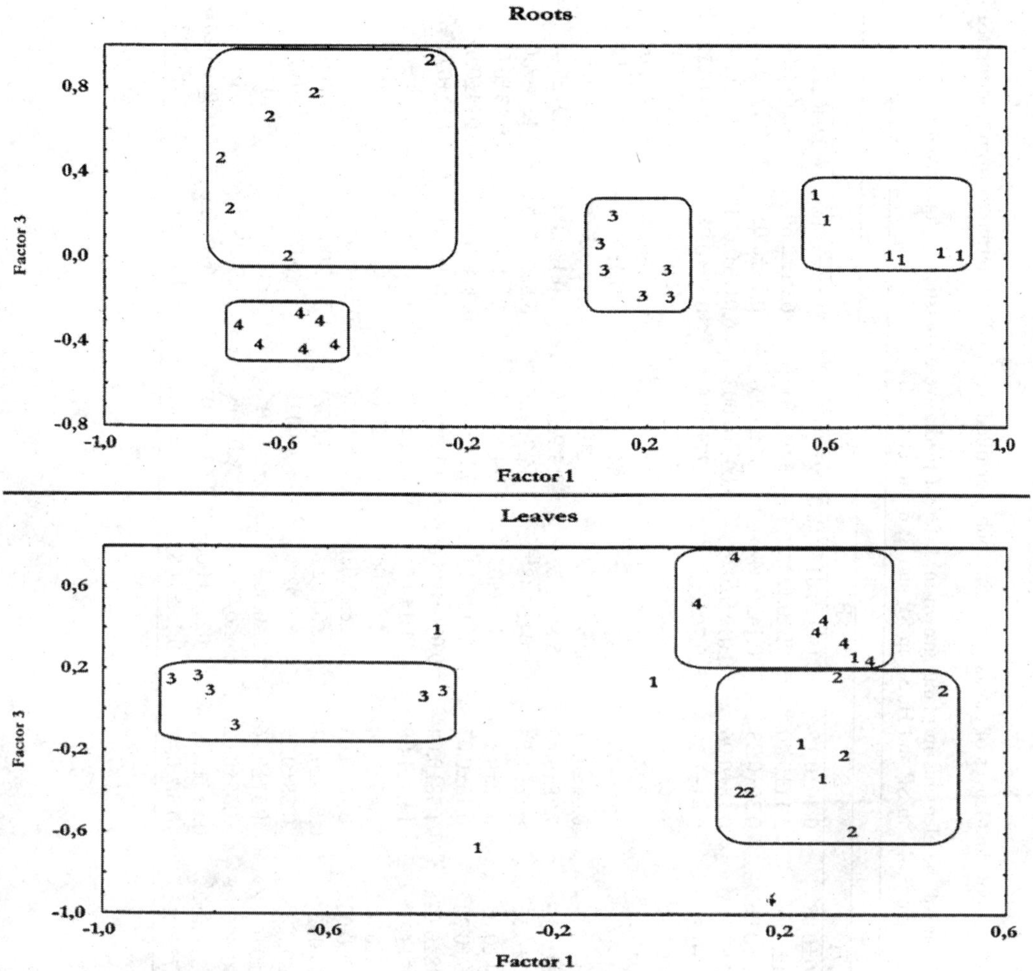

Fig. 2. Score plot of the first and third principal components for roots and leaves. 1 - plants were grown in soil, 2 - in spring water, 3 - in nutrient solution, 4 - in double distilled water

evidence of the fact that influence of environmental conditions on uptake of ions is more important for roots than for leaves.

CONCLUSIONS

The experiments were carried out to determine the influence of environmental media (soil and different culture solutions) on uptake of elements by roots and translocation of the elements to leaves. In many cases concentration of elements in experimental media and different parts of plants as well as relationships between the elements were quite different. Roots were the most suffering from environmental stress part of the plants. Elemental composition of the roots varied more considerably with environmental conditions than elemental composition of leaves. Nevertheless, in spite of very different concentrations of elements in the experimental media, plants keep their elemental composition at the determined level.

Table 1. Mean concentrations ± SD of elements (mg kg^{-1} in roots and leaves of plants grown in doubly distilled water (B), water taken from a spring (SP), nutrient solution of Hoagland (H) and soil (S).
The numbers show that differences between concentrations of the elements in plants grown in different media were statistically significant (P<0.01). **1** - plants were grown in B and SP; **2** - in B and H; **3** - in B and S; **4** - in SP and H; **5** - in SP and S; **6** - in H and S

	Roots				Leaves			
Elem.	B	Sp	H	S	B	Sp	H	S
Na, %	$0.48±0.09^{1,3}$	$0.77±0.19^{4,5}$	0.40±0.08	0.40±0.15	$0.049±0.013^{1}$	$0.080±0.011^{4,5}$	0.058±0.009	0.046±0.020
K, %	$0.45±0.09^{1,2,3}$	$1.94±0.88^{4}$	$4.11±1.18^{6}$	2.11±0.40	$1.29±0.12^{1,2,3}$	3.60±1.91	6.01±2.69	4.24±1.56
Ca, %	$0.24±0.08^{1}$	0.40±0.11	0.29±0.06	0.30±0.23	0.14±0.04	0.16±0.06	0.18±0.05	0.17±0.10
Sc	$0.024±0.013^{1,3}$	$0.053±0.011^{4,5}$	$0.028±0.006^{6}$	0.18±0.06	0.019±0.026	$0.009±0.003^{5}$	$0.010±0.001^{6}$	0.013±0.002
Cr	$1.82±0.46^{3}$	$1.54±0.49^{5}$	$1.82±0.50^{6}$	9.41±1.39	0.70±0.15	0.77±0.18	0.89±0.28	0.84±0.22
Fe	$377±144^{1,2,3}$	$569±133^{4,5}$	$2050±754^{6}$	686±199	135±19	$134±9^{4}$	159±20	161±30
Co	$0.33±0.03^{1,2,3}$	$1.26±0.53^{4,5}$	$0.65±0.15^{6}$	0.29±0.08	$0.038±0.008^{1}$	$0.051±0.012^{4,5}$	0.038±0.006	0.039±0.005
Zn	$379±59^{2,3}$	$328±69^{4}$	$70.4±14.2^{6}$	231±99	$62.4±2.4^{2}$	64.8±2.8	70.7±7.5	61.2±11.8
As	≤0.2	0.46±0.13	≤0.2	0.64±0.31	0.098±0.038	≤0.08	≤0.03	≤0.02
Br	10.5±8.1	15.9±12.6	13.0±10.3	7.78±6.43	8.89±2.27	13.9±7.6	10.1±3.0	7.97±6.41
Rb	$4.97±1.73^{1,3}$	$14.0±7.8^{4}$	$3.87±0.50^{6}$	8.16±1.54	$6.04±0.48^{1,3}$	$15.3±9.3^{4}$	$6.50±0.76^{6}$	13.4±4.0
Ag	0.44±0.21	$0.35±0.03^{5}$	$0.31±0.06^{6}$	0.12±0.11	0.15±0.03	0.19±0.06	0.14±0.04	0.13±0.02
Sb	$0.15±0.04^{2}$	0.77±0.95	$0.37±0.20^{6}$	0.12±0.07	0.11±0.14	$0.11±0.03^{4}$	0.064±0.023	0.074±0.027
Cs	0.067±0.050	0.050±0.021	0.038±0.018	0.058±0.044	0.016±0.006	0.019±0.006	0.019±0.007	0.012±0.008
Ba	$20.2±4.4^{2}$	$26.3±6.1^{4}$	$475±192^{6}$	14.1±14.8	$11.1±4.0^{2}$	$12.2±4.9^{4}$	$55.4±28.3^{6}$	12.8±4.1
La	$0.29±0.12^{2,3}$	$0.25±0.06^{5}$	$0.26±0.03^{6}$	1.49±0.30	0.13±0.06	0.15±0.04	0.16±0.02	0.12±0.05
Sm	$0.010±0.015^{3}$	≤0.03	$0.017±0.005^{6}$	0.16±0.03	0.016±0.020	0.006±0.003	0.004±0.002	≤0.009
Eu	≤0.004	≤0.004	≤0.002	0.049±0.010	≤0.002	≤0.002	≤0.001	≤0.002
Yb	≤0.03	≤0.1	≤0.03	0.38±0.09	≤0.02	≤0.02	≤0.03	≤0.1
Hf	≤0.03	≤0.01	≤0.02	1.57±0.26	≤0.009	≤0.01	≤0.007	≤0.01
Ta	≤0.006	≤0.01	≤0.006	0.065±0.021	≤0.004	≤0.008	≤0.004	≤0.007
Au	$0.044±0.015^{3}$	0.051±0.025	$0.033±0.012^{6}$	0.015±0.003	0.022±0.006	$0.033±0.012^{4}$	0.016±0.005	0.029±0.020
Th	$0.047±0.022^{3}$	0.25±0.45	0.775±1.15	0.56±0.44	0.15±0.08	0.12±0.12	0.13±0.11	0.098±0.027

ACKNOWLEDGEMENTS

This work was supported by NATO C&L grant N EST.CLG.977448. The authors appreciate F. Carrot for the ion chromatography measurements.

REFERENCES

Briat J.F., Lebrun M. Plant responses to metal toxicity. Comptes Rendus de l'Academie des Sciences Serie 111-Sciences de la Vie-Life Sciences, 1999, 322 (1), 43-54.

Brownell P.F. Sodium as an essential micronutrient for plants and its possible role in metabolism. Adv. Bot. Res., 1979, 7, 117-224.

Jiang W., Liu D., Hou W. Hyperaccumulation of lead by roots, hypocotyls and shoots of *Brassica juncea. Biologia Plantarum*, 2000, 43 (4), 603-606.

Karley A.J., Leigh R.A., Sanders D. Where do all the ions go? The cellular basis of differential ion accumulation in leaf cells. Trends in Plant Science, 2000, 5 (11), 465-470.

Kuiper P.J.C., Walton C.S., Greenway H. Effect of hypoxia on ion uptake by nodal and seminal wheat roots. Plant Physiology and Biochemistry, 1994, 32 (2), 267-276.

Mackay A.D., Barber S.A. Comparison of root and root hair growth in solution and soil culture. J. of Plant Nutrition, 1984, 7 (12), 1745-1757.

Schachtman D., Weihong L. Molecular pieces to the puzzle of the interaction between potassium and sodium in plants. Trends in Plant Science, 1999, 4 (7), 281-287.

Shabala S. Ionic and osmotic components of salt stress specifically modulate net ion fluxes from bean leaf mesophyll. Plant, Cell and Environment, 2000, 23, 825-837.

Shtangeeva I.V. Behaviour of chemical elements in plants and soils. *Chemistry and Ecology*, 1995, 11, 85-95.

Vasconcelos M.T.S.D., Leal M.F.C. Adsorption and uptake of Cu by *Emilania huxleyi* in natural seawater. *Env. Sci. Techn.*, 2001, 35, 508-515.

Metal Ions in Biology and Medicine: vol. 7. Eds L. Khassanova, Ph. Collery, I. Maymard, Z. Khassanova, J.-C. Étienne.
John Libbey Eurotext, Paris © 2002 pp. 296-299.

Algal resistance to heavy metals

V. I. Ipatova[1], V. Yu. Prokhotskaia[2]

[1,2]*Moscow State University, 1-12, Leninskie Gory, 119899, Moscow, Russia.*

ABSTRACT

There are many experimental data about algal adaptation to heavy metals. The result of adaptation is increasing of resistance to toxicants during the time. In the case of chronic intoxication this process can develop by selection of already existent forms in genetically heterogeneous population (genetic adaptation) or by forming of resistant cells of algae within population (biochemical or phenotypic adaptation). It is important to know the limits of algal population resistance to long-term high intensive toxic effects for hydrosphere monitoring. In present work we estimated resistance of laboratory population of green chlorococcal alga *Scenedesmus quadricauda* (Turp.) Breb. to potassium dichromate ($K_2Cr_2O_7$) as a model toxicant. After the pre-adaptation to relatively low concentrations of toxicant the algae were re-inoculated twice in medium with 10.0 mg/l $K_2Cr_2O_7$ (excluding control culture). It was revealed that the number of cells depended on initial $K_2Cr_2O_7$ concentration: the higher the toxicant concentration was, the less cell number was. So, in a 30 days maximum cell number was in pre-adapted with 0.1 mg/l $K_2Cr_2O_7$ culture. In another 30 days the cell number was the same in all samples (5-6% of initial cell number). During the experiment photosynthetic efficiency decreased accordingly to toxicant concentration: the higher concentration was, the lower photosynthetic activity was. Living cells were found out in cultures after three-time intoxication. They had size spectra and functional characteristics the same than a control ones. The resistant cells are suggested to relate to their constant presence in population or are the result of selection. It is need of special research for clarification of this phenomenon.

The laboratory microalgal cultures are optimal models for investigation of toxic action on population and cell levels [1, 2]. They have short-term life cycle and, therefore, can produce many daughter cells during short period of time. The using of algal cultures in toxicological experiments gives an opportunity; first, to research effects of toxicants on functional and morphological characteristics of plant cell and, second, to estimate their action on microalgal model population. The intensity of toxic action should be evaluated by the cell number changing during the cultivation. Variability of cell number is reported to reflect the algal population state adequately [3, 4].

There is vast experimental information about chemical waste effects on plants, including algal adaptation to heavy metal action [5-7]. The result of adaptation is increasing of resistance to toxicants during the time. In the case of chronic intoxication this process can develop by selection of already existent forms in genetically heterogeneous population (genetic adaptation) or by forming of resistant cells of algae within population (biochemical or phenotypic adaptation). The limits of algal cells resistance to long-term high intensive toxic effects determine survival of population as whole. Thus, research of algal adaptation to adverse conditions is of primary importance for hydrosphere monitoring. We were shown earlier that the algal culture previously growing in the presence of low concentrations of toxicants attained resistance to higher toxic concentrations [10].

The aim of present investigation was to estimate resistance of laboratory population of green chlorococcal alga *Scenedesmus quadricauda* (Turp.) Breb. to potassium dichromate ($K_2Cr_2O_7$) as a model toxicant.

MATERIALS AND METHODS

A pure laboratory culture of green chlorococcal alga *Scenedesmus qudricauda* (Turp.) Breb. was obtained from the collection of the Department of Microbiology, Biological Faculty, Moscow State University (DMMSU, strain S-3). The alga was grown in Uspenskii medium no1 (composition, g/l: 0.025 KNO_3, 0.025 $MgSO_4$, 0.1 KH_2PO_4, 0.025 $Ca(NO_3)_2$, 0.0345 K_2CO_3, 0.002 $Fe_2(SO_4)_3$; pH 7.0-7.3) in conical flasks in luminostat under periodic illumination (12 h/day). Cells were counted with a Goryaev's hemocytometer under a light microscope. Number of dead cells was counted with a MLD-1 (LOMO, Russia) fluorescent microscope. Under illumination of UV and blue light dead cells emit green, whereas living cells emit red light. The functional state of the photosynthetic apparatus of the alga was characterized by *in vivo* measuring of delayed fluorescence of chlorophyll *a*. The toxicant potassium dichromate ($K_2Cr_2O_7$) was added to the alga cultures (initial cell number was about 2×10^5 cells/ml) at a logarithmic phase. The concentrations of the toxicant were obtained after dilution of initial solution 1 mg/ml. All measurements for each sample were repeated three times.

RESULTS AND DISCUSSION

We worked out a program of "step by step" experiment, which has been carried out to develop Cr-tolerant cells of algae through previous exposure at various concentrations of $K_2Cr_2O_7$ 0.1; 1.0; 3.0; 10.0 mg/l during 30 days. Then the alga was re-inoculated twice in medium with 10.0 mg/l $K_2Cr_2O_7$.or in the medium without toxicant. The re-inoculation did by following manner: after intoxication during 30 days the algae were infiltrated via membrane filters NN 4 and 5, washed by distilled water and transferred to the Uspenskii medium with or without toxicant. The three experiments were conducted according to this scheme. After step II before re-inoculation the control cultures were diluted by Uspenskii medium to the initial cell number 2×10^5 cells/ml. At the end of step III of experiment the algae were re-inoculated in the Uspenskii medium without $K_2Cr_2O_7$. adding (step IV) for estimation of population restore possibility. In spring we did additional experiment with $K_2Cr_2O_7$. After 30-day exposure with 10 mg/l the algae were we re-inoculated twice in Uspenskii medium with 10 mg/l $K_2Cr_2O_7$. Then, the algae were transferred in the pure medium without toxicant.

The scheme of the experiment ($K_2Cr_2O_7$ concentrations, mg/l which the algae were inoculated in consecutively are indicated):

The cultures were pre-adapted to the $K_2Cr_2O_7$ action because of growing with various concentrations of the toxicant. Then the algae were transferred to the Uspenskii medium with 10 mg/l $K_2Cr_2O_7$ (step I).

In the end of step II pre-adaptation has led to the following results. The number of living algal cells was different: the more the initial $K_2Cr_2O_7$ concentration was, the less the cell number was. The maximal cell number was 73000 cells/ml in the culture previously exposed in the presence of $K_2Cr_2O_7$ 0.1 mg/l (seemingly inactive concentration, "dead zone" on the dose-response curve). We suggested that the increasing of algal response to the toxic action is a result of initial exposition at this concentration.

After the re-inoculation in $K_2Cr_2O_7$ 10 mg/l we revealed that the cell number was 6000-8000 cells/ml in all samples (step III). In sample, pre-adapted with 10 mg/l $K_2Cr_2O_7$, living cells were not found in winter, but in spring the cell number was 5000 cells/ml (variant 5). It means that metal-resistance changes in the course of year. During the experiments efficiency of photosynthesis decreased accordingly to $K_2Cr_2O_7$ concentrations: the higher concentration was, the lower efficiency of photosynthesis was.

Thus, in spite of the long-term exposition with toxicant some algal cells remained alive. Their number was 5-6% of initial cell number. We analyzed the size-age population structure and

Steps and terms their carrying out (excluding variant 5)	Variants of the experiment ($K_2Cr_2O_7$, mg/l)					
	Control	1	2	3	4	5 (14.03-26.07.95)
I 3.10.94-1.11.94	0	0.1	1	3	10	10
II 1.11.94-6.12.94	↓ 0	↓ 10	↓ 10	↓ 10	↓ 10	↓ 10
III 6.12.94-10.01.95	↓ 0	↓ 10	↓ 10	↓ 10	↓ 10	↓ 10
IV 10.01.95-14.02.95	↓ 0	↓ 0	↓ 0	↓ 0	-	↓ 0

Note: "0" - culture without toxicant;
"-" - algae were not re-inoculated.

photosynthetic activity in control cultures and after treatment. The result of size-age cell distribution demonstrated in *figure 1*. It is clear that cell size spectrum is the rather same than control one. It indicates that after toxic exposure the normal algal cells remain in population. The photosynthetic activity of these cells was the same than control one, too. The number of these cells (5-6%) corresponds with frequency of mutation for unicellular algae, fungi and bacteria in nature.

The resistant cells cause quick population restoration after the intoxication arrest. For example, the growth rate of the cells, which were pre-adapted with 3.0 mg/l $K_2Cr_2O_7$ and re-inoculated twice to the medium with 10 mg/l $K_2Cr_2O_7$, was ten times as many as that of the control.

The presence of resistant cells can be related to their constant presence in population or is the result of selection. It is need of special research for clarification of this phenomenon.

CONCLUSION

Thus, we were shown the possibility of adaptation of unicellular algal laboratory population *S. quadricauda* to the potassium dichromate action with using of "step by step" experiment. The maximal resistance of the algae to $K_2Cr_2O_7$ was revealed in spring-summer, the minimal resistance - in winter. The functional and structural characteristics of the algal population after toxic treatment were not different from that of the control. We count the size of pool of resistant cells in heterogeneity algal population (5-6% of initial cell number). These cells are able to restore the algal population after the stop of toxic treatment.

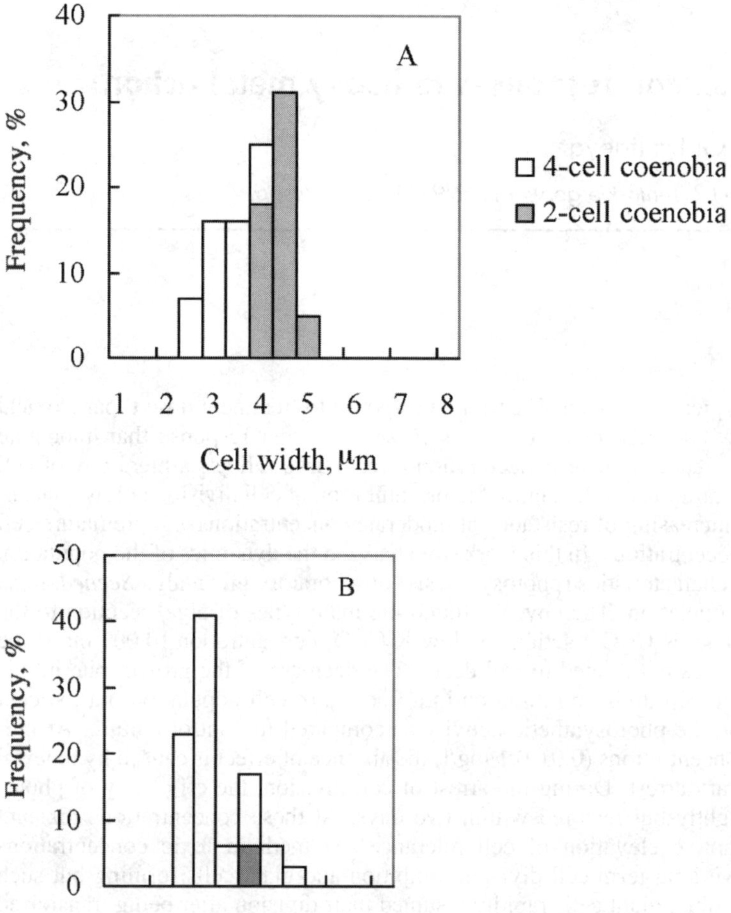

Fig. 1. Cell width distribution in the cultures of *S. quadricauda*.
A - control cultures; B - after the triple intoxication of $K_2Cr_2O_7$ 10 mg/l

REFERENCES

1. Trainor F. R. Indicator algal assay: laboratory and field approaches//Algae as ecological indicator. 1984.
2. Biomonitoring of coastal waters and estuaries. 1995.
3. Schafer H., Hettler H., Fritsche U., Pitzen G., Rodoger G., Wenzel A. Biotests using unicellular algae and ciliates for predicting long-term effects of toxicants. *Ecotoxicol Environ Saf.* 1994, 27, 1, 64-81.
4. van Leeuwen C. J., Luttmar W. J., Griffioen P. S. The use of cohorts and populations in chronic toxicity studies with Daphnia magna: a cadmium example. *Ecotoxicol Environ Saf.* 1985, 9, 1, 26-39.
5. Stockner J. L., Antia N. J. Phytoplankton adaptation to environmental stresses from toxicants, nutrients and pollutants - a warning. *J Fish Res Board Canada*. 1976, 33, 2089-2096.
6. Hall J. L. Cellular mechanisms for heavy metal detoxification and tolerance. *J Exp Bot*. 2002. 53, 366, 1-11.
7. Lasat M. M. Phytoextraction of toxic metals. *J Environ Quality*. 2002, 31, 109-120.

*Metal Ions in Biology and Medicine: vol. 7.*Eds L. Khassanova, Ph. Collery, I. Maymard, Z. Khassanova, J.-C. Étienne.
John Libbey Eurotext, Paris © 2002 pp. 300-305.

Types of algal population responses to heavy metal action

V. Yu. Prokhotskaia[1], A. G. Dmitrieva[2]

[1,2]*Moscow State University, 1-12, Leninskie gory, 119899, Moscow, Russia.*

ABSTRACT

A usual behavior of biological systems is three-phase response to treatment dose ("paradoxical reaction"). It means that low and high doses of factor cause a stronger response than moderate ones. We were shown earlier that nonlinear concentration-effect curve reflects a hierarchy of cell responses to increasing concentrations of damaging factor: inhibition of cell division at low concentrations; stress and adaptive increasing of resistance at moderate concentrations and premature cell division and death at high concentrations. In this work, we analyzed the dynamics of the population structure and cell functional characteristics (photosynthesis) of laboratory green alga *Scenedesmus quadricauda* (Turp.) Breb. population. Thereby, we found the main types of algal reaction to the toxicant (potassium dichromate, $K_2Cr_2O_7$) action. At low $K_2Cr_2O_7$ concentration (0.001 mg/l) the total cell number decreasing was not related to cell death. The decrease of the growth rate in this case was due to the long-term cell division inhibition in a fraction of cell population but toxicant did not have strong effect on the photosynthetic activity as compared to control culture. At moderate, seemingly inactive concentrations (0.01-0.1 mg/l) the absence of effect is caused by renewal of cell division after temporary arrest. During the arrest of cell division, the efficiency of photosynthesis decreased only slightly but restored within two days. At these concentrations toxicant induced cell stress and adaptive elevation of cell tolerance. At medium toxic concentrations (1.0-3.0 mg/l) we can observe long-term cell division inhibition and giant cells forming but such a state of the algae was reversible: giant cells rapidly resumed their division after being transferred to a toxicant-free medium. Sublethal $K_2Cr_2O_7$ concentrations did not significantly inhibit photosynthesis. At lethal concentration (10.0 mg/l) the cell division is stimulated and the small immature cells predominated in the beginning of intoxication. The toxicant caused reducing photosynthetic efficiency to a double as compared to control level. It indicated irreversible cell damage. Thus, changes of the population structure characterize its state. We can recommend using described types of reaction to the toxic action for risk assessment and biotesting.

It is known that the population number changes by complex way in increasing of damaging factors intensity. A usual behavior of biological systems is three-phase response to treatment dose ("paradoxical reaction"). It means that low and high doses of factor cause a stronger response than moderate ones [1, 2]. We have shown earlier that nonlinear concentration response curve of cell survival reflects of hierarchy of cell responses to increasing concentration of fungicide imazalil sulfate: cell division inhibition in low doses, stress and adaptive tolerance increasing in moderate doses and immature cell division and cell death in high doses [3]. The purpose of our research was to reveal the main principles of algal population response to the toxicant (potassium dichromate, $K_2Cr_2O_7$). Moreover, we display a possibility of their using as a diagnostic tool for biotesting.

MATERIALS AND METHODS

A pure laboratory culture of green chlorococcal alga *Scenedesmus qudricauda* (Turp.) Breb. was obtained from the collection of the Department of Microbiology, Biological Faculty, Moscow State University (DMMSU, strain S-3). The alga was grown in Uspenskii medium no1 (composition, g/l: 0.025 KNO_3, 0.025 $MgSO_4$, 0.1 KH_2PO_4, 0.025 $Ca(NO_3)_2$, 0.0345 K_2CO_3, 0.002 $Fe_2(SO_4)_3$; pH 7.0-7.3) in conical flasks in luminostat under periodic illumination (12 h/day). Cells were counted with a Goryaev's hemocytometer under a light microscope. Number of dead cells were counted with a luminescent microscope. Cell width was measured with a calibrated occular micrometer (no less than 80 cells in each sample) with an accuracy of 0.1 μm. The functional state of the photosynthetic apparatus of the alga was characterized by *in vivo* measuring of delayed fluorescence of chlorophyll *a*. The toxicant potassium dichromate ($K_2Cr_2O_7$) was added to the alga cultures (initial cell number was about 2×10^5 cells/ml) at a logarithmic phase to a final concentration of 0.001- 10.0 mg/l.

RESULTS

Two cell groups are mainly presented in culture. A small, daugther cells had width 3.0 μm and formed 4-cellular coenobia, predominated after cell division; large, mature cells had width 4.5 μm and formed 2-cellular coenobia, predominated before cell division. Efficiency of photosynthesis of small cells was higher than that of large cells. Thus, appearance in the culture small cells is a marker of cell division and increasing cell number.

The total cell number changed in a complicated pattern in a presence of $K_2Cr_2O_7$. At low and high concentrations of the toxicant the number of cells was less than in the control culture, whereas at moderate concentrations toxicant had no visual effect *(fig. 1*, curve *1)*. Such three-phase concentration-effect relationship we observed beginning from 3-4 days after toxicant adding to the end of experiment (30 days).

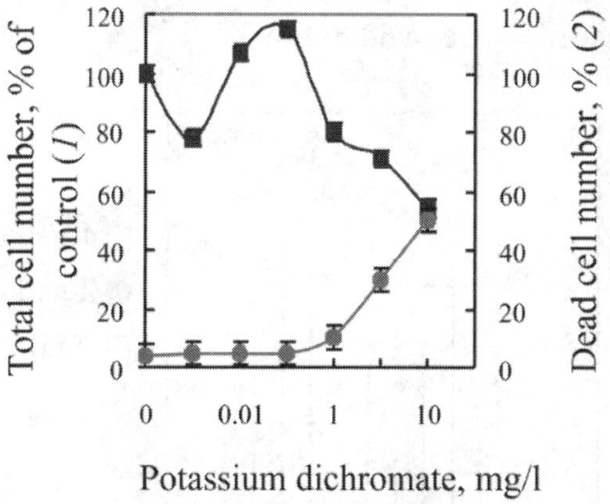

Fig. 1. Changes of the total cell number *(1)* and dead cell number *(2)* in the *S. quadricauda* culture as a function of $K_2Cr_2O_7$ concentration on the 4th-7th days of treatment

The number of dead cells in the culture increased only at high $K_2Cr_2O_7$ concentrations *(fig. 1*, curve *2)*. Therefore, the change in the relative cell number at low concentrations cannot be explained by the simple summation of the cell division and cell death.

We were shown earlier that the lack of effect at the moderate toxicant concentrations ("dead zone") was caused by renewal of cell division after temporary inhibition [3]. At these concentrations the toxicant initiated cell transfer to the state of nonspecific resistance (stress) and the cell reparative mechanisms were activated.

We analyzed changes size-age distribution, coenobial composition and functional (photosynthesis) cell characteristics in $K_2Cr_2O_7$ presence.

At low concentration 0.001 mg/l $K_2Cr_2O_7$ we observed slowdown population growth as compared to the control culture starting from 3th-4th days. Analysis of size-age distribution showed the appearance of large cells (width 4.5-5.5 μm) in 2-cell coenobia *(fig. 2, A)*. It was seemingly caused by cell division inhibition. Later, the size of these cells increased to 6.0-6.5 μm, they became single and formed 50% of population *(fig. 2, B)*. The size distribution of cells had two maxima: the first wide maximum included proliferating cells, united in 2- and 4-cell coenobia and the second maximum was comprised by large single cells. By the 25th day of experiment, large cells transformed into single round "giant cells". Thus, the reason of population growth delay was the arrest of proliferation of some cells rather than deceleration of cell cycle in all cells. Toxicant had not strong effect on the photosynthetic activity as compared to the control level.

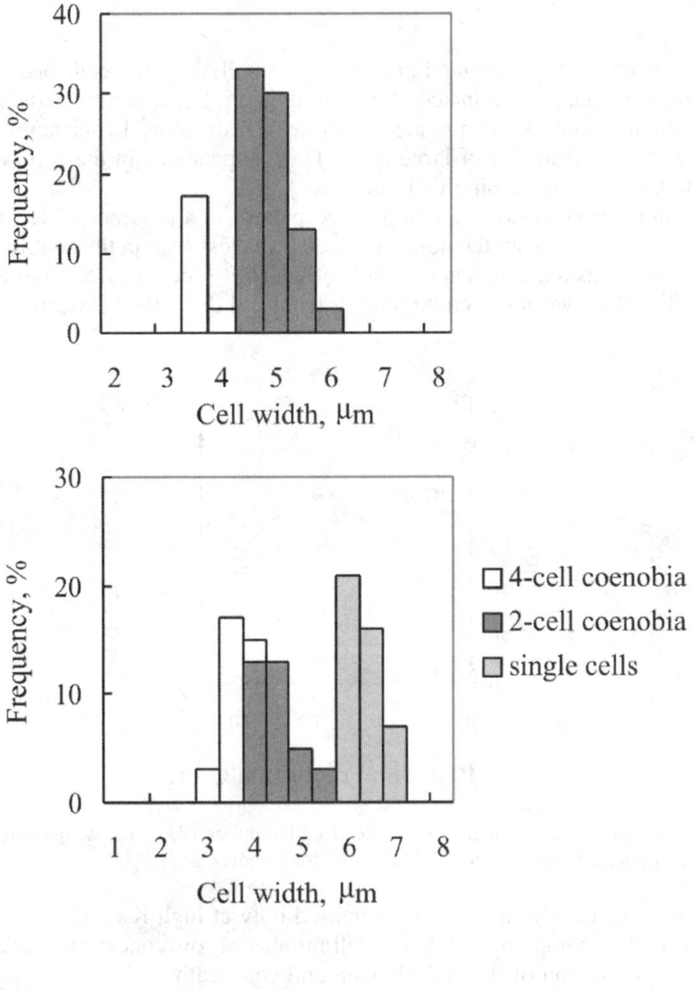

Fig. 2. Size-age and coenobial structure of *S. quadricauda* population after $K_2Cr_2O_7$ 0.001 mg/l incubation: A - 4th day, B - 30th day

At moderate doses 0.01-0.1 mg/l $K_2Cr_2O_7$ cell division was stopped during two days, and size both large and small cells increased. Then, cell division was restored synchronously and cell number was only slightly differed from control level. The efficiency of photosynthesis was the same as the control value.

At sublethal concentration $K_2Cr_2O_7$ 1.0 mg/l we observed 70% total cell number decreasing by the 30th day of experiment. Number of dead cells was not more than 15%. During 3th-21th days the large cells (width 6.5-7.0 µm) appeared in both 2- and 4-cell coenobia *(fig. 3, A)*. They did not divide and had only one nucleus. Later, (21th-30th days) size distribution was the same as the control one with maxima 3.5 and 5.0 µm *(fig. 3, B)*. It means that initial cell division arrest was reversible and the usual cell cycle was restored. These $K_2Cr_2O_7$ concentration did not significant inhibit photosynthetic efficiency, too.

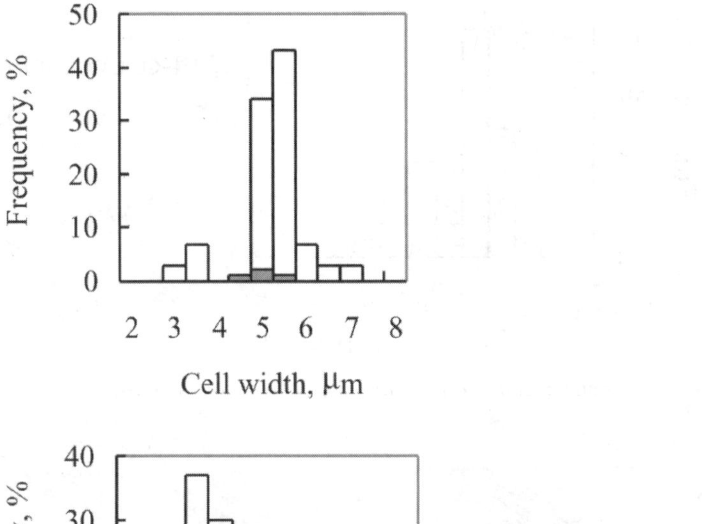

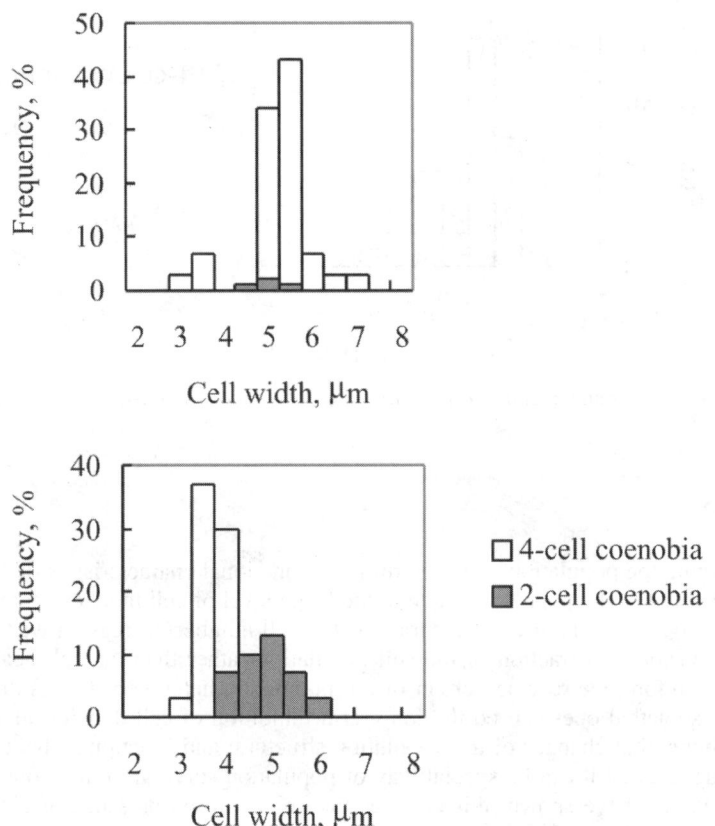

Fig. 3. Size-age and coenobial structure of *S. quadricauda* population after $K_2Cr_2O_7$ 1.0 mg/l incubation: A - 4th day, B - 30th day

At the concentration 3.0 mg/l the cell number was the same as initial one during the experiment. Analysis of size-age structure and functional characteristics of the cells showed that there were at least two reasons: delay of cell division of one cells and division and death of others. By the 2th-4th days we observed both undividing large cells (width 6.0 µm) in 2-cell coenobia and small proliferating cells in 4-cell coenobia. Then, (4th-7th days) 2-cell coenobia with cells (width 3.5-4.0 µm) which were smaller than control ones appeared. It means that $K_2Cr_2O_7$ disturbed coenobial wall integrity caused their breakdown. Beginning from 15th day size-age structure was the same as control one again.

At the lethal concentration 10.0 mg/l the cell number decreasing was caused by their death, but during the first day of cultivation the cell number did not change. The very small cells (width 2.0-2.5 µm) in 4-cell coenobia appeared within population *(fig. 4)*. Therefore, toxicant first initiated cell division in all cells, including those that had not attained the mature cell size. In the normal culture cells divided after attaining about 4.5 µm in diameter, whereas in the presence of toxicant they divided after attaining the size of 3.5 µm. Since the total cell number did not change, it is clear that a certain part of cells died. Therefore, the analysis of size-age population structure can find out the lethal effect earlier than counting of cell number. After 1-day incubation $K_2Cr_2O_7$ the photosynthetic activity was reduced to a double as compared to the control level. It implies that cell damage was irreversible.

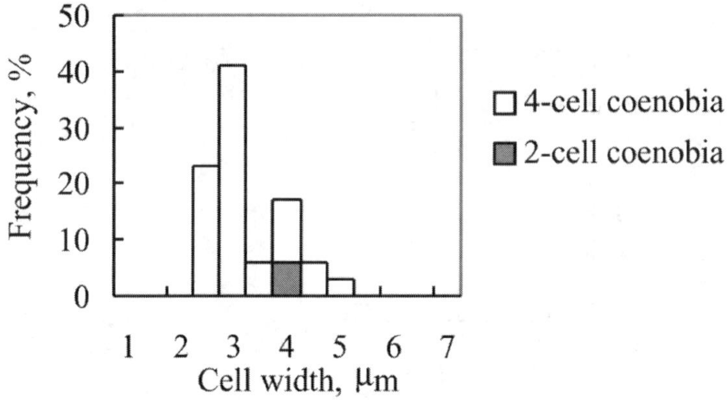

Fig. 4. Size-age and coenobial structure of *S. quadricauda* population after 1-day $K_2Cr_2O_7$ 10.0 mg/l incubation

DISCUSSION

Thus, analyzing the population cell spectrum and functional characteristics of cells we showed the differences between the population state at the same level of cell number decreasing under the toxicant action *(fig. 1)*. At the low concentrations the cell number decreasing was caused by inhibition of cell division in a fraction of the cell population rather than the cell death or cell cycle deceleration. Therefore, we revealed effect of the population heterogeneity. As distinct from low concentrations, sublethal ones caused the long-term inhibition of cell division in all cells.

We were shown that changes of the population structural and functional characteristics reflect its state. We suggest that it can be special way of population survival in unfavourable conditions. As our investigations have shown, this changes take place in the presence of different toxicants [3]. That is why it can be used as a tool for estimation of toxicant dangerous for water ecosystems. By this method we could do qualitative analysis of population reaction to the toxicants: appearance of large and giant cells denotes possible presence of sublethal toxicant concentrations, appearance of very small cells as the result of premature cell division means lethal effect of toxicant.

CONCLUSION

The concentration-response curve of cell survival reflects a hierarchy of cell responses to increasing concentration of the toxicants. On the base of structural and functional population characteristics analysis we suggest to appropriate the following types of population reaction to the toxicant action: at low concentrations (0.01 mg/l) the decreasing of cell number is the result of cell division arrest; at moderate, seemingly inactive concentrations (0.01-0.1 mg/l) the absence of effect is caused by renewal of cell division after temporary arrest; at medium concentrations (1.0-3.0 mg/l) we can observe long-term cell division inhibition and giant cells forming; at lethal concentration (10.0 mg/l) the cell division is stimulated and the small immature cells predominated at the beginning of intoxication. We offer using described types of reaction to the toxic action for risk assessment and biotesting.

REFERENCES

1. Alexandrov V. Ya. Reactivnost' kletok i belki (Cell reactivity and proteins). Leningrad: Nauka, 1985.
2. Burlakova E. V. Effect of extra low doses. *Vestn. Akad. Nauk*, 1994, 64, 425-431.
3. Prokhotskaia V. Yu., Veselovskii V. A., Veselova T. V., Dmitrieva A. G., Artyukhova V. I. On the nature of the three-phase response of Scenedesmus quadricuda populations to the action of imazalil sulfate. *Russian J Plant Physiol*, 2000, 6, 772-778.

Metal Ions in Biology and Medicine: vol. 7. Eds L. Khassanova, Ph. Collery, I. Maymard, Z. Khassanova, J.-C. Étienne.
John Libbey Eurotext, Paris © 2002 pp. 306-311.

The role of cations in the functioning of glutamine synthetase from *Azospirillum brasilense*

Victoria E. Smirnova[1], Lyudmila P. Antonyuk[1]*, Alexander A. Kamnev[1], Leonid A. Kulikov[2] and Yurii D. Perfiliev[2]

[1]*Laboratory of Biochemistry of Plant-Bacterial Symbioses, Institute of Biochemistry and Physiology of Plants and Microorganisms, Russian Academy of Sciences, 13 Prosp. Entuziastov, 410015 Saratov, Russia.*
[2]*Laboratory of Nuclear Chemistry Techniques, Department of Radiochemistry, Faculty of Chemistry, Moscow State University, 119899 Moscow, Russia.*
* Corresponding author. Tel./Fax: +7-(8452)-947303; E-mail: micbio@ibppm.saratov.su.

ABSTRACT

In the present work, we studied *(i)* the effect of divalent cations on the secondary structure of glutamine synthetase (GS) isolated from *Azospirillum brasilense* Sp245, and *(ii)* binding of Co^{2+} at the GS active sites using emission ^{57}Co Mössbauer spectroscopy (EMS). The native enzyme contained strongly bound metal ions, which could not be removed by dialysis. Circular dichroism (CD) analysis of the enzyme showed the *A. brasilense* GS to be a highly structured protein (59% of the residues as α-helices and 13% as β-strands). Adding divalent metal cations (1 mM Mg^{2+}, Mn^{2+}, or Co^{2+}) to the native enzyme caused only slight alterations in the CD spectra of the enzyme. On the contrary, a 30-min incubation of the *A. brasilense* GS with 5 mM EDTA leading to the removal of the cations from the protein molecule caused some noticeable changes in the CD spectrum. Calculations showed that the proportion of α-helix in the cation-free enzyme diminished as compared to that in the native enzyme, whereas the proportion of β-strands increased. EMS study of binding of $^{57}Co^{II}$ at the GS active sites revealed that *(i)* the *A. brasilense* GS has two divalent cation-binding sites per active site of the enzyme and *(ii)* the affinity of one metal ion site is higher than that of the other.

Key words: glutamine synthetase; *Azospirillum brasilense*; divalent cations; active sites, circular dichroism (CD) spectroscopy; emission (^{57}Co) Mössbauer spectroscopy.

INTRODUCTION

Plant growth-promoting rhizobacterium *Azospirillum brasilense* Sp245 belongs to facultative endophytes, i.e. to bacteria which not only colonise the exterior of the host plant root system, but can penetrate into plant cells and tissues. Nitrogen metabolism of azospirilla has not as yet been adequately explored; nevertheless, in the case of *A. brasilense* Sp245 it may have some specific features, since strain Sp245 as a simbiont is able to meet the needs of the host plant in bound nitrogen [1]. In this relation, investigation of the *A. brasilense* Sp245 glutamine synthetase (GS), a key enzyme of its nitrogen metabolism, is of great interest.

GS (EC 6.3.1.2.) catalyses the ATP-dependent and metal-ion dependent synthesis of L-glutamine from L-glutamate and ammonia [2]. It is well documented that GS, together with glutamate synthase (GltS), forms the main pathway for ammonia assimilation in bacteria. In addition to its assimilatory function, GS is involved in metabolic regulation and is recognised as a key enzyme of nitrogen metabolism in many organisms including bacteria. Regulation of activity and synthesis of bacterial GSs is very complex and has so far been investigated in detail for enteric bacteria only. GS activity in many bacteria, including *Azospirillum brasilense*, is modulated by reversible adenylylation in response to the cellular N-status. The enzyme is maintained in a top-active unadenylylated or slightly adenylylated form under nitrogen-limiting conditions, and the adenylylation level increases under the conditions of ammonia abundance [2].

In the present views, bacterial GS molecules are dodecamers formed from two face-to-face hexameric rings of subunits with 12 active sites formed between the monomers. Divalent cations (commonly Mg^{2+}, Mn^{2+} or Co^{2+}) are absolutely necessary for the activity of all known bacterial GSs both in the case of glutamine synthesis and for the transferase reaction catalysed by the enzyme. The X-ray crystallographic study of the *Salmonella typhimurium* GS revealed that each active site of the enzyme has two divalent cation binding sites, n1 and n2; both of them must be saturated for the activity to be expressed (see [2] and references therein). Judging from the earlier data obtained with $^{54}Mn^{2+}$ [3], along with the n1 and n2 sites, GS of enteric bacteria has 48 additional metal binding sites per oligomer. These additional cation binding sites have relatively low affinity [3]; they are considered to be arranged outside the active sites of the enzyme and to be important for the conformational stability of the molecule.

Previous research from our laboratory gave the first data on the enzymology of the *A. brasilense* GS [4-6]. A number of kinetic properties appeared to depend strongly on the nature of the divalent cation used for GS activation [5-6]. The aim of this work was to study *(i)* the effect of divalent cations on the secondary structure of the *A. brasilense* GS, and *(ii)* binding of Co^{2+} at the GS active sites using emission ^{57}Co Mössbauer spectroscopy (EMS).

MATERIALS AND METHODS

A. brasilense Sp245 was grown in a minimal salt medium (MSM) supplemented with 0.5% (w/v) sodium succinate and 5 mM L-glutamate, from which fully non-adenylylated GS was isolated and purified [6]. In the case of obtaining of slightly adenylylated GS, the MSM was supplemented with 5 mM NH_4Cl as a nitrogen source and 0.6% sodium malate as a carbon source [7]. Both enzyme preparations were isolated from frozen cells disrupted by passing through a French press at 60 MPa. To obtain the slightly adenylylated A. brasilense GS, the extract was treated with 0.25 mg/ml DNAse I (Sigma) and 2 mg/ml streptomycin sulphate, centrifuged, and the supernatant was subjected, first, to ion-exchange chromatography in 50 mM Tris-HCl buffer (pH 7.0) with a linear NaCl gradient from 0 to 1 M using DEAE-Toyopearl 650M (Toyo Soda), and then to FPLC on a Mono-Q column (Pharmacia Biotech) with the same buffer and NaCl gradient. In the case of fully non-adenylylated GS, the purification protopol was somewhat different, as described in detail earlier [6]. Each of the preparations obtained was electrophoretically homogeneous, contained divalent metal cations (see Results) and was referred to as "native GS". GS activity was estimated in the transferase reaction or in the biosynthetic reaction as described previously [5].

To obtain metal ion-free enzyme for non-spectroscopic experiments, EDTA was added up to the final concentration of 5 mM to A. brasilense Sp245 GS solution, and after a 30-min incubation period the preparation was dialysed against 200 volumes of 50 mM Tris-HCl buffer, pH 7.0 (for 24 h at 4°C), to remove the EDTA together with the chelated cations. For EMS, the slightly adenylylated A. brasilense Sp245 GS preparation was dialysed against 1000 volumes of 50 mM KCl (pH 7.0) at 4°C to separate the enzyme from EDTA-metal chelates. Carrier-free $^{57}CoCl_2$ (1 mCi) was added to 1 ml of aqueous solution of the resulting cation-free enzyme up to a ratio of ca. 24 $^{57}Co^{2+}$ ions per enzyme molecule (see below). This solution was then incubated for 1 h at ambient temperature, and finally rapidly frozen in liquid nitrogen. EMS measurements were performed by placing the ^{57}Co-containing sample (source) in a cryostat filled with liquid nitrogen (at ca. 80 K) using a conventional constant-acceleration Mössbauer spectrometer (absorber $K_4[Fe(CN)_6].3H_2O$) combined with a PC-operated multichannel analyser. Standard PC-based statistical analysis consisted of fitting the experimental data obtained (converted into a form compatible with that of absorption ^{57}Fe Mössbauer measurements) as the sum of Lorentzians using a least squares minimisation procedure, which enabled determination of the isomer shift (δ; relative to α-Fe), quadrupole splitting (Δ), linewidth (i.e. full width at half maximum, Γ) and relative areas of spectral components (S_r). Circular dichroism (CD) spectra were recorded on a JASCO J-500C spectropolarimeter (Japan) using a solution with a protein concentration of 0.8 mg/ml. Calculation of the content of the GS secondary structure elements was performed using the CONTIN program [6].

RESULTS AND DISCUSSION

Efficiency of divalent cations in supporting the GS activity. Native and cation-free GSs

The homogeneous preparations of GS from *A. brasilense* Sp245 contained divalent metal cations, and dialysis (the final purification step) failed to remove them from the enzyme. This conclusion is evident from the fact that the enzyme exhibited activity in the biosynthetic reaction when divalent cations were omitted from the assay mixture *(table 1)*. The GS preparation obtained was referred to as "native GS". In order to remove divalent metal cations from the enzyme, we used the treatment of the enzyme with 5 mM EDTA (see Materials and Methods). After such a treatment, the *A. brasilense* Sp245 GS was inactive in the absence of divalent cations in the assay mixture *(table 1)*. The results obtained suggest that the native unadenylylated GS contains metal ions at their active sites. All the three tested cations, Mn^{2+}, Mg^{2+} and Co^{2+}, were effective in supporting the biosynthetic activity of the unadenylylated GS from *A. brasilense* Sp245, as well as with the native GS and with the metal-ion free enzyme *(table 1)*.

Table 1. Biosynthetic activity of the native and metal-free non-adenylylated glutamine synthetase (GS) from *Azospirillum brasilense* Sp245

Metals in the assay mixture, 1 mM	Biosynthetic activity (mg P_i per min per mg protein) in:	
	native GS	metal-free GS
-	9.5	0
Mn^{2+}	11.0	6.9
Mg^{2+}	24.5	8.5
Co^{2+}	24.5	19.1

The effect of divalent cations on the secondary structure of A. brasilense GS

To study the effect of cations on the secondary structure of the enzyme, CD spectra were measured of the native non-adenylylated GS, to which 1 mM Mg^{2+}, Mg^{2+}, Co^{2+} or EDTA was added. Besides that, a CD spectrum of the native GS after treatment with 5 mM EDTA was recorded. The spectra obtained (not shown) were used for calculating the content of secondary structure elements in the enzyme *(table 2)*.

The native non-adenylylated *A. brasilense* Sp245 GS (bearing bound metal cations, as was determined in the experiments described above) contained a high proportion of α-helix (59% of the enzyme polypeptide chain, *table* 2). The proportion of β-strands appeared to be lower (13%). As a whole, the native unadenylylated *A. brasilense* Sp245 GS appears to be a highly structured protein: according to the calculations, 72% of the molecule is structured (α-helices and β-strands) while only 28% is unordered. As follows from *table 2*, introducing additional divalent cations to

Table 2. Content of secondary structure elements of non-adenylylated glutamine synthetase (GS) from *Azospirillum brasilense* Sp245 calculated from circular dichroism spectra

Variants of treatment	Secondary structure elements, %		
	α-helices	**β-strands**	**Unordered structure**
GS (native enzyme)	59±2	13±5	28±3
GS + 1 mM $CoCl_2$	57±2	21±4	22±4
GS + 1 mM $MgCl_2$	52±4	21±7	27±7
GS + 1 mM $MnCl_2$	56±2	13±4	31±3
GS + 1 mM EDTA	43±1	24±3	32±2
GS + 5 mM EDTA	38±2	32±4	30±3

the system (besides those already bound to the protein) did not cause significant changes in the content of the secondary structure elements of the protein. Conversely, the removal of the metal ions resulted in alterations in the enzyme secondary structure. Calculations showed that the proportion of α-helix in the metal ion-free enzyme is significantly diminished as compared to that in the native enzyme, whereas the proportion of β-strands is increased (see *table 2*).

EMS study of ^{57}Co-activated glutamine synthetase from A. brasilense Sp245

The EMS spectrum of cation-free *A. brasilense* GS incubated for 1 h with $^{57}Co^{2+}$ measured in rapidly frozen solution *(fig. 1)* contains two components corresponding to two high-spin daughter Fe^{II} forms *(table 3)* with different coordination. It may be supposed that the Fe^{II} form with Δ=2.39 mm/s corresponds to the site with more symmetrically coordinated O-donor ligands. Note that, in case of *Salmonella typhimurium* GS, the cation in site n1 (with a much higher affinity) is coordinated by three glutamate (Glu) residues (i.e., 3 carboxylic groups), whereas in site n2 by one histidine (His) and two Glu residues (i.e., one N-donor atom of the His heterocycle and two carboxyls) [2]. If such is the case with *A. brasilense* GS, the higher Δ value (3.08 mm/s) may reflect a lower coordination symmetry owing to different donor atoms (the isomer shift (δ) values are close in these cases). Similar parameters were reported for Fe^{II}-DNA complex: δ=1.07 mm/s and Δ=2.98 mm/s at 300 K ([8], p. 382); note that in going from 300 K to 80 K, most quadrupole splitting values for Fe^{II} compounds tend to slightly increase. In neutral DNA solutions, purine nitrogen atoms may coordinate Fe^{II} [8], and it is noteworthy that the heterocycle of the His residue is similar to the 5-member heterocycle in purine. This conclusion is also in line with a noticeably higher proportion (60%) of the Fe^{II} component with Δ=2.39 mm/s (see *table 3*) which may thus correspond to the site of the type n1 (with a higher affinity to cations) in *A. brasilense* GS.

Note also that the yield of the stabilised daughter $^{57}Fe^{III}$ component resulting from after-effects is much lower (22%; see *table 3*) than for very dilute, rapidly frozen (80 K) aqueous solutions of $^{57}Co^{2+}$ (see [7] and references therein). This corresponds to a more firm binding of parent $^{57}Co^{II}$ to the aforementioned bioligands with a higher electron-donor properties in the inter-subunit spaces of the two GS hexamers [2].

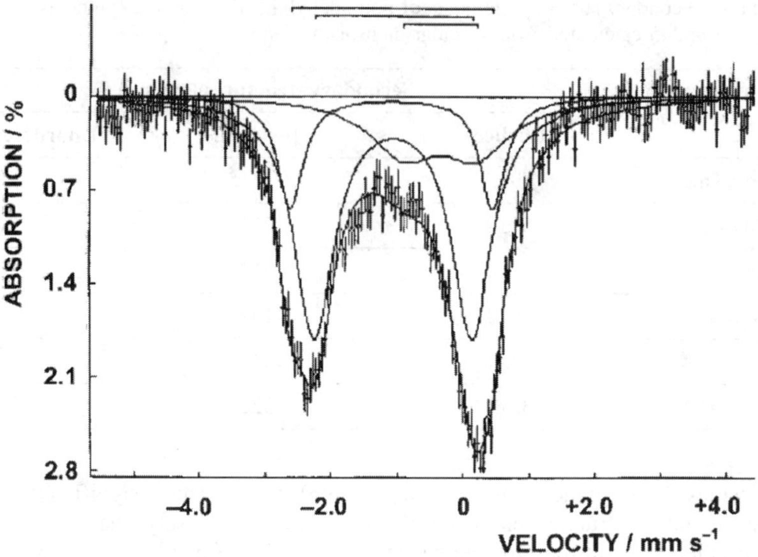

Fig. 1. Emission Mössbauer spectrum of cation-free glutamine synthetase from *Azospirillum brasilense* Sp245 activated by $^{57}Co^{2+}$ (frozen solution; T = 80 K)

Table 3. EMS parameters[a] for $^{57}Co^{II}$-activated glutamine synthetase (GS) isolated from *Azospirillum brasilense* Sp245 [7] (T = 80 K)

Oxidation state[b]	δ,[c] mm/s	Δ,[d] mm/s	Γ,[e] mm/s	S_r,[f] %
+2	1.08(2)	3.08(8)	0.48(5)	18(1)
+2	1.05(2)	2.39(6)	0.75(8)	60(1)
+3	0.34(10)	1.12(20)	1.25(30)	22(1)

[a]Errors (in the last digits) are given in parentheses. [b]For the nucleogenic ^{57}Fe components stabilised after nuclear decay of the parent $^{57}Co^{II}$. [c]Isomer shift (relative to α-Fe; converted to the normal absorption convention). [d]Quadrupole splitting. [e]Full line width at half maximum. [f]Relative areas of spectral components (represent relative contents of the corresponding Fe forms assuming a common recoilless fraction for all forms in a sample).

ACKNOWLEDGEMENTS

This work was supported in parts by INTAS (EC, Brussels, Belgium), NATO (Grant LST.CLG.977664) and the Russian Academy of Sciences' Commission (Grant No. 205 under the 6th Competition-Expertise of research projects).

REFERENCES

1. Döbereiner J., Pedrosa F.O. Nitrogen-fixing bacteria in nonleguminous crop plants, Berlin: Springer Verlag (1987).
2. Eisenberg D., Gill H.S., Pfluegl G.M.U., Rotstein S.H. Structure-function relationships of glutamine synthetases. *Biochim Biophys Acta*, **1477**, 122-145 (2000).
3. Denton M.D., Ginsburg A. Conformational changes in glutamine synthetase from *Escherichia coli*. I. The binding of Mn^{2+} in relation to some aspects of the enzyme structure and activity. *Biochemistry*, **8**, 1714-1725 (1969).

4. Bespalova L.A., Korshunova V.E., Antonyuk L.P., Ignatov V.V. Isolation, purification, and some kinetic properties of moderately adenylylated glutamine synthetase from *Azospirillum brasilense* Sp245. *Biochemistry* (Moscow), **59**, 41-45 (1994).
5. Bespalova L.A., Antonyuk L.P., Ignatov V.V. *Azospirillum brasilense* glutamine synthetase: influence of activating metal ions on the enzyme properties. *BioMetals*, **12**, 115-121 (1999).
6. Antonyuk L.P., Smirnova V.E. Kamnev A.A., Serebrennikova O.B., Vanoni M.A., Zanetti G., Kudelina I.A., Sokolov O.I., Ignatov V.V. Influence of divalent cations on the catalytic properties and secondary structure of unadenylylated glutamine synthetase from *Azospirillum brasilense*. *BioMetals*, **14**, 13-22 (2001).
7. Kamnev A.A., Antonyuk L.P., Smirnova V.E., Serebrennikova O.B., Kulikov L.A., Perfiliev Yu.D. Trace cobalt speciation in bacteria and at enzymic active sites using emission Mössbauer spectroscopy. *Anal. Bioanal. Chem.*, **372**, 431-435 (2002).
8. Vértes A., Korecz L., Burger K. Mössbauer Spectroscopy, Elsevier, Amsterdam (1979).

*Metal Ions in Biology and Medicine: vol. 7.*Eds L. Khassanova, Ph. Collery, I. Maymard, Z. Khassanova, J.-C. Étienne.
John Libbey Eurotext, Paris © 2002 p. 312.

The influence of environment polluted with some metal ions on generative sphere of plants (palynoteratical data from the Leningrad atomic power station area)

G.M. Levkovskaya*, O.A. Gavrilova**, L.A. Karzeva**

**Institute of History of Material Culture, RAS, 18 Dvortsovaya emb., St.-Petersburg, 191186, Russia.*
***Komarov Botanical Institute. RAS, 2 Popova st., St.-Petersburg, 197376, Russia.*

The rate of heavy - and radioactive metal ions (except copper and zinc) and benz(a)-pirene in the Leningrad Atomic Power Station (APS) industrial zone is not high, though the quantity of Mn, Co, Ni, As and Pb ions in separate points exceeds normal concentrations (Dzuba et al., 2001). The level of chemical contamination in the town of Sosnovyi Bor is higher than at the APS.

A new statistical palynoteratical method (G.M. Levkovskaya. Palynoteratical Complexes as Indicators of Ecological Stress, Past and Present// Proceedings of 5th European Palaeobotanical and Palynological Conference. Acta Palaeobotanica 2, Krakow, 643-648, 1999) allows to use complexes with domination of morphologically abnormal pollen of different types as indicators of normal, stressed and catastrophic states of generative sphere of single plants and even their societies.

A specific feature of pollen complexes of soil samples from Sosnovyi Bor is high percentage of palynoterates with divergence from the norm in several morphological characters simultaneously which was proved by statistical and SEM data.

This type of palynoteratical complex is a "response" of the plants' generative sphere on an anrtropogenic stress. It was a result of polluting the area with heavy and radioactive metal ions.

A new method of environmental control could be worked out as a result of joint research projects of the author of the palynoteratical method (Levkovskaya, 1999; see e-mail for contacts) and specialists on metal ions.

The author is grateful to the Russian Foundation for Basic Research (grant: 01-06-80-222) and to St. Petersburg Scientific Center for financial support of some palynoteratical researches.

Metal Ions in Biology and Medicine: vol. 7.Eds L. Khassanova, Ph. Collery, I. Maymard, Z. Khassanova, J.-C. Étienne.
John Libbey Eurotext, Paris © 2002 pp. 313-316.

Evidence for heavy metal inhibition of algal growth in Lake Pamvotis (Greece)

Kagalou I.[1], Beza P.[1], Papagiannis I.[2], and V. Kalfakakou[2]

[1]*Technological Educational Institute, Dept. of Ichthyology and Fisheries, Laboratory of Aqu. Ecosystems, P.O Box 152, 46100, Igoumenitsa, Greece.*
[2]*Laboratory of Exp. Physiology, Medical School, University of Ioannina, 45110, Ioannina, Greece.*
Corresponding author: Dr. Ifigenia Kagalou.

ABSTRACT

Heavy metals from natural and anthropogenic sources are continually released into aquatic ecosystems and they are a serious threat because of their toxicity, long persistence, bioaccumulation and biomagnification in the food-chain.

In the present study the effect of Cu and Pb, on Chlorophyll-a production as a net growth rate of algal biomass index, in Lake Pamvotis (Greece) was investigated. Lake water samples collected from two sampling-stations were metal treated to 0.05, 0.1 and 5 μg/l final concentrations, in three replicates each one, for both Cu and Pb correspondingly. Samples as well as controls were incubated for 48h under simulated in situ conditions of temperature, light level and photoperiod. Upon termination of incubation, Chlorophyll-a was, spectrophotometrically, determined after the acetone extraction of the pigment.

Additional limnological parameters (included temperature, pH, dissolved oxygen and nutrients) were measured. Promoting effects on algal growth were observed in the treatment with the highest Cu conc. whereas stimulatory and inhibitory effects were recorded in Pb treatment.

Key words: Heavy metals, copper, lead, chlorophyll-a, lake Pamvotis.

INTRODUCTION

Regular discharge of toxic materials from anthropogenic point pollution sources as industries, urban settlements, as well from non point pollution sources (i.e run-off from agricultural fields) results in considerable degradation of water quality in aquatic ecosystems (Whitton, 1970). That trace elements have a function in biological matter has been appreciated only recently even though experimental evidence of their importance was postulated many years ago (Gonzalez-Davilla, 1995).

Heavy metals exert multiple inhibitory or promoting effects on photosynthesis at different structural and metabolic levels (Giardi et al., 1997). Biochemical interactions between algae and metals have been studied in marine environments, but fresh water ecosystems have received less attention. Moreover most work has involved laboratory experiments rather than field studies (Nosher et al., 2000).

The aim of the present study was to investigate the effects of copper and lead on the fresh water algal biomass of lake Pamvotis by use of bioassay techniques.

MATERIALS AND METHODS

The study area

Lake Pamvotis is located in the N-W part of Greece *(fig. 1)*. It's a shallow (mean depth < 7m) lake and occupies an area of 22.8 km^2. It has a great recreational value and also it supports local agriculture, tourism and fisheries. During recent decades since the 1980s, ecosystem of Pamvotis sustained many activities such as irrigation, input of domestic sewages, sediment deposit, causing a serious degradation problem. The main sources for metal pollution of Pamvotis lake are the following: garages, motor workshops, car washing stations and agricultural effluents are activities releasing mainly Pb but also Cu and Zn (Kalfakakou et al. 2000).

Methodology

In September and December 2001, lake water samples were collected at two sampling stations selected as representative of eutrophic and hypertrophic conditions (st.1, st.2 correspondigly, *fig. 1*). At each sampling station 12 1-l polyethylene bottles were filled with surface lake water. Samples were treated immediately with Cu and Pb addition using stock solutions (CuSO4.5H2O and $Pb(NO_3)_2$) in final metal concentrations of 0.05, 0.1 and 5 μg/l. Three replicates were prepared along with three control samples (no addition). Experimental bottles were incubated for 48-h under simulated conditions of photoperiod, temperature and light level. Upon termination of the incubation two 250 ml aliquots were removed from each one bottle and the concentration of Chlorophyll-a was determined spectrophotometrically after the acetone extraction of the pigment (APHA, 1986). Additional limnological parameters were studied included pH, diss.oxygen, phosphates and nitrates acccording APHA methodology.

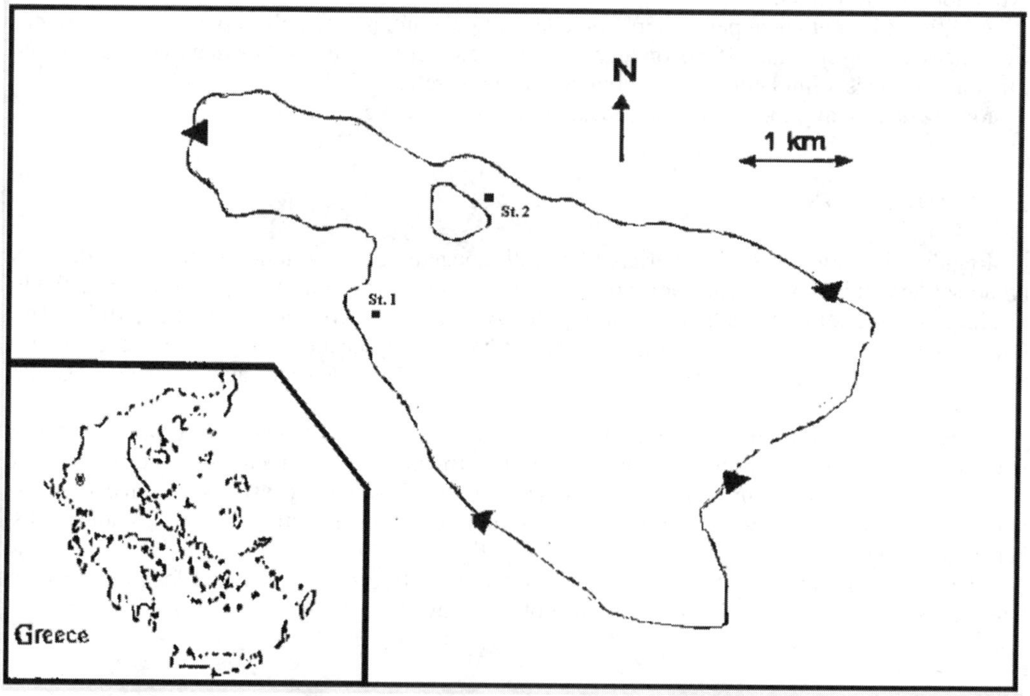

Fig. 1. Lake Pamvotis (NW Greece) map showing the sampling sites

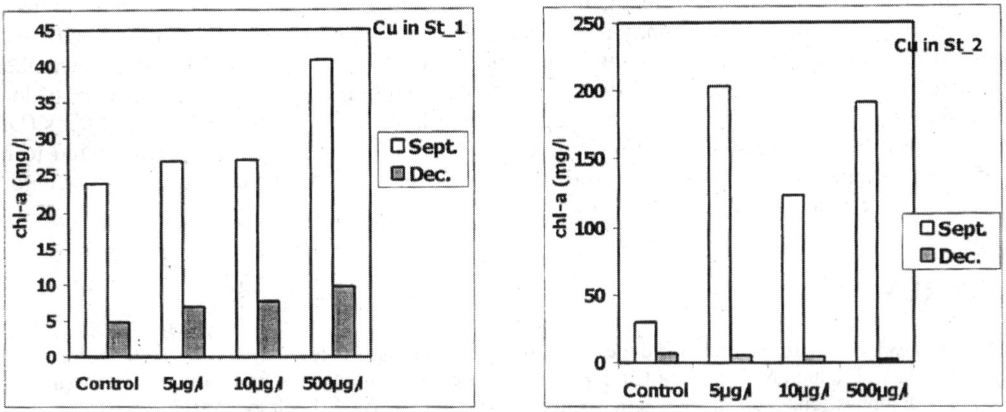

Fig. 2. **a.** Concentrations of chl-a in station 1, **b.** Concentrations of chl-a in station 2

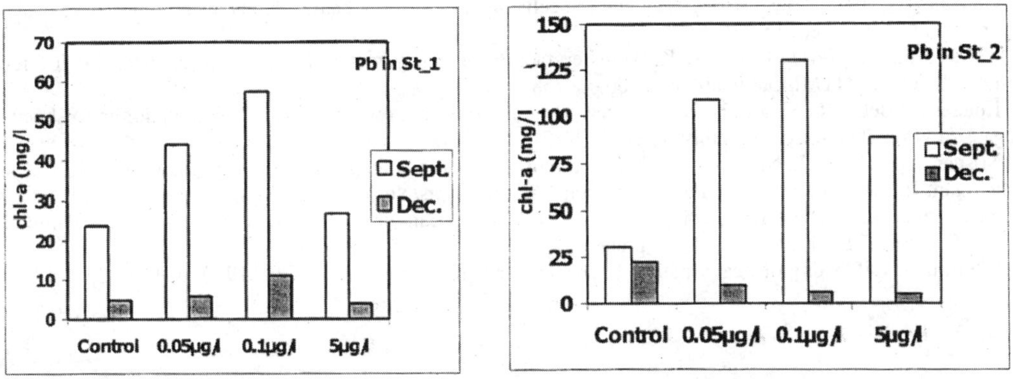

Fig. 3. **a.** Concentrations of chl-a in station 1, **b.** Concentrations of chl-a in station 2

RESULTS AND DISCUSSION

Experimental results concerning the chlorophyll-a concentrations during the exposure period, in various concentrations of Cu and Pb are shown in the *figures 2 (a & b)* and *3 (a & b)*.

In September, at both stations (st.1, st.2) Cu treatments with 0.05 and 0.1 μg/l (final concentrations) showed no effects on chlorophyll-a growth, relative to control. With increasing Cu concentrations (5 μg/l) algal biomass was enhanced. At the same period limnological properties at the two stations are promoting algal biomass (i.e at st.1: water temp.: 22°C, pH: 9.13, Nitrates-Nitrogen: 1.4 mg/l, Phosphates: 1.84 mg/l. At st.2: water temp.: 23°C, pH: 9.44, Nitrates-Nitrogen: 2.2 mg/l, Phosphates: 12,5 mg/l).

Results from December showed that Cu addition did not influenced chlorophyll-a biomass relative to control. The values of limnological parameters were coincided with low algal production (st.1: water temp.: 17°C, pH: 9.07, Nitrates-Nitrogen: 1.9 mg/l, Phosphates: 0.23 mg/l. At st.2: water temp.: 18°C, pH: 9.17, Nitrates-Nitrogen: 1.5 mg/l, Phosphates: 0.03 mg/l). According our previous study (Papagiannis et al., 2001) the mean Cu conc. in lake Pamvotis is 0.14 μg/l. The present finding suggests that copper concentrations between 0.1-5 μg/l, above ambient could serve as a growth factor where the other basic algal requirements (temperature, nutrients etc.) are available. According Knauer et al. (1997) copper has been implicated as a limiting factor for algae in eutrophic lakes. Regarding at Pb treatments, it was found that during September concentrations of 0.05 and 0.1 μg/l promoted chlorophyll-a biomass while the highest concentration (of 5 μg/l) showed a stimulatory effect on algal growth. The same profile was produced at st.1 during December. No differences were observed at st.2 during December. Inhibition or stimulatory capacity of lead on different physiological parameters and algal growth have been detected in earlier studies (Danilov, 2001). In the case study of lake Pamvotis where Pb mean concentration was found to be 0.05 μg/l (Papagiannis et al., 2001), treatment with 5 μg/l Pb can be concluded to have stimulatory effects on freshwater algae.

REFERENCES

- APHA, 1985. Standard methods for the examination of water and wastewater, 16th ed. New York.
- Danilov R.A., Ekelund N. Effects of Cu, Ni, Pb, Zn and pentachlorophenol on photosynthesis and motility in Chlamydomonas reinhardii in short term exposure experiments. BMC Ecology, 2001, 1, 1-10.
- Giardi M.T., Masojidek J., Godde D. Effects of abiotic stress on the turnover of the D1 reaction centre II protein. Physiol. Plant., 1997, 101, 635-642.
- Gonzalez-Davila M. The role of phytoplankton cells on the control of heavy metal concentration in seawater. Mar. Chem., 1995, 48, 215-236.
- Kalfakakou V., Akrida-Demertzi K. Transfer factors of heavy metals in aquatic organisms of different trophic levels. In HTML publications, 2000, 1, 768-778.
- Knauer K., Behra R., Sigg L. Effects of free Cu and Zn ions on growth and metal accumulation in freshwater algae. Environ. Toxic. And Chemistry, 1996, 16(2), 220-229.
- Nosher T., Lechman J., Nriagu J. Evidence for copper inhibition of algal growth in SaginawBay-LakeHuron, Michigan. In www.sph.umich.edu/eih/heavymetals/Manuscripts/NosherT.htm.
- Papagiannis J., Kagalou I., Paleologos E., Kalfakakou V. Karayiannis M. Heavy metals in lake Pamvotis (N-W Greece). Fresen. Envir. Bull. (in press).
- Whitton B.A. Toxicity of heavy metals to freshwater algae: a review. Phykos, 1970, 9, 116-125.

Metal Ions in Biology and Medicine: vol. 7. Eds L. Khassanova, Ph. Collery, I. Maymard, Z. Khassanova, J.-C. Étienne.
John Libbey Eurotext, Paris © 2002 pp. 317-321.

Isolation, purification and characterization of a chromate-reductase from an ochrobactrum SPP. 5BVL-1

Romeu Francisco[1,2], Paula V. Morais[1,2], Paula Veríssimo[2] and M. C. Alpoim[1,2]

1. *Instituto do Ambiente e Vida, 3004-517 Coimbra, Portugal.*
2. *Departamento de Bioquímica, Universidade de Coimbra, 3001-401 Coimbra, Portugal.*

ABSTRACT

Hexavalent chromium (Cr(VI)) is a strong oxidant and a toxic pollutant. Bacterial reduction of Cr(VI) to the less toxic and less water soluble Cr(III) has previously been reported and there is accumulated evidence that bacterial reduction of chromate can occur under both aerobic and anaerobic conditions. Furthermore, microorganisms can be a potential useful tool in the treatment of contaminated soils and waters.

Recently, we isolated several Cr(VI)-resistant bacteria strains from a Cr(VI)-contaminated activated sludge. One of them, belonging to the species *Ochrobactrum tritici* (strain 5bvl-1), was found to show both high Cr(VI)-resistance and reduction capacities. The isolation and characterization of the enzyme(s) responsible for reducing Cr(VI) would afford information to implement or improve bioremediation strategies.

The cells of the strain 5bvl-1 were disrupted by sonication and each fraction obtained was tested for chromate-reductase activity. Several preliminary tests were realized on the fraction with Cr(VI) reduction capacity, preceding our attempt to isolate, purify and characterize the chromate-reductase. The preliminary work performed showed that the enzyme chromate-reductase has different physiological characteristics than the ones previously described. The role this chromate reductase(s) plays in bacteria physiology has not been explored yet.

INTRODUCTION

Hexavalent chromium (Cr(VI)) is a strong oxidant and a toxic pollutant. Bacterial reduction of Cr(VI) to the less toxic and less water soluble Cr(III) has previously been reported [1] and there is accumulated evidence that bacterial reduction of chromate can occur under both aerobic and anaerobic conditions [1, 2]. Serious concerns about the toxicity of Cr(VI) compounds necessitates recovery and reuse of chromium from industrial wastes or at least rendering it to a less toxic form. The use of bacteria in bioremediation of Cr(VI)-contaminated soils and ground waters has achieved growing attention, because the biological reduction of Cr(VI) usually generates an insignificant quantity of chemical sludge. Therefore microorganisms can be a potential useful tool in the treatment of contaminated soils and waters.Recently, we isolated several Cr(VI)-resistant bacteria strains from a Cr(VI)-contaminated activated sludge, belonging to (β-Proteobacteria, high G+C Gram-positive bacteria, and to the genus *Acinetobacter* [3] One of the isolated strains, belonging to the genus *Ochrobactrum* (strain 5bvl-1), was found to show both high Cr(VI)-resistance and reduction capacities [4] in aerobic conditions. The isolation and characterization of the enzyme(s) responsible for reducing Cr(VI) would afford information to implement or improve bioremediation strategies. Work was done with the bacterium in order to obtain the fraction(s) with Cr(VI) reduction capacity, preceding our attempt to isolate, purify and characterize the chromate-reductase.

MATERIALS AND METHODS

Cellular fractionation and extract preparation

The strain 5-bvl-1 was grown for 36±2 hours (late exponential phase) at 30°C in 3-4 liters of buffered mineral medium (pH 7.0) as previously described [4] containing 0.25 mM $Na_2Cr_2O_7$ and 0,5% glucose (w/v). The cells were centrifuged at 4 500 g during 10 minutes and the pellet washed 3 times with NaCl 0.85%. The pellet containing the cells was re-suspended in phosphate buffer 50 mM pH 7.0. The cell suspension was sonicated with 15W pulses, during a total time of 5 minutes. The suspension was centrifuged at 12 000 g during 10 minutes to remove the unbroken cells and debris **(P1)**, resulting in crude extract **(S1)**. That extract was recovered and centrifuged at 150 000 g during 1 hour at 4°C, obtaining a light-brown to yellow soluble extract **(S2)** and a brown and translucid pellet **(P2)**, containing cell membranes and membrane-associated material. The soluble extract was recovered and kept at -20°C, and the pellet P2 was re-suspended in phosphate buffer before being kept at the same temperature.

Chromate-reductase purification

To achieve partial purification of the chromate-reductase present in the fraction S2, the fraction was first dialyzed and lyophilized, and then concentrated by molecular exclusion chromatography with a Hiload 16/60 superdex 200 XK16 column and anion-exchange chromatography with a 1 ml Hitrap Q sepharose HP column. The elution flow rate in both chromatographies was of 1 ml/min and in both cases the solution used was phosphate buffer 50 mM with 0.8 mM $MgSO_4.7H_2O$. The protein quantity eluted in the molecular exclusion chromatography was of 10 mg per injection. In the anion-exchange chromatography, a NaCl gradient was created in a way that strongly bounded proteins would unbound at a concentration of 1 M NaCl and moderately bound ones unbound under a linear gradient of up to 0.5 M. The fractions obtained were analysed by sodium dodecyl sulfate polyacrylamide gel electrophoresis (SDS-PAGE) and tested for chromate-reductase activity. In the final of the process, an enriched fraction showing chromate-reductase activity was dialyzed and lyophilized.

Cr(VI) reduction experiments

The experiments were performed at 30°C, with a protein extract concentration of 3 mg/ml, in phosphate buffer 50 mM with 0,8 mM $MgSO_4.7H_2O$. The Cr(VI) reduction capacity was estimated using the diphenylcarbazide method [3]. When using whole-cells, the cells were removed by centrifugation before Cr(VI) reduction determination. The Cr(VI) reduction when necessary was also continuously evaluated following the absorbance at 410 nm, with a protein extract concentration of 0,3 mg/ml.

RESULTS

Chromate-reductase activity was found in the crude and in the soluble extracts and not in the fraction containing cell membranes *(fig. 1)*. The cell extract S2 showed a high chromate-reductase activity at 20°C and a drastic decrease of activity above 40°C. The optimum initial Cr(VI)-concentration was of 2 mM, and the cell extract showed a decrease in the Cr(VI)-reduction activity for higher concentrations. Using cell extracts, the enzyme activity was dependent on NADH *(fig. 2)* and the cell extract showed maximum activity at a NADH concentration of 4 mM. At optimum conditions described above, the chromate-reductase activity using cell extracts was of 60,9 µM/min/mg. The presence of sulfate did not affect the chromate-reductase activity of the fraction.

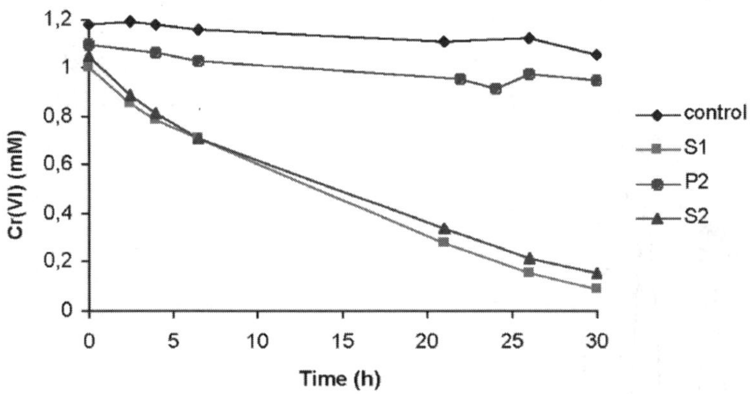

Fig. 1. Cr(VI) reduction of crude extract (S1), soluble extract (S2) and cell membranes and membrane-associated material fraction (P2), in phosphate buffer with a total protein concentration of 3 mg/ml, 1 mM Cr(VI) and 0,2 mM NADH. The control was performed in the same conditions as the extract assays without protein

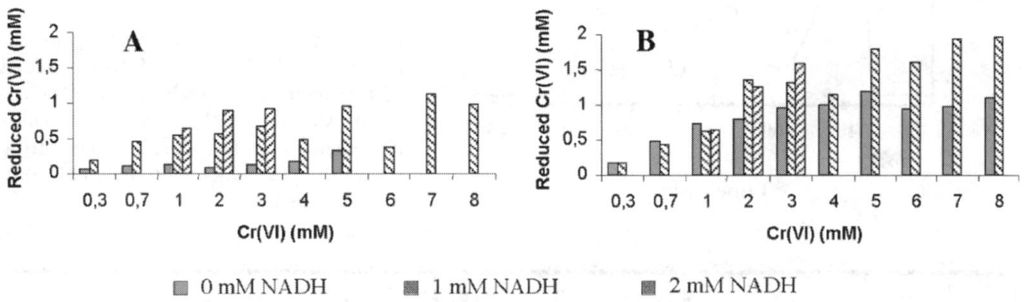

Fig. 2. Cr(VI) reduction using the soluble extract (S2) at increasing Cr(VI) concentrations in the absence of NADH and in the presence of 1 mM NADH. Experiments were performed with 2 mM NADH at Cr(VI) concentrations ranging from 1 to 3 mM Cr(VI). Results are showed at 4 h 30m (A) and at 48 h (B) incubation

After molecular exclusion chromatography of the soluble fraction (S2) four peaks were obtained. Chromate reductase activity was found on the second peak *(fig. 3)*. The second chromatogram was performed and an enriched fraction showing chromate reductase activity was obtained *(fig. 4)* after elution with buffer containing a high NaCl concentration. The eluate was concentrated and analyzed SDS-PAGE. A single band with molecular weight between 37 kDa and 50 kDa was visible. An additional fraction, eluted as abroad peak, that did not bind to the column showed chromate reductase activity but further purification was not performed and the fraction was abandoned.

DISCUSSION

The preliminary work performed on the chromate reductase purified in this study is from *Ochrobactrum tritici* strain 5-bvl-1 suggests that it could be a new enzyme. It is soluble as the reductases from *Pseudomonas putida* MK1 and PRS2000 [5, 7] while that of *Enterobacter cloacae* is membrane bound [6]. For example, the MK1 enzyme is optimally active at 80°C and *P. ambigua* enzyme is 50°C while ours has an optimal temperature between 20°C and 40°C. The elution pattern of the enzyme from strain 5-bvl-1 points to the fact that the PI is lower than 7.0 as it is found in PRS2000 while MK1 enzyme PI is higher than 7.0.

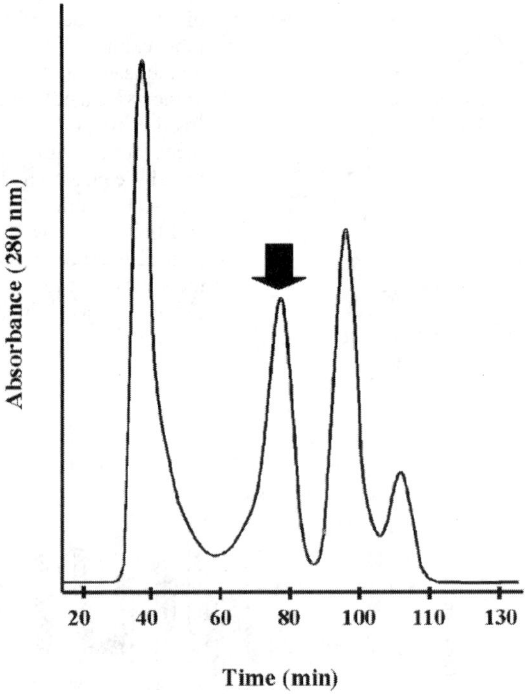

Fig. 3. Molecular exclusion chromatography of the soluble fraction (S2) on a Hiload 16/60 superdex 200 XK16 column. The column was pre-equilibrated with 50 mM phosphate buffer (pH 7.0). The arrow shows the peak where chromate-reductase activity was found

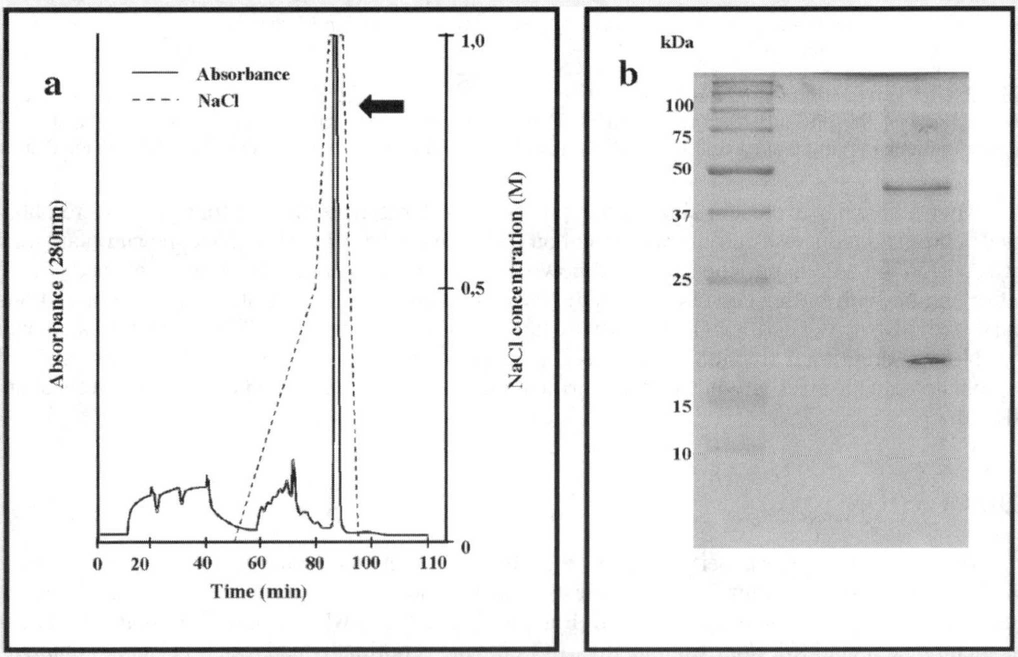

Fig. 4. Anion-exchange chromatography of peak two sample *(fig. 3)* using on a Hitrap Q sepharose HP column (A). At the side (B), an SDS-PAGE gel was run with the fraction eluted as a sharp peak in A containing chromate reduction activity (indicated by an arrow). A single band was evident

The 5-bvl-1 gave a molecular weight between 37 kDa and 50 kDa on SDS-PAGE, that seems to be close of the molecular weight of the native protein, as is suggested by the analyses of the molecular exclusion chromatogram.

The question of the role that chromate reductase(s) plays in the bacterium physiology have not been explored but will be the aim of further studies. However, independently of the physiological role of the enzyme(s), chromate reductase activity is useful in bioremediation.

REFERENCES

1. McLean J., Beveridge T. Chromate reduction by Pseudomonad isolated from a site contaminated with chromated copper arsenate. *Applied and Environmental Microbiology*, 2001, 67, 1076-1084.
2. Rege M., Petersen J., Johnstone D., Turick C., Yonge D., Apel W. Bacterial reduction of hexavalent chromium by Enterobacter cloacae strain HO1 grown on sucrose. *Biotechnology Letters*, 1997, 19, 691-694.
3. Francisco R., Alpoim M.C., Morais, M.P. Diversity of chromium-resistant and reducing bacteria in a chromium-contaminated activated sludge. *J. Applied Microbiology*. In Press.
4. Branco R., Alpoim M.C., Morais P.V. A new bacteria resistant to several metal ions and able to reduce hexavalent chromium. *Metal Ions in Biology and Medicine*, 2000, 6, 661-663.
5. Park C., Keyhan M., Wielinga B., Fendorf S., Matin A. Purification to homogeneity and characterization of a novel *Pseudomonas putida* chromate reductaase. *Applied and Environmental Microbiology*, 2000, 66, 1788-1795.
6. Wang, P., Mori T., Komori K., Sasatsu M., Toda K., Ohtake H. Isolation and characterization of an *Enterobacter cloacae* strain that reduces hexavalent chromium under anaerobic conditions. *Applied and Environmental Microbiology*, 1989, 55, 1665-1669.
7. Ishibashi Y., Cervantes, C., Silver, S. Chromium reduction in Pseudomonas putida. *Applied and Environmental Microbiology*, 1990, 56, 2268-2270.

*Metal Ions in Biology and Medicine: vol. 7.*Eds L. Khassanova, Ph. Collery, I. Maymard, Z. Khassanova, J.-C. Étienne.
John Libbey Eurotext, Paris © 2002 pp. 322-325.

Study of heavy metal stress on plants by hyphenated techniques

Gyula Záray[1,3], Victor G. Mihucz[1], Eniko Tatár[2], Tibor Gasparics[3] and Anita Varga[1]

[1]Research Group of Environmental and Macromolecular Chemistry of the Hungarian Academy of Sciences, PO Box 32, H-1518 Budapest; [2]Department of Inorganic and Analytical Chemistry, L. Eötvös University, Budapest, PO Box 32, H-1518 Budapest; [3]Department of Chemical Technology and Environmental Chemistry, L. Eötvös University, Budapest, PO Box 32, H-1518 Budapest, Hungary

ABSTRACT

Cucumber plants grown in nutrient solutions and artificially contaminated with nickel, were chosen as model edible plants for the investigations focusing on the determination of the chemical transport of essential and toxic heavy metals in their xylem vessels, which aimed the elucidation of the mechanism of phytotoxicity. For this purpose, "off-line" size exclusion high performance liquid chromatographic - graphite furnace atomic absorption spectrometric (SE-HPLC - GF-AAS), "off-line" SE-HPLC - total reflection X-ray fluorescence spectrometric (TXRFS) and "on-line" SE-HPLC - inductively coupled plasma atomic emission spectrometric (ICP-AES) methods were developed. By employing these methods for standard solutions containing citric acid in various concentrations (100 - 500 $\mu g/cm^3$) and nickel in concentration of 2 $\mu g/cm^3$ as well as for xylem sap samples, citric acid seemed to be involved in the nickel transport within the plants. Also, the role of citric acid in the Fe and Mn transport in plants could not be discarded.

INTRODUCTION

Up to the eighties, there was no evidence on the essentiality of nickel in plant metabolism, although some works [1] reporting the beneficial effects of Ni on plant growth indicated that this metal might be of plant physiological importance. In plants, nickel may be bounded to compounds of molecular weight ranging from 1 to 10 kDa [2, 3]. It has been reported that nickel also forms stable complexes with nicotianamine [4, 5], cysteine and citrate ions [6], which may be found in xylem saps, roots and other plant parts. Depending on the number of ligands and of a series of experimental factors like temperature, ionic strength, the logarithm values of the stability constants of nickel citrate range between 3.11 and 7.87 [7]. The evidence of the binding capability of nickel by citric acid in xylem sap of tomato was demonstrated by *in vitro* electrophoretic experiments by White et al. [8]. Kersten and his coworkers [9] stated, by analyzing the nickel content of leaf extracts previously passed through a gel filtration column by GF-AAS and originating from nickel accumulating plants, that nickel was encountered in the samples together with citric acid.

The arising difficulties in justifying the existence of nickel - citric acid complex compound in biological samples are the following: i) the lack of commercially available nickel(II) citrate compound; ii) the developing of a chromatographic method that both allows separation at the physiological pH of the samples and whose separation principle does not alter the samples species; iii) the low metal concentration of the samples; iv) the multi-fold dilution that occurs during the separation, and consequently the analytical capability of the spectrometric detection methods. Regarding the choice of the chromatographic method, the least sample denaturing technique is size-exclusion chromatography [10].

MATERIAL AND METHODS

Plant growth and sampling of the xylem fluid

Cucumber plants were grown for one month in modified Hoagland nutrient solutions artificially contaminated with nickel ions in concentration of 10^{-5} mol/dm^3 and having nitrate ions or urea as sole nitrogen form. Iron(III) was added to the nutrient solutions as Fe(III) citrate in concentration of 10 µM. After a growth period of one month, the plants were decapitated 5 mm above the root collar and the bleeding xylem sap was collected in PE vials with micropipettes for one-hour. The sap samples were stored at -20°C before their analysis. The samples were directly injected onto the column without any previous sample preparation. In case of the experiments investigating the effect of citric acid on the Ni signal, 400 µl xylem sap samples were spiked with 20 µl citric acid in concentration of 10 g/dm^3.

Reagents and solutions

HPLC purity grade citric acid and p.a. inorganic nickel(II) salts were supplied by ACROS ORGANICS (New Jersey, USA) and Reanal Kft. (Hungary, Budapest), respectively. The ICP standard solution of Ni in concentration of 1 g/dm^3 was purchased from Merck (Darmstadt, Germany). Throughout the experiments, bidistilled water was used. Ammonium acetate and acetic acid of HPLC purity grade used for the preparation of the mobile phase were produced by SCHARLAU Chemie S. A. (Barcelona, Spain). HPLC purity grade methanol was purchased from Fisher Scientific (UK). The mobile phases were filtered through a NALGENE nylon membrane filter of 0.45 µm pore size supplied by NALGE Co. (Rochester, USA). The standard solutions as well as the samples were filtered through a Minisart RC 15 membrane filter of 0.2 µm pore size supplied by SARTORIUS (Göttingen, Germany).

Analytical procedures

For the SE-HPLC investigations, silica based and diol layer containing 250 mm × 4.6 mm 7 µm Macrosphere GPC 60 PEEK coated analytical column (ALLTECH, Deerfield, IL, USA) connected with a 7.5 × 4.6 mm guard cartridge of the same packaging were used. The mobile phase was a 50 mM ammonium acetate (NH_4OAc) solution of pH 5.6 adjusted with glacial acetic acid (AcOH). The employed flow rate was 0.5 cm^3/min. The fractions eluting from the column were analyzed by using total-reflection X-ray fluorescence spectrometer (Oberschleissheim, Germany) after freeze-drying, graphite furnace atomic absorption spectrometer (Varian, Australia) and "on-line" by employing inductively coupled plasma atomic emission spectrometer (Plasmalab, Australia). In case of GF-AAS measurements, Ni was measured at 232.0 nm, meanwhile for the ICP-AES detection the 221.647 nm analytical line was used. For TXRFS investigations, Mo anode X-ray tube and Si(Li) detector were used. The analytical line used for Ni detection was 7.471 keV.

RESULTS AND DISCUSSION

A SE-HPLC method - due to its unique capability of assuring the least denaturing conditions for organic compounds during a chromatographic separation - was used for the elucidation of the possible role of citric acid and in the transport of nickel in the xylem vessels. For these experiments nickel containing model solutions were used. The concentration of Ni in those solutions was 2 µg/cm^3, which was a typical concentration of the xylem sap samples. The nickel concentrations of the collected fractions, determined "off-line" by means of GF-AAS, indicated that, in case of 250:1 and 125:1 organic acid- nickel mass ratios, citric acid might carry nickel. Those results could also be demonstrated by having analyzed the identical solutions by coupling the HPLC system to an ICP-AES device by using hydraulic high-pressure nebulizer (HHPN) *(fig. 1 and 2)*. Nebulization

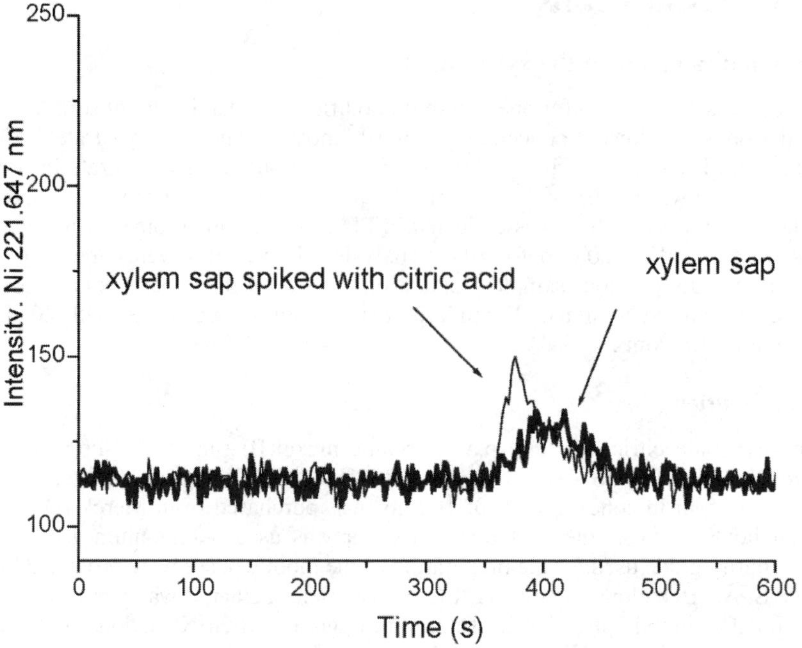

Fig. 1. Influence of citric acid addition on the Ni signal recorded at 221.647 nm with the SE-HPLC-HHPN-ICP-AES system of xylem sap originating from nickel contaminated plants whose nitrogen source was nitrate salt; 400 μl xylem saps were spiked with 20 μl citric acid in concentration of 10 g/dm^3

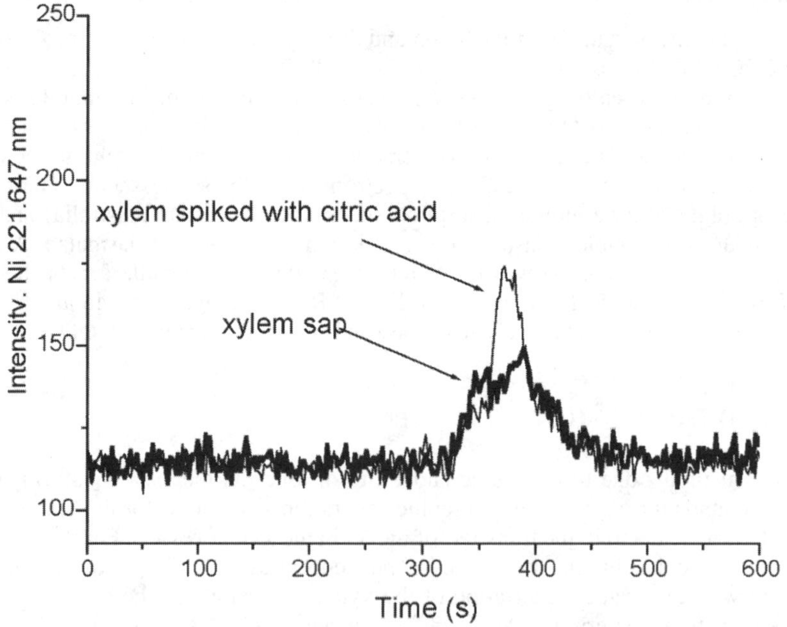

Fig. 2. Influence of citric acid addition on the Ni signal recorded at 221.647 nm with the SE-HPLC-HHPN-ICP-AES system of xylem sap originating from nickel contaminated plants whose nitrogen source was urea; 400 μl xylem saps were spiked with 20 μl citric acid in concentration of 10 g/dm^3.

was enhanced by pumping methanol from another HPLC pump. Moreover, by spiking xylem sap samples with citric acid, sharper Ni peaks were obtained slightly moved toward the retention time of citric acid detected separately by UV detection at 210 nm (not shown). The proof of these results could not be accurately justified in case of xylem exudate samples due to the low nickel and organic acid and high nitrate ion concentrations in the saps. In case of plants whose sole nitrogen source was urea, after pooling 4 times the SE-HPLC fractions of 0.5 cm^3 that underwent to freeze-drying and a 10-fold concentration, the fraction corresponding to organic acids also contained Ni and Fe in detectable concentration by applying a TXRFS method. Also, because of having similar size, the separation of the organic acids of the sap samples (malic, citric and fumaric acids) previously identified by using a reversed-phase HPLC method, could not be achieved by SE-HPLC. However, this method is promising for the characterization of the heavy metal transport by the organic acid molecules present in biological matrices.

CONCLUSIONS

The hyphenation of SE-HPLC with GF-AAS, ICP-AES and TXRFS was investigated for studying the binding capability of Ni^{2+} by citric acid, which might be useful to elucidate the role of citric acid in the heavy metal transport within the plants. The proposed method was applied for real xylem sap samples coming from cucumber plants artificially contaminated with nickel ions. The results obtained by these hyphenated techniques are promising in heavy metal speciation studies.

ACKNOWLEDGEMENTS

The financial support of the Hungarian National Research Foundation (OTKA) through Grant T030845 is gratefully acknowledged.

REFERENCES

1. Mengel K., Kirkby E. A. in: Principles of Plant Nutrition, International Potash Institute, Worblaufen, Bern, 1978, p. 593-598.
2. Cataldo D.A., Garland T.R. Wildung R.E. Nickel in plants. II. Distribution and chemical form in soybean plants. *Plant Physiology*, 1978, 62, 563.
3. Wiersma D, Van Goor B.J. Chemical forms of nickel and cobalt in phloem of *Ricinus comunis*. *Physiologia Plantarum*, 1979, 45, 440-445.
4. Stephan U.W., Scholz G. Nicotianamine: mediator of transport of iron and heavy metals in the phloem? *Physiologia Plantarum*, 1993, 88, 522-529.
5. Stephan U.W., Schmidke I., Stephan V.W., Scholz G. The nicotianamine molecule is made-to-measure for complexation of metal micronutrients in plants. *Biometals*, 1996, 9, 84-90.
6. Thauer R.K., Diekert G., Schönheit P. Biological role of nickel. *Trends in Biochemical Sciences*, 1980, 5, 304-306.
7. Martell A.E., Smith R.M. Critical Stability Constants, Vol. 2., Plenum Press, New York, 1974, p. 115, p. 126, p. 163.
8. White M.C., Chaney R., Decker M.A. Metal complexation in xylem fluid. III. Electrophoretic evidence. *Plant Physiology*, 1981, 67, 311.
9. Kersten W.J., Brooks R.R., Reeves R.D., Jaffré T. Nature of nickel complexes in Psychotria Douareei and other nickel accumulating plants. *Phytochemistry*, 1980, 19, 1963-1965.
10. Snyder L.R., Kirkland J.J., Glajch J.L. in Size-exclusion chromatography.

*Metal Ions in Biology and Medicine: vol. 7.*Eds L. Khassanova, Ph. Collery, I. Maymard, Z. Khassanova, J.-C. Étienne.
John Libbey Eurotext, Paris © 2002 pp. 326-329.

Effect of hydrofluoric acid on recovery of selected trace elements in acid digests of plant and peat materials

Michael Krachler[1], Hendrik Emons[2] and William Shotyk[1]

[1]*Institute of Environmental Geochemistry, University of Heidelberg, Heidelberg, Germany, E-mail: krachler@ugc.uni-heidelberg.de,* [2]*Institute of Phytospheric Research, Research Centre Juelich, Juelich, Germany.*

INTRODUCTION

There is a growing interest in the use of peat bogs as archives of atmospheric metal deposition. However, accurate and precise analytical data are needed to provide reconstructions of atmospheric metal deposition rates. Thus, reliable analytical procedures for trace element analyses are a critical first step in any such study. Almost all analytical techniques for the determination of trace elements require solid samples to be dissolved. This digestion is commonly achieved by heating the samples with acids. Hydrofluoric acid (HF) is normally used in the acid mixture to attack silicates which are present in peat and plant samples mainly in the form of microscopic soil dust. However, many elements such as Ca and rare earth elements (REE) form insoluble fluorides that easily precipitate and result in poor recoveries of these elements. Additionally, elements such as arsenic form volatile element fluorides that evaporate during open vessel digestion with acid mixtures containing HF. Therefore, we studied the use of tetrafluoroboric acid (HBF_4) as a possible replacement for HF in the digestion solutions in three closed pressurized digestion procedures.

Inductively coupled plasma mass spectrometry (ICP-MS), a powerful multi-element technique with low detection limits (for many elements < ng/L), is frequently used for the determination of trace elements in a variety of environmental matrices. However, several spectral interferences originating from major matrix constituents, acids used for digestion and from argon used as plasma gas hamper this endeavour. Therefore, the accurate and precise determination of trace elements in environmental samples by ICP-MS requires carefully optimized operating procedures.

The current investigation was undertaken to develop a simple, robust and reliable analytical procedure for the dissolution of silicates in plant and peat materials followed by quantification of selected trace elements in the digestion solutions by ICP-MS.

MATERIAL AND METHODS

Reagents

For the preparation of all solutions, high purity water (18.2 MΩ cm) from a MilliQ-system (Millipore, Milford, MA, USA) was used. Nitric acid (65%, analytical-reagent grade, Merck, Darmstadt, Germany) was further purified by sub-boiling distillation (MWS Vertriebs GmbH, Leutkirch, Germany). Other acids for digestions were hydrofluoric acid (40%, suprapur®, Merck), perchloric acid (70%, suprapur®, Merck), hydrochloric acid (32%, Riedel-de Haen, Seelze, Germany) and tetrafluoroboric acid solution (HBF_4, ~50%, purum, Fluka, Buchs, Switzerland). Microwave assisted digestion in the autoclave additionally required hydrogen peroxide (30%, Baker analysed, J.T. Baker, Deventer, Holland).

Digestion procedures

Three different pressurized digestion approaches were tested for the dissolution of 250 mg aliquots of peat and plant samples, namely i) closed vessel acid digestion on a hotplate at 180 °C, ii) digestion in a microwave high pressure autoclave at temperatures of 240 °C and iii) high pressure ashing (HPA) at temperatures of 300 °C. Acid mixtures for digestion contained concentrated nitric acid (3-5 ml) alone or additions of hydrofluoric acid (HF) or tetrafluoroboric acid (HBF_4) at volumes of 0.05 ml to 1.0 ml.

Determination of trace elements

Trace elements were subsequently determined in the diluted digests by inductively coupled plasma mass spectrometry (ICP-MS). Beside the determination of REE, we also quantified 16 other elements (Ag, Ba, Ca, Cd, Co, Cu. Cr, Mn, Ni, Pb, Rb, Th, Tl, U, V and Zn) in peat and plant materials. Elements whose determination was hampered by spectral interferences or which were present at low concentration levels in plant and peat samples were aspirated with an ultrasonic nebulizer with membrane desolvation into the plasma of the ICP-MS. Other elements were introduced into the ICP-MS via pneumatic nebulization.

Quality control

Adequate certified plant reference materials have been analyzed with every batch of samples to ensure the accuracy of the results. Additionally, selected peat samples had previously been characterized by instrumental neutron activation analysis (INAA), largely confirming ICP-MS results obtained in the present study.

RESULTS AND DISCUSSION

Experiments revealed that the appropriate choice of internal standards that are used to correct for the instrumental instability and fluctuations of the ICP-MS signals of the elements to be determined, is of utmost importance to obtain accurate and precise results. After testing several elements to be used as potential internal standards, the two elements rhodium and rhenium were successfully employed for this purpose.

The temperature applied during digestion was not the decisive factor of the entire analytical procedure, but rather the composition of the digestion mixture. In particular, the presence or absence of HF was found to play a crucial role. As depicted in *table 1*, digestions carried out in the microwave autoclave at a maximum temperature of 240 °C using different acid mixtures yielded strongly differing results for various elements. Generally, highest concentrations were obtained when small amounts of HBF_4 were present in the digestion mixture. When only nitric acid was used for sample dissolution, elements were not fully liberated from the siliceous matrix in almost all cases. However, some elements such as Cd were readily available from those HNO_3 digestion solutions *(table 1)*. The addition of HF to the digestion mixture, in turn, provoked the formation of precipitated fluorides which negatively affected the recoveries of REE, Ca and Th. These poor recoveries of REE improved with their increasing atomic numbers, the reason for which is related to the well known effect called lanthanide contraction.

The replacement of HF by HBF_4 efficiently helped to solve this problem and is therefore recommended for dissolution of samples containing silicates. Although the use of HBF_4 largely prevents the precipitation of fluorides, the formation of precipitates in the digestion solutions can be observed with increasing volumes of added HBF_4. Therefore, the amount of HBF_4 required, which obviously also depends on the composition (i.e. silicate content) and the mass of the sample to be digested has to be considered. It should be emphasized, however, that the negative effects of fluorides present in digestion solutions were distinctly less pronounced when HBF_4 was used in comparison to HF, for example, when 1 ml HF was added to the digestion mixture, recoveries for REE amounted to only about 1%. Using the same amount of HBF_4, concentrations of REE in the worst case amounted to about 60% relative to digestion mixtures containing only HNO_3.

Table 1. Influence of the composition of the acid mixture on the results of selected elements (mean ± standard deviation) in a bulk peat sample as determined by the microwave high pressure autoclave/ICP-MS approach

		Digestion mixture		
		3 ml HNO_3	**3 ml HNO_3 + 0.1 ml HF**	**3 ml HNO_3 + 0.1 ml HBF_4**
Isotope for ICP-MS	**Unit**	**N = 8**	**N = 8**	**N = 8**
^{43}Ca	mg/g	1.67±0.08	1.64±0.09	2.66±0.24
^{52}Cr	µg/g	3.62±0.54	4.06±0.39	4.84±0.13
^{85}Rb	µg/g	1.77±0.06	1.80±0.06	2.12±0.09
^{114}Cd	µg/g	0.074±0.009	0.077±0.005	0.071±0.004
^{140}Ce	µg/g	2.94±0.16	2.19±0.33	3.44±0.15
^{141}Pr	ng/g	334±17	263±34	387±18
^{153}Eu	ng/g	46±3	38±3	51±2
^{175}Lu	ng/g	9.4±0.4	12.1±0.8	14.3±0.4
^{232}Th	µg/g	0.303±0.016	0.221±0.043	0.371±0.015
^{238}U	µg/g	0.075±0.002	0.134±0.007	0.122±0.003

Among the three closed pressurized digestion procedures evaluated in this study, the high pressure microwave autoclave provided the most efficient approach. Forty samples can be simultaneously digested in one run (approximately 2 h) under identical experimental conditions, thus contributing to both productivity and reproducibility of the results. Temperatures up to 300 °C and pressures up to 200 bar can be applied to the samples and guarantee fast and efficient dissolution of even difficult-to-digest samples.

A detailed description of the development, optimization and evaluation of the digestion as well as ICP-MS procedures can be found in the references cited below. Given the advantages of the microwave assisted high pressure autoclave and the use of HBF_4 to dissolve silicates, the following analytical procedure is recommended. Aliquots of 200-250 mg of peat and plant samples should be digested with 3 ml HNO_3 and 0.1 ml HBF_4 at a temperature of 240 °C. When operating conditions for ICP-MS measurements are chosen carefully, REE and further 16 elements of environmental interest can be successfully determined in the diluted digests by this powerful multi-element technique.

CONCLUSIONS

For many, if not for most trace elements of environmental interest, HNO_3 alone cannot liberate 100% of all elements during digestion of silicate-containing peat and plant matrices. The conventional approach which is to add HF, dissolves siliceous matter reasonably well, but forms insoluble metal fluorides that negatively affect the quantification of many elements. The addition of HBF_4

to the digestion mixture efficiently solves this problem by simultaneously attacking the silicates while preventing the precipitation of metal fluorides.

Carefully optimized and thoroughly validated digestion and ICP-MS procedures allow the accurate and precise determination of at least 30 elements in the diluted digestion solutions of peat and plant samples by ICP-MS. The developed analytical procedure therefore represents a powerful approach for environmental studies involving trace elements in plant and peat materials.

REFERENCES

1. Krachler M., Burow M. and Emons H. Biomonitoring of antimony in environmental matrices from terrestrial and limnic ecosystems. *J Environ Monit* 1: 477-481, 1999.
2. Krachler M., Shotyk W. and Emons H. Digestion procedures for the determination of antimony and arsenic in small amounts of peat samples by flow injection hydride generation atomic absorption spectrometry. *Anal Chim Acta* 432: 307-314, 2001.
3. Krachler M., Emons H., Barbante C., Cozzi G., Cescon P. and Shotyk W. Inter-method comparison for the determination of antimony and arsenic in peat samples. *Anal Chim Acta*, in press.
4. Krachler M., Mohl C., Emons H and Shotyk W. Influence of digestion procedures on the determination of rare earth elements in peat and plant samples by USN-ICP-MS. *J Anal At Spectrom*, submitted.
5. Krachler M., Mohl C., Emons H and Shotyk W. Analytical procedures for the determination of selected trace elements in peat and plant samples by inductively coupled plasma mass spectrometry. *Spectrochim Acta B*, submitted.

VII TOXICOLOGY

Metal Ions in Biology and Medicine: vol. 7.Eds L. Khassanova, Ph. Collery, I. Maymard, Z. Khassanova, J.-C. Étienne.
John Libbey Eurotext, Paris © 2002 pp. 333-335.

Influence of lead and cadmium on the electrolyte exchange and morphological characteristics in rat's male gametes

I. Andrusishina

Institute for Occupational Health AMS of Ukraine, 75, Saksaganskogo, Kyiv 01033, Ukraine

ABSTRACT

Presented are the experimental data about disturbances in Ca, Mg a turnover and morphological parameters in the gonadal cells of mature male rats against the background of administration of 1/200 LD_{50} lead acetate and cadmium chloride in experiments of 48-days.

Administration of lead and cadmium to adult rats resulted in an increase of its level in blood and spermatozoa, alteration in the contents of intracellular Ca, Mg. The disturbance of turnover of electrolytes was assumed to result in impairment of biochemical processes in spermatozoa and negative influence on their fertility.

Key words: lead, cadmium, male gametes, electrolyte.

INTRODUCTION

In last years saw the increase of heavy metals pressure, especially lead and cadmium on the population of Ukraine. However, it is know, that the toxicity of lead and cadmium continues to cause health problem [1-2]. In many publications has been shown that lead and cadmium upset the balance in the exchange in various body compartments. The importance on electrolyte for the normal function of male gametes is well established [3-4]. There are a number of diseases with abnormal concentration of electrolytes in male gametes has damaging effect on fertility. One of their high levels lead and cadmium in male gametes.

MATERIAL AND METHODS

Experiments were performed on 190 adult male rats (weight 150-200 g). For a period of 48-days the animals were intra-abdominally administred by lead acetate (in doses 1/200 LD_{50} salt of lead) and cadmium chloride (in doses 1/200 LD_{50} salt of cadmium). The control rats were affected by the distilled water. The spermatozoon were separated out of epididimises in saline solution. For determination in the blood heavy metals the samples were prepared by means of wet mineralization for blend acids [5]. The spermatozoon samples were cultured at concentration 20×10^6 cells/ml and treated by 2% saline azotic acid for AAS-analysis [6].

The blood and male gametes lead and cadmium levels were determined using the AAS-flame and electrotermal method. Flame AAS can be used for the detection of calcium and magnesium too. The phisiological activity of sperm was measured using light microscopy under ×600. Was defined total number and alive number. All data obtained from control and lead and cadmium poisoned animals were compared using Student's-test for upaired means.

RESULTS

The heavy metal levels in blood and sex cells were different. Experimental data are shown in *table 1-2*. The Pb levels in blood (Pb-B) was increased to 1 from 7 weeks for 4,79 one level Pb-B in control. The Cd levels in blood (Cd-B) was increased in these period too (for 9,13 one big of control), but only 1 week it was bigest. The Pb and Cd levels in male gametes (Pb-Sp, Cd-Sp) was increased in investigated period for 4,03 and 2,88 ones bigest corespond of control. Probably the way of collection of lead and cadmium of male gametes and blood different. The results of electrolytes exchange are presentation in *table 3*. In has been established that Pb-exposure was increased levels of Ca in male gametes to 2 for 5 weeks on 1,9 ones, Cd-exposure was increased levels of Ca too to 2 for 5 weeks on 2,0 ones. The Mg levels under the Pb influence it was unchanged. Influence of Cd was increased levels of Mg in male gametes to 1-2, 7 weeks (to 7 weeks level was big of control on 70,3%.

Table 1. Contents of lead and cadmium in blood of rats during experiments of 48-day's (M ± m, mkM/l)

Conditional of experiments	Element	1 week	2 weeks	5 weeks	7 weeks
Control	Pb	1,25±0,31	1,20±0,46	1,23±0,51	1,27±0,30
1/200 LD_{50} Pb	Pb	3,38±0,49*	3,69±0,42*	4,0±0,40*	5,13±0,22*
Control	Cd	0,37±0,12	0,40±0,14	0,43±0,10	0,49±0,10
1/200 LD_{50} Cd	Cd	3,38±0,62*	2,14±0,65*	2,06±0,27*	2,20±0,70*

The difference from control is valid: *-P ≤ 0,05

Table 2. Llead and cadmium concentration in male gametes during experiments of 48-day's (M±m, nM/10^6 cells)

Condition of experiment	Element	1 week	2 week	5 week	7 week
Control	Pb	0,22±0,05	0,23±0,05	0,24±0,06	0,29±0,04
1/200 LD_{50}Pb	Pb	0,16±0,05	0,53±0,23	0,67±0,17*	1,17±0,17*
Control	Cd	0,04±0,01	0,06±0,007	0,25±0,03	0,25±0,03
1/200 LD_{50}Cd	Cd	0,09±0,03	0,15±0,04*	0,39±0,04*	0,72±0,16*

Table 3. Calcium and magnesium concentration in male gametes during the experiment of 48-day's (M±m, nM/10^6 cells)

Conditional of experiment	Electrolytes	1 week	2 week	5 week	7 week
Control	Ca^{2+}	35,62±6,78	27,81±3,34	22,58±1,51	34,75±1,81
	Mg^{2+}	5,05±0,76	2,30±0,34	4,59±0,87	4,44±0,29
1/200 LD_{50}Pb	Ca^{2+}	35,58±3,67	31,30±4,48*	40,20±1,23*	31,50±1,73
	Mg^{2+}	3,77±0,68	4,77±1,74	5,57±0,99	5,05±0,80
1/200 LD_{50}Cd	Ca^{2+}	41,55±4,80	48,70±3,20*	40,92±5,05*	35,60±2,75
	Mg^{2+}	8,07±0,70*	11,10±4,11*	6,27±1,47	7,56±0,71*

Data about morphological characteristics are shown in *table 4*. Was shown that total number and number of moving cells was decreased as Pb-exposure since Cd exposure. Alterations of total number, movement of gametes and rise intracellular Ca and Mg as leading heavy metal damage of rat's male gametes are reveated. The disturbance of turnover of electrolytes was assumed to result in impairment of biochemical processes in spermatozoa and probably negative influence on their fertility. The result obtained can be used as markers for early diagnosis of reproductive health disturbances in case of lead and cadmium intoxication.

Table 4. Morphological male gametes during the experiment of 48-day's (M±m, 10^6 cells)

Condition of experiments	Indicator	1 week	2 week	5 week	7 week
Control	Total number (10^6)	76,60±6,70	88,00±9,94	60,20±5,22	92,26±1,74
	Number of moving (10^6)	54,40±4,73	75,50±9,10	33,53±2,09	67,06±2,02
1/200 $LD_{50}Pb$	Total number (10^6)	39,40±3,28*	63,30±11,27	47,20±7,54	61,40±1,86*
	Number of moving (10^6)	43,40±1,86*	56,00±11,27	41,20±8,09	45,40±3,48*
1/200 $LD_{50}Cd$	Total number (10^6)	38,00±1,15*	44,00±6,03*	48,00±1,00*	70,66±1,20*
	Number of moving (10^6)	20,00±1,00*	33,34±3,71*	42,70±2,03*	58,00±1,53*

REFERENCES

1. Recommended health based limites in occupational exposure to heavy metals. Geneva: WHO, 1995-p. 96.
2. Environmental Health Criteria Human Ewxposure Assessment/ World Health Organization (IPCS), Geneva, 2000-378 p.
3. Gavriliuk I., Gavriliuk M., Samchyk O., Chanek G. Diagnosis of diseaces epididimus gland about bielectrolytes data //Medical efficient 1990-N 2.-p. 75-77.
4. Evdocimov V., Erasova B., Evseev P. Moviment of spermatozoa and contents Ca and Mg in eiyacylat of man /Mat. IV Vses.Ymp. Urologov M., 1990-p. 506.
5. Dobrovolsky L., Vitte P., Belashova et al Blood lead monitoring studies in Chernobyl region in 1992 //Abstracts Symp. Trace Elem. In Man and Animals TEMA-8 (Dresden, May 16-21 1993)-Dresden: Friadrich-Shiller University, 1993.-p. 140.
6. Andrusishina I., Dudko I. Experimental data about the influence of lead and cadmium on male gametes // Abstract V Inte. Symp. "Metal Ions Biology and Medicine" v.5 Munich, (Germany May 8-10 1998)-Munich, 1998.-v.-5.-p. 282-292.

*Metal Ions in Biology and Medicine: vol. 7.*Eds L. Khassanova, Ph. Collery, I. Maymard, Z. Khassanova, J.-C. Étienne.
John Libbey Eurotext, Paris © 2002 pp. 336-341.

Estimation of DNA-protein cross-links, abnormal sperm heads and micronuclei in mice continuously exposed to heavy metals and gamma-radiation at low doses

A.N. Osipov[1], M.D. Pomerantseva[2], L.K. Ramaiya[2], V.D. Sypin[1], V.A. Shevchenko[2]

[1]*Moscow Scientific and Industrial Association "Radon", 2/14, 7th Rostovsky lane, 119121 Moscow, Russia; E-mail: aosipov@radon.ru;* [2]*N.I. Vavilov Institute of General Genetics, Russian Academy of Science, Gubkin str. 3, 117809 GSP-1, Moscow B-333, Russia*

ABSTRACT

The present study gives an estimation of DNA-protein cross-links (DPC), abnormal sperm heads (ASH) and micronuclei (MN) in mice continuously exposed to heavy metals and/or gamma-radiation at low doses. CBA/lac male-mice 6-7 weeks old were exposed to low dose-rate gamma-radiation (0.07 cGy/day) and/or heavy metals (lead and cadmium) with drinking water (0,3 mg Pb2+/l; 0.01 mg Cd2+/l) for different time (20-80 days) and compared with control mice. The DPC level was determined by K/SDS assay. The percentage of ASH and the percentage of peripheral blood normochromatic erythrocytes (NCE) with MN were evaluated by standard cytogenetic tests. The results of our studies have demonstrated that the dependence of the DPC level on the total dose (exposure time) of gamma-radiation and/or heavy metals is nonlinear. The pattern of changes of DPC level was similar for different agents. The exposure to cadmium or γ-radiation or the combined exposure to gamma-radiation and lead caused an increase in the ASH frequency at 80-th day of the experiment. No increase was found in the frequencies of MN in NCE and reciprocal translocations in spermatocytes. Summing up the results obtained it can be concluded that action of gamma-radiation and/or heavy metals at the applied doses do not result in pronounced cytogenetic effects. Non-linear shape of DPC dose-response dependencies is strongly support the idea that DPC participate in overall response of the exposed cells to heavy metals and ionizing radiation at low doses.

INTRODUCTION

The global pollution of the environment has led to the fact that humans and other living organisms are frequency exposed to the low dose action of various technogenic factors, including ionizing radiation and heavy metals. In recent years the results of numerous studies have been published demonstrating that the action of even small doses of ionizing radiation and heavy metals in a wide diversity of living organisms may lead to significant biological effects.

In the present work the genetic aspects of the action of chronic low-dose irradiation and heavy metals (lead and cadmium) on the organism of laboratory mice are considered.

The criteria for estimating the effect of radiation and cadmium have been DNA-protein cross-links (DPC) in mouse thymus and spleen lymphocytes, the number of micronuclei (MN) in normochromatic erythrocytes (NCE) of peripheral blood, the number of abnormal sperm heads (ASH) and the number of reciprocal translocations (RT) in spermatocytes. The choice of these tests is determined by their high sensitivity to the action of the damaging agents, on the one hand, and by a crucial importance of the given systems (immune, hemopoietic and reproductive) for the existence of both individual organisms and a species as a whole, on the other hand.

In choosing the dose rate of gamma-radiation and the concentration of heavy metals we were guided by the data on the contamination by radionuclides and/or heavy metals of different regions of Russia.

MATERIALS AND METHODS

Mice and treatments

The experiments were done using CBA/lac male mice weighing 16-18 g (6-7 weeks of age) purchased from the nursery Stolbovaya of the Russian Academy of Medical Science. The distribution of animals in experimental and control groups was done randomly. The animals were housed in plastic cages and were acclimated for 7 days prior to treatment. The mice received a dry standard vivarium ration and water *ad libitum*. ?he control and exposure groups were kept under standard conditions (20±2°C, 50±10% relative humidity, 12-hr light/dark cycle).

The animals were exposed to radiation from a UOG-1 unit with a cesium-137 source for 20, 40 and 80 days at a dose rate of 0.07 cGy/day. The radiation dose rate was regularly controlled with a DRG-01T radiometer (Russia). Irradiation was 24 hours a day with a short break (10-15 min) for taking care of the animals (change of a bedding, giving fodder, etc.). The total doses received by the animals made up 1.4, 2.8 and 5.6 cGy, respectively. Dosimetry was made with the use of thermoluminescence detectors on the basis of lithium fluoride TLD-100 (Sweden) and DTG-4 (Russia). Measurements were made with RE-1 (RADOS, Finland) and DTF-01 (Russia) devices.

Cadmium chloride or lead acetate was dissolved in drinking water up to a final concentration of 10 μg/l cadmium ions or 300 μg/l lead ions. The solutions were given to animals. According to our calculations, one mouse drank 3 to 5 ml of the solution a day.

Quantitative measurement of DPC

The DPC level was determined by the K/SDS assay [1]. The method utilizes harsh treatment by SDS to dissociate non-covalent DNA-protein complexes and selectively precipitate cross-linked complexes by adding of KCl. The level of DPC was determined as a ratio of the amount of DNA in the supernatant to the total DNA in the sample.

The amount of DNA was measured with the Hoechst 33258 reagent using a FL-2110 fluorimetric analyzer (Solar, Belarus) with 365 nm excitation and 460 nm emission.

Cytogenetic Methods

The MN frequency in NCE was determined according to a routine procedure [2]. 2000 NCE were analyzed for each animal. The ASH frequency was registered in air-dry preparations from the contents of epidermis [3]. 500 spermatozoa from each male were analyzed. The RT frequency was determined in spermatocytes at the stage of diakinesis-metaphase of the 1st meiotic division [4].

Statistics

Data are presented as mean ± SEM. Statistic processing of the experimental results was made using the one-tailed Student's t-test. The samples from each animal were treated separately. The results are a mean for a group of 10-15 mice.

RESULTS

DNA-Protein Cross-links

For quantitative estimation of DPC, the coefficient k_{DPC} was proposed as the ratio between the DPC level in cells of the experimental animals and that in the control animals. It is supposed that the existence of a background level of DPC in the norm is the consequence of cell metabolism [5].

The results of the studies have demonstrated that at the used dose rate of γ-radiation and concentration of heavy metal ions in water the corresponding relationships of k_{DPC} on the total dose (exposure time) in cells of thymus and spleen for each type of exposure are nonlinear *(figs. 1, 2)*. Both for spleen lymphocytes and for thymocytes the maximal values of k_{DPC} upon long-term exposure to γ-radiation or heavy metals were recorded on the 40-th day of the experiment (*figs. 1, 2*, curves 1, 2 and 4). Under the combined action of radiation and cadmium (*figs. 1, 2*, curve 3) the maximal values of k_{DPC} in cells of each of the studied lymphoid organs were recorded on the 20-th day. On the 80-th day of the experiment the DPC level in cells of each of the organs returns to the control values and in thymocytes upon the combined treatment with radiation and cadmium the DPC level even decreases below the control values (*fig. 2*, curve 3). The combined action of lead and γ-radiation did not induce an increase throughout the whole all time of the experiment (*fig. 2*, curve 5), and an increase of the k_{DPC} value in splenocytes on the 40 day was not statistically significant (*fig. 1*, curve 5).

Micronucleus test

The mutagenic effect of heavy metals and γ-radiation was judged from the frequency of MN in NCE. MN are known to represent delayed chromosomes and acentric fragments persisting throughout the whole life of an erythrocyte. No significant alteration in the MN frequency was observed in the experimental groups as compared to the control (data not shown).

Abnormal sperm heads and reciprocal translocations

The ASH frequency is a rather sensitive index in determining the gonadotropic and mutagenic effects of different physical and chemical factors. It was shown earlier that upon acute exposure to radiation a significant increase in the ASH frequency was observed within 35-40 days after irradiation [6].

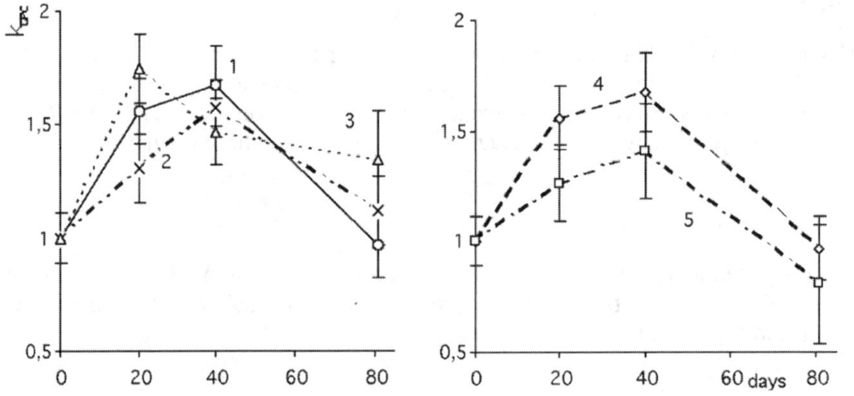

Fig. 1. Changes in k_{DPC} values for splenocytes depending on the time of exposure (dose); γ-radiation (1), cadmium (2), cadmium and γ-radiation (3), lead (4), lead and γ-radiation (5)

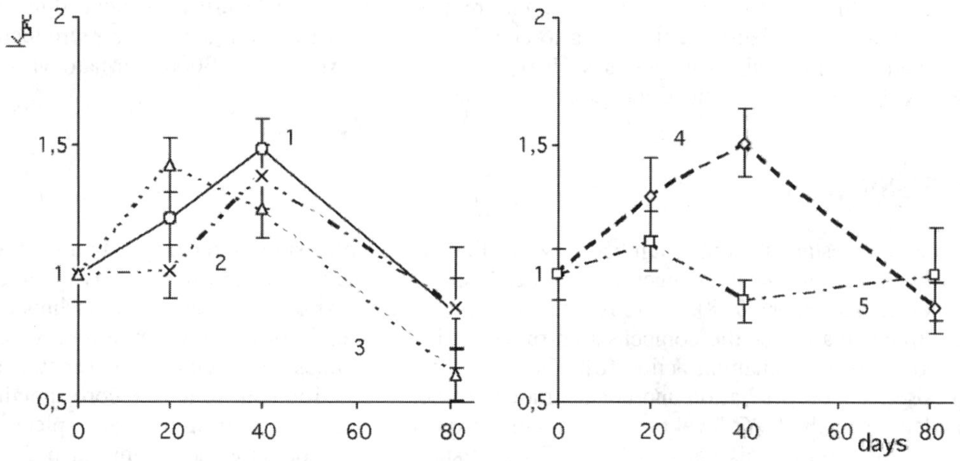

Fig. 2. Changes in k_{DPC} values for thymocytes depending on the time of exposure (dose); γ-radiation (1), cadmium (2), cadmium and γ-radiation (3), lead (4), lead and γ-radiation (5)

Figure 3 presents the data on the ASH frequency in the case of chronic exposure to cadmium, lead and γ-radiation. It has been found that under the action of cadmium a significant increase in the ASH frequency as compared to the control occurs on the 80-th day of the experiment. Irradiation also causes an increase in the ASH frequency on the 80-th day but the effect is less pronounced as compared to the action of cadmium. Upon combined exposure to radiation and cadmium the ASH frequency is lower than upon exposure to cadmium alone. Upon exposure to lead an increase in the ASH frequency was not significant. The combined exposure to radiation and lead causes an increase in the ASH frequency.

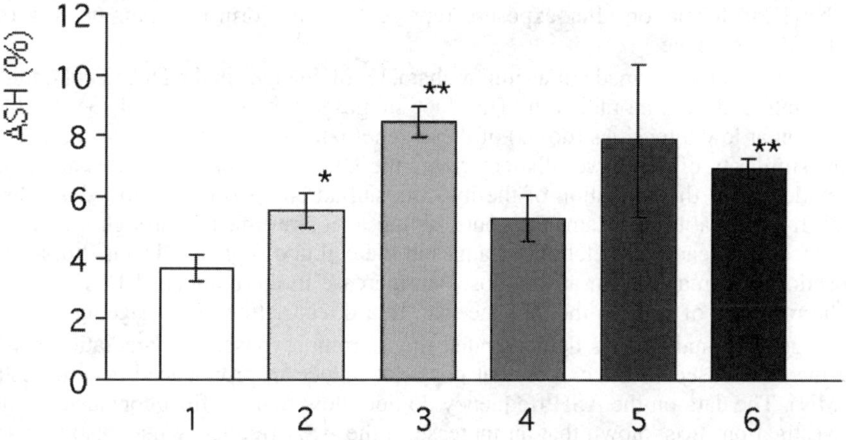

Fig. 3. The percentage content of ASH's on the 80th day of the experiment:
1 - control, 2 - γ-radiation, 3 - cadmium, 4 - cadmium and γ-radiation, 5 - lead, 6 - lead and γ-radiation (* - $p<0,05$, ** - $p<0,01$)

It should be noted that the testis weight index (testis mass to body mass ratio) was the same in all variants of the experiment.

The earlier described induced genetic disturbances are not transmited to the next cell generations being eliminated with cell death. Of particular importance for assessing genetic consequences from the action of different factors are the data on the frequency of genetic disturbances induced in stem spermatogonia. Disturbances not leading to cell death remain throughout the whole reproductive period and are inherited. Damages of such type include RT. Analysis of 3000 spermatocytes revealed no RTs in any variant of the experiment (data not shown).

DISCUSSION

It has been established in a number of works that chronic exposure to ionizing radiation at low doses leads to non-linear changes in the metabolic parameters of cells (in the activity of cell enzymes, in particular) [7, 8]. There is a belief that the non-linear dose-response relationships may reflect the activation of the compensation-restoration processes in order to preserve homeostasis in response to the damaging action [8]. The results of our studies have demonstrated that at the used dose rates of γ-radiation and concentrations of heavy metal ions in water the corresponding dependencies of the DPC level on the total dose (exposure time) in cells of thymus and spleen for each type of exposure also are nonlinear. These results are in good agreement with our previous data [9, 10].

In present time is not clear whether additionally-induced DPC are directly involved in the formations of mutations and present the factor of risk. It should be realized that determined DPC are no uniform type lesions. Untreated cells always have proteins that are covalently bound with DNA due to normal cell metabolism [5], for instance topoisomerases I and II [11]. Under the action of ionizing radiation and heavy metals other proteins (nuclear matrix proteins in particular) covalently bind with DNA [12, 13]. The non-repairable damage-induced DPC disturb the normal functions of nuclear chromatin, such as replication and transcription, and may lead to serious genetic consequeNCE (mutagenesis, carcinogenesis, etc). However, the yield of first-induced DPC is very low (~ 150 DPC per Gy per cell) [5] During the post-exposure period, an additional formation of DPC to take place [14]. Subramanian et al. [11] shows that topoisomerase I is are the proteins involved in the formation of DPC during incubation following the treatment. It is suggest, that DPC formation after exposure represents some form of repair process or cellular response to DNA damage.

Special notice should be made of a similar character of changes in the DPC level upon exposure to heavy metals and gamma-radiation. This fact suggests non-specificity of the mechanisms of DPC formation at low intensities (doses) of these genotoxic agents. This conclusion is also confirmed by the similarity of the curves characterizing the DPC level for thymocytes and splenocytes. There is evidence for the activation of the hypophysial-adrenal system by low-dose chronic irradiation [7]. It is probable that damage agents, acting as a stress-factor through the neurohumoral system, induces a cascade of metabolic changes at the cell and organism levels. So, in the case of continue action of damage agent at low doses an increase in the number of DPC is due rather to a metabolic response of cells to the damage than to a direct action of damage agent.

The cytogenetic analysis has demonstrated that chronic exposure to γ-radiation and/or heavy metals at the doses used in the experiment does not induce in pronounced genetic disturbaNCE (RT and MN). The data on the ASH frequency do not allow making final conclusions and call for further investigation. It is known that an increase in the ASH frequency may also be a result of a metabolic response of the organism to unfavorable factors [7]. It is most likely that the increase in the ASH frequency on the 80-th day of the experiment is related to this fact.

CONCLUSIONS

Summing up the results obtained it can be concluded that action of gamma-radiation and/or heavy metals at the applied doses do not result in pronounced cytogenetic effects. Non-linear shape of DPC dose-response relationships is strongly support the idea that DPC participate in overall response of the exposed cells to low doses of heavy metals and/or ionizing radiation.

REFERENCES

1. Zhitkovich A., Costa M. A simple, sensitive assay to detect DNA-proten cross-links in intact cells and in vivo. *Carcinogenesis*, 1992, 13, 1485-1489.
2. Schlegel R., MacGregor J.T. The persistence of micronuclei in peripheral blood erythrocytes: detection of chronic chromosome breakage in mice. *Mutation Res.*, 1982, 104, 367-369.
3. Wyrobec A.J., Heddle J.A., Bruce W.R. Chromosomal abnormalities and the morphology of male mouse sperm heads. *Can. J. Genet. Cytol.*, 1975, 17, 675-681.
4. Evans E.P., Breckon G., Ford C.E. An air-drying method for meiotic preparations from mammalian testes. *Cytogenetics*, 1964, 3, 289-294.
5. Oleinick N.L., Chiu S.M., Ramakrishnan N., Xue L.Y. The formation, identification, and significance of DNA-protein cross-links in mammalian cells. *Br. J. Cancer*, 1987, 55, 135-140.
6. Pomerantseva M.D., Ramaiya L.K., Vilkina G.A. Comparative efficiency of three different tests for estimation of mutagenicity of some factors in mammals: II. The frequency of abnormal sperm head in mice treated with different factors. *Genetika*, 1980, 16, 1397-1403 (in Russian).
7. Alesina M.Yu. Radiobiological effects formation in experimental animals under low dose chronic internal and external irradiation. *Intern. Journ. Radiat. Medicine*, 1999, 2, 92-99.
8. Slozhenikina L.V., Fialkovskaya L.A., Kolomiytseva I.K. Ornithine decardoxylase in organs of rats following gamma-irradiation at low dose-rate. *Intern. Journ. Radiat. Biol.*, 1999, 75, 195-199.
9. Osipov A.N., Grigoryev M.V., Sypin V.D., Pomerantzeva M.D., Ramaiya L.K., Shevchenko V.A. The influence of chronic exposure to cadmium and γ-radiation at low doses on genetic structures of mice. *Radiats. Biol. Radioecol.*, 2000, 40, 373-377 (in Russian).
10. Osipov A.N., Sypin V.D., Puchkov P.V., Razumova ?.S., ?uznetsova E.M. Changes in the level of DNA-protein cross-links in spleen lymphocytes of mice exposed to low-intensity γ-radiation at low doses. *Radiats. Biol. Radioecol.*, 2000, 40, 516-519 (in Russian).
11. Subramanian D., Rosenstein B.S., Muller M.T. Ultraviolet-induced DNA damage stimulates topoisomerase I-DNA complex. Formation in vivo: possible relationship with DNA repair. *Cancer Res.*, 1998, 58, 976-984.
12. Chiu S.M., Friedman L.R., Sokany N.M., Xue L.Y., Oleinick N.L. Nuclear matrix proteins are crosslinked to transcriptionally active gene sequeNCE by ionizing radiation. *Radiat. Res.*, 1986, 107, 24-38.
13. Wedrychowski A., Schmidt W.N., Hnilica L.S. The in vivo cross-linking of proteins and DNA by heavy metals. *The Journal of Biological Chemistry*. 1986, 261(7), 3370-3376.
14. Osipov A.N., Kolomijtseva G.Ya. Post-irradiation change of DNA-protein cross-links and single-strand DNA breaks in different organs of γ-irradiated rats. *Biochemistry (Moscow)*. 1996, 61, 667-670.

Metal Ions in Biology and Medicine: vol. 7. Eds L. Khassanova, Ph. Collery, I. Maymard, Z. Khassanova, J.-C. Étienne.
John Libbey Eurotext, Paris © 2002 pp. 342-345.

Altered open-field performance in depleted uranium exposed rats

Wayne Briner

Department of Psychology University of Nebraska at Kearney, Kearney, NE, 68849 USA

ABSTRACT

The administration of depleted uranium in drinking water produced significant behavior changes in rats in a manner that was dependent on the length of administration. The behavioral changes reflected changes in overall activity level rather than altering the emotional responsiveness of the animals. These data suggest that DU is a cumulative toxin.

Key words: depleted uranium, behavior, rat.

INTRODUCTION

Depleted uranium (DU) is a heavy metal with suspected toxic effects on the central nervous system (CNS) (Briner & Davis, 2002; Briner & Abboud, 2002). If DU proves to have CNS toxicity its effects may resemble those of other heavy metals, especially lead. The effects of lead are known to be dose dependent, in the classic pharmacologic sense, as well as time dependent due to its accumulation in the body. Dose dependency of DU toxicity has been demonstrated earlier (Briner & Davis, 2002; Briner & Abboud, 2002), in this article we describe DU toxicity being influenced by the temporal domain.

MATERIALS AND METHODS

Male and female Long-Evans rats were raised under standard laboratory conditions. Experimental groups consisted of age-matched animals that were administered either tap water or water containing DU acetate (75 mg/L) for either 2 weeks or 6 months. After the appropriate length of time animals were tested for 5 minutes in the open-field maze where line crossing, rearing, grooming, and boli were measured. After testing the animals were sacrificed and body and brain weights taken.

RESULTS

There were time dependent increases in line crossing ($F(2,53)=2.66$, $p=.08$) and significantly greater rearing activity for both genders with increased length of DU exposure ($F(2,53)=3.37$, $p=.04$; *fig. 1*). Boli deposition was also had a trend effect in a time and gender dependent manner (F(2.70. p=.08). There were no significant differences for urination and grooming *(fig. 2)*.

The general health of the animals was good and there were no significant differences in brain and body weight due to DU exposure *(fig. 3)*. This suggests that whatever effect DU had was because of a specific CNS toxicity and not due to some general systemic effect.

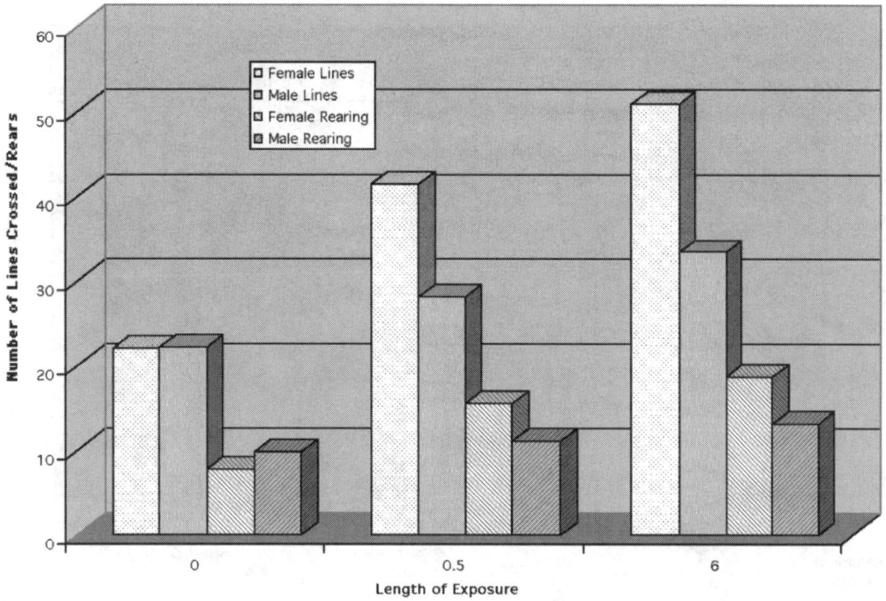

Fig. 1. Line Crossing and Rearing Behavior

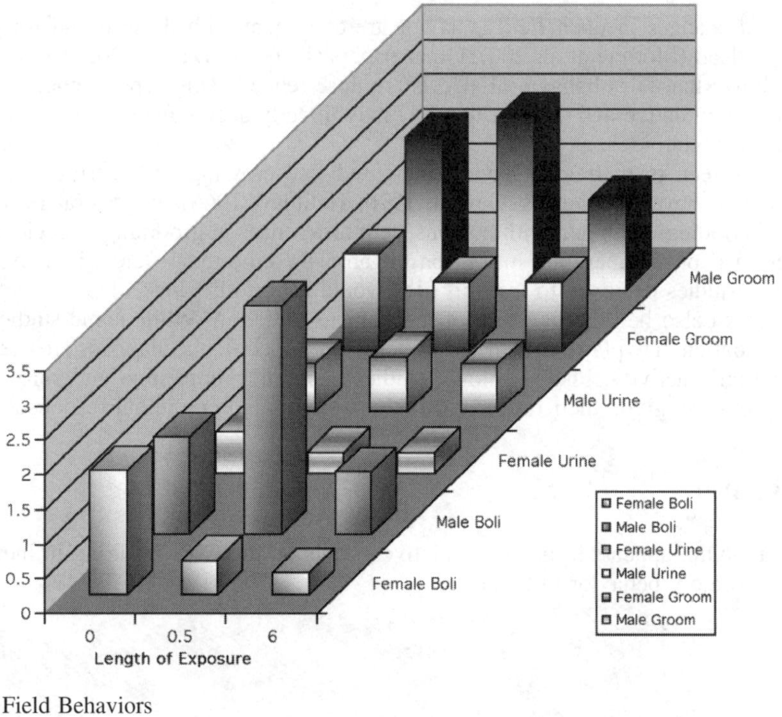

Fig. 2. Open Field Behaviors

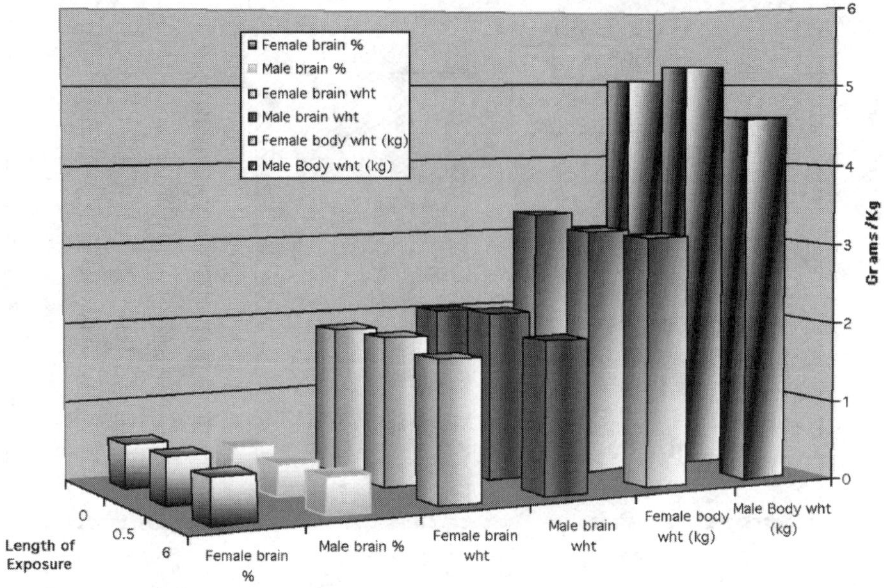

Fig. 3. Brain and Body Weights

DISCUSSION

DU induced changes in open field behavior are consistent with those found for other heavy metals, such as lead (Morgan et al, 200). Other papers (Briner & Davis, 2002; Briner & Abboud, 2002) have shown that DUs behavioral effects are dose related. This paper demonstrates that the effects are also cumulative and suggest that DU may directly accumulate in tissue or at least that the effects are cumulative.

The changes seen open field behavior, increased line crossing and rearing, reflects overall changes in activity level (Barnett & Cowan, 1976; Halliday, 1966). In general fewer boli were deposited. This finding combined with no consistent differences in grooming behavior or urination suggests there was only a marginal impact on the animals' emotional state (Halliday, 1968). However, specific studies designed to test this idea would need to be done.

These findings also beg the question of mechanism of action. While some studies have suggested lipid oxidation may play a role in DU toxicity more specific mechanisms are need to explain changes in specific activity. Pharmacologic studies examining dopamine and serotonin activity would be appropriate, given their role in regulating activity and emotional responses.

CONCLUSIONS

Depleted uranium appears to have cumulative effects on the CNS resulting in increased locomotor and exploratory behavior in adult rats.

REFERENCES

Barnett S.A. & Cowan P.E. Activity, exploration, curiosity and fear: an ethological study. *Interdisciplinary Science Reviews*, 1976 1, 43-62.
Briner W. & Davis D. Lipid oxidation and behavior are correlated in depleted uranium exposed mice. *This volume*.
Briner W. & Abboud B. Behavior of juvenile mice chronically exposed to depleted uranium. *This volume.*
Halliday M.S. Exploration and fear in the rat. *Symposium of the Zoological Society of London*, 1966, 18, 45-54.
Halliday M.S. (). Exploratory behavior. *In Analysis of Behavioral Change* (Ed. L. Weiskrantz). 1968, NY, Harper & Row.
Morgan R.E., Levitsky D.A., Strupp B.J. Effects of chronic lead exposure on learning and reaction time in a visual discrimination task. *Neurotoxicology & Teratology*, 2000, 22, 337-345.

Metal Ions in Biology and Medicine: vol. 7. Eds L. Khassanova, Ph. Collery, I. Maymard, Z. Khassanova, J.-C. Étienne.
John Libbey Eurotext, Paris © 2002 pp. 346-348.

A new approach to risk assessment of lead intoxication

T.K. Larionova, R.D. Tukayev, N.I. Simonova, R.G. Nafikov

Research Institute of Occupational Medicine and Human Ecology, 94, Kuvikina street, 450106 Ufa, Russia

ABSTRACT

In the process of dynamic clinic-experimental studies we have revealed the effect of a number of nonspecific therapeutic methods (hypnotherapy, acupuncture as well as an intake of ascorbic acid or adaptogens) on urinary excretion of lead.

The content of lead in blood and urine was determined using the method of atomic-absorption spectrometry.

The method of lead intoxication risk assessment depending of lead content in urine following either hypnotherapeutic effects, or acupuncture has been developed on the basis of data obtained.

INTRODUCTION

Lead is one of basic technogenic toxicants being constantly released into the environment via wastes discharged by high temperature technologic processes, automobile emissions, severs from metal extracting and processing factories. It is known to pose the most toxic hazard to humans. They are exposed to lead mainly via air and food stuffs.

When making evaluations of risks to humans, the development and introduction of new effective analytical methods for revealing interconnections between lead content in the body biological media, its cumulation level in the body and clinical signs of intoxication.

Medication methods of lead intoxication therapy do not exhaust all therapeutic possibilities. The study of the body biological media is in one of the perspective trends in this context. With an increase in sensitivity and selection of chemical analysis these media may serve as a safe biologic indicator reflecting "health level" of both man and his environment.

The aim of present investigation was to study sanogenic mechanisms of hypnotherapy and reflexo-therapy and to reveal their possibilities in the diagnosis of damage rate and excretion of cumulating ecotoxicants such as lead from the body.

MATERIALS AND METHODS

The main group (40 individuals 15 woman at a mean age of 32 years and 25 men at a mean age of 34 years) comprised workers of radioelectronic industry who were exposed to lead in the process of soldering and were its carriers. Hypnotherapy and acupuncture reflexotherapy were used in complex therapy of this patient group.

The control group consisted of patients of the clinic of occupational diseases who also were in contact with lead but undergo neither hypnotherapy nor reflexotherapy during the treatment. In the process of hypnotherapy or reflexotherapy, capillary and venous blood samples were drawn from the patients of the main group for clinical, immunological and biochemical tests. Blood analysis test was performed four times: 1) before the first sitting; 2) after the first sitting; 3) before the final sitting in the treatment course; 4) after the final sitting. Besides, the content of heavy metals in blood and urine was determined by atomic-absorption method.

RESULTS AND THEIR DISCUSSIONS

The data obtained have allowed to reveal effects typical of both hypno- and reflexo-therapy including immunologic and detoxication ones, protein, bilirubin and cholesterol metabolism activation as well as dynamic effects on the kinetics of heavy metal metabolism.

Immunologic effects include complete reaction of both links of the immune system: cellular and humoral. Detoxication effects are manifested by an active excretion of metabolic residues from the organs and tissues into blood flow with their subsequent renal excretion from the body. Nonspecific therapy being a powerful detoxication means leads to a significant decrease in autointoxicant content in the body. In the process of hypno- and reflexo-therapy protein metabolism and hence, corresponding enzymatic systems of the body are activated. At the same time a significant decrease in the content of blood bilirubin apparently connected with the activation of its excretion from the body and based on an increase in protein metabolism occurs. Normalization of cholesterol metabolism has also been revealed. This may testify about restoration of cellular membrane activity.

Biologic effects of hypnotherapy and reflexotherapy on the body have been found to coincide. Early detected neutrophilia, lymphopenia, eosinophilia become less marked by the end of the treatment. In accordance with the factor analysis data, their functional value is decreased. This is accompanied by significant changes in biological homeostasis at hematologic, immunologic, biochemical levels.

In the process of dynamic clinico-experimental studies we have revealed the effect of hypnotherapy and reflexotherapy on urinary excretion of lead, that is nonspecific in regard to toxic metals.

According to the investigation results we have picked out three types of dynamics of urinary excretion of lead in response to hypnotherapy: 1) an increase in excretion; 2) the absence of excretion dynamics; 3) a decrease in excretion.

The type of dynamics of lead excretion is determined by its initial content in the body which in its turn depends on lead cumulation level in the body.

Initially low concentrations of lead in blood and urine as a result of nonspecific readaptation therapy, a significant increase in its content in blood and urinary excretion from the body occurs.

In threshold indices of lead concentration in urine as a result of nonspecific readaptation therapy the content of lead both in blood and urine is not authentically changed.

In initially high concentrations of lead in blood and urine that is in clinical signs of lead intoxication, a significant (almost a twofold) decrease in urinary excretion of lead from the body with its marked increase in blood occurs as a result of nonspecific readaptational therapy.

CONCLUSIONS

It has been established that:

- if lead content in urine is increased to more than 10% as compared to the initial one the risk of lead intoxication development is evaluated as low and human adaptive capacity is evaluated as safe or satisfactory;
- if lead content is decreased to more than 10% the risk is high and adaptive capacity of the body is exhausted;
- the absence of significant dynamics of lead content (its concentration in urine is not changed) testifies about moderate risk and adaptive capacity of the body is strained.

Therapeutic efficiency of hypno- and reflexotherapy in the body detoxications in based on mechanisms of hypnogenic stress which produces a complicated, multicomponent, powerful biologic effect leading to optimizing, readaptative alterations at the level of the entire body. The study of biological grounds of therapeutic effects of hypno- and reflexotherapy regarding occupational diseases contributes not only to a better understanding of their mechanisms but a better detection of mechanisms of natural sanogenesis.

*Metal Ions in Biology and Medicine: vol. 7.*Eds L. Khassanova, Ph. Collery, I. Maymard, Z. Khassanova, J.-C. Étienne.
John Libbey Eurotext, Paris © 2002 pp. 349-352.

Effect of metal ions on the liver nuclei

N. Kralj-Klobucar, S. Vecek-Šimunovic

Department of Biology, Faculty of Science, Rooseveltov trg 6, 1000 Zagreb, Croatia

ABSTRACT

The influence of metal ions Mg (0,1 mg/L), Pb (0,3 mg/L) and Se (0,1 mg/L) on liver cells of carps (*Cyprinus carpio* L.) was investigated during the period of 80 days. Already after 7 days of treatment a major increase in number of nuclei with two or three nucleoli was observed. After this initial increase oscillations followed but with a reduction tendency of nucleoli number to untreated group level. Nucleolus is the place of ribosomal RNA production and is especially well developed in cells active in protein synthesis. A larger number of nucleoli suggest metal inducted increase of protein synthesis, probably connected to cells detoxification mechanism. Apart from the nucleolus significant changes were found in nucleus chromatin formation. In all cases metal contamination led to the presence of small heterochromatin grains circularly arranged around nucleus. During prolonged period of contamination grains become bigger and can mask the nucleolus. Heterochromatin present in karyoplasm is regarded to be inactive chromatin, and the increase in number and density of chromatin grains suggests metal inhibitory effect on some genetic activity. It is possible that prolonged contamination causes cells polyploidia that is manifested by the presence of big nuclei.

INTRODUCTION

Aquatic system pollution with metal ions is one of the key ecological problems today. There are signs that exposure to heavy metals has genotoxic effects (Babich et al., 1985). In this context it is very important to know what effects does the low level concentration of metals have in conditions of long term exposure. In our researches we have investigated the effects of Pb, as a representative of toxic non essential elements, Se as toxic but essential element, and Mg as one of the leading essential elements that must be present in organism.

Pb induce histopathological changes in the tissue (Fantin at al. 1992; Kralj-Klobucar at al. 1996). In the cell it binds with protein which is synthesised in cytoplasm and forms a complex which is transported to nucleus, where it is polymerised in fibrils which build intranuclear inclusions (Pounds at al. 1982).

Mg in nucleus modulates DNA and RNA polymerase activity, and the processes of DNA synthesis and its duplication, transcription and degradation (Ghanbi at al. 1989).

Se exerts a strong protection action against the poisoning effects of many heavy metals, but in higher concentration, beside other effects, shows mutagenous characteristics and brings to the increase of chromosome aberrations elevated DNA repair synthesis and inhibition of mitosis (Wilber, 1980; Liu, 1995).

MATERIALS AND METHODS

One year old carps (*Cyprinus carpio* L.) were kept in aquariums with constant concentration level of elements. During the 80 days of research following concentrations were used: 0.1 mg Mg/L ($MgCl_2 \times 6\ H_2O$), 0,3 mg Pb/L ($Pb\ (NO_3)_2$) and 0.1 mg Se/L ($Na_2SeO_3 \times 5\ H_2O$). Animals were sacrificed in 7-day intervals. Liver tissue was used for the analysis; it was fixated in Bouin fixative, enbedded in paraffin and coloured with hematoxylin-eosin.

RESULTS

Largely increased number of nuclei with two or three nucleoli was found after 7 days of treatment in all treated animals *(fig. 1)*. Increase was the biggest in case of Pb treatment, it was slightly less in case of Mg treatment and the least in case of Se. After an initial increase a decrease followed in number of nuclei with two nucleoli, and then again a significant peak between day 28 and day 35 of the treatment. During further treatment, oscillations became smaller and smaller, and after 80 days the number of nuclei with two or more nucleoli was similar to the untreated group of animals.

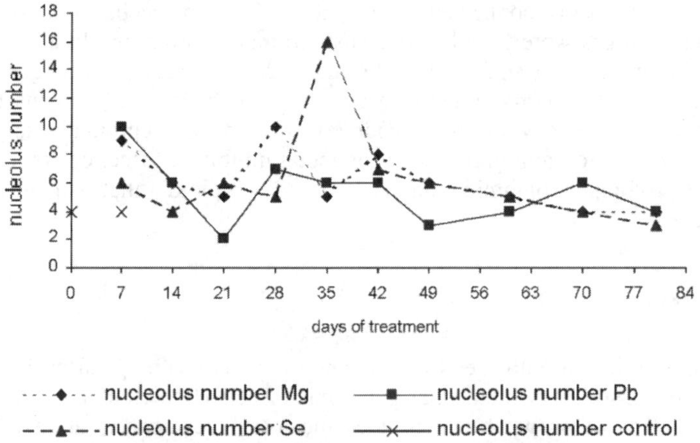

Fig. 1. Changes in number of nucleoli depending of metal contamination duration

Nucleus of untreated carps contains pale homogenous karyoplasm with slight heterochromatin edge that marks nuclear membrane and one, or very rarely two, dark coloured nucleolus *(fig. 2a)*.

7-day treatment with Mg did not cause any significant changes in the look of the nuclei *(fig. 2e)*. But starting from day 14 the presence of chromatin in the form of small grains circularly arranged around the nucleolus *(fig. 2f)*, was observed. After 70 days *(fig. 2g)* in many nuclei chromatin was coarsely grained and after 80 days of treatment small number of cells with untypical enlarged nucleus could be found *(fig. 2h)*.

In case of Pb, changes, in a form of small chromatin grains near the nuclear membrane or around nucleolus, were present already after 7 days of treatment *(fig. 2b)*. After 42 days of treatment grains became larger *(fig. 2c)*, and in this interval small number of cells with large nucleus and coarse cloddy chromatin could be found. Further treatment increased the number of cells with large nucleus *(fig. 2d)*, increased basophile karyoplasm and roughly grained chromatin that masks the nucleolus.

Se treatment causes enlargement of nuclei and presence of small chromatin grains alredy after 7 days *(fig. 2i)*. From day 14 *(fig. 2k)* some nuclei are present with larger chromatin grains. From day 42 *(fig. 2m)* some nuclei are significantly enlarged and chromatin is collected in a large number of big coarse cloddy that completely mask the nucleolus. Similar nuclei are also found throughout the further treatment. Most of the nuclei contain nucleolus with fine-grained chromatin, but much larger nuclei with a few nucleoli and many chromatin grains unevenly spread across karyoplasm are also present *(fig. 2n)*.

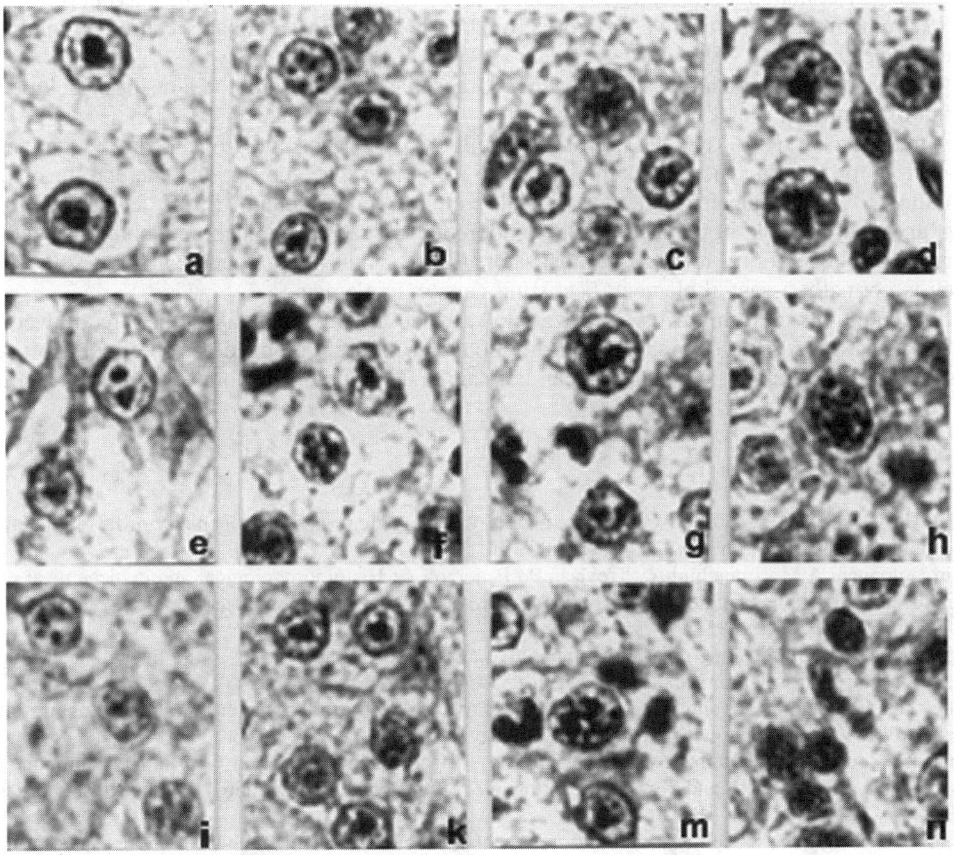

Fig. 2. Differences in the nucleus morphology during metal treatment: a) untreated group, b) Pb 7 days, c) Pb 42 days, d) Pb 70 days, e) Mg 7 days, f) Mg 14 days, g) Mg 70 days, h) Mg 80 days, i) Se 7 days, k) Se 14 days, m) Se 42 days, n) Se 80 days. ×1250

DISCUSSION

It is known that metals cause genotoxic effect in the cell. Already one week of treatment resulted in significant increase in number of nucleolus in nuclei and that obviously represent the result of the increased activity of genes responsible for rRNA synthesis. The increase of rRNA is in a connection with ribosome activity and with the increased protein synthesis in cytoplasm. Probably is a question of metallothionein synthesis that protects the cell by binding the metal to itself. This explains reduction in nucleoli number after starting insult, because plasma proteins protect nucleus from metal effect. This also explains the oscillations in number of nucleolus during the treatment. After prolonged contamination, excess of metal accumulated in amounts that surpass the protein binding ability, repeatedly stimulates synthetic activity in a nucleus that creates new RNA. Continuous long contamination as a final result shows reduction tendency concerning the number of nuclei with two nucleoli, and that in the end results in numbers that are similar to the numbers of untreated group of animals.

This may mean that genes responsible for RNA synthesis are inactivated which results in reduction of plasma protein production, and with reduced cells defence ability, or in other words with manifestation of the toxic effect of the metal.

Prolonged contamination leads to forming of heterochromatin grains in karyoplasm. Since heterochromatin presents inactive form of chromatin, its presence is probably caused by inhibitory effect of metal.

Appearance of untypical cells with volume larger than normal could be a result of polyploidia as sign of genotoxic metal effect.

Stimulation of rRNA synthesis, which is built into the ribosome from the nucleolus and is responsible for protein synthesis, is the answer to metal presence and serves to defend the cell. Pb is binding to protein plasma and forms metaloprotein, while selenium forms selenoprotein and plays an important role in defence against oxidoreductive radicals. When contamination with Se is prolonged, gene ability to make compensatory RNA synthesis is reduced which results in Se, which is normally essential element, becoming genotoxic element.

Concentrations used in our research were low, inside the maximal values that are allowed and defined as harmless. But, even these concentrations during the prolonged period of contamination cause effects that are not pathohistological, but can be evidented in a more sensitive genetic area.

REFERENCES

Babich H., Devanas M.A., Stotzky G. The mediation of mutagenicity and clastogenicity of heavy metals by physicochemical factors. *Environ. Res.*, 1985, 37, 253-286.

Fantin Bolognani A.M., Franchini A., Trevisan P., Pederzoli A. Histomorphological and cytochemical changes induced in the liver of goldfish *Carassius carassius var. auratus* by short-term exposure to lead. *Acta histochem.*, 1992, 92, 228-235.

Kralj-Klobucar N., Stunja A., Kopjar N. Comparative analysis of the lysosomes in the liver treated with the lead, magnesium and selenium. *Acta Histochem. Cytochem.*, 1996, 29, 523-524.

Liu D., Zhai L., Jiang W., Wang W. Effect of Mg^{2+}, Co^{2+}, and Hg^{2+} on the nucleus and nucleolus in root tip cells of *Allium cepa. Bull. Environ. Contam. Toxicol.*, 1995, 55, 779-787.

Ghanbi Z., Rouabhia M., Othmane O., Deschaux P.A. Effects of metal ions on Cyprinid fish immune response: In vivo effects of Zn^{2+} and Mg^{2+} on the mitogenic response of carp pronephros lymphocytes. *Ecotoxicol. Environ. Saf.*, 1989, 17, 183-189.

Pounds J.G., Wright R., Kodell R.L. Cellular metabolism of lead: A kinetic analysis in the isolated rat hepatocyte. *Toxicol. Appl. Pharmacol.*, 1982, 66, 88-101.

Wilber C.G. Toxicology of selenium: A Review, *Clin. Toxicol.*, 1980, 17 (2), 171-230.

Metal Ions in Biology and Medicine: vol. 7. Eds L. Khassanova, Ph. Collery, I. Maymard, Z. Khassanova, J.-C. Étienne.
John Libbey Eurotext, Paris © 2002 pp. 353-356.

Behavior of juvenile mice chronically exposed to depleted uranium

Wayne Briner & Bridget Abboud

Department of Psychology University of Nebraska at Kearney, Kearney, NE, 68849 USA

ABSTRACT

Mice exposed to depleted uranium during development show disturbed development and behavior. Modulation of responses to environmental stimuli appears to be effected as well as learning and exploration. Lipid oxidation is evident in the brains of DU exposed animals and may be one of DU's mechanisms of action.

Key words: Depleted uranium, development, brain.

BACKGROUND

Uranium is a heavy metal with toxic potential outside of radioactivity. Depleted uranium (DU) use by the military is escalating with uranium entering the environment in increasing amounts. Use of DU in industrial applications has also increased. Previous studies in mice have found that uranium is a reproductive toxin producing fetal death, smaller pups, and fewer litters at 25 or 50 mg/kg of uranium acetate (oral administration) depending on length and timing of exposure [Domingo et al., 1989a; Paternain et al., 1989]. Skeletal malformations have been found in mouse pups when pregnant dams were exposed to 25-50 mg/kg of uranium acetate [Domingo et al., 1989b]. Injections of as little as 0.5 mg/kg of uranium acetate can produce discernable effects on pregnant mice or their offspring [Bosque et al., 1993]. We have previously presented work demonstrating that mice exposed to DU effected their development on a number of measures [Briner & Byrd, 2000]. Collectively, these previous findings demonstrated an impaired ability of the animal to regulate its response to the environment. We now report additional observations based on a larger sample size and biochemical studies of CNS tissues.

MATERIAL AND METHODS

Female Swiss-Webster mice, housed under standard laboratory conditions, were exposed to 0, 19, 37, or 75 mg/L of uranium acetate in drinking water for two weeks, and then mated. Exposure of dams and pups continued until sacrifice. Access to food and water was ad-lib. Mice were assessed using the Fox Developmental Scale [Fox, 1965] until age 21 days. At 21 days of age the pups were assessed with a Functional Observation Battery [O'Donoghue, 1996] after-which the brains were removed for study.

After behavioral assessment the animals were killed and the brains removed. Lipid oxidation of brain tissue was measured using the thiobarbituric acid (TBA) assay [Ohkawa et al, 1976]. Protein content of the sample will be determined using the commassie blue method and lipid oxidation expressed as a ratio to total protein content.

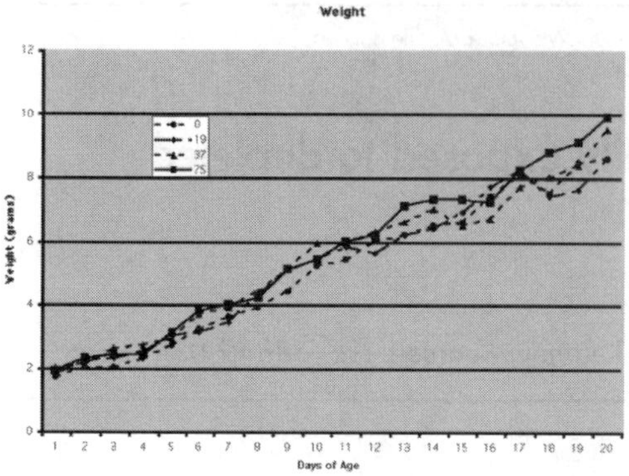

Fig. 1.

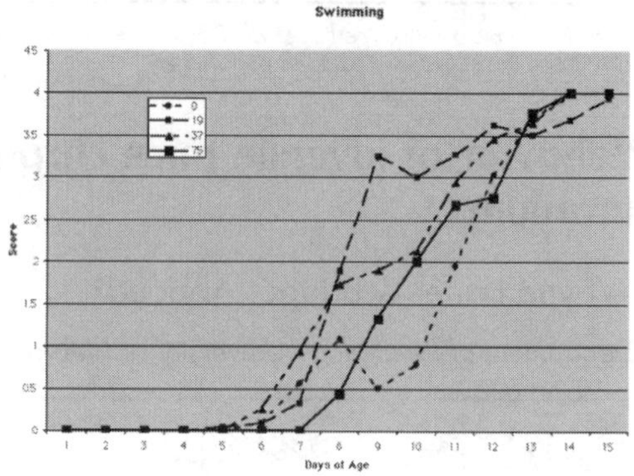

Fig. 2.

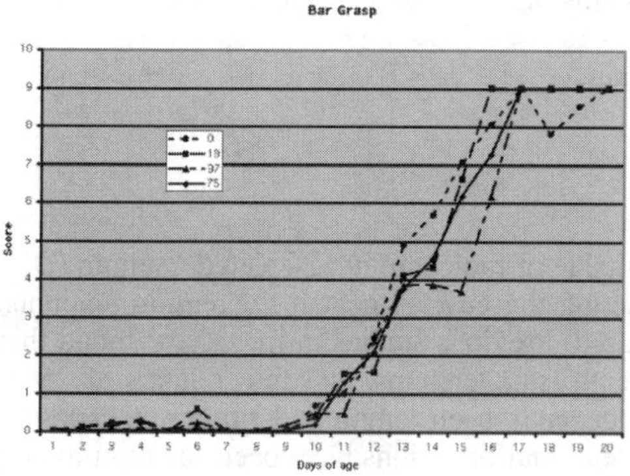

Fig. 3.

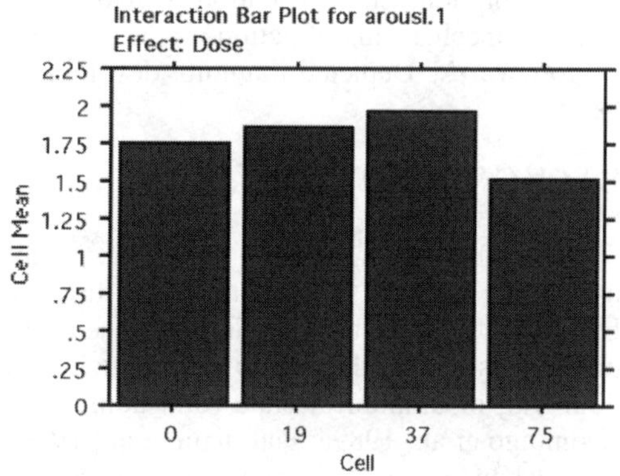

Fig. 4. Interaction Bar Plot for arousl.1
Effect: Dose

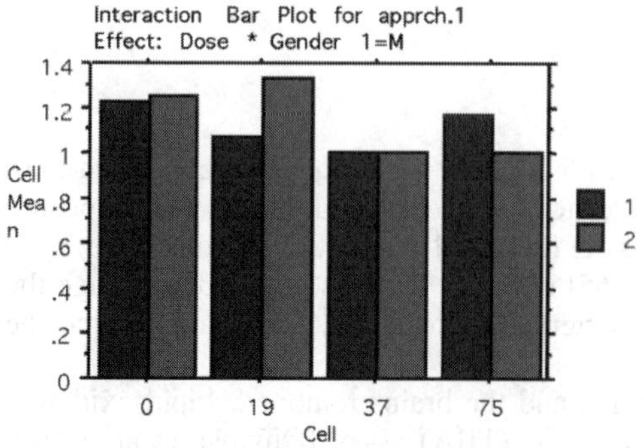

Fig. 5. Interaction Bar Plot for apprch.1
Effect: Dose * Gender 1 = M

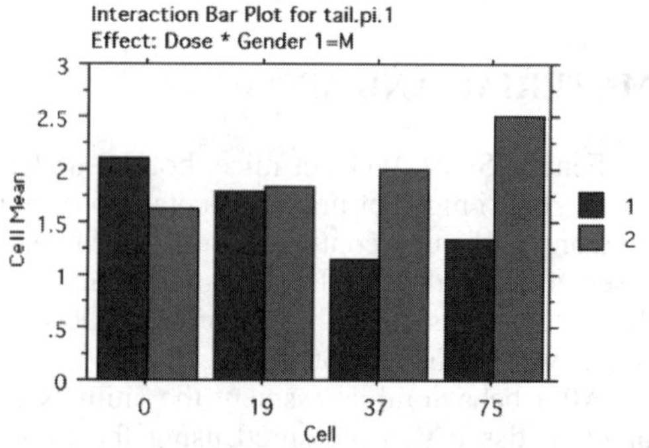

Fig. 6. Interaction Bar Plot for tail.pi.1
Effect: Dose * Gender 1 = M

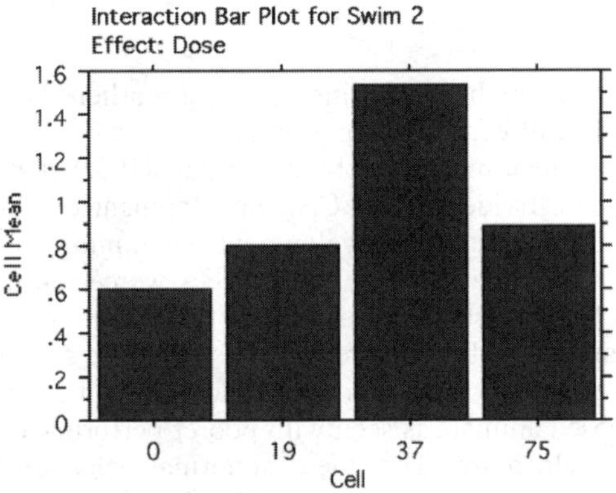

Fig. 7. Interaction Bar Plot for Swim 2
Effect: Dose

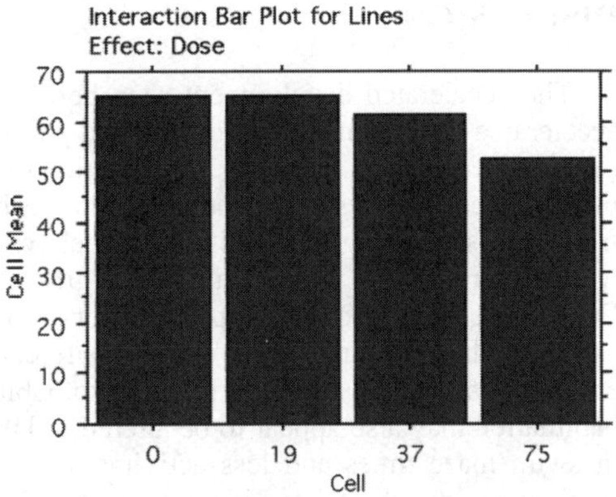

Fig. 8. Interaction Bar Plot for Lines
Effect: Dose

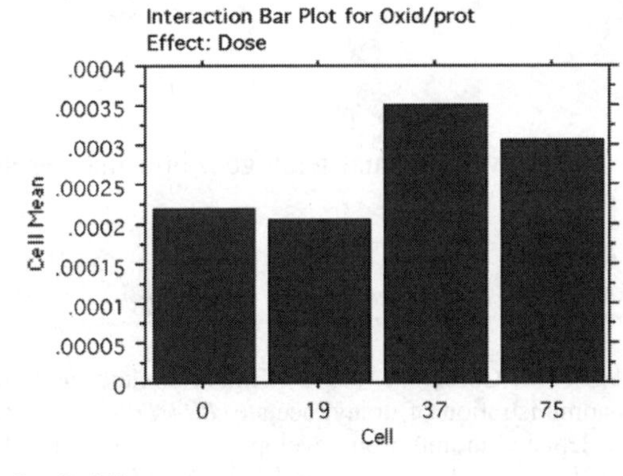

Fig. 9. Interaction Bar Plot for Oxid/prot
Effect: Dose

RESULTS

All of the pups were in good health with no discernable abnormalities at birth. Neurodevelopmental progress differed between the groups on three measures including; weight *(fig. 1)* and development of swimming behavior *(fig. 2)* with the DU exposed pups gaining weight more rapidly and developing swimming behavior at an accelerated rate. However, bar grasp behavior was delayed in DU exposed animals *(fig. 3)*.

Behavior at 21 days of age also differed for DU exposed animals. Arousal was significantly different from controls in what appears to be an inverted U-shaped response *(fig. 4)*. Approach response was altered in a dose dependent fashion interacting with gender *(fig. 5)*, as did the tail-pinch response *(fig. 6)*. Swim maze learning behavior was also effected by DU, with DU exposed animals taking longer to solve the task and showing less improvement on the second trial *(fig. 7)*. DU exposed animals also demonstrated less activity in the open field *(fig. 8)*. DU exposed animals also had larger bodies, reflected in smaller brains as a percentage of body weight. DU exposed animals also demonstrated greater lipid oxidation levels in their brains *(fig. 9)*.

DISCUSSION

The accelerated development seen here is consistent with the findings reported earlier. This accelerated development may be due to a pharmacologic effect of DU on growth hormone or some other related system. However, acceleration of development may disrupt the coordinated development of neuronal systems producing altered behaviors. Evidence that CNS development comes from data suggesting that while some behaviors are accelerated in development (swimming and weight gain), others were delayed (bar grasp). Later, at 21 days of age, differences were seen in DU exposed pups Arousal, approach, and tail pinch responses were abnormal for DU exposed animals. These findings argue that DU affects the animal's ability to regulate its response to the environment, perhaps through inability to habituate or alterations in emotional state. Learning and habituation may also appear to be altered in DU exposed animals as seen with poorer performance in swim maze times and less activity in the open field maze. That these abnormal behavioral findings are due to DU exposure, and not an extraneous variable, is supported by the presence of oxidized lipids and small brain sizes (as percentage of body weight) seen in DU exposed pups.

This study raises the possibility that DU exposure can produce subtle alterations in brain structure or function that produce alterations in behavioral development at doses far lower than those needed to produce blatantly toxic effects. While the mechanism by which DU produces its toxicity is unknown, lipid oxidation and alteration in growth physiology may be important.

CONCLUSIONS

Depleted uranium may be a developmental teratogen, in a manner similar to lead and other heavy metals.

REFERENCES

Bosque MA, Domingo JL, Llobet JM, Corbella J. (1993). Embryotoxicity and teratogenicity of uranium in mice following subcutaneous administration of uranyl acetate. *Biol Trace Elem Res* 1993; 36: 109-118.

Briner W & Byrd K. Effect of depleted uranium on development of the mouse. In Centeno JA, Collery PH, Vemet G, Finkelman RB, Gibb H, Etienne JC (Eds). *Metal Ions in Biology and Medicine*, 2000, 6th, 459-461.

Domingo JL, Ortega A, Paternain JL, Corbella J. Evaluation of the perinatal and postnatal effects of uranium in mice upon oral administration. *Archv Environ Health* 1989a; 44: 395-398.

Domingo JL, Paternain JL, Llobet JM, Corbella J. The developmental toxicity of uranium in mice. *Toxicol* 1989b; 55: 143-152.

Fox WM. Reflex-ontogeny and behavioral development of the mouse. *Anim Behav* 1965; 13: 234-241.

Ohkawa H, Ohishi N & Tagi K. Assay for lipid peroxides in animal tissues by thiobarbituric acid reaction. *Anal. Biochem.*, 1979, 95, 351-358.

O'Donoghue JL. Clinical neurologic indicies of toxicity in animals. *Environ. Health Perspect.*, 1996, 104 (suppl. 2), 323-330.

Paternain JL, Domingo JL, Ortega A, Llobet JM. The effects of uranium on reproduction, gestation, and postnatal survival in mice. *Ecotoxi Environ Safety* 1989; 17: 291-296.

*Metal Ions in Biology and Medicine: vol. 7.*Eds L. Khassanova, Ph. Collery, I. Maymard, Z. Khassanova, J.-C. Étienne.
John Libbey Eurotext, Paris © 2002 pp. 357-362.

Modulation in hepatic parameters of carp (*cyprinus carpio* L.) induced by copper and chitosan

Claire Dautremepuits[1,2], Stéphane Betoulle[1,2], Sylvie Biagianti[1,2], Jean-Claude Etienne[2] and Guy Vernet[1,2]

[1]*Laboratory of Eco-Toxicology;* [2]*International Research Institute on Metal Ions, University of Reims Champagne-Ardenne, BP 1039, 51687 Reims cedex 2, France*

ABSTRACT

Copper is used in treatment mixtures to control fungal diseases in vineyards and concentrations are relatively high in some aquatic ecosystems collecting vineyard runoff water. Better awareness of the ecological and health problems associated with copper and its potential accumulation through the food chain has prompted the demand for alternative used of low toxic molecules in plant protection and for the removal of heavy metal ions from waste water.

Chitosan is the N-deacetylated derivative of chitin, a natural abundant polysaccharide. The numerous properties of chitosan (especially antifungal, and heavy metal ions chelating agent) offer a wide range of applications. On the one hand, chitosan could be used as a natural agri-bioactive substance controlling fungal diseases and contributing to reduce the use of copper in plant protection. On the other hand, chitosan is a metal ion chelator already used in waste water treatment. It is then necessary to investigate the potential toxicity of chitosan for aquatic animal health, alone and associated with copper.

In this study, Carp were exposed to sublethal concentrations of chitosan (0.5% and 1%) or copper (0.1, 0.25 $mg.L^{-1}$) or chitosan and copper (0.5% and 0.1 $mg.L^{-1}$ respectively) for 5 and 10 days.

In chitosan-exposed carp, hepatic catalase and glutathione S-transferase activities were significantly increased in comparison with control. An increase of these parameters was also observed in copper-exposed carp. Moreover, physiological response of carp exposed to chitosan/copper mixture was modulated too. This study showed that chitosan is potentially noxious molecule for fish and its industrial and/or future agricultural uses will have to take care of this problem.

INTRODUCTION

Numerous chemicals called contaminants are introduced deliberately or accidentally into the aquatic ecosystem, impairing the quality of the water and making it unfavourable for aquatic life. They can produce an adverse response (effect) in a biological system, seriously damaging its structure or function or producing death (Sörensen, 1991).

Copper is used in agriculture especially to control fungal diseases in vineyards. High concentrations of copper were detected in some aquatic ecosystems collecting vineyard runoff water (GERBE, 1996; Teisseire, 1999). This heavy metal is already known to be accumulated in organs of aquatic animals (Rójik *et al*, 1983; Bàlint *et al*, 1997). It can lead to redox reactions generating free radicals and, therefore, may change some physiological processes (Varanka *et al*, 2001; Paris-Palacios *et al*, 2000). The mechanisms of its toxicity for fish are still poorly understood and the health problems associated to the copper uses has prompted the demand for an alternative strategy using low toxic molecules in plant protection against pathogens such as chitosan.

Chitosan is the deacetylated form of chitin and is extracted primarily from shells of crustaceans such as shrimps and crabs, and from squid pens. Chitosan is used as a non toxic cationic flocculent in treatment of wastewaters from sewage, sludge or breweries and as a chelator of heavy metal (Sandford, 1989). Agriculture applications include coatings for seeds and fruit preservation associated with fungistatic properties (Ravi Kumar, 2000). We postulate that chitosan could be used as a natural agri-bioactive substance controlling fungal diseases contributing to reduce the use of copper in plant protection. It is then necessary to investigate the potential toxicity of chitosan alone and associated with copper for aquatic animal health.

Carp (*Cyprinus carpio* L.) is one of the most common freshwater fish species. It gives a good opportunity to study responses and possible adaptations of local fish populations to distinct treatment mixtures used to control diseases in vineyard.

Like mammals, fish also possesses defense mechanisms to counteract the impact of toxics. Taking account that enzymatic and non-enzymatic antioxidants serve as an important biological defense against environmental oxidative stress induced by metal ions, we investigate the effect of copper or chitosan alone and their combination on some physiological parameters induced in carp liver in controlling conditions. The purpose of this work was to determine the potential toxicity of chitosan alone or associated with copper in term of ecotoxicological risk assessment.

MATERIALS AND METHODS

Fish

Immature common carp (*Cyprinus carpio* L.) (22 ± 4 g and 100 ± 10 mm) were obtained from a local hatchery (Ets "Au Vairon", Reims, France). Fish were kept in 20L tanks containing aerated spring water following standard fish maintenance procedure. They were acclimated 10 days before the start of the experiment. Water quality parameters such as dissolved oxygen (75 ± 5%), pH (8 ± 0.5), temperature (20 ± 1 °C) were determined daily until the end of the experiment. Group of 6 fish were used for each treatment. Carp were maintained in water containing different copper concentrations, i.e. 0.1 or 0.25 mg/L ($CuSO_4$, Prolabo, France), chitosan concentrations, i.e. 0.5 or 1% (Chitogel®, Aber Tech, France), or copper (0.1 mg/L) + chitosan (0.5%). All experiments were realized in duplicates. The median lethal concentration (LC_{50}) of copper for carp after 96 hours exposure was obtained by Shariff *et al* (2001) (0.53 mg/L) and the median lethal concentration of chitosan for carp after 96 hours exposure was determined in a previous experiment (2 ± 0.12%). Copper concentrations used here corresponded approximatively to 20 and 50% of LC_{50}-96 h and chitosan concentrations corresponded to 25 and 50% of LC_{50}-96 h.

Sampling of tissue

Samples of carp liver were prepared by suspended 0.2g of liver in 3 ml of Phosphate Buffer Saline (PBS) (Life Technologies, France) and homogenised using a potter-Servodyne homogeniser (Cole-Parmer Instrument Co, Niles, Illinois). Homogenates were centrifuged at 30.000 g for 15 min at 4 °C. The supernatants were collected; pools of 4 fishes were made and stored at -80 °C until further analysis.

Total protein assay

Proteins of fish liver were determined by the method of Bradford (1976) using Bovine Serum album (Sigma, France) in as a standard.

Catalase activity

Catalase activity was assayed by the method of Teisseire *et al* (1998) as described by Giri *et al* (1996). The assay mixture consisted of 1.95 ml phosphate buffer (0.05 M, pH 7.0), 1 ml hydrogen peroxide (19 mM) and 0.05 ml of sample in a final volume of 3 ml. Change in absorbance was recorded at 240 nm. Catalase activity was calculated in terms of nmol H_2O_2 consumed/min/mg protein.

Glutathione s-transferase activity

Glutathione s-transferase (GST) activity was determined by the method of Regoli *et al.* (1997) with some modifications. The reaction mixture consisted of 1 ml phosphate buffer (0.1 M, pH 6.5), 0.5 ml reduced glutathione (1 mM), 0.3 ml H_2O, 50 µl 1-chloro-2,4-D nitro-benzene (CDNB) (1 mM) and 150 µl of sample in a total volume of 2 ml. The change in absorbance was recorded at 340 nm and the enzyme activity calculated as µmol CDNB conjugate formed/min/mg protein using a molar extinction coefficient of $9.6.10^3$ $M^{-1}.cm^{-1}$.

RESULTS

GST and catalase activities were significantly increased in carp maintained in chitosan-containing water for 5 days *(fig. 1a)*. The catalase activity remained significantly high at 10 days in chitosan-exposed carp in comparison with control *(fig. 1a)*. However, at the same time, GST activity was decreased and lower than control. There was no difference in total plasma protein content between control fish and fish maintained in chitosan-containing water *(fig. 1a)*. Concerning the effects of copper on the same parameters, the total plasma protein were significantly increased in fish exposed to 0.25 mg/L copper for 5 days in comparison with control whereas no difference was observed after 10 days of exposure *(fig. 1b)*. Moreover, GST activity was increased only in fish exposed to 0.1 mg/L of copper for 10 days. Carp exposed to 0.25 mg/L copper for 5 days had a decrease in catalase activity in comparison with control. At 10 days, this activity was increased in copper exposed fish whatever the concentration may be. Concerning the fish exposed to copper and chitosan *(fig. 2)*, their total plasma proteins were significantly increased at 5 days of exposure in comparison with control. GST and catalase activities were significantly decreased in carp exposed to the mixture copper and chitosan for 5 days whereas no effect on enzymatic activities was observed at 10 days of exposure *(fig. 2)*.

DISCUSSION

Antioxidant enzymes such as catalase or glutathione-S transferase are induced to deal with reactive oxygen species. Exposure of carp to copper or chitosan or copper and chitosan resulted in induction or suppression of enzymatic antioxidants in fish. Carp exposed to chitosan had an increased catalase activity whatever the exposure duration may be whereas in copper exposed fish this activity decreased firstly at 5 days and secondly increased after 10 days of exposure time. The inhibition of catalase activity by high concentrations of copper has already been reported by Radi and Matkovics (1988). This inhibition was related to an oxidative stress. A longest time of exposure (10 days) was associated with an increase in catalase activity probably corresponding to an adaptative mechanism. This enzymatic activity was also increased in chitosan-exposed fish. Chitosan then improved the fish antioxidant potential related to catalase activity.

Moreover, GST activity was increased in fish exposed to chitosan for 5 days but not at 10 days where an inhibition was observed. The GST in liver has been reported as a bio indicator for assessing the environmental impact of xenobiotics that generate oxidative stress (Livingstone, 1998). As GST is involved in xenobiotic detoxification and excretion of xenobiotics and their

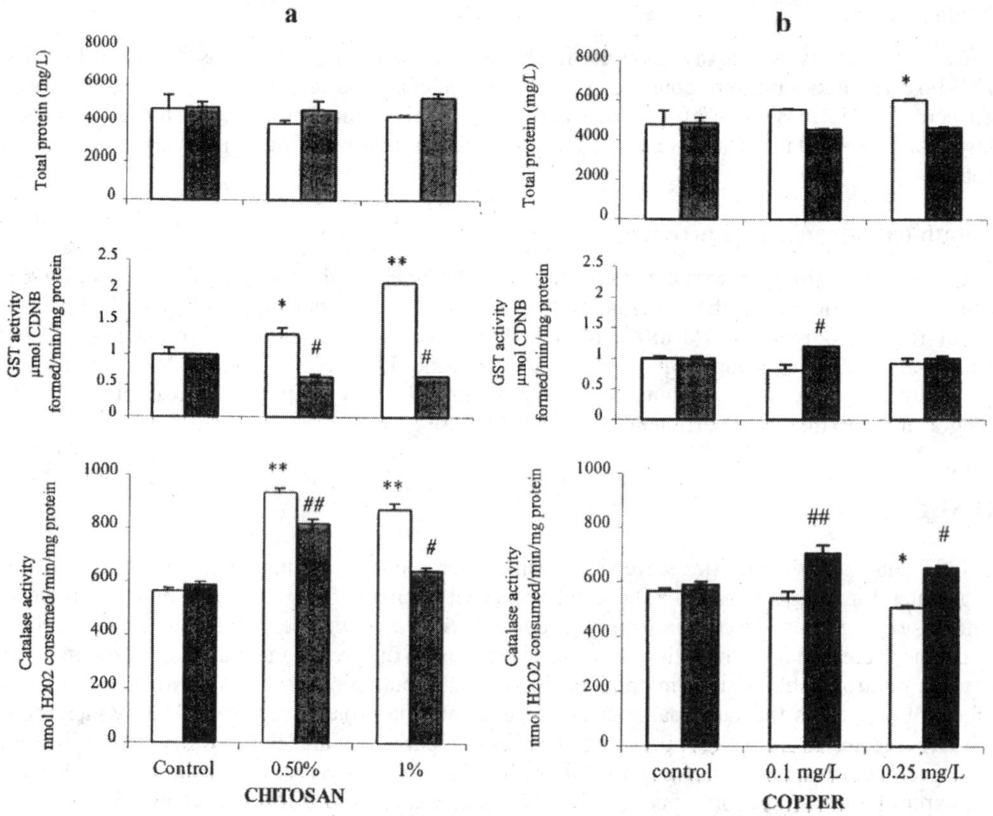

Fig. 1. Total protein content, glutathione S-Transferase and catalase activities of fish liver after 5 (□) and 10 (■) days of exposure to 0.5 and 1% of chitosan (a) and 0.1 and 0.25 $mg.L^{-1}$ of copper (b). Values are means ± S.E. For each treatment, means obtained at 5 or 10 days were compared to their respective control: for 5 days, *$p<0.05$ and **$p<0.01$; for 10 days, #$p<0.05$, ##$p<0.01$.

metabolites, the increased activity in liver observed at 5 days of carp exposure to chitosan may indicate development of a defensive mechanism of fish to eliminate naturally this molecule. Chitosan could then improve the antioxidant capacity of fish but liver reactions corresponding to xenobiotic metabolism were comparable to those observed for other organic chemicals (Ahmad *et al.*, 2000).

Although some studies indicate a decrease in total protein content during heavy metal exposure (De Smet and Blust, 2001), a low increase was observed only in 0.25 mg/L copper-exposed fish for 5 days.

When carp were treated with copper/chitosan mixture, there was no significant modulation of hepatic GST activity excepted a relatively low decrease observed in fish contaminated for 5 days. An identical response was observed for catalase activity. Fish treated with 0.1 mg/L copper or with 0.5% chitosan had a high hepatic catalase activity in comparison with control whereas carp contaminated with copper associated to chitosan presented no real modulation of this activity. These responses were probably due to the chelating activity of chitosan molecule. Chitosan seems to inhibit the hepatic impact of copper particularly on enzymatic antioxidant system. Moreover, effects of chitosan alone on catalase and GST activities were modulated when fish were simultaneously contaminated by copper. No long-term modulation in antioxidant enzymatic activities was observed in fish treated with chitosan + copper

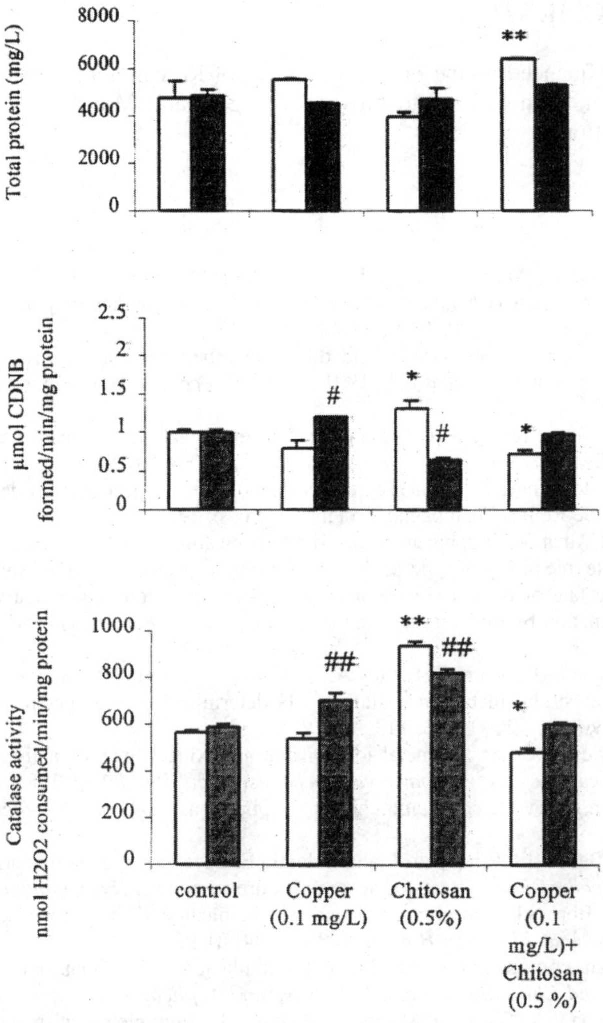

Fig. 2. Total protein content, glutathione S-Transferase and catalase activities of fish liver after 5 (□) and 10 (■) days of exposure to copper (0.1 mg/L), chitosan (0.5%) or copper + chitosan (0.1 mg/L and 0.5% respectively). Values are means ± S.E. For each treatment, means obtained at 5 or 10 days were compared to their respective control: for 5 days, *p<0.05 and **p<0.01; for 10 days, #p<0.05, ##p<0.01.

CONCLUSIONS

Thus, chitosan significantly increased the antioxidant enzymes activities in carp liver and contributed to decrease the oxidative stress potentially induced by the metal ion copper. However, this chitin derivate reduced the GST activity after 10 days of fish exposure. In perspective to be used as an agricultural pesticide, chitosan seems to be a relatively low toxicant after a short-term exposure (5 days) whereas its long term toxicity (as shown in hepatic GST activity) may be more pronounced as shown after 10 days of fish exposure. These observations open perspectives for their used to control fungal diseases in vineyards.

ACKNOWLEDGMENTS

This study was financed by the French Ministery of Research and Europol' Agro Foundation. Author thanks Aber Tech Society to provide chitogel® and Mrs A. Conreux for her precious technical collaboration.

REFERENCES

Ahmad I., Hamid T., Fatima M., Chand H.S., Jain S.K., Athar M. and Raisuddin S. Induction of hepatic antioxidants in freshwater catfish (*Channa punctatus* Bloch) is a biomarker of paper mill effluent exposure. *Biochimica et Biophysica Acta*, 2000, 1519, 37-48.

Bálint T., Ferenczy J., Kátia F. Similarities and differences between the massive eel (*Angilla angilla* L.) devastations that occurred on lake Balaton in 1991 and 1995. *Ecotoxicology Environmental Safety*, 1997, 37, 17-23.

De Smet H., Blust R. Stress response and changes in protein metabolism in carp *Cyprinus carpio* L. during cadmium exposure. *Ecotoxicology and Environmental Safety* 2001, 48, 255-262.

GERBE. Programme de recherche "Toxicologie, Ecotoxicologie des pesticides et des métaux lourds". Europol' Agro Université de Reims Champagne-Ardenne. 1996, p. 39.

Giri U., Iqbal M. and Athar M. Porphyrin-mediated photosensibilization has a weak tumor promoting effect in mouse skin: possible role of *in situ* generated reactive oxygen species. *Carcinogenesis*, 1996, 17, 2023-2028.

Livingstone D.R. The fate of organic xenobiotics in aquatic ecosystems: quantitative and qualitative differences in biotransformation by invertebrates and fish. *Comparative Biochemistry and Physiology* 1998, 120, 43-49.

Paris-Palacios S., Biagianti-Risbourg S, Fouley A., and Vernet G. Metallothioneins in liver of *Rutilus rutilus* exposed to Cu^{2+}. Analysis by metal summatation, SH determination and spectrofluorimetry. *Comparative Biochemistry and Physiology* 2000, 126, 113-122.

Radi A.A.R., Matkovics B. Effects of metal ions on the antioxidant enzyme activities, protein content and lipid peroxidation of carp tissues. *Comparative Biochemistry and Physiology* 1988, 90, 69-72.

Ravi Kumar M.N.V. a review of chitin and chitosan applications. *Reactive & Functionnal Polymers*, 2000, 46, 1-27.

Regoli F., Nigro M., Bertoli E., Principato J. and Orlando E. Defences against oxidative stress in the Antartic scallop *Adamussium colbecki* and effects of acute exposure to metals. *Hydrobiologia*, 1997, 355, 139-144.

Rójik I., Nemcsók J., Boross L. Morphological and Biochemical studies on liver, kidney and gill of fishes affected by pesticides. *Acta Biologica Hungarica* 1983, 34, 81-92.

Sandford P.A. chitosan: commercial uses and potential applications. In: Skjak-braek G., Anthonsen T., Sanford P. (Eds), *Chitin and Chitosan: Sources, Properties and Applications*, Elsevier, London, 1989, 51-69.

Shariff M., Jayawardena P.A., Yusoff F.M. and Subasinghe R. Immunological parameters of Javanese carp *Puntius gonionotus* (Bleeker) exposed to copper and challenged with *Aeromonas hydrophila. Fish & Shellfish Immunology*, 2000, 11, 281-291.

Sörensen E.M. Cadmium. In: *Metal Poisoning in fish* CRC Press, Boston, 1991, 175-234.

Stripped R.A. and Trombetta L.D. Antioxidant enzyme activity as biomarkers of acid stress in natural aquatic ecosystems. In: Muller R., Lloyd D. (Eds), *Sublethal and Chronic Effects of Pollutants on Freshwater Fish*, Fishing New Book. Blackwell Science, Oxford, 1994, 249-261.

Teisseire H., Couderchet M. and Vernet G. Toxic responses and catalase activity of *Lemna minor* L. exposed to folpel, copper and their combination. *Ecotoxicology Environmental Safety*, 1998, 41, 194-200.

Teisseire H. Toxicologie et ecotoxicologie des pesticides et des métaux lourds susceptibles d'être présents dans le vignoble champenois: étude de leur impact physiologique et biochimique sur *Lemna minor. Thèse Université de Reims Champagne Ardenne* 1999, p. 272.

Varanka Z., Rojik I., Varanka I., Nemcsok J., Abraham M. Biochemical and morphological changes in carp (*Cyprinus carpio* L.) liver following exposure to copper sulfate and tannic acid. *Comparative Biochemistry and Physiology* 2001, 128, 467-478.

VIII ENVIRONMENT

*Metal Ions in Biology and Medicine: vol. 7.*Eds L. Khassanova, Ph. Collery, I. Maymard, Z. Khassanova, J.-C. Étienne.
John Libbey Eurotext, Paris © 2002 pp. 365-370.

Health risk assessment and management of arsenic toxicity and carcinogenesis

Paul B. Tchounwou

Molecular Toxicology Research Laboratory, NIH-Center for Environmental Health, School of Science and Technology, Jackson State University, Jackson, MS 39217, USA

ABSTRACT

Acute and chronic exposure to arsenic has been reported in several countries of the world, with major outbreaks of arsenosis occurring in Argentina, Bangladesh, India, Mexico, Thailand, and Taiwan, where a large proportion of drinking water (groundwater) is contaminated with high concentrations of arsenic. Hence, arsenic poisoning appears to be one of the major public health problems of pandemic nature. A comprehensive analysis of published data indicates that arsenic exposure induces cardiovascular diseases, developmental abnormalities, neurologic and neurobehavioral disorders, diabetes, hearing loss, heamatologic disorders, and various types of cancer, including skin, and lung neoplasms. Recent epidemiologic studies have demonstrated a strong correlation between arsenic exposure and the increase in incidence of human cancers. Research has also pointed out significantly higher standardized mortality rates for cancers of the bladder, kidney, skin, liver, and colon in many areas of arsenic pollution. There is therefore a great need for developing a comprehensive risk assessment (RA) model, to be used in the management of health risks associated with arsenic exposure. The development of such model requires a thorough understanding of the physical and chemical properties of arsenic, its production and use, fate and transport, toxicokinetics, systemic and carcinogenic health effects, regulatory and health guidelines, analytical methods, and treatment technologies. This paper aims at using the National Academy of Science's RA paradigm as a guide; to present a conceptual RA framework for managing the potential risks of toxicity and carcinogenesis associated with arsenic exposure.

INTRODUCTION

Arsenic is released to the environment from natural sources as a result of natural phenomena such as erosion of mineral deposits and volcanoes, but releases from human activities such as metal smelting, coal combustion, chemical production and use, and waste disposal can lead to substantial contamination of the environment [1]. The National Academy of Science estimates that about 75,000 to 100,000 tons of arsenic are produced annually on a global scale [2]. More than 80% of arsenic compounds are used to manufacture products with agricultural applications such as insecticides, herbicides, fungicides, algicides, sheep dips, wood preservatives, dyestuffs, and medicines for the eradication of tapeworms in sheep and cattle. Arsenic compounds have been used for at least a century in the treatment of syphilis, yaws, amoebic dysentery, and trypanosomiasis [2]. Arsenical drugs are still used in treating certain tropical diseases such as African sleeping sickness and amoebic dysentery, and in veterinary medicine to treat parasitic diseases, including filariasis in dogs and black head in turkeys and chickens [2]. Recently, arsenic has been used as an anticancer agent in the treatment of acute promeylocytic leukemia, and its therapeutic action has been attributed to the induction of programmed cell death (apoptosis) in leukemia cells [3].

In recent years, arsenic contamination of natural resources (ground water) has emerged as one of the major environmental health issues in several countries of the world. Outbreaks of arsenosis and other health effects associated with ground water contamination in Argentina, Bangaldesh, Chili, China, Mexico, India, Thailand, and Taiwan are examples of such concerns [1]. Arsenic exposure has lead to a significant number of health concerns including hyperkeratosis, jaundice, vascular diseases, and cancer of various organs/tissues including the skin, liver, lung and bladder [4].

The mechanism by which arsenic exerts its toxic effect is through impairment of cellular respiration by the inhibition of various mitochondrial enzymes, and the uncoupling of oxidative phosphorylation. Most toxicity of arsenic results from its ability to interact with sulphydryl groups of proteins and enzymes, and to substitute phosphorus in a variety of biochemical reactions. Although the evidence of carcinogenicity of arsenic in humans seems strong, the mechanisms by which it produces tumors in humans is not completely understood. In contrast to most other human carcinogens, it has been difficult to confirm the carcinogenicity of arsenic in experimental animals. Research evaluating the chronic effects in laboratory animals exposed to inorganic and organic arsenic compounds by the oral route or skin contact, has not shown any potential for initiation or promotion of carcinogenicity [5].

Based on its high degree of toxicity to humans, and the non-threshold dose-response assumption, a zero level exposure is being recommended for arsenic even though it is practically non attainable [1]. As with other human carcinogens, the development of a comprehensive risk assessment (RA) and risk management (RM) protocol for arsenic requires a thorough understanding of the four components (hazard identification, dose-response assessment, exposure assessment, and risk characterization) of the RA paradigm, and an evaluation of RM options related to arsenic contamination [2].

HAZARD IDENTIFICATION

There are many case reports of death in humans due to ingestion of high doses of arsenic. The clinical manifestations of arsenic poisoning depend on the type of arsenical involved and on the duration of exposure. Symptoms of acute intoxication usually occur within 30 minutes of ingestion but may be delayed if arsenic is taken with food. In nearly all cases, the most immediate effects are severe nausea and vomiting, colicky abdominal pain, profuse diarrhea with rice stools, gastrointestinal hemorrhage and death may ensue from fluid loss and circulatory collapse. Drowsiness and confusion are often seen along the development of psychosis associated with paranoid delusions, hallucinations and delirium. Finally, seizures, coma and death, usually due to shock, may ensue [6]. Cardiac manifestations include acute cardiomyopathy, subendocardial hemorrhages, and electro-cardiographic changes. The pathological lesions described in patients with rapidly fatal arsenic intoxication are fatty degeneration of the liver, hyperemia and hemorrhages of the gastrointestinal tract, renal tubular necrosis, and demyelination of peripheral nerves [6]. Chronic exposure to arsenic affects the gastrointestinal tract, circulatory system, skin, liver kidneys, nervous system and heart. There is clear evidence from epidemiological studies that exposure to inorganic arsenic increases the risk of cancer [7]. When exposure occurs by the oral route, the main carcinogenic effect is increased risk of skin cancer. In addition to skin cancer, increased risk of other internal tumors (mainly of liver, kidney, lung, and bladder) have been reported with arsenic exposure [8, 9]. *Table 1* presents a synopsis of clinical and pathological manifestations in acute and chronic arsenic poisoning [6].

Experimentally, arsenicals are fetotoxic and teratogenic in laboratory animals. The common developmental effects seen include malformations of the brain, urogenital organs, skeleton, ear, as well as small or missing eyes. Generally, these are only seen at doses that also result in maternal toxicity. *In vitro* experiments with many arsenicals have shown that they are powerful clastogens

Table 1. Clinical and pathological manifestations of acute and chronic arsenic poisoning

Organ/Tissue Level	Acute Effects	Chronic Effects
• Dermatologic	• Capillary flush • Contact dermatitis • Folliculitis • Hair loss	• Melanosis • Bowen's disease • Facial edema • Palmoplantar hyperkeratosis • Cutaneous malignancies • Hyperpigmentation • Desquamation
• Neurologic	• Hyperpyrexia • Convulsions • Tremor / Coma • Disorientation	• Encephalopathy • Headache • Peripheral neuropathy • Axonal degeneration
• Gastrointestinal/Hepatic	• Abdominal pain • Dysphagia • Vomiting • Blood/rice water diarrhea • Garlicky odor to breath • Mucosal erosions • Fatty liver • Cholangitis • Cholecystitis	• Nausea • Vomiting • Diarrhea • Anorexia • Weight loss • Hepatomegaly • Jaundice • Pancreatitis • Cirrhosis • Liver cancer
• Renal	• Tubular damage • Glomerular damage • Oligura • Uremia	• Nephritis • Proteinuria
• Hematologic	• Anemia • Thrombocytopenia	• Bone marrow hypoplasia • Anemia • Thrombocytopenia • Basophilic stippling • Karyorrhexis
• Cardiovascular	• Ventricular fibrillation • Tachycardia	• Arrhythmias • Pericarditis • Acrocyanosis • Raynaud's, and gangrene
• Respiratory	• Pulmonary edema • Bronchial pneumonia • Tracheobronchitis	• Cough • Pulmonary fibrosis • Lung cancer

in many cell types. Tests for genotoxicity have indicated that arsenic compounds inhibit DNA repair, and induce chromosomal aberrations, and sister chromatid exchanges [4]. There are also several epidemiological studies reporting an association between exposure to inorganic arsenic and increased risk of adverse developmental effects such as congenital malformations, low birth weight, and spontaneous abortion [10].

DOSE-RESPONSE ASSESSMENT

It has been demonstrated that the risk of arsenic intoxication increases as a function of exposure level and duration. The National Research Council of Canada reported that 9, 16 and 44% incidence of symptoms of arsenic poisoning are observed at drinking water arsenic concentrations of 50, 50-100, and >100 ug/L, respectively [11]. The frequencies of skin cancer associated with arsenic-contaminated water have been reported by the U.S. EPA as 0.26% and 2.14% at 290 and 600 ug/L, respectively. Also, cancer risks of 10^{-5}, 10^{-6}, and 10^{-7} have been estimated for drinking water containing 0.022, 0.0022, and 0.00022 ug As/L, or for eating aquatic organisms living in contaminated water containing 0.175, 0.0175, and 0.00175 ug As/L [12]. In a study of cancer prevalence in patients treated with Fowler's solution (1% potassium arsenite concoction), a dose-response relationship was found between the ingested dose of Fowler's solution and the incidence of skin cancer. In patients who ingested 200 to 800 mL of Fowler's solution (1.6 to 6 g of arsenic), the prevalence of skin cancer was 5 to 10%, while patients treated with the Fowler's solution at an equivalent arsenic dose of 7.6 g had a prevalence of 20% [13]. A strong dose-response relationship between arsenic concentration in drinking water and skin cancer mortality has been documented in the endemic area of blackfoot disease in Taiwan. Similar correlations have been reported between arsenic levels in drinking water and age-adjusted mortality rates for cancers of the lung, liver, bladder, kidney, and colon [14].

EXPOSURE ASSESSMENT

A very large number of people are exposed to arsenic chronically throughout the world. Exposure to arsenic occurs via the oral route (ingestion), inhalation, dermal contact, and the parenteral route to some extent. For most people, the diet is the largest source of arsenic exposure, with an average intake of about 50 ug per day. Intake from air, water and soil are usually much smaller, but exposure from these media may become significant in areas of arsenic contamination. People who produce or use arsenic compounds in occupations such as non-ferrous metal smelting, pesticide manufacturing and application, wood preservation, semiconductor manufacturing, or glass production can be exposed by substantially higher levels of arsenic [4, 6, 11, 14]. As stated earlier, different species of arsenic have different degrees of toxicity, with arsine and the trivalent arsenic causing the most injury. In addition to the exposure route and the arsenic species involved, the toxicity also depends on the exposure dose, frequency, duration, the biological species, age, and gender, as well as on individual susceptibilities, genetic and nutritional factors [4, 14].

RISK CHARACTERIZATION

A review of biological properties and toxic effects of arsenic indicates that this chemical is a systemic toxicant capable of causing a significant number of health effects including: cardiovascular disease, peripheral vascular disease, developmental effects, neurologic and neurobehavioral effects, diabetes, hearing loss, portal fibrosis of the liver, lung fibrosis, hematologic effects (anemia, leukopenia, and eosinophilia), and carcinogenic effects [14, 15]. Studies in Taiwan have reported that in areas where blackfoot disease is endemic, the standardized and cumulative mortality rates were significantly higher for cancer of the bladder, kidney, skin, liver, lung and colon. These studies have also pointed out that the lifetime risks for developing cancer due to an arsenic intake of 10 ug/kg/day were 0.012, 0.0043, 0.012, and 0.0042 in male residents for cancers of the lung, liver, bladder, and kidney, respectively [14]. The corresponding figures for female residents were 0.013, 0.0036, 0.017, and 0.0043 [14]. The International Agency for Research on Cancer, and the

U.S. EPA classify arsenic in Group 1/A-known human carcinogens [15]. The oral reference dose (RfD) is $3x10^{-4}$ mg/kg/day, and the cancer potency factor is 1.75 $(mg/kg/day)^{-1}$ [16].

RISK MANAGEMENT

Because of its high potential to cause adverse effects in exposed persons, a number of regulations and guidelines have been established for various inorganic and organic forms of arsenic by international, federal, and state agencies. In 1976, the permissible limit for arsenic in drinking water (maximum contaminant level) was fixed at 50 ug/L [17]. In 1999, the National Research Council of the National Academy of Science recommended a standard of 10 ug/L [18]; a standard that has just been adopted by the Bush Administration. The World Health Organization?s tolerable daily intake for inorganic arsenic is 2 ug/kg BW [19]. The action level for arsenic in the air is 5 ug/m^3. The permissible exposure limit-total weighted average (PEL-TWA) is 10 ug/m^3 for inorganic arsenicals, and 500 ug/m^3 for organic arsenicals [20]. British Anti-Lewisite (BAL or 2,3-Dimercaptopropanol) has been used to treat acute dermatitis, and the pulmonary symptoms associated with arsenic exposure. However, because of the side effects associated with BAL, other agents such as sodium 2,3-dimercapto-1-propane sulfonate (DMPS or Dimaval), and dimercaptosuccinic acid (DMSA) are being tested for the chelation therapy of arsenic poisoning [5]. The atomic absorption spectrophotometry (AAS) appears to be the most common analytical method used for arsenic analysis; more sophisticated techniques, however, such as HPLC-FGAAS, and ICP-MS, are highly valuable when arsenic speciation is required [1]. Removal of arsenic from drinking water by appropriate technologies is one of the most important control and management strategies. Several treatment methods including chemical precipitation (coagulation processes), ion exchange, reverse osmosis/electrodialysis, use of activated alumina or carbon, and oxidation, have therefore been recommended for arsenic removal in water [21].

ACKNOWLEDGMENTS

This research was financially supported by a grant from the National Institutes of Health (No. 1G12RR13459). The author thanks Dr. A. Mohamed, Dean of the JSU School of Science and Technology for his technical advise on this project.

REFERENCES

1. Tchounwou P.B., Wilson B., Ishaque A. Important considerations in the development of public health advisories for arsenic and arsenic-containing compounds in drinking water. *Rev Environ Hlth*, 1999, 14, 1-19.
2. NAS. Arsenic. National Academy of Science. Washington D C, 1977.
3. Rousselot P., Laboume S., Marolleau J.P., Larghero T., Noguera M.L., Brouet J.C., Fermand J.P. Arsenic trioxide and melarsoprol induce apoptosis in plasma cell lines and in plasma cells from myeloma patients. *Cancer Res*, 1999, 59, 1041-1048.
4. ASTDR. Toxicological Profile for Arsenic TP-92/09. Agency for Toxic Substances and Disease Registry. Center for Disease Control, Atlanta, GA, 1993.
5. Goyer R.A. Toxic effects of metals. In: Klaassen CD ed. Cassarett & Doull=s Toxicology-The Basic Science of Poisons. McGraw Hill. New York, NY, 1996; 691-736.
6. Gorby MS. Arsenic in human medicine. In: Nriagu JO, ed. *Arsenic in the Environment; Part II: Human Health and Ecosystem Effects*. New York, NY: John Wiley & Sons, Inc., 1994, 1-16.
7. Wu M.M., Kuo T.L., Hwang Y.H. Dose-response relation between arsenic concentration in well water and mortality from cancers and vascular diseases. *Am J Epidemiol*, 1989, 130, 1123-1132.

8. Tseng W.P., Chu H.M., How S.W., Fong J.M., Lin C.S., Yeh S. Prevalence of skin Cancer in an endemic area of chronic arsenicism in Taiwan. *J Natl Cancer Inst*, 1968, 40, 453-463.
9. Chen C.J., Chen C.W., Wu M.M., Kuo T.L. Cancer potential in liver lung, bladder and kidney due to ingested inorganic arsenic in drinking water. *Br J Cancer*, 1992, 66, 888-892.
10. Aschengran A., Zierler S., Cohen A. Quality of community drinking water and the occurrence of spontaneous abortion. *Arch Environ Hlth*,1989, 44, 283-290.
11. NRCC. Effects of arsenic in the environment. National Research Council of Canada. *Natl Res Counc Can Publ*, 1978, 1-349.
12. U.S. EPA. Ambient Water Quality Criteria for Arsenic. EPA 400/5-80-021. Environmental Protection Agency. Washington DC, 1980.
13. Fierz U. Catamnestic investigations of the side effects of therapy of skin diseases with inorganic arsenic. *Dermatologia*, 1965, 131, 41-58.
14. Chen C.J., Lin L.J. Human carcinogenicity and atherogenicity induced by chronic exposure to inorganic arsenic. In: Nriagu JO, ed. *Arsenic in the Environment; Part II: Human Health and Ecosystem Effects*. New York, NY: John Wiley & Sons, Inc., 1994, 109-131.
15. IARC. Monographs on the Evaluation of Carcinogenic Risks of Chemicals to Humans. Supplement F. Overall Evaluation of Carcinogenicity. International Agency for Research on Cancer. World Health Organization. Lyon, France, 1987, 29-57.
16. IRIS. Integrated Risk Information System. U.S. EPA. Washington, DC, 1992.
17. U.S. EPA. National primary drinking water regulations. *Fed Regist*, 1985, 50, 46931-47022.
18. National Research Council. Arsenic in drinking water. 2001 update. On line at: http://www.nap.edu/books/0309076293/html/
19. WHO. Guidelines for Drinking Water Quality. World Health Organization, Geneva, Switzerland, 1984.
20. OSHA. Occupational Safety and Health Administration. *Fed Reg*, 1989, 54, 2332-2335.
21. Jekel MR. Removal of arsenic in drinking water treatment. In Nriagu JO ed. *Arsenic in the Environment, Part I: Cycling and Characterization*. New York, NY: John Wiley & Sons, Inc. 1994, 119-132.

*Metal Ions in Biology and Medicine: vol. 7.*Eds L. Khassanova, Ph. Collery, I. Maymard, Z. Khassanova, J.-C. Étienne.
John Libbey Eurotext, Paris © 2002 pp. 371-374.

Safety assessment for children from beryllium contained in atmospheric dust

Mayorova Oxana A., Ginzburg Lev N.

Institute of Mineralogy, Geochemistry & Crystal Chemistry of Rare Elements (IMGRE), 15 Veresaeva str., 121357, Moscow, Russia

ABSTRACT

Environment pollution in Moscow City with plenty of pollution sources is the real danger for people health especially for children health. The study of environment pollution influence on human organism and people health was provided in South-Eastern Administrative District (SEAD) of Moscow City. The atmospheric air of all environment mediums in cities affects directly the human organism. The hazard quotient (HQ) from Be intaking in human organism with inhaled dust was assessed using snow dust sampling data. It is more preferable to use the snow sampling data instead of atmospheric air sampling data in assessment of environment state. Beryllium contained in dust intaking in human organism by inhalation route is dangerous for population health on 65% of SEAD territories. The hazard quotient from inhaled Be correlates with Be concentration in children hair. From the medical-ecological point of view human is the final link of metals migration in environment. Negative influences of man-made geochemical load on human organism were revealed with the help of abnormal contents of metal in children hair.

INTRODUCTION

South-Eastern Administrative District (SEAD) of Moscow City was chosen as a model for study of environment pollution influence on human organism and people health. District area is more than 100 km^2. Population is about 1 million. More than 30% of the territory is occupied by industrial objects, 40% of the population work at them. There are factories of metallurgical, oil-processing, chemical etc. industry, aeration stations and filtration fields on the territory of district. The main pollution sources are factories and road transport.

The beryllium influence on children health in SEAD of Moscow is considered in the manuscript. The main sources of Be discharges is road transport, oil-processing, metallurgical and other factories. The main stable form of Be in atmospheric air is BeO [1]. Be has toxic and probable carcinogen influence on the organism. Be and its compounds have ability to penetrate into all organs, cells and their nucleuses, to injure cells membranes. Penetrating into cell nucleuses Be reduces activity of synthesis DNA ferments etc. Be slows down transport of Mg and Ca, and also reduces activity of ferments enlivened by Mo. Be causes the chronic beryllium disease (CBD) in human organism. Initial symptoms of early cases of CBD typically include dyspnea, cough, fatigue, weight loss, chest pain etc. CBD is characterized by the symptoms including bibasilar crackles, clubbing of the fingers and skin lesions, heart failure, and an enlarged liver or spleen. Prominent diagnostic findings are diminished vital capacity, diffuse infiltrates, and radiographically visible hilar adenopathy. Fibrosis occurs at late stages of the disease. Granulomatous inflammation also has been reported in extrapulmonary organs, such as extrapulmonary lymph nodes, skin, liver, spleen, kidney, bone, myocardium, central nervous system, and skeletal muscle [2]. The main Be exposure route to human organism is inhalation, other routes are of smaller importance [3].

MATERIALS AND METHODS

The study of atmospheric air pollution with Be was carried out on the data of snow cover sampling in SEAD of Moscow. Sampling scale is 1:50.000. The dust fraction with the size less than 0.25 mm was analyzed by emission spectrum analysis. The average content of Be is 4.0 mg/kg, standard deviation is 2.4 mg/kg.

Average daily concentration of Be in atmospheric dust in the winter period [mg/m^3]: **$?_{Be}=(P_n*C)/(h*24)$**, where **P_n** - winter dust load, [$mg/(m^2*day)$]; **C** - Be concentration in dust, [%]; **h** $\approx$ 50 m - height of the lower layer of troposphere which contains the main dust quantity; **24** - quantity of hours in day, because the dust particles are in air during 1 hour, and then settle on surface.

For the hazard assessment for children health from influence of Be, intaking in organism with inhaled dust, hazard quotient (**HQ**) was accounted: **HQ=E/RfC**, where **E** - exposure is quantified as the amount of an agent available at the exchange boundaries of the organism (lungs), [mg/m^3]; **RfC** - is an estimate of a daily inhalation exposure of human population (including sensitive subgroups like children) that is likely to be without an appreciable risk of deleterious effect during a lifetime. RfC Be = 0.00002 mg/m^3 [2].

The data of official medical statistics and individual medical documentation of children from SEAD of Moscow were used. Age, residing terms of family in the SEAD, and health conditions of a child were taken into account in selective formation of children group for research of Be contents in hair. Hair was sampled from healthy children without deviations in physical and psycho-emotional development (control group), and patients with various chronic diseases outside of aggravation stage. The spatial distribution of control group covered all the territory of SEAD. All children were selected from families constantly living in the SEAD territory not less of 5 years. Hair samples of 215 children were investigated. The Be content in a hair was analyzed by local laser emission method. The interrelation of the Be contents in hair with the Be contents in atmospheric dust was revealed by cartographical method.

RESULTS

Average value C_{Be} in atmospheric dust is equal to 0.00004 mg/m^3. Increased daily concentration of Be are typical of zones of factories emissions influence of oil-processing, metallurgy and mechanical engineering, and also territories located along large automobile highways.

The value of exposure calculated for estimation of toxic inhalation influence of Be on child organism (with the averaging period of exposure 6 years) varies from 0.00001 to 0.0002 mg/m^3 with average value 0.00003 mg/m^3.

The HQ value exceeding 1 is typical of 65% of territory. Staying of people in territories with HQ>1 is dangerous for their health *(fig. 1)*. The maximal excesses of HQ value (from 3 to 9 times) are typical of zones of factories emissions influence of oil-processing, metallurgy and mechanical engineering, and also territories located along large automobile highways.

The average Be concentration in hair of healthy children is equal to 0.062 mg/kg, standard deviations being 0.055 mg/kg. The excess of the average Be concentration in hair is characteristic for children with the diagnoses: blood diseases, bronchial asthma. The increased Be concentration in hair are typical of children living in zones of factories emissions influence of oil-processing, metallurgy and mechanical engineering, and also territories located along large automobile highways *(fig. 2)*.

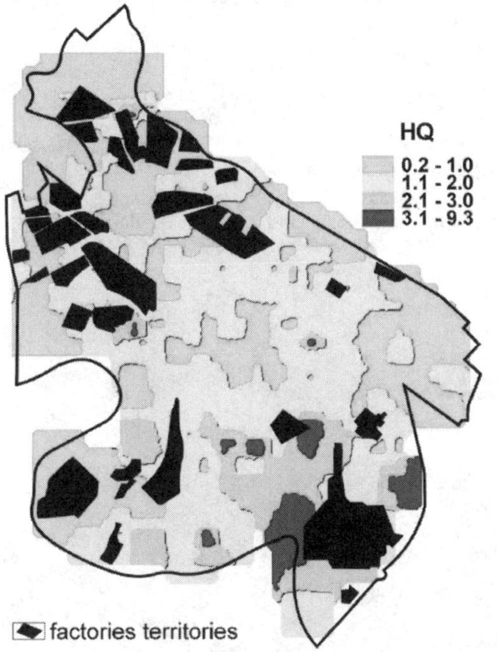

Fig. 1. The map-scheme of HQ in SEAD of Moscow City scale 1:20000

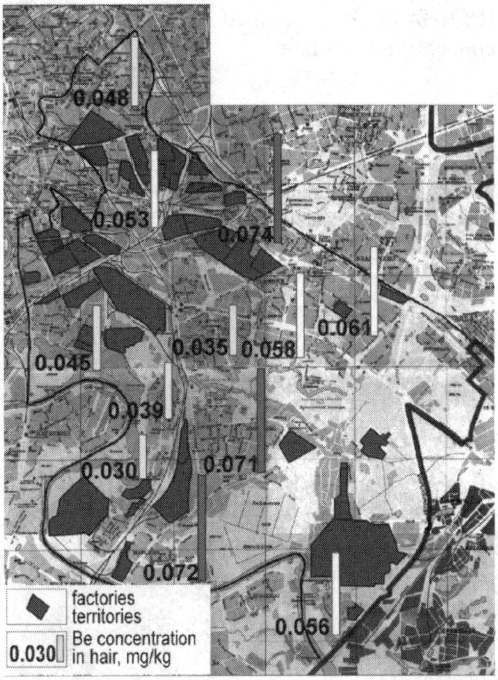

Fig. 2. The map-scheme of Be concentration in children hair in SEAD of Moscow scale 1:20000

DISCUSSION

The main influence of environment on a human in cities is by inhalation route. However direct study of Be concentration in atmospheric air is problematic. The air is a very dynamical transporting medium. However all metal emissions from the factories and transport in the winter period are accumulated in snow cover. The anomalies of increased concentration of elements in snow cover correlate with location of pollution sources better than with anomalies in atmospheric air. The snow sampling data are more reproducable than atmospheric air sampling data. Therefore pollution research of atmospheric air and assessment of average daily dozes of metals intaken to human organism by inhalation route are more preferable to be studied using the data of snow cover sampling.

Hair is one of human organism biosubstances accumulating metals contained in atmospheric air. Hair sampling in comparison with other methods is most simple and painless. The microelements concentrations in hair substantially correspond to their accumulation by other human organism biosubstances. Hair quickly accumulates metals and keeps them for a long time. The results of researches have shown, that the hair sampling data reflect the load of environment on human organism and correspond to assessment of environment hazard for the population.

CONCLUSIONS

Application of various approaches to estimation of environment state and influence of environment on children health in SEAD of Moscow has shown that the ecological norms are exceeded. Spatial correlations were revealed in SEAD of Moscow between: halos of increased hazard quotient

(HQ) from Be, contained in inhaled air dust, halos of children inhabitance with increased Be concentration in hair.

REFERENCES

1. Ivanov V.V. Ecological geochemistry of elements. 1994, v.1, p. 304.
2. http://www.epa.gov/iris/subst/0012.htm
3. Keyzer C.A. Beryllium. Toxicology, hygiene, prevention, diagnostics and cure of Be-damage. 1985, p. 160.

Metal Ions in Biology and Medicine: vol. 7. Eds L. Khassanova, Ph. Collery, I. Maymard, Z. Khassanova, J.-C. Étienne.
John Libbey Eurotext, Paris © 2002 pp. 375-381.

The low Volga basin's microorganisms capable to accumulate iron compounds

N. V. Chertov

Astrakhan State Technical University, 16 Taticheva Street, 414025 Astrakhan, Russia

ABSTRACT

Processes of bacteria interaction with iron compounds have been attracting microbiologists attention for more than 100 years. It was established not long ago that several bacteria are able to accumulate ferromagnetic minerals - magnetite, greigite in environment and also in their cells - the last case stipulate their capability to magnetotaxis. The purpose of the research became the investigation of natural water basins in the area of the Low Volga for availability of magnetotactic bacteria (MTB). The following tasks were established: to study physical and chemical parameters that help compounds' biomineralization; to investigate the species content of water and silt microecosystems that have magnetic characteristics; to single out accumulating culture of MTB, determine its magnetic characteristics and establish the species belonging of determined MTB. In the research the following methods were used: photometrical, atomic and absorbing method, chromate-mass-spectrometric, phylogenetic and also analytical scanning and enlightening electronic microscopy. Investigation showed that there can be iron compounds' biomineralization in the Low Volga basin in the condition of low content of diluted organic substance and higher iron concentration both in water and silt and in low alkaline environment. Silt microorganisms with high parameters of magnetic repletion are presented by 23 species of bacteria that belong to 19 families. Accumulating culture of MTB was determined in the model system of Lake Belyachnoye that is characterized by low alkaline environment, low content of diluted organic substance, higher content of general iron and also by high magnetic silt repletion. It was established that the determined MTB have magnetic characteristics: capability to magnitotaxis and biomineralization of intercellular iron compounds. The carried phylogenetic analysis proved that the investigated bacteria belongs to (-subclass of proteobacteria. Inside this group the bacteria was most close to representatives of *Magnetospirillum* family showing not less than 96% similarity of sequences. The investigated bacteria belongs to phylogenetic subclass that contains the typical species *Magnetospirillum magnetotacticum*. The level of similarity in gene sequence 16s rRNA of determined bacteria and typical species *Magnetospirillum magnetotacticum* was quite high (99.6%), which proves its belonging to the species *M. magnetotacticum*.

INTRODUCTION

The processes of bacteria interaction with iron compounds have been attracting microbiologists' attention for some time already. The iron necessity for most organisms is not random. It is determined by the metal peculiarities, which allow this metal to form a big amount of life-important iron compounds. The capability of living organisms to create different minerals has been known for a long time. Complexes consisting of organic and mineral substances, created either by living organisms or with the help of their participation, are called biominerals [Korago, 1992], and the process of their formation that takes place with the help of living organisms and unites a wide rank of mechanisms, is called biogenesis [Lowenstam, 1981].

Recently it has been established, that some bacteria are able to accumulate ferromagnetic minerals in the environment or in the cells. The ferromagnetic minerals - magnetite, greigite, pirrotin - stipulate bacteria capability to magnetotaxis [Blakemore, 1975; Bazylinski et al., 1988; Kirshvinka et al., 1989; Heywood et al., 1990; Sparks et al., 1990]. These minerals can be formed as a result of two different biogenesis ways: "Bioindicated mineralization" (BIM) [Lowenstam, 1981] and "Biocontroled mineralization" (BCM) [Lowenstam, 1981; Mann, 1986]. Magnetotactic bacteria (MTB) are capable to in-cell synthesis of magneto regulating iron compounds (BCM way) and are capable to react on geomagnetic field's changes moving along its power lines (magnetotaxis) [Blakemore, 1975, 1979; Bazylinski et al., 1997]. Biominerals, created in BCM process, can be determined by a strictly specified crystallographic shape, cleanness and specific structure. They can act as compass arrows making determined movements along the lines of Earth's magnetic field [Frankel, Blakemore, 1980]. Nowadays a few MTB species singled out as a pure culture are known.

Despite considerable interest that biogenic minerals represent from practical point of view, there are a lot of unclear moments that deal with physiological aspects of their formation, environmental topics related to MTB and roles of separate representatives of physiological groups of microorganisms that help to carry this process out in nature.

The main purpose of the research became the investigations of the Low Volga Basins to determine the presence of MTB. According to that the following goals were addressed: to study physical and chemical parameters that help iron compounds' biomineralization in the model systems based on natural material from the Low Volga basins; to investigate the species content of water and silt microecosystems that have magnetic characteristics; to single out accumulating culture of MTB, determine its magnetic characteristics and establish the species belonging of determined MTB.

MATERIALS AND METHODS

As materials the investigation results of model systems were used. Model systems were based natural material collected from the Low Volga basins: the Volga and Akhtuba rivers, north of Caspian Sea, Lake Belyachnoye, small tributaries Uliankin and Sazaniy. The water and bottom samples (sand, silt etc) were collected in the northern part of natural water basins, which are exposed to human activities, in combination 4:1 and then were placed in hermetically sealed flasks (0.7L.), that were kept without light to create gradient oxidizing and reducing conditions. To the outside of one the flask side at the phase division border "silt-water" a magnet by a south pole was attached to receive a population of MTB [Mann et al., 1990; Verhovtseva, 1992; Filina, 1998]. In general, 24 model systems were investigated.

The measuring of environmental reaction data (pH) and oxidizing and reducing potential (Eh) were done on pH-meter pH121. The content of diluted organic substance (DOS) was measured photometrically [Pannikov et al., 1988]. General iron amount was determined by atom and adsorption spectroscopic methods by Ivanov and Lerner (1974). The measuring of magnet parameters (χ_{par} - paramagnetic component of magnetic receptivity and I_{mar} - magnetizing repletion) of natural silt and bacteria biomass samples were done by using Faraday-Seksmith method [Babanin et al., 1995]. The measuring of fatty and acidic content of silt microsystems' summary biomass was done on chromate-mass-spectrometer QP-2000 (Shimadzu, Japan). To analyze the minor lipid components Folch method of fragment graphical mass was used [Folch et al., 1957]. Quantity and quality of microbe systems was analyzed by "EXCEL" with the help of developed counting algorithm [Turova, Dsipov, 1996]. The initial culturing of MTB took place in synthetic environment CMT-2 [Schleifer et al., 1991] modified by N. U. Filina (1998). Nutrition environment developed by Maratea, Blakemore (1981) was used to get the pure MTB culture. Electronic and microscopic analysis of in-cell crystals was done by analytical scanning and enlightening electronic microscopic system [Gritsaenko et al., 1969; Scott, Love, 1971; Sergeeva, 1977]. To determine phylogenetic MTB placement, its 16s ribosome gene sequence analysis was done. Phylogenetic tree of studied bacteria was built using methods from program package TREECON [Van de Peer, De Wachter, 1994]. Statistic analysis was done on Microsoft Excel 7.0 for Windows 95.

RESULTS AND DISCUSSION

During the arrangements of model systems (MS) the Low Volga Basins were chosen according to different hydrological, hydrochemical and hydrobiological parameters: North of Caspian Sea (microecosystems (1-4), running water - the Volga river (5-8), Akhtuba river (9-12), stable water - Belyachnoe lake (13-16), small tributaries Uliankin (17-20) and Sazaniy (21-24). As physical and chemical MS parameters that stipulate the presence of MTB in microbe systems were determined.

The analysis of data showed that by DOS, iron content, pH, χ_{par}, I_{mar} in water and silt it is possible to determine MS that are capable to form and accumulate iron containing minerals. In those systems together with neutral- low acid environmental reaction considerable amount of iron was found in the water-up to 1,7 mg/l and in silt - up to 7400 mg/l. Investigations of magnet parameters showed that high I_{mar} is usually peculiar to systems N° 9, 15, 18, 22 where its values constitute 59,7; 72,3; 67,8; 66,2 Gs(sm^3/g accordingly. Thus, in the condition of low organic substance amount and higher iron content with low alkaline values of pH MS can lead to formation and accumulation of magnetic regulating iron compounds in MTB cells. In general, it leads to "magnetizing" MS bottom silt and "attraction" of magnetic fraction to magnet.

Based on experimental data analysis it was established, that heterogeneity of nutrition types was a feature of studied microsystems. The differences of quantity and diversity of MS microsystems were established at that point. The most impressive can be called those MS microsystem of the Akhtuba river where 26 bacteria species were found, including 3 actinomicetes, 1 yeast species, 4 micromicetes species, protozoa and microalgae. As the least impressive the MS microsystem of Uliankin tributary can be approved, which includes 7 bacteria species belonging to 6 families. Micromicetes and microalgae were found in all studied MS but yeast and protozoa were not found in Uliankin and Sazaniy tributaries of MS. In all described systems the bacteria content of micromicetes and actinomicetes was practically similar. But at the same time, the content of associates and quantity of particular species in MS of Belyachnoye Lake is quite different from other MS.

The community of microorganisms from magnet area MS N° 15 of Belyachnoye Lake's silt was presented by 23 bacteria species belonging to 19 families. It contains micromicetes (*Fusarium sp., Aspergillus sp., Mucor sp., Penicillum sp.*); similar to actinomicetes (Maduromycetes); budding yeast fungi (*Caulobacter sp.*), algae and protozoa. Among associates of microsystem there are saprophyte organisms (*Candida albicans, Fusarium sp., Acetobacter diazotrophicus, Bacillus subtilis, Bacteroides ruminicola, Clostridium difficile, Cl. perfringens, Cl. putrificum, Corynebacterium sp., Cytophaga sp., Desulfovibrio vulgaris, Eubacterium sp., Flavobacterium sp., Leptothrix sp.*, Maduromycetes, *Pseudomonas sp., P. cepacia, Rhodococcus sp., Sphingobacterium sp.*), enterobacteria (f. *Klebsiella*), oxygenic photoautotrophic cyanic bacteria (*Gloebacter sp.*) and facultative lithoautotrophic bacteria (*Nitrobacter sp.*). From quantity point of view the dominating species are saprophyte H_2S-creating bacteria *Cl. putrificum* and budding bacteria of *Caulobacter* family that are capable to carry out the nitrification of phase 2. The microsystem includes obligate aerobes (*A. diazotrophicus, Caulobacter sp.*), microaerophylic bacteria (*B. ruminicola*), strict anaerobes (*D. vulgaris, Eubacterium sp.*). In the microsystem there are: proteolythic (*B. subtilis, Cl. difficile*), nitrogen fixing (*A. diazotrophicus*), denitrificating (*Corynebacterium sp., Eubacterium sp.*, f. *Klebsiella, Pseudomonas sp., P. putida*), nitrificating (*Caulobacter sp., Nitrobacter sp.*), cellulose destroying (*B. ruminicola, Cytophaga sp.*), sulfate reducing (*D. vulgaris*) bacteria. It is known that increased amount of saprophyte bacteria producing H_2S as living product, helps biomineralization process of iron compounds.

With the help of gas chromatographic-mass-spectrometric methods it was established that there is a presence of hemorganic heterotrophic bacteria *Aquaspirillum sp.* (synonym to *Magnetospirillum sp.*), capable to biomineralization of iron compounds. Those bacteria were not found in other systems. Bacteria of *Aquaspirillum* family belong to aerofilic and microaerofilic species capable to denitrification. Besides, bacteria *Leptothrix sp.*, that are capable to form specific cover filled up with oxidized iron and manganese.

The investigations of dry biomass of microsystems' cells showed that high values of magnetic parameters were pointed out in MS N° 15 of Belaychnoye Lake, which contained bacteria of *Aquaspirillum* family capable to accumulate iron compounds in cell. Thus, the value χ_{par} that shows concentration of paramagnetic compounds in the sample, for this particular MS it was 5-8 times as high as the values in MS based on the natural material of Akhtuba river and Uliankin and Sazaniy tributaries, and constitutes $28{,}25\times10^{-6}$ sm^3/g. The value of another magnetic parameter - I_{mar} allows to judge about relative amount of magnetic regulating compounds in the substance. The results showed that I_{mar} of cell biomass of MS?15 is 3-4 times as high and constitutes $43{,}16 \leftrightarrow 10^{-3}$ $Gs \leftrightarrow sm^3/g$. From this particular MS the MTB pure culture was obtained. After a series of sequent replantations of obtained accumulating culture on nutrient environment Maratea and Blakemore (1981), a pure MTB culture was extracted, which cells had specific shape with round ends, 10-35 mcm long, 3.05-5.5 mcm wide, very dynamic and contain in-cell elements (from 18 up to 21 cristals), forming a chain. The possibility of iron compound's biomineralization was studied to determine the magnetic qualities of bacteria pure culture.

Electronic and microscopic investigations of in-cell crystals of the samples showed the presence of two types of iron formations, which can be determined as silicates and iron oxidizes. The silicate mineral iron fractions are located as accumulations; their spectrum is presented as several peaks of iron, silicon and carbon. The presence of sodium, potassium, phosphorus, nickel and sulphur are found in very small amounts. Inconsiderable amount of sulphur, for sure, is stipulated by organic mass content, which allows to conclude that there is no iron in the studied form of sulphide sample. The sizes of most oxidized iron fractions are from 2 up to 3 mcm, sometimes they reach up to 5 mcm. The shape of the fractions is round but sometimes they have angles reminding edges. The content spectrum of such fractions represented by iron, oxygen, sodium and carbon peaks.

The studied on the scanning enlightening microscope sample was extracted from surface of graphite disk, transferred into test-tube and processed be ultrasound in distillate water. Firm round formations with the diameter from 1.5 up to 3 mcm, most part of which is, probably, electronically amorphous, are found in large amounts in the samples. Some of those determined fractions form unregulated pointed microdiffraction pictures and are supposed to belong to iron silicates. From peripheral departments of fractions circled and discreet - circled microdiffraction pictures conforming to magnetite: $FeO\times Fe_2O_3 \approx Fe_3O_4$ and gematite: Fe_2O_3 in the form of "balls" and "disks" were received. It was noticed that not all of the "balls" are going into diffractions. This event provides the opportunity to suppose that even in the cases when diffraction picture was received from "ball's" peripheral part, its central part doesn't remain crystallized and, thus, the initial form of iron oxidize is electronically amorphous, unstructured or partially non-crystallized as a type of thin dispersion magnetite. In general, the magnetite qualities of extracted MTB are stipulated by biomineralization of in-cell iron compounds in the form of magnetite (Fe_3O_4) or gematite (Fe_2O_3). It is proved, that magnetite is the initial mineral form of in-cell compounds. After oxidizing it becomes gematite. This phenomenon is known in nature and has been discussed in literature [Chuhrov et al., 1975].

To establish taxonomic status of obtained bacteria Sengera (1977) method was used. As a result a considerable part of gene sequence of 16S rRNA (1408 nucleotides) of obtained group, this was correlating with positions from 47 up to 1510 on *E.coli* nomenclature [Brosius et al., 1978].

Phylogenic analysis proved that the studied group belonged to α-subclass of proteobacteria, to phylogenic group which unites *Rhodospirillum* family and close families to be exact. Inside the larger group the particular group was most close to representatives of *Magnetospirillum* family showing about 96% of similarity. With the other families' representatives of this phylogenic group the level of similarity didn't increase 92%. Nowadays two species belong to *Magnetospirillum*: *M. magnetotacticum* and *M. gryphiswaldense* [Schleifer et al., 1991]. These two species constitute cluster on the built phylogenic tree *(pic.1)*.

The studied species with the given name *Magnetospirillum magnetotacticum*, species *N7*, is included into phylogenic sub-cluster containing type species *M. magnetotacticum (pic.1)*. Several unidentified magnetospiral species also belong to this phylogenic cluster [Burgess et al., 1993]. The level of sequence similarity of gene 16S rRNA species *N7* and type species *M. magnetotacticum* is rather high (99.6%), that allows considering it to be *M. magnetotacticum* [Stackebrandt, Goebel, 1994].

Thus, the investigations on establishing taxonomy status of determined bacteria showed that they belong to magnetospirals. The phylogenic analysis proved that the new magnetotactic bacteria group - *M. magnetotacticum* species *N7*.

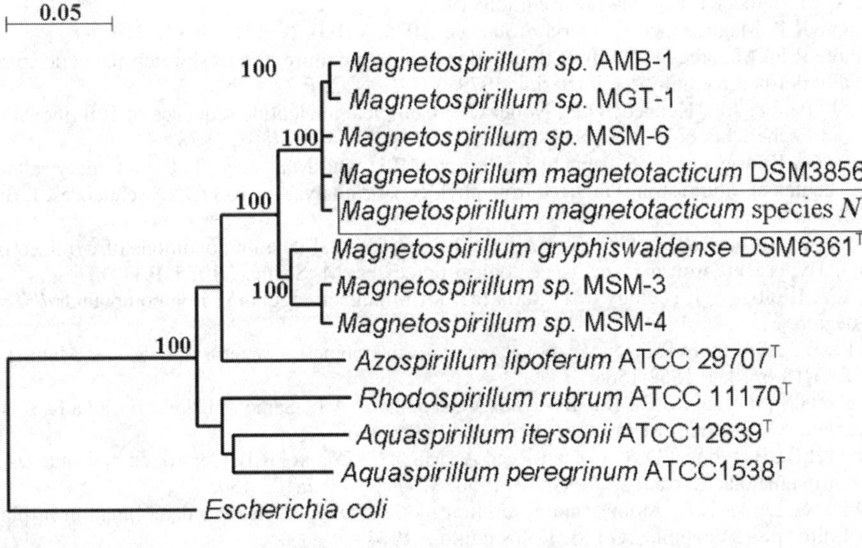

Picture 1. Phylogenic tree of magnetospirals and several close to them bacteria families

CONCLUSIONS

1. The analysis of physical and chemical parameters of studied MS showed that they don't change within wide ranks and don't always assist biomineralization process of iron compounds due to low concentration of general iron amount both in water and silt and due to low magnetizing ability of silt.

2. Microorganisms community obtained from natural material of Belyachnoye Lake, which is capable to provide high magnetic parameters, consists of 23 bacteria species including the representatives of *Aquaspirillum* family (synonym to *Magnetospirillum*), capable to in-cell biomineralization of iron compounds.

3. The process of iron compound biomineralization happens only in Belyachnoye Lake stipulated by low content of diluted organic substance, increased iron concentrations (0,5-1,7 mg/l in

water and 1205-7400 mg/kg in silt), higher magnetic parameters (χ_{par} 54,2 ↔ 10^{-6} sm^3/g, I_{mar}72,3 ↔ 10^{-3} Gs ↔ sm^3/g) and presence of MTB.

4. The MTB capability to in-cell biomineralization of iron compounds on the example of magnetite (Fe_3O_4) and gematite (Fe_2O_3) was shown.

5. As a result of phylogenic analysis the obtained MTB were identified as *Magnetospirillum magnetotacticum* species *N7*.

REFERENCES

1. Babanin V.F., Truhin V.I., Karpachevskiy L.O., Ivanov A.V., Morozov V.V. Soil magnetizings. Yaroslavl: YSTU edition. 1995. 223 p.
2. Bazylinski D.A., Moskowitz B.M. Microbial biomineralization of magnetic iron minerals: microbiology, magnetism and environmental significance // Review in Mineralogy. Ed. By Jill Banheld, K.N. Nealson. V.35. P.181-223.
3. Bazylinski D.A., Frankel R.B., Jannasch H.W. Anaerobic magnetite production by a marine magnetotactic bacterium // Nature. 1988. V.334. N° 6182. P.518-519.
4. Biogenic magnetite and magnetic reception. New in biomagnetizing: 2v. // Editors: J. Kirshvinka, D. Johns, B. Mcfadden. Moscow: Mir edition. 1989.
5. Blakemore R.P. Magnetotactic bacteria // Science. 1975. V.190. N° 4212. P.377-379.
6. Blakemore R.P., Maratea D., Wolfe R.S. Isolation and pure culture of a freshwater magnetic spirillum in chemically defined medium // J. Bacteriol. 1979. V.140. N° 2. P.720-729.
7. Brosius J., Palmer J.L., Kennedy H.P., Noller H.F. Complete nucleotide sequence of 16S ribosomal RNA gene from Escherichia coli. Proc. Natl. Acad. Sci. USA 75, 4801-4805, 1978.
8. Burgess J.G., Kawaguchi R., Sakaguchi T., Thornhill R.H. and Matsunaga T. Evolutionary relationships among Magnetospirillum strains inferred from phylogenetic analysis of 16S rDNA sequences. J. Bacteriol. 175 (20), 6689-6694, 1993.
9. Chuhrov F.V., Ermilova L.I., Gorshkov A.I. et al. Experimental data on conditions of oxidized iron formation // Hypergenic iron oxidizes in geological processes. M. Science. 1975. P.11-33.
10. Filina N.Y. Biology and ecology of bacteria that form magnetoregulating iron compounds // Thesis for Ph.D. degree. M. 1998. P.24.
11. Frankel R.B., Blakemore R.P. Navigational compass in magnetic bacteria // J. Magn. Magnet. Mater. 1980. V.15-18. p.III. P.1562-1564.
12. Gricaenko G.S., Zvyagin B.B., Boyarskaya R.V., Gorshkov A.I., Samotoin N.N., Frolova K.Y. Methods of electronic mineral microscopy // M. Science. 1969. 309p.
13. Heywood R.B., Bazylinski D.A., Garratt-Reed A., Mann S., Frankel R.B. Controlled biosynthesis greigite (Fe_3S_4) in magnetotactic bacteria // Nature. 1990. V.77. N° 11. P.133-136.
14. Ivanov D.,N., Lerner L.A. Atomic and absorbtion method of microelement determination in plants and soils // Editor prof. Vazhinina G.I.M. Kolos edition. 1974.
15. Korago A.A. Introduction to biominerology. St. Petersburg: Nedra edition. 1992. 280 p.
16. Lowenstam H.A. Minerals formed by organisms // Science. 1981. V.211. P.1126-1131.
17. Mann S. On the nature of boundare-organised biomineralization // J. Inorg. Chem. 1986. N° 28. P.263-971.
18. Mann S., Sparks N.H.C., Board R.G. Magnetotactic bacteria: microbiology, biomineralization, paleomagnetism, and biotechnology // Advan. In Microb. Physiol. 1990a. V.31. P.125-181.
19. Maratea D., Blakemore R.P. Aquaspirillum magnetotacticum sp. nov. a magnetic spirillum // Internat. j of System. Bacteriol. 1981. V.31. N° 4. P.452-455.
20. Quantity electronic and zond microanalysis. Editor Scott V., G. Love // M. Mir. 1971. 260 p.
21. Schleifer K.H., Schuler D., Spring S., Weizenegger M., Amann R., Ludwig W., and Kohler M. The genus *Magnetospirillum* gen. nov. description of *Magnetospirillum gryphiswaldense* sp. nov. and transfer of *Aquaspirillum magnetotacticum* to *Magnetospirillum magnetotacticum* comb. nov. *Syst. Appl. Microbiol.*, 1991, 14, P.379-385.
22. Sergeeva N.E. introduction to electronic microscopy of minerals // M. MSU edition. 1977. 144 p.
23. Sparks N.H.C., Mann S., Bazylinski D.A., Lovley D.R., Jannasch H.W., Frankel R.B. Structure and morphology of magnetite anaerobically-produced by a marine magnetotactic bacterium and a dissimilatory iron-reducing bacterium // Earth Planet Sci. Lett. 1990. V.98. P.14-22.
24. Stackebrandt E., Goebel B.M. Taxonomic note: A place for DNA-DNA reassociation and 16S rRNA

sequence analysis in the present species definition in bacteriology.// Int. J. Syst. Bacteriol. 1994, V. 44. N 4. P.846-849.
25. Turova E.S., Osipov G.A. The studies of microbe community, active in bio transformation of iron minerals in caoline // Microbiology. 1996. V.65. N° 5. P.682-689.
26. Van de Peer Y., De Wachter R. TREECON for Windows: a software package for the construction and drawing of evolutionary trees for the Microsoft Windows environment // Comput. Applic. Biosci. 1994. V. 10. P.569-570.
27. Verhovceva N.V. Formation of magnetite by bacteria and magnetotaxis //Microbiology success. M.: Science. 1992. V.25. pp. 51-79.

*Metal Ions in Biology and Medicine: vol. 7.*Eds L. Khassanova, Ph. Collery, I. Maymard, Z. Khassanova, J.-C. Étienne.
John Libbey Eurotext, Paris © 2002 pp. 382-388.

The relationship between fluorine in drinking water and dental health of oppidans in P. R. China

B.S. Zheng[1], B.B. Wang[1], C. Zhai[1], H.Y. Wang[2], X.J. Liu[1], Y.K. Ping[3], D.M. Liu[3], J. Li[3], L.Z. Sun[3], Y.H Tao[4], S.M. Yang[4], H.S. Wang[4], X. Cheng[4], S.T Qiao[4], H.P. Zhang[4], X.X. Bai[5], G.W. Zhu[6], Z.X. Yun[7], P.Z. Chen[7]

[1]The State Key Laboratory of Environmental Geochemistry, Geochemistry Institute of Chinese Academy of Science; [2]Prevention Medical Department, Stomatology College of Beijing University; [3]Stomatology College of Hebei University of Medicine; [4]Endemic Prevention Institute of Hebei Province; [5]Sanitation & Epidemic Prevention Station of Sichuan Province; [6]Resources & Environment College of Zhejiang University; [7]Endemic Prevention Institute of Shandong Province, P. R. China

ABSTRACT

In this study, the relation of intake fluorine (including drinking water and main foodstuff) and the fluorine content in urine was evaluated. The incidence of DMF (mean number of decayed, missing and filled teeth), the ratio of dental caries and the index of dental fluorosis was evaluated and compared with the data of fluorine content in drinking water and in main foodstuff from 30 cities in China, where dental epidemiological investigation has been available since 1995. At concentration ranges between 1.0mg/L and 4.0mg/L of fluorine content in drinking water, the relative incidence of DMF and dental fluorosis was studied. Samples of drinking water and main foodstuff in 50 largest cities in China were selected and fluorine levels determined.

INTRODUCTION

The study on the relationship between fluorine in drinking water and dental caries has lasted for more than 5 decades. In 1942, Dean [2] found that with the increasing of fluorine concentration in drinking water from 0.1 mg/l to 1.0 mg/l, DMFT decreases from 7.0 to 3.5. When more fluorine added, DMFT decreases but slightly. This result was conformed by a great deal of following researches. USPHS (The U.S. Public Health Service) [3] has comprehensively summarized this issue. In 1945, USPHS decided to conform the efficacy of fluorine on decreasing the ratio of dental caries by carrying out experiments in 3 cities, which approved that the ratio of dental caries among children in the city of Grand Rapid decreased to half of the original value when the content of the fluorine in drinking water was up to 1.0 mg/L. Similar studies conducted in different regions in Europe, the U.S. and Canada received the same results (USPHS 1991, Murray et al. 1991) [3] [4] [5] [6]. The social and economic benefits of drinking water fluoridation have been well accepted in the western world. There had been 145 million people living in the biggest 42 cities in the U.S. using fluoride drinking water in by the year of 1993 (Hinman et al. 1996) [7]. The project drinking water fluoridation was also carried out in Australia, Brazil, Canada, Colombia, Hong Kong, Ireland, Malaysia, New Zealand, Singapore, Spain and The Great Britain (WHO 1986) [8].

In China, the efficacy of drinking water fluoridation needs reevaluation because of the Chinese different food customs from western countries. Despite some experiments being conducted in some areas in China in the past 4 decades, the confirmation of the efficacy of drinking water fluoridation still lacks comprehensive evaluations. An experiment of drinking water fluoridation was carried out in Guangzhou in 1965 [11] but didn't receive the liable result because of the disordered administration during the Culture Revolution and this experiment was stopped in 1983 [12].

During the experiment in Guangzhou, the verity of the ratio of dental caries and the fluorosis of the local people were accurately consistent with the changes of no fluoridation, superfluous fluoridation, well-balanced fluoridation, discontinues fluoridation and stop fluoridation. The ratio of dental caries decreased 30%-70% in the period of drinking water fluoridation and rebounded when fluoridation was broken. The ratio of fluorosis was slight prevalence due to the superfluous fluoridation in the early-days of the experiment and disappeared when the concentration of fluorine was reduced [13-16]. From the statistics of dental epidemiological investigation conducting by China Organization On Dental Disease Prevention in 1995, the ratio of dental caries of 18 year-old people in 3 cities in Guangdong province was 7.95%, two times of the ratio of the dental caries of the whole country. When in the period of break fluoridation, the teeth of this age grope are in the procedure of mineralization. In this study, the comprehensive evaluation and the accurate confirmation of the efficacy of drinking water fluoridation are expected through the investigations on drinking water and oral health in the biggest 34 cities in China.

METHOD

Subjects

Most subjects in the study are come from 10 of the 30 cities where an investigation on the epidemic dental health have been taken in 1995. These cities are Shenyang, Hangzhou, Linyi, Ningbo, Jinhua, Dalian, Jinan, Chaoyang, Shanghai and Qingdao. There are also some subjects coming from some towns where the public water supply system with high fluorine content in Bazhou city of Hebei Province.

Investigation Items

The concentration of the fluorine in drinking water and in the urine of the 16-19 year old subjects of the investigation cities and areas are determined. The examination on dental health including DMF (Mean number of decayed, missing and filled teeth.), the ratio of dental caries and the index of dental fluorosis of 5-year-old group, 12-year-old group, 15-year-old group and 16-19 year old group coming from the mentioned cities and areas are taken by local dentists (WHO) [17].

Investigation Method

Drinking water samples are from taps. Water is collected into a 300ml clean and dry plastic bottle after being flowed for several minutes. Every subject collects urine sample to a dry clean plastic bottle of 100ml in the morning. The average volume of urine sample should be 50ml. It is approved that the fluorine concentrations of the different periods of one day have slight difference but which is not distinct, so when there is not a good condition for collecting urine in the morning, a random urine sample is instead. The sample is examined at the same day of the collection by using fluorine electrodes. The record of each dental examination by local dentist is filled in a special questionnaire. The standardization of the investigation on dental decay is according to the basic method on dental investigation established by the World Health Organization, and the standardization of the investigation on dental fluorosis is according to Dean's standard method.

WORK COMPLETED

Water sample selection, determination of fluorine in drinking water, main foodstuff and urine in 12 of 30 cities and the area where fluorine content in drinking water ringing between 1.0-4.0mg/L has been completed. In addition, the selection of water samples from another 10 cities was conducted. The investigation of 29214 person's dens and the determination of fluorine in tap waters from 22 cities and 1931 urine samples have been carried out.

PRELIMINARY RESULTS

The Relation Of Fluorine Between Drinking Water And Urine

From *fig. 1*, at fluorine levels below 4 mg/L, the fluorine content in adult's urine increases in the proportion of fluorine content in drinking water. Based on these findings, the following conclusions are presented:

a. Fluorine content in drinking water can represent the total fluorine intake of human being. It shows a very compact relationship between fluorine in drinking water and urine, which indicates that drinking water is the main source of fluorine intake of human being.

b. Regression equation of fluorine content in drinking water and urine is: $Y=1.3146\times +0.3136$ ($r=0.9709$)

Y: average concentration of fluorine in urine

X: average concentration of fluorine in drinking water

The average fluorine content in urine is 0.3136mg/L when no fluorine is present in drinking water. Considering the average quantity of excreted urine per person per day as 1L, the average fluorine intake from main foodstuff and inhalation per person per day is approximately 0.3mg.

The above expression can also be used to demonstrate that the fluorine content in foodstuff and air of different zones in China doesn't have a significance difference.

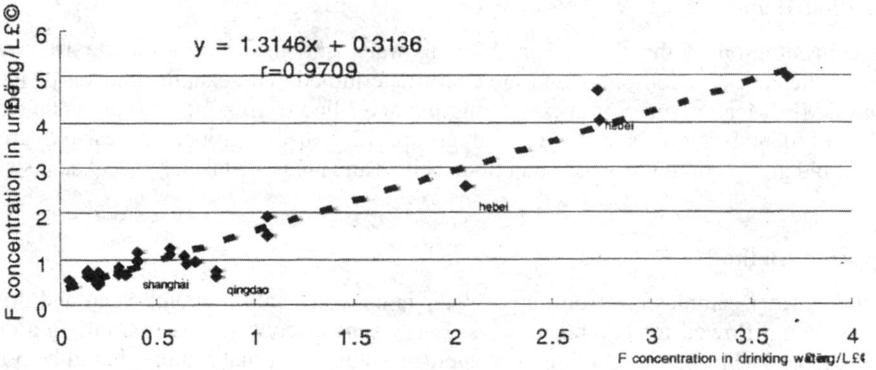

Fig. 1. The relation of fluorine concentration in drinking water and in urine of 16-19 year-old group of some area in china

The Relation Between Fluorine Concentration In Drinking Water And The Ratio Of Dental Caries

Fig. 2-5 show that the ratios of dental caries of different age groups all slightly decrease with the increasing of fluorine content in drinking water. In the above figures, the even trend lines indicate the relationship between fluorine concentration in drinking water and the ratio of dental caries is not very sensitive. The ratio of dental caries cannot alone accurately indicate the dental damage. So it's not the best parameter for conforming the function of fluorine to dental health.

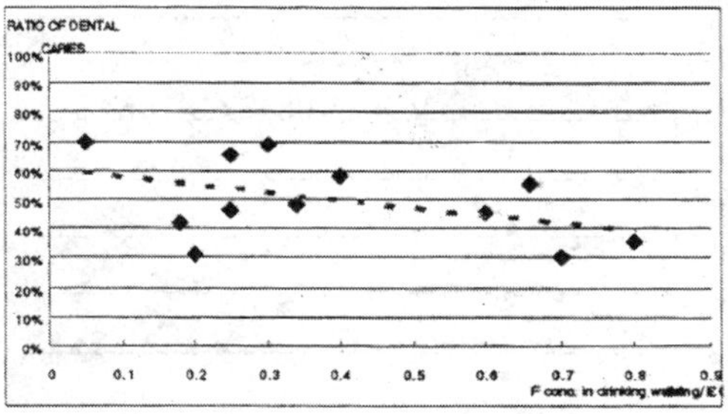

Fig. 2. The relation of Fluorine concentration in drinking water and the ratio of dental caries of local 5-year-group in some cities and areas in China

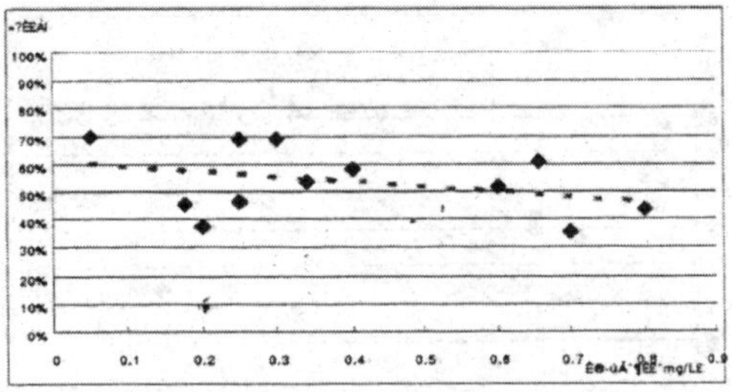

Fig. 3. The relation of Fluorine concentration in drinking water and the ratio of dental caries of local 12-year-group in some cities and areas in China

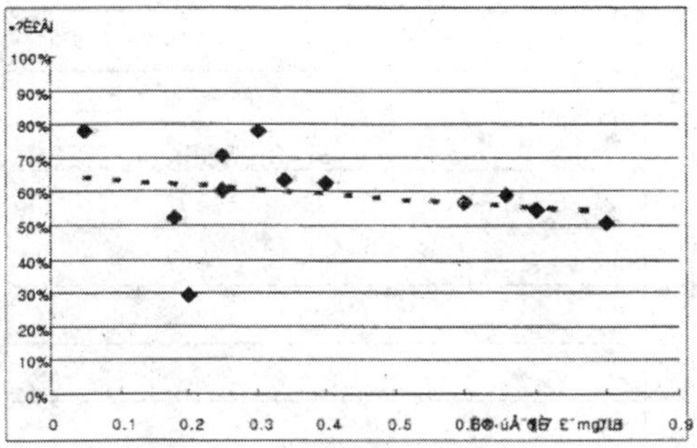

Fig. 4. The relation of Fluorine concentration in drinking water and the ratio of dental caries of local 15-year-group in some cities and areas in China

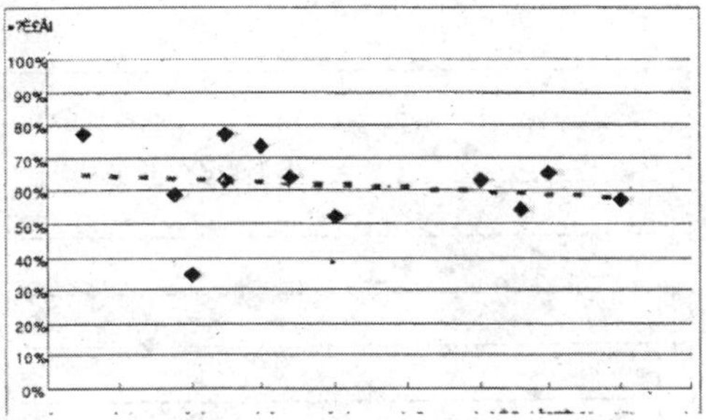

Fig. 5. The relation of Fluorine concentration in drinking water and the ratio of of dental caries of local 18-year-group in some cities and areas in China

The Relation Between Fluorine Concentration In Drinking Water And DMF

In the following Figs, it shows that DMF of all different age groups decrease with the increasing of fluorine content in drinking water.

The more sharp trend lines showed in *fig. 6-9* indicate that DMF has closer relation with the concentration of fluorine in drinking water than the ratio of dental caries has. When fluorine concentration in drinking water increasing 0.5mg/L, DMF averagely decreases of 2 milk teeth in 5-year-old group and 0.5 tooth in 18 year-old-group. In *fig. 10-13*, DMF decreases with the increasing of fluorine content in drinking water when the fluorine concentration in drinking water below 2.0mg/L, and increases with the increasing of fluorine content in drinking water when fluorine concentration in the drinking water exceeding 2.0mg/L, which indicate that high ratio of DMF exists with the excessively high or low fluorine content in drinking water.

The parabolas of the trend lines in *fig. 10-13* indicate that there has a minimum DMF with a certain concentration of fluorine in drinking water. In different age groups, the concentration of fluorine in drinking water which corresponding to the minimum value of DMF are different, but all within the range of 1.5-2.5mg/ml. From this point, the most appropriate range of fluorine concentration in drinking water, concerning DFM, is 1.5-2.5 mg/ml.

The Relation Between Fluorine Concentration In Drinking Water And Index Of Dental Fluorosis

From *fig. 14*, it is showed that the index of dental fluorosis increases with the increasing of the fluorine concentration in drinking water. It is known that when the index of dental fluorine is

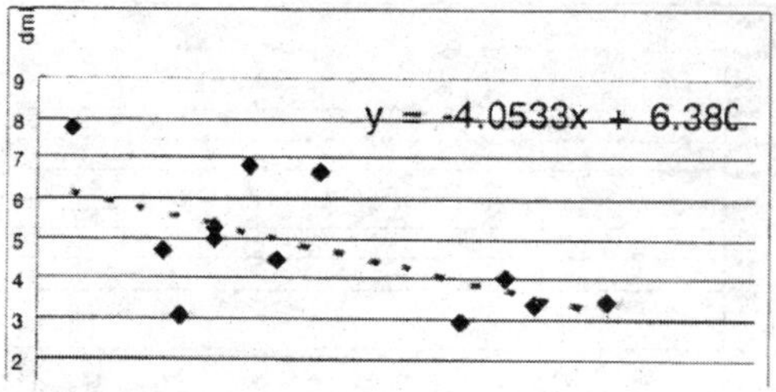

Fig. 6. The relation of Fluorine concentration in drinking water and DMF of local 5-year-group in some cities and areas in China

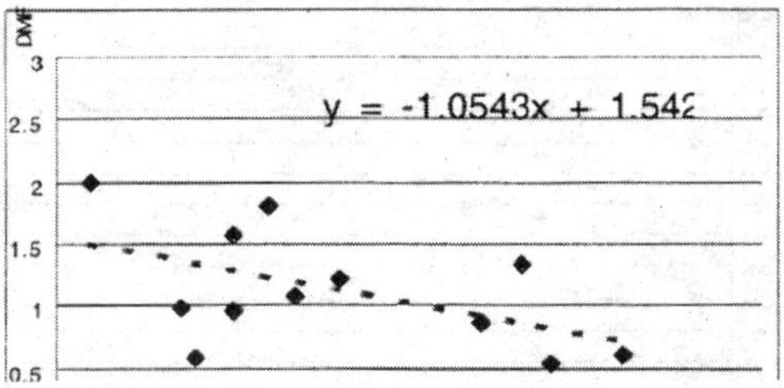

Fig. 7. The relation of Fluorine concentration in drinking water and DMF of local 12-year-group in some cities and areas in China

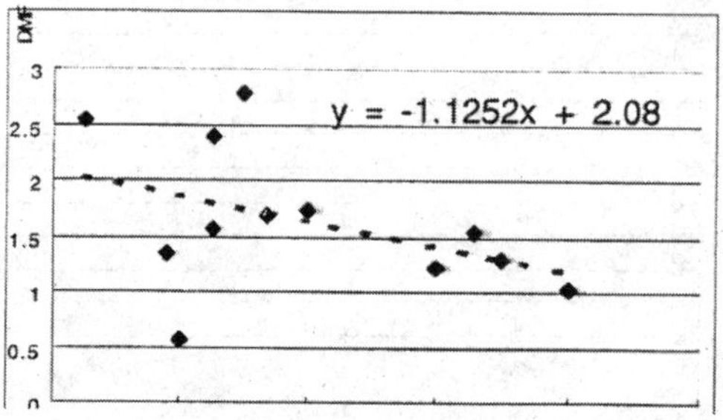

Fig. 8. The relation of Fluorine concentration in drinking water and DMF of local 15-year-group in some cities and areas in China

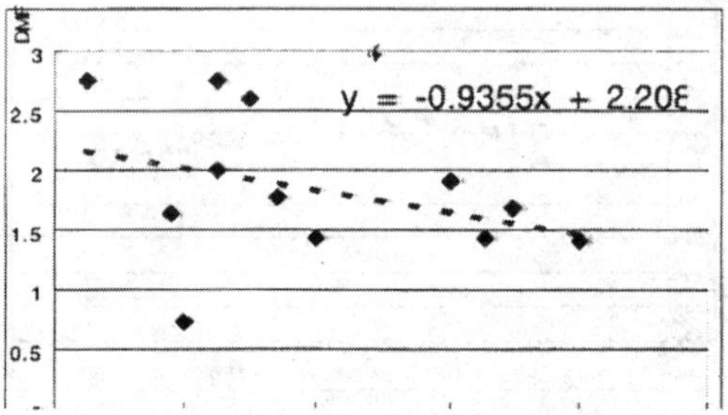

Fig. 9. The relation of Fluorine concentration in drinking water and DMF of local 18-year-group in some cities and areas in China

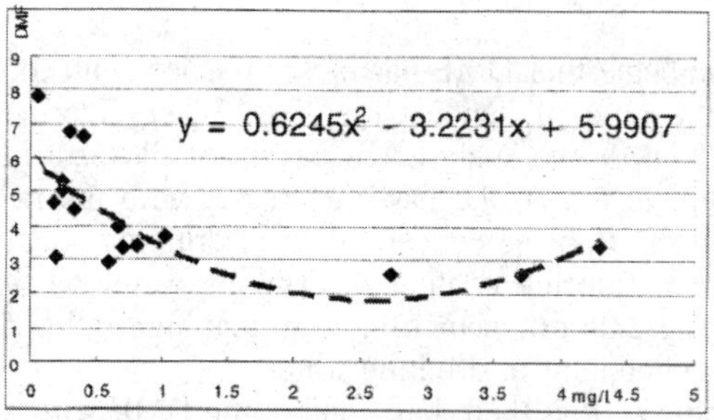

Fig. 10. The relation of Fluorine concentration in drinking water and the ratio of dental caries of local 5-year-group in some cities and areas in China

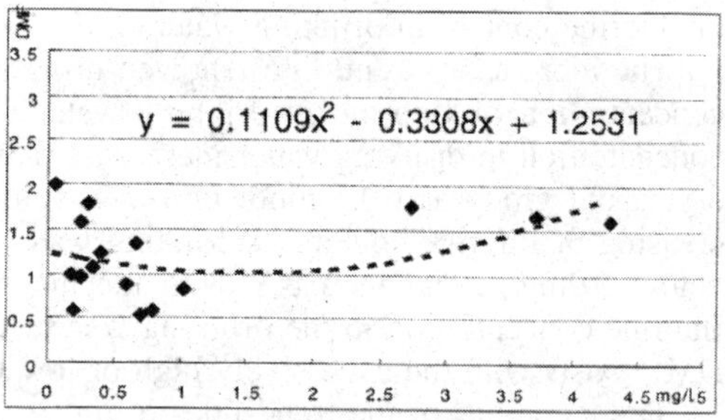

Fig. 11. The relation of Fluorine concentration in drinking water and the ratio of dental caries of local 12-year-group in some cities and areas in China

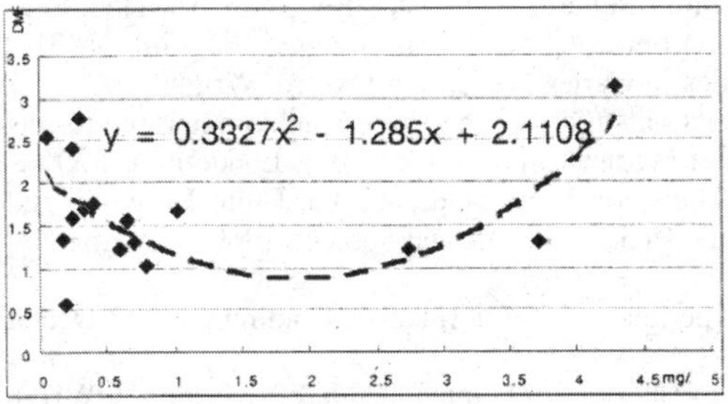

Fig. 12. The relation of Fluorine concentration in drinking water and the ratio of dental caries of local 15-year-group in some cities and areas in China

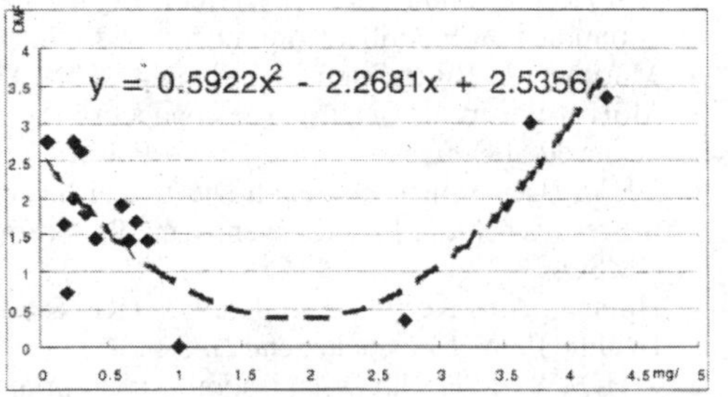

Fig. 13. The relation of Fluorine concentration in drinking water and the ratio of dental caries of local 18-year-group in some cities and areas in China

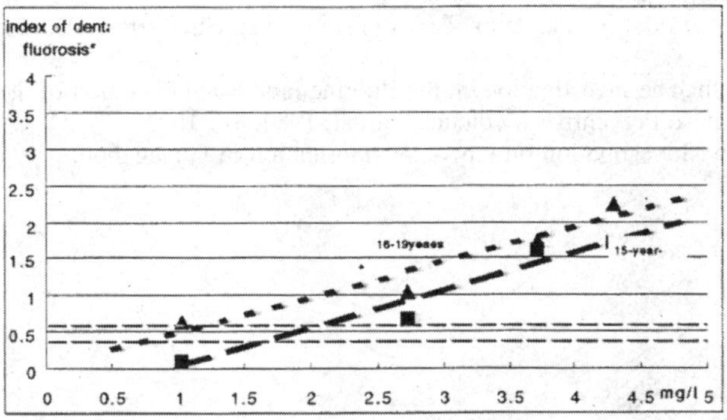

Fig. 14. The relation of Fluorine concentration in drinking water and the index of dental fluorosis of different age groups in some cities and areas in China

* index < 0.4, no dental fluorosis prevalent; 0.4 < index < 0.6, suspected; index > 0.6, slight prevalence of dental fluorosis

higher than 0.6, a slight dental fluorosis will be prevalent. In *fig. 14*, the different minimum values of fluorine concentration in drinking water that inducing dental fluorosis prevalent are 2.129mg/L in 15-year-old group and 1.260mg/g in 16-19 year-old group. Therefore, the maximum fluorine content in fluoride drinking water should not higher than 1.260mg/L.

DISCUSSION

In this study, some big cities and areas of china have been investigated, and the influence of verity of temperatures, drinking and eating habits and etc in the individual cities are not involved in the study for the time being. Different temperatures of respective areas can influence the amount of daily water intake. Different drinking and eating habits and the large scale of using fluoride toothpastes can also influence the result. Therefore the conclusions here need Additional studies to adjust, substantiate and evaluate and these studies are currently in progress.

REFERENCES

1. The Second Epidemiological Investigation On Dental Health In China. 1999, China Organization On Dental Disease Prevention, People's Sanitation Press, Beijing.
2. Dean, H.T. The investigation of physiological effects by the epidemiological method. In: Moulton, R.F (Ed), Fluorine and dental health. Washington D.C., American Association for Advancement of science, 1942.

3. USPHS (US Public Health Service). Review of Fluorine: Benefits and Risks: Report of Ad. Hoc. Sub-committee on Fluoride. Committee to Co-ordinate environmental Health and Related Programs, 1991.
4. Murray, J.J., Rugg-Gunn, A.J., Jenkins, G.N. Fluorides in Caries Prevention Oxford, Wright, 1991.
5. MoH (Ministry of Health). The conduct of the fluoridation studies in the Unite Kingdom and the results achieved after five years. Report on Public Health and Medical Subject 1962, 105. London: HMSO.
6. DHSS (Department of Health and Social Security) The fluoridation studies in the Unite Kingdom and the results achieved after eleven years. Report on Public Health and Medical Subject, 1969, 122. London: HMSO.
7. Hinman, A.R.,Sterritt, G.R., Reeves, T.G. The US experience with fluoridation. Community and Dental Health, 1996, 13 (supplement 2): 3-5.
8. WHO (World Health Organization). Appropriate use of fluoride for Human health, ed. Murray JJ, WHO, Geneva, 1986.
9. Guangzhou Tape Water Fluoridation and Dental Caries Prevention Group. The evaluation of discontinuous water fluoridation in Guangzhou. 1985.
10. Yangcheng Evening paper. 18 years tap water fluoridation stops. Oct 2rd, 1983.
11. Guangzhou Tape Water Fluoridation and Dental Caries Prevention Group. The efficacy of tap water fluoridation in Guangzhou. Chinese medicine journal, 1973, 53: 632.
12. Liang Shaoren. Summarize on tap water fluoridation and dental caries prevention in Guangzhou. Journal of Chinese Stomatology, 1983, 18: 159.
13. Shen Yanmin, Jiang Junrong and Li lanxin. The investigation on the fluorine intake and excretion of the residents in fluoride area. Journal of Chinese Preventive Medicine Journal, 1984, 18: 162.
14. Guangdong Student Dental Caries Survey. Investigation on tap water fluoridation in Guangzhou.
15. WHO Oral Health Survey third ed.

* This research supported by the State Key Lab. of Environmental Geochemistry, the Chinese Academy of Sciences and the Foundation of Chinese Oral Health. The authors would like to thank Dr. Jose Centino for his editing and advice.

*Metal Ions in Biology and Medicine: vol. 7.*Eds L. Khassanova, Ph. Collery, I. Maymard, Z. Khassanova, J.-C. Étienne.
John Libbey Eurotext, Paris © 2002 pp. 389-396.

Mercury Toxicity: A Clinical Perspective

Jeff Longacre, MD

Uniformed Services University of the Health Sciences, Bethesda, Maryland, USA

ABSTRACT

"Patients, a truce to the disgust which may be caused by this remedy! For if it is disgusting, the disease (syphilis) is still more so. Besides, your cure is at this price. So without hesitation, spread this mixture on your body and cover with it your entire skin, with the exception of the head and of the precordial region. Then, carefully wrap yourself in wool and tow; then get into bed, load yourself with bed covering and thus await until a sweat bathes your limbs with an impure dew. Ten days in succession renew this treatment, for ten entire days you are to undergo this cruel trial whose beneficial effect will not cause you to wait.

As a matter of fact, very soon an infallible presage will announce to you the hour of your freedom. Very soon you will feel the ferments of the disease dissolve themselves in your mouth in a disgusting flow of saliva, and you will see the virus, even the virus, evacuate itself at your feet in rivers of saliva.

If during the course of this treatment, small ulcers develop in your mouth, have a care to fight them with gargles of milk or by a decoction of pomegranate privet. This treatment being completed, you may then, without fear, recall Bacchus to your table and enjoy in full liberty the generous nectars of Phetia, of Falcernum, and of Chios."

Girolamo Fracastoro, "Syphilis, sive Morbus Gallicus", 1530 A.D.

Throughout recorded time, mercury has held the dubious distinction as both a valuable resource in medical, agricultural, industrial, and household uses, as well as a toxin causing immense morbidity and mortality. The above passage by Fracastorius not only described and named the scourge of "syphilis" as it swept across Europe, but also vividly illustrated the poisonous nature of mercury (note the salivation, sweating, and mucosal ulcers) even as a medicinal cure.

The clinical presentation of excessive mercury exposure can vary tremendously, depending on the form of mercury, route of exposure, dose, duration of exposure, and susceptibility of the exposed individual. Since treatment options are limited, prevention remains the mainstay of therapy.

INTRODUCTION

Mercury occurs in three forms: elemental (Hg^0); inorganic (mercurous, Hg^{1+} or mercuric, Hg^{2+}); and organic (methyl-, ethyl-, and phenylmercury).

ELEMENTAL MERCURY

Sources

Elemental mercury (also known as quicksilver) can be found in the environment as either a liquid or vapor form. In its liquid form, it is most commonly known for its use in thermometers (and in other measuring tools such as sphygmomanometers, thermostats, and barometers). Some large fluorescent light bulbs are manufactured with an inner coating of mercury. Disc batteries, such as watch batteries, can contain mercury. Most recently, much attention has been focused on dental amalgam. Typical "silver" dental amalgams contain nearly 50% mercury by weight.

The vapor form of mercury can be spewed naturally by erupting volcanoes. The largest source, however, is attributed to the burning of fossil fuels, such as high-sulfur coal. It also has been found as an effluent of plastics and chloralkali processing plants (used in electrolysis in producing sodium hydroxide, hydrogen chloride, chlorine, and bleach). Residual mercury waste can be created during mercury mining, processing, or smelting, or found as part of the waste at incinerators and crematoriums.

EXPOSURE PATHWAYS

Elemental mercury is most harmful when inhaled as an odorless, colorless vapor. This vapor easily passes through the lungs and enters the blood stream, where it has its greatest effect on the blood and nervous system. Much of the inhaled mercury vapor (97%) enters the brain (in the unoxidized form) dissolved in the blood, easily crossing the blood-brain barrier. In the brain tissue, it is oxidized and amassed. The mercury vapor also oxidizes in the red blood cell and becomes Hg^{2+}. This oxidized mercury has great affinity for sulphhydryl components of protein and glutathione. Elemental mercury is minimally absorbed through the gastrointestinal tract or through the skin, yet has been shown to cross into the placenta and accumulate in the developing fetus.

CLINICAL CONSEQUENCES

Acute inhalation of large concentrations of mercury vapor can cause bronchitis and pneumonitis, ultimately leading to respiratory failure. This has occurred in poorly ventilated homes where mercury was heated, and in neonatal incubators with faulty mercury switches. Chronic exposure to mercury vapor has been shown to primarily affect the central nervous system. The initial clinical presentation is difficult to distinguish from other medical and psychiatric conditions, since the findings of decreased appetite, absent-mindedness, and difficulty sleeping are found in so many other afflictions. Significant exposure, whether prolonged or acute, can lead to salivation, sweating, and erethism (a syndrome characterized by red palms, emotional lability, and memory impairment). The kidney is the primary target organ for intoxication since the body attempts to clear the toxin through this mechanism. Renal toxicity may be manifested as proteinuria, when mercury accumulates in the kidney.

CLINICAL CONTROVERSIES

Newer and safer resources are presently phasing out the widespread use of liquid mercury in objects such as thermometers. This form of mercury is not very toxic if swallowed or handled. However, because of its ability to vaporize at room temperature, it can be toxic via the respiratory tract if not recovered and disposed of properly.

Mercury-containing dental amalgams have been investigated recently in having a role in the onset of several nervous system conditions, from demyelinating disorders to multiple sclerosis. Although an association does exist with placement of dental amalgams and slightly elevated urinary mercury excretion, multiple reviews (including an expert National Institute of Health panel) have not revealed any association with recognizable clinical conditions. Consequently, current public health guidance recommends that metal amalgams remain in place.

DIAGNOSTIC RESOURCES

Mercury vapor concentrations in exhaled air and urinary mercury excretion can be measured, yet the clinical correlation of elevated levels and clinical signs and symptoms remains poorly understood.

INORGANIC MERCURY

Sources

Inorganic mercury salts exist in several forms. They have been used as antibacterials, antiseptics, cathartics, diuretics, and numerous folk remedies. One of the most notorious compounds is mercurous chloride, also known as calomel, which was used as a teething powder in children. Mercuric chloride, known as corrosive sublimate, and mercuric oxide both have industrial and commercial uses.

EXPOSURE PATHWAYS

Although only a fraction of inorganic mercury salts are absorbed through the gastrointestinal tract and skin, resulting in little systemic absorption, these salts can be quite caustic and irritating to the area of contact, creating significant local damage. The kidney is by far the organ most affected by accumulation of mercuric mercury. Nephropathy is characterized by severe damage of the proximal tubule. Glomerular damage can occur at a lesser degree. Absolute injury to kidney cells occurs with acute exposures, while chronic exposure can induce an immunologic response within the glomeruli. Biliary excretion is considerably lower than that found in organic mercury compounds.

CLINICAL CONSEQUENCES

Although absorption is small, accidental or intentional ingestion or mucosal contact can have devastating consequences. Mercurous chloride (calomel) used in children as a teething powder has resulted in acrodynia, and overuse as a laxative has resulted in renal failure, colitis, and dementia. If ingested orally, the gastrointestinal tract can become compromised, breaking down its protective barrier and thus advancing the toxic absorption, which can lead to severe hemodynamic and renal collapse. Prolonged exposure may also result in neuropathy and hypersensitivity reactions.

DIAGNOSTIC RESOURCES

Inorganic mercury levels can be identified and measured in the urine. Spot values can provide useful information, but the optimal method is using a 24-hour urine collection. Standards for elevated levels do exist. However, the urinary concentration does not necessarily correlate with the duration or degree of exposure, especially if it has been intermittent. Blood levels can also be measured (using standardized normal values), but these have a tendency to quickly normalize when exposure is discontinued.

ORGANIC MERCURY

Sources

Organic mercury compounds (methyl-, ethyl-, and phenylmercury) have been produced primarily as biocides, such as pesticides and antiseptics. Methylmercury is ubiquitous in the environment, since microorganisms methylate elemental mercury deposited in the air or discharged into the watershed from both natural and manmade sources. This can then accumulate in the food chain to a toxic level, as seen in certain fish species. It has also been used as a fungicide on crops, causing toxic ingestions. Ethylmercury, in the form of thimerosal, has been used previously as an antiseptic and vaccine preservative. Prior to 1999, thimerosal was an ingredient in many of the routine childhood immunizations. Phenylmercury has been used in latex paints to prevent mildew and as a preservative to prevent discoloration from microorganism growth. Phenylmercury and ethylmercury are still used as bacteriostatics in topical therapeutic compounds.

EXPOSURE PATHWAYS

Since the organic mercury compounds are lipid soluble, they are nearly completely absorbed in the gastrointestinal tract. They easily cross the blood-brain barrier and the placenta, attaining significant levels in the breast milk and fetal circulation. Additionally, because the phenyl-mercury bond is unstable, this can result in a release of elemental mercury, which can then be inhaled through the lungs. Not only does methylmercury possess a predilection for the nervous system, it seems to have an affinity for very localized central nervous system cells (visual cortex and cerebellum).

CLINICAL CONSEQUENCES

Acute toxic exposure can present as a spectrum of complaints, to include weakness, paresthesia, visual and auditory deficits, tremor, ataxia, coma, and death.

Methylmercury has been determined to be a teratogen in the fetal brain, disrupting the organization and migration of neurons. Several toxicologic catastrophes (e.g., Iraq and Japan) have provided evidence that mercury exposure in children and fetuses results in widespread neurotoxicity (seizures, psychomotor retardation, and visual-auditory impairment) despite outwardly normal appearance at birth.

CLINICAL CONTROVERSIES

In recent years, much discussion and investigation have revolved around the potentially toxic levels of methylmercury in fish. Efficient absorption of this lipid soluble organic mercury presents a potential hazard in the methylmercury-laden fish. Predator fish, such as swordfish, shark, king mackerel, and large tuna, have been found to accumulate levels of methylmercury via the food chain. Studies have documented moderate elevations in mercury levels in those individuals who consume large quantities of predator fish. Consequently, the United States Food and Drug Administration has recommended that pregnant or nursing mothers and young children completely refrain from ingesting these fishes, while others should practice moderation. One of the most dramatic examples of mercury intoxication from fish consumption was in Minamata Bay, Japan, where entire villages were poisoned by consuming contaminated fish from by-products of an industrial plant. Similarly, the Cree Indians of Quebec were poisoned by eating contaminated fish from industrial discharge.

The Iraqi experience of exposing a population to toxic levels of an organic mercury fungicide suggested an increased incidence of psychomotor retardation. Two prospective studies were subsequently undertaken to assess the effects of methylmercury exposure in geographically separate coastal populations. The Faroe Island cohort (off the coast of Iceland) was known to consume up to three meals of cod each week and have occasional feasts of pilot whale. The cod possess very little mercury, while the whales have significantly more. These children were discovered to have deficits in language, attention, and memory correlating with increased maternal hair mercury levels. In contrast, the Seychelles cohort (in the Indian Ocean) consumed 12 meals of fish weekly, which have a comparatively low level of organic mercury. The cumulative mean mercury level in the Seychelles was higher than the Faroe cohort. Despite this higher mean (6.8 ppm vs 4.3 ppm), there was no evidence of neurodevelopmental deficits in the Seychelles cohort. This leaves speculation as to whether intermittent elevated mercury levels at crucial developmental stages are more harmful than constant mildly elevated levels.

Thimerosal has likewise become a very controversial compound in the past several years. In addition to hypersensitivity reactions, doses of intramuscular injections, omphalocele painting, or ear canal irrigation have resulted in significant toxicity. A recent study consisting of medical record reviews suggested a link between the presence of thimerosal in vaccines and neurodevelopmental delays. Although not proven, this prompted further review and removal of this preservative from childhood immunizations within the United States.

DIAGNOSTIC RESOURCES

Since organic mercury compounds accumulate in red blood cells, levels in the blood are considered an accurate reflection of exposure. Methylmercury also accumulates in hair, enabling a noninvasive method of determining chronic exposure, assuming appropriate hair collection and processing occurs (e.g., is free of superficial mercury contamination).

MERCURY POISONING TREATMENT OPTIONS

Elemental and Inorganic Mercury

Chelation can decrease the mercury stores by increasing urinary excretion (succimer, d-penicillamine) or hepatic and urinary excretion (BAL). However, it is still unclear whether chelation actually decreases the toxic effects of mercury poisoning. Consequently, the strict indications for chelation theory are yet to be determined.

Organic Mercury

There is no recognized chelation regimen for victims of organic mercury poisoning. Although chelation has been used in severe cases (succimer), chelation with some agents (Dimercaprol) can be harmful (may increase brain mercury concentrations). N-acetyl cysteine (Mucomyst) is now receiving attention as a chelator.

Since treatment options are limited, it is paramount to recognize the importance of prevention in mercury poisoning.

CLINICAL SYNDROMES OF MERCURY POISONING

Acrodynia (also known as Pink Disease, Swift Disease, Feer Disease,
Erythredema Polyneuropathy, and Dematopolyneuritis)

The Greek derivation of acrodynia refers to painful extremities. Initially, the ends of the fingers, toes, and nose develop a pink color, to be followed by the hands and feet becoming a dusky pink with patches of ischemia. This discoloration fades at the ankles and wrists. Often, the cheeks and nose develop a scarlet coloration.

As the condition progresses, the fingers and toes swell, exhibiting continuous itching and excruciating pain or burning in the hands and feet. The nails become dark and slough, occasionally developing gangrene or ulcers on the hands and feet. Sweat glands enlarge and sweating is markedly increased. Photophobia is dramatic, with young children often burying their heads in pillows. Severe hypotonia allows the children to assume unique resting positions not typically possible.

As the poisoning severity worsens, the gums become swollen and red, teeth may fall out, and the mandible may necrose. Salivation is profuse, as is sweating. Diarrhea is a common complication.

The effects of severe mercury toxicity ultimately have devastating effects on the central nervous system. The child who is exposed to chronic excesses of mercury will usually present with apathy, hypoactivity, lethargy, and irritability. Deep tendon reflexes eventually disappear and muscle wasting occurs due to lack of use. "There is no time when a child with acrodynia appears happy or comfortable; the child does not play or smile, but appears dejected and melancholic, a picture of abject misery."

Minamata Disease

This constellation of signs and symptoms was exhibited in villagers living on Minamata Bay, Japan after years of consuming mercury-contaminated fish from a local industrial plant. Adults, children, and fetuses were all tragically affected with devastating and irreversible neurological damage. Congenital exposure resulted in significant mental and physical developmental delay, to include abnormal motor movements. Examination of the brain revealed noteworthy cortical atrophy and microcephalus. The infantile form included visual deficits, speech delays, abnormalities of gait and coordination, tremors, and seizures. The damage inflicted was irreversible, requiring extensive chronic care, to include remedial education and cognitive therapy, as well as intensive physical rehabilitation.

CONCLUSION

For thousands of years, the unique and useful characteristics of mercury have been exploited. Time and science continue to tip the scales in a direction of cautious application, balancing the relationship between man and metals in the environment. A solid knowledge base and awareness of the toxicity of metals, an index of suspicion, and a careful and comprehensive environmental and occupational history remain the key components of appropriate diagnostic and therapeutic interventions that may reduce the detrimental effects of this potentially toxic metal.

REFERENCES

Agency for Toxic Substances and Disease Registry. Toxicological Profile for Mercury. Atlanta, Georgia: United States Public Health Service; 1999.

Akabane T. Mercury. In: Nelson Textbook of Pediatrics, 12th ed. Behrman RE and Vaughan CV, eds. WB Saunders, Philadelphia; 1983: 1797-1800.

American Academy of Pediatrics, Committee on Environmental Health. Mercury. Handbook of Pediatric Environmental Health. American Academy of Pediatrics, Elk Grove Village, IL; 1999: 145-154.

American Academy of Pediatrics, Committee on Environmental Health. Technical Report: Mercury in the Environment: Implications for Pediatricians. Pediatrics. 2001; 108: 197-205.

American Academy of Pediatrics, Committee on Infectious Diseases and Committee on Environmental Health. Thimerosal in vaccines - an interim report to clinicians. Pediatrics. 1999; 104: 570-574.

Amin-Zaki L et al. Intrauterine methylmercury poisoning in Iraq. Pediatrics. 1974; 54: 587-595.

Davidson PW et al. Effects of prenatal and postnatal methylmercury exposure from fish consumption on neurodevelopment: outcomes at 66 months of age in the Seychelles Child Development Study. JAMA. 1998; 280: 701-707.

Goldfrank LR, Bresnitz EA, Howland MA, Weisman RS. Mercury. In: Goldfrank's Toxicologic Emergencies, Goldfrank LR et al., eds. Appleton-Century-Crofts, Norwalk, CT; 1986: 640-650.

Kales SN and Goldman RH. Mercury Exposure: Current Concepts, Controversies, and a Clinic's Experience. J Occup Environ Med. 2002; 44: 143-154.

Miura K et al. Mercury Toxicity. In: Toxicology of Metals: Biochemical Aspects, Goyer RA and Cherian MG, eds. Springer-Verlag, Berlin; 1995: 163-188.

National Academy of Sciences, Committee on the Toxocological Effects of Methylmercury. Toxicological Effects of Methylmercury. National Academy Press: Washington, DC; 2000.

Source Book of Medical History: Compiled with Notes by Logan Clendening. Fracastorius. Dover Publications Inc., New York; 1960: 119-120.

Steuerwald U et al. Maternal seafood diet, methylmercury exposure, and neonatal neurological function. J Peds. 2000; 136: 599-605.

United States Environmental Protection Agency. Mercury Study Report to Congress. United States Environmental Protection Agency: Washington DC; 1997.

United States Public Health Service, Committee to Coordinate Environmental Health and Related Programs, Subcommittee on Risk Management. Dental Amalgam: A Scientific Review and Recommended Public Health Service Strategy for Research, Education, and Regulation. United States Public Health Service: Washington, DC; 1993.

IX EPIDEMIOLOGY

Metal Ions in Biology and Medicine: vol. 7. Eds L. Khassanova, Ph. Collery, I. Maymard, Z. Khassanova, J.-C. Étienne.
John Libbey Eurotext, Paris © 2002 pp. 399-401.

Parkinson disease and hereditary hemochromatosis mutations in a Swedish sample

Sharon Moalem[1,2], Martin J. Somerville[3], Lisa O'Brien[1,2], Ada Ng[3], Shelagh Haase[3], and Maire E. Percy[1,2]

[1]*Department of Physiology, University of Toronto, Toronto ON, Canada, M5S 1A8;* [2]*Neurogenetics Laboratory, Surrey Place Centre, 2 Surrey Place, Toronto ON Canada M5S 2C2;* [3]*Department of Medical Genetics, 839 Medical Sciences Building, University of Alberta, Edmonton ALTA, Canada, T6G 2H7.*

INTRODUCTION

Hereditary hemochromatosis is one of the most common single gene disorders in individuals of Celtic or northern European ancestry [1]. Mutations in the gene HFE results in excessive iron uptake and deposition in many organs and tissues [2]. As continuous iron uptake into the brain occurs normally, the possibility that a defect in iron metabolism could influence brain iron uptake must be considered [3]. A rather complex association between familial Alzheimer disease (AD) and the HFE mutations C282Y and H63D, the absence of these mutations appearing protective against familial AD in males lacking the apolipoprotein E4 allele, has been previously reported [4]. Also, presence of the H63D is associated with earlier age at onset of sporadic AD in a dose dependent fashion [5]. As there is evidence for dysmetabolism of iron in Parkinson disease (PD) as well as in AD [6], there is a strong rationale for studying the involvement of HFE mutations in PD. Since the frequencies of HFE mutations differ among different populations [1], designing a case/control study to evaluate the role of HFE mutations in PD can be difficult in countries with great ethnic diversity. Because Sweden is a country with relatively low ethnic diversity, and relatively high frequencies of HFE mutations [7] and PD [8], the present study was carried out on a Swedish sample.

AIMS

In order to study the involvement of HFE mutations in PD, the frequencies of the two most common HFE mutations (C282Y and H63D) were determined and compared in individuals with PD and in healthy normal individuals, with/without consideration of sex and subject age.

METHODS

Swedish Caucasians were recruited through the Karolinska Institute. Blood samples were obtained from 86 PD patients (43% female, mean age 70 years [range 50 to 90 years] and 57% male, mean age 63 years [range 43 to 89 years]) and 127 geographically matched controls (50% female and 50% male, mean age 38 years [range 20 to 65 years]). Patients were diagnosed according to the "brain bank clinical diagnostic criteria" [9], with the exception of a few who had more than one affected relative, and were classified as idiopathic or with a family history of PD. To avoid misdiagnosis, DNAs extracted from blood samples were independently screened for C282Y and H63D using two different published procedures [10, 11]. Associations were evaluated using Chi-square analysis.

RESULTS

The results are summarized in *table I*. Overall, the frequency of individuals with one or two H63D mutations was greater ($p = 0.19$), and of individuals with one or two C282Y mutations somewhat lower ($p = 0.46$), in PD patients than in the controls. With stratification for sex, the overrepresentation of H63D was apparent in PD males ($p = 0.20$) and the underrepresentation of C282Y in PD females ($p = 0.29$). In PD patients under the age of 55 years, the frequency of those with one or two H63D mutations was significantly greater than in those over age 55 ($p = 0.037$), and significantly greater than in the controls ($p = 0.01$). These effects were apparent in patients with or without a family history of PD.

Table 1. Percentages of individuals with Parkinson disease and normal individuals with one or more H63D or C282Y mutations: effects of sex, subject age and family history of disease

	H63D		C282Y	
	N	**Percentage with one or more mutation**	**N**	**Percentage with one or more mutation**
C (all)*	127	18.1	127	12.6
PD (all)	86	25.6	86	9.3
C (m)	50	16.0	50	12.0
C (f)	48	27.1	48	14.6
PD (m)	49	26.5	49	12.2
PD (f)	37	24.3	37	5.4
PD (all, >55)	65	20.0	65	10.8
PD (all, <55)	21	42.9**	21	4.8
fPD (<55)	6	50.0	6	0
Idiopathic PD (<55)	15	40.0	15	6.7

Legend: N: total number of individuals in group; C: control; PD: Parkinson disease; fPD: PD with family history of disease; m: male; f: female; <55: under 55 years of age; >55: over 55 years of age.
* The gender was not recorded for 29 control individuals.
** Significantly greater than in PD patients >55 or than in controls. See also text.

DISCUSSION

This study is the first to suggest an involvement of HFE mutations in PD that is complex and dependent upon sex and subject age. The data suggest that H63D might be a predisposing risk factor for PD especially in males under age 55 and possibly mildly protective in females. It is of interest that like the C282Y and H63D mutations which can result in hemochromatosis [1], the rates of PD are reported to be highest among populations of European origin [12]. Furthermore, PD occurs more frequently in males than in females [13] as does hemochromatosis [1, 2] Aside from increasing chronological age, the extent to which environmental and genetic factors contribute to PD is controversial. Linkages of PD to mutations in alpha-synuclein, parkin, tau and other loci have previously been reported [14]. Nevertheless, on the basis of our findings, further investigation of the involvement of HFE mutations in PD, especially for those under the age of 55 years and separately for males and females, would seem warranted. Aside from the issues of whether or not HFE mutations are susceptibility factors for PD or its complications, clinicians should be aware that a predisposition for hereditary hemochromatosis can occur in individuals with PD and their families, as well as in the general population, and that established procedures exist for preventing or reducing body iron overload that can occur in individuals with HFE mutations. For treatment to be effective, this must be implemented before tissue damage occurs [15].

ACKNOWLEDGMENTS

Andrea Carmine, Silvia Buervenich, Lars Olson, Ann-Christin Thelander and Olof Sydow donated the DNA samples and assisted with the data analysis. Supported by the Swedish Parkinson Foundation, the Swedish Research Council, Karolinska Institutet funds, the Scottish Rite Charitable Foundation of Canada and the Alzheimer Society of Canada.

REFERENCES

1. Merryweather-Clarke AT, Pointon JJ, Jouanolle AM, Rochette J, Robson KJ. 2000. Geography of HFE C282Y and H63D mutations. Genet Test 4(2):183-198.
2. Feder JN, Gnirke A, Thomas W, Tsuchihashi Z, Ruddy DA, Basava A, Dormishian F, Domingo Jr. R, Ellis MC, Fullan A, Hinton LM, Jones NL, Kimmel BE, Kronmal GS, Lauer P, Lee VK, Loeb DB, Mapa FA, McClelland E, Meyer NC, Mintier GA, Moeller N, Moore T, Morikang E, Wolff RK and et al. 1996. A novel MHC class I-like gene is mutated in patients with hereditary haemochromatosis. Nat Genet 13(4): 399-408.
3. Connor JR, Milward EA, Moalem S, Sampietro M, Boyer P, Percy ME, Vergani C, Scott RJ, Chorney M. 2001. Is hemochromatosis a risk factor for Alzheimer's disease? Journal of Alzheimer Disease 5(3): 471-477.
4. Moalem S, Percy ME, Andrews DF, Kruck TP, Wong S, Dalton AJ, Mehta P, Fedor B, Warren AC. 2000. Are hereditary hemochromatosis mutations involved in Alzheimer disease? Am J Med Genet 93(1): 58-66.
5. Sampietro M, Caputo L, Casatta A, Meregalli M, Pellagatti A, agliabue J, Annoni G, Vergani C. 2001. The hemochromatosis gene affects the age at onset of sporadic Alzheimer's disease. Neurobiol Aging 22: 563-568.
6. Berg D, Gerlach M, Youdim MB, Double KL, Zecca L, Riederer P, Becker G. 2001. Brain iron pathways and their relevance to Parkinson's disease. J Neurochem 79(2): 225-236.
7. Cardoso EM, Stal P, Hagen K, Cabeda JM, Esin S, de Sousa M, Hultcrantz R. 1998. HFE mutations in patients with hereditary haemochromatosis in Sweden. J Intern Med 243(3): 203-208.
8. Zhang ZK, Roman GC. 1993. Worldwide occurrence of Parkinson's disease: an updated review. Neuroepidemiology 12(4): 195-208.
9. Daniel SE, Lees AJ. 1993. Parkinson's Disease Society Brain Bank, London: overview and research. J Neural Transm Suppl 39: 165-172.
10. Somerville MJ, Sprysak KA, Hicks M, Elyas BG, Vicen-Wyhony L. 1999. An HFE intronic variant promotes misdiagnosis of hereditary hemochromatosis. Am J Hum Genet 65(3): 924-926.
11. Stott MK, Fellowes AP, Upton JD, Burt MJ, George PM. 1999. Simple multiplex PCR for the simultaneous detection of the C282Y and H63D hemochromatosis (HFE) gene mutations. Clin Chem 45(3): 426-428.
12. Rosati G, Granieri E, Pinna L, Aiello I, Tola R, De Bastiani P, Pirisi A, Devoto MC. 1980. The risk of Parkinson disease in Mediterranean people. Neurology 30(3): 250-255.
13. Baldereschi M, Di Carlo A, Rocca WA, Vanni P, Maggi S, Perissinnnotto E, Grigoletto F, Amaducci L, Inzitari D. 2000. Parkinson's disease and parkinsonism in a longitudinal study: two-fold higher incidence in men. ISLA Working Group. Italian Longitudinal Study on Aging. Neurology 55(9): 1358-63.
14. Martin ER, Scott WK, Nance MA, Watts RL, Hubble JP, Koller WC, Lyons K, Pahwa R, Stern MB, Colcher A, Hiner BC, Jankovic J, Ondo WG, Allen FH Jr, Goetz CG, Small GW, Masterman D, Mastaglia F, Laing NG, Stajich JM, Ribble RC, Booze MW,Rogala A, Hauser MA, Zhang F, Gibson RA, Middleton LT, Roses AD, Haines JL, Scott BL, Pericak-Vance MA, Vance JM. 2001. Association of single-nucleotide polymorphisms of the tau gene with late-onset Parkinson disease. JAMA 286(18): 2245-2250.
15. Camaschella C, Piperno A. 1997. Hereditary hemochromatosis: recent advances in molecular genetics and clinical management. Haematologica 82(1): 77-84.

Metal Ions in Biology and Medicine: vol. 7. Eds L. Khassanova, Ph. Collery, I. Maymard, Z. Khassanova, J.-C. Étienne.
John Libbey Eurotext, Paris © 2002 pp. 402-404.

Biological monitoring of occupational exposure to cytostatic drugs with platinum

F. Deschamps[1], V. Marinutti-Liberge[1], D. Lamiable[1]

[1]*Department of Occupational Health, Faculté de Médecine, 51 rue Cognacq-Jay, 51100 Reims, France.*

ABSTRACT

Background: the past 20 years has been an explosion in the introduction of effective cytostatic agents for treating malignant disease. Several cytostatic drugs have been shown to be mutagenic, teratogenic and carcinogenic in experimental systems. Medical staff is potentially exposed to a wide spectrum of antineoplastic drugs in combinations not used in therapy althrough in significantly lower concentration.

The aim of the study was to investigate whether oncology workers are occupationally exposed to antineoplastic drug by measuring platinum urinary concentration used as marker.

Patients and methods: the exposed group consisted of nurses, pharmacy technicians and pharmacists, handling antineoplastic drugs for at least one month. The control group consisted of medical workers who had never been occupationally exposed to known carcinogenic substances. Each person was interviewed. The questions covered a detail occupational, medical and family history, and adverse effects outbreaks. Frequency of handling cytostatic drugs was evaluated. Last informations include the type of protective measures used.

Results: platinum was included in two of the four most frequently used drugs. An elevated level of urinary platinum (0.63 µg/g creat.) was found in one exposed worker, and none in the control group. No relationships were observed between the excretion rate and the kind of activity, the frequency or the duration of handling cytostatic drugs, or the excretion rates and the use of gloves or masks.

Conclusion: occupational surveillance of health care workers is often inconsistent. The biomonitoring using platinum or other methods is simple, fast and selective. Since even very low exposure levels may result in a health hazard, high sensitivity by metal dosage is required.

INTRODUCTION

Over the past five decades, there has been a marked increase in the use of cytostatic drugs (C.S.) for cancer treatment. The classification of agents is based on their action on the target cell. Their application depends on the type of tumour and the known resistance of interaction of applied drugs.

Although improved safety standards for the handling of anticancer drugs have been established [1], significant incorporation of trace amounts of these agents has been found in hospital pharmacy personal [2, 3, 4]). Many C.S. are mutagens, carcinogens and teratogens [5]. Considering the mechanisms of the C.S. that are used, it is not surprising that many persons involved in health care, especially nurses are worried about the effects that these drugs may have on their health. But during occupational activities, such as preparation and administration of these drugs, acute toxic effects have not been observed, except for accident in the course of which workers have been exposed to large amounts of spilled drugs [6]. This raised questions about the relative efficacy of current prevention measures and pointed to the need for useful methods of monitoring. The assessment of exposure and health effects is complicated by the fact that nurses, pharmacists and pharmacy technicians are exposed to a wide spectrum of these agents in subtherapeutic concentrations with unknown biological consequences.

Institutions have used various methods to monitor occupation exposure to C.S., including work history records, drug logs, registries of drug spills, and punctures, and industrial hygiene measures such as air sampling [7].

Exposure to chemical agents can be assessed by measuring some biological variables. Biological monitoring takes into account the fact that the exposure is not always constant and also absorption by other routes of exposure than the lungs. Many chemicals can enter the body absorption through the skin of the intestinal tract [8].

In order to monitoring exposure levels and to measure possible uptake of C.S., it is necessary to use sensitive and compound specific detection methods. The aim of our study was to carry out internal dose monitoring of nurses and pharmacy personal regularly handling anticancer drugs.

MATERIALS AND METHODS

The study population consisted of a group of pharmacists and nurses who reconstitute C.S. They have been employed in an oncology department for at least one month. The control group consisted of medical workers (1 control for 5 exposed workers) who had never been occupationally exposed to known carcinogenic substances. All the workers completed a questionnaire from which we were able to assess drug taking, which C.S. were handled and how often, and the type of protective measures used when handling the drugs. Questions included demographic data, general health perception, and smoking habits. A physical examination was administered. We tried to identify adverse effects including irregular menstrual period, hair loss, chromosomal abnormalities, decreased white blood cells, liver damage, and spontaneous abortion, as well as light headedness, dizziness, nausea, headache, cough and skin / mucous membrane and allergic reactions.

Among the large number of currently employed C.S., platinum is a promising tracer of the contamination levels. Biological monitoring of internal dose implies the assessment of the parent compound or its metabolites in biological samples. The media most commonly used is urine. Total urine was collected in portion during the period of about 24 hours starting from the beginning of the working day. The amounts excreted and the excretion periods were registered. To assess platinum urinary level, a standard HP 4500 benchtop ICP-MS, fitted with a V-groove nebulizer and a glass spray chamber was used. The samples were prepared by diluting 500 µl of urine with 4500 µl of an aqueous diluent containing 1% HNO3 (Merck, suprapur), 0.1% of Triton X 100 (Jansen) and 10 µg/l of Bismuth (Merck 1 µg/l in 5% nitric acid solution used as internal standard. The calibration graph used in routine conditions was obtained from a platinum standard solution (Aldrich), containing 980 µg/l of platinum in 5% Hcl solution. The calibration range was 0 to 1 µg/l. The blank and the three standards (0.2-0.5 and 1 µg/l of platinum) were diluted with the same diluent as the samples.

RESULTS

Thirty three technicians and nurses at seven different departments participated in this study. Preparation of the drugs occurred in the clinical pharmacy department and in the preparation room of the out patient department. The group of drugs handlers was 88% female, had a mean age of 36 years, and had worked in their profession an average of 8 years. The oldest exposed worker was age 59 years and the youngest age 24 years. The control group which included 6 medical workers was 50% female, had a mean age of 41 years. Thus the drug handlers and controls were very similar on these key demographic variables. During occupational activities, such as preparation and administration of these drugs, acute toxic effects have not been observed. The physical examination, medical history and laboratory findings of 39 study participants, were unremarkable. An elevated level of urinary platinum (0.63 µg/g creat.) was found in one exposed worker and none in the control group.

The method used showed a good linearity in the calibration range. The detection limit calculated as being equivalent in concentration at three standard derivations of the back ground signal was 0.011 µg/l. The within day reproducibility (n = 10) was determined under routine condition at 1 and 0.1 µg/l of platinum, was 9.33% and 1.83% respectively.

DISCUSSION

Urinary platinum was significantly increased in one urine sample of healthcare workers compared with a non exposed control group. The more recently developed analytical methods are developed and validated to establish occupational exposure. The sensitive method (ICP-MS) for the detection of exposure to C.S., including platinum, has shown its suitability and applicability in the detection of low exposure levels. It is striking that despite the introduction of safety guidelines and protective measures, healthcare workers are still exposed to these toxic drugs. In fact, spillage fractions represent between 1-5% of the C.S. used. Urinary platinum measured (with ICP-MS) as a marker for exposure to chemotherapeutic agents seems to be one of the best methods to identify C.S. contamination.

CONCLUSION

To summarise the proposed biomonitoring is simple, fast and selective. Since even very low exposure levels may result in an healthcare hazard, high sensitivity is required. This procedure can be successfully used to improve the safety of personal handling C.S.

REFERENCES

1. Valanis B., Vollmer W., Labuhn K., Glass A., Corelle C. Antineoplastic drug profession, handling activity and work site. *J. Occup. Med.*, 1992, 34, 149-155.
2. Sessink P.J.M., Boer K.A., Scheefhals A.P.H., Anzion R.B.M., Bos R.P. Occupational exposure to antineoplastic agents at several departments in a hospital. *Int. Arch. Occup. Environ. Health*, 1992, 64, 105-112.
3. Ensslin A.S., Stoll Y., Pethran A., Pfaller A., Römmelt H., Fruhmann G. Biological monitoring of cyclophosphamide and ifosfamide in urine of hospital personnel occupationally exposed to cytostatic drugs. *Occup. Environ. Med.*, 1994, 51, 229-233.
4. Mader R.M., Rizovski B., Steger G.G., Wachter A., Kotz R., Rainer H. Exposure of oncology nurses to methotrexate in the treatment of osteosarcoma. *Arch. Environ. Health*, 1996, 5, 310-314.
5. Burgaz S., Karahalil B., Bayrak P. et al. Urinary cyclophosphamide excretion and micronuclei frequencies in peripheral lymphocytes and in exfoliated buccal epithelial cells of nurses handling antineoplastics. *Mutat. Res.*, 1999, 439, 97-104.
6. Sessink P.J.M., Bos R.P. Drugs hazardous to healthcare workers. *Drug safety*, 1999, 20, 347-359.
7. Labuhn K., Valanis B., Schoeny R., Loveday K., Vollmer W.M. Nurses' and pharmacists' exposure to antineoplastic drugs: findings from industrial hygiene scans and urine mutagenicity test. *Cancer Nurs.*, 1998, 21, 79-89.
8. Hoet P., Haufroid V. Biological monitoring: state of the art. *Occup. Environ. Med.*, 1997, 54, 361-366.

Metal Ions in Biology and Medicine: vol. 7. Eds L. Khassanova, Ph. Collery, I. Maymard, Z. Khassanova, J.-C. Étienne.
John Libbey Eurotext, Paris © 2002 pp. 405-408.

Strontium concentrations in bone, whole blood and urine from osteoporotic subjects submitted to femur head repair surgery

M. Burguera[1], J.L. Burguera[1], M.L. Di Bernardo[1], O.M. Alarcón[1], E. Nieto[2], J.R. Salinas[2] and E. Burguera[3]

[1]*Faculty of Sciences (Venezuelan, Andean Institute for Chemical Research); [2]Faculty of Medicine, (Metabolic Bone Research Group); [3]Faculty of Odontology (Department of Preventive and Social Odontology), Los Andes University, Mérida 5101-A, Venezuela.*

ABSTRACT

A total of 23 subjects (17 female and 6 male) with ages between 47 and 89 years (69.39 ± 11.23) submitted to femur head fracture repair surgery were recruited for this study. The Sr content in bone, whole blood and urine had an inverse relationship with the age, being that in urine the most representative (49% compared to 17% and 14% for bone and blood, respectively. These weak associations might be attributed to the narrow age range studied. The concentration of Sr in bone (28.00 ± 7.68 μg g^{-1}) has a highly positive correlation (94%) with its content in blood (25.17 ± 6.28 μg l^{-1}) and less marked (only 14%) with that in urine (87.43 ± 18.94 μg l^{-1}), indicating that the Sr depleted from bone deposits circulates in the blood stream and then it is eventually excreted. The determination of Sr in blood could be useful for diagnosing osteoporosis, while its concentration in urine could be related to the decreased renal function in elderly population.

INTRODUCTION

The deposition of strontium (Sr) in bones is believed to be a result of two distinct processes: i) a rapid incorporation of the element attributed to ionic exchange, surface adsorption and/or by preosseous protein binding and ii) a slow incorporation of strontium into the lattice structure of the crystals during bone formation [1]. In old bones, the Sr is almost exclusively taken up onto the surface of crystals and therefore its depletion due to bone mass loss is easier [2]. These processes depend mainly on the strontium concentration in blood as well as on its urinary excretion. The accurate determination of Sr in complex matrixes gained in interest during the last decade in view of this element's potential to resemble calcium (Ca) absorption and so to better understand its metabolism, renal clearance and interaction with important biological processes in the human body [3]. Our research team recently developed sensitive, accurate and precise analytical procedures which allow routine determination of Sr in bone [4], blood and urine [5].

The purpose of this work was to show the correlation between strontium concentrations in samples of different types (bone, blood and urine) from osteoporotic patients who undergone surgery due to femur head fracture.

METHODS

Sr was determined in bone samples by atomic absorption spectrometry (AAS) with nitrous oxide/acetylene flame [4] while in blood and urine by AAS with electrothermal atomization [5].

Statistical evaluation of the data was carried out with Statistical Analysis System (SAS) software operated under UNIX using ANOVA and paired t-Student tests for a confidence level of 95% ($p = 0.05$). Sampling and samples processing have been described in the recent literature [4, 5].

RESULTS

Experimental results expressed as mean ± Standard Deviation (SD) are summarized in *table 1*. A total of 23 subjects, 17 postmenopausal female (F) and 6 male (M), all submitted to femur head fracture repair surgery were recruited for this study.

Table 1. Simple statistics for the experimental data

Variable	Male and Female*	Male*	Female*	Min.++	Max.++
Age	69.39±11.23	65.17±6.79	70.88±12.24	52.00	89.00
Sr in bone	28.00±7.68	25.33±5.38	28.94±8.26	21.00	52.00
Sr in blood	25.17±6.28	23.50±4.72	25.76±6.77	18.00	46.00
Sr in urine	87.43±18.94	89.50±24.68	85.53±15.52	58.00	135.00
n+++	23	6	17		

Mean ± Standard Deviation; ++ Min. and Max. refer to minimum and maximum Values for all patients; +++ number of subjects.

The bone mineral density (BMD) was assessed by dual energy X-ray absorptiometry at femoral trochanter and at L1-L4 lumbar spine vertebrae. Independent of the measurement site, the BMD values were below - 2.5 (t-score) for all patients, indicating osteoporosis [6]. In a recent publication [7] we reported that age is significantly related with the decreased Ca and Sr values in bone. This effect was significant over 55 and 45 years of age for male and female respectively. The subjects studied in this work were all over 47 years of age, thus only a weak correlation between the Sr concentration and their age might be expected. As indicated by the Pearson correlation coefficients *(table 2)* the Sr content in the three matrixes had an inverse relationship with the age, being that in urine the most representative (49%). The weak association of Sr in bone (17%) and in blood (14%) with the age could be attributed to the narrow age range studied this time.

Table 2. Pearson correlation coefficients*

	Age	Bone	Blood	Urine
Age	1.000			
Bone	-0.166 -0.242 -0.078	1.000		
Blood	-0.144 -0.274 0.521	0.939 0.972 0.707	1.000	
Urine	-0.488 -0.541 -0.308	0.123 0.239 0.213	0.138 0.262 0.123	1.000

* In each column, the numbers in the first row correspond to all subjects, second row to female and last row to male.

Sr in bone was positively and significantly correlated with its content in blood (r = 0.939). Obviously the Sr depleted from bone deposits will be circulating in blood *(fig. 1)* and its level in this matrix reflects the element's body burden. Therefore, the determination of Sr in blood may be useful for diagnosis of diseases related to its deficiency, like osteomalacia or osteoporosis, and also to get a better insight into its metabolism.

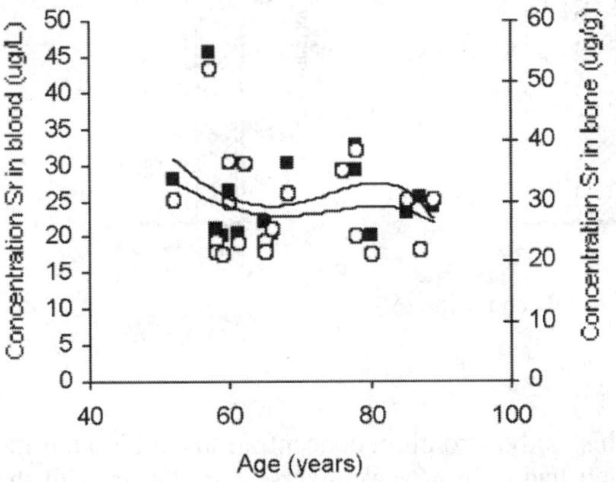

Fig. 1. Age-related blood (o) and bone (■) Sr concentrations

On the contrary, the concentrations of Sr in bone *(fig. 2A)* and in blood *(fig. 2B)*, although were not significantly correlated with those in urine (r = 0.138 and 0.123, respectively), showed an antagonic behavior. These findings may reflect an inefficient renal function in elderly population, although a stronger correlation might be found if 24 h period collection of urine is implemented.

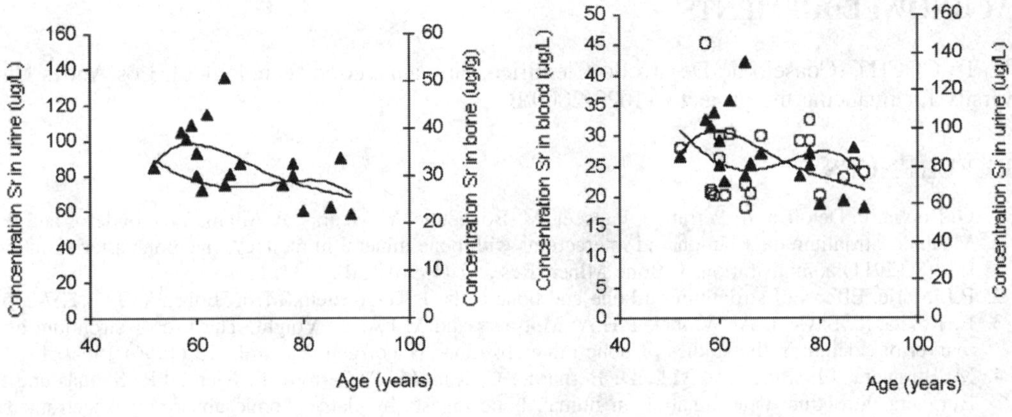

Fig. 2. Tendencies of bone (A □) and blood (B o) Sr concentrations with respect to its content in urine (▲)

Although Sr concentrations were not significantly influenced by sex *(fig. 3)* in either matrix, it is interesting to state that most of the osteoporotic population is represented by women (17 female compared to 6 male). As Sr resembles Ca in many biochemical processes, their depletion from bone, accentuated by the low hormonal activity might be responsible for the deterioration of bone quality in elderly population.

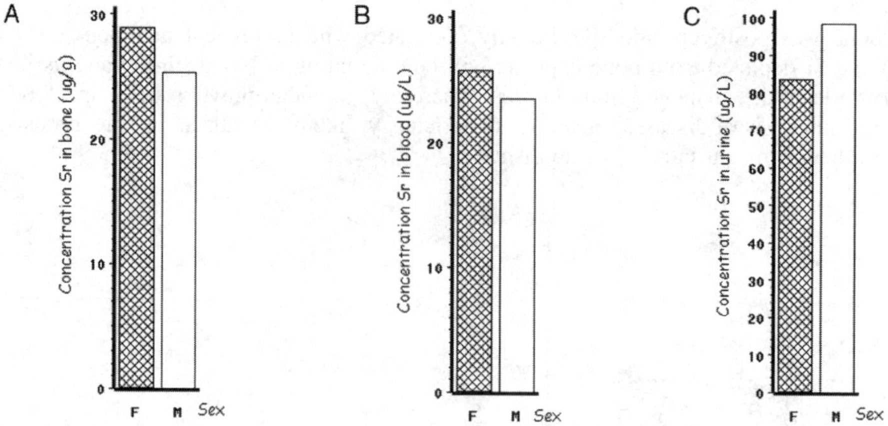

Fig. 3. Sex-related Sr concentrations in bone (**A**), blood (**B**) and urine (**C**)

CONCLUSION

For the osteoporotic population selected for this study, strontium concentrations in blood, urine and bones do not depend significantly on sex and had only a weak inverse correlation with the age probably due to the narrow age range of the subjects selected for this study. There is a highly positive statistical correlation between strontium levels in bones and blood assuring the homeostatic condition thus the determination of Sr in blood could be useful for diagnosis purposes. The behavior of Sr in urine might be related with the decreased renal function in old people. Extending this study to a more numerous elderly population and sampling 24 h period urine might be useful to: a) investigate the interactions of Sr with other elements (e.g. Ca) and b) get a better insight into the renal clearance in osteoporotic subjects.

ACKNOWLEDGEMENTS

To CDCHT (Consejo de Desarrollo Cientifico, Humanistico y Tecnologico), Los Andes University for financing the project C-1023-2000/B.

REFERENCES

1. G. Boivin, P. Deloffre, B. Perrat, G. Panczer, M. Boudeulle, Y. Mauras, P. Allain, Y. Tsouderos and P.J. Meunier, Strontium distribution and interactions with bone mineral in monkey iliac bone after strontium salt (S 12911) administration. J. Bone Miner. Res., 11 (1996) 1302-1311.
2. P.J. Marie, Effects of strontium on bone and bone cells, J. Trace Elem. Med. Biol., 9 (1995) 227-235.
3. E. Rokita, T. Savicki, A. Wrobel, P.H.A. Mutsaers and M.J.A. de Voight, The use of strontium as a marker of calcium in the studies of bone mineralization. Trace Elem Electrol., 13 (1996) 155-161.
4. M. Burguera, J.L. Burguera, M.L. Di Bernardo, C. Rondón, P. Carrero, E. Nieto, J.R. Salinas and E. Burguera. Strontium determination in human bone digests by flame atomic absorption spectrometry. Quim. Anal. 18 (1999) 305-312.
5. M. Burguera, J.L. Burguera, C. Rondón, M.L. Di Bernardo, M. Gallignani, E. Nieto and J.R. Salinas. Appraisal of different electrothermal atomic absorption spectrometric methods for the determination of strontium in biological samples. Spectrochim. Acta Part B, 54 (1999) 805-818.
6. G. Vezzoli, I. Baragetti, S. Zerbi, A. Caumo, L. Soldati, P. Bellinzoni, A. Centemero, A. Rubinacci, G. Moro and G. Bianchi. Strontium absorption and excretion in normocalciuric subjects: relation to calcium metabolism. Clin. Chem., 44 (1998) 586-590.
7. M. Burguera, J.L. Burguera, M.L. Di Bernardo, O.M. Alarcón, E. Nieto, J.R. Salinas and E. Burguera. Age- and sex-related calcium and strontium concentrations in different types of human bones. Trace Elem. Electol., (2001).

Metal Ions in Biology and Medicine: vol. 7. Eds L. Khassanova, Ph. Collery, I. Maymard, Z. Khassanova, J.-C. Étienne.
John Libbey Eurotext, Paris © 2002 pp. 409-411.

Iron-deficiency anemia and its genetic markers

G.Sh. Safuanova[1], A.A. Morozova[1], E.K. Khusnutdinova[2], Z.M. Sultanaeva[1], T.V. Victorova[2]

[1]*Bashkir State Medical University; Ufa, Russian Federation;*
[2]*Institute of Biochemistry and Genetics; Ufa, Russian Federation*

ABSTRACT

The polymorphism of genes cytochrome P-4501A1 (CYP1A1), glutathione S-transferase M1 (GSTM1), N-acetyltransferase 2 (Nat2), angiotensin-converting enzyme (ACE) and activator of plasminogene (AP) has been investigated among 102 patients with iron-deficiency anemia and 105 practically healthy persons (controls). The increased frequency of mutant form of gene CYP1A1 and homozygous deletion of gene AP in patients with iron-deficiency anemia showed significant difference from that in healthy controls. There were detected combinations of genotypes on polymorph systems CYP1A1, GSTM1, Nat2, ACE and AP that are characterized for iron-deficiency anemia patients. It's possible to use them as genetic markers of disease.

Key words: iron-deficiency anemia, genetic, genes, risk of disease, predisposition, combinations of genotypes.

INTRODUCTION

Iron is one of the most obligatory elements in human body and is inserted in erythrocariocytes and erythrocytes, lots of enzymes, ferritin and transferrin. Iron decrease determines progress of iron-deficiency anemia. This pathology is one of the most widespread diseases in the world [2, 5]. This problem is actual in Bashkortostan as a region with developed oil-chemistry industry [6]. There was established that environmental factors influence on developing of iron-deficiency anemia [3] mostly on genetic predisposition population [1].

The aim of our investigation was determination of genetic markers in iron-deficiency anemia patients.

METHODS AND MATERIAL

The material was DNA that was extracted from blood's lymphocytes by using method of phenol-chloroform extraction [4]. The polymorphism of genes cytochrome P-4501A1 (CYP1A1, genotypes: Ile/Ile, Ile/Val, Val/Val) glutathione S-transferase M1 (GSTM1, genotypes: 0/0, +/+, +/0), N-acetyltransferase 2 (Nat2, genotypes: S/S, S/R, R/R), angiotensin-converting enzyme (ACE, genotypes: I/I, I/D, D/D) and activator of plasminogene (AP, genotypes: I/I, I/D, D/D) has been performed among 102 patients with iron-deficiency anemia and 105 practically healthy persons (controls) by PCR method with using specific oligonucleotide primers.

RESULTS

We have performed molecular-genetic investigation Ile/Val polymorphism of gene CYP1A1 in iron-deficiency anemia patients and control individuals. Polymorphism of gene CYP1A1 consists in oligonucleotide amino acid's replacement from isoleucine to valin that leads to appearance of restriction site to restrictase HincII. During PCR fragment of gene CYP1A1 is amplificated by length of 139 p.n. If restriction site is lost this fragment would be changed (allele Ile). When restriction site is present this DNA locus would be hydrolyzed with appearance fragments length of 120 and 19 p.n. (allele Val). The increased frequency of Val allele of gene CYP1A1 in patients with iron-deficiency anemia (10,21%) has showed significant difference from that in healthy controls (4,29%, p<0,05, OR=3,17) *(table 1)*. The frequency of Ile/Val genotypes of CYP1A1 gene in patients with iron deficiency anemia (16.12%) was also statistically significant compared with the controls (5,71%, p<0,05, OR=2,51).

Table 1. Frequency of genotypes and alleles gene CYP1A1 in iron-deficiency anemia patients and controls

Examined groups	N	Frequency of genotypes gene CYP1A1			Frequency of alleles gene CYP1A1	
		Ile/Ile n (%)	Ile/Val n (%)	Val/Val n (%)	Ile n (%)	Val n (%)
IDA patients	93	76 (81,72)	15 (16,12)	2 (2,15)	167 (89,79)	19 (10,21)
controls	70	65 (92,86)	4 (5,71)	1 (1,43)	134 (95,71)	6 (4,29)
χ^2		1,42	5,57	1,93	1,78	6,95

We have studied frequency of homozygous on deletion of gene GSTM1 by length approximately 10 th.p.n. that leads to formation null genotype GSTM1 0/0. Homo- and heterozygous of normal allele (genotypes GSTM1 +/+, +/0) was determined by present of amplification's fragment length of 271 p.n. Analysis of frequencies genotypes GSTM1 gene in iron-deficiency anemia patients and controls showed tendency of increase frequency of normal allele in iron-deficiency anemia patients (62,24%; 49,44%; p>0,05).

Table 2. Frequency of genotype's combinations on polymorph systems CYP1A1, GSTM1, Nat2, ACE and AP in iron-deficiency anemia patients and controls

Genotype's combinations on polymorph systems CYP1A1, GSTM1, Nat2, ACE and AP	IDA patients (%)	Controls (%)	Odds Ratio
Ile/Ile,0/0,S/R,I/D,I/D	4,17	0	9,75
Ile/Ile,+/+,S/S,I/D,I/D	4,17	0	9,75
Ile/Ile,+/+,S/R,I/I,I/D	5,56	0	12,84
Ile/Ile,+/+,S/R,D/D,I/I	5,56	0	12,84
Ile/Val,0/0,R/R,I/I,I/D	4,17	0	9,75

Polymorphism of gene Nat2 was studied. Amplificated fragment by length of 547 p.n. had three polymorph sites that were determined by restrictases KpnI, TagI and BamH. Loss one of the restriction sites leads to formation mutant allele (S-form) of gene Nat2. Dominant S-allele have as iron-deficiency anemia patients (55,7%) as controls (57,7%, χ^2=0,001, p=1,00).

Gene's polymorphisms of ACE and AP are determined by present (insertion) or loss (deletion) of Alu-repetition. Investigation these gene's polymorphisms showed us no significant differences between iron-deficiency anemia patients and controls.

We have combined genotypes of five investigated genes and have detected combinations of genotypes on polymorph systems CYP1A1, GSTM1, Nat2, ACE and AP that are characterized for iron-deficiency anemia patients. They are: Ile/Ile,+/+,S/R,D/D,I/I, (OR=12,84); Ile/Ile,+/+,S/R,I/I,I/D, (OR=12,84); Ile/Val,0/0,R/R,I/I,I/D, (OR=9,75); Ile/Ile,0/0,S/R,I/D,I/D (OR=9,75); Ile/Ile,+/+,S/S,I/D,I/D (OR=9,75) *(table 2)*.

CONCLUSIONS

1. Mutant form of gene CYP1A1 is characterized for iron-deficiency anemia patients.
2. Combinations of genotypes Ile/Ile,+/+,S/R,D/D,I/I; Ile/Ile,+/+,S/R,I/I,I/D; Ile/Val,0/0, R/R,I/I,I/D; Ile/Ile,0/0,S/R,I/D,I/D; Ile/Ile,+/+,S/S,I/D,I/D have possibility to be used as genetic markers of iron-deficiency anemia.

REFERENCES

1. Baranov V.S., Barnova E.V., Ivachenko T.E., Aseev M.V. Human genome and "predisposition" genes: Introduction to predictive medicine.-SPb.: "Intermedica", 2000. - 271 p.
2. Cmunt E., Umlaufova A., Tichva I. et al. Normal blood values in the adult population in the Czech Republic / Cas. Lec. Cesh. - 1995. - Vol. 134. - N 19. - P. 630-633.
3. Kozinets G.I., Novodergkina A.I. Ecological factors and constant of blood formation / Actual Questions of Hematology and Transfusiology: Abstracts of the third Russian conference of hematologists and transfusiologists. - M.: Altus, 1996. - P. 9.
4. Mathew C. C. The isolation of high molecular weight eucariotic DNA // Methods in Molecular Biology / Eds. Walker J. M. - N.Y.: Human Press, 1984. - Vol. 2. - P. 31-34.
5. Milmon N., Kirchhoff. Iron stores in 1359, 30 - to 60 year old. Danish women: evaluation by serum ferritin and haemoglobin / Ann. Haematolog. - 1992. - Vol. 64. - N 1. - P. 22-27.
6. Nikulicheva V.I. Iron-deficiency anemies. - Ufa, 1993. - 201 p.

*Metal Ions in Biology and Medicine: vol. 7.*Eds L. Khassanova, Ph. Collery, I. Maymard, Z. Khassanova, J.-C. Étienne.
John Libbey Eurotext, Paris © 2002 pp. 412-414.

Selenium and haemodialysis

RM. Pérez-Beriain[1], A. García de Jalón[1], JF. Escanero[2], MD. Zapatero[1], M. Guerra[2], ML. Calvo Ruata[1], T. Pérez-Beriain[3]

[1]Biochemistry department. Miguel Servet Universitary Hospital. Pº Isabel la Católica 1-3. 50009 Zaragoza, Spain. [2]Phisiology department of University. C/ Domingo Miral s/n. 50006 Zaragoza, Spain. [3]Veterinary School, C/ Miguel Servet 177, 50013 Zaragoza, Spain.
The mail will be remitted to:
Dr. Ángel García de Jalón Comet.
Jefe de Servicio de Bioquímica Clínica
Hospital Universitario Miguel Servet
Paseo Isabel la Católica 1-3
50009 ZARAGOZA (Spain)
E-mail: jcomet@teleline.es

ABSTRACT

In the bibliografy, some studies show lower selenium levels in patients on regular hemodialysis treatment compared to healthy controls.

Patients and methods : 466 chronically dialyzed patients and 256 age-matched healthy controls were enrolled in the study. All patients were divided into 3 groups: 1) patients treated by 1-24 months hemodialysis, 2) patients treated by 25-49 months hemodialysis, and, 3) patients treated by 50 months or more.

Serum selenium levels were measured by atomic absorption spectrophotometry (AAS) with a graphite furnace and Zeeman background corrector.

The statistical test used has been the analysis of variance (ANOVA).

Results and conclusions : In all patients groups, selenium concentrations were significantly lower (group 1 = 57.04 µg/l; group 2 = 56.11 µg/l; group 3 = 52.90 µg/l) than in healthy subjets group (75.02 µg/l). The differences between group 1 and group 3 was statistically significant.

The findings of this study suggest that selenium levels in hemodialysis patients in lower than healthy population, and serum selenium decrease during hemodialysis. Supplementation with this trace element may be indicated.

BACKGROUND

Epidemiological, animal and human studies have indicated that selenium deficiency is a risk factor for death from malignant diseases. The mechanisms that could modify selenium status may, therefore, be of particular interest in hemodialysis patients, considering their high cancer mortality rates [1].

Some studies have shown that haemodialysis is connected with increased free radicals production. Oxidative damage due to free radical production is increased in uraemic patients and has been suggested as a possible factor contributing to the anaemia of chronic renal failure and the pathogenesis of atherosclerosis [2].

One factor contributing to this reduction of antioxidative capacity is selenium deficiency [3].

In the bibliografy, some studies show lower selenium levels in patients on regular hemodialysis treatment compared to healthy controls [4-6].

The question about supplementation of antioxidants in hemodialysis patients is open although are some positive data regarding the use of moderate and safe selenium supplementation in hemodialysis patients [7].

AIMS

Our objetive, was to investigate the existence of an alterared selenium balance in patients on regular hemodialysis treatment, considering the duration of the treatment.

MATERIALS AND METHODS

Subjects

466 chronically dialyzed patients and 256 age-matched healthy controls were enrolled in the study.

All patients were divided into 3 groups:

1) patients treated by 1-24 months hemodialysis,
2) patients treated by 25-49 months hemodialysis, and,
3) patients treated by 50 months or more.

Instrumentation

Serum selenium levels were measured by atomic absorption spectrophotometry (AAS) with a graphite furnace and Zeeman background corrector *(Perkin Elmer 4110 ZL)*, using $Pd(NO_3)_2$ solution as matrix modifier.

Statistical analysis

The statistical calculations were carried out using SPSS statistics program. The statistical test used has been the analysis of variance (ANOVA). The Multiple Range Test used was the LSD test with significance level of 0.05.

RESULTS

The results of the serum concentratios of selenium in the groups *(table 1)* were as follows:

Table 1.

	N	Selenium (µg/l)	C.I. (95%)
Control group	256	75.02	73.33-76.72
Hemodialyzed patients:			
• treated by 1-24 months	260	57.04	55.40-58.69
• treated by 25-49 months	96	56.11	53.41-58.82
• treated by 50 months or more	110	52.91	50.17-55.65

In all patients groups, selenium concentrations were significantly lower (group 1 = 57.04 µg/l; group 2 = 56.11 µg/l; group 3 = 52.90 µg/l) than in healthy subjets group (75.02 µg/l), with a $p<0.001$.

The differences between group 1 and group 3 *(fig. 1)* was statistically significant too.

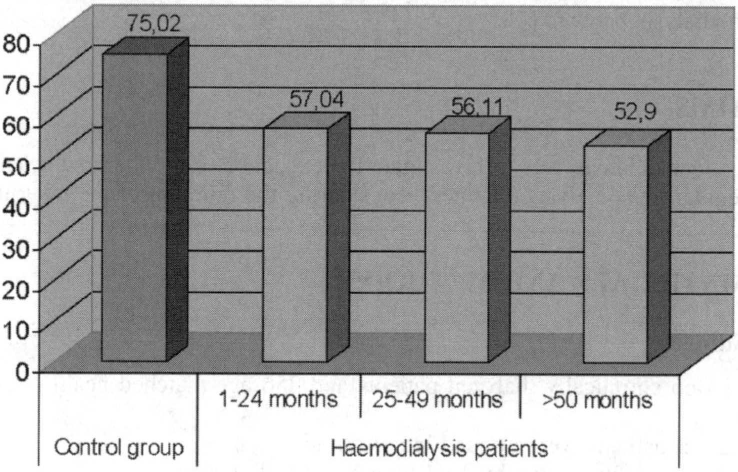

Fig. 1.

CONCLUSIONS

The findings of this study suggest that selenium levels in hemodialysis patients in lower than healthy population, and serum selenium decrease during hemodialysis.

Selenium deficiency is to be suspected in dialyzed patients and selenium supplementation may be beneficial (increasing glutathione peroxidase activity, cardioprotective effect, inmunostimulatory properties) for chronic renal failure patients.

Supplementation with a trace element may be indicated when its depletion was unequivocally documented and when there is evidence of the positive effects of this element on the quality of life of the hemodialyzed patients.

REFERENCES

1. Bogye G, Tompos G, Alfhan G. Selenium depletion in hemodialysis patients treated with polysulfone membranes. Nephron. 2000 Feb. 84(2). P 119-23.
2. McGrath LT, Douglas AF, McClean E, Brown JH, Doherty CC, Johnston GD, Archbold GP. Oxidative stress and erythrocyte membrane fluidity in patients undergoing regular dialysis. Clin Chim Acta. 1995 Mar 31. 235(2). P 179-88.
3. Koening JS, Fisher M, Bulant E, Tiran B, Elmadfa I, Druml W. Antioxidant status in patients on chronic hemodialysis therapy: impact of parenteral selenium supplementation. Wien Klin Wochenschr. 1997 Jan 17. 109(1). P 13-9.
4. Zima T, Mestek O, Nemecek K, Bartova V, Fialova J, Tesar V, Suchanek M. Trace elements in hemodialysis and continuous ambulatory peritoneal dialysis patients. Blood Purif. 1998. 16(5). P 253-60.
5. Lin TH, Chen JG, Liaw JM, Juang JG. Trace elements and lipid peroxidation in uremic patients on hemodialysis. Biol Trace Elem Res. 1996 Mar. 51(3). P 277-83.
6. Bonomini M, Forster S, Manfrini V, De Risio F, Steiner M, Vidovich MI, Klinkmann H, Ivanovich P, Albertazzi A. Geographic factors and plasma selenium in uremia and dialysis. Nephron. 1996. 72(2). P 197-204.
7. Zima T, Janebova M, Nemececk K, Bartova V. Retinol and alpha-tocopherol in hemodialysis patients. Ren Fail. 1998 May. 20(3). P 505-12.

Metal Ions in Biology and Medicine: vol. 7. Eds L. Khassanova, Ph. Collery, I. Maymard, Z. Khassanova, J.-C. Étienne.
John Libbey Eurotext, Paris © 2002 pp. 415-417.

Could epidemic pathogenic selection (EPS) be an explanation for hereditary hemochromatosis?

Sharon Moalem[1,2], Maire E. Percy[1,2,3], Theo Kruck[1] and Richard R. Gelbart[4]

[1]*Department of Physiology, University of Toronto, Toronto, Ontario, Canada M5S 1A8;* [2]*Neurogenetics Laboratory, Surrey Place Centre, 2 Surrey Place, Toronto, Ontario, Canada M5S 2C2;* [3]*Department of Obstetrics & Gynaecology, University of Toronto, Ontario, Canada;* [4]*Department of Languages, Literatures and Linguistics, York University, Toronto, Canada.*

ABSTRACT

Using a newly developed metod by HPLC with ultraviolet detection we measured plasma coenzyme Q_{10} (CoQ_{10}) level in group of 43 children (19 femals and 24 males: ages 1 month - 9 years) with an acute inflammatory process. The results for coenzyme Q_{10} were expressed as molar concentration (µmol/l plasma). Our study confirmed that CoQ_{10} concentration (median - 0.8 µmol/l) was independent of sex, and we established which biochemical parameters influence on ubiquinone levels. The results indicate that CoQ_{10} concentration is connected with leukocytosis, calcium and magnesium levels. These findings suggested that transferin, amylase and serum glutamic transaminase may also determine the CoQ_{10} levels plasma.

Key words: coenzyme Q_{10} level, pediatric population, acute inflammatory process.

INTRODUCTION

Mutations were recently discovered in individuals with hereditary hemochromatosis (HH), an iron metabolism disorder that leads to excessive iron uptake and deposition. Two HH mutations initially described result in a cysteine-to-tyrosine substitution (C282Y) and a histidine-to-aspartic acid substitution (H63D) in a protein, coded by the gene now named HFE, on chromosome 6p21.3 [1, 2]. The C282Y mutation is causally associated with HH, but the role of the other mutations in HH, including H63D, is still controversial [3]. The C282Y mutation prevents HFE protein from associating with β_2-microglobulin and from being predominantly expressed on the cell surface. Thus cells expressing C282Y mutation protein might be protected from iron seeking organisms or organisms that might require HFE antigen for cellular recognition, adhesion and/or entry. As many pathogens are iron-requiring organisms, they may have receptors that acquire iron from transferrin or from transferrin receptor-HFE complexes on cells.

The HFE gene product is a major histocompatibility complex (MHC) class I-like protein and is highly expressed in epithelial cells, particularly the crypt enterocytes of the small intestine [3, 4, 5]. The latter observation is important, since HFE is involved in iron uptake, metabolism and regulation, and absorption of dietary iron occurs in crypt cells [4]. Iron seeking microorganisms may well target crypt cells as a source of iron and concomitantly as a point of entry into the host. Mutations in HFE, which increase clearance of iron from the intestinal lumen, should reduce the pathogenicity of infective agents that depend upon the availability of iron to proliferate and infect the human host.

Another hallmark of HH is very low iron levels in macrophages and in reticuloendothelial cells in spite of body iron overload. This phenomenon is reversed when the wild type HFE gene is inserted back into the macrophages [6]. This finding is interesting because while there is iron loading in parenchymal cells with HFE mutations, the reverse is true for macrophages. The reason for the paradox may be explained by considering the fact that as the level of iron increases within macrophages so does their susceptibility to some pathogens [7]. Furthermore, mutations in the gene NRAMP 1, which involves iron transport and affects macrophage function, seem to be responsible for susceptibility to intracellular pathogens such as the aforementioned *Salmonella typhi* and *Mycobacterium tuberculosis* as well as other pathogenic organisms [8]. Thus a reduction in iron within macrophages and not just increased iron clearance from the intestinal lumen might result in the selection for HFE mutations by pathogens. Overall, decreased susceptibility might be conferred either by limiting the availability of iron, or removal of the presentation of HFE as an antigen.

POPULATION DISTRIBUTIONS OF HFE MUTATIONS

HH is the most common inherited single gene disorder in people of northern and western European descent. Although H63D is present in most populations studied so far, the highest frequencies still are in Europeans and, like C282Y, especially in those of northern and western European descent [9]. These observations invite the conclusion that a very significant and specific type of genetic selection for C282Y and possibly H63D must have occurred.

HFE MUTATIONS AND THE PLAGUE

Researchers have estimated, using linkage-disequilibrium, that the C282Y mutation originated 60-70 generations ago [10]. Assuming a generation time of 15-20 years, this event dates to 600-1100 A.D. Given this time frame, we examined catastrophic events in the last 1500 years that might result in selection and amplification of C282Y and/or H63D. The one event, beginning possibly as early 6th century, that had a striking effect on the population of Europe, was the Black Death - also known as the bubonic plague, that is thought to have been caused by the pathogen *Yersinia pestis* [11]. Almost one-third of the European population succumbed to this generally fatal disease. According to some scholars, the plagues of the 14th century reduced the population of England by as much as 40 percent [12]. Some of the regions that suffered the highest mortality rates during the initial plague phases of the 14th century - Northern Europe, Belgium, England and France - are today among the countries that have the highest prevalence of HFE mutations [9].

EPIDEMIC PATHOGENIC SELECTION (EPS)

During disease epidemics, some forms of class I and class II MHC molecules are known to stimulate T cell responses that favour better survival, but which ones they are depends upon the particular type of infection [13]. Thus, our suggestion that plague epidemics or other pathogens might have altered the frequency of the C282Y, and possibly also the H63D mutation, is not without precedent. Yet, if C282Y is protective should we not also find it in areas, such as China, that have been historically endemic for the plague? One may think so, but the mutation would have to be initially present for a selective factor to have influence. Because epidemics of different pathogens affect survival in different ways, we suggest that any pathogen-host interactions that alter the frequency of a particular genetic marker, or result in disequilibrium between two or more genetic markers, be called Epidemic Pathogenic Selection (EPS). As outlined, the concept of EPS can be used to help identify a particular pathogen that selects for a mutation and changes its prevalence in a population's history, to determine when this selection occurs, and to glean knowledge about the selection mechanism. We have suggested that certain mutations in HFE (e.g., C282Y) hinder pathogens from entering the epithelium, in the small intestine and/or in the lung, thus reducing their pathogenicity. Decreased susceptibility might be conferred either by limiting the availability of iron to infective agents, and/or possibly by removal of the presentation of HFE as an antigen which a pathogen might require for cellular recognition, adhesion and/or entry. Based on reasoned evidence, we hypothesize that mutations in the HFE gene are involved in defense against pathogenic infection.

ACKNOWLEDGMENTS

This work was supported, in part, by grants to Sharon Moalem from the Alzheimer Society of Canada and the Margaret and Howard Gamble Research Grant. Maire E. Percy was supported by grants from the Alzheimer Society of Canada and the Queen Elizabeth Hospital Research Institute. The authors are grateful to Surrey Place Centre for infrastructure support and to Dr. John Percy for helpful discussions and comments.

REFERENCES

1. Kaplan, F., *Tay-Sachs disease carrier screening: a model for prevention of genetic disease*. Genet Test, 1998. **2**(4): p. 271-92.
2. Feder, J.N., et al., *A novel MHC class I-like gene is mutated in patients with hereditary haemochromatosis*. Nat Genet, 1996. **13**(4): p. 399-408.
3. Jazwinska, E.C., *Hemochromatosis: a genetic defect in iron metabolism*. Bioessays, 1998. **20**(7): p. 562-8.
4. Blumberg, R.S., *Current concepts in mucosal immunity. II. One size fits all: nonclassical MHC molecules fulfill multiple roles in epithelial cell function*. Am J Physiol, 1998. **274**(2 Pt 1): p. G227-31.
5. Parkkila, S., et al., *Association of the transferrin receptor in human placenta with HFE, the protein defective in hereditary hemochromatosis*. Proc Natl Acad Sci USA, 1997. **94**(24): p. 13198-202.
6. Montosi, G., et al., *Wild-type HFE protein normalizes transferrin iron accumulation in macrophages from subjects with hereditary hemochromatosis*. Blood, 2000. **96**(3): p. 1125-9.
7. Barton, C.H., et al., *Nramp1: a link between intracellular iron transport and innate resistance to intracellular pathogens*. J Leukoc Biol, 1999. **66**(5): p. 757-62.
8. Canonne-Hergaux, F., et al., *The Nramp1 protein and its role in resistance to infection and macrophage function*. Proc Assoc Am Physicians, 1999. **111**(4): p. 283-9.
9. Merryweather-Clarke, A.T., et al., *Global prevalence of putative haemochromatosis mutations*. J Med Genet, 1997. **34**(4): p. 275-8.
10. Ajioka, R.S., et al., *Haplotype analysis of hemochromatosis: evaluation of different linkage- disequilibrium approaches and evolution of disease chromosomes*. Am J Hum Genet, 1997. **60**(6): p. 1439-47.
11. McNeill, *Plagues and People*. 1998, Garden City (NY): Doubleday.
12. Poos, L.R., *Plague mortality and demographic depression in later medieval England*. Yale J Biol Med, 1981. **54**(3): p. 227-34.
13. Parnham, P., *Immunogenetics: soaring costs in defense*. Nature, 1999. **401**: p. 870-871.

This work was supported by the grant BW264476 from Medical Academy of Bialystok.

Metal Ions in Biology and Medicine: vol. 7. Eds L. Khassanova, Ph. Collery, I. Maymard, Z. Khassanova, J.-C. Étienne.
John Libbey Eurotext, Paris © 2002 pp. 418-420.

The content of zinc in brain in cases of suicide

Krystyna Sadlik[1], Bernadeta Szewczyk[2], Wojciech Piekoszewski[1,3], Franciszek Trela[4], Gabriel Nowak[2,5]

[1]Institute of Forensic Research, Kraków, Poland; [2]Institute of Pharmacology, Polish Academy of Sciences, Kraków, Poland; [3]Department of Clinical and Industrial Toxicology, Collegium Medicum, Jagiellonian University, Kraków, Poland; [4]Department of Forensic Medicine, Collegium Medicum, Jagiellonian University, Kraków, Poland; [5]Laboratory of Radioligand Research, Collegium Medicum, Jagiellonian University, Kraków, Poland

ABSTRACT

Zinc plays an important role in neurotransmission and in other systems of living organisms. In recent years, the role of zinc in depression has attracted the attention of researchers.

The aim of the work was to determine the content of zinc in samples of brain (frontal cortex, hippocampus, cerebellum) taken from the bodies of persons who have committed suicide, and then to assess whether this disorder (depression) influences the content and distribution of this element in the studied anatomical structures of the brain.

The work is a preliminary investigation. 14 persons were studied, of whom 8 had committed suicide and 6 had died of other causes (control group). Determination of zinc was carried out by the method of flame atomic absorption spectrometry following microwave digestion.

The mean content of zinc was lower in persons who had committed suicide (the greatest difference was ascertained in the cerebellum). The mean content of zinc [micrograms per gram of wet weight] in the control and studied group were in the cerebellum - 12.1 and 9.7; in the frontal cortex - 10.8 and 9.4; in the hippocampus - 11.0 and 10.7; respectively.

The obtained results may indicate the possible role of zinc in the psychopathology of suicide (including depression). This conclusion should, however, be regarded as preliminary, because, amongst other reasons, both the group of studied persons and the control group was small.

INTRODUCTION

Amongst persons who die as a result of suicide (commit suicide), a significant percentage (est. 50-70%) of cases is made up of persons suffering from depression. In recent years a marked, progressive increase in the incidence of depression has been noted. It is estimated that various forms of depression and depressive symptoms occur in about 10% of the population (many depressive symptoms are considered normal and that is why depression is rarely recognized and treated). The mechanisms of the pathophysiology and therapy of depression are not precisely known at present. In recent years attention has turned towards a link between the nervous and immunological system and the homeostasis of zinc in depression. This element plays a fundamental role in many biochemical processes in the living organism; amongst others roles, it is a modulator in the central nervous system. Several authors have found a lowered level of zinc in the blood of persons suffering from depression [2, 3, 4, 7]. At the current stage of research it is difficult to ascertain whether this drop in the level of zinc precedes the occurrence of depression, or whether it is a result of it [6]. After successful anti-depressive therapy, the lowered level of zinc returned to normal [2, 8]. In animals, with experimentally induced depression, the antidepressant activity of zinc has also been demonstrated [1].

THE AIM

The aim of the work was:

- to define the content of zinc in segments of the frontal lobe cortex, hippocampus and cerebellum taken from persons who had committed suicide and persons who had died of other causes (control group),
- to assess, on the basis of results obtained, whether there exists a difference in content of this element between the two groups.

MATERIAL AND METHODS

14 men aged 19 to 67 were studied. This group included 8 persons (aged 19 to 67) who had died as a result of suicide by hanging and 6 persons (aged 24 to 46) who had died of other causes.

Fragments of the brain (frontal cortex, hippocampus and cerebellum) were collected during autopsy at the Forensic Medicine Department, Collegium Medicum of Jagiellonian University. Before analysis of each sample it was divided in two, weighed (0.3 g - 0.6 g) and wet-digested with nitric acid and hydrogen peroxide (microwave digestion, Milestone MLS-1200 Mega Microwave Digestion System). Determination of zinc was carried out by flame atomic absorption spectrometry. The equipment used was a Pye Unicam SP-9 800 AA Spectrophotometer with deuterium background correction (air flow was 4.5 L/min, acetylene flow - 1.1 L/min, analytical wavelength - 213.9 nm). Relative standard deviation (RSD) of the method (the whole analytical procedure: digestion + zinc determination) did not exceed 2.4%. Mean recovery of zinc was 99% (SD 0.78).

RESULTS

Results obtained - mean values and standard deviations (SD) and value ranges - are shown in *table 1*.

Table 1. Content [micrograms/g of wet weight] of zinc in the studied material

Material	Examined group (N=8)		Control group (N=6)	
	Mean (SD)	Range	Mean (SD)	Range
Frontal cortex	9.4 (1.77)	7.8 - 13.3	10.8 (0.90)	10.1 - 12.3
Hippocampus	10.7 (2.13)	8.7 - 14.3	11.0 (0.85)	10.2 - 12.4
Cerebellum	9.7 (1.55)	8.7 - 12.7	12.1 (1.70)	8.8 - 13.2

In particular cases the highest zinc content was observed in the hippocampus. The mean zinc content in the studied group was lower than in the control group. The greatest difference was ascertained in the cerebellum - the mean zinc content in the studied group relative to the control group was lower by about 20%. In the frontal cortex the mean content of zinc was lower by about 10%. A significant difference in mean content of this element in the hippocampus was not ascertained between groups.

CONCLUSIONS

The presented results may indicate that zinc has a role in the psychopathology of suicides (including depression). This conclusion should, however, be regarded as preliminary, because, amongst other reasons, the group of studied persons was small, as was the control group, and, furthermore, other factors, such as age, diet and other illnesses, may also influence the content of zinc. In order to verify the above conclusion further studies are required.

REFERENCES

1. Kroczka B., Branski P., Palucha A. et al.: "Antidepressant-like properties of zinc in rodent forced swim test", Brain Research Bulletin, 55, 297-300, 2001.
2. Maes M., D'Haese P. C., Scharpe S. et al.: "Hypozincemia in depression", Journal of Affective Disorders, 31, 135-140, 1994.
3. Maes M., Vandoolaeghe E., Neels H. et al.: "Lower serum zinc in major depression is a sensitive marker of treatment resistance and of the immune/inflamatory response in that illness", Biological Psychiatry 42(5), 394-58, 1997.
4. Narang R. L., Gupta K., Narang A. P., Singh R.: "Levels of copper and zinc in depression" Indian Journal of Physiology & Pharmacology, 35, 272-274, 1991.
5. Nowak G.: "Alterations in zinc homeostasis in depression and antidepressant therapy", Polish Journal of Pharmacology, 50, 1-4, 1998.
6. Nowak G., Kubera M., Maes M.: "Neuroimmunological aspects of the alterations in zinc homeostasis in the pathophysiology and treatment of depression", Acta Neuropsychiatrica, 12, 49-53, 2000.
7. Nowak G., Zieba A., Dudek D. et al.: "Serum trace elements in animal models and human depression. Part I. Zinc", Human Psychopharmacology Clinical and Experimental, 14, 83-86, 1999.
8. Schlegel-Zawadzka M., Zieba A., Dudek D. et al.: "Effect of depression and antidepressant therapy on serum zinc levels - a preliminary clinical study" in: Trace Elements in Man and Animals 10, Kluwer Academic Plenum Press, 607-610, 2000.

*Metal Ions in Biology and Medicine: vol. 7.*Eds L. Khassanova, Ph. Collery, I. Maymard, Z. Khassanova, J.-C. Étienne.
John Libbey Eurotext, Paris © 2002 pp. 421-426.

Coenzyme Q_{10} level in plasma of children with inflammatory process

Bozena Mikoluc[a], Joanna Karpinska[b]*, Radoslaw Motkowski[a], Janina Piotrowska - Jastrzebska[a]

[a]*Department of Propedeutics of Pediatrics,
Medical Academy, Bialystok, ul. J. Waszyngtona 17
15-274 Bialystok, Poland.*
[b]*University of Bialystok, ul. J. Pilsudskiego 11/4
15-447 Bialystok, Poland*

ABSTRACT

Using a newly developed metod by HPLC with ultraviolet detection we measured plasma coenzyme Q_{10} (CoQ_{10}) level in group of 43 children (19 femals and 24 males: ages 1 month - 9 years) with an acute inflammatory process. The results for coenzyme Q_{10} were expressed as molar concentration (µmol/l plasma). Our study confirmed that CoQ_{10} concentration (median - 0.8 µmol/l) was independent of sex, and we established which biochemical parameters influence on ubiquinone levels. The results indicate that CoQ_{10} concentration is connected with leukocytosis, calcium and magnesium levels. These findings suggested that transferin, amylase and serum glutamic transaminase may also determine the CoQ_{10} levels plasma.

Key words: coenzyme Q_{10} level, pediatric population, acute inflammatory process.

INTRODUCTION

The coenzyme Q_{10} (ubiquinone) plays an important role in bioenergetic conditions in nearly all human tissues. It participated in electron transport and ATP synthesis related to the mitochondrial respiratory chain. The reduced form of ubiquinone protects cells from peroxidative damage [1]. In blood coenzyme Q_{10} (CoQ_{10}) is transported by lipoproteins and prevent free radical damage caused by neutrophils and oxidative injury by endothelial cells in ischemia - reperfusion. Adequate serum concentrations of CoQ_{10} seem necessary to prevent peroxidative damage [1]. The studies of [2] showed the hematopoetic activites of CoQ_{10} in animal models and children having kwashiorkor. It seems to be interesting that patients with AIDS showed a "striking" clinical response to therapy with CoQ_{10}. The macrophage potentiating activity of CoQ_{10} and CoQ_{10} significantly increased the levels of IgG in patients. Coenzyme Q_{10} has been promoted as an effective agent for reducing the deleterious effects of septic shock by acting as an oxygen free radical scavenger and thus stabilizing mitochondrial membranes and by inhibiting the arachidonic acid metabolic pathway and the formation of various prostaglandis [3].

The last years studies have contributed to a better knowledge of the function of the CoQ_{10} in many diseases like cancer [4], various thyroid disorders [5], diabetes mellitus [6], cardiovascular [7] and mitochondrial diseases [8]. Little is known about the CoQ_{10} concentrations in plasma in pediatric disorders. The studies of [9] showed that normal fetal CoQ_{10} plasma levels are lower than 0,3 mg/ml. [10] established serum reference values for the next two groups of pediatric population. The first group aged, 1 month - 7 years - median CoQ_{10} levels was - 0,8 µmol/l and second

8-18 years - median CoQ_{10} levels was - 0,57 μmol/l. They revealed that CoQ_{10} concentration was independent of sex, decreased significantly with age and correlated with cholesterol concentrations.

Taking into consideration the above findings we measured the plasma CoQ_{10} concentration during the first 24 hours of acute inflammatory process in pediatric population.

MATERIAL AND METHODS

Patients

Forty three children with acute inflammatory process were included into the study. In this group of children was 19 femals and 24 males ages 1 month - 9 years.

Chemicals

All reagents used were of chromatographic grade. Coenzyme Q_{10} - Sigma-Aldrich, Germany. The standard solution of Q_{10} (100 ppm) was prepared by dissolving appropriate amount of compound in n-hexane. Working solutions were prepared by the appropriate dilutions of the stock standard solution with n-hexane. The working solutions were prepared freshly every day. The stock solution was stored in brown bottles at 4°C.

METHODS

Apparatus

Chromatographic analyses were performing using the HPLC system, Merck, Germany, consisted of: pump type L-6200A Inteligent Pump, a reversed-phase analytical column, Lichrospher 100 RP-18 250×4 mm (5 μm) with a guard column 4×4 mm (5 μm) (Merck, Germany), detector DAD, type L-4500 (Merck, Germany). For manual injection a injection valve with 20 μl sample loop was used. All system was managed by interfejs HPLC-Manager, D-6500. The results were recorded by PC-computer supplied with DryLab Windows G/plus programme (LCResourcesinc). The mobile phase used consisted 72% of methanol and 28% of n-hexane at flow rate 1.5 ml/min.

Assay of coenzyme Q_{10} in plasma samples

0.5 ml of pooled human plasma was pipetted in triplicate into 2 ml polypropylene centrifuge tubes. Next 0.1 ml of methanol and 0.1 ml of hexane and n-propanol were added. The tubes were mixed for 1 minute and centrifuged for 15 minutes at 2000g to separate the layers. The n-hexane phase was transferred to small vial and extraction was repeated with 0.1 ml portion of n-hexane alone. The hexane extracts were dried under nitrogen atmosphere at room temperature. The residue was dissolved in 0.1 ml of n-hexane.

STATISTICS

The results were statistically analysed by the multiple regression analysis. The arithmetic mean and standard deviation were calculated for each measurable parameter. Next the correlation coefficients between coenzyme Q_{10} level and selected parameters were computed. The significance were analysed by t-Student's test at P=0.05.

RESULTS

At the beginning of presented research the HPLC method was optimalized in order to obtain maximum speed of analysis and good reproducibility. For this purpose four mobile phases were examined using standard solutions of 1 μg Q_{10}/ml in hexane. The first tested eluent: methanol-water appeared not suitable because of extremely (35 min) long retention time of coenzyme. The use of methanol alone resulted in slight reduction of retention time but the shape of obtained peaks was distorted. The mixture of methanol-chloroform was also unsuitable due to high UV-absorption of chloroform. The mobile phase, according to reference [11] consisted of n-hexane in methanol was found the most useful and mixture made up of 72% of methanol and 28% of n-hexane was employed for further experiments. Such composed eluent assured acceptable short retention time 6.65%±0.22 s) with good reproducibility at flow rate 1.5 ml/min. In order to verify the linearity of UV-detection the calibration curve was constructed. A linear relationship between concentration and peak area was confirmed within the range up to 1.5 μg Q_{10}/ml with the following regression equation: y=99519x +51.414 (r=0.9992); where y denotes concentration in μg Q_{10}/ml and x - peak area. The method is characterised by good reproducibility. The values of RSD (%) at 0.5 μg Q_{10}/ml is 0.2% and at 1.5 μg Q_{10}/ml is 0.04%. The recoveries for the calibration plot were in the range 94-100% (in all cases three determination were averaged).

DISCUSSION

The biochemical parameters characterised observed group of patients are assembled in *table 1.*

Table 1. The biochemical characteristic of investigated group

	Unit	Mean	Std Dev	Minimum	Maximum	Valid N
CoQ_{10}	μmole/l	0.80	0.58	0.0366	2.7530	43
Magnesium	mg/100ml	2.32	0.25	1.7000	2.7200	37
Calcium	mmole/l	2.42	0.15	2.0300	2.7100	38
α_1-globulins	g%	3.09	0.66	1.4500	4.3000	38
Phosphorus	mg/100ml	4.98	1.18	2.1000	6.9000	33
Protein	g%	6.90	0.69	5.2000	8.3000	40
β-globulin	g%	9.19	1.96	5.6000	17.3000	38
γ-globulin	g%	10.71	3.37	4.7000	24.6000	37
α_2-globulins	g%	12.56	2.23	8.1000	17.6000	38
Age	Months	23.72	23.50	1.0000	108.0000	43
GPT	IU	41.60	57.71	10.0000	332.0000	43
Iron	μg/100ml	54.23	32.21	15.0000	138.0000	28
GOT	IU	56.81	53.35	23.0000	325.0000	43
Albumin	g%	63.67	5.28	47.0000	73.3000	39
Protein C-reactive	g%	78.88	38.13	6.0000	102.0000	34
IgE	IU	121.8	159.19	1.0000	453.0000	20
Cholesterol	Mg/100ml	128.33	27.98	81.0000	189.0000	15
UIBC	μg/100ml	262.82	77.66	26.0000	380.0000	28
FA	IU	273.00	135.85	132.000	687.0000	23
Leukocytes	Number of cells/μl	13499.07	5970.04	5100.00	28900.00	43

The plasma CoQ_{10} concentrations was determined individually for each patient and analysed in accordance to others biochemical parameters. The mean level of CoQ_{10} for investigated group of patients is 0.788 μmole/ml.

It has been found which biochemical parameters influences on ubiquinone levels at the first hours of inflammatory process. Using the multiple regression analysis the correlation was found between leukocytes, Mg, Ca, UIBC, FA, GPT levels and coenzyme Q_{10} concentration. The results are as sembled in *table 2*.

The results indicate that CoQ_{10} concentration is connected with leukocytosis (9,4%), calcium (18,5%) and magnesium (29,8%) levels. The statistic analysis has revealed that transferrin (44.1%), amylase (51.4%) and serum glutamic transaminase (60%) may determine also the CoQ_{10} levels plasma *(fig.)*.

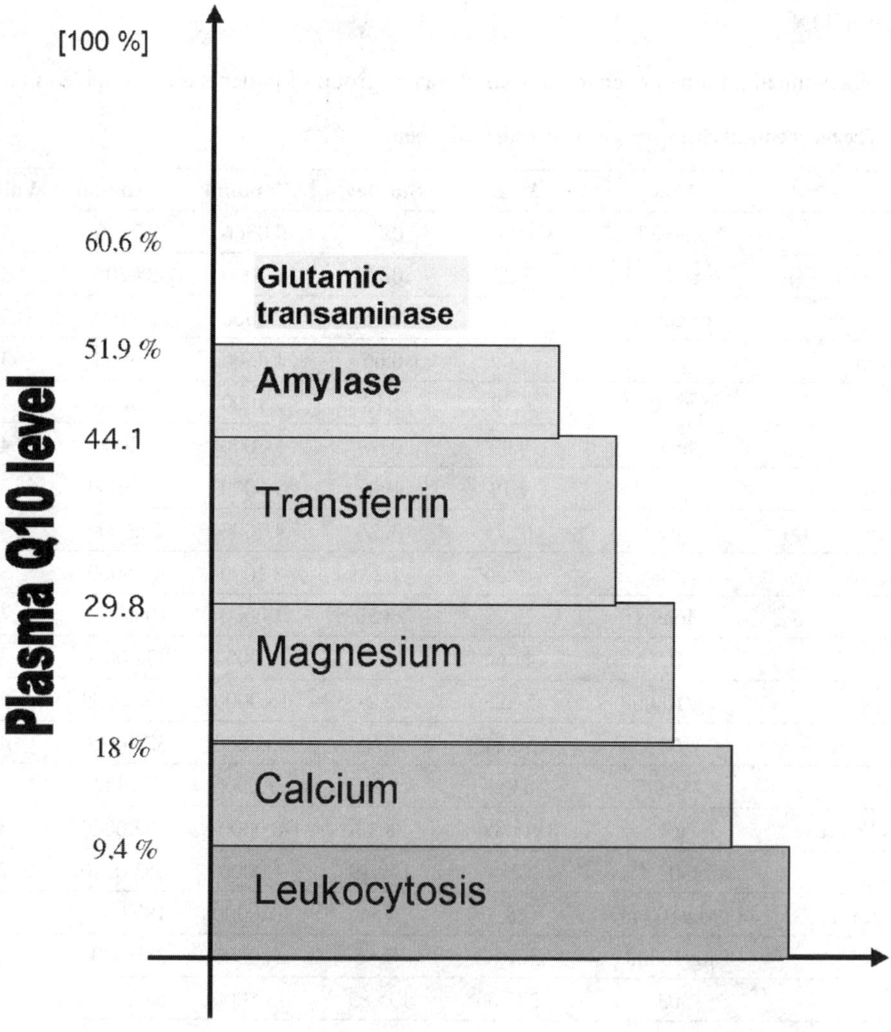

Figure. The graph of correlation between some biochemical parameters and coenzyme Q_{10} level

Table 2. The results of multiple regression analysis

Analysed parameter	Regression coefficient	Coefficient of determination
Lu	0.000028	0.0949
Ca	-1.236907	0.1859
Mg	-0.818773	0.2983
UIBC	0.003261	0.4413
FA	0.03874	0.5141
GPT	0.004309	0.6060

According to the present state of knowledge the correlation's between plasma CoQ_{10} levels and findings biochemical parameters is still difficult to explain but it seems to be important that CoQ_{10} levels does not changes during first hours of inflammatory process. In our study of CoQ_{10} its mean concentrations in plasma was determined as 0,788 μmole/l, like in healthy children. The same serum reference values established [10] for a healthy pediatric group. Our results indicated that increase in plasma leucocytosis, transferrin, amylase and glutamic transaminase levels caused an increased in CoQ_{10} concentrations. Plasma level of coenzyme Q_{10} decreased after calcium and magnesium lowering.

CONCLUSION

Our study confirmed that CoQ_{10} concentration (median 0.788 μmol/l) was independent of sex and the acute inflammatory process did not change ubiquinone levels. We established which biochemical parameters may have influence on the CoQ_{10} levels plasma.

REFERENCES

1. Ernster L., Dallner G. Biochemical, physiological and medical aspects of ubiquinone function. Biochim Biophys Acta 1995; 1271: 195-204.
2. Folkers K., Brown R., Judy W.V., Morita M. Survival of cancer patients on therapy with coenzyme Q_{10}. Biochem. Biophys. Res. Commun 1993; 192: 241-245.
3. Lelli J.L., Drongowski R.A., Sastman B., Remick D.G., Coran A.G. Effects of coenzyme Q_{10} on the mediator cascade of sepsis. Circ. Shock 1993; 39: 178-87.
4. Jolliet P., Simon N., Barre J., Pons J.Y., Boukef M., Paniel B.J., Tillement J.P. Plasma coenzyme Q_{10} concentrations in breast cancer: prognosis and therapeutic consequences. Int. J. Clin. Pharmacol. Ther 1998; 36: 506-509.
5. Mano T., Iwase K., Hayashi R., Hayakawa N., Uchimura K., Makino M., Nagata M., Sawai Y., Oda N., Hamada M., Aono T., Nakai A., Nagasaka A., Itoh M. Vitamin E and coenzyme Q concentrations in the thyroid tissues of patients with various thyroid disorders. Am. J. Med. Sci. 1998; 315: 230-232.
6. Suzuki S., Hinokio Y., Ohtomo M., Hirai M., Hirai A., Chiba A., Kasuga S., Sotoh Y., Akai H., Toyota T. The effects of coenzyme Q_{10} treatment on maternally inherited diabetes mellitus and deafness and mitochondrial DNA 3243(A to G) mutations. Diabetologie 1998; 41: 584-588.
7. Mortensen S.A., Perspectives on therapy of cardiovascular diseases with coenzyme Q_{10} (ubiquinone). Clin. Investig. 1993; 71 (Suppl 8): 116-23.
8. Gvozdjakova A., Kucharska J., Gvozdjak J., Redox therapy in mitochondrial diseases using coenzyme Q_{10}. Bratisl. Lek Listy 1994; 95: 443-451.

9. Noia G., Romano D., De Santis H., Mariorenzi S., Caruso A., Mancuso S. Coenzyme Q_{10} fetal plasma levels. Fetal Diagn Ther 1998; 13:. 127-130.
10. Artuch R., Moreno J., Quintana M., Puig R.M., Vilaseca M.A. Serum ubiquinone-10 in a pediatric population. Clin. Chem. 1998; 44: 2378-2379.
11. Kaplan P., Sebastianova N., Turiakova J., Kucera I. Determination of Coezyme Q_{10} in human plasma. Physiol. Res. 1996; 45: 39-45.

This work was supported by the grant BW264476 from Medical Academy of Bialystok.

*Metal Ions in Biology and Medicine: vol. 7.*Eds L. Khassanova, Ph. Collery, I. Maymard, Z. Khassanova, J.-C. Étienne.
John Libbey Eurotext, Paris © 2002 pp. 427-432.

The content of the chosen trace elements in hair of children with food allergy

M.J. Piotrowska-Depta[1], J.D. Piotrowska-Jastrzebska[1], M. Borawska[2], M. Kaczmarski[3], R. Markiewicz[1]

[1] *Department of Auxology Medical Academy in Bialystok, 15-274 Waszyngton Street 17, Poland;* [2]*Department of Bromatology Medical Academy in Bialystok;* [3]*The IIIrd Department of Paediatrics Medical Academy in Bialystok, 15-274 Waszyngton Street 17, Poland.*

ABSTRACT

A well-balanced diet, rich in all indispensable nutrients is one of the factors responsible for the right functioning of a child's organism. Food hypersensitivity has a chronic course and requires the application of individually selected elimination diet.

It refers not only to the main nutrients but also vitamins and trace elements.

The aim of the study is to denote the content of the chosen trace elements (Zn, Cu, Se) in hair of children with food allergy at its acute stage.

The research covers the group of 26 children aged from one month to three years; mean age - 15 months.

According to the type of clinical manifestation of the illness, the children were divided into three subgroups, I - where the skin manifestation dominated [15], II - with the prevailing digestive manifestation [6], and III - with the predominance of respiratory system manifestation [5].

The control group consisted of 21 healthy children of the same age. The content of trace elements in the hair of children was denoted with the usage of atomic absorption spectrophotometry method.

The results of the research - it was proved that the average content of zinc in the hair of 26 tested children was 146.19 μg/g as compared to the control group - 128.8 μg/g.

The average content of cooper in the hair children with food allergy was 18.815 μg/g and selenium was 0.68 μg/g. The average content of cooper in the hair of control group was 11.92 μg/g ($p<0,0001$) and selenium was 0.410 μg/g ($p<0,0001$).

Comparative analysis of the data concerning various types of clinical manifestation of allergy (subgroups I, II, III) indicated the lowest content of zinc and copper for the group of children with prevailing respiratory system manifestation (subgroup III) and the lowest content of selenium in the hair children with enteral manifestation of food allergy (subgroup II).

Conclusions: The content of copper and selenium in the hair of children with allergy was higher with reference to the control group.

Key words: zinc, copper, selenium, children, food allergy.

Among the factors responsible for the regular functions of human organism a special role should be assigned to a well-balanced diet which guarantees a proper supply of proteins, carbohydrates and fats. But along with these basic nutrients, also vitamins and mineral elements have their significant role here [1, 2].

Because of intensive growth processes taking place in the initial stage of human development a properly composed diet is especially important for children and adolescents who should receive more nutrients (including mineral elements) as compared to adults. Deficiency of trace elements and excess of toxic elements results in various metabolic disorders.

Zinc and copper are indispensable in regular functioning of a living organism. As important constituents and activators of numerous enzymes, such as carbonic anhydrase or alkaline phosphatase, zinc and copper are responsible for a number of biochemical processes, especially during the early stage of intensive growth of human organism.

Zinc is a fundamental constituent of DNA and RNA polymerase, enzymes responsible for transcription and replication of genetic material [4, 5, 6, 7]. Deficiency of zinc can manifest itself in growth delay and retardation of skeletal age, underdevelopment of gonads, anaemia [8], disorders of nervous system [9], impaired twilight vision [10, 11], lack of appetite, dysfunction of taste sensation [12, 13, 14], skin anomalies, predisposition to candidiasis infections. Zinc can also have a protective role in lead intoxication as it has been proved that higher concentration of zinc in food is related to slower absorption of lead [15, 16, 17, 18].

The main function of copper is connected with its role in erythropoesis, generation of hem, maturation of erythroblasts and production of thrombocytes. Copper is also indispensable in the osteogenesis processes, production of myelin, collagen and elastin synthesis [19, 20].

Another trace element considered asfundamental in proper functioning of living organisms is selenium. For a long time it was regarded as a toxic element [21, 22] but in the second half of the last century it was discovered that selenium has an essential role in life processes.

Selenium has a significant role in the process of removing of free radicals which can lead to a damage of DNA and disturbances in regular life processes. Selenium is also known for its role as a factor restraining production and development of carcinogenesis [23, 22]. It also has many other positive attributes, such as antiatherogenic, immunomodulative or protective properties towards some heavy metals [22, 24]. The role of selenium in the treatment of some allergic diseases is pointed out in the publications issued in recent years [22, 25, 26, 27, 28, 29].

The content and role of trace elements in the group of children with allergy is a subject of quite little research studies and the results of these analyses are divergent [33, 34, 35, 36, 37].

Thus the examination of the content of the above mentioned trace elements in hair of children with food allergy seems to be a vital concern [30, 31, 32].

MATERIAL AND METHOD

The research covered a group of 26 children (17 girls and 9 boys) aged from 2 months to 3 years (the average 15 months) with diagnosed food allergy and undergoing treatment in the IIIrd Department of Paediatrics of Medical Academy in Bialystok.

Clinical diagnostics was established on the grounds of a thorough medical and family history of the patients, physical examination as well as complex allergological and immunological testing with respect to the established criteria. The children were selected in the initial stage of illness, with various clinical manifestations, before the application of elimination diet. According to a clinical picture of the illness the children were divided into 3 subgroups: I - where the skin manifestation dominated (16 children), II - with the prevailing digestive tract manifestation (7 children), III - with the predominance of respiratory system manifestation (3 children). The control group consisted of 21 healthy children of the same age, selected during preventive examination in regional outpatient clinics in Bialystok.

The content of zinc, copper and selenium in hair was determined for each child. The examination was conducted in the Bromatology Department of the Medical Academy in Bialystok.

Calibration solutions of proper concentration were prepared from standard solutions by Fluka Company (concentration 1 mg/ml). 3-4 cm-long segments of hair (from the head skin) were taken for analysis.

The hair was subjected to the process of washing in accordance with the standard procedures of International Atomic Energy Agency. The samples were mineralised in nitric acid in a microwave mineralizer UniClever, Plazmatronika. The content of zinc, copper and selenium in hair was determined by atomic absorption spectrometry method with the usage of a spectrophotometer Z-5000 with Zeeman background correction, Hitachi. The accuracy method was tested with reference material - GBW 09101 (human hair).

Statistical analysis was conducted by using t-Student test.

RESULTS AND ANALYSIS

The results of the research demonstrated the decrease of average content of zinc, copper and selenium in hair of children with food allergy as compared to the control group. The differences between the content of zinc in the group of healthy children and the group of children with allergy were not statistically significant. The results of comparison of the content of cooper and selenium in the two mentioned groups were statistically significant ($p < 0{,}0001$). The average content of zinc, copper and selenium in hair of the studied children and control group is presented in the *table 1.*

Table 1. Average level of zinc, copper and selenium in hair of children with food allergy and healthy children

Group of tested children	Average level of zinc (Zn) in hair (μg/g)	Scope of the level	Average level of cooper (Cu) in hair (μg/g)	Scope of the level		Average level of selenium (Se)in hair (μg/g)	Scope of the level	
Children with food allergy n=26	146,19	min. 44,129 max. 251,51	18,815	min. 8,074 max. 29,564	p<0,0001	0,680	min. 0,293 max. 0,965	p<0,0001
Healthy children n=21	128,79	min. 59,98 max. 203,20	11,915	min. 7,943 max. 13,434		0,410	min. 0,128 max. 0,791	

Table 2. Level of copper, zinc and selenium in hair of children with food allergy in relation to a clinical manifestation of allergy

Manifestation of allergy	Level of zinc in hair (μg/g)	Scope of the level	Level of copper in hair (μg/g)	Scope of the level	Level of selenium in hair (μg/g)	Scope of the level
I - skin manifestation n = 16	146,56	min. 71,164 max. 251,51	17,91	min. 8,074 max. 29,564	0,740	min. 0,405 max. 0,970
II - enteral manifestation n = 7	152,100	min. 62,827 max. 225,69	22,41	min. 9,317 max. 31,092	0,530	min. 0,293 max. 0,780
III - respiratory system manifestation n = 3	130,320	min. 44,129 max. 205,14	15,24	min. 8,206 max. 20,81	0,700	min. 0,480 max. 0,920

The average content of zinc in hair of all the children with food allergy was 146.19 μg/g and was higher in relation to the healthy children (128.79 μg/g).

Comparative analysis of the data concerning various types of clinical manifestation of allergy (subgroups I, II, III) indicated the lowest content of zinc for the group of children with prevailing respiratory system manifestation (subgroup III) *(table 2)*. Moreover, the cases of zinc level lower than the acknowledged deficiency level of 70 μg/g, were observed in all the types of clinical manifestation of allergy [38].

The average content of copper in hair of the ill children was 18.815 μg/g and it was significant higher in comparison with the group of healthy children (11.915 μg/g). The lowest levels were observed in the subgroup of children with prevailing respiratory system manifestation (subgroup III). The highest level of copper in hair (31.092 μg/g) was noted in the case of a child who, at the same time, had a low content of zinc in hair (62.827 μg/g). Such an observation proves a well-known statment that zinc and copper compete for protein carriers resulting from the fact that zinc bonds metallothioneine easier than zinc [39, 40].

The average content of selenium in hair of the children with food allergy was 0.680 μg/g and it was significantly higher ($p < 0,0001$) than in the case of healthy children (0.410 μg/g). The lowest levels (0.530 μg/g) were observed in the group of children with predominant digestive system symptoms (subgroup II).

DISCUSSION

There are only few publications (both Polish and foreign) which deal with the problem of zinc content in hair of children with allergy, especially food allergy. The conclusions of the studies described in these sources are ofen divergent.

The research conducted by Di Toro et al. [33] indicated the lower zinc level in hair of children with allergy in relation to healthy children. The level of copper was considerably higher in hair of children with allergy as compared to healthy children. Similar results were achieved in the research conducted by David et al. [35], Goldey et al. [36] and Hambridge et al. [41]. All these publications showed the lower zinc level in hair of children with allergy.

Slighter differences were observed by Krosniak et al. [42] who examined the Polish population. Krosniak found lower zinc level in hair of children with symptoms of allergy, in comparison with healthy children. However, after the group of studied children was divided into two subgroups - one with asthma, and another - with symptoms of atopic dermatitis, the results of analysis indicated a considerably lower level of zinc for the children with asthma, and higher level of zinc in the group of children with atopic dermatitis, in comparison with the group of healthy children. In our group of children with allergy the cases of children with atopic dermatitis were predominant, which can explain the divergence of the obtained results. Piechota et al. point out that in the case of moderate deficiency of zinc the rate of hair growth drops, while the content of zinc remains unchanged, but in the case of severe deficiency the decrease of the rate of hair growth can even result in the increase of zinc content in hair. Different results (in relation to the ones given above) were obtained by Steckiewicz et al. [34], however that study dealt with blood plasma. Steckiewicz detected a considerably higher level of zinc in blood plasma of children with allergy and suffering from recurrent infections, in comparison with healthy children.

The increase of copper level in children with allergy in comparison with healthy children, was also detected by Sinha and Gabrieli [44]. Similar results were also obtained in our study. Different conclusions were drawn by Godley et al. who did not observe any considerable differences between copper content in children with allergy and healthy children.

The known sources concerned with the content of selenium in children with allergy concentrate on the question of selenium content in serum or whole blood and erythrocytes [28, 29, 45, 46], and they mostly deal with children with inhalatory type of allergy. Most of these publications indicate the decrease of selenium level in blood or erythrocytes of children with allergy [22, 26, 28, 29].

In our study we denoted the content of selenium only in hair of children.

As the mentioned data indicate, the results concerning content of zinc, copper and selenium in hair of children with allergy are divergent. Thus, the subject seems worth dealing with.

CONCLUSIONS

The research proved the higher content of zinc, copper and selenium in hair of children with food allergy in comparison with healthy children. The obtained differences were statistically significant for copper and selenium.

REFERENCES

1. Kulikowska E., Moniuszko-Jakoniuk J., Miniuk K.: The role of zinc in physiological and pathological processes, Pol. Tyg. Lek., 1991, (46), 24-26; 470.
2. Chan S., Gerson B., Subramaniam S: the role of coper, molybdenum, selenium, and zinc in nutrition and health, Clin. Lab. Med., 1998, Dec. 18(4): 673-85.
3. Mertz W.: The essential trace elements. Science, 1991, 213/40514, 1332-1338.
4. Obserleas D., Prasad A.S.: Effect of zinc on thymidine kinase activityand DNA metabolism w Trace elements metabolism in animals (red. W.G. Hoestra). University Park Press, Baltimore 1974.
5. Auld D.S. et al.: Reverse transcriptase from avian myeloblastosis virus. Zinc metalloenzyme. Biochem. Biophys. Res. Commun., 1974, 57, 967.
6. Duncan J.R., Dreostil J.E.: The effect of zinc deficiency on the tining of deoxyribonucleic and synthesis in regenerating rat liver; S. Afr. Med. J., 1974, 48, 1697.
7. Sullivan V.K., Cousins R.J.: Competitive reverse transcriptase-polymerase chain reaction shaws that dietary zinc supplementation in humans increases monocyte metallothionein mRNA levels; J. Nutr., 1997 May, 127(5): 694.
8. Oelshlegel F.J. et. al.: Studies on theinteractionof zinc with human hemoglobin. Arch. Biochem. Biophys., 1974, 163, 742.
9. Prenland J.G., Sandstead H.H. et al.: A preliminary report: effects of zinc and micronutrient repletion on growth and neuropsychological function of urban Chinese children, J. Am. Coll. Nutr., 1997, Jun., 16(3); 268-72.
10. Brown E.D. et al.: Vitamin A metabolism during the repletionof zinc dificient rats. J. Nutr., 1976, 106, 563.
11. Christian P., West K.P.Jr: Interactions between zinc and vitamin A: an update, Am. J. Clin. Nutr., 1998 Aug.; 68 (2 Suppl.); 435S-441S.
12. Bystron-Prajener U., Trojanowska D.: Sense of taste of patients with acute renal insufficiency. Pol. Arch. Med. Wew., 1983, 69, 97.
13. Henkin R.J. et al.: A zinc protein isolated from human parotid saliva. Proc. Nat. Acad. Sci. USA, 1975, 72, 488.
14. Schechter P.J.: Idiopathic hypogeusia, a description of the syndrome and a single-blind study with zinc sulphate w Neurobiology of trace metals zinc and cupper (red. C.C. Pfeiffer). Academic press, New York 1976.
15. Das J.: Effects of zinc deficienty on ethanol metabolism and alkohol and aldehyde dehydrogenase activities. J. Lab. Clin. Mmed, 1984, 104, 610.
16. Lisiewicz J., Moszczynski P.: The influence of lead on haematopoetic system including the role of enviromental and work exposition. Postepy Hig. Med. Dosw., 1986, 40, 45.
17. Krol B.: The exposition to lead and simultaneous application of copper. Pamietnik III KrajowegoZjazdu Pol. Tow. Med. Pracy. Lódz 9-11.11.1978, Lódz 1981.
18. Markowitz M.E., Rosen J.F.: Zinc and copper metabolism in CaEDTA-treated children with plumbism, Pediatr. Res., 1981, 15, 635.

19. Kozielec T., Kaszczyk-Kaczmarek K., Kotkowiak L.: Copper deficiency in children and adolescents from Szczecin. Biul. Mag. 1994, 5/4, 105-108.
20. Graczyk A., Radomska K., Konarski J.: Copper. Mag. Med. 1993, 4/10, 24-33.
21. Fan W.: Pathogenic factors of Keshan disease in the grains cultivated in endemic areas, medicine (edit. G.F. Combs, J.E. Spallholz, O.A. Levander, J.E. Oldifield). In: Selenium in biology and medicine. Van Nostrand Reinhold Company. New York. 1987.
22. Kalita B., Nowak P., Slimok M., Sikora A., Szkilnik R., Obuchowicz A., Sulej J., Sabat D.: Analysis of selenium concentration in serum of children with food allergy. Pol. Merk. Lak. 2001, 10, 60: 411-413.
23. Combs G.F.: Selenium as a cancer-protective agent. Bull. Selenium-Telurium develop. Assoc., 1997, 22, 1-4.
24. Zbikowska H.M.: Metabolism of selenium in human cells and organism. Post. Biol. Kom., 1997, 24, 303-313.
25. Fairris G.M.: The effect on atopic dermatitis of supplementation with selenium and vitamin E. Acta Derm. Venorol., 1989, 69, 359-362.
26. Hasselmark L., Malgren R., Zefferstrom O., Unge G.: Selenium supplementation in intristic asthma. Allergy, 1993, 48 (1), 30-36.
27. Kadrabova J.: Selenium status is decreased in patients with intristic asthma. Biol. Trace Elem. Res., 1996, 52, 241-248.
28. Misso N.L., Powers K.A., Gillon R.L., Stewart G.A., Thompson P.J.: Reduced plateled glutatione peroxidase activity and serum selenium concentration in atopic asthmatic patients. Clin. Exp. Allergy, 1996, 26, 838-847.
29. Ward K.R., Arthur J.R., Russell G., Aggett PJ.: Blood selenium content and glutathione peroxidase activity in children with cystic fibrosis, celiac disease, asthma, and epilepsy. Eur. J. Pediatr., 1984 Apr., 142 (1), 21-24.
30. Piotrowska-Jastrzebska J.: Auksological examination of children with cow milk protein intolerance treated with elimination diet therapy. Ann. Acad. Med. Bial. 1995; 40, 573-579.
31. Piotrowska-Jastrzebska J: Auxological and densitometric examination of children with food oversensitivity treated with elimination diet therapy. The thesis for the degree of assistant professor Wydawnictwo Uczelniane. Bialystok 1998.
32. Korol D.: Food hypersensitivity of children and adolescents as seen through retro- and prospective analysis. Doctoral thesis. Akademia Medyczna w Bialmstoku 1997.
33. Di Toro, Galdo Capotorti G., Gialanella G., Miraglia del Giudice M., Moro R., Perrone L.: Zinc and copper status of allergic Children; Acta Paediatr. Scand. 1987 Jul, 76(4), 612-617.
34. Steckiewicz W., Suplacz J., Luszczanska A.: Zinc level in blood of children with allergy and recurrent infections. Wiad. Lek. 1987, 15, 5; 315-318.
35. David T.J., Wells F.E., Sharpe T.C., Gibbs A.C.: Low serum zinc in children with atopic eczema; Br. J. Dermatol. 1984 Nov., 111(5): 597-601.
36. Goldey D.H., Mansmann H.C., Rasmussen A.I.: Zinc ststus of astmathic, prednisone-trated asthmatic, and non-asthmatic children. J. Am. Diet. Assoc. 1984; 84: 157-63.
37. Ojuowo A., Lindley KJ., Milla PJ.: Serum zinc, selenium and cooper concentration in children with allergic colitis. East. Afr. Med. J., 1996, 73 (4), 236-238.
38. Hambidge KM.: Low levels of zinc in hair, anorexia, poor growth and hypogeusia in children, Pediatr. Res., 1972, 6: 868-74.
39. Hall A.C.: J. Organic Biochem., 1979, 11, 57.
40. Piletz J.E., Ganshow R.E.: Am. J. Clin. Nutrir., 1979, 32, 275.
41. Hambridge K.M.: Hair analysis. Pediatr. Clin. North. Am. 1981; 27: 855-60.
42. Krosniak M.: Ca, Zn and Mg levels in hair of children with some allergic diseases; Metal Ions in Biology and Medicine, vol. 4, Paris, 1996, 547-549.
43. Piechota W.: The role of zinc in physiology and metabolic disorders Diagn. Lab. 1983, 19, 5-6.
44. Sinha S.N., Gabrieli E.R.: Serum cooper and zinc levels in various pathologic conditions. Am. J. Clin. Pathol. 1970; 54: 570-77.
45. Micetic-Turk M.D., Turk Z., Radiolli L.: Serum selenium Values in healthy children aged 1-18 years in NE Slovenia. Eur. J. Clin. Nutr., 1996, 50, 1920-1924.
46. Wasowicz W., Kocur J.: Selenium concentration in the blood of children and adults. Enviromental and clinical aspects. Pol. J. Environ. Study, 1997, 6, 192-194.

Metal Ions in Biology and Medicine: vol. 7. Eds L. Khassanova, Ph. Collery, I. Maymard, Z. Khassanova, J.-C. Étienne.
John Libbey Eurotext, Paris © 2002 pp. 433-436.

Maternal selenium and fetal locomotor system malformations

K. Hukalowicz, J. Popko*, M. Borawska, R. Markiewicz

Department of Bromatology,
**Department of Orthopaedic Children Surgery,*
Medical Academy, 1 Kilinski St, 15-230 Bialystok, Poland

ABSTRACT

The aim of this study was to determine blood levels and hair selenium content in mothers of children with locomotor system malformations (LSM). The level of selenium in mothers of children with LSM was significantly lower than in control group. Low selenium status in pregnant women may be cause of locomotor system malformations in their children.

INTRODUCTION

Selenium is an essential nutrient. The key role of selenium in human metabolism is attributed to the presence of four selenocysteine residues in the enzyme glutathione peroxidase. It plays a role in the stabilization of hydrogen and lipid peroxides (Takahashi et al., 1987). This enzyme protects against heart disease through catalytic reduction of hydrogen peroxide and fatty acid hydroperoxides that would otherwise damage cell membranes (Avissar et al., 1989). Selenium also plays an important role in the control of thyroid hormone metabolism. The iodothyronine deiodinases, which are responsible for the conversion of thyroxine (T4) to its active form triiodothyronine (T3), are selenoenzymes (Arthur et al., 1996). Selenium may also protect against cancer through other mechanisms including inhibition of cell proliferation and stimulation of the immune system (Medina, 1986). Deficiency of selenium may be cause of recurrent spontaneous abortion (Kocak et al., 1999). Areas of China where the soil is extremely low in selenium are associated with clear selenium deficiency diseases - an endemic cardiomyopathy - Keshan disease and a deforming arthritis - Kashin-Beck disease (Lockitch, 1989). It is an osteoarthropathy of the hands and fingers, elbows, knees, and ankles in children and adolescents. It is characterized by necrosis of growth-plate and epiphyseal chondrocytes and proliferation of surrounding chondrocytes, and it leads to a hypertrophic osteoarthropathy and, in some subject, to short stature (Yang et al., 1988, Ge and Yang, 1993). It has been suggested that the locomotor system malformations in children may be explained by the selenium status in their mothers.

The aim of this study was to compare blood levels and hair selenium content in mothers of healthy children and mothers of children with locomotor system malformations (LSM).

MATERIALS AND METHODS

Mothers, aged 26-42 years (mean age 27,7), who gave birth to children in the obstetric teaching hospitals in Bialystok in 1999-2001, were examined. The selenium in the blood and hair was determined in 46 mothers of healthy children and 49 mothers of children with locomotor system malformations (LSM). The newborns were clinically examined with ultrasound screening for hip dysplasia, performed by orthopaedic surgeons, usually 2-6 days after birth. According to the International Classification of Diseases (ICD-10) we found newborns with LSM (limb reduction defects, clubfoot, hip dislocation, syndactylia, polydactylia).

The selenium in the blood and hair of mothers was determined in the first weeks after delivery and in the blood in the year after delivery.

We obtained the consent of the local Committee of Ethics to perform the examinations.

The blood and hair samples (after the washing procedure recommended by the International Atomic Energy Agency) were decomposed with concentrated nitric acid in a microwave mineralizer BM - 1z instrument UniClever (Plazmatronika, Poland). The selenium content in the blood and hair was analyzed by electrothermal atomic absorption spectrometry (ETAAS) on a Z-5000 instrument (Hitachi, Japan). Certified reference materials - Seronorm Trace Elements (Nycomed) for whole blood and GBW 09101 for human hair (from the Chinese Academy of Science) - were used to test the accuracy of this method. Statistical analyses were performed using Statistica v. 5'97 software.

RESULTS AND DISCUSSION

In mothers of children with LSM the mean selenium blood concentration was 57.4 µg/l, which was significantly lower than in the control group - 69.9 µg/l *(fig. 1)*.

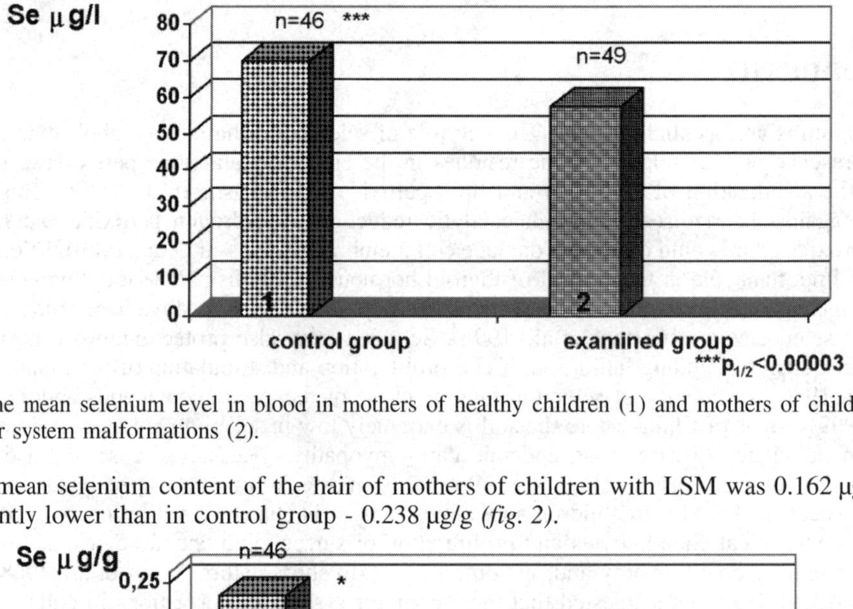

Fig. 1. The mean selenium level in blood in mothers of healthy children (1) and mothers of children with locomotor system malformations (2).

The mean selenium content of the hair of mothers of children with LSM was 0.162 µg/g also significantly lower than in control group - 0.238 µg/g *(fig. 2)*.

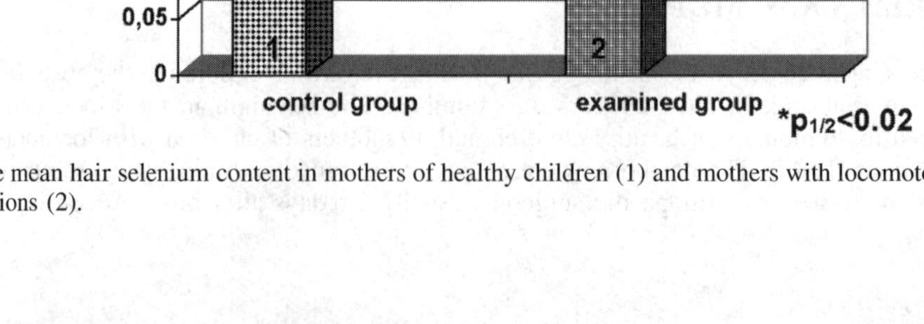

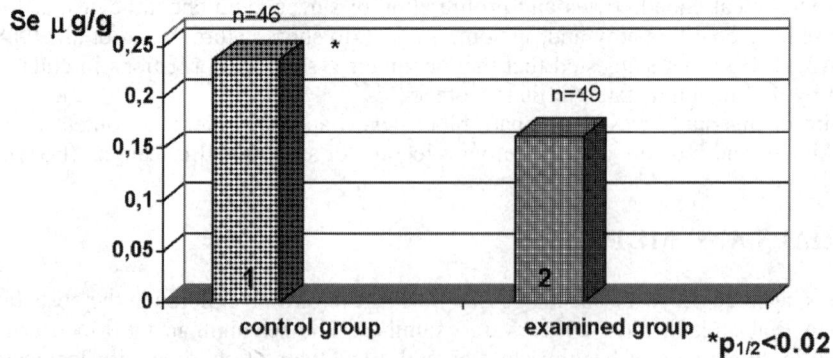

Fig. 2. The mean hair selenium content in mothers of healthy children (1) and mothers with locomotor system malformations (2).

In mothers of healthy children a significant correlation (correlation coefficient = 0.36) between the blood and hair was noted *(fig. 3)*.

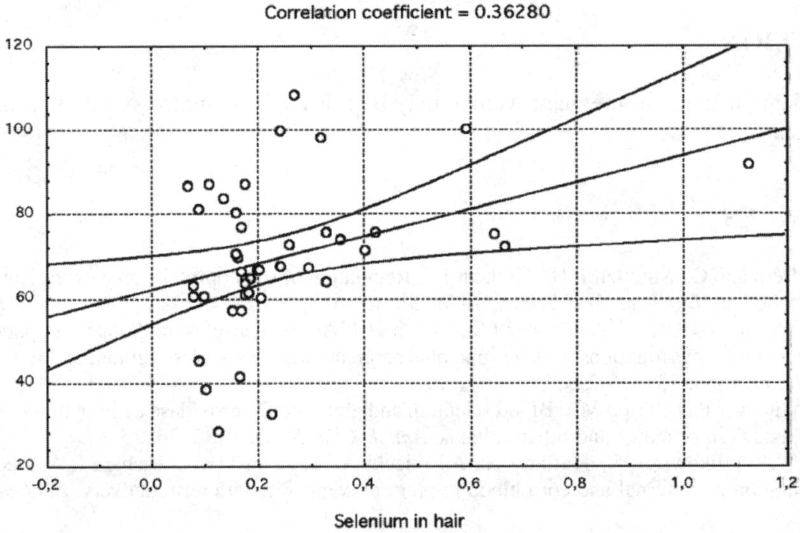

Fig. 3. The correlation coefficient between selenium in blood and selenium in hair in mothers of healthy children.

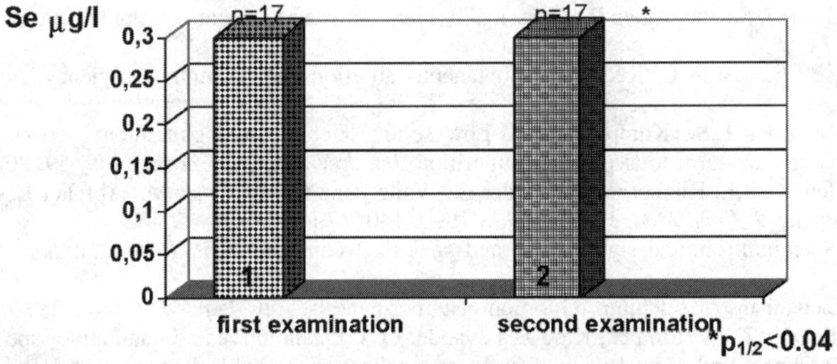

Fig. 4. The mean selenium level in blood of mothers of children with locomotor system malformations examined in 1999-2000 (1) and in the year after delivery (2).

In the year after delivery the mean level selenium in the blood of mothers of children with LSM increased and we are not observed significantly differences between the control group *(fig. 4)*.

The selenium concentration in the blood and hair of mothers participating in this study is rather low. The mean blood level of selenium in mothers of healthy children in our study (69.9 µg/l) is similar than mean Se concentration in whole blood of mothers at term delivery noted by Dobrzynski - 72,3 µg/l (Dobrzynski et al., 1998).

The hair selenium content in women after delivery has not been described. The mean hair selenium content in mothers of healthy children (0,24 µg/g) is about 30% lower than mean selenium content in healthy people from Poznan (Kaniewski et al., 1997).

Several workers have shown that in healthy pregnant women both whole blood and plasma Se concentrations decreased linearly with the progress of pregnancy, reached a nadir just before delivery (Zachara et al., 1993, Butler et al., 1982, Swanson et al, 1983, Lechner et al., 1990). The requirement for Se of pregnant women increases as a result of Se transport to the foetus (Korpela et al., 1989, Kasik and Rice, 1995).

CONCLUSIONS

Low selenium status in pregnant women may be cause of locomotor system malformations in their children.

REFERENCES

Arthur J.R., Bermano G., Mitchell J. H., Hesketh J.E. Regulation of selenoprotein gene expression and thyroid hormone metabolism. *Biochem. Soc. Trans.*, 1996, 24, 384-388.

Avissar N., Whitin N., Allen P.Z., Palmer I.S., Cohen H.J. Antihuman plasma glutathione peroxidase antibodies: immunologic investigations to determine plasma glutathione peroxidase protein and selenium content in plasma. *Blood*, 1989, 73, 318-323.

Butler J.A., Whanger P.D., Tripp M.J. Blood selenium and glutathione peroxidase activity in pregnant women: comparative assays in primates and other animals. *Am. J. Clin. Nutr.*, 1982, 36, 15-23.

Dobrzynski W., Trafikowska U., Trafikowska A., Pilecki A., Szymanski W., Zachara B.A. Decreased selenium concentration in maternal and cord blood in preterm compared with term delivery. *Analyst*, 1998, 123, 93-97.

Ge K., Yang G., The epidemiology of selenium deficiency in the etiological study of endemic diseases in China. Am. J. Clin. Nutr., 1993, 57, Suppl: 259S-263S.

Kaniewski A., Takagi Y., Marcinkowski J.T. Zawartosc 21 pierwiastków we wlosach mieszkanców Poznania. *Probl. Med. Spol.*, 1997, 32, 391-396.

Kasik J.W., Rice E.J. Selenoprotein P expression in liver, uterus and placenta during late pregnancy. *Placenta*, 1995, 16, 67-74.

Kocak I., Aksoy E., Ustun C. Recurrent spontaneous abortion and selenium deficiency. *Int. J. Gynaecol. Obstet.*, 1999, 65, 79-80.

Korpela H., Nuutinen L.S., Kumpulainen J. Low serum selenium and glutathione peroxidase activity in patients receiving short-term total parenteral nutrition. *Int. J. Vitam. Nutr. Res.*, 1989, 59, 80-84.

Lechner W., Jenewein I., Ritzberger G., Solder E., Waitz-Penz H., Schirmer M., Abfelter E. Selenium deficiency in pregnancy? *Wien. Med. Wochenschr.*, 1990, 140, 370-372.

Lockitch G. Selenium: clinical significance and analytical concepts. *Crit. Rev. Clin. Lab. Sci.*, 1989, 27, 483-541.

Medina D. Mechanisms of selenium inhibition of tumorigenesis. *Adv. Exp. Med. Biol.*, 1986, 206, 465-472.

Swanson C., Reamer D., Veillon C., King J., Levander O.A. Quantitative and qualitative aspects of selenium utilization in pregnant and nonpregnant women: an application of stable isotope metodology. *Am. J. Clin. Nutr.*, 1983, 38, 169-180.

Takahashi K., Avissar N., Whitin J.C., Allen P.Z., Cohen H. Purification and characterization of human plasma glutathione peroxidase: a selenoglycoprotein distinct from the known cellular enzyme. *Arch. Biochem. Biophys.*, 1987, 256, 677-686.

Yang G.Q., Ge K.Y., Chen J.S., Chen X.S. Selenium-related endemic diseases and the daily selenium requirement of humans. *World Rev. Nutr. Diet.*, 1988, 55, 98-152.

Zachara B.A., Wardak C., Didkowski W., Maciag A., Marchaluk E. Changes in blood selenium and glutathione concentrations and glutathione peroxidase activity in human pregnancy. *Gynecol. Obstet. Invest.*, 1993, 35, 12-17.

*Metal Ions in Biology and Medicine: vol. 7.*Eds L. Khassanova, Ph. Collery, I. Maymard, Z. Khassanova, J.-C. Étienne.
John Libbey Eurotext, Paris © 2002 pp. 437-439.

Lead contents in mothers and newborns with locomotor system malformations

Maria H. Borawska, Janusz Popko*, Slawomir Olszewski*, Katarzyna Hukalowicz, Renata Markiewicz

*Department of Bromatology, *Department of Orthopaedic Children Surgery, Medical Academy, 1 Kilinski St, 15-230 Bialystok, Poland*

SUMMARY

The aim of the study was to evaluate concentration of lead in the blood and in the hair of 46 women after the delivery of healthy newborns and 48 women whose newborns were born with locomotor system malformations (LSM).

The samples of blood and hair were collected from the mothers during the first week after delivery. Lead levels were determined by electrothermal atomic absorption spectrometry with a Zeeman background correction. Mothers of healthy newborns had a lower lead concentration in their blood - mean 6.30 µg/dL and hair - mean 0.77 µg/g; in comparison to the mothers of newborns with LSM, 14.08 µg/dL and 1.99 µg/g respectively. In mothers from control group, birth weights of newborns were higher by a mean of 417g in comparison of children born with LSM. In our studies the increase of lead concentrations in the blood and the hair in mother of newborns with LSM does not appear to be influenced by the nutritional habits of a pregnant woman.

Our results reflect the negative influence of lead on intrauterine fetal development.

INTRODUCTION

There have reports on individual cases of significant lead intoxication in pregnant women. The case of a girl whose mother in the 8th week of pregnancy as a result of occupational exposure had a lead concentration in the blood of 62 µg/dl was described [1]. At birth, the girl exhibit numerous congenital defects, known as VACTERL-syndrome [1]. Its causes have not yet been discovered. However, some authors [2, 3] have denied the existence of a relationship between lead and congenital malformations; therefore the problem requires further investigation. The aim of the present study is to evaluate the relationship between lead content in the blood and hair mothers and congenital malformations of locomotor organs in their newborns.

MATERIALS AND METHODS

During the 3-year period 1997-1999, there were 4 519 live births at the teaching hospital. The newborns were clinically examined with ultrasound screening for hip dysplasia, performed by orthopaedic surgeons, usually 2-6 days after birth. During these examinations we found 48 newborns with locomotor system malformations. According to International Classification of Diseases (ICD-10), the locomotor system malformations in the newborns were limb reduction defects - 4, clubfoot - 21, hip dislocation - 14, syndactylia - 4, polydactylia - 5 cases. Their mothers, aged 16-37, mean age 26.2 years, gave consent to allow measurement of lead concentrations in their blood and hair. 46 mothers, aged 17-42 years, mean 28.1, who gave birth to healthy children during the same 3 - year period, and who voluntarily gave consent to allow the same measurement of lead concentrations in their blood and hair, served as our control group. All data relating to lenght of pregnancy, birth weights of newborns, pre-natal environmental factors, work history, alcohol and drug use, tobacco, and the data relating to dietary preferences, for all 94 mothers was gathered in on interviev-based questionnaire. They did not take special medicines exept for vitamin and mineral supplements. We obtained the consent of the local Committee of Ethics to perform the examinations.

They consented to give 3 ml of venal blood and 0.3 g of the hair. The lead content in the blood was determined in deproteinated whole blood with 1 mol/l nitric acid and 1% Triton X-100. To remove exogenous contamination from the hair samples, the washing procedure recommended by the International Atomic Energy Agency was used. Hair samples were decomposed with concentrated nitric acid in a microwave mineralizer, BM-1z instrument UniClever (Plazmatronika). The lead content in the blood and the hair was analyzed by electrothermal atomic absorption spectrometry on a Z5000 instrument (Hitachi, Japan)with Zeeman correction. Certified reference materials - Seronorm Trace Elements for whole blood and GBW 09101 for human hair - were used to confirm the accuracy of this method. The department participates in a quality control program of the National Institute of Hygiene.

For statistical analysis a STATISTICA for Windows, v. 5'97, StatSoft Ink. pack was used. Statistical analysis was performed with a non-parametric test. Differences between independent groups were tested by the Mann-Whitney U-test. Correlation was calculated and tested by the Sperman rank test. When not otherwise stated, values of $p<0.05$ were considered significantly different.

RESULTS

In mothers of children with locomotor system malformations, the mean lead blood concentration was 14.08 μg/dL, which was significantly higher than in the control group - 6.30 μg/dL. The lead content of hair in this group was 1.99 μg/g while in control mothers - only 0.77 μg/g *(table. 1)*.

Table 1. The mean lead concentration in blood and hair in mother of children with locomotor system malformations (1) and mother from control group (2) and the birth weight of newborns

No	Mothers of:	Concentration of Pb in mother		Birth weight of newborns g
		whole blood μg/dL	hair μg/g	
1.	children with locomotor system malformations	14.08 (1.48-30.08)	1.99 (0.001-10.69)	3047.2 (900-4150)
2.	healthy children	6.30 (0.68-15.16)	0.77 (0.001-2.35)	3464 (2350-4550)

There was a correlation between the lead concentration in blood and hair in all mothers; the correlation coefficient was $r=0.24$; $p<0.02$ and reflected that hair will be a non-invasive material for study of concentrations of lead in pregnant women.

The birth weight of newborns with locomotor system malformations *(table 1)* were significantly lower with a mean value of 3047.2 g while in healthy newborns - 3464g. There were no correlations between lead content in maternal blood ($r=0.07$) or hair ($r=0.08$) and the weights of newborns or the duration of pregnancy.

DISCUSSION

Taking into account that lead penetrates fetal circulation via the umbilical cord [4, 5], our results suggest the negative influence of lead on intrauterine fetal development. Our results have highlighted the fact that the lead concentration in the blood of mothers whose children had defects was twice as high as in mothers of healthy children. Furthermore, lead content in hair was approximately 2.6 times higher than in the control group. Needleman et al. [6] have detected the correlation between lead concentrations in umbilical blood, and slight anomalies such as hemangiomas, lymphangiomas, hydroceles, cryptorchism, and skin anomalies. Investigations conducted so far [7, 8] have shown a high correlation coefficient between the lead concentration in maternal blood and the newborn umbilical blood. The mean birth weights of newborns with locomotor system malformations were lower by 417 g compared to healthy newborns *(table 1)*. Bellinger et al. [9] have shown that birth weights of newborns with lead concentrations in the umbilical blood $\geq$ 15µg/dl were lower by 80-100g than birth weights of newborns with lead concentrations <15µg/dl.

However, we have only examined lead concentrations in maternal blood (the source of lead for a fetus) and hair. Information gathered in a survery concerning environmental factors, coffee drinking, alcohol, and cigarette smoking as well as the nutritional habits of pregnant woman, cannot help to find the causes of this phenomenon. Lead accumulates in bones and has a long-life period of 20-30 years. During pregnancy, as a result of hormonal changes, the mobilization of lead from the bones may occur where it had accumulated during prior exposure [10].

CONCLUSIONS

Our results reflect the negative influence of lead on intrauterine fetal development. Studies have shown too that noninvasive examinations of lead contents in hair might be an important source of information concerning a mother - to - be previously exposed to that toxic element.

REFERENCES

1. Levine F., Muenke M. VACTERL association with high prenatal lead exposure: similarities to animal models of lead teratogenicity. *Pediatrics*, 1991, 87, 390-392.
2. Macdonell J.E., Campbell H., Stone D.H. Lead levels in domestic water supplies and neural tube defects in Glasgow. *Arch Dis Child*, 2000, 82, 50-53.
3. Sallmèn M., Lindbohm M.L., Antilla A., Taskinen H., Hemminki K. Paternal occupational lead exposure and congenital malformations. *J Epidemiol Community Health*, 1992, 46, 519-22.
4. Clark A.R. Placental transfer of lead and its effects on the newborn. *Postgrad Med J*, 1977, 53, 674-678.
5. Korpela H., Loueniva R., Yrjänheikki E., Kauppila A. Lead and cadmium concentrations in maternal and umbilical cord blood, amniotic fluid, placenta, and amniotic membranes. *Am J Obstet Gynecol*, 1986, 155, 1086-1089.
6. Needleman H.L., Rabinowitz M., Leviton A., Linn S., Schoenbaum S. The relationship between prenatal exposure to lead and congenital anomalies. *JAMA*, 1984, 251, 2956-2959.
7. Saxena D.K., Singh C., Murthy R.C., Mathur N., Chandra S.V. Blood and placental lead levels in an Indian city: a preliminary report. *Arch Environ Health*, 1994, 49, 106-110.
8. Zareba A., Strugala-Stawik H., Rudkowski Z., Dembicka D., Pastuszek B., Jedrzejczyk-Musiak A. Blood lead levels in mothers and their neonates - ecological factors and influence on fetus growth. *Ped Pol*, 1996, 4 (suppl), 87-94.
9. Bellinger D., Leviton A., Rabinowitz M., Allred E., Needleman H., Schoenbaum S. Weight gain and maturity in fetuses exposed to low levels of lead. *Environ Res*, 1991, 54, 151-158.
10. Kasznia-Kocot J., Latusek B. Lead intoxication and prenatal child development. *Przegl Ped*, 1999, 29, 17-21.

*Metal Ions in Biology and Medicine: vol. 7.*Eds L. Khassanova, Ph. Collery, I. Maymard, Z. Khassanova, J.-C. Étienne.
John Libbey Eurotext, Paris © 2002 pp. 440-444.

Zinc deficiency among the Mayan children of Guatemala

Donald Oberleas[1] and Barbara F. Harland[2]

[1]*Texas Tech University (Emeritus), Lubbock, Texas and* [2]*Howard University, Washington, D.C., USA*

ABSTRACT

The Maya are the indigenous population of Guatemala who live in the mountains of western Guatemala, Quiche Province. The economy is mostly agricultural and the lifestyle is primarily a survival existence. Sixty-seven of these children (34 male and 33 female) were subjects of this study. Anthropometrical measurements were made of the children and food frequency data were collected from their parents. Samples of the foodstuffs were purchased from the local market and analyzed for phytate and zinc. The major dietary components were maize tortillas and refried black beans eaten routinely three times a day; in addition some fruit, vegetables and much candy were also consumed. Meat, usually chicken, was consumed about once each week as a stew, soup or gruel. The phytate:zinc molar ratio necessary to sustain homeostasis is 10 or less. A maize tortilla, dried before analysis, had a phytate:zinc molar ratio of 33, commercially canned refried black beans 13, mature black beans 41, hybrid whole corn 21, open pollinated black corn 35, and open pollinated white corn 33. Thus the dietary composition alone would indicate that zinc deficiency is present among the children of this population. The height and weight of both male and female children, compared with a 50th percentile standard for the U.S., confirm that both height and weight of the children studied were compromised. The confirmatory evidence indicates that zinc deficiency is a universal problem among the children of this population. Since the adults consume these same diets, it may be concluded that zinc deficiency is by far the most prevalent deficiency among the Maya population of Guatemala.

MATERIALS AND METHODS

This study was done as a part of a medical missionary project to the area of Guatemala in close proximity to Chichicastenango, Quiche Province, Guatemala. The climate of Guatemala provides for a 12-month growing season with more rainfall recorded from May through November. The native population is predominantly Maya, the historical native population of southern Mexico and much of Central America. Medical clinics are established in various locations on weekdays and the people of the outlying villages are invited to the clinic location to be examined by physicians and dentists. In an effort to learn about the zinc status of this population, children whose parents had a complaint of poor appetite for their children were referred for evaluation of zinc status. Anthropometrical measurements were made, including height, weight, head-circumference, mid upper arm circumference and arm span. No invasive techniques were made, nor were any necessary. A Spanish/English qualitative food frequency questionnaire was completed by a parent for not more than one child per family. Each child was given a 100-day supply of zinc gluconate (25 mg zinc/day).

The economics of the population dictate that most of the food consumed is locally grown. Corn is the most prominent crop in this area. A black bean, Phaseolus genus, is also grown in the area. Though there was some variation in food consumption patterns between families, the typical diet consists of maize tortillas and refried black beans with some fruit or vegetable three times each day. Candy is also consumed 1 to 3 times each day. Meat, usually chicken, is consumed about once each week.

Several varieties of corn (maize) and beans were purchased from the local market and returned to the laboratory, ground to 20 mesh, and a sample analyzed by an HPLC method for phytate [7]. A maize tortilla was also placed in a plastic bag, dried and analyzed. Samples were also prepared and analyzed for zinc by atomic absorption [1, 3].

RESULTS AND DISCUSSION

The analyses of the foodstuffs for phytate and zinc are shown in *table 1*. A phytate:zinc molar ratio larger than 10 compromises zinc homeostasis [4]. The mechanism for the phytate compromise of zinc homeostasis has been published and confirmed [5, 8]; and a human model has been reported [6]. These analytical data for the foodstuffs projected over the diets alone is sufficient to demonstrate zinc deficiency in this population.

Table 1. Analysis of Foods from Guatemala

Foodstuff[1]	Phytate mmol/kg	Zinc mmol/kg	Phytate:Zinc molar ratio
Hybrid yellow corn	6.92	0.34	21
Black corn	9.25	0.27	35
White corn	15.69	0.47	33
Mature black bean	22.31	0.54	41
Maize tortilla	13.63	0.41	33
Refried black beans	5.43	0.42	13

[1] Foodstuffs were air dried except refried black beans
[2] Black and white corn were open pollinated
[3] Refried black beans were commercial canned product, non-dried

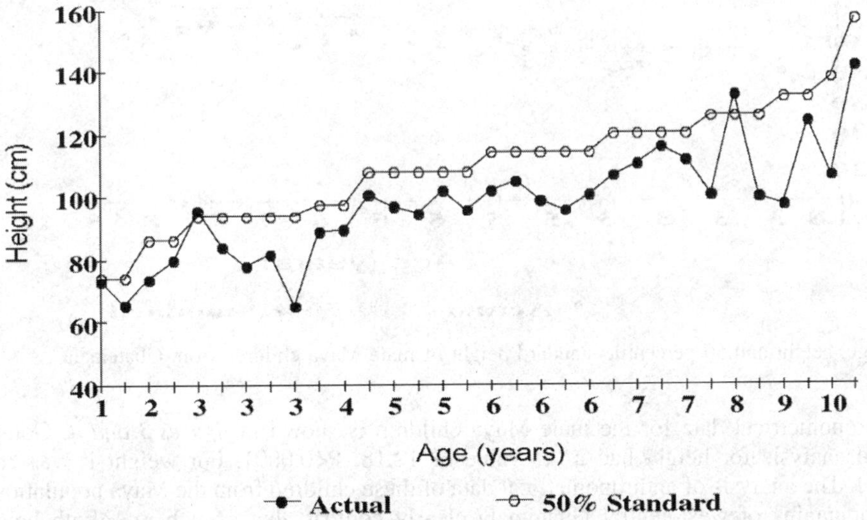

Fig. 1. Age, height and 50 percentile standard heights of female Maya children from Guatemala

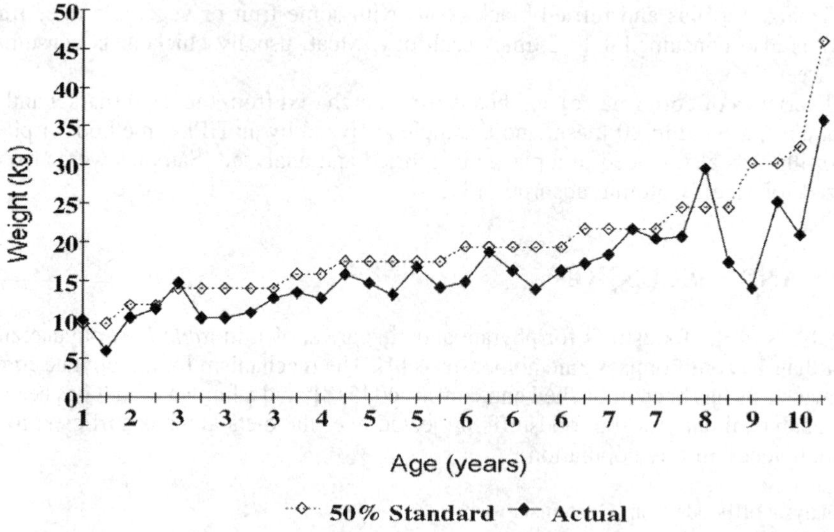

Fig. 2. Age, weight and the 50 percentile standard weight of female Maya children from Guatemala

The anthropometrical data for the female children are shown in *figures 1 and 2*. The Maya population is inherently shorter than many populations of the world. No appropriate standard for height and weight is available that is totally adequate for use with the Maya population. For comparison therefore, the 50 percentile standard established by the National Center for Health Statistics was selected for statistical comparison [2]. The actual height was measured to the nearest centimeter and weight to the nearest 0.5 pound and converted to kilograms. A paired-"t" analysis was made against the age corrected standard. In each analysis, the actual height and weight was consistently below the comparable 50 percentile standard for all ages. The analysis for height of female children had a "t" of 8.24, P<0.0001. For weight, similar analysis was "t" 5.08, P<0.0001.

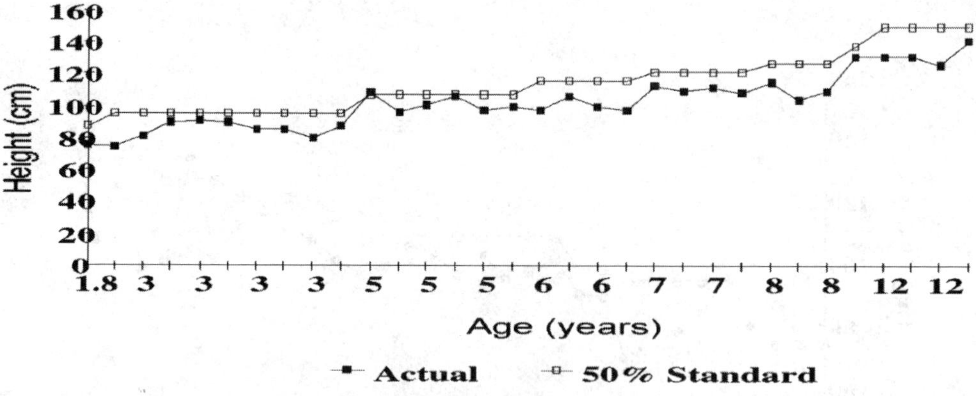

Fig. 3. Age, height and 50 percentile standard height of male Maya children from Guatemala

Anthropometrical data for the male Maya children is shown in *figures 3 and 4*. Comparable statistical analysis for height had a "t" value of 12.18, P<0.0001. For weight it was "t" 8.58, P<0.0001. The analysis of anthropometrical data of these children from the Maya population living in the mountains of west-central Guatemala clearly confirm that growth rate, both height and weight, is compromised. Since the most prevalent symptom of zinc deficiency is a restriction of growth rate, these data provide strong evidence for zinc deficiency being the primary nutritional deficiency.

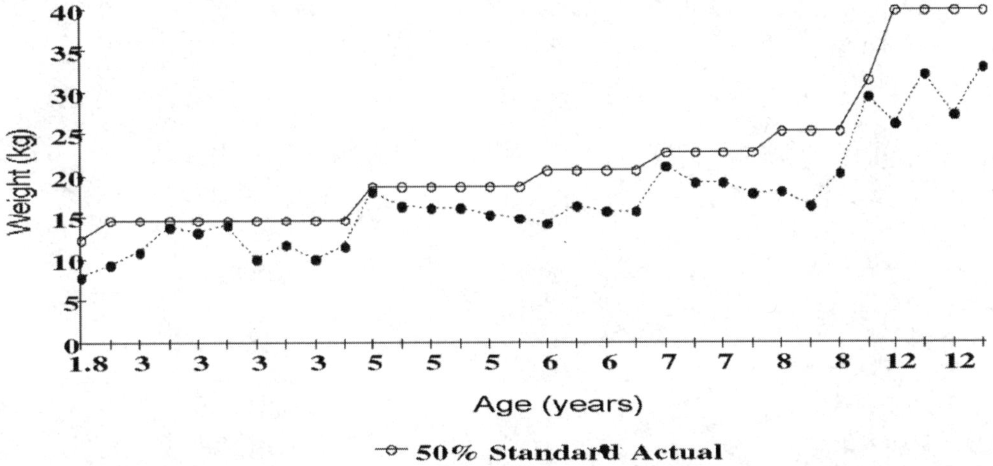

Fig. 4. Age, weight and 50 percentile standard weight of male Maya children from Guatemala

CONCLUSION

Since nutritional zinc deficiency is caused exclusively by the dietary phytate, some knowledge of the food consumption pattern in any population and the phytate and zinc analysis of the major foods consumed by that population are the necessary components to establish the zinc status of that population. This is clearly demonstrated in the current study. The anthropometrical data collected from a sample of children from that population confirms that zinc deficiency is universal among the Maya of west central Guatemala. These later comparisons were made against an age corrected 50 percentile standard available for the U.S. population. Since the adult population also consumes the same diet as the children, it may be inferred that zinc deficiency is prevalent among the entire Maya population.

REFERENCES

1. Anderson, RA, Bryen, NA, Polansky, MM. (1985) Serum chromium of human subjects: effects of chromium supplementation and glucose. Am. J. Clin. Nutr. 41: 571-577.
2. Hamill, PVV; Drizd, TA; Johnson, CL, Reed, RB; Roche, AF; Moore, WM (1979): Physical growth: National Center for Health Statistics percentiles. Am. J. Clin. Nutr. 32, 607-629.
3. Jackson, MJ, Jones, DA Edward, RHT, Swainbank, IG, Coleman, ML. (1984) Zinc homeostasis in man: studies using a new stable isotope-dilution technique. Br. J. Nutr. 51: 199-208.
4. Lo, GS; Settle, SL; Steinke, FH; Hopkins, DT (1981): Effect of phytate:zinc molar ratio and isolated soybean protein on zinc bioavailability. J. Nutr. 111, 2223-2235.
5. Oberleas, D (1996): Mechanism of Zinc Homeostasis. J. Inorg. Biochem. 62, 231-241.
6. Oberleas, D (1999): In search of a human model for zinc homeostasis. Proceedings of the 2nd International Symposium on Trace Element in Human: New Perspectives, October 7-9, 1999, 651-659.
7. Oberleas, D; Harland, BF (2001): Newer methods for phytate analysis. In: CRC Handbook of Dietary Fiber in Human Nutrition. 3rd ed. (Ed: Spiller, GA) CRC Press, Boca Raton, 113-126.
8. Oberleas, D; Kwun, IS (1998): Confirmation of the mechanism of zinc homeostasis. Metal Ions in Biology and Medicine 5, 140-145.

X NUTRITION

*Metal Ions in Biology and Medicine: vol. 7.*Eds L. Khassanova, Ph. Collery, I. Maymard, Z. Khassanova, J.-C. Étienne.
John Libbey Eurotext, Paris © 2002 pp. 447-453.

Copper nutrition and copper metabolism in rat newborns

L.V. Puchkova(#*), T.V. Zhivulko(*), B.S. Mishenko(*), E.A. Zhiguleva(#), N.K. Bichevaya(#), N.A. Platonova(#), N.V. Tsymbalenko(#), N. Guolikhandanova(*), A. Vasin(#), R. Povalihin(#)

Institute for Experimental Medicine, 12 Pavlova Street, 197376 St. Petersburg (#), St. Petersburg State Technical University(*), 29 Polytechnicheskaya, 195251 St. Petersburg, Russia

The work was supported by RFBR (grants No. 98-04-49790 and No. 01-04-49597), by Russian Program "Leading Scientific Schools" (grant No. 96-15-97742) and by Russian Federal Programs "Russian Universities - Fundamental Research" (grant No. 1316), "Human Genome" (grant No. 74-98).

ABSTRACT

Background. Copper (Cu) ability to easy change of states Cu(I)?Cu(II) in redox cycle is utilized by Cu-containing enzymes. This property makes it toxic to biological systems. The special copper metabolic systems (CMS) keep up copper homeostasis. The main source of Cu for newborns is tissue specific milk ceruloplasmin (Cp). The level of Cu in milk is controlled by Cp gene activity. The newborns, fed with baby formula (BF), take up the excess of Cu "not packaged" in the protein envelope. Theoretically it is able to fail the copper homeostasis.

The aim of the work is to study the effect of nutrition Cu on the CMS gene expression in newborns.

The methods of velocity and equilibrium sedimentation, RNA/DNA dot- and Northern-blot hybridization, immunoblotting, rocket immunoelectrophoresis, enzymatic assays of Cp, atomic absorption spectrometry as well as the solid-phase method of peptide synthesis were used.

Results. The newborn rats were fed with BF during 8 days (experimental group). In serum Cu and Cp levels were increased in 3 times. Simultaneously the liver Cu concentration was decreased in 2 times, comparing to 8-day old rats fed by rat females (control group). The liver Cp-mRNA content was increased and Wilson ATPase gene expression appeared in experimental group. The brain Cu concentration was not changed, but in cerebrospinal fluid Cp and Cu levels were increased in 7 times. During the first 12 days of life Cu is progressively accumulated in the liver of the control group. The major part of this Cu was found in lysosomes and nuclei. The cellular Cu distribution was similar in both groups. However the drop of Cu concentration has been observed in rats from the experimental group already in 5 days. The liver starts copper accumulation during embryonic development, but the yolk sac (YS) cells do not accumulate it. Two Cp-mRNA forms as well as full length Menkes ATPase and Wilson ATPase are expressed in YS cells.

Conclusions: 1) newborn CMS is adapted to Cp as a source of Cu; 2) nutrition Cu that is "not packed" into Cp changes the activity of CMS genes in newborns.

INTRODUCTION

Simultaneous essentiality and toxicity of copper are overpassed by living organisms using an elaborate appropriate mechanism (copper metabolic system, CMS) for uptake, distribution and elimination of copper as well as preventing its accumulation to toxic level [1]. In mammals CMS is changed during ontogenesis: the embryonic copper metabolic type (ECMT) is replaced by adult copper metabolic type (ACMT) [2]. The characteristic properties of the embryonic type are the low copper concentration in blood and brain as well as its progressive accumulation in liver. The change to the ACMT consists of copper redistribution between blood and liver and the subsequent rise of brain copper concentration [3]. These quantitative changes are expressed by qualitative tissue specific alterations of CMS genes expression on the molecular-genetic level. So ceruloplasmin (Cp, a main extracellular copper transporting glycoprotein with 6-9 copper atoms per one molecule [4]) gene as well as Wilson ATPase (copper transporting ATPase P1 type) gene are activated in liver of adult rats [5]. The mechanism of switching of metabolic types remains unknown so far. This problem is of the main interest today because the nutrition copper excess or deficiency can result in copper associated diseases [6].

In breast milk copper is almost (~75%) revealed in whey [7]. It is included into the tissue specific milk Cp [8]. During the first 20 days of lactation the Cp maintenance is decreased almost 5-fold [9]. It guarantees the maintenance of nutrition copper on physiological level in newborn [10]. The electrophoretically homogeneous milk $[^{125}I]$-Cp preparation administered *per os* to six-day old rats (ECMT) was transferred across the intestinal mucosa, entered the bloodstream and tissue-specifically distributed round body. Alternatively, radiolabeled milk Cp administered *per os* to the 33-day old rats (ACMT) become degraded in the digestive tract into the low molecular weight fragments [11]. Baby formulas (BF) are prepared without taking these data in consideration and so newborns fed by BF take up an excess of copper that, in addition, is "not packed" into Cp globule [7, 10]. This work is the first attempt to research the effect of BF on newborn copper metabolism. Such data can be value for understanding of the character of embryonic copper metabolic type, and besides this, for preparation of BF corresponding with breast milk in copper concentration and "wrapping".

MATERIALS AND METHODS

The work was carried out on embryos and newborns of outbred albino rats. Tissues were homogenized in 0.25M sucrose with 100 mM KCl, 10 mM $MgCl_2$, 5 mM β-mercaptoethanol, 50 mM tris-HCl buffer, pH7.4, and 1.5 μl/ml protease inhibitor cocktail ("Sigma", USA). Subcellular fractions of liver were isolated by velocity and equilibrium sedimentation. To identify the lysosomes the acid phosphatase activity was determined in subcellular fractions. The isolation of total cell RNA, RNA/DNA dot- and Northern-blot hybridization and Western blotting were made in compliance with handbook on gene cloning [12]. The 16- and 13-mers peptides having the sequences coinciding with fragments Menkes ATPase (^{968}?-?^{983}) and Wilson ATPase (^{951}Q-Q^{963}) respectively were synthesized by the solid-phase method. The conjugates of P16 and P13 with crab hemocyanin were used to obtain the rabbit antibodies. Their specificity and the absence of cross-reactions were proved in binding tests with the BSA-P16 and BSA-P13 conjugates. To measure copper concentration in brain and liver the tissues were dried to constant weight in vacuum under P_2O_5 and then burned in mixture of $HClO_4$:H_2SO_4:H_2O taken as 100:25:125, correspondingly. The copper concentration was measured by the atomic-absorption spectrometry with electrothermal atomization and the Zeemann correction of nonselective absorption on a 4100ZL Perkin-Elmer spectrometer (United States). Cp content was determined by Ravin's methods [13] and rocket immunoelectrophoresis using antibodies to rat Cp. Student's *t*-criterion was used for the evaluation of the meaning of results.

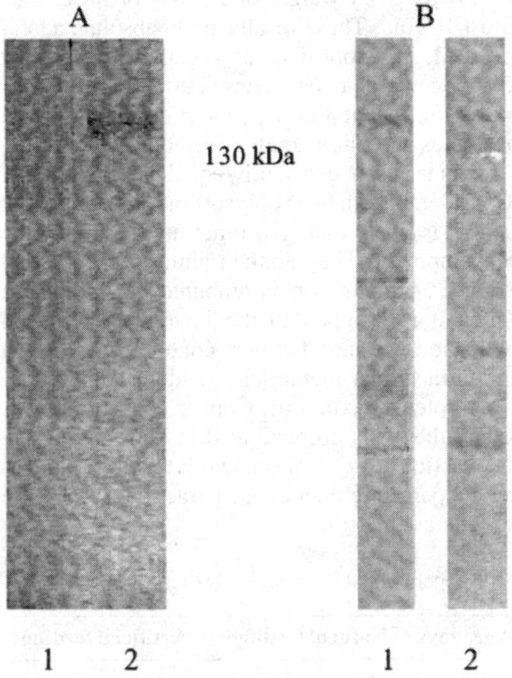

Fig. 1A. Immunoblotting of the liver membrane proteins of 2 day-old rats fed by rat females (1) and with BF (2) with using antibodies to P13 (Wilson ATPase)
Fig. 1B. Immunoblotting of rat yolk sac membrane proteins on 19th days of embryonic development 1 - antibodies to P16 (Menkes ATPase), and 2 - antibodies to P13 (Wilson ATPase)

RESULTS AND DISCUSSION

Newborn rats from different females were combined into one group and then random segregated into two equal groups. One of them was given back to rat females (control groups) while another group (experimental group) was fed by BF during 8 days after birth. The protein content of BF prepared for nutrition corresponded to protein content of rat milk (about 20 mg/ml), copper concentration was 0.4 (µg/ml). At the end of the experiment cerebrospinal fluid (CSF), blood serum, liver and brain of both rat groups were collected and Cp content and copper concentration were measured then. The results of *table 1* show that Cp serum level and copper concentration of the rat fed by BF are in coordination and significantly increased. Moreover in liver of experimental rats Cp-mRNA content defined by dot-hybridization is raised about 2-fold (data are not shown). Simultaneously copper concentration of experimental rats liver is decreased more than 2-fold (4.58+0.04 µg/organ, n=7, of control rats versus 1,91+1.21 (g/organ, n=7, of experimental rats). The brain copper concentration does not change and is equal to 1.71+0.04 and 1.72+0.062 µg/organ in control and experimental groups, respectively. In CSF the both copper and Cp levels remain low during ontogenesis (~0.20 µg of Cu/L and ~0.3 mg of Cp/100 ml in CSF of 6-, 14- and 160-day old rats). Cp and copper concentrations of CSF do not depend on their fluctuations in serum during development (200, 240 and 1040 µg of Cu/L and 6.1, 13.1 and 51.2 mg of Cp/100 ml in serum of 6-, 14- and 160-day old rats respectively). On the contrary, CSF Cp content is increased 7-fold (2.3 versus 0.3 mg of Cp/100 ml) in 8 day-old rats fed with BF. The copper redistribution and serum Cp increase have been already observed in 48 hours after beginning of BF feeding. So Cp content is increased about 1.5-fold (9.6 mg of Cp/100 ml versus 15.06 mg of Cp/100 ml in control and experimental rats respectively). Besides Wilson ATPase gene are activated *(fig. 1)*.

The copper accumulation in liver of control rats occurs progressively during embryogenesis as well as after birth till 12 days of life (from 10.37 to 31.05 µg/g dry weights of 19th and 22nd days of embryonic development, respectively, and to 121.37 μg/g dry weight of 12th day of life). Then copper concentration in liver sharply decreases almost 10-fold. These results are in absolute agreement with the earlier data from different laboratories [2, 3]. More detailed research showed that copper is non-uniformly distributed among cell components of hepatocytes. Furthermore the intracellular distribution of copper is changed during of ECMT and is dependent on feeding type. Thus it turns out that copper concentration in peroxisomes and mitochondria is not changed during ontogenesis. It does not dependent on the type of feeding and is constantly equal to 0.025+0.001 and 0.051+0.0021 μg/mg of protein, respectively. Considerable changes of lysosome copper content were observed during the first 21 days of life *(fig. 2)*. It is likely that lysosomes are the one of the main copper containing organell in this stage of development. They contain almost 60% of total cellular copper after 7th day of life. Lysosome copper concentration is amounted to maximum 1.25 μg/mg of protein to 10-11th days of life. Then it is dropped to the lowest level in cell (0.014 μg/mg of protein). The same curves of growth and lowering of copper content are observed in cytoplasm (0.40 μg/mg of protein maximum) and intracellular membranes (endoplasmic reticulum and Golgi complex membranes, 0.37 μg/mg of protein maximum). Copper concentration is increased in both organells till the 12th day of life and then it is dropped to 0.15 and 0.04 μg/mg of protein, respectively. It can be suggested that the fractions of cytoplasm and intracellular membranes are contaminated by lysosomes so they keep copper more than another fractions. The more detail research work would be carried out to check this assumption.

Table 1. The Cp and copper contents in serum of control and experimental rat groups

Parameters	Age, days	Natural feeding	Artificial feeding
Serum: 1) oxidase activity Cp, mg/100 ml	8	14,1±2,13 (6)*	38,47±4,46 (6)
2) antigen content Cp, mm² of rocket (2,5 μl of serum)	8	56±2 (6)	192±16 (7)
Cu, μg/L	8	300±50 (6)	940+71 (7)
Cu atoms per 1 Cp molecule	8	4,5	5,05

* - Number of rats

In opposite, the different curve depicts the dynamics of copper content alteration in nuclei *(fig. 2)*. The maximum of the nuclear copper concentration is reached to 5th day of life. It drops to about 3-fold and keeps almost the same level to 12th day. After that it is sharply reduced. In rats fed with BF the intracellular copper distribution and concentration correspond to ACMT (15-day old rats) already to 6th days of life *(fig. 2)*.

These data, taken together, show that newborns are adapted to uptake copper from Cp. And moreover the "copper packing" has more meaning for copper homeostasis of the newborns than copper excess. Perhaps it is concerned with the mechanism of copper safeguard of fetus. It was shown that during embryonic development a fetus received copper from maternal Cp [14]. And it is also known that peptide moiety of maternal Cp is not absorbed by fetus tissues [15]. It is likely that maternal Cp delivers copper ions into cells of fetus barrier tissue. This data are well agreed with the fact that yolk sac cells synthesize Cp, which is secreted to embryo [16]. It is interesting that yolk sac copper concentration is not changed during embryonic development (15.8 μg/g dry weight on 19th and 22nd days of embryonic development). It is equal to copper concentration of adult rat liver (near 15 μg/g dry weight [3]). During the same period the embryonic liver copper content is sharply increased (see above). It is possible that yolk sac is a central organ supporting copper homeostasis in embryos in conditions when fetal liver Cp gene activity is low and fetal tissues need escalating quantity of copper. It partly confirms with set forth below data showing that yolk sac cells contain the main proteins of CMS. So the immunoreactive polypeptides of Menkes ATPase and Wilson ATPase were found in plasma membrane-enriched fraction of the

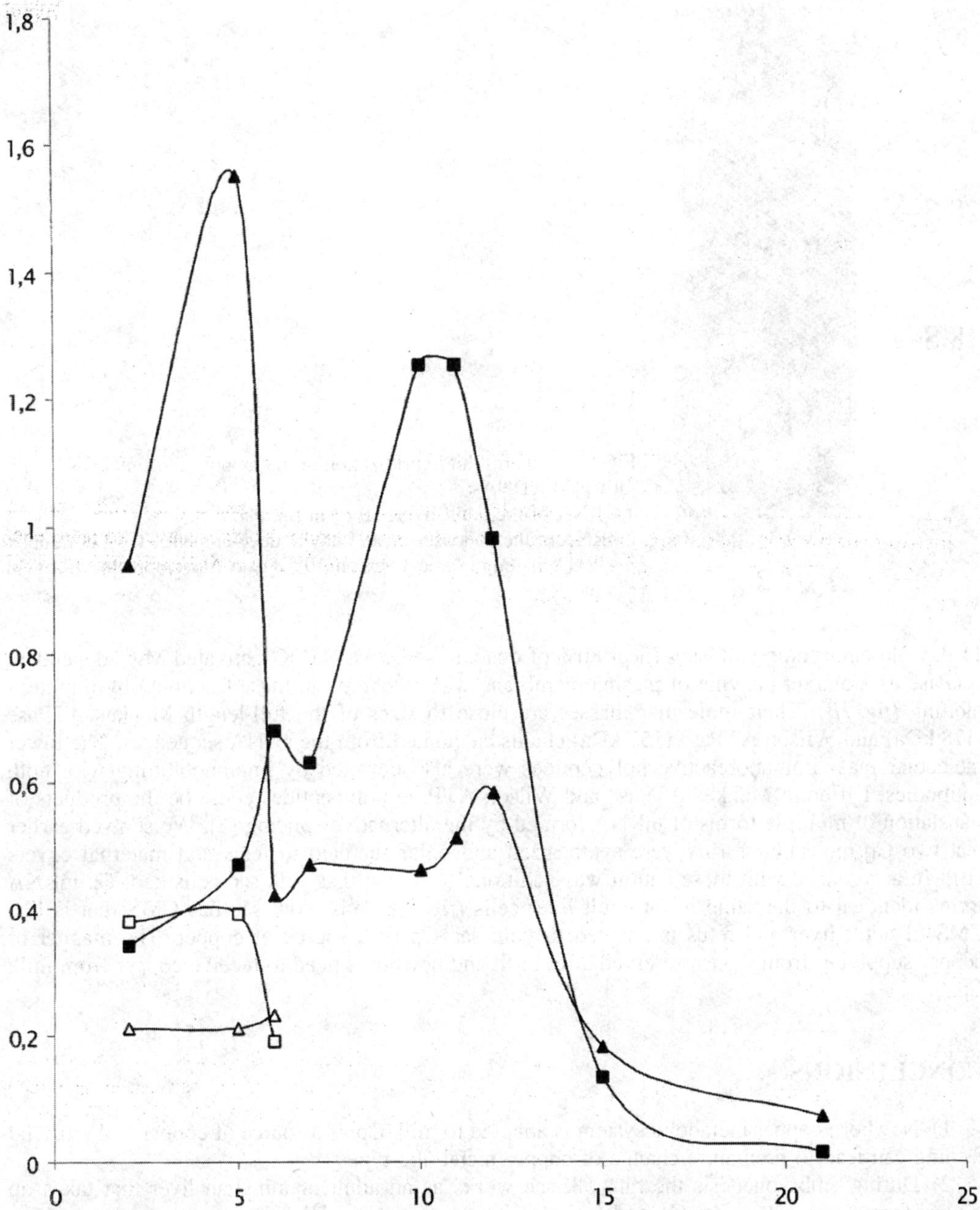

Fig. 2. The change of the intracellular copper distribution in newborn rat liver during development
Each point is a middle means from two experiments. For a one measurement was taken one g liver tissue (3-10 livers depending on age of rats)
Abscissa - age of rat, days; ordinate - Cu, µg/mg protein;
- - lysosomes of rats fed by females; - - lysomes of rats fed with BF;
- - nuclei of rats fed by females; - - nuclei of rats fed with BF

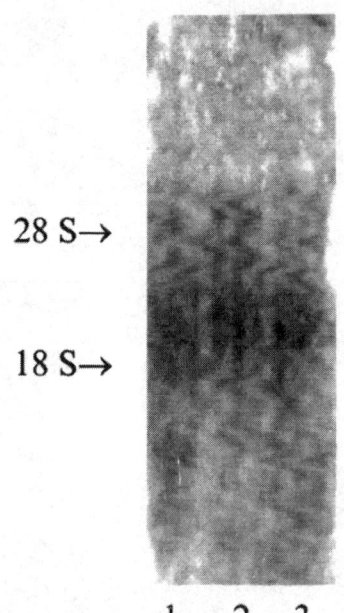

Fig. 3. Northern-blot hybridization of the rat polyribosomal RNA with rat Cp[^{32}P]cDNA
1 - RNA of the adult liver, 70 (g in sample;
2 - RNA of the yolk sacs on 19th day of development, 70 μg in sample;
3 - RNA of embryonic liver on 19th day of development, 70 μg in sample

11-day old embryonic yolk sacs (near 80% of ouabain-sensitive Na^+/K^+-activated Mg^{2+}-dependent ATPase as a marker enzyme of plasma membrane was recovered in these fractions) by immuno-blotting *(fig. 1B)*. Their molecular masses are close to sizes of the full-length Menkes ATPase (178 kDa) and Wilson ATPase (157 kDa) chains calculated from the cDNA sequences. The lower molecular mass immunoreactive polypeptides were also detected by immunoblotting with both antibodies. Different Menkes ATPase and Wilson ATPase polypeptides could be the products of translation of multiple forms of mRNA formed by the alternative splicing [5]. We showed earlier that two Cp molecular forms were synthesized and polar secreted to fetus and maternal covers [16]. In accordance with these data it was additionally shown that yolk sac cells kept Cp-mRNA forms identical to the same in rat adult liver cells *(fig. 3)*. Thus, yolk sac has CMS that is like CMS of adult liver and fetus is adapted to yolk sac Cp as a source of copper. The manner of copper supplying from Cp is preserved after birth and newborns need to receive copper from milk Cp.

CONCLUSIONS

1) Newborn copper metabolic system is adapted to milk Cp as a source of copper and artificial feeding can lead to premature change of copper metabolic types.

2) During embryogenesis the rat yolk sac works as an adult mammalian liver that takes up copper from maternal serum Cp and includes it into own Cp, which transports copper to fetus tissues then.

REFERENCES

1. Harris E.D. Cellular copper transport and metabolism. *Annu Rev Nutr*, 2000, 20, 291-310.
2. Hurley L.S., Keen C.L., Lonnerdal B. Copper in fetal and neonatal development. *Hunt Ciba Foundation Symp.*, 1980, 80: 227-245.

3. Mason K.E. A conspectus of research on copper metabolism and requirements of man. *J. Nutr.*, 1979, 109, 1979-2066.
4. Zaitseva I., Zaitsev V., Card G. et al. The nature of the copper centers in human ceruloplasmin. *J. Biol. Inorg. Chem.*, 1996, 1, 15-23.
5. Bull P.C., Cox D.W. Wilson disease and Menkes disease: new handles on heavy-metal transport. *TIG*, 1994, 10, 246-252.
6. Muller T., Muller W., Feichtinger H. Idiopatic copper toxicosis. *Am. J. Clin. Nutr.*, 1998 (suppl), 67, 1082S-1086S.
7. Lonnerdal B. Copper nutrition during infancy and childhood. *Am. J. Clin. Nutr.*, 1998 (suppl). 67, 1046S-1053S.
8. Puchkova L.V., Aleinikova T.D., Tsymbalenko N.V. et al. Biosynthesis and secretion of ceruloplasmin by rat mammary cells during lactation. *Biokhimiia (Rus)*, 1994, 59, 296-303.
9. Puchkova L.V., Zakharova E.T., Aleinikova T.D. et al. Comparative analysis of the molecular heterogeneity of ceruloplasmin from human blood and breast milk. *Biochemistry (Mosc)*, 1997, 62, 928-930.
10. Puchkova L.V., Alelnikova T.D., Zakharova E.T. et al. Content of ceruloplasmin as a source of copper in breast milk at different stages of lactation. *Vopr Pitan (Rus)*, 1997, 4, 19-22.
11. Puchkova L.V., Aleinikova T.D., Bichevaia N.K. et al. A comparative study of the transport dynamics of the peptide moiety of the milk ceruloplasmin molecule in the body of rats with embryonic and adult types of copper metabolism. *Ontogenez*, 1999, 30, 31-39.
12. Sambrook J., Fritch I.F., Maniatis T. Molecular cloning. Laboratory manual. *Cold Spring Harbor, Laboratory Press*, 1991.
13. Ravin H.A. Rapid test for hepatolenticular degeneration. *Lancet*, 1956, 1, 7267-7271.
14. Shavlovski M.M., Chebotar N.A., Konopistseva L.A. et al. Embryotoxicity of silver ions is diminished by ceruloplasmin-futher evidence for its role in the transport of copper. *BioMetals*. 1995, 8, 122-128.
15. Gorbunova V.N., Baranov V.S. The character of rat ceruloplasmin biosynthesis during embryogenesis and early postnatal period. *Ontogenez*, 1984, 15, 63-72.
16. Puchkova L.V., Zhiguleva E.A., Mokshina S.V. et al. The role of yolk sac in copper metabolism during rat embryogenesis. *Ontogenez*, 2001, 32, 204-211.

Metal Ions in Biology and Medicine: vol. 7. Eds L. Khassanova, Ph. Collery, I. Maymard, Z. Khassanova, J.-C. Étienne.
John Libbey Eurotext, Paris © 2002 pp. 454-459.

Metabolism of fatty acids in broiler chicken is affected by chromium

Kuryl Tomasz, Lipko Maciej, Debski Bogdan

Faculty of Veterinary Medicine, Div. Anim. Biochemistry, SGGW, ul. Nowoursynowska 159, 02-787 Warsaw, Poland

INTRODUCTION

Chromium was recognized as a factor engaged in insulin action on glucose metabolism. Broiler chicken fed a diet supplemented with chromium exhibits elevated protein concentration in liver and muscles as well as lower fat and cholesterol content in these tissues [1]. Lipid metabolism is one of the main energetic processes. Disorders in this metabolic pathway should effect among others development of miopathies and atherosclerosis [2].

The enzymatic reactions of mammalian beta-oxidation in mitochondria were elucidated some 40 years ago. In steps leading to this process are engaged a great number of enzymes i.e. membrane-associated acyl-CoA dehydrogenase and carnitine palmitoyl transferases [3]. Inherited beta-oxidation disorders should manifest clinically as usual by recurrent Reye-like episodes (acute hepatopathy and encephalopathy, myopathies). The findings at the tissue level are dominated by steatosis of organs with high beta-oxidation level (liver, kidney, heart and skeletal muscles) and toxic action of metabolites of beta-oxidation process [4]. Additional effect of it is obesity, associated with metabolic disorders related to alterations in lipid mobilization and oxidation [5]. Lipogenesis is stimulated by a high carbohydrate diet, whereas is inhibited by polyunsaturated fatty acids and by fasting [6]. Lipid metabolism in flying bird is a source not only of ATP but also of water. In human determination of total activity of beta-oxidation of fatty acids is important in medical diagnostics [7, 8, 9]. This problem was not studied in animal systems yet. These studies were undertaken with to transfer methodology from human to animal systems. The aim of these studies was to optimize the reaction parameters for broiler system and to check if chromium, nutritional Cr (Cr-yeast) supplementation may influence beta-oxidation process in broilers' lymphocytes.

METHODS

Broilers were fed 4 weeks with standard diet and diet supplemented with 0,5 ppm of chromium in form of chromium enriched yeast. After the nutritional period of experiment, the blood was collected from brachial vein on standard EDTA. The samples of the blood were transferred to the laboratory without cooling and lymphocytes were isolated by centrifugation on FICOLL-HYPAQUE 1077 (2000 rpm at 18° for 20 min.). The lymphocyte layer was collected, washed with phosphate buffered saline (PBS) without Ca^{+2} and Mg^{+2} ions. Cell suspension was centrifuged again for 10 minutes at 2000 rpm at 18^0C. Obtained pellet was washed third time with PBS and transferred to plastic 4 ml tube, mixed and centrifuged as above. Final pellet was resuspended in PBS at a final concentration of protein 1-1,5 mg of protein per 1 ml. Method of Manning et al. [7] and Olpin et al. [8] in own modification [10] for human system was used.

Beta-oxidation of fatty acids was estimated in reaction mixture containing lymphocytes (25 µl) and substrate - palmitic acid supplemented with radiolabelled [9,10-3H] palmitic acid (200 µl) (Amersham) dissolved in Hank's Balanced Salt Solution (HBSS) containing 0,5 mg of fatty acids free albumin (Sigma) per 1 ml of HBSS. The samples were incubated in water bath with shaker for 1 hour at 38°C. After incubation 200 µl of 10% trichloroacetic acid and 100 µl of HBSS were given into reaction mixture. Tubes were vortexed and after 5 min at room temperature centrifuged at 400 rpm for 10 min. Supernatant was transferred to new tube, alkalized with 100 µl of 2 M NaOH and mixed. This solution was applied on the column of Bio-Rex AG 1×8 (8 x 20 mm) and effluent was collected straight into scintillation vial. The column was washed with 1 ml of water and combined effluents were supplemented with 5 ml HiSafe scintillation cocktail (Pharmacia) and radioactivity was measured in Packard TriCarb counter. Blank samples containing only substrate solution were treated as experimental tubes. The results were expressed as pmoles of decomposed palmitic acid per 1 minute per 1 mg of lymphocyte protein.

Protein content was assayed by the method of Lowry et al. using fatty acid free albumin as a standard.

Table 1. The effect of substrate concentration on beta-oxidation 0f [9,10-3h] palmitic acid by broiler lymphocytes

Susbtrate (pmoles/sample)	Beta-oxidation activity (pmoles/min/mg protein) [Mean±S.D.: n=24]
10	26,303 ± 1,768
25	40,157 ± 5,962
50	56,365 ± 4,839
100	103,370 ± 6,067
200	105,963 ± 8,915

Table 2. The effect of temperature of incubation on the activity of beta-oxidation of [9,10-^{3}H] palmitic acid by broiler lymphocytes

Temperature [°C]	**Activity of beta-oxidation (pmoles/min/mg of protein)* [Mean±S.D.: n=9]**
30°	28,532 ± 2,260
37°	36,628 ± 1,768
42°	37,025 ± 2,844
48°	28,307 ± 2,844

* - protein amount - 14,49 µg/sample
- substrate - 26,2 pmoles/sample

Table 3. The effect of chromium yeast suuplementation on the activity of beta-oxidation of [9,10-^{3}H] palmitic acid by broiler lymphocytes

Experimental variant	**Beta-oxidation activity (pmoles/min/mg protein) [Mean ±S.D.; n=12]**
Control	40,151 ± 5,961
+ 0,5 ppm Chromium yeast	85,883 ± 5,812
2 weeks without chromium	68,175 ± 2,181
6 weeks without chromium	40,738 ± 2,619
20 weeks without chromium	40,096 ± 0,698
50 weeks without chromium	32,906

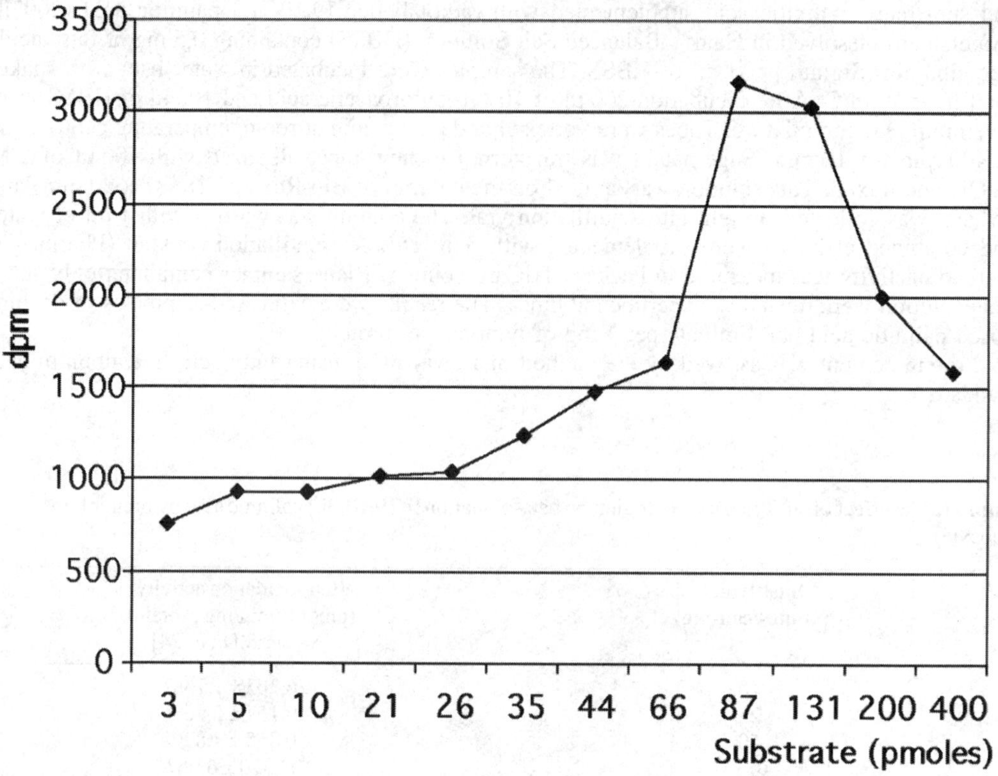

Fig. 1. The effect of substrate concentration on the beta-oxidation activity of palmitate in broiler lymphocytes

RESULTS

Lymphocytes from broiler chicken metabolize palmitic acid and should be used in studies on fatty acid metabolism. In control system the activity of β-oxidation of ^{14}C-palmitic acid was 40,151±5,961 units (pmol/min/mg of protein) in the presence of 26,2 pmol of the substrate (as in human system). The question was if chicken lymphocytes required reaction conditions different from those from human? Experiments were done with changing amounts of lymphocyte protein (at constant substrate concentration) as well as the effect of protein amount on palmitic acid degradation. As can be seen from *fig. 1*, at very low concentration of the substrate (2,62 to 20,96 pmoles) degradation of palmitic acid was very low and only very small differences between samples were observed (in the range 754±280 dpm and 1017±157 dpm). On the other hand, the best activity of beta-oxidation took place at concentration of substrate in the range 87,33-131 pmoles (3169±122 and 3029±341 dpm, respectively). The activity of beta-oxidation in chicken lymphocytes was concentration dependent up to almost 100 pmoles of palmitic acid *(table 1)*. It changed from 26,303±1,768 to 40,157±5,962 and 56,365±4,839 pmoles/min/mg for 10, 25 and 50 pmoles of palmitic acid, respectively. For concentration of substrate of 100 and 200 pmoles degradation of palmitic acid was the same - 103,370±6,067 and 105,963±8,915 pmoles/min/mg of protein. These results led us to conclude, that in chicken system the best concentration of palmitic acid as a substrate for beta-oxidation studies is 100 pmoles per sample.

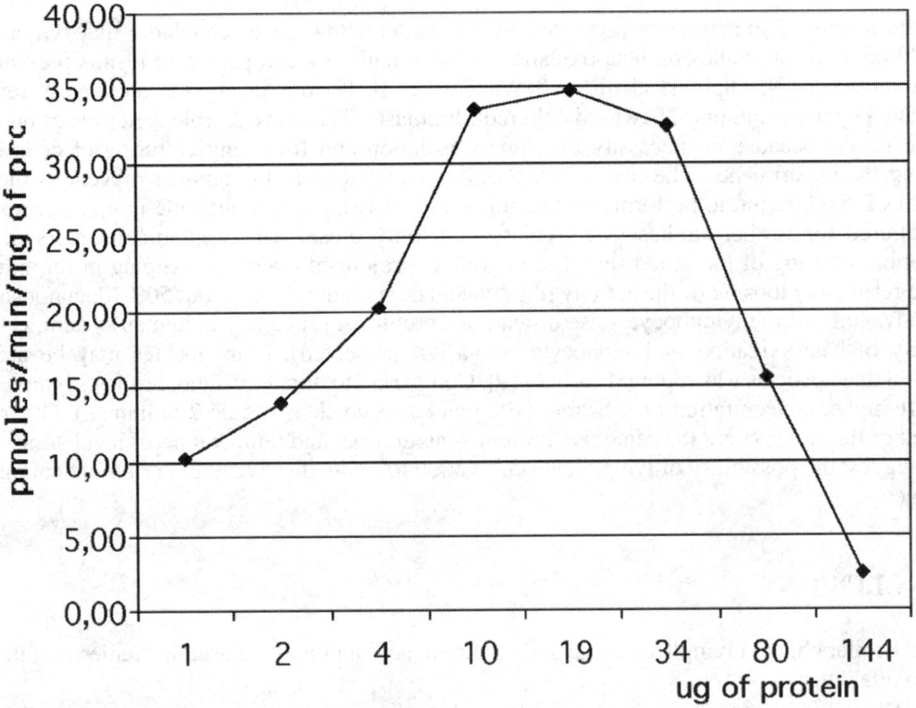

Fig. 2. The effect of protein concentration on the activity of palmitic acid beta-oxidation

Activity of fatty acids degradation in chicken lymphocytes strongly depends on protein amount in the reaction mixture. These experiments were done with 26,2 pmole of palmitic acid in the reaction mixture. As can be seen from *fig. 2* the highest degradation of substrate occurs at protein content ranging from 10 to 35 µg of lymphocytes protein.

Temperature dependency of the fatty acids degradation by broiler lymphocytes showed plateau between 37°C and 42°C *(table 2)*. However, at lowest (30°C) and highest (48°C) beta-oxidation of palmitic acid in the cells studied was decreased only by at about 25%.

All these experiments led us to conclude, that the optimal conditions for beta-oxidation studies in chicken lymphocytes are: substrate concentration - 100 pmoles of palmitic acid per sample, protein amount of 10 to 35 µg per sample and 1 hour incubation at 42°C.

The effect of chromium supplementation was studied in experiment with the parameters of the reaction as in human system. The results are presented in *table 3*. In those preliminary experiments substrate at the concentration of 26,2 pmole/sample was used. This resulted in lower activity of fatty acids degradation. Anyway, 2-weeks supplementation of broilers with 0,5 ppm of chromium yeast to the feed elevated the activity from 40,151±5,961 to 85,883±5,812 pmoles/min/mg of protein. 2, 6 and 20 weeks after termination of chromium supplementation the activity of fatty acids beta-oxidation lowered to 68,175±2,181, 40,738±2,619 and 40,096±0,698 pmoles/min/mg, respectively. Experiment with only one broiler after 1 year after without chromium supplementation exhibits the activity of beta-oxidation of 32,906 pmoles/min/mg of protein.

DISCUSSION

The results of experiments presented in this paper allow us to conclude, that lymphocytes should be used for studies on beta-oxidation of fatty acids. As compare with results presented by Manning et al. [7], Olpin et al. [8] and Wanders et al. [9] our reaction conditions differ from described by these authors. They used cultured fibroblasts. This caused prolonged time of obtaining of the results because on necessity of primary isolation and following culturing of cells before starting the experiments. The use of lymphocytes isolated from the amount of venous blood as small as 2-5 ml permit to perform all experiment in just 1 day. The remaining lymphocytes should be cultured for further studies as fibroblasts, but only those, which exhibit disorders of beta-oxidation. Cooling of the blood samples, as well as prolonged overnight keeping in room temperature resulted in loosing of the activity of fatty acid degradation by at about 50% (data not shown). The advantage of the lymphocytes use instead of fibroblasts, especially in human system, is higher activity of beta-oxidation in lymphocytes (data not presented). Lymphocytes may be used for medical diagnostic as was reported earlier [10]. Comparing to human system, broiler's lymphocytes require higher concentration of substrate (100 pmoles in chicken and 26,2 in human). Other parameters of the reaction are the same, i.e. protein content, time and temperature of incubation. These data suggest the possibility of lymphocyte cultivation to omite disadvantages of biological material storage.

CONCLUSIONS

1. Broiler chicken lymphocytes degrade palmitic acid and may be used in studies on fatty acid beta-oxidation.
2. Chicken given 0,5 ppm two-weeks long chromium yeast supplementation metabolize fatty acids almost 2-times faster than control birds.
3. The influence of different compounds on beta-oxidation may be studied using lymphocytes.

REFERENCES

1. Debski B., Zalewski W., Gralak M.A., Effect of dietary chromium sources on broilers glycogen and glucose level., in: Anke M., Muller R., Schafer U. (ed.) Mineralstoffe, Mengen-, Spuren- und Ultraelemente in der Prevention WBA Wissenschaftliche Verlagsgesellschaft mbH Stuttgart 2001, 302-308.
2. Saudubray J.M., Martin D., de Lonneley O.P. et al., Recognition and management of fatty acid oxidation defects: a series of 107 patients., J. Inher. Metab. Dis., 22 (1999), 488-502.
3. Eaton S., Barlett K., Pourfarzam M., Mammalian mitochondrial beta-oxidation., Biochem. J. 320 (1996), 345-347.
4. Elleder M., Kmoch S., Hejtmankowa M., Bouska I., Autopsy and biopsy findings in disorders of mitochondrial beta-oxidation of fatty acids. Role of the pathologists in the diagnostic process., Cesk. Patol. 35(3) (1999), 98-105.
5. Horowitz J.F., Regulation of lipid mobilization and oxidation during exercise in obesity., Exerc. Sport Sci. Rev. 29 (1) (2001), 42-46.
6. Kersten S., Mechanisms of nutritional and hormonal regulation of lipogenesis., EMBO Rep. 2(4) (2001), 282-286.
7. Manning N.J., Olpin S.E., Pollitt R.J., Webley J., A comparison of [9,10-3H] palmitic acid and [9,10-3H] myristic acid for the detection of defects of fatty acids oxidation in intact cultured fibroblasts., J. Inher. Metab. Dis., 13 (1990), 58-68.
8. Olpin S.E., Manning N.J., Pollitt R.J., Bonham J.R., Downing M., Clark S., The use of [9,10-3H] palmitate and [9,10-3H] oleate for the detection and diagnosis of medium and long-chain fatty acid oxidation disorders in intact cultured fibroblasts., Adv. Exp. Med. Biol., 466 (1999), 321-325.

9. Wanders R.J.A., Vraken M.E.J., den Boer F.A. et al., Disorders of mitochondrial fatty acyl-CoA beta-oxidation., J. Inher. Metab. Dis., 22 (1999), 442-487.
10. Kuryl T., Adamowicz M., Debski B., Bertrandt J., Martinik K., Degradation of [9,10-3H] myristic acid by lymphocytes. Screening test of inherited disorders of activation, transport and mitochondrial oxidation of fatty acids., Atherosclerosa, 23 (2001), 23-26.

Metal Ions in Biology and Medicine: vol. 7. Eds L. Khassanova, Ph. Collery, I. Maymard, Z. Khassanova, J.-C. Étienne.
John Libbey Eurotext, Paris © 2002 pp. 460-463.

Selenium and diabetic elders

JF. Escanero[1], A. García de Jalón[2], RM. Pérez-Beriain[2], ML. Calvo Ruata[2], A. Sanz-París[3], J. Abanto[4], A. Cabeza[2]

[1]Phisiology department of University. C/ Domingo Miral s/n. 50006 Zaragoza, Spain.
[2]Biochemistry department. [3]Endocrinology and Nutrition department. Miguel Servet University Hospital. Pº Isabel la Católica 1-3. 50009 Zaragoza, Spain
The mail will be remitted to:
Dr. Ángel García de Jalón Comet.
Jefe de Servicio de Bioquímica Clínica
Hospital Universitario Miguel Servet
Paseo Isabel la Católica 1-3
50009 Zaragoza (Spain)
E-mail: jcomet@teleline.es

ABSTRACT

Many studies conclude that mean serum selenium concentrations in patients with diabetes are lower than in controls. Our objetive was to study the relationship between diabetes and serum concentrations of selenium in elders.

Materials and methods : Samples were analyzed from 47 NIDDM elders, interned in a long-term care old people's home, comparing to a control group (n=49) of similar characteristics not diabetic.

Serum concentrations of selenium were analyzed by atomic absorption spectrophotometry with a graphite furnace and a Zeeman background corrector.

Results and conclusions : The results of the serum concentration of selenium in the diabetic group was of 61.19 µg/l. In the control group the mean of the selenium levels was of 68.99µg/l. The means comparison using the Student T test shows a statistically significant difference with a $p<0.001$.

The findings of this study suggest that the diabetic elders showed lower selenium concentrations than the normal elders.

BACKGROUND

Approximately 11% of the population is aged 65 years or older, with a projected increase to 20% by the year 2030 [1].The Census Bureau predicts that by the year 2020 the average life expectancy for women will be 82 years and for men 74.4 years, and that the population aged 65 and older will outnumber those under the age of 15 [2].

Nutrition and aging are inseparably connected as eating patterns affect the progress of many degenerative diseases associated with aging. In turn, the nutritional status of elderly may be adversely affected by a number of factors associated either directly or indirectly with aging [3].

Plasma glucose levels increase with advancing age as the result of progressive changes in insulin action and metabolism, and the incidence of non-insulin-dependent diabetes mellitus (NIDDM) increases with age and increased body fat [4].

Knowledge of mineral element status and requirements in older people is incomplete. There is a general lack of information about mineral status and metabolism in very old people, and specially in diabetic elders.

Some studies supports the theory that free radical stress has a role in the causation of non-insulin dependent diabetes mellitus [5]. Other studies concluded that mean serum selenium concentrations in patients with diabetes mellitus were lower than those determined in the control group [6].

AIMS

To study the relationship between diabetes and serum concentrations of selenium in elders subjets.

MATERIALS AND METHODS

Subjects

Samples were analyzed from 47 non-insulin-dependent diabetic elders, interned in a long-term care old people's home, aged between 66 and 95 (mean of 81.62 years).

As a control group, we selected the serums of 49 apparently healthy elders interned in the same long-term care old people's home, who were aged between 70 and 91 (mean of 79.36 years) *(fig. 1)*.

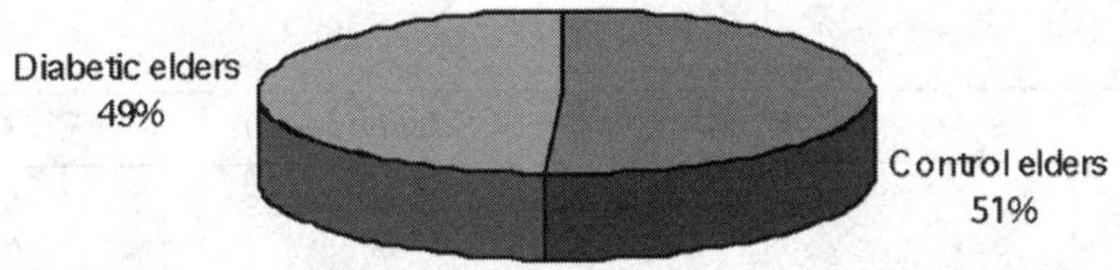

Fig. 1.

Instrumentation

Serum concentrations of selenium were analyzed by atomic absorption spectrophotometry with a graphite furnace and a Zeeman background corrector (Perkin Elmer 4110 ZL) *(fig. 2)*, using Pd(NO3)2 solution as matrix modifier.

Statistic method

The statistical calculations were carried out using SPSS statistics program. The statistical test used, was the Student T-test for equality of means.

RESULTS

According to age, the differences are not statistically significant. The results *(table 1)* of the serum concentration of selenium in the diabetic group was of 61.19 μg/l. In the control group the mean of the selenium levels was of 68.99 μg/l.

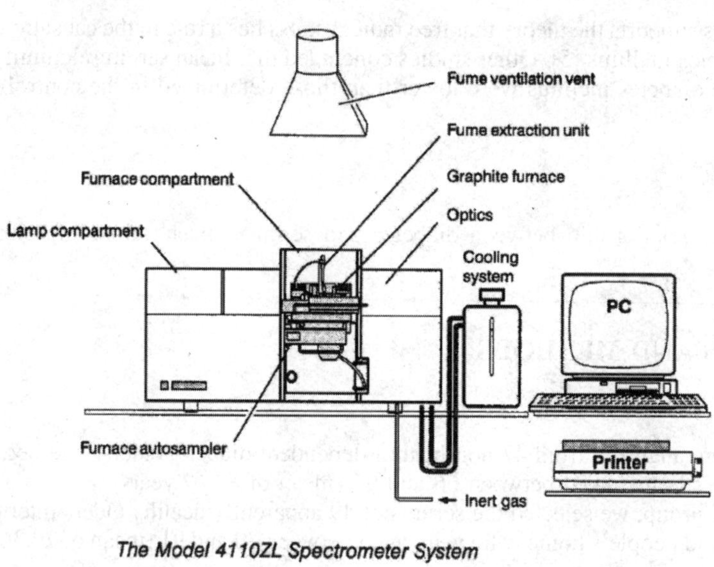

Fig. 2.

Table 1.

	N	Selenium (μg/l)	C.I. (95%)
Diabetic elders	47	61.19	58.35-64.03
Control elders	49	68.99	66.14-71.83

The distribution of selenium levels in the differents groups are defined to the following illustration *(fig. 3)*:

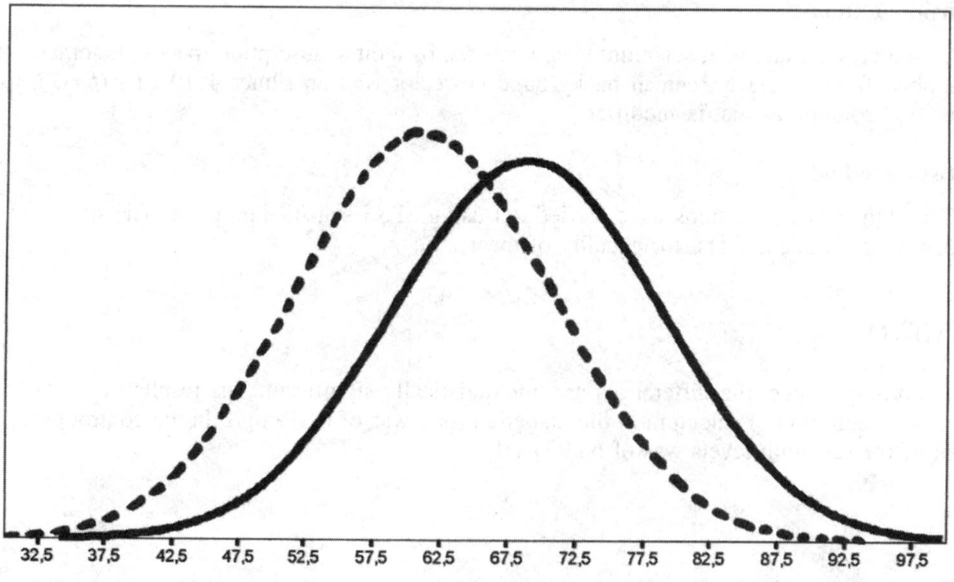

Fig. 3.

The means comparison using the Student T test shows a statistically significant difference with a $p<0.001$.

CONCLUSIONS

The findings of this study suggest that the diabetic elders showed lower selenium concentrations than the normal elders.

Selenium supplementation may be beneficial (increasing glutathione peroxidase activity, cardioprotective effect, inmunostimulatory properties) for diabetic patients. Supplementation with a trace element may be indicated when its depletion was unequivocally documented and when there is evidence of the positive effects of this element on the quality of life of the diabetic patients.

REFERENCES

1. Kendrich ZV, Nelson-Steen S, Scafidi K. Exercise, aging and nutrition. Southern Medical Journal. 1994 May. 87(5). P 550-60.
2. U.S. Bureau of the Census. Statistical abstracts of the United States, 1991. 3rd ed. Washington, DC: U.S. Government Printing Office, 1991.
3. Buchowski MS, Sun M. Nutrition in minority elders: current problems and future directions. Journal of Health Care for the Poor & Underserved. 1996 Aug. 7(3). P 184-209.
4. Lum OM. Health status of Asians and Pacific Islanders. Clin Geriatr Med. 1995 Feb. 11(1). P 53-67.
5. Salonen JT, Nyyssÿonen K, Tuomainen TP et al. Increased risk of non-insulin dependent diabetes mellitus at low plasma vitamin E concentrations: a four year follow up study in men. BMJ. 1995 Oct 28. 311(7013). P 1124-7.
6. Navarro-Alarcón M, López-G de la Serrana H, Pérez-Valero V, López-Martínez C. Serum and urine selenium concentrations as indicators of body status in patients with diabetes mellitus. Sci Total Environ. 1999 Mar 22. 228(1). P 79-85.

*Metal Ions in Biology and Medicine: vol. 7.*Eds L. Khassanova, Ph. Collery, I. Maymard, Z. Khassanova, J.-C. Étienne.
John Libbey Eurotext, Paris © 2002 pp. 464-467.

Zinc, copper and selenium content in the diets of children with food intolerance

Renata Markiewicz, Maria H. Borawska, Janina D. Piotrowska-Jastrzebska*, Maria J. Piotrowska-Depta*, Katarzyna Hukalowicz

Department of Bromatology, Medical Academy, 1 Kilinski St., 15-230 Bialystok, Poland,
**Department of Axiology, Medical Academy, 17 Waszyngton St., 15-274 Bialystok, Poland*

ABSTRACT

Zinc, copper and selenium level in food rations of children with food allergy was compared to the level of these elements in rations of healthy children.

The level of the analysed trace elements in 40 full-day food rations was determined using atomic absorption spectrometry. The analysed food rations did not provide the recommended amount of zinc and copper for children. However, the food rations prepared in the Children's Home were more variegated and with lower deficiency of zinc and copper. The food rations from Children's Home were rich in selenium.

INTRODUCTION

Incidence of allergic diseases is high and it has increased since last decade (Kaczmarski,1997). Manifestation of food allergy in children below the age of three can take a form of dyspepsia, rhinitis, asthma, hives, or atopic skin inflammation (Cantani 1999, Rance et al., 1999). Mild symptoms can pass unnoticed, while acute cases are not always diagnosed as allergy. Thus an early diagnosis and application of an appropriate elimination diet therapy are extremely important (Kaczmarski, 1993). Because of the long-lasting, recurrent character of the disease, diet therapy is often applied for months or even years. Then, especially in the case of the youngest children, a prolonged elimination diet should be put under careful control respecting its influence on children's health state and their developmental processes.

Epidemiological investigation of the relationship between supply of nutrients and health state indicates a fundamental role of mineral elements, especially for children (Kierst, 1989).

The function of zinc, copper and selenium has been described in numerous publications which point out multidirectional biological activity of zinc, its role in infant's and children's development and functioning of immune system. Zinc is also applied in therapy of Wilson's disease, acute and prolonged diarrhoea in infants and children, or even in treatment of colds (Sandstead, 1991).

Copper has a significant role in the development and functioning of nervous system, it affects metabolism of lipids, connective and elastic tissue in membranes of blood vessels, plays role in hardening processes of collagen, hair keratinization and synthesis of melanin (Kabata-Pendias and Pendias, 1993, Lewis, 1984). Copper ad zinc ions activate superoxide dismutase, an enzyme which protects cell membranes from oxidative damage by H_2O_2 (Graczyk et al., 1993).

Selenium is an integral part of the enzyme glutatione peroxidase, which assists in intracellular defense mechanisms against oxidative damage. It is also present in type I 5'-iodothyronine deiodinase (Arthur et al., 1990), selenoprotein P and other selenium proteins (Sunde, 1990), and has recently been implicated in the immune response (Kiremidjian-Schumacher and Stotzky, 1987).

Selenium deficiency can lead to cardiomyopathy called Keshan disease and osteoarthropathy - Kashin-Back disease (Litov and Combs, 1991).

The aim of this study was to evaluate content of zinc, copper and selenium in food rations of children undergoing elimination diet therapy, and to compare it to the content of this minerals in food rations of healthy children.

MATERIALS AND METHODS

The research material included full-day rations of children (1-3-years old) with food intolerance undergoing treatment in IIIrd Department of Paediatrics Medical Academy of Bialystok, and food rations of healthy children from Children's Home No. 2 in Bialystok.

The meals were prepared in June and July, 2001. Randomly chosen full-day rations were weighted, homogenised and decomposed with concentrated nitric acid in a microwave mineralizer BM-1z instrument UniClever (Plazmatronika, Poland), following appropriate minaralization procedure.

The content of zinc and copper was determined using flame atomic absorption spectrometry method, and the selenium content was analysed by electrothermal atomic absorption spectrometry (ETAAS) on a Z-5000 instrument (Hitachi, Japan) with a Zeeman background correction. Certified reference material - Simulated diet A - was used to test the accuracy of this method.

The obtained results were compared to the recommended dietary allowances. Statistical analyses were performed using Statistica v. 5'97 software.

RESULTS AND DISCUSSION

Twenty food rations from IIIrd Department of Paediatric and twenty rations from the Children's Home were examined. The content of zinc, copper and selenium is presented in *tables 1 and 2*.

The average weight of the rations from the hospital was 2.289 kg, while in the Children's Home it was considerably lower - 2.221 kg.

Table 1. Zinc, copper and selenium content in the daily food rations of children

No	Element	Unit	The mean content in rations A ± SD (min-max)	The mean content in rations B ± SD (min-max)
1.	zinc	mg/kg	1.98 ± 0.28 (1.52 - 2.67)	2.64 ± 0.75 (1.57 - 4.99) $p_{(1A/1B)}<0,0008$
2.	copper	mg/kg	0.23 ± 0.03 (0.19 - 0.31)	0.30 ± 0.07 (0.22 - 0.47) $p_{(2A/2B)}<0,0004$
3.	selenium	µg/kg	6.48 ± 0.50 (5.72 - 7.77)	19.78 ± 10.93 (2.21 - 48.27) $p_{(3A/3B)}<0,000003$

A) Hospital; B) Children's Home

Table 2. Zinc, copper and selenium intake with the daily food rations of children

No	Element	Unit	Average intake ± SD (min - max) A	**Average intake ± SD (min - max) B**
1.	zinc	mg/day	5.31 ± 0.81 (4.10 - 7.40)	5.81 ± 1.48 (3.50 - 9.70)
2.	copper	mg/day	0.63 ± 0.10 (0.46 - 0.86)	0.67 ± 0.15 (0.44 - 1.03)
3.	selenium	µg/day	17.39 ± 1.56 (14.80 - 19.70)	43.84 ± 24.21 (5.00 - 101.50) $p_{3A/3B}$<0,00002

A) Hospital; B) Children's Home

The level of zinc in the hospital rations was between 1.52 and 2.67 mg/kg, while in the rations from the Children's Home it was more unstable - from 1.57 to 4.99 mg/kg. The mean content of zinc in the hospital rations (1.98 mg/kg) was significantly lower than the mean content of zinc in the rations from the Children's Home (2.64 mg/kg).

The level of copper in hospital rations ranged from 0.19 to 0.31 mg/kg; the average - 0.23 mg/kg; in the ration's from the Children's Home it was between 0.22 and 0.47, the average - 0.30 mg/kg and it was significantly higher.

In the hospital rations selenium level was between 5.72 and 7.77 µg/kg, while in the rations from the Children's Home it was between 2.21 and 48.27 µg/kg.

It has been observed the average selenium level in the hospital rations (6.48 µg/kg) was significantly lower than the level of this element in the rations from the Children's Home (19.78 µg/kg).

The average supply of zinc in the hospital food rations was 5.31 mg/day and it was similar to the average supply of zinc in the rations from the Children's Home - 5.81 mg/day.

Selenium level was considerably lower in the rations from the hospital - 17.39 µg/day than in the rations from the Children's Home - 43.84 µg/day.

The research proved zinc and copper deficiency in the analysed food rations and it was higher in the meals served in the hospital. Requirement for zinc of children at the age of 1-3 (10 mg/day) was supplied in 53.1-58.1%; the requirement for copper (0.7-1 mg/day) (Ziemlanski et al., 1998) - in 63.0-67.0%. The problem of zinc deficiency in diverse populations was also considered by other authors (Ilow et al., 1999, Switoniak, 1999).

According to quantitative analysis the average selenium supply in the hospital rations was 17.39 µg/day, which supplied the requirement in 87.0%. The rations from the Children's Home contained 2.2 times more selenium than the recommended level - 20 µg/day (Ziemlanski et al., 1998).

Up to now the problem of selenium consumption in Poland has not been satisfactorily researched and described. Zagrodzki (Zagrodzki, 2000) points out that the research results provided by some authors deal with the chosen populations only, such as students - 417 µg Se/day (Skorkowska-Zieleniewska et al., 1983), or pensioners - 176.14 g/day (Klos and Bertrandt, 1998), and these results seem exceedingly overestimated.

Reliable data concerning consumption of mineral elements, especially in children at a stage of intensive growth and development, can be helpful in preventing insufficient supply of these elements. The appropriate level of the mineral elements in food is indispensable in regular functioning of human organism.

CONCLUSIONS

1. Food rations of the children undergoing elimination diet therapy in the hospital did not supply recommended daily intake for zinc, copper and selenium.

2. Full-day rations prepared in the Children's Home were more variegated - rich in selenium and with lower zinc and copper deficiency.

REFERENCES

Arthur J.R., Nicol F., Beckett F.J. Hepatic iodothyronine deiodinase: the role of selenium. Biochem. J., 1990, 272, 537-540.

Cantani A. The growing genetic links and the early onset of atopic diseases in children stress the unique role of the atopic march: a meta-analysis. J. Investig. Allergol. Clin. Immunol., 1999, 9, 314-320. Graczyk A., Radomska K., Kowalski J. Miedz, Mag. Med., 1993, 4/10, 24-33.

Ilow R., Regulska-Ilow B., Szymczak J. Ocena sposobu zywienia dziewczat ze szkól srednich z Glogowa i Lublina. Cz. II: Ocena ilosciowa. Bromat. Chem. Toksykol., 1999, 33, 27-33.

Kabata-Pendias A., Pendias H. Biogeochemia pierwiastków sladowych. PWN, Warszawa, 1993.

Kaczmarski M. Alergia i nietolerancje pokarmowe - mleko, cukry, soja. Sanmedia, Warszawa, 1993.

Kaczmarski M. Alergia pokarmowa u dzieci. Postepy w alergologii II. Med-press, Warszawa, 1997, 120-129.

Kierst W. Nauka o zywieniu czlowieka zdrowego i chorego. PZWL, Warszawa, 1989.

Kiremidjian-Schumacher L., Stotzky G., Reviev. Selenium and imune response. Envir. Res., 42, 277-303.

Klos A., Bertrandt J. Zawartosc selenu w racjach pokarmowych stosowanych w zywieniu pensjonariuszy Domu Emeryta Wojskowego. W: Materialy konferencji: Witaminy i mikroelementy w zywieniu czlowieka - biodostepnosc i stan odzywienia. Warszawa, 1998, 186.

Lewis A.J. The role of copper in inflammatory disorders. Agent Action, 1984, 15, 513-519.

Litov R.E., Combs G.F.Jr. Selenium in pediatric nutrition. Pediatrics. 1991, 87, 339-351.

Rance F., Kanny G., Dutau G., Moneret-Vautrin D.A. Food allergens in children. Arch. Pediatr., 1999, 6, suppl 1, 61-66.

Sandstead H.H. Zinc deficiency: A public health problem? Am. J. Dis. Child., 1991, 145, 835-859.

Skorkowska-Zieleniewska J., Brzozowska A., Józefowicz T. Ocena spozycia selenu na tle zlozonych wspól-zaleznosci metabolicznych tego pierwiastka. Roczn. PZH. 1983, 469-474.

Sunde R.A. Molecular biology of selenoproteins. Ann. Rev. Nutr., 1990, 10, 451-474.

Switoniak T. Sposób zywienia studentów w warunkach gospodarki rynkowej. Bromat. Chem. Toksykol., 1999, 33, 55-61.

Zagrodzki P. Selen w zywieniu czlowieka. Cz. I: Zawartosc selenu w zywnosci, zalecane i rzeczywiste spo-zycie selenu. Bromat. Chem. Toksykol., 2000, 33, 209-214.

Ziemlanski S., Bulhak-Jachymczyk B., Budzynska-Topolowska J., Panczenko-Kresowska B., Wartano-wicz M. Normy zywienia dla ludnosci w Polsce. Nowa Medycyna, 1998, 4, 1-27.

Metal Ions in Biology and Medicine: vol. 7. Eds L. Khassanova, Ph. Collery, I. Maymard, Z. Khassanova, J.-C. Étienne.
John Libbey Eurotext, Paris © 2002 pp. 468-472.

Dietary selenium (Se) intake of breast-fed and formula-fed young infants in Hungary

Ibolya Sziklai-László[1], Mária Ágnes Cser[2], Judit Cholnoky[3]

[1]*KFKI Atomic Energy Research Institute, H-1525 Budapest, P.O. Box 49.,* [2]*Bethesda Children's Hospital, H-1146 Budapest, Bethesda Street 3.,* [3]*Nestlé Hungária Ltd., H-1525 Budapest, Hungary*

ABSTRACT

Background: Selenium is a cofactor of several antioxidant seleno - enzymes such as glutathione peroxidase and therefore is essential for the proper function of cellular defense mechanisms. Sufficient supply of Se for infants is necessary for normal growth and development.

The aim of the study was to measure Se concentrations in human milk and milk-based formulas and to calculate Se intake for infants fed on human milk or milk-based formulas.

Method: Se concentrations were measured by Instrumental Neutron Activation Analysis (INAA). The age of the babies in both groups ranged from 4-8 weeks.

Results: human milk (mature milk, n=30) gave a mean of 13.7 ± 2.3 mcg Se/L with a range from 10.4 to 16.6 mcg/L. The Se content of infant formulas ranged from 4.4 to 9.9 mcg/L and found to be significantly lower in comparison with human milk. The daily Se intakes were calculated from the milk volumes (mean: 721±83 mL/d) consumed, determined by the weight gain of the babies after feedings. This resulted in 10.7 ± 2.7 mcg daily Se intake in babies fed exclusively on human milk. The Se content of starter formulas ranged from 4.4 to 8.4 mcg/L, follow up formulas varied from 7.2 to 9.9 mcg/L. The lowest Se intake was obtained in the group of infants fed locally produced formula: 3.2 ± 0.4 mcg/d (n=9), whereas 6.9 ± 0.8 mcg/d were observed in babies (n=9) fed on imported formulas. The Se content of formulas positively correlated with its protein content ($r=0.76$, $p<0.01$) and this association was even stronger in the human milk ($r=0.99$, $p<0.001$).

Conclusion: the formulas provide significantly less Se than the 10 mcg/day corresponding to the Recommended Daily Allowances (USA, 1989) for infants aged 0 to 6 months. All infant formulas provided significantly less Se intake ($p<0.001$) compared to human milk. Locally produced milk-based formulas had the lowest Se concentrations. The Se intake of breast-fed infants was within the range of the recommended values. Infant formulas have lower Se concentrations than human milk. Therefore babies fed on formulas may be at increased risk for inadequate Se intake induced diseases.

Key words: Selenium, human milk, infant formula.

This work was supported by grant from KFKI Condensed Matter Research Centre (KFKI-CMRC, No. ICA1-CT-2000-70029).

INTRODUCTION

Selenium (Se) is a cofactor of several antioxidant seleno-enzymes such as glutathione peroxidase and therefore is essential for the proper function of cellular defense mechanisms. Se is of particular interest in infant nutrition, Se deficiency may cause increased sensitivity to oxidant stress (Rudolph). In neonatal and infant periode human milk is the optimal food since it provides optimal nutritional, immunological, physiological benefit. [IOM]. Adequate Se intake is necessary for normal infant growth and development [Brätter]. The concentrations of microelements and the composition of human milk, however, change during the lactation time and in each individual. In general, the Se content is highest in colostrum [Ellis, Tamari, Hojo, Smith] and decreases significantly within 1 to 2 weeks of parturition and remains relatively constant during the first year of lactation in mature milk [Debski, Funk]. The Se concentration in human milk is influenced by the maternal dietary Se intake, which depends on the food Se content reflecting geographically different soils [Kumpulainen, Yang]. A significant correlation between Se in mature human milk and maternal Se intake was found in different regions of the world [Brätter, Sanz Alaejos, Mannan]. Consequently, the Se status of the exclusively breast-fed infants depends strongly on the nutritional status of the nursing mother. The Se availability of human milk significantly higher compared to that of cow milk [Shen]. The Se content of infant formulas based on cow's milk is much lower than that of breast milk, consequently the Se supply for formula-fed infants is lower [Hojo, Lombeck, Smith, Stockhausen, Tamari]. The aim of the study was to measure the Se content of human milk during the 4 to 8 weeks of lactation, as well as concentrations of Se in milk-based formulas available in Hungary and to calculate the daily Se intake for infants fed on human milk or milk-based infant formulas.

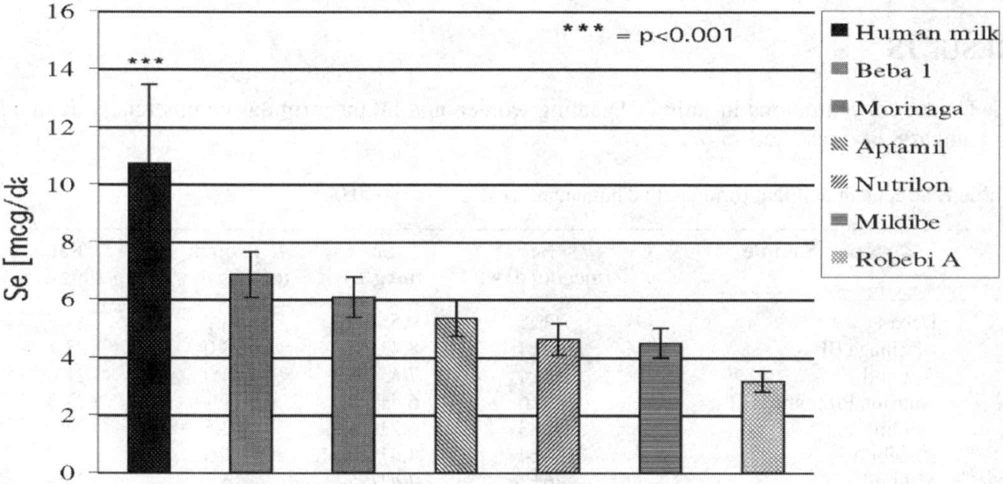

Fig. 1. Daily Se intake of infants from human milk and starter infant formulas

MATERIALS AND METHOD

Human milk samples were collected from ten healthy lactating mothers (4 to 8 weeks after birth) at the Bethesda Children's Hospital in Budapest. Milk samples were obtained by manual pressure from the breast and were directly collected into polyethylene tubes and stored deep-frozen until analysis. A known volume of milk (about 5 ml) was freeze-dried, the dry weight of each sample was determined. The Se intake of the breast-fed infants was calculated from the amount of milk consumed, determined by the weight gain of the babies after feedings, and from the analysed Se concentration in milk samples. In formula-fed babies the daily volume consumed was directly measured by the bottles, Se concentrations were analysed in duplicates from several commercially available powdered cow's milk-based infant formulas. Neutron irradiation and gamma-ray spectrometry: For neutron irradiation 100-300 mg samples were sealed in irradiation vials of high purity quartz (Suprasyl). The samples together with Se standards (Merck) were irradiated with identical geometry to that of the samples, for 24 to 48 hours at a neutron flux density of 8 ×

10^{13} n.cm^{-2}.s^{-1}. Irradiations were carried out at the WWR-M type 10 MW nuclear research reactor of the Budapest Neutron Centre (BNC, KFKI Atomic Energy Research Institute, Budapest, Hungary). Prior to the gamma counting the irradiated quartz vials were surface cleaned by etching in 10% hydrogen fluoride acid solution for 5 minutes and washed with double distilled water and dried. Gamma-spectrometric measurements were performed with the aid of a PC based Canberra type spectrometer with a Canberra HPGe detector with energy resolution of 1.74 keV and efficiency of 13.6% for the 1332.5 keV ^{60}Co line, associated linear electronics, an 8K ADC and an ACCUSPEC/B type 16K MCA board. Counting losses were corrected with a Westphal-type Loss-Free Counting (LFC) module with dual spectrum storage. The gamma-spectra were measured for 5, 15 and 20 hours, after a decay time of 40-90 days. The gamma-ray spectra were evaluated by the program HYPERMET-PC, involving automatic peak search, energy calibration, net peak counts computation and isotope identification. The Se content was determined by measuring the gamma rays of ^{75}Se isotope ($T_{1/2}$:120 d, $E\gamma$:136.0, keV, 264.7 keV, the main gamma-lines). The accuracy and precision of the Se determinations by INAA were tested by co-analyses of biological standard reference material NBS SRM 1549 Non-Fat Milk Powder: the value found 105±4 ng Se/g (n=6), agreed very well with the certified value of 110±10 ng Se/g (deviation was 4.5%). For statistical significance analysis Student t-test was used.

RESULTS

The Se concentrations in milk of lactating women and infant formulas commercially available in Hungary are presented in *table 1*.

Table 1. Se content of infant formulas and human milk (M ± SD) *=p<0.001

No.	Sample	Se (mcg/kg d. w.)	Se (mcg/L)	Protein (g/100g d. w.)	Fat (g/100g d. w.)
1	Beba 1	73±5	9.5±0.7	13.9	27.5
2	Morinaga BF	65±4	8.4±0.5	13.0	27.1
3	Aptamil	53±4	7.4±0.5	14.0	21.0
4	Nutrilon Premium	49±6	6.4±0.7	11.1	21.3
5	Mildibé	48±3	6.2±0.4	12.5	26.5
6	Robébi A	34±4	4.4±0.5	11.6	25.9
7	Milumil	46±3	7.4±0.5	16.0	23.3
8	Beba 2	67±11	9.9±1.6	16.0	24.3
9	Robébi B	56±6	7.2±0.8	14.6	23.6
	Human milk	108±18*	13.7±2.3*	11.2	33.9

The human milk samples of Hungarian mothers yielded a mean Se concentration of 13.7 ± 2.3 mcg/L with a range from 10.4 to 16.6 mcg/L was significantly higher, than that of the starters (samples 1-6) or follow up formulas (samples 7-9). Se content calculated for dry weight gave a mean of 108 mcg/kg with a range from 78 to 135 mcg/kg in the human milk, indicating to twice to four times higher Se concentration in the breast milk compared to that in formulas. There was a positive correlation between the protein and Se content of the formulas (r=0.76, n=27, p<0.01), even stronger in the human milk, where the coefficient was 0.99 (n=18, p, p<0.001). The Se supply in the human milk correlated with the age of the babies (r=0.97). Fat content did not associate with Se intakes. The daily Se intake of the babies directly related to the Se concentration in breast milk or formulas consumed and ranged from 6 to 15 mcg.

The Se content of formulas based on cow's milk varied between 34 to 73 mcg/kg (d. w.) and 4.4 to 9.9 mcg/L and found to be significantly lower in comparison with human milk. Correspondingly the Se supply of formula-fed infants was less than that of the breast-fed babies. Three follow up formulas were analysed and their Se content varied from 7.2 to 9.9 mcg/L. Six starter formulas had Se content between 4.4-8.4 mcg/L. The lowest Se intake was calculated in the group of infants fed with locally produced formula, prepared from Hungarian cow milk: 3.2 ± 0.4 mcg Se/day or 0.76 ± 0.05 mcg Se/kg/b. w., whereas higher intakes were observed in babies fed on imported formulas 6.9 ± 0.8 mcg Se/day. The Se content of both the starter and the follow up formulas manufactured in Hungary was less than 30% of that of human milk. All formulas of foreign products reached a Se content, which corresponded in average 60% of the breast milk Se. The highest Se was measured in a Danish formula.

DISCUSSION

It is known that the Se concentration of human milk may vary considerably, depending on the geographical region or the type of local food products consumed by breast-feeding mothers and depends on physiological, pathological, environmental factors. Se concentration in mature human milk varies widely, from as low as 2.6 mcg/L in the Keshan area of China, trough 18 mcg/ L with a range of 15-22 mcg/L in Canada and the United States [Levander, Mannan, Smith] and Japan and Korea [Tamari and Kim], to as high as 49-90 mcg/L in Venezuela [Brätter] and is directly related to the Se intake by lactating mothers. Differences found in the human milk Se contents from various countries and geographical regions are mostly due to differences in maternal Se intake and status. Selenium levels in wheat, different grain crops and some basic carbohydrate nutrients found to be low in comparison with other European countries indicating low Se supply in Hungary [Sziklai-László]. Plasma Se concentrations of Hungarian women aged between 20-35 years were 0.76 ± 0.11 mcmol/L much lower than the European mean values [Cser]. In our study human milk values were close to those reported by others in Europe and were lower than the ones from the USA [Levander], Japan and Korea [Tamari and Kim]. In our study the average Se content of mature human milk from healthy Hungarian mothers was lower than in Germany [Brätter], comparable with those in Austria [Li F., Rossipal], Belgium [Robberecht], prior to Se fertilisation in Finland [Kumpulainen], Sweden [Walivaara], Poland [Debski, Trafikowska, Zachara], Serbia [Mandic] and in Spain [Torres] but significantly lower than the ones from different locations in the USA, Japan and Korea. Similar to our present results, low Se values on Hungarian human milk were reported a decade ago by Parr in a WHO/IAEA study including Hungary (range: 6.4-21 mcg Se/L with a median value of 13.9 ± 0.4 mcg/L) indicating the similar eating patterns of the Hungarian women as well as similar natural Se supply in this country resulting in similar Se intakes. A range of 7.5-15 mcg Se/day were estimated as the general Se intake for babies world-wide. Daily Se intakes of babies ranged from 6 to 15 mcg with an average of 10.7 ± 2.7 mcg were still within the recommended safe and adequate daily intake for infants (10-40 mcg), but it was close to the lower limit of the RDA (National Research Council, USA, 1989) for infants of 0 to 6 months of age and to the recommended 5-15 mcg/day for infants of 0- to 4 months of age (German Society of Nutrition 1991). Despite the limited number of cases, our data showed that the absolute amount of daily Se intake was significantly higher ($p<0.001$) in breast-fed infants than in formula-fed babies. When Se intake was calculated for body weight this difference remained significant ($p<0.001$). The fact, that the breast milk gave higher Se supply for the babies, than formulas, was also observed by many authors. Most of these studies, however, investigated only one or the other group separately, and only a few of them investigated the two feeding groups simultaneously [Torres]. Babies fed on formulas, receiving less than optimal Se amount from the cow's milk based preparations may be at increased risk for inadequate Se intake-induced diseases later on life.

REFERENCES

Brätter P., Negretti de Brätter V.E., Rosick U., von Stockhausen HB. 1991. Selenium in the nutrition of infants. Chandra R.K. ed., Nestle Nutrition Workshop Series Vol. 23, pp. 79-90. Nestle Ltd.

Cser MÁ., Sziklai-László I., Menzel H., Lombeck I. 1996. Se and glutathione peroxidase activity in Hungarian children and adults. J. Ped. Hun. 47: 384-394.

Debski B., Zarski T.P., Milner JA. 1992. Se content and distribution in bovine and human milk from different region s of Poland. J. Vet. Med. Assoc. 39: 313-318.

Deutsch Gesellsch. Eraehrung (DGE). Empfehlungen fur die Nach., Ic Tab 25, 5; Frankfurt, Germany, 1991.

Ellis L., Picciano MF., Smith AM., Hamosh M., Mehta NR. 1990. The impact of gestational lenght on human milk Se concentration and GSH-Px activity. Pediatr Res. 27: 32-35.

Funk MA., Hamlin L., Picciano MF., Prentice A., Milner JA. 1990. Milk Se of rural African women: Influence of maternal nutrition, parity and lenght of lactation. Am J Clin Nutr. 51: 220-224.

Hojo Y. 1986. Sequential study on GSH-Px and Se contents of human milk. Sci. Total Environ, 52: 83-91.

IOM (Institute of Medicine). 1991. Nutrition During Lactation. Washington, DC: National Academy Press.

Kumpulainen J. 1989. Se: requirement and supplementation. Acta Paediatr. Scand. Suppl., 351: 114-117.

Levander OA, Moser PB., Morris VC. 1987. Dietary Se intake and Se conc. of plasma, erythrocytes, and breast milk in pregnant and postpartum lactating and nonlactating women. Am J Clin Nutr. 46: 694-8.

Li F., Rossipal E., Irgolic K.J. 1999. Determination of Se in human milk by hydride cold-trapping atomic absorption spectrometry and calculation of daily Se intake. J Agric Food Chem., 47: 3265-8.

Lombeck I., Kasperek K., Bonnermann B., Feinendegen LE., Brener HJ. 1978. Selenium content of human milk, cow's milk and cow's infant formulas. Eur. J Pediatr. 129: 139-145.

Mandic Z., Mandic M.L., Grgic J., Hasenay D., Grgic Z. 1995. Se content of human milk. Z. Lebensm. Unters. Forsch., 201 (3): 209-212.

Mannan S., Piccano MF. 1987. Influence of maternal selenium status on human milk selenium concentration and glutathione peroxidase acticity. Am J Clin Nutr 46: 95-100.

Parr R.M., De Maeyer E.M., Iyengar V.G., Byrne A.R. Kirkbright G.F., et al., 1991. Minor and trace elements in human milk from Guatemala, Hungary, Nigeria, Philippines, Sweden and Zaire. Results from a WHO/IAEA Joint Project. Biol. Trace Elem. Res., 29: 51-75.

Robberecht H., Roekens E., Van Caillie-Bertrand M., Deelstra H., Clara L. 1985. Longitudinal study of the Se content in human human milk in Belgium. Acta Paediatr. Scand., 74: 254-258.

Rudolph N., Wong S.L. 1978. Selenium and glutathione peroxidase activity in maternal and cord blood plasma and red cells. Pediatr. Res. 12: 789-792.

Sanz Alaejos M., Diaz Romero C. 1995. Se in human lactation. Nutr. Rev. 53:6, 159-66.

Shen L., Van Dael P., Luten J., Deelstra H. 1996. Estimation of selenium bioavailability from human, cow's, goat and sheep milk by an in vitro method. Int J Food Sci. Nutr. 47(1): 75-81.

Smith AM., Picciano MF., Milner JA. 1982. Se intakes and status of human milk and formula fed infants. Am. J. Clin Nutr 35: 521-526.

Von Stockhausen H.B., Rösick U., Negretti de Brätter V.E., Brätter P. 1988. In: Berger H., ed. Vitamins and Minerals in pregnancy and lactation. Nestle Nutrition Workshop Series Vol. 16, pp. 49-50. Nestle Ltd.

Sziklai-László I., Cser MÁ., Snyder P., Snyder RD. 2000. Selenium in diet samples in Hungary. Metal Ions in Biol. And Medicine Eds. JA. Centeno et al., J. Libbey Eurotext Paris, 6: 273-275.

Tamari Y., Kim ES. 1999. Longitudinal study of the dietary Se intake of exclusively breast-fed infants during early lactation in Korea and Japan. J. Trace Elem. Med. Biol., 13: 129-133.

Torres M.A., Verdoy J., Alegria A., Barbera R., Farre R., Lagarda M.J. 1999. Se contents of human milk and infant formulas in Spain. Sci Total Environ., 228: (2-3) 185-92.

Trafikowska U., Sobkowiak E., Butler J.A., Whanger P.D., Zachara B.A. 1998. Organic and inorganic Se supplementation to lactating mothers increase the blood and milk Se concentrations and Se intake by breast-fed infants. J Trace Elem Med Biol., 12 (2): 77-85.

Walivaara R., Jansson L., Akesson B. 1983. Se content of human milk sampled in 1978 and 1983 in Sweden. Acta Paediatr. Scand., 75: 236-239.

Yang G. 1986. Research on Se-related problems in human health in China. In: Proc. Of the 3d Int. Symp. on Selenium in Med. and Biol. Avi Publishing Co. 9.

Zachara B.A., Pilecki A. 2000. Selenium concentration in the milk of breast-feeding mothers and its geographic distribution. Environ Health Perspect 108 (11) 1043-6.

Metal Ions in Biology and Medicine: vol. 7.Eds L. Khassanova, Ph. Collery, I. Maymard, Z. Khassanova, J.-C. Étienne.
John Libbey Eurotext, Paris © 2002 pp. 473-475.

The intake of selected minerals in daily food rations taken by Polish adolescents

J. Przyslawski[1], M. Schlegel-Zawadzka[2], L. Wadolowska[3], E. Babicz-Zielinska[4]

[1]Chair and Department of Bromatology and Human Nutrition, Medical Academy, 42 Marcelinska, 60-345 Poznan, [2]Department of Food Chemistry and Nutrition, and Department of Continuing Education and Postgraduate Studies Pharmaceutical Faculty, Jagiellonian University, 9 Medyczna, 30-688 Cracow, [3]Institute of Human Nutrition University of Warmia and Mazury, 44a Sloneczna, 10-718 Olsztyn, [4]Chair and Department of Nutrition Sciences, Gdynia Martime Academy, 81-87 Morska, 81-225 Gdynia; Poland

ABSTRACT

The objective of the study was to evaluate the intake of calcium, phosphorus and magnesium, respectively in daily food rations taken by Polish girls and boys aged between 13 and 15 years. The survey was conducted in the year 2000. A random sample from the general population consisted of 443 girls and 321 boys. Food intake in daily food rations (DFR's) was assessed by the 24-h recall method. The qualitative and quantitative assessments of consumed food products were based on photo-models of various foods and their portion size. The study demonstrated that the intake of energy, calcium, phosphorus and magnesium were 2104-2521 kcal, 741-850 mg, 1057-1252 mg, 256-285 mg (girls vs. boys), respectively. The nutritional value of DFR's taken by healthy Polish adolescents was far from the recommended value. It was clear in the case of calcium and phosphorus. From the epidemiological point of view this may lead to disturbances in the development of bone structure of adolescent girls and boys and with may time aggravate the risk of osteoporosis.

INTRODUCTION

It is well known that one of the factors which determine good health is a proper nutrition, the key to which is the balance of nutrients in daily food rations (WHO, 1990). Studies carried out by many Polish research centers, assessing the dietary habits of various population groups, demonstrated an inadequate health-promoting nutrition pattern. Such patterns may result in a high frequency of several diseases i.e., obesity, cardiovascular disease, colon cancer, osteoporosis (Maly Rocznik Statystyczny, 2000). The special concern should be taken for the nutrition of young people. In this group of population inadequate food intake may lead to disturbances in their development (Ziegler and Filer, 1996).

The objective of the study was to evaluate the intake of calcium, phosphorus and magnesium, respectively in daily food rations taken by Polish girls and boys aged between 13 and 15 years.

MATERIAL AND METHODS

The survey was conducted in the year 2000. A random sample from the general population consisted of 443 girls and 321 boys. Food intake in daily food rations (DFR's) was assessed by the 24-h recall method (Mann and Truswell, 1998). The qualitative and quantitative assessments of consumed food products were based on photo-models of various foods and their portion size

(Szczyglowa et al., 1991). All nutritional factors were expressed as mean (X) + standard deviation (SD). The statistical differences between selected groups were calculated with the use of Mann-Whitney test with a significance level of $p<0.05$.

The results of the questionnaire studies were analyzed using computer database for Microsoft Access 7.0 prepared on the basis of tables for the composition and nutrition value of food products. All statistical analyses were investigated with the statistical software Statistica 5.5. StatSoft firm Tulsa USA.

RESULTS

The study demonstrated that the intake of energy ranged from 2104±673 kcal to 2521±607 kcal in DFR's of girls and boys, respectively. The calcium intake level of the total daily food rations for both groups was below the recommended one. It was found to be 741±392 mg and 850±515 mg DFR's for girls and boys, respectively. Inadequate consumption of calcium was magnified by the high level of intake of phosphorus (1057±397 mg - girls and 1252±425 mg - boys). The intake of magnesium was congenial with the recommended daily allowances; particularly for boys. The values were found to be 285±99.4 mg and 256±99.2 mg for boys and girls, respectively. The degree of realizations of recommended daily allowances had the values of 100% (energy intake), 62% to 71% (Ca), 117% to 139% (P), and 85% to 95% (Mg). The nutrient density per 4.184 MJ (1000 kcal) of minerals analysed in DFR's of girls were 363±175 mg (Ca), 509±126 mg (P), 124±35.2 mg (Mg) and 336±185 mg, 498±128 mg, 113±29.7 mg in DFR's taken by boys, respectively. The differences in the nutrient density between daily food rations were found to be statistically significant in case of calcium ($p<0.01743$) and magnesium ($p<0.00009$).

Table 1. The intake level of energy, calcium, magnesium and phosphorus in the daily rations of boys and girls groups studied [Nutrient density per 4.184 MJ (1000 kcal)]

Parameter	Boys X±SD	Nutrient density	Girls X±SD	Nutrient density
Energy [kcal]	**2521±607**	-	2104±673	-
Ca [mg]	**850±515**	336±185*	741±392	363±175*
P [mg]	**1252±425**	498±128	1057±397	509±126
Mg [mg]	**285±99.4**	113±29.7*	256±99.2	124±35.2*

* p<0.05 - difference statistically significant

Table 2. Polish recommended values for energy, calcium, magnesium and phosphorus for girls and boys aged 13-15 years and their percentage of realization (Ziemlanski et al., 1994)

Gender	Calcium (mg/person)	Phosphorus (mg/person)	Magnesium (mg/person)
Boys			
Safe level of intake	1200	900	300
The realization percentage of recommended daily allowances	71	139	95
Girls			
Safe level of intake	1200	900	300
The realization percentage of recommended daily allowances	62	117	85

DISCUSSION

The studies done by the same authors regarding food preferences among Polish youth indicate that more attention should be paid to food products that are rich source of fiber, vitamins, and some minerals (calcium, magnesium and zinc). Polish girls and boys underestimate the healthy role of milk and milk products as a main source of easy available calcium. They didn't correlate the healthy quality of milk products with their protein and calcium content. They choose food to eat taking into account sensory factors such as freshness and taste (Wadolowska et al., 2001). Bad food habits, among that a frequency of drinking milk less than 3-4 times a week obviously correlate with the level of calcium and magnesium intake (Babicz-Zielinska et al., 2000). The content of these two minerals in the youth daily diets was unsatisfactory and was below the recommended value, both in boys and girls (Ziemlanski et al., 1994). Anxiety may arise regarding the reversion of the calcium to phosphorus physiological ratio, which should be 1.0-1.3:1.0.

CONCLUSIONS

The nutritional value of DFR's taken by healthy Polish adolescents was far from the recommended value. It was clear in the case of calcium and phosphorus. From the epidemiological point of view this may lead to disturbances in the development of bone structure of adolescent girls and boys and with time may aggravate the risk of osteoporosis.

REFERENCES

Babicz-Zielinska E., Wadolowska L., Schlegel-Zawadka M., Przyslawski J. Preferencje i spozycie mleka i jego przetworow wsrod mlodziezy szkolnej in: Konsument zywnosci i jego zachowania rynkowe. Wydawnictwo SGGW, Warszawa, 2000, 254-257.

Maly Rocznik Statystyczny. PWE, Warszawa, 2000.

Mann J., Truswell A.S. Essentials of human nutrition. Oxford University Press, New York, 1998.

Szczyglowa H., Szczepanska A., Ners A. et al. Album porcji produktów i potraw. Instytut Zywnosci i Zywienia, Warszawa, 1991.

Wadolowska L., Schlegel-Zawadzka M., Babicz-Zielinska E., Przyslawski J. Factors influencing food choice among Polish youth. Ann. Nutr. Metab., 2001, 45 (supl. 1), 160.

WHO. Diet, nutrition and the prevention of chronic diseses. Report of a WHO Study Group. Technical Report Series 797, WHO, Geneva, chapter 3-4, 1990.

Ziegler E.E., Filer L.J. Present knowledge in nutrition. ILSI Washington DC, 1996.

Ziemlanski S., Bulhak-Jachymczyk B., Budzynska-Topolowska J., Panczenko-Kresowska B., Wartanowicz M. Normy zywienia dla ludnosci w Polsce. Zyw. Czlow. Metab.1994, 4, 303-338.

Metal Ions in Biology and Medicine: vol. 7. Eds L. Khassanova, Ph. Collery, I. Maymard, Z. Khassanova, J.-C. Étienne.
John Libbey Eurotext, Paris © 2002 pp. 476-479.

The intake of Ca and Mg in the Polish habitual diet and their relation to socio-economic factors

M. Schlegel-Zawadzka[1], J. Przyslawski[2], J. Bertrandt[3], A. Klos[3]

[1]*Department of Food Chemistry and Nutrition, and Department of Continuing Education and Postgraduate Studies Pharmaceutical Faculty, Jagiellonian University, 9 Medyczna, 30-688 Cracow,* [2]*Chair and Department of Bromatology and Human Nutrition, Medical Academy, 42 Marcelinska, 60-345 Poznan,* [3]*Military Institute of Hygiene & Epidemiology,* [4]*Kozielska, 01-163 Warsaw; Poland*

ABSTRACT

The objective of the study was to evaluate the intake of calcium and magnesium occurring in the habitual diet of healthy men and women in the region of Mid-West Poland.

The survey was conducted in the year 1999 and 2000. A sample from the general population consisted of 631 women and 522 men with the age range between 36 and 41. The nutritional assessment was evaluated by a quantitative method consisted of 24-hour recalls of actual food intake. The qualitative and quantitative assessments of consumed food products were based on photo-models of various food and their portion size. The realization percentage of recommended values for calcium and magnesium were 66% and 84% respectively.

INTRODUCTION

Proper, conformable with the necessaries of physiology, nutrition habits are one of the elements in the "way of life" of every man. Under this idea, the supply of energy, protein, fat carbohydrates, vitamins and minerals, adequate to the age, gender and the physiological state of the organism becomes understood (Ziegler and Filer, 1996). Many studies from nutritional epidemiology, supported by clinical experiments and prospective studies, provide sufficient scientific evidence of the nutritional role in the promotion and/or the prevention of such diseases as: osteoporosis, coronary ischemic disease, brain stroke, hypertension, obesity, and some form of cancer (WHO, 1990).

The objective of the study was to evaluate the intake of calcium and magnesium occurring in the habitual diet of healthy men and women in the region of Mid-West Poland.

MATERIALS AND METHODS

The survey was conducted in the year 1999 and 2000. A sample from the general population consisted of 631 women and 522 men with the age range between 36 and 41. The nutritional assessment was evaluated by a quantitative method consisted of 24-hour recalls of actual food intake (Mann and Truswell, 1998). The qualitative and quantitative assessments of consumed food products were based on photo-models of various food and their portion size (Szczyglowa et al., 1991). To Make things easier for the subjects questioned, a correctly filled in questionnaire (with full details of meals) was completed with an enclosure including a list of the most often consumed food items along with their quantity in grams and some suited examples of domestic measures.

Computer databases prepared with Mirosoft Access 7.0 on the basis of the food composition and nutritive value tables were used to analyse results of the total ration diet questionnaire study of qualitative and quantitative composition. The recommended standards as optimal values, which not only protect against nutritional deficiency, but may also have prophylactic meaning in the prevention of some civilization-related diseases were used to establish a degree of standard realization. Moreover, it was assumed that moderate physical activity characterized the studied male and female groups.

Statistical analysis

The intake of calcium and magnesium respectively were expressed as median (Me) ± quartile deviation (QD), because of analyzed factors distribution deviated from the normal. In addition the average intake level of components analysed for comparative targets was expressed using classical measure as arithmetic average mean (Xaa).

The Mann-Whitney U test was used to test the hypothesis of significant differences between average values on the level of significance set at $p<0.05$. The Chi2 test was employed for qualitative variables at the same level of significance.

RESULTS

A general anthropometrical characteristic of the male and female groups studied is presented in *table 1* and chosen socio-economic features (place of residence, type of education, financial situation, cigarette smoking, light alcohol intake, strong alcohol intake) in *table 2*. Body mass indexes of group studied are lower than the first step of obesity (25-26). The percentage of women who declared strong alcohol drinking is significantly higher than in the men group. The intake of calcium and magnesium occurring in the total daily rations of the male and female groups is presented in *table 3*. The calcium intake was similar in the male and female food rations (about 600 mg), however magnesium intake was about 30% higher in the male total daily rations.

Table 1. General anthropometrical characteristics of the male and female groups studied

Gender	Statistical parameter	Analysed factor			
		Age (year)	Height (cm)	Body mass (kg)	BMI
Men N=522	Me+QD Xaa V%	36.0±3.0 35.3 8.33	179±3.0 178 1.67	79.0±5.0 78.8 6.32	24.5±1.52 24.6 6.22
Women N=631	Me±QD Xaa V%	36.0±2.5 35.6 6.94	165±4.50 165 2.72	60.0±5.5 61.2 9.16	21.7±2.00 22.4 9.23

Me - median; QD - quartile deviation; N - number of subjects; Xaa - arithmetic average mean; V% - coefficient of variation, BMI - body mass index

DISCUSSION

The integrity of this estimation is a derivative of the precision of the food composition and nutritive value Table, which was used to prepare the computer databases. It is not possible to eliminate totally a subjective estimation related to the health state as well as the intake quantity of food and dishes, either from a respond or questioner. In spite of these faults, the questionnaire

Table 2. The chosen socio-economic feature of male and female groups studied (% from a line)

Gender	Analysed factor												
	Place of residence		Type of education			Financial situation		Cigarette smoking		Light alcohol intake		Strong alcohol intake	
	ST	BT	G	S	H	Good	Bad	Yes	No	Yes	No	Yes	No
Men N=522	58.6	41.4	31.4	39.4	28.9	92.5	7.47	48.8	51.1	36.6	63.4	32.8	67.1
Women N=631	54.8	45.2	27.6	56.4	15.8	90.0	9.98	61.2	38.8	39.3	60.7	73.4	26.5

ST - small town; BT - big town; G - basic; S - intermediate; H -high school

Table 3. The intake level of calcium and magnesium in the daily rations of male and female groups studied

Mineral	Male group		Female group	
	Me±QD	Xaa	Me±QD	Xaa
Calcium (mg)	621±254	704	574±202	634
The realization percentage of recommended values	69%		63%	
Magnesium (mg)	319±62.3	333	249±52.9	262
The realization percentage of recommended values	86%		83%	

Me - median; QD - quartile deviation, Xaa - arithmetic average mean

study methods in nutritional sciences are recognized and generally used. The precision of this estimation is also sufficient for nutritional epidemiological purposes.

Returning to the discussion results relating to some socio-economic factors and mineral intakes the data suggest a possible factors influence the increase rate of osteoporosis, depression and colon cancer in Polish population (Maly Rocznik Statystyczny, 2000). An interesting result concerned cigarette smoking and the intake of light and strong alcohol - especially in the female group. It follows from analysis that women smoked more cigarettes than men. Also women drank more strong alcohol.

The level of mineral intake was unsatisfactory. The intake level of calcium and magnesium was below recommended values in both male and female (Ziemlanski et al., 1994). Calcium and magnesium cations take part in the passing on nervous stimuli (Katzman and Pappius, 1973). Their deficits in the diet may lead to changes in their distribution in the brain and in consequence to behavior changes (Katzman and Pappius, 1973; Saris et al., 2000).

There are many evidence too, of correlation between skeletal mass and bone loss, bone remodeling and colon cancer with the low calcium intake.

CONCLUSIONS

The nutritional value of calcium and magnesium in DFR's taken by healthy Polish adults was far from the recommended. And together with bad behavior habits (smoking, alcohol intake) may lead to increase incidences of osteoporosis, colon cancer and depression in Polish population.

REFERENCES

Maly Rocznik Statystyczny. PWE, Warszawa, 2000.
Mann J., Truswell A.S. Essentials of human nutrition. Oxford University Press, New York, 1998.
Katzman R., Pappius H.M. Brain electrolytes and fluid metabolism. Williams&Wilkins, Baltimore, 1973.
Saris N.E.L., Mervaala E., Karppanen H., Khawaja J.A., Lewenstam A. Magnesium. An update on physiological, clinical and analytical aspects. Clin. Chim. Acta 294, 1-26, 2000.
Szczyglowa H., Szczepanska A., Ners A. et al. Album porcji produktów i potraw. Instytut Zywnosci i Zywienia, Warszawa, 1991.
WHO. Diet, nutrition and the prevention of chronic diseses. Report of a WHO Study Group. Technical Report Series 797, WHO, Geneva, chapter 3-4, 1990.
Ziegler E.E., Filer L.J. Present knowledge in nutrition. ILSI Washington DC, 1996.
Ziemlanski S., Bulhak-Jachymczyk B., Budzynska-Topolowska J., Panczenko-Kresowska B., Wartanowicz M. Normy zywienia dla ludnosci w Polsce. Zyw. Czlow. Metab. 4, 303-338, 1994.

Metal Ions in Biology and Medicine: vol. 7. Eds L. Khassanova, Ph. Collery, I. Maymard, Z. Khassanova, J.-C. Étienne.
John Libbey Eurotext, Paris © 2002 pp. 480-485.

Silicon and fluoride in water for human consumption and bone mineralisation

L. Fabiani[1], M. Scatigna[1], G. Mosca[1], G. Farello[1], V. Leoni[2]

[1]*Università degli Studi dell'Aquila, Dipartimento di Medicina Interna e Sanità Pubblica, via San Sisto 22/E, 67100 L'Aquila, Italy*
[2]*Università di Roma "La Sapienza", Dipartimento di Scienze di Sanità Pubblica, piazzale Aldo Moro 5, 00185 Roma, Italy*

ABSTRACT

The level of bone mineral density in adolescents depends on numerous factors including the intake of calcium in the diet, the stage of puberty, physical activity and genetic characteristics. Study by means of ultrasound made it possible to carry out a cross-sectional epidemiological survey aimed at analysing the effect of environmental factors such as the mineral composition of the water intended for human consumption on two groups of adolescents aged 10 to 14 resident: one in an area supplied with oligomineral water, and the other with water containing fluoride in optimal quantity as well as a high level of silicon.

The following were noted: anthropometric parameters, bone ultrasonometry of the proximal phalanxes of the non-dominant hand, and oral health. Analysis of the results shows evidence of a greater bone mineralisation in the subjects resident in the area served with water having a high silicon content and optimal fluoride content compared to those resident in the area with oligomineral water.

INTRODUCTION

The measurement of bone mineral density in adolescents is justified, in addition to any conditions requiring a diagnostic investigation, by the need to characterise the factors fostering the attainment of maximum peak levels [1, 2]. In fact bone mass in adults depends on accumulation in the stage of growth and on the speed of loss, which is greater in women and following the menopause. In the relatively short period of puberty considerable increases in bone mass may be observed, so that the possibility of fully understanding the factors responsible for a maximum accumulation would be extremely important for the promotion of health in advanced age. The impact of osteoporosis on present-day society has greatly stimulated studies relative to the diagnostic and predictive capacity of the various techniques of measurement of bone mass and, albeit to a lesser extent, studies on the factors of promotion of optimal mineralisation.

The use of ultrasound to evaluate bone mineral density has numerous advantages since it is a relatively inexpensive technique which is portable, easy to apply and does not emit ionising radiations. Bone ultrasonometry was employed in subjects of pediatric age on various parts of the body: tibia [2], heel [3] and proximal phalanxes [4]. Evaluation of the bone with ultrasounds uses different measurements: the speed of sound (SoS), bone broad band ultrasound attenuation (BUA) and other derived measurements.

Many studies were carried out in order to evaluate the accuracy and the reproducibility of the osteosonographic measurements, above all compared with the Dual X-ray absorptiometry - DXA

- [5, 6] and the peripheral Quantitative Computed Tomography - pQCT - [7]. Nevertheless, would appear not to be important in epidemiological investigations comparing different groups of subjects in relation to environmental and behavioural variables, while in such studies what appears fundamental is the reproducibility of the measurement considered as good, in all of the studies.

Table 1. Mineral content of the water distributed for human consumption in the two areas considered

	L'Aquila	Two comunes of the province of Caserta	
Hardness °F	12,0	10,0	5,2
pH	7,1	7,5	8,0
SiO_2 (mg/L)	3,8	76,6	61,5
Ca^{2+} (mg/L)	24,5	32,8	14,0
F^- (mg/L)	<0,05	1,7	1,2
Fixed residue at 180°	128,9	123,0	77,5

Table 2. Multiple linear regression of the stage of puberty, age and percentage of ideal weight on ADSOS° (m/s)

	Coeff.	p>|t|	[95% confidence interval]	
Chronological age	6.068816	0.049	0.0275528	12.11008
StPuP*	13.137490	0.007	3.6040690	22.67092
StPuG*	9.726403	0.042	0.3399182	19.11289
Percent of weight**	- 1.134033	0.000	- 1.3995610	- 0.868505

°: "Adjusted Speed of Sound" (m/s)
*: The "Stage of puberty" is classified in 5 classes both with reference to the pubic hairs (StPuP) and to the gonads or the breast (StPuG)
**: The "Percent of weight" is percent ratio between real on the ideal weight and represent the excess weight of the subject. It's negatively correlate to ADSOS

Table 3. Average value and SD of ADSOS by area of residence, sex and stage of puberty

	L'AQUILA			CASERTA		
StPuP	**N°**	**ADSOS** (m/s)	**SD**	**N°**	**ADSOS** (m/s)	**SD**
			MALES			
1	44	1891.38	86.12	56	1920.27*	75.89
2	43	1905.32	73.32	49	1929.80*	55.76
3	21	1929.06	40.55	26	1957.62*	44.50
4	18	1937.64	52.03	26	1973.65*	63.12
5	6	2003.00	168.29	9	1970.78	52.96
			FEMALES			
1	12	1935.00	61.28	13	1959.15	53.73
2	12	1925.75	53.17	24	1964.38*	58.99
3	43	1939.95	71.59	45	1990.60*	50.96
4	22	1962.55	64.15	41	2013.66*	74.14
5	16	2004.54	64.71	25	2010.20	60.52

*: difference in the two areas is significant at the level of p>0.05

The level of bone mineralisation achieved in the different phases of growth is also determined by genetic factors [8, 9].

Another factor that modifies the bone mineralisation in a directly proportional way is physical activity [10].

The intake of nutrients plays a fundamental role in achieving an adequate bone mineralisation in adolescence and is more and more frequently indicated as an opportunity for fostering the skeletal health and preventing osteoporosis and fractures at an advanced age. In particular the role of calcium is well known, even if there is less agreement on the quantities that should be taken (which some consider should be high [11] in order to guarantee optimal mineralisation [12, 13]. Other elements too would appear to be particularly important in growth and mineralisation of the skeleton [14]. A central role has on various occasions been suggested for fluoride [15, 16, 17, 18]. The effect of fluoride in relation to the dose taken shows a U shaped curve with a window of optimal activity in the prevention of dental caries and in the promotion of bone mineralisation at between 1 and 4 mg per day [15]. Silicon has been shown to be essential for the normal growth of the skeleton, above all of the long bones and the skull, but very little is known on the recommended levels of its intake [14, 19]. The intake of calcium is based above all on dairy products, however the overall intake may be integrated if the water used in the diet contains a high quantity of it. Water is the main source of fluoride in the diet. The water distributed for human consumption affects the overall mineral content of the diet, since it is used not only as a beverage but also for cooking food in, and vegetables especially may be impoverished or enriched by the different mineral species according to the composition of the water. The hypothesis according to which the use of a water rich in minerals may foster a better mineralisation during adolescence lies at the base of our work.

The noting of bone mineral density in adolescents should consider also in mind the stage of puberty. In fact there is not always a constant correspondence between chronological age and puberal development and mineralisation is more closely correlated with the state of puberty than with age itself [20, 21].

Finally the values of the SoS and of BUA are lower in relation to the subject's weight, and accordingly it is necessary to check for any situation of overweight in the girls and boys examined [22]. Thus in the study we analysed bone mineral density on the basis of the composition of the water supplied for human consumption, checking over the modifying and confusing factors referred to earlier.

MATERIAL AND METHODS

The water distributed for human consumption in the Comune of L'Aquila and in two Comunes of the Province of Caserta were analysed in double in two laboratories at different times of the year according to the official methods. Specifically the parameters relative to the mineralisation and to the components considered were evaluated: hardness, fixed residue, calcium, fluoride and silicon.

In the two areas a cross-sectional epidemiological study was carried out, enrolling 745 girls and boys attending the 6° to 9° grade school (10-14 years old). The young people were contacted with the collaboration of the School Medical services that carry out routine screening activities and ensure the promotion of health in schools, and parents were asked to consent to their participation in the study.

The young people for whom the parents' consent had been obtained were subjected to an auxological visit and ultrasonographic investigation and dental examination, which will be reported on in a separate note. The measurement of Adjusted Speed of Sound (ADSOS) on the proximal phalanxes of the last four fingers of the non-dominant hand was carried out using a DBM Sonic 1200 (IGEA, Italy) and was recorded as an average value. The auxological visit provided the opportunity for noting the main anthropometric parameters using the Harpenden instruments installed and calibrated in the school infirmary and the definition of stage of puberty recorded in accordance with the Tanner criteria [23], in five stages with reference to the development of the gonads (males) and of the mammary gland (females) and of pubic hair (males and females). With a view to controlling the effect of weight on ADSOS, weight was recorded also as a percentage of the ideal weight in relation to real height (percent w).

The data were analysed using the Stata - 6 statistical package.

Study of the results of the ADSOS was correlated with the determinant parameters in the entire sample: height, stage of puberty. An evaluation was likewise carried out of the difference in the two groups distinguished by the water distributed for consumption, of the average values of ADSOS achieved. Finally a multivariate analysis was carried out to highlight the independent contribution of the different variables in the determination of ADSOS.

RESULTS AND DISCUSSION

Table 1 shows the average results of the analyses of the water distributed in the two areas. In the Province of Caserta the average content of fluoride is optimal, the content of calcium more or less the same as that from the other area, while the content in silicon is undoubtedly high. In the water distributed at L'Aquila fluoride is in a concentration close to the limits of sensitivity of the method, calcium in a concentration of 24.5 mg/l and silicon is less than 4 mg/l. Then the water employed for alimentary purposes in the two areas shows clearly different characteristics as regards fluoride and silicon. As far as calcium is concerned, the difference is smaller. Fluoride has been reported to have positive effect on bone strength both fluoride forming fluorohydroxiapatite (that is a less soluble crystal) and increasing osteoblast number and bone formation). Furthermore the lifelong exposure to water fluoride level of 1-1,5 mg/L (that is the level in the Caserta water) seems to lower hip fracture incidence (15-18) in the elderly.

Silicon has been considered has essential nutrient for normal bone matrix formation and also for bone mineralisation [24] It has a positive effect on the formation of the cross-links between collagen and proteoglycans during bone growth [14]. In experimental studies silicon deficiency resulted in growth retardation and abnormalities in the skull and bone architecture [25]. The intake of silicon is 60-120 mg/day for the Caserta teenagers versus less than 8 mg/day for L'Aquila teenagers.

The multiple linear regression given in *table 2* shows that the stage of puberty - both with reference to pubic hair and with reference to the gonads and the breast - chronological age and in the opposite way excess weight are significant determinants independent of the bone mineralisation inferred as ADSOS. The results confirmed what was already known regarding the decisive factors, although in the multiple linear regression age and stage of puberty completely absorb correlation with height; which proved inversely to be correlated with percentage overweight.

The subjects from Caserta have a higher average age by about 4 months and a greater asymmetry in their distribution by age, which goes up to 17. Accordingly evaluation of the differences between the two groups was limited to the subjects aged 11 to 14 years of age.

In the comparison between the two areas there emerged in a significant way though to a limited extent a greater bone mineralisation in the residents from the area of Caserta with its more mineralised water. In fact the average values of ADSOS in the two groups, as stratified by sex and by stage of puberty, were found to be significantly greater in the group using water with a higher concentration of minerals. The differences between the level of mineralisation noted in the two areas are significant at the level of $p<0.05$ in all groups except for the females at stages 1 and 5 and the males at 5, according to the pattern recorded in *table 3*. The groups in which significance is not reached are also the least numerous ones.

CONCLUSIONS

The adolescents in the Caserta zone are lightly more advancial in their growth than the subjects of same age in L'Aquila and the correlation between age and height and bone mineralisation is closer suggesting a better or easier bone growth.

In conclusion therefore the adolescents with a diet comprising water having a higher concentration of fluoride and of silicon show a greater bone mineralisation, as inferred by ultrasonometry of the proximal phalanxes of the hand.

REFERENCES

1. Gilsanz V. 1998 Bone density in children: a review of the available techniques and indications. Eur J Radiol 26: 177-182.
2. Lequin M.H., van Rijn R.R., Robben S.G.F., Hop W.C.J., van Kuijk C. 2000 Normal values for tibial quantitative ultrasonometry in Caucasian children and adolescents (Aged 6 to 19 years). Calcif Tissue Int 67: 101-105.
3. Mughal M.Z., Ward K., Qayyum N., Langton C.M. 1997 Assessment of bone status using the contact ultrasound bone analyser. Arch Dis Child 76: 535-536.
4. Baroncelli G.I., Federico G., Bertelloni S., de Terlizzi F., Cadossi R., Saggese G. 2001 Bone quality assessment by quantitative ultrasound of proximal phalanxes of the hand in healthy subjects aged 3-21 years. Pediatr Res 49: 713-718.
5. Moris M., Peretz A., Tjeka R., Negaban N., Wouters M., Bergmann P. 1995 Quantitative ultrasound bone measurements: normal values and comparison with bone mineral density by dual X-ray absorptiometry. Calcif Tissue Int 57: 6-10.
6. Blanckaert F., Cortet P., Coquerelle P., Filipo R.M., Duquesnoy B., Delcambre B. 1999 Ultrasound velocity through the phalanges in normal and osteoporotic patients. Calcif Tissue Int 64:28-33.
7. Louis O., Moreels X., Osteaux M. 1998 Reproducibility of phalanx osteosonography and relation with forearm peripheral quantitative computed tomography: single finger versus average measurement on the last four finger. Eu J Radiol 28:270-275.
8. Howard GM, Nguyen T.V., Harris M., Kelly P.J., Eisman J.A. 1998 Genetic and environmental contributions to the association between quantitative ultrasound and bone mineral density measurements: a twin study. J Bone Miner Res 13: 1318-1327.
9. Daly R.M., Rich P.A., Klein R. 1997 Influence of high impact loading on ultrasound bone measurements in children: a cross-sectional report. Calcif Tissue Int 60: 401-404.
10. Prentice A. 2001 The relative contribution of diet and genotype to bone development. Proc Nutr Soc 60: 45-52.
11. NIH Consensus Development Panel on Optimal Calcium Intake 1994 Optimal Calcium Intake. JAMA 272: 1942-1948.
12. Lloyd T., Andon M.B., Rollings N., Martel J.K., Landis R., Demers L.M., Eggli D.F., Kieselhorst K., Kulin H.E. 1993 calcium supplementation and bone mineral density in adolescent girls 270: 841-844.
13. Anderson J.J. 2001 Calcium requirements during adolescence to maximize bone health. J Am Coll Nutr 20: 186S-191S.
14. Cashman K, Flynn A. Trace Elements and Bone Metabolism. In: Sandström B, Walter P. Role of Trace Elements for Health Promotion and Disease Prevention. Basel: Karger, 1998: 150-164.
15. Allolio B., Lehmann R. 1999 Drinking water fluoridation and bone. Exp Clin Endocrinol Diabetes 107: 12-20.
16. Phipps K.R., Orwoll E.S., Bevan L. 1998 The association between water-borne fluoride and bone mineral density in older adults. J Dent Res 77: 1739-1748.
17. Lehmann R., Wapniarz M., Hofmann B., Pieper B., Haubitz I., Allolio B. 1998 Drinking water fluoridation: bone mineral density and hip fracture incidence. Bone 22: 273-278.
18. Fabiani L., Leoni V., Vitali M. 1999 Bone fracture incidence rate in two italian regions with different fluoride concentration level in drinking water. J Trace Elements Med Biol 13: 232-237.
19. Trumbo P, Yates AA, Schlicker S, Poos M. Dietary reference intakes: vitamin A, vitamin K, Arsenic,

Boron, Chromium, Copper, Iodine, Iron, Manganese, Molibdenum, Nickel, Silicon, Vanadium and Zinc. J Am Diet Assoc 2001; 101 (3): 294-301.
20. Southard R.N., Morris J.D. Mahan J.D., Hayes J.R., Torch M.A., Sommer A., Zipf W.B. 1991 Bone mass in healthy children: measurement with quantitative DXA. Radiology 179: 735-738.
21. Farello G. Tollis G., Pozone M., Fabiani L., de Matteis F. 1997 The evaluation of bone mineralisation in adolescence: the relathionship between anthropometric variables and bone mineral content in the distal radius. 25° European Symposium on calcified Tissues. Bone 20 (4S): 39.
22. Rico H., Gòmez M., Aguado F., Villa L.F., Hernàndez E.R., Cortés J. 1999 Impact of weight in obese subjects on bone speed of sound. Invest Radiol 34: 596-599.
23. Tanner J.M. 1962 Growth at adolescence. 2nd ed. Oxford, England, Blackwell Scientific.
24. Carlisle EM. Silicon as a trace nutrient. Sci Total Environ 1988; 73: 96-106.
25. Carlisle EM. Silicon an essential element for the chick. Science 1972; 178: 619-621.

XI INFLAMMATION, IMMUNOLOGY

*Metal Ions in Biology and Medicine: vol. 7.*Eds L. Khassanova, Ph. Collery, I. Maymard, Z. Khassanova, J.-C. Étienne.
John Libbey Eurotext, Paris © 2002 pp. 489-491.

Rhenium cluster compounds as POL regulators in human RBC

I.I. Patalakh, N.I. Shtemenko, O.P. Baydal

Dnipropetrovsk National University,Department of Biophysics and Biochemistry, 13 Naukoviy by-St., 49050, Dnipropetrovsk, Ukraine

ABSTRACT

According to our recent investigations some organic binuclear rhenium compounds (RC) with a quadruple metal-metal bond like $Re_2(C_3H_6COOH)_4Cl_2$ (I) or $Re_2(GABA)_2Cl_6$ (II)) reveal an anti-hemolytic and antitoxic effects. Therefore lipid peroxidation (POL) intensity and glutathione-dependent defensive system: glutathione-peroxidase (GP) - glutathione-reductase (GR) were examined during red blood cells (RBC) treatment with rhenium organic compounds I and II *in vitro* (1-hour incubation with 10^{-4} - 10^{-12} mol/L).

RC I has demonstrated strong antioxidant effect with POL inhibition on 15-35% (maximal effect was found for 10^{-8} M concentration) while complex II effectively inhibited POL only in range of low concentrations (10^{-10} - 10^{-12} M).

These results prompted our investigations to elucidate the mechanism of action of compounds I and II in enzyme antioxidant systems. Our data indicate that compound II lowered activity of GP and GR whereas compound I diminished GP activity and considerably stimulated GR activity. Thus, it seems that compound I can be used as an effective scavenger of free radicals so what it rather influence enzymes activity on substrate inhibition mechanism whereas compound II has shown certain enzymes modulation properties.

INTRODUCTION

According to our recent investigations some organic binuclear rhenium compounds (RC) with a quadruple metal-metal bond reveal cell-stabilizing [1] and antioxidative [2, 3] abilities. Especially two of them, $[Re_2(i\text{-}C_3H_6COOH)_4]Cl_2$ (I) and $[Re_2(NH_3^+CH_2CH_2CH_2COO)_2Cl_5(H_2O)_2Cl^-H_2O$ (II)) have taken dose-dependent antihemolytic effects on human red blood cells (RBC) and neutralized phenylhydrazine intoxication of rats considerably. Proceed from the assumption of their antioxidative properties we aimed to analyze which points of cell defensive system RC can potentate. Three sources of peroxidation processes are known to exist in all living cells: 1) free radicals such as reactive oxidant species (ROS) and secondary organic radicals; 2) enzymatic prooxidants (majority of oxidoreductases and monooxigenases, enzymes of electron transport chain); 3) non-enzymatic prooxidants like some ions of transition metals. So RC might be a direct scavenger of free radicals or potentate efficiency of certain cell enzymatic and non-enzymatic antioxidants.

Hence we tested rhenium compounds I and II as possible modulators of lipid peroxidation process (POL) produced in the course of normal cell metabolism. Since that may result in perturbation of the cellular redox state we also evaluated some changes of glutathione-peroxidase (GPO) and glutathione-reductase (GR) activity.

MATERIALS AND METHODS

Preparation of human erythrocytes and hemolysates. Fresh human RBC were prepared from blood of healthy volunteers. Blood was collected in tubes containing EDTA as anticoagulant. Blood cells were washed at 1,500 *g* in four volumes of cold saline solution (pH 7.4) and the supernatants were removed after three additional washes. After 1-hour incubation with CR in range of 10^{-4} - 10^{-12} mol/L at 37°C RBS were washed again. Hemolysates were obtained on hypotonic lysis. Hemoglobin content were determined with hemoglobincyan color reaction [4].

POL intensity. POL was evaluated by measuring MDA (malondialdehyde) level through thiobarbituric acid reaction [5]. Spontaneous POL was initiated by adding to hemolysates 1 mM $FeSO_4$, enzymatic POL - with 10 ADP and 2,4 NADPH, non-enzymatic POL - with 1,7 ascorbic acid and 2,4 NADPH.

GPO activity. GPO activity was determined through GSH peroxidation in reaction initiated by H_2O_2. Content of GSSG was evaluated with spectrophotometric method on 260 nm [6].

GR activity. GR activity was determined as rate of NADPH oxidizing during GSSG reduction with spectrophotometric method on 340 nm [6].

Statistical analysis. All data are presented as mean ± standard error of mean (SEM). Analysis of variance (one-way ANOVA) was used for comparing experimental group with control value. If global test for differences was significant, pair-wise tests for differences between groups were applied (Student's *t*-test for paired data). *P* value < 0.05 was considered statistically significant.

RESULTS AND DISCUSSION

The results present at *figure 1* indicate some differences in POL affection by RC I and RC II. Although both I and II have demonstrated considerable reduction of spontaneous processes in POL initiation about 30% in range of 10^{-8} - 10^{-12} M there were essential specific features in mechanisms of POL affection. So the most drastic changes were initiated by I in spontaneous POL intensity whereas II influenced enzyme-dependent POL component most of all. An opposite effects were demonstrated for enzymatic POL component: inhibition with II after 1 hour incubation RBC at subnanomolar concentrations whilst I enhanced enzyme-dependent ROS production which was activated by 25-43% in relation to control.

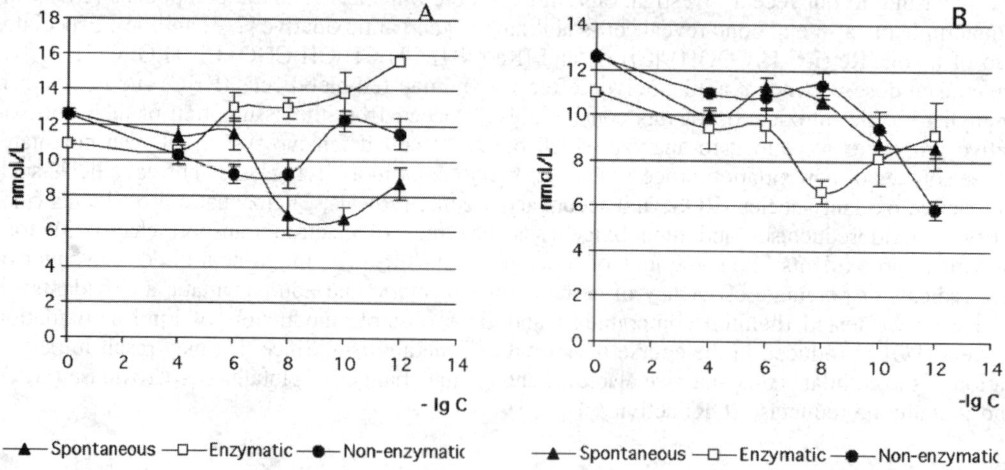

Fig. 1. POL levels in RBC dependent on concentration and chemical structure of rhenium compounds: A - isobutiric ligand (RC I), B - GABA ligand (RC II)

GR and GPO are present in the erythrocyte cytoplasm. An experimental model was designed to determine their activity in hemolysate mimicking the ability of erythrocyte cytoplasm to o restore GSH level and decompose H_2O_2. So as it is shown at *figure 2* opposite effects was observed for glutathion-reductase activity that was impaired after low concentrations of I influence but was in excess of average 50% after RBC treatment with II. It was surprising that both RC demonstrated the inhibition of glutation-peroxidase activity by 20%-50% but statistically significant differences could be marked only in range of low concentrations: 10^{-8} - 10^{-12} M for I and 10^{-9} - 10^{-10} M for II.

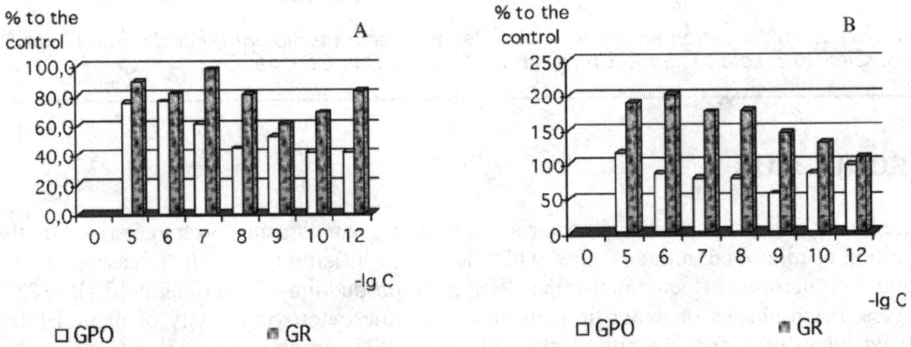

Fig. 2. Changes in activity of glutathion-dependent enzymes in RBC incubated with A - RC I, $[Re_2(C_3H_6COOH)_4]Cl_2$ and B - RC II, $[Re_2(GABA)_2]Cl_6$

Experiments with RBC incubation in range of rhenium compounds concentrations allowed us to study a possibility of RC penetration into intracellular space. We suggest the better transport properties for RC rather than RC II because of more essential affection of total cytosolic enzyme-oxidative cytosolic activity. According to obtained results we would suggest the main role of cluster rhenium compounds as a direct scavengers of the intracellular free radicals. Being better scavenger RC I is possible to diminished H_2O_2 level that has caused more strong substrate-dependent inhibition of GPO. We also can suggest an explanation of non-enzymatic POL decreasing: obviously RC I and RC II have a possibility to potentate such non-enzymatic antioxidants as ascorbat and GSH. GSH was directly involved in organic peroxides reduction and was oxidized to GSSG. In that case it has become clear that GR activity was enchanted because of involving in process of GSSG reduction. In case of RC II there was a good correspondence between the level of non-enzymatic POL inhibition and GR activation.

Thus, it seems that compound I can be used as an effective scavenger of free radicals so what it rather influence enzymes activity on substrate inhibition mechanism whereas compound II has shown certain enzymes modulation properties.

REFERENCES

1. Shtemenko N.I., Zelenyuk M.A., Patalakh I.I. Influence of the liposomal forms of rhenium compounds on human erythrocytes hemolysis. *Vestnik DNU (Biology and ecology)*, 9, 24-28.
2. Shtemenko N.I., Oliynik S.A., Patalakh I.I. Antioxidant enzymes activity in conditions of hemolytic anemia and rhenium cluster compounds application. *Proceedings of scientific conference at the honor of acad. Bulankin I.N.*, 2001, 97-99.
3. Patalakh I.I., Shtemenko N.I., Shtemenko A.V., Rhenium cluster compounds as modulators of RBC redox state. *Abstracts of International symposium intracellular signaling in plant and animal systems (ISPAS)*, 2001, Kyiv, Ukraine, 87.
4. Goryachkovskiy A.M. Handbook on clinical biochemistry, 1994.
5. UshIjama M., Mihara M. *Analyt. Biochem.* 1978, 86, 271-278.
6. Vlasova S.N., Shabunina.E.I., Pereslegina I.A. *Laboratornoe delo*, 1990, 8, 19-21.

Metal Ions in Biology and Medicine: vol. 7. Eds L. Khassanova, Ph. Collery, I. Maymard, Z. Khassanova, J.-C. Étienne.
John Libbey Eurotext, Paris © 2002 pp. 492-494.

Zinc effects over IL-12 gene expression and IL-12 protein secretion in mice macrophages

MD Lastra, AE Aguilar, K Humanez, R Hernandez, Saldivar L, R Pastelin

Laboratorio de Investigación en Inmunología, Departamento de Biología, Facultad de Química, UNAM. Circuito Escolar, Ciudad Universitaria, México, D.F. CP 04510

INTRODUCTION

The innate immune response to bacteria and to protozoan parasites is characterized by the rapid recognition of microbial antigens, after which activated inflammatory cells release soluble mediators and antimicrobial effector molecules. The early production of interleukin-12 (IL-12) by granulocytes, macrophages or dendritic cells stimulates the cytotoxic activity of natural killer cells (NK) and enhances their release of interferon-gamma (IFN gamma), effectors in these mechanisms.

Interleukin 12 thus, plays a pivotal role in the immune response and provides a crucial link between the innate and the adaptive immune responses to a broad spectrum of pathogens.

With this in mind we continue to explore the effect of zinc supplementation over the macrophages immune functions by studying IL-12 secretion in peritoneal macrophages from BALB/c mice supplemented with zinc from gestation to lactation.

MATERIALS AND METHODS

BALB/c mice were divided into groups according to zinc oral supplementation (500 mg/L) and periods of administration: from gestation to lactation (6 wks, group I), and from gestation to weaning (9 wks, group II), as formerly reported.

Semicuantitative reverse transcriptase polymerase chain reaction (RT-PCR) was applied, cells being prepared as previously described (Lastra 2001). Total ribonucleic acid (RNA) was extracted by the Chomczynski & Sacchi method with a guanidinethiocyanate/phenol solution. After isolation the RNAs were reverse transcribed and amplifications were performed with the rTTh desoxiribonucleic acid (DNA) polymerase enzyme and a commercial kit (Gene Amp PCR - Perkin-Elmer). IL-12 amplification was carried out (35 cycles) in a Perkin-Elmer 2400 thermocycler. PCR products were separated by gel electrophoresis in 1.5% agarose with ethidium bromide. Polaroid photographs were taken. Cells were adjusted to $1x10^6$ cells/mL in the same medium. Macrophages ($5x10^6$ per well) were stimulated with 1 microgram lipopolysaccharide (LPS) in polystirene microplates; microplates were incubated in a 5% CO2 atmosphere at 37°C; supernatants were harvested at 14 h. Interleukin-12 analysis was performed with an ELISA (Immunoassay kit Mouse interleukin 12 BioSource International CA USA).

The following primers were used:

5'primer CCA CTC ACA TCT GCT GCT CCA CAA

3'primer CAG TTC AAT GGG CAG GGT CTC CTC

Macrophages were drawn from mice peritoneal cavity in sterile conditions after a 5 mL RPMI intraperitoneal injection, for IL-12 evaluation. Macrophages ($5x10^6$ per well) were stimulated with 1 microgram lipopolysaccharide (LPS) in polystyrene microplates; microplates were incubated in a 5% CO_2 atmosphere at 37°C; supernatants were harvested at 14 h. Interleukin-12 analysis was performed with an ELISA (Immunoassay kit Mouse interleukin 12 BioSource International CA USA). Statistical analysis was carried out by ANOVA.

RESULTS

IL-12 gene expression: Gene expression evaluated by RT-PCR messenger RNA (mRNA) 45 min after LPS cell in vitro stimulation, showed a 3.6 fold increase relative to controls in mice group I (6 wks treatment). In mice group II (9 wks zinc *in vivo* treatment), the mRNA presented a significant 5.8 fold increase *(fig. 1.)*

IL-12 in serum: IL-12 was assayed on zinc orally supplemented mice (500 mg/mL) by an enzyme linked immunosorbent assay (ELISA). The assay evaluates the dual expression of the p70 form and the p40 subunit in the LPS stimulated mice serum. IL-12 serum concentration had a significant increase in mice with 9 wks in vivo zinc treatment. IL-12 concentration went from 288 pg/mL to 335 pg/mL *(fig. 2).*

DISCUSSION AND CONCLUSIONS

Infectious diseases remain the biggest killer of children and young adults worldwide despite scientific advances. Poor nutritional status increases susceptibility to infectious diseases and severe micronutrient deficiencies lead to specific syndromes. Zinc status directly affects the monocyte - macrophage cells. Thus the changes in pro-inflammatory (IL-12, interleukin 1 and TNF, etc.) cytokine secretion should be addressed, emphasizing the importance of studying the implications of zinc supplementation on cell interactions as well as the consequential effect on cytokines secretion.

IL-12 has vital roles in both innate immunity and later adaptive immune response and as previously stated, the most distinctive of IL-12 activity is its ability to regulate the balance between Th1 and Th2 cells. Therefore the necessity of studying the influence of zinc supplement in the IL-12 secretion.

The IL-12 gene expression evaluated by RT-PCR (mRNA) produced an important increase relative to controls in the 6 wks zinc treatment group (group I), and of almost double in the 9 wks zinc treatment group (group II), which are consistent with the serum protein concentrations.

The zinc *in vivo* administration is measured by the effects it produces on the cells (macrophages) which is remarkable. We are working now with these same systems assessing the intracellular zinc concentrations and the interleukin concentrations. We have observed (preliminary data) that zinc *in vivo* administration increases the IL-12 intracellular concentrations in mice with 9 wks supplementation.

The IL-12 serum concentration in mice supplemented with zinc, from gestation to weaning (group II), resulted in a significant increase, relative to controls. This could provoke an augmentation in the IL-12 effects through the stimulation of Th1 cells and the consequent production of IFN gamma and IL-2 which would result in beneficial effects on the animals, and are a consequence of zinc *in vivo* administration.

If the expression and production of these cytokines is influenced by supplemented zinc as well as by dietary zinc, this could affect how we view the relationship between the zinc nutritional status and immune functions particularly as observed during the perinatal periods.

REFERENCES

1. Fragoso G, Lastra MD, Aguilar AE, Pastelin R, Rosas G, Meneses G, Sciutto E, Lamoyi E (2001) Effect of oral zinc supplementation upon Taenia crassiceps murine cysticercosis *J Parasitol* 87: 1034-1039.

2. Frankenberger M, Sternsdorsf T, Pechumer H, Pforte A, Ziengler Heitrock HW (1996) Differential cytokine expression in human blood monocyte subpopulations: a polymerase chain reaction analysis *Blood* 87:373-377
3. Lastra MD, Pastelin R, Camacho A, Monroy B, Aguilar AE (2001) Zinc intervention on macrophages and lymphocytes response *J Trace Elem Med Biol* 15: 5-10.
4. Prasad AS (2000). Effects of zinc deficiency on Th1 and Th2 cytokine shifts *J Infect Dis* 182: S62-S68.
5. Salgueiro MJ, Zubillaga M, Lysioneck A, Cremaschi G, Goldman CG, Caro R, De Paoli T, Hager A, Weill R, Boccio J (2000). Zinc status and immune system relationship: a review *Biol Trace Elem Res* 76: 193-205.
6. Shankar AH and Prasad AS (1998). Zinc and immune function: the biological basis of altered resistance to infection *Am J Clin Nutr* 68: 447S-463S.

*Metal Ions in Biology and Medicine: vol. 7.*Eds L. Khassanova, Ph. Collery, I. Maymard, Z. Khassanova, J.-C. Étienne.
John Libbey Eurotext, Paris © 2002 pp. 495-497.

Effect of metal ions on adrenaline autoxidation coupled with the formation of superoxide radicals

T. V. Sirota, N. V. Chunderyakova, and M. N. Kondrashova

Institute of Theoretical and Experimental Biophysics, Russian Academy of Sciences, Pushchino, Moscow region, 142290 Russia

It was found that ions of bivalent metals regulate the reaction of adrenaline autoxidation in alkaline medium coupled with the formation of superoxide radicals, which serves as a model of the quinoid pathway of adrenaline conversion in the organism. The stimulating (prooxidant) effect of the ions investigated at the initial stage of the reaction was stated for: $??^{2+}$, Sr^{2+}, Mg^{2+}, Mn^{2+}. The most stimulatory was 2.5-50 μM Mn^{2+}; 25-500 μM Ca^{2+} exhibited a lower prooxidant dose-dependent effect. A similar effect was induced by low (50-100 μM) concentrations of Sr^{2+}. However, its higher concentrations (250 μM) induced an inhibition instead of activation. The effect of Mg^{2+} was close to that of high Sr^{2+} concentrations: an activation of the initial stage with the following inhibition. The regulation by physiological cations can modulate the effect of adrenaline in the organism.

One of the pathways of adrenaline conversion to adrenochrome in the organism is the quinoid oxidation [1]. It was found that, at the intermediate stages of this transformation (from adrenaline to adrenaline semiquinone and then to adrenaline quinone) in the aqueous phase under aerobic conditions, the formation of superoxide anion radicals (O_2^-•) occurs [2-5]. The formation of O_2^-• can also occur during the reverse reaction, the reduction of adrenochrome by the action of enzymes DT-diaphorase and NADPH-cytochrome P-405 reductase [3, 5]. The quinoid pathway of oxidation of catecholamines (adrenaline), which was described 40 years ago, became currently again a subject of wide discussion in view of its close relation with the formation of reactive radicals (semiquinones and active oxygen species) [3-5, 8, 9]. It is believed that the cardiotoxic action of catecholamines is associated just with active oxygen species rather than adrenochrome [7].

The *in vitro* quinoid conversion of adrenaline with the formation of adrenochrome without involvement of enzymes occurs in the alkaline medium. Intramolecular rearrangements of adrenaline occurring in a buffer solution of low H^+ concentration initiate a chain process of its autoxidation. In biochemistry, this reaction is known as a model superoxide-generating system and also as a superoxide-detecting reaction because it is inhibited by superoxide dismutase (SOD) [10]. The activity of SOD is determined from its ability to inhibit the accumulation of the adrenaline oxidation product adrenochrome (by absorbance at 480 nm) [10]. Earlier it was found in our laboratory that autoxidation of adrenaline in alkaline medium and, consequently, the generation of O_2^-•, can be detected not only at 480 nm but also at 347 nm, approximately in the middle of the absorption shoulder (320-380 nm). This assay is more sensitive since the intensity of the increment in optical density in this spectrum region is substantially higher than at 480 nm. The formation of adrenochrome measured at 347 nm is SOD-sensitive [11, 12].

Here we studied the effect of physiologically active metal ions (Ca^{2+}, Mg^{2+}, Mn^{2+} and Sr^{2+}, an analogue of Ca^{2+}) with permanent and variable valence on the rate of generation of superoxide radicals during adrenaline autoxidation in alkaline medium. It is known that Ca^{2+} and Mg^{2+} are involved in the regulation of O_2^-• formation by various superoxide-generating systems (e. g., blood cells, such as neutrophils, eosinophils, macrophages, etc.). Mn^{2+} is known as a cofactor of mitochondrial SOD; in addition, it can serve as a classical catalyst of chain reactions.

MATERIALS AND METHODS

Absorption spectra and the kinetics of adrenaline autoxidation to adrenochrome were measured for 5 min at 22° in a thermostated cuvette in 0.1 M carbonate buffer using a Uvikon-923 spectrophotometer (Italy) in the time Driver mode at 347 nm under continuous stirring. The concentration of adrenaline was 2.6×10^{-4} M. Solutions of metal salts were added to the buffer prior to introducing adrenaline. The addition of the metal salts did not change the pH value. The data presented are the average of several independent experiments, with 3-6 parallel measurements in each experiment.

RESULTS AND DISCUSSION

As it follows from *fig. 1*, calcium ions induced a dose-dependent activating effect on the accumulation of adrenochrome, i. e., enhanced the generation of O_2^{-}•. The minimal effective concentration of Ca^{2+} under the conditions of the experiment was 25 μM (*fig. 1*, curve 2). The latent period of Ca^{2+}-activated adrenaline autoxidation was considerably shorter than that of the control reaction, and at a concentration of 0.5 mM the curve has a clearly pronounced sigmoidal shape (*fig. 1*, curve 4). Strontium ions, which are chemically close to calcium ions, also activated the reaction at a concentration of 50-100 μ M; however, the shape of the kinetic curves of adrenaline activation was different. At a 50 μM concentration of Sr^{2+}, the shape of the curve was similar to that of the control sample (*fig. 2*, curve 2 and 1); the activating effect of Ca^{2+} at this concentration, however, was much stronger, and the lag period was substantially shorter (*fig. 2*, curve 3). At higher concentrations (250 μM), the difference between the effects of Ca^{2+} and Sr^{2+} was still more pronounced *(fig. 3)*. The effect of strontium ions was strictly biphasic (*fig. 3*, curve 1): the initial low-amplitude short activation with the oxidation rate higher than in the presence of Ca^{2+} was followed (after a 2-min interval) by a decrease of activation with the oxidation rate lower than in the presence of Ca^{2+}.

The curve illustrating the effect of magnesium ions (100 and 200 μM) is shown in *fig. 4*. The magnitude of the effect can be considered as intermediate between those produced by Ca^{2+} and Sr^{2+}; however, after a 2-min interval and at high concentrations, magnesium ions have an inhibitory effect. At a concentration of 100 μM, Mg^{2+} induces a pronounced activation without a lag period (*fig. 4*, curve 2), which, however, is followed by a decrease in the oxidation rate, which is the greater the higher is Mg^{2+} concentration (*fig. 4*, curves 2, 3).

Manganese and copper ions constitute the prosthetic groups of the mitochondrial and cytoplasmic SOD, respectively. However, these ions exert different effects. Manganese ions (in the form of both chloride and sulfate) at a concentration of 2.5-50 μM activated the reaction to a maximum degree *(fig. 5)*. Such a strong effect is inherent in ions of metals with variable valence [13, 14], which are known to be initiators and catalysts of chain processes. However, as it was shown previously, copper ions at low concentrations (2.5 and 5 μM) inhibited the generation of O_2^{-}• [11, 12]. There is evidence that Cu^{2+}, a metal of variable valence, induces the break of chains upon oxidation of alcohols and amines and inhibits chain reactions, i. e., acts as an antioxidant [13]. Previously we also showed that the classical activator of chain processes Fe^{2+}, an ion with variable valence, has an activating action on adrenaline autoxidation [11].

Thus, we showed that Ca^{2+}, Sr^{2+}, Mg^{2+} and Mn^{2+} enhanced to a different extent the generation of superoxide radicals at the initial stage of adrenaline oxidation; however, the further course of the reaction substantially differed and depended on both the concentration of the ion and its nature:

some ions retained their activating effect (Ca^{2+}), whereas in the presence of the other ions (Mg^{2+} and high concentrations of Sr^{2+}) the inhibition of the reaction was observed. Such a biphasic kinetics is typical for chain free radical processes [13, 14].

These metal ions may affect the generation of superoxide radicals in the living organism if the adrenaline oxidation proceeds via the quinoid pathway of conversion. Free radical processes occurring during the oxidation of catecholamines and reduction of aminochromes, which are accompanied by the generation of superoxide radicals, are observed in the pathogenesis of some diseases (schizophrenia [8, 15, 16] and acute form of hypertension [9]). A biochemical prerequisite for the development of the manic depressive form of schizophrenia is known to be related to the oxidation of adrenaline via the quinoid oxidation pathway: in the urine of these patients, pathological metabolites of indole and its derivatives are found [16]. The products of adrenaline oxidation are also found in the serum and plasma of these patients [16].

The study was supported by the Russian Federation for Basic Research, "Leading Scientific Schools" (project no. 00-15-97847).

REFERENCES

1. Men'shikov V.V., Bol'shkova T.D. In: Adrenline and noradrenaline, M. Nauka, 1964, 284-293.
2. Uemura T., Chiesara E., Cova D., Mol. Pharmacol., 1977, **13**, 196-215.
3. Marques F., Duarte R. O., Moura J. J., Bicho M. P., Biopl. Signals., 1996, **5**, 275-282.
4. Polewski K., Biochim. Biophys. Acta., 2000, **1523**, 56-64.
5. Baez S., Segura-Aguilar J., Biochem. Mol. Med., 1995, **56**, 37-44.
6. Bindoli A., Deeble D. J., Rigobello M. P., Galzigna L., Biochem. Biophys. Acta., 1990, **1016**, 349-356.
7. Rump A.F., Klaus W., Pharmacol. Toxicol., 1995, **77**, 102-105.
8. Smythies J. R., Schizophr. Res., 1997, **24**, 357-364.
9. Zhou Q., Hulea S., Kummerow F. A., Res. Commun. Pathol. Pharmacol., 1995, **89**, 111-126.
10. Misra H. P., Fridovich I., J. Biol. Chem., 1972, **247**, 3170-3175.
11. Sirota T. V., Vopr. Med. Chem. (Russia), 1999, **45**, 263-272.
12. Sirota ?.V., Lange N. V., Kosjakova N. I., Vanichkin A.V., Kondrashova M. N., Current Topics in Biophysics, 2000, **24**, 185-189.
13. Emanuel N.M., Denisov E.T., Maizus Z.K. Chain oxidation reactions of hydrocarbons in the liquid phase, M. Nauka, 1965, PP. 13, 203-204.
14. Denisov E.T., Sarkisov O.M., Lichtenstein G.I., Chemical kinetics., Acta Khimiya, 2000, PP. 397-413, 510-512.
15. Pogodaev K.I. Epileptology and brain pathology, M.: Meditsina, 1986, PP. 50, 57-65.
16. Hoffer A. In: Chemical concepts of Psychosis, M. Gos. Izd. Med. Lit., 1963, PP. 112-124.

*Metal Ions in Biology and Medicine: vol. 7.*Eds L. Khassanova, Ph. Collery, I. Maymard, Z. Khassanova, J.-C. Étienne.
John Libbey Eurotext, Paris © 2002 pp. 498-501.

Role of mercury in lipid peroxidation

Marjatta Kantola[1], Kari Seppänen[1], Pasi Soininen[1], Jukka T. Salonen[2], Simo Lötjönen[1], Reino Laatikainen[1]

[1]*Department of Chemistry and* [2]*Research Institute of Community Health and General Practice, University of Kuopio, P.O. Box 1627, FIN-70211 Kuopio, Finland*

ABSTRACT

The effects of mercury, copper and iron on the non-enzymatic oxidation of LDL and the enzymatic oxidation of dihydroxyfumaric acid were investigated. Mercury did not cause non-enzymatic oxidation of LDL but copper appeared to activate peroxidase by 93%. We conclude that mercury and methylmercury do not directly cause or promote the non-enzymatic radical peroxidation of lipids but, that a simultaneous exposure to copper and mercury, the former via acceleration of both enzymatic and non-enzymatic peroxidation, and the latter via the inhibition of glutathione peroxidase, promote lipid peroxidation in biological systems, and thus may promote development of atherosclerosis.

INTRODUCTION

There is a strong epidemiological evidence for an association between a high body mercury concentration and accelerated atherosclerosis [1-4]. Mercury has also been reported to accelerate lipid peroxidation in the rat [5]. Previously, iron and copper have been shown to be associated to atherosclerosis via direct lipid peroxidation [6, 7] and, thus, it has been proposed that also mercury promotes lipid peroxidation in the same way [1]. This explanation is not convincing because the redox and coordination properties of mercury differ greatly from those of the two others. On the other hand, the inverse association between coronary heart disease and selenium [8] and the high affinity of mercury to selenium in the glutathione peroxidase [9-11], a key enzyme of the peroxidation protection system, suggest that the balance of mercury, selenium, iron and copper, play a role behind the association between mercury and lipid peroxidation.

AIMS

The aim of this study was to assess the role of mercury, copper and iron in lipid peroxidation using native LDL (low density lipoprotein), horseradish peroxidase and dihydroxyfumaric acid as the model systems.

MATERIALS AND METHODS

Materials

The horseradish peroxidase type VI-A (Donor:hydrogen peroxide oxidoreductase, EC 1.11.1.7), HEPES (N-[2-hydroxyethyl]piperazine-N'-[2-ethanesulfonic acid]), β-NADH and low density lipoprotein (native LDL) were from Sigma Chemical Co. (St. Louis, USA). $CuCl_2$, $FeCl_3$, D_2O,

$CDCl_3$ and NaCl were from Merck (Darmstadt, Germany). Dihydroxyfumaric acid was from ICN Biomedicals Inc. (Ohio, USA). H_2O_2 and $HgCl_2$ were from Riedel-deHaen AG (Seelze, Germany) and CH_3HgCl from Johnson Matthey GmbH (Karlsruhe, Germany). All the used reagents were of analytical grade unless indicated otherwise.

NMR studies

For the NMR studies LDL was diluted to the final concentration of 0.40 mg ml^{-1} by 0.15 M NaCl in H_2O:D_2O (9:1). $CuCl_2$, $HgCl_2$ or CH_3HgCl was added so that its final concentration was 350 µM. Also 5 µl of 30% H_2O_2 was added to the samples. The EDTA concentration arising from the LDL preparate was 5 µM. After 6 hour incubation at 27°C, the preparates were extracted with $CDCl_3$. The spectra were recorded by a Bruker Avance 500-spectrometer (Bruker, Karlsruhe, Germany) in routine way. The signal areas were integrated using the deconvolution method with the PERCH software [12].

Peroxidase activity

The peroxidase activity measurement was based on the reaction of dihydroxyfumaric acid to dioxosuccinic acid. Long-chain fatty acids were not used because they dissolve poorly into the water. The peroxidase activity was measured in a system which contained 30 µg ml^{-1} horseradish peroxidase (specific activity 1310 units(mg^{-1}) in 0.9 ml of 100 mM HEPES buffer pH 7.5. Final concentrations in the reaction mixture (2.0 ml) were 0.07 mM NADH, 2.0 µM dihydroxyfumaric acid and 10 µM $CuCl_2$, $FeCl_3$ or $HgCl_2$. The reaction was started by 40 µl of 500 µM H_2O_2. The mixture was incubated at 37°C for 30 min. The enzyme activity (decrease of NADH nmol(min^{-1}l^{-1}) was calculated using 0.07 mM NADH-standard.

Statistical analyses

Each group of the study was compared to the group of the pure enzyme. The equality of the means of the activities between two different groups was tested by the non-parametric Mann-Whitney test. The statistical significance was based on two-sided tests.

RESULTS

The NMR method [13] was applied to follow the reactions of LDL in presence of mercury and methylmercury chlorides, using the oxidation reaction by copper as the reference reaction. The copper catalyzed oxidation of the LDL lipids is seen in a decrease of the linoleic and arachidonic acid signals at 2.75 ppm *(fig. 1)*. In our experiments with mercury or methylmercury no oxidation or decomposition of the LDL particles was seen.

In the enzymatic experiments *(table 1)* Cu^{2+} almost doubled the activity of peroxidase. Iron did not affect the activity, Hg^{2+} seems to act as a weak inhibitor of peroxidase.

Table 1. Enzyme activities of peroxidase without and with different activators and inhibitors

Enzyme and activator or inhibitor	n	Enzyme activity[a] nmol · min^{-1}l^{-1}	p
		Decrease of NADH	
None	18	327 ± 41	
Copper (Cu^{2+})	15	631 ± 102	<0.001
Iron (Fe^{3+})	15	355 ± 61	0.051
Mercury (Hg^{2+})	20	268 ± 42	<0.001
Copper (Cu^{2+}) and mercury (Hg^{2+})	5	652 ± 25	0.001
Iron (Fe^{3+}) and mercury (Hg^{2+})	5	232 ± 5	0.001

[a]Mean ± SD

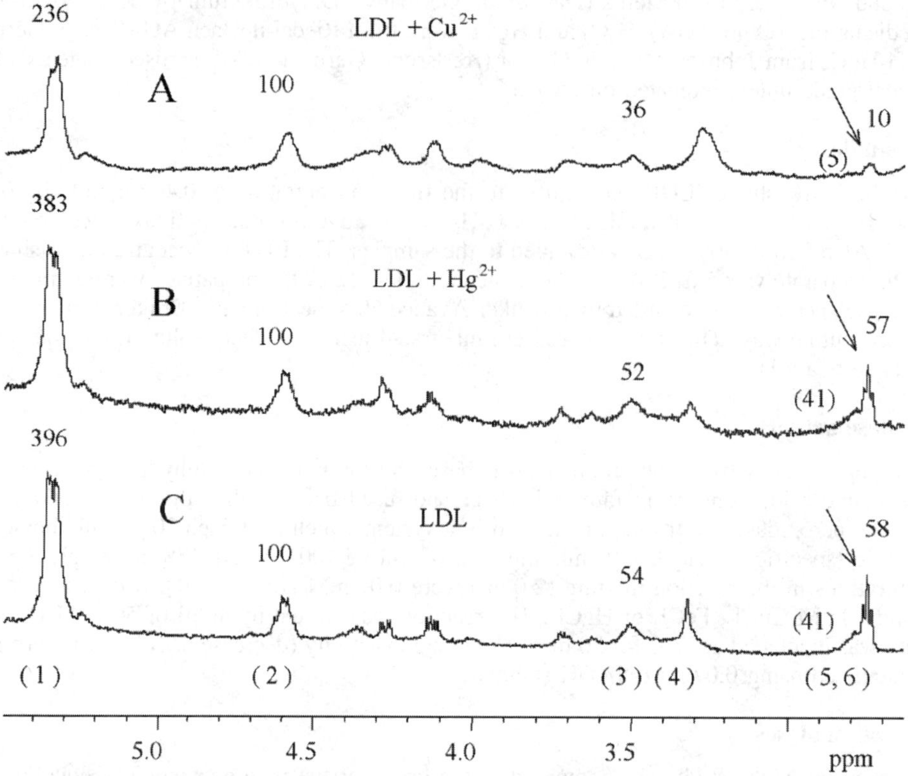

Fig. 1. A part of 500 MHz ^{1}H NMR spectrum of LDL after 6 hours incubation in presence of a trace amount of H_2O_2. (A) LDL+ Cu^{2+}, (B) LDL + Hg^{2+}, (C) Native LDL. The signals: (1) CH=CH, (2) cholesterol ester C-3, (3) cholesterol C-3, (4) $(N(CH_3)_3)^+$ and (5,6) arachidonic and linoleic acid proton signals. The numbers above the signals give their areas when the signal (2) is set to 100

DISCUSSION

Lipid peroxidation is thought to occur through the radical mechanism [14]. The radicals can be produced by peroxidase enzymes [8] or non-enzymatically by the Fenton ($H_2O_2 + M^{n+} = OH^- + {}^{\cdot}OH + M^{(n+1)+}$) and Haber-Weiss ($H_2O_2 + O_2^{\cdot -} + M^{n+} = O_2 + {}^{\cdot}OH + OH^- + M^{(n+1)+}$) reactions. The analogous reactions can be written also for the Hg_2^{2+}/Hg^{2+}-pair. However, the reactions do not easily occur between copper and double bonds in normal water or methanol solutions, as also confirmed in some NMR experiments with arachidonic acid. Our NMR experiments *(fig. 1)* indicate now that mercury is not able to induce the oxidation of lipids in LDL.

Enzymes such as superoxide dismutase, catalase and glutathione peroxidase scavenge reactive oxygen species and protect the cell against the oxidative stress [15]. From the point of view of the protection mechanism, the glutathione peroxidase is considered most important [14-16]. A most significant point in glutathione peroxidases is that they contain selenium in active site and that, as previously well-known [9-11], mercuric ion is a strong inhibitor for glutathione peroxidase.

In our studies copper acted as an activator of peroxidase. Previously it has been suggested that Cu^{2+} inhibits the peroxidase-catalyzed reactions [17]. Although the active peroxidase enzyme contains Fe^{2+} or Fe^{3+} ion as a co-factor (18), in our experiments Fe^{3+} did not activate the enzyme, likely due to iron in the enzyme preparate. However, preliminary results *in vivo* with LDLR-deficient mice show that the combination of mercury and iron has 3-fold atherogenic effect compared with iron alone [19].

CONCLUSIONS

On the basis of the above results, we conclude that mercury is not able to induce the oxidation of lipids in LDL nor, probably, in the vascular intima via the direct radical mechanism. However, because the mercuric ion is a strong inhibitor for glutathione peroxidase and copper increases the activity of peroxidase, it is possible that the balance of copper, iron, selenium, mercury and the peroxidase-glutathione peroxidase system may have a role in lipid peroxidation in biological systems. Horseradish peroxidase is widely used as a model enzyme in lipid peroxidation because it catalyzes same reactions as human peroxidases. However, the enzyme system of lipid peroxidation needs more examination because plant enzyme differs very much from human peroxidases. Anyway, our results propose that metal ions may have significant effects on enzyme reactions in lipid peroxidation.

REFERENCES

1. Salonen, J.T., Seppänen, K., Nyyssönen, K., Korpela, H., Kauhanen, J., Kantola, M., Tuomilehto, J., Esterbauer, H., Tatzber, F., and Salonen, R. (1995) *Circulation* **91**, 645-655.
2. Guallar, E., Sanz-Gallardo, M.I., van's Veer, P., Martin-Moreno, J.M., and Kok, F.J. (1999) *7th Nordic Symposium on Trace Elements in Human Health and Disease*, Espoo, Finland.
3. Salonen, J.T., Seppänen, K., Lakka, T.A., Salonen, R., and Kaplan, G.A. (2000) *Atherosclerosis* **148**, 265-273.
4. Rissanen, T., Voutilainen, S., Nyyssönen, K., Lakka, T.A. and Salonen, J.T. (2000) *Circulation*, **28**, 2677-2679.
5. Huang, Y.L., Cheng, S.L., and Lin, T.H. (1996) *Biol. Trace Elem. Res.* **52**, 193-206.
6. Salonen, J.T., Salonen, R., Seppänen, K., Kantola, M., Suntioinen, S., and Korpela, H. (1991) *Br. Med. J.* **302**, 756-760.
7. Salonen, J.T., Nyyssönen, K., Korpela, H., Tuomilehto, J., Seppänen, R., and Salonen, R. (1992). *Circulation* **86**, 803-811.
8. Salonen, J.T., Alfthan, G., Huttunen, J.K., Pikkarainen, J., and Puska, P. (1982) *Lancet* **2**, 175-179.
9. Splittgerber, A.G., and Tappel, A.L. (1979) *Arch. Biochem. Biophys.* **197**, 534-542.
10. Bem, E.M., Mailer, K., and Elson, C.M. (1985) *Can. J. Biochem. Cell Biol.* **63**, 1212-1216.
11. Hirota, Y. (1986) *Am. Ind. Hyg. Assoc. J.* **47**, 556-558.
12. Laatikainen, R., Niemitz, M., Malaisse, W.J., Biesemans, M., and Willem, R. (1996) *Magn. Reson. Med.* **36**, 359-365.
13. Lodge, J.K., Sadler, P.J., Kus, M.L.n and Winyard, P. (1995) *Biochim. Biophys. Acta* **1256**, 130-140.
14. Kanner, J., German, J.B., and Kinsella, J.E. (1987) *Crit. Rev. Food Sci. Nutr.* **25**, 317-364.
15. Takane, T., Asayama, K., Kodera, K., Hayashibe, H., Uchida, N., and Nakazawa, S. (1998) *Free Radical Biol. Med.* **25**, 504-511.
16. Raderecht, H.J. (1998) *Clin. Lab.* **44**, 33-50.
17. Halliwell, B. (1977). *Biochem. J.* **163**, 441-448.
18. Lippard, S., and Berg, J. (1994) *Principles of Bioinorganic Chemistry*. Ed., University Science Books, California, Mill Valley, California.
19. Leppänen, P., Seppänen, K., Salonen, J.T., and Ylä-Herttuala, S. (2000) *XIIth International Symposium on Atherosclerosis*, Stockholm, Sweden.

*Metal Ions in Biology and Medicine: vol. 7.*Eds L. Khassanova, Ph. Collery, I. Maymard, Z. Khassanova, J.-C. Étienne.
John Libbey Eurotext, Paris © 2002 pp. 502-505.

Strontium effects on lipid peroxidation due to $FeCl_3$ and ascorbic acid in rat synaptosomes

S. Millán-Plano, J.J. García, E. Martínez-Ballarín, S. Ortega-Gutiérrez, J.L. Allué, R.M. Lázaro, J.F. Escanero

Department of Pharmacology and Physiology. University of Zaragoza, Spain

To whom correspondence should be addressed:
Professor Jesús F. Escanero
Departamento de Farmacología y Fisiología
Facultad de Medicina. Universidad de Zaragoza
c) Domingo Miral s/n
50009 Zaragoza, Spain
Phone + (34) 976 76 16 99
Fax + (34) 976 76 17 00

ABSTRACT

The objective of this work was to test the effects of several strontium (Sr) concentrations in the lipid peroxidation initiated by $FeCl_3$ and ascorbic acid in rat synaptosomal membranes. In addition, it was tested the combination of Sr with pinoline, an antioxidant that reduces efficiently iron mediated lipid peroxidation in biological membranes. Aliquots of membranes were incubated at 37°C in the presence or absence of either $FeCl_3$, ascorbic acid, $SrCl_2$ and pinoline. Lipid peroxidation was estimated by measuring malondialdehyde (MDA) and 4-hydroxyalkenals (4-HDA) concentrations. Under these experimental conditions, Sr did not have significative influence on lipid peroxidation initiated by $FeCl_3$ and ascorbic acid. Moreover, Sr did not improve the antioxidant ability of pinoline against iron-induced lipid peroxidation.

Key Words: Strontium, Iron, Pinoline, Free Radical, Lipid Peroxidation, Synaptosome.

INTRODUCTION

Free radicals are responsible of the oxidative damage in the cells. They react with the phospholipids of biological membranes and initiate a chain reaction known as lipid peroxidation [1]. This chemical process determines loss or modification of some membrane-dependant functions and cell death [2].

Strontium (Sr) is a divalent cation included in the Alkaline Earth metals of the Periodic Table of the Elements. Sr shows a similar behavior with calcium, and they share the same metabolic pathways [3, 4]. Sr is clinically used with diagnostic and therapeutic purposes. Several authors have proposed its indication in the treatment of osteoporosis [5] and to try to remove intoxications by ^{90}Sr [6]. Previous studies have shown that in animals fed with Sr-rich diets, their muscular and hearth mitochondria where protected against the swelling caused by calcium, thyroxine, oleate, phosphate and free fatty acids [7]. Possibly these effects may be explained because Sr reduces oxidative damage in the mitochondria. Other Alkalin Earth metals, *e.g.*, calcium and magnesium, have been related to oxidative stress in biological membranes [8, 9]. Thus, the aim of this work was to test the Sr effect on iron-lipid peroxidation of membranes isolated from rat brain.

MATERIALS AND METHODS

Female Sprague-Dawley rats weighing 225±25 g were used to isolate synaptosomes. Animals were sacrificed by decapitation. Their brains were quickly removed, washed in saline solution (0.9% NaCl) and homogenized 1:10 (w/v) in cold 0.32 M sucrose. The resulting suspension was centrifuged at 1,000×*g* for 10 min at 4°C to remove cell debris and nuclei. The supernatant was centrifuged at 20,000×*g* for 20 min at 4°C. The pellet obtained was resuspended in cold water and centrifuged at 8,000×*g* for 20 min at 4°C. The supernatant, which contained the membranes, was recentrifuged at 48,000×*g* for 20 min at 4°C. The resulting pellet was washed 2 times and the final pellet was suspended 1/2 (v/v) in 20 mM Tris-HCl buffer (pH = 7.4) and frozen at -30°C until assay.

Aliquots of membranes (0.5 mg protein/ml) were incubated at 37°C for 60 min in the presence or absence of either 0.1 mM $FeCl_3$, 0.1 mM ascorbic acid, $SrCl_2$ (0.001, 0.01, 0.1, 1, 10 mM) and pinoline 0.01 mM. Lipid peroxidation was stopped by placing the synaptosomal suspensions into ice-cold water for 10 min. MDA+4-HDA concentrations, an index of lipid peroxidation, were measured using a colorimetric commercial assay kit (Caymen Chemical). In this assay, MDA+4-HDA react with a chromogenic reagent at 45°C, yielding a stable chromophore with maximal absorbance at 586 nm wavelength [10]. Protein concentrations were determined by the Bradford method using bovine albumin as a standard [11]. Statistical analysis was done using the Student's two-tailed t-test ($p<0.05$).

RESULTS AND DISCUSSION

When synaptosomes were incubated with Sr in absence of $FeCl_3$ and ascorbic acid, no significant differences in MDA+4-HDA levels were found between control samples and Sr-treated membranes *(fig. 1A)*. Iron and ascorbic acid are frequently used to induce oxidative stress in biological membranes [12]. Our data are in agreement with these previous observations, since MDA+4-HDA concentrations raised in the presence of $FeCl_3$ and ascorbic acid *(figure 1B)*.

The addition of Sr in the oxidative model did not change MDA+4-HDA levels, relative to control samples exposed to iron and ascorbic acid alone. In rats subjected to exhaustive physical exercise, Sr seems to protect the energy production in the mitochondria, perhaps because it reduces the swelling and the alteration of its internal structure [6], which suggests a reduction of oxidative damage. Our results of MDA+4-HDA levels reflect absence of lipid peroxidation due to Sr. However, it is possible that Sr may determine changes in the phospholipid dynamic of the bilayer. Some metals, *e.g.*, aluminum, bind to the phospholipids and modify the order in the membrane [13]. Disturbation in the phospholipid motions influence the susceptibility of biological membranes to free radicals.

Finally, we investigated the Sr influence on the antioxidant ability of pinoline against iron-lipid peroxidation. Pinoline is a tricyclic β-carboline isolated from the pineal gland and other tissues [14, 15]. Although its physiological role remains unknown, pinoline increases serotonin levels in the brain and modulates the 3H-imipramine recognition site [16]. Recently, we showed the antioxidant ability of pinoline [17]. The presence of Sr did not improve the protective effect of this β-carboline *(fig. 1C)*.

In conclusion, under the experimental conditions above indicated, Sr did not influence lipid peroxidation due to $FeCl_3$ and ascorbic acid in synaptosomal membranes. Moreover, Sr did not modify the antioxidant ability of pinoline in protecting against iron-lipid peroxidation in membranes isolated from rat brains.

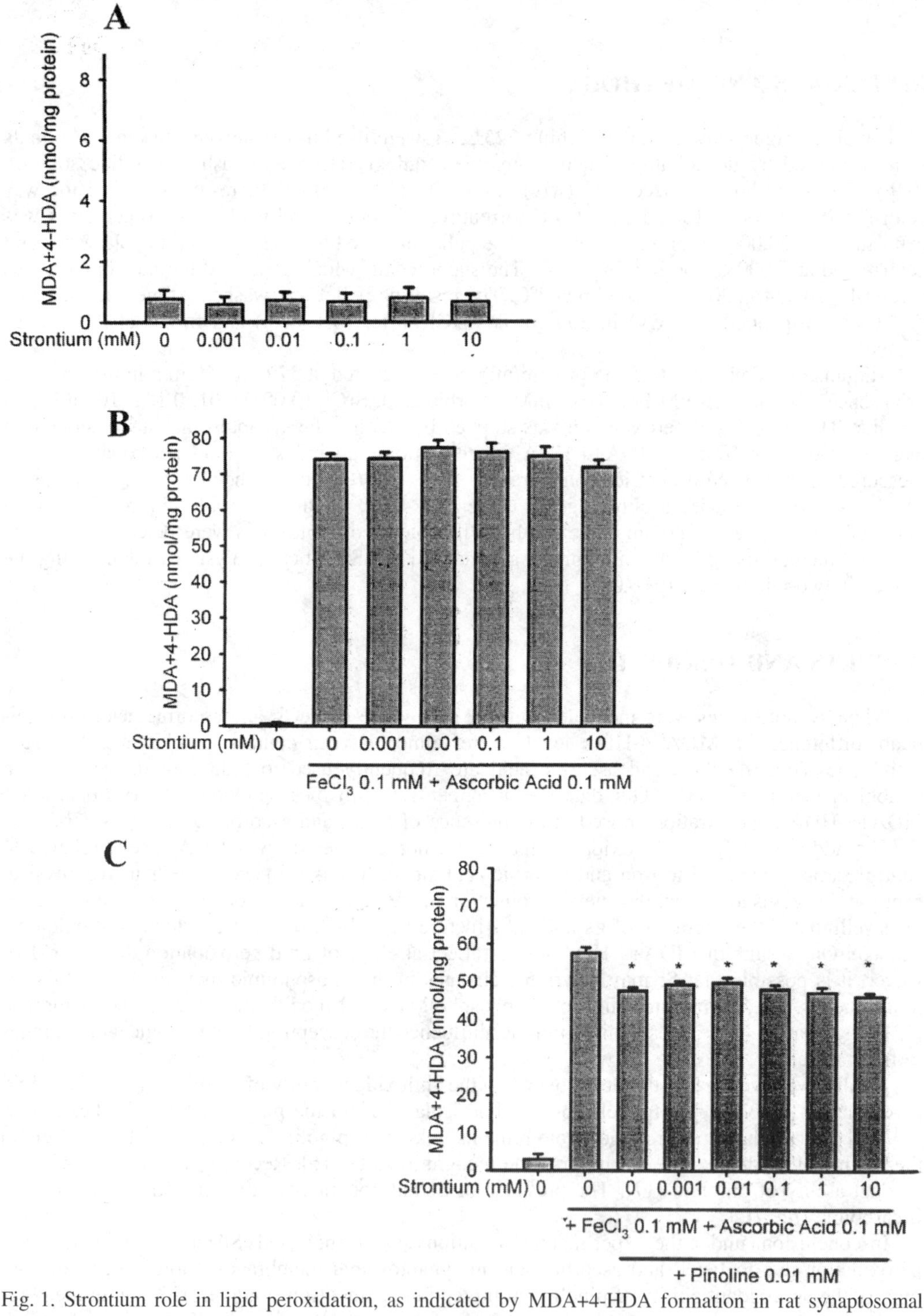

Fig. 1. Strontium role in lipid peroxidation, as indicated by MDA+4-HDA formation in rat synaptosomal membranes exposed to $FeCl_3$ and ascorbic acid. **A:** Effect of Sr (0.001, 0.01, 0.1, 1 and 10 mM) in absence of the oxidant reagents; **B:** Effect of Sr (Same concentrations) + oxidant reagents (0.1 mM $FeCl_3$ + 0.1 mM ascorbic acid); **C:** Effect of Sr on the antioxidant ability of 0.01 mM pinoline in the iron-lipid peroxidation. Values are expressed as means ± standard errors obtained in 5 independent experiments. *: $p<0.05$ vs. synaptosomes treated with $FeCl_3$ and ascorbic acid.

REFERENCES

1. Kanner, J., German J.B. & Kinsella, J.E. Initiation of lipid peroxidation in biological systems. *Crit. Rev. Food Sci.*, 1987, **25**, 317-364.
2. García, J.J., Reiter, R.J., Guerrero, J.M., Escames, G., Yu, B.P., Oh, C.S. & Muñoz Hoyos, A. Melatonin prevents changes in microsomal membrane fluidity during induced lipid peroxidation. *FEBS Lett.*, 1997, **408**, 297-300.
3. Escanero, J.F., Carre, M. & Miravet, L. Effets des différents métabolites de la vit. D_3 et de la concentration calcique sur l'absorption intestinale de strontium. *C. R. Soc. Biol.*, 1976, **170**, 47-53.
4. Escanero, J.F. & Cordova, A. Effect of glucagon on serum calcium, strontium and magnesium levels in rats. *Miner. Electrol. Metab.*, 1991, **17**, 190-193.
5. Morohashi, T., Sano, T. & Yamada, S. Effects of strontium on calcium metabolism in rats. I. A distinction between the pharmacological and toxic doses. *Jpn. J. Pharmacol.*, 1994, **64**, 155-162.
6. Skorina, S.C. & Fuskova, M. Effects of stable strontium supplementation. In: *Handbook of stable strontium*. Ed. S.C. Skorina. New York: Plenum Press, 1981, 593-617.
7. Caplan, A.I. & Carafoli, E. The effect of Sr on swelling and ATP-linked contraction of mitochondria. *Biochim. Biophys. Acta*, 1965, **97**, 99-106.
8. Günther, T., Vormann, J. & Höllriegl, V. Effects of magnesium and iron on lipid peroxidation in cultures hepatocytes. *Mol. Cell. Biochem.*, 1995, **144**, 141-145.
9. Maciel, E.N., Vercesi, A.E., Castilho, R.F. Oxidative stress in Ca^{2+}-induced membrane permeability transition in brain mitochondria. *Neurochem.*, 2001, **79**, 1237-1245.
10. Reiter, R.J., Tan, D.X., Kim, S.J., Manchester, L.C., Qi, W., García, J.J., Cabrera, J. & Rouvier-Garay, V. Augmentation of indices of oxidative damage in life-long melatonin-deficient rats. *Mech. Ageing Dev.*, 1999, **110**, 157-173.
11. Bradford, M.M. A rapid and sensitive method for the quantification of microgram quantities of protein utilizing the principle of protein-dye binding. *Anal. Biochem.*, 1976, **72**, 248-254.
12. Andorn, A.C., Britton, R.S. & Bacon, B.R. Ascorbate-stimulated lipid peroxidation in human brain is dependent on iron but not on hydroxyl radical. *J. Neurochem.*, 1996, **67**, 717-722.
13. Gutteridge, J.M.C., Quinlan, G.J., Clark, I. & Halliwell, B. Aluminium salts accelerate peroxidation of membrane lipids stimulated by iron salts. *Biochim. Biophys. Acta*, 1985, **835**, 441-447.
14. Shoemaker, D.W., Cummins, J.T. & Bidder, T.G. β-carbolines in rat arcuate nucleus. *Neuroscience*, 1976, **3**, 233-239.
15. Kari, I. 6-Methoxy-1,2,3,4-tetrahydro-β-carboline in pineal gland of chicken and cock. *FEBS Lett.*, 1981, **127**, 277-280.
16. Langer, S.Z., Lee, C.R., Segonzac, A., Tateishi, T., Esnaud, H., Schoemaker, H. & Winblad, B. Possible endocrine role of the pineal gland for 6-methoxytetrahydro-β-carboline, a putative endogenous neuromodulator of the ^{3}H-imipramine recognition site. *Eur. J. Pharmacol.*, 1984, **102**, 379-380.
17. Pless, G., Frederiksen, T.J.P., García, J.J. & Reiter, R.J. Pharmacological aspects of N-acetyl-5-methxytryptamine (melatonin) and 6-methoxy-1,2,3,4-tetrahydro-β-carboline (pinoline) as antioxidants: Reduction of oxidative damage in brain region homogenates. *J. Pineal Res.*, 1999, **26**, 236-246.

Metal Ions in Biology and Medicine: vol. 7. Eds L. Khassanova, Ph. Collery, I. Maymard, Z. Khassanova, J.-C. Étienne.
John Libbey Eurotext, Paris © 2002 pp. 506-511.

Immunological-related perturbations induced by a sublethal concentration of gallium in carp (*Cyprinus Carpio* L.)

Stéphane Betoulle[1,2], Claire Dautremepuits[1,2], Sylvie Biagianti[1,2] Jean-Claude Etienne[2] and Guy Vernet[1,2]

[1]Laboratory of Eco-Toxicology; [2]International Research Institute on Metal Ions, University of Reims Champagne-Ardenne, BP 1039, 51687 Reims cedex 2, France

ABSTRACT

As the manufacture of semiconductor technology using gallium can produce negative consequences for the environment especially for aquatic ecosystems, it is useful to assess and predict the potential risk of gallium for aquatic animal health. Fish is one of the most important indicators of environmental contamination of water and its immunophysiological system is particularly sensitive to water quality. In term of ecotoxicological risk assessment, we reported here the sublethal impact of gallium on immune parameters of carp. Fish were exposed to a sublethal concentration of gallium (chosen as 5% of 96 h LC_{50} in carp) for 96 hours. Selected parameters [*e.g.*, phagocyte killing activity, total plasma immunoglobulins and proteins, plasma lysozyme and ceruloplasmin activities] were evaluated at 24, 48 and 96 hours of exposure. The phagocyte killing and plasma lysozyme activities were not altered by exposure to gallium. Total plasma immunoglobulin was decreased after 48 h exposure. A same result was observed in level of total proteins in plasma of fish exposed to gallium from 48 h of exposure while the ceruloplasmin activity was significantly increased at the same time. In case of an acute environmental contamination gallium could be implicated as a relatively low immunopathological metal ion in fish. However, such aquatic gallium contamination of carp may induced an inflammatory process where the non specific humoral immunity should be the first immune function of fish implicated. Changes in immune parameters observed can reflecte a decrease in fish health (*e.g.* inflammatory response) associated with stressfull conditions and could then increase the sensibility of fish to pathogens.

INTRODUCTION

In many years, gallium has seen increased used for integrated circuits in semiconductor industry. Numerous gallium compounds exist including gallium arsenide or gallium antimonide and gallium dust emissions are relatively high (Fowler and Dobrota, 1989). According to Chepesiuk (1999), the manufacture of semiconductor technology using gallium can produce negative consequences for the environment. Toxicological effects of semiconductor components were evaluated in humans (Kuroda *et al.*, 1991) and animals (Omura *et al.*, 1996). They reported testicular toxicity produced by gallium arsenide to hamsters and found that gallium concentration was 32 times higher than arsenic in GaAs-treated hamsters. Exposure to gallium arsenide results in systemic arsenic intoxication, although the gallium moiety may also produce toxicity (Hayes, 1988). Gallium alone is a metallic ion inducing many biomedical and toxicological effects in mammals (Domingo, 1994). It showed specific immunomodulatory activities in mammals (Bernstein, 1998). Actually, no levels of tolerance have been established for gallium and the extent of environmental pollution caused by this industry is not yet evaluated (Hayes, 1988). It is then useful to assess and predict the potential risk of gallium for aquatic ecosystems. Fish are important indicators of environmental contamination of water. Their physiological functions can be used as effective bioindicators of water quality (Fournier *et al.*, 2000). The integrity of the immune system is fundamental to good defence against a variety of pathogenic agents in the environment. Metal ions released into aquatic systems have been recognised as stressors causing immunosuppression in fish resulting in increased succeptibility to infectious diseases (Zelikoff, 1993). The immunological alterations observed in fish can be used as indicators of environmental metal exposure (Zelikoff, 1994). The present study reports the sublethal effects of gallium on some of immune parameters of the freshwater carp in term of ecotoxicological risk assessment.

MATERIALS AND METHODS

Common carp weighing 25 ± 3 g were purchased from a local hatchery (Ets "Au Vairon" - Reims, France) and were placed in aerated spring water (Aurele, Ardennes, France) with a photoperiod 12 h: 12 h. Fish were allowed to acclimatize for 3 weeks before the start of the experiment. In tanks, the water quality parameters (dissolved oxygen, temperature, pH,...) were determined daily until the end of the experiment according to Standard Methods for the Examination of Water and Wastewater (APHA, 1989). For experiments, gallium nitrate ($Ga(NO_3)_3$, Sigma, France) was dissolved in water and added in tank to the dilution water to produce the test concentration. For determination of gallium effects on immune related parameters, we chose a sublethal (5 mg/L) exposure levels of gallium. It was calculated as 5% of estimated 96 h LC_{50} value of gallium to carp (96 h LC_{50} = 95.6 ± 14.3 mg/L) (Betoulle *et al.*, 2002). For each treatment, twenty fish were stocked in 80 l tank. The acute intoxication experiment was run for 96 h under static conditions to examine the mechanisms underlying the acclimatation to the toxic effects of the metal. At different times (0, 24, 48 or 96 h), fish were sacrified and bled from the caudal vein using an heparinized syringe to isolate plasma. At the same time, phagocytes were isolated from head kidneys (Secombes, 1990). The Lowry *et al.* (1951) method was followed to measure the total plasma protein. The immunoglobulin levels was evaluated by separating immunoglobulins from the plasma by precipitation with polyethylenic glycol and a turbidimetric assay was used to determine plasma lysozyme activity with procedures of Anderson and Siwicki (1993). The plasma ceruloplasmin activity was measured as p-phenylenediamine oxidase activity, an assay based on the method described by Pelgrom (1995). The potential killing activity of head kidney phagocytes was evaluated with the chemiluminescence assay (Secombes, 1990) by estimating reactive oxygen species production of cells with a liquid scintillation biocounter (Packard, France). Data were collected in terms of counts per minute (cpm) corresponding to potential killing activity of phagocytes. For statistical analysis, differences in means were analyzed by one-way ANOVA in conjuntion with Dunnet's test using the software SigmaStat for MS Windows version 2.03.

RESULTS

The phagocyte killing activity of carp maintained in gallium-containing water was not significantly different of control maintained in normal springwater as observed in *table 1.* A same result was observed for plasma lysozyme activity which was not modulated in fish exposed to gallium in comparison with control *(table 3).* However, the total plasma immunoglobulin was significantly reduced in carp living in gallium-containing water for 48 or 96 hours *(table 2).* This result was associated with low plasma protein levels in comparison with control fish *(table 2).* The ceruloplasmin activity of gallium-exposed carp was modulated too but measured activities were significantly higher in plasma of carp maintained in gallium-containing water in comparison with control *(table 3).* No change in ceruloplasmin activity was observed for fish maintained in gallium-containing water for 24 hours. The immunological parameters were modulated in fish only after 48 hours of exposure duration.

Table 1. Killing activity of phagocytes isolated from gallium-exposed carp.
Values are means ± SE of maximal killing activity measured in cpm (n = 10). (*) For each exposure duration, significant differences from the respective control ($[Ga^{3+}] = 0$) were estimated at $p<0.05$

	Phagocyte killing activity (cpm)	
Time (hours)	**$[Ga^{3+}] = 0$ (control)**	**$[Ga^{3+}] = 5$ mg/L**
0	81555 ± 12846	88130 ± 3723
24	87775 ± 12766	84497 ± 12026
48	82840 ± 7497	75310 ± 14632
96	88220 ± 4599	74128 ± 9650

Table 2. Total plasma protein and immunoglobulin measured in gallium-exposed carp. Values are means ± SE of total plasma protein or immunoglobulin measured in mg/L (n = 10). (*) For each exposure duration, significant differences from the respective control ($[Ga^{3+}] = 0$) were estimated at $p<0.05$

	Total plasma protein (mg/L)	
Time (hours)	**$[Ga^{3+}] = 0$ (control)**	**$[Ga^{3+}] = 5$ mg/L**
0	27.3 ± 3.6	25.3 ± 3.2
24	25.2 ± 3.5	22.4 ± 2.8
48	26.3 ± 2.4	17.4 ± 0.9 (*)
96	28.0 ± 2.3	15.3 ± 1.8 (*)
	Total plasma immunoglobulin (mg/L)	
Time (hours)	**$[Ga^{3+}] = 0$ (control)**	**$[Ga^{3+}] = 5$ mg/L**
0	4.2 ± 1.3	3.2 ± 1.3
24	3.3 ± 0.9	2.9 ± 1.5
48	4.4 ± 1.1	1.05 ± 0.8 (*)
96	4.9 ± 1.4	0.9 ± 0.2 (*)

Table 3. Plasma lysozyme and ceruloplasmin activities measured in gallium-exposed carp. Values are means ± SE of plasma lysozyme activity measured in μg/mL and plasma ceruloplasmin activity measured in absorbance at 550 nm (n = 10). (*) For each exposure duration, significant differences from the respective control ($[Ga^{3+}] = 0$) were estimated at $p<0.05$.

	Plasma lysozyme activity (μg/mL)	
Timc (hours)	**$[Ga^{3+}] = 0$ (control)**	**$[Ga^{3+}] = 5$ mg/L**
0	0.5 ± 0.05	0.43 ± 0.03
24	0.47 ± 0.04	0.51 ± 0.07
48	0.49 ± 0.1	0.43 ± 0.08
96	0.55 ± 0.02	0.42 ± 0.1
	Plasma ceruloplasmin activity (A_{550}-azide blank)	
Time (hours)	**$[Ga^{3+}] = 0$ (control)**	**$[Ga^{3+}] = 5$ mg/L**
0	0.3 ± 0.02	0.33 ± 0.013
24	0.32 ± 0.021	0.37 ± 0.025
48	0.29 ± 0.04	0.46 ± 0.07 (*)
96	0.25 ± 0.015	0.55 ± 0.14 (*)

DISCUSSION

No data were available concerning effects of gallium on fish and especially on their immune system. Head kidney phagocytes have an important role in regulating nonspecific and specific immune responses in fish (Secombes and Fletcher, 1992). No modulation of their killing activity was observed for fish exposed to 5 mg/L gallium whatever the exposure duration might be *(table 1)*. However, in mammals, there are many examples of the immunomodulating action of gallium (especially cytotoxicity) particularly against tumorals cells (Bernstein, 1998). Gallium can modulate fish immune defence. As a matter of fact, we showed here that total plasma protein and immunoglobulin levels were significantly decreased in plasma of gallium-exposed carp for 48 or 96 hours. However, no effect was observed on plasmatic lysozyme activity. Thus, in case of an acute environmental contamination, high gallium levels in water can be then implicated as an immunosuppressive metal ion in fish as already observed for other metal ions (Zelikoff, 1993). But, in carp, gallium seems to have no impact on some important nonspecific immunological parameters such as phagocyte killing activity and plasma lysozyme activity whereas the specific immune response is modulated. This modulation was observed only after 48 h of exposure time. Gallium is then a low potential toxicant for carp in comparison with other classical metal ions. As a matter of fact, the mean LC_{50} of gallium for carp after 96 h exposure were determined as 95.6 ± 14.3 mg.l^{-1}. For comparison 96 h LC_{50} for Zn^{2+} and for Cu^{2+} were 25 mg.l^{-1} and 1 mg.l^{-1} respectively in cyprinidae (Leland and Kuwabara, 1985).

Gallium reduced total plasma immunoglobulin levels in carp from 48 h of exposure time whereas an increase in ceruloplasmin activity was observed at the same time. Ceruloplasmin is a part of the non cellular and nonspecific defence mechanisms of fish (Alexander and Ingran, 1992). This acute phase protein complexes with copper and other metal cations (Beisel, 1976). The protein acts as an oxidase and converts molecular oxygen into water and not hydrogen peroxide. It is also called ferroxidase because it oxidises ferrous iron to ferric which can then bind to transferrin. The defence role of ceruloplasmin appear to be more through its ferroxidase activity because in converting ferrous to ferric iron from the environment and therefore decreases its availability to microorganisms. Thus, in this case, the increase in ceruloplasmin levels in gallium-contaminated fish would improve the protection of fish against pathogens. But, the ceruloplasmin modulation observed in our study, is more certainly due to the ionic perturbations induced by expositions of fish to the toxic metal ion, associated with stressfull conditions. Moreover, changes in immune parameters observed are probably due to decrease fish health and not to the direct action of gallium upon the immune system. As other metal ions, gallium can modulate the fish mineral balance. Metal ions affect gill functions, thus inducing the loss of critical ions (Reid and Mc Donald, 1991). Cortisol, in addition to being a glucocorticoid, is a mineralocorticoid, therefore cortisol levels rise to balance ion loss in the gills (Richman and Zaugg, 1987) This rising cortisol then affects immune functions (Maule and Schreck, 1990). Cortisol can stimulate or inhibit such activity, depending on the dose and the exposure duration. The immunotoxic effects observed in gallium-exposed carp were probably due to the indirect effect of the chemical according to this endocrine hypothesis. But these effects can be due to a direct action of gallium on immune response as observed in mammals where gallium directly reduces some immune functions too (Bernstein, 1998). For example, at high concentrations, it directly inhibits macrophage activation and suppresses cytokine secretion by these cells (Makkonen *et al.*, 1995).

CONCLUSIONS

The acute toxicity test showed here that gallium is a relatively low toxic chemical for carp in comparison with other metal ions. But as an immunotoxicological point of view, the sublethal effects of gallium on some immune parameters, especially plasma immunoglobulin levels and ceruloplasmin activity, were significant and were able to modulate the immune response against pathogens (Anderson, 1990). This study conducted under short-term exposure conditions gave informations on the acute immunotoxic effects of gallium to carp. However, the effects of this metal ion on fish immunity might be important in case of chronic environmental contaminations and should be studied.

REFERENCES

Alexander J.B., Ingram G.A. Noncellular nonspecific defence mechanisms of fish. *Annual Review of Fish Diseases*. 1992, 249-279.

American Public Health Association (APHA) Standard Methods for the examination of Water and Wastewater (17th ed.). Washington, DC. 1989.

Anderson D.P. Immunological indicators: effects of environmental stress on immune protection and diseases outbreaks. *American Fish Society Symposium* 1990, 8, 38-50.

Anderson D.P., Siwicki A.K. Basic hematology and serology for fish health programs. In: Stolen JS, Fletcher TC, Anderson DP, Roberson BS, Van Muiswinkel WB (eds) *Techniques in fish immunology*, SOS publications, Fair Haven, 1993. p 113.

Beisel W.R. Trace elements in infectious processes. *Med.. Clin. North America*. 1976, 60, 831-849.

Bernstein L.R. Mechanisms for therapeutic activity for gallium. *Pharmacological Review* 1998, 50, 665-682.

Betoulle S., Etienne J.C., Vernet G. *In vivo* and *in vitro* modulation of carp (*Cyprinus carpio* L.) phagocyte oxidative burst activity by gallium. *Journal of Toxicology and Environmental Health* 2002, 65, 101-113.

Chepesiuk R. Where the chips fall: environmental health in the semiconductor industry. *Environment Health Perspectives* 1999, 107, 1-8.

Domingo J.L. Metal-induced developmental toxicity in mammals: a review. *Journal of Toxicology and Environmental Health* 1994, 42, 123-141.

Fournier M., Cyr D.G., Brousseau P., Tryphonas H. Biomarkers in immunotoxicology. In: Guillette L, Crain D (eds) *Environmental endocrine disruptors*, Taylor and Francis Publishers, New York, 2000, p. 182.

Fowler B.A., Dobrota M. Toxicity of semiconductor metal compounds. Robens Institute meeting on toxicity and therapeutics of newer metals and organometallic compounds. Guildford, Surrey, U.K. 1989.

Hayes R.L. Gallium. In: Seiler H, Sigel, A (eds) *Handbook of toxicity of inorganic compounds*, Marcel Dekker, New York, 1988, p. 297.

Kuroda K., Endo G., Okamoto A., Yoo Y.S., Horiguchi S. Genotoxicity of beryllium, gallium and antimony in short-term assays. *Mutat. Res.* 1991, 264, 163-170.

Leland H.V., Kuwabara J.S. Trace metals. In: Rand GM, Petrocelli R (eds *Fundamentals in aquatic toxicology*, Hemisphere Publishing Corporation, London. 1985, p. 374.

Lowry O.H., Rosebrough N.J., Farr A.L., Randall R. Protein measurements with the folin phenol reagent. *Journal of Biological Chemistry* 1951, 193, 265-275.

Makkonen N., Hirvonen M.R., Savolainen K., Lapinjovski S., Mönkönnen J. The effect of free gallium and gallium in liposomes cytokine and nitric acid secretion from macrophage like cells *in vitro. Inflammation Research* 1995, 44, 523-528.

Maule A.G., Schreck C.B. Changes in numbers of leucocytes in immune organs of juvenile coho salmon after acute stress or cortisol treatment. *Journal of Aquatic Animal Health* 1990, 2, 298-304.

Omura M., Hirata M., Tanaka A., Zhao M., Makita Y., Inoue N., Gotoh K., Ishinishi, N. Testicular toxicity evaluation of arsenic-containing binary compound semiconductors, gallium arsenide and indium arsenide, in hamsters. *Toxicology Letters* 1996, 89, 123-129.

Pelgrom S.M.G.J., Lock R.A.C., Balm P.H.M., Wendelaar Bonga S.E. Integrated physiological response of tilapia *(Oreochromis mossambicus)* to sublethal copper exposure. *Aquatic Toxicology* 1995, 32, 303-320.

Reid S.D., Mc Donald D.G. Metal binding activity of the gills of the rainbow trout *(Oncorhynchus mykiss). Canadian Journal of Fish Aquatic Sciences* 1991, 48, 1061-1068.

Richman N.H., Zaugg W.S. Effects of cortisol and growth hormone in osmoregulation in pre- and de- smoltified coho salmon *(Oncorhynchus kisutch). General and Comparative Endocrinology* 1987, 67, 194-201.

Secombes C.J. Isolation of salmonid macrophages and analysis of their killing activity. In: Stolen JS, Fletcher TC, Anderson DP, Roberson BS, Van Muiswinkel WB (eds) *Techniques in fish immunology*, SOS publications, Fair Haven, 1990, p. 137.

Secombes C.J., Fletcher T.C. The role of phagocytes in the protective mechanisms of fish. *Annual Review of Fish Diseases* 1992, 2, 53-71.
Zelikoff J.T. Metal pollution-induced immunomodulation in fish. *Annual Review of Fish Diseases* 1993, 3, 305-325.
Zelikoff J.T. Immunological alterations as indicators of environmental metal exposure. *Modulators of Fish Immune Responses* 1994, 1, 101-109.

*Metal Ions in Biology and Medicine: vol. 7.*Eds L. Khassanova, Ph. Collery, I. Maymard, Z. Khassanova, J.-C. Étienne.
John Libbey Eurotext, Paris © 2002 pp. 512-516.

Selenium (Se) status and inflammation markers in diseases influenced by air pollution

Ilona Kovács[1], Anna Stocker, Ibolya Sziklai-László[2], Nóra Adányi[3] Mária Ágnes Cser[1]

[1]Bethesda Children's Hospital, 1146 Budapest, Bethesda St.3., [2]KFKI Atomic Energy Research Institute, 1525 Budapest 114. P.O.B.49, [3]Central Food Research Institute, 1537 Budapest, P.O.B.393

ABSTRACT

Background: the incidence of allergic and respiratory diseases are increasing not only in Europe, but all over the world. The prevalenece of asthma bronchiale remarkably increased in Hungary similarly to European, overseas as well as to developing countries. Hungarian soil is low in Se and Se intake is lower than the Recommended Daily allowance.

Our aim was to investigate the possible associations between blood Se status, clinical symptomes and inflammation markers in diseases strongly influenced by air pollution.

Patients: asthmatic children (n=64), and adults (n=23), and children with allergic rhinitis and or atopic dermatitis (n=40) were investigated. Se and immune parameters were compared to healthy, age and sex mached controls (n1=81, n2=46, n3=27 respectively).

Method: Se concentrations in erythrocytes, whole blood and plasma were determined by atomic absorption spectrophotometry (AAS). Granulocyte macrophage colonic stimulating factor (GM-CSF), intracellular adhesion molecules (ICAM 1) by ELISA.

Results: patients with chronic rhinitis and atopic dermatitis had plasma and whole blood Se levels similar to the healthy children, but Se in their erythrocytes decreased. Asthmatic children had less erythrocyte Se ($p<0.001$) and in adult asthmatics Se content significantly decreased in all blood compartments ($p<0.001$). Total IgE increased in asthmatics irrespective of the severity of symptoms. 20 fold increase of IgE has been observed in status asthmaticus. GM-CSF levels increased in chronic rhinitis and asthma in active phases of the diseases. ICAM-1 concentration was higher in rhinitis and asthma patients but those with severe viral infections had a 300% increase compared to values of healthy children.

In conclusion we could say, that in atopic diseases the blood Se parameters associated with the severity of clinical symptomes and with an altered synthesis of immune proteins, inflammation markers.

Key words: selenium, serum GM- CSF, serum ICAM-1, IgE, asthma, allergic rhinitis, atopic dermatitis, children, adults.

This work was supported by the grant Ministry Public Health 031/2000 and partly supported by KFKI-CMRC (No.ICA1-2000-70029).

INTRODUCTION

Over the last few years a vast amount of new information accumulated on the field of atopic diseases especially in allergy and asthma [1, 2]. Allergens reaching the respiratory mucosa, taken by B cells, after transformation by the TH2 cells. Lymphokines produced and activate the antigen specific B cells. As a result immunoglobuline isotype change occurs, inducing IgE production.

The granulocyte macrophage stimulating factor (GM-CSF) induces eosinophil proliferation and histamin release. Intracellular adhesion molecule-1 (ICAM-1) helps to establish contact between the antigen-presenting B cells and T-lymphocytes, and its serum level is increased in allergic diseases [3]. It has been suggested, that ICAM-1 could be the major human rhinovirus receptor [4]. The inflammatory cells and cytokines cause airway damage that perpetuate the inflammatory process [5]. It is well known, that Selenium (Se) plays an important role in immune functions[6, 7]. In asthma the inflammation procedure increases the reactive oxygen species production [8, 9] and association has been found between the increased ROS and decreased Se status. The incidence of allergic and respiratory diseases are increasing not only in Europe, but all over the world [10]. The prevalence of asthma remarkably increased in Hungary [11] similarly to other European and overseas [12] as well as in developing countries [13]. The aim of the study was to investigate the possible associations between clinical symptomes, inflammation markers and Se status in diseases influenced by allergens.

PATIENTS

Se was measured in asthmatic, 2-18 years old children (n=64) and adults (n=23) and in children with allergic rhinitis and or atopic dermatitis (n=4o) Immune parameters were determined in 37 symptome free asthmatics in 67 asthmatics (stageII:asthma attacks severeal times per week-stageIII: symptomes occure daily), in allergic rhinitis (n=19) and atopic dermatitis (n=12) patients compared to healthy, age and sex matched controls. Protocols were approved by the Hospital Research and Ethical Committee, informed consent was obtained from the parents of the children and of each child. Blood was taken in fasting state, plasma and serum alliquots kept at -20°C until measurements.

METHODS

Se was measured by AAS [14] Total IgE levels were determined by ELISA (Abbott, Hungary Kft.). GM-CSF and ICAM-1 were measured by human ELISA (Endogen Inc. Woburn, Ma,. USA). Coefficients of variations were below 8%. Data analysis: SPSS™, p values<05 were considered statistically significant.

RESULTS

Plasma Se respresenting the acute Se supply was similar in all children *(Fig. 1.)*, but erythrocyte

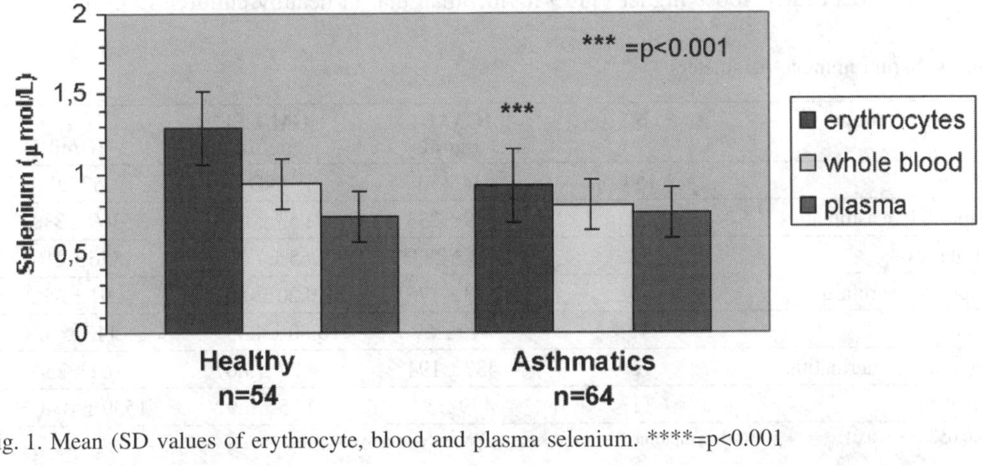

Fig. 1. Mean (SD values of erythrocyte, blood and plasma selenium. ****=p<0.001

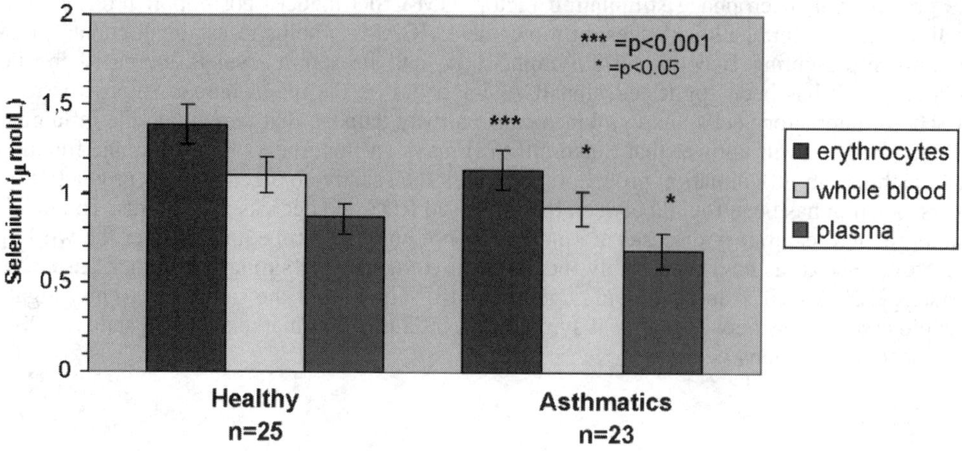

Fig. 2. Mean(SD values of erythrocyte, blood, plasma selenium,*=p<0.05,**=p<0.01, ****=p<0.001

Se content representing the long term Se supply of the individuum was less in asthmatics compared to the healthy group of children *(Fig. 1)*. However in adult asthmatics all blood Se parameters were lower, than in the healthy population *(Fig. 2)*.

There was a negativ correlation between the erythrocyte Se levels and the severity of asthma (r = 0.75) or the duration of asthma (r = 0.62). Total IgE levels were higher in all groups of patients compared to healthy controls. Total IgE levels were highest in children with atopic dermatitis. At the time of exacerbation, presenting severe clinical symptomes, their IgE values were 16 times higher than that of healthy controls and seven times higher than that in atopic patients in a silent, symptom free periode. There was a strong, positive correlation between the magnitude of serum IgE level and the severity of symptomes both in asthmatics and others with dermatitis. (r = 0.87). Serum GM-CSF levels were not detectable in healthy children, none of their values reached 1 pg/ml. In contrast levels were increased in five groups of patient except in children with simple rhinitis. These cytokine levels were similarly high in acute phases of asthma and rhinitis, reaching individual levels as high as 12-15 pg/ml. In atopic dermatitis GM-CSF levels were significantly higher than that of healthy children, but lower than in the the acutely ill patients *(table 1)*. Serum ICAM-1 levels were not different in healthy or ill patients regardless of their clinical status and symptomes. However in a group of children with acute RS viral infection ICAM-1 dramatically increased, it was nearly 400% higher (1129 ng/ml) than that in healthy children (345 ng/ml).

Table 1. Serum immune parameters

Group	N	ICAM-1 ng/ml	GM-CSF pg/ml	IgE IU/ml
Healthy	154	345(50	ND	52(20
Symptom free asthmatics	29	549 ± 238	2.53 ± 1.43	369 ± 338
Asthmatics	52	623 ± 283*	11.65 ± 7.23***	856 ± 418
Symptom free rhinitis	15	443 ± 175	0.20 ± 0.01	91 ± 35
Rhinitis	31	479 ± 59	10.80 ± 4.33***	113 ± 46
Symptom free dermatitis	16	487 ± 194	4.27 ± 3.07	167 ± 250
Dermatitis	11	479 ± 59	7.75 ± 0.92	1539 ± 1040**

*=p<0.05, **=p<0.01, ****=p<0.001 symptoms free compared to patients with symptoms

DISCUSSION

Our data are in agreement with others, who found reduced Se status in asthmatic patients [15, 16, 17, 18]. These were observed in different countries, like U.K. with a medium Se supply [16, 18] and New Zelaland [15] known of extreme low Se supply and blood Se status. Decreased glutathione peroxidase activity indicating low Se status was observed in asthmatic patients [19], and decreased dietary Se intake associated with bronchial asthma [20]. Hungary has low soil Se content and even the healthy population has lower Se status than that in many other European countries [21]. Se intake in Hungary is lower, than the RDA given by WHO [22]. The fact, that adult asthmatics had a low Se status in both short and long terms in contrast to young asthmatics, who had low erythrocyte Se content indicates, that Se status is strongly associated with the outcome, severity of their disease. It has been accepted, that the antibody immunglobulin E is responsible for allergic reaction and IgE production is different to healthy the atopic patients and asthmatics have raised values [23]. In healthy individuals IgE production is generally suppressed unless they have a viral infection [24], therefore our healthy children were carefully selected to be suitable controls, which could be proved by the less than 60 IU/ml IgE levels. The association found between the increased IgE production and the severity of asthma was in accordance with findings in young adults [25]. The association between the increased IgE production and severity of atopic dermatitis was not unexpected, our observations are in agreement with result of Croatian authors [26]. *In vitro* sensitized mononuclear cells from asthmatic children induced higher levels of eosinophil chemotaxins such as IL-3, IL-5 and GM-CSF [27]. Hashimoto and his coworkers [28] reported on elevated ICAM-1 levels in sera from adult asthmatics, they observed that ICAM-1 concentrations were higher during asthma attacks than those in stable conditions. Similarly increased ICAM-1 values were observed in childhood asthma both of serum [3] and bronchoalveolar lavage fluid [29], in atopic dermatitis [30] and allergic rhinitis [31, 32]. Selenium supplementation reduced ICAM-1 production *in vitro* epithelial cell culture [33] suggesting a Se can modulate the immunological mechanism of asthma. In conclusion we have found that the the monitoring of cytokines were useful markers in the evaluation of activity of atopic diseases. The high ICAM-1 expression during infection suggest that the prevention of virus infection could diminish the exacerbations of the disease. The correlations between Se status and the severity of asthma and immunglobulin synthesis indicate, that Se supplementation could improve the life of asthmatics

REFERENCES

1. Leung, DYM Role of inflammation in allergic diseases Pediatric Research, 1997, 42, 559-568.
2. Holgate ST, Church M.K.: Allergy in Allergy Ed. Holgate S.T. 1993, Raven Press, New York.
3. Carlsen KH: Markers of airway inflammation in preschool wheezers. Arch Chest Dis 1997:52:455-460 4.
4. Bianco A, Sethi SK, Wilton S, Spiteri M Factors modifying host susceptibility of the airways to rhinovirus infection. Clin Exp Immunol 2000, 10, 112-115.
5. Björnsdottir U.S., Cypcar D.M.: Asthma:an inflammatory mediator sup. Allergy 1999, 54, 55-61.
6. Kiremidijan-Schumacher L., Stoczky G.: Se and immune responses. Envir.Res. 1987,42,277-303.
7. Spallholz J.E.:Selenium and glutathione peroxidase:essential nutrient and antioxidant component of the immune system. Adv.Exp.Med.Biol. 1990, 62, 145-158.
8. Flohé L, Günzler A, Loschen G. The glutathione peroxidase reaction: A key to understand the Se requirements of mammals. In: Kharasch N ed. Trace metals in health and disease. NY Raven Press, 1979:263-85.
9. Burk RF. Biological activity of selenium. Ann Rev Nutr. 1983, 3, 53-70.
10. Gergen PJ, Weiss KB Changing patterns of asthma hospitalization among children JAMA 1990, 264, 1688-1692.
11. Kovács I, Cser MÁ Epidemiology of allergic diseases of Hungarian children aged 6 to 14 years, Pediatric Research 1998,44, 456.

12. Knorr B, Matz J, Bernstein JA, Nguyen Ha, Seidenberg BC, Reiss TF, Becker A Montelucast for chronic asthma in 6- to 14 year-old children. JAMA, 1998,279,1181-1186.
13. Cerveri I, Locatelli F, Zoia MC, Accordini S, de Marco R: International variations in asthma treatment. Eur Respir J 1999,14, 288-294.
14. Adányi N., Sziklai-László I., Váradi M., Snyder P., Snyder R.D., Cser M.Á.: Determination of Se balance in healthy children by AAS-hydride generation and INAA technique. Acta Alim. 2002 in press.
15. Flatt A., Pearse N., Thomson C.D., Sears M.R., Robinson M.F., Beasley R.: Reduced selenium in astmatic subjects in New Zealand. Thorax 1990, 45, 95-99.
16. Stone J, Hinks LJ, Beasley R, Holgate ST, Clayton BA. Reduced selenium status of patients with asthma. Clin.Sci 1989,77,495-500.
17. Pearson DJ, Suartz-Mendez VJ, Day JP, Miller PF. Selenium status in relation to reduced glutathione peroxidase activity in aspirin-sensitive asthma. Clin.Exp.Allergy 1991,21,203-208.
18. Kadrabova J., Madaric A., Lovacikova Z.-, Podivinsky F., Ginter E., Gazdik F.:Selenium status is decreased in patients with intrinsic asthma Biol.Trace Elem.Res. 1996,52, 241-248.
19. Misso NL, Powers KA, Gillon RL, Stewart GA, Thompson PJ. Reduced platelet glutathione peroxidase activity and serum Se concentration in atopic asthmatic patients. Clin.Exp.Allergy 1996; 26,838-847.
20. Picado C., R. Deulofeu, R. Leonart, M. Agusti, J. Mullol, L. Quinto, M. Torra Dietary micronutrients/antioxidants and their relationship with bronchial asthma severity Allergy 2001,56,43-49.
21. Cser M.Á., Sziklai-László I., Menzel H., Lombeck I.:Selenium and glutathione peroxcidase activity in Hungarian children. J.Trace Elem.Med.Biol. 1996,10,163-173.
22. Cser M.Á., Sziklai-László I., Adányi N., Snyder P., Snyder R.D.: Selenium balance in healthy American and Hungarian children living in Budapest, Hungary. In Metal Ions in Biology and Medicine,Vol.6., eds: Centeno J.A.,et a. John Libbey Eurotext Paris, 2000, 248-250.
23. Johansson SGO, Berglund A, Kjelmann NIM Comparison of IgE values as determined by different solid plate radioimmunoassay methods Clin Allergy 1976, 6, 91-98.
24. Frick OJ Effect of respiratory and other virus infection on IgE immunoregulation J Allergy Clin Immunol 1986,5, 1013-1018.
25. Peat JK, Toelle BG, Dermand J,van den berg R, Britton WJ, Woolcock AJ Serum IgE levels, atopy and asthma in young adults: results from a longitudinal cohort study Allergy 1996,51,804-810.
26. Lugovic L, Lipozencic J Are respiratory allergic diseases related to atopic dermatitis Coll Antropol 2000,24, 335-345.
27. Lantero S, Sacco O, Scala C, Rossi GA Stimulation of blood mononuclear cells of atopic children with the relevant allergen induces the release of eosinophil chemotaxins such as IL-3, IL-5 and GM-CSF J Asthma 1997,34, 141-152.
28. Hashimoto S, Imai K, Kobayashi T Elevated levels of soluble ICAM-1 in sera from patients with bronchial asthma Allergy 1993,48, 370-372.
29. Marguet C, Dean TP, Warner JO Soluble intracellular adhesion molecule-1 (sICAM-1) and interferon gamma in bronchoalveolar lavage fluid from children with airway diseases Am J Respir Crit Care Med 2000,162, 1016-1022.
30. Halmerbauer G, Frischer T, Koller DY Monitoring of diseases activity by measurement of inflammatory markers in atopic dermatitis. Allergy 1997,52,765-769.
31. Ciprandi G, Tosca M, Ricca V. Cetirizine treatment of rhinitis in children with pollen allergy, evidence of its antiallergic activity Clin Exp Allergy 1997, 27, 1160-1166.
32. Fasce L, Ciprandi G, Pronzato C, Cozzani S, Tosca MA, Grimaldi I, Canonica GW Cetirizine reduces ICAM-1 on epithelial cells during nasal minimal persistent inflammation in asymptomatic children with mite-allergic asthma Int Arc Allergy Immunol 1996,109, 272- 276.
33. Horvathova M., Jahnova E., Gazdik F.: Effect of selenium supplementation in asthmatic subjects on the expression of endothelial cell adhesion molecules on culture. Biol.Trace Elem. Res. 1999, 69, 15-26.

*Metal Ions in Biology and Medicine: vol. 7.*Eds L. Khassanova, Ph. Collery, I. Maymard, Z. Khassanova, J.-C. Étienne.
John Libbey Eurotext, Paris © 2002 pp. 517-522.

Lactoferrin and albumin protect ceruloplasmin against Cu-mediated degradation induced by H_2O_2

K.V.Solovyov, M.O. Pulina, A.V. Sokolov, E.T. Zakharova, M.M. Shavlovski and V.B. Vasilyev

Institute for Experimental Medicine, 12 Pavlov street, Saint-Petersburg 197376, Russia

Address for correspondence: Vadim B. Vasilyev, Department of Molecular Genetics, Institute for Experimental Medicine, 12 Pavlov St., Saint Petersburg 197376 Russia

Abbreviations: CP - ceruloplasmin (EC 1.16.3.1); LF - lactoferrin; LYZ - lysozyme from chicken egg white (EC 3.2.1.17); HSA - human serum albumin; DMSO - dimethyl sulfoxide; PAG - polyacrylamide gel.

ABSTRACT

It is known that EDTA inhibits pro-oxidative effect of copper ions of ceruloplasmin (CP) exerted as the latter is peroxidized. We previously demonstrated that CP is able to form a specific complex with lactoferrin (LF). LF is known as a natural chelator, so we studied its ability to prevent CP peroxidation by scavenging the copper ions released from CP. Degradation of peroxidized CP molecule was followed by electrophoresis in PAG and by changes of the oxidase activity of the enzyme. Apo-LF more efficiently protects CP against peroxidative destruction as compared to EDTA. Like LF, serum albumin no less efficiently protects CP from peroxidation, which may be explained by its ability to scavenge copper ions. In contrast, lyzozyme has no protective effect, as it is unable to bind copper ions. When pH dropped to 5.5, both LF and albumin lost their protective effect. We suggest that LF that is normally secreted in its apo-form can serve as a natural chelator of highly pro-oxidant copper ions, thus preventing the deleterious reactions involving oxygen radical species. Protective function of LF and albumin in the foci of inflammation where pH is low must be questioned in view of the results obtained.

INTRODUCTION

We have shown previously [1] that copper protein CP selectively interacts with an iron protein LF both *in vitro* and *in vivo*. CP/LF complex is found in various biological fluids. Some physiological meaning may underlie this interaction of these two proteins. CP is an acute-phase reactant, whose concentration in plasma (ca. 300 μg/ml) noticeably increases upon inflammation [2]. Normally LF is found in serum at low concentrations (ca. 1 μg/ml), the latter increasing seriously upon activation of neutrophils (up to 200 μg/ml in case of sepsis) [3]. Hence, in case of acute-phase reactions concentrations of both proteins in plasma increases.

It was shown that CP is destroyed in the presence of peroxinitrite [4] and when affected by H_2O_2 [5]. Copper ions that are released become pro-oxidant interacting with hydrogen peroxide in reactions of Fenton chemistry, generating reactive hydroxyl radicals [6, 7].

We were interested to learn whether LF predominantly synthesized as apo-protein [8], is able to bind copper ions, thus inhibiting the reactions of peroxidation.

MATERIALS AND METHODS

Reagents were purchased from Merck, Pharmacia, Sigma, Serva and Reanal.

CP was purified as described [9]. Apo-LF was purified from breast milk according to Zakharova *et al.* [1].

Reactions with CP were carried out in PBS (10 mM phosphate buffers containing 0.15 M NaCl) pH 7.4-5.5.

All solutions were run through Chelex-100 to eliminate traces of heavy metals.

CP was oxidized in 1 ml of buffer solution, containing 2.7 μM CP. H_2O_2 was added to final concentrations in the range 0.1-15 mM. Five hours of incubation at 37°C under stirring followed. The remaining hydrogen peroxide was neutralized by adding 10 μg of catalase. This oxidative mixture was used with different amounts of LF and of other substances to evaluate the protective effect of the latter. Upon the end of oxidation CP was sampled and its denaturing was visualized by SDS-containing [10] and SDS-free PAG electrophoresis [11]. Relative oxidase activity in the samples of CP was also measured in reaction with p-phenylene diamine according to Ravin [12].

Copper incorporation into LF was checked spectroscopically using Beckman DU-50 spectrophotometer.

Experiments were done in triplicate; standard errors are shown.

RESULTS

Our preliminary results demonstrated that 5 hr of incubation of purified CP with H_2O_2 result in decrease of the enzyme's activity and its degradation *(fig. 1)*. Gradual increase of hydrogen peroxide concentration is accompanied by destruction of the protein moiety of CP as evidenced by disappearance of the protein band in SDS-free PAG electrophoresis *(fig. 1,A)*. Concomitantly CP loses its oxidase activity *(fig. 1,B)*. Five hours of incubation of CP with 10 mM H_2O_2 causes the drop of its oxidase activity to zero *(fig. 2, B)*. Enzymatic activity of CP is mainly lost within the first three hours of incubation, its decrease being linearly time-dependent. SDS-free PAG electrophoresis showed time-dependent fading of CP band *(fig. 2,A)*.

In all experiments that followed, H_2O_2 concentration was 10 mM and incubation time was 5 hours. Under such conditions neither 100 mM mannitol, nor 2% DMSO had any protective effect on CP, as judged by the results of PAG electrophoresis. At the same time 20 mM histidine added to the reaction mixture preserved 75% and Chelex-100 - 100% of CP oxidase activity. EDTA in concentrations up to 10 mM preserved not more than 20% of oxidase activity of CP. No further increase of EDTA content improved the protection (data not shown).

Upon adding LF to the reaction mixture (CP + H_2O_2, pH=7.4) destruction of CP became less pronounced *(fig. 3, 4)*. After 5 hr CP preserved 80% of its oxidase activity. HSA has similar effect, while LYZ did not protect CP at all.

Effect of LF and HSA was most pronounced at pH=7.4. At pH=5.5 the proteins did not protect CP. Upon decrease of pH the protective effect of these proteins weakens. A decrease of pH by one unit (pH 7.4 - pH 6.5) resulted in a drop of protective effect of LF by 50% *(fig. 5)*.

LF and HSA themselves are not destroyed upon incubation with 10 mM H_2O_2, as judged by PAG electrophoresis (data not shown).

Incubation of CP and LF under oxidative conditions (pH - 7.4) resulted in loss by CP its absorption at 610 nm (type I copper) with concomitant appearance of absorption band at 435 nm ascribed to Cu-LF [13] (data not shown).

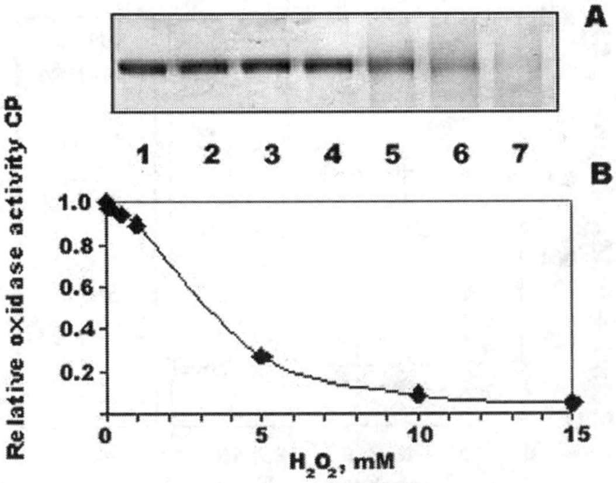

Fig. 1. CP degradation dependent on hydrogen peroxide concentration.
Panel A - SDS-free PAG electrophoresis. 3 μM of CP was loaded on each lane.
Lanes: 1 - CP; 2 - CP + 0.1 mM H_2O_2; 3 - CP + 0.5 mM H_2O_2; 4 - CP + 1 mM H_2O_2; 5 - CP + 5 mM H_2O_2; 6 - CP + 10 mM H_2O_2; 7 - CP + 15 mM H_2O_2
Panel B - Dependence of CP relative oxidase activity (ordinate) upon H_2O_2 concentration (abscissa)

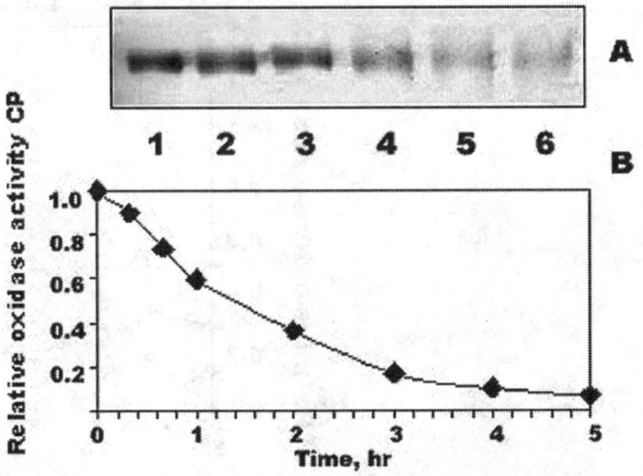

Fig. 2. CP degradation in the presence of H_2O_2 dependent on incubation time.
Panel A - SDS-free PAG electrophoresis. 3 μM of CP was loaded on each lane. All samples, except for number 1, contained 10 mM H_2O_2.
Lanes: 1 - CP; 2 - 1 hr; 3 - 2 hr; 4 - 3 hr; 5 - 4hr; 6 - 5hr.
Panel B - Dependence of CP relative oxidase activity (ordinate) upon the time of incubation with H_2O_2 (abscissa)

DISCUSSION

Our results showing that CP can get destroyed by hydrogen peroxide are in line with the data of another group [5]. These authors suggest that copper ions released from CP favour the protein destruction. Our data also show that copper ions released at the initial stages of oxidation trigger the destruction of CP. The absence of protection of CP by mannitol and DMSO may be explained by their inability to bind copper ions and to remove them from the metal-binding sites in CP, even when the binding is relaxed, which is also corroborated by the concept of site-specific nature of Fenton chemistry, when relatively large molecules are unable to reach the metal-binding sites

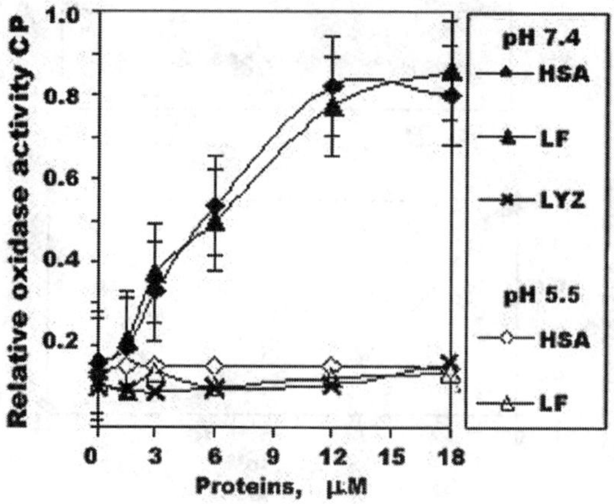

Fig. 3. Effect of pH upon the capacity of LF, HSA and LYZ to protect CP from destruction by H_2O_2. Relative oxidase activity is plotted by ordinate; protein concentrations are plotted by abscissa

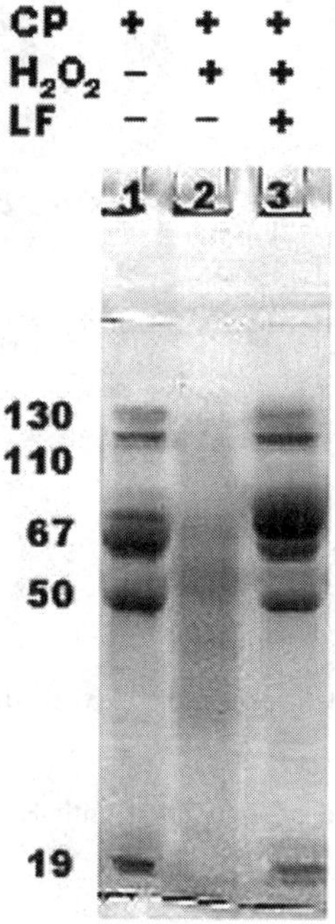

Fig. 4. SDS-PAG electrophoresis of CP samples affected by H_2O_2 in the presence and in the absence of LF.
1 - CP; 2 - CP + H_2O_2; 3 - CP + H_2O_2 + LF

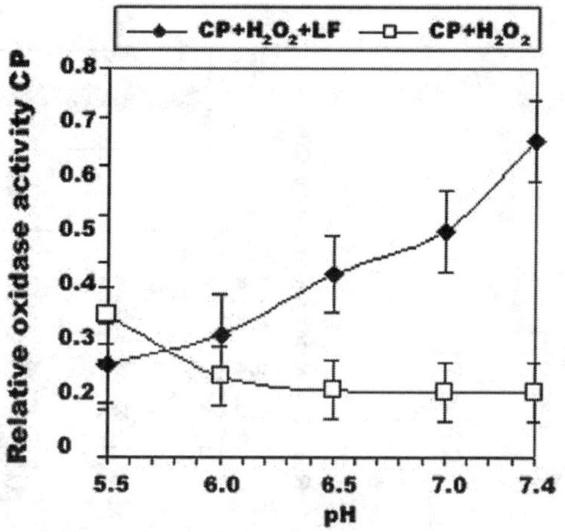

Fig. 5. Effect of pH upon the capacity of LF to protect CP against peroxidation.
Molar ratio LF: CP is 4:1. Control experiment (plotted in blanc squares) is done with the mixture of CP and H_2O_2. Black squares are showing the level of oxidase activity in the presence of LF

where reactive oxygen species originate [7]. Moderate protective effect of EDTA (20%) can be explained by the fact that this chelator does not prevent penetration of water into the coordination sphere of metal ions. Consequently, redox reactions are still possible upon chelation of a metal ion by EDTA [14]. In contrast, histidine chelates copper at the same time scavenging reactive oxygen species [6]. This explains its higher efficiency (75%) in protection of CP against oxidative destruction.

All chelators in our study could protect CP against H_2O_2. The most efficient protector was Chelex-100 that binds all the copper ions in solution. This is an argument in favour of an important role of these ions in degradation of CP.

Proteins that are able to bind copper ions (*e.g.* LF and HSA) also protect CP against destruction with hydrogen peroxide.

Decrease of pH to 5.5 diminishes the protective effect of both LF and HSA.

This can be explained by facilitated release of CP copper at low pH i.e. more ions enter the reactions of Fenton chemistry, the products of which damage CP.

We suggest that inhibition of oxidation of CP by H_2O_2 in our experiments resulted from inactivation of copper ions by metal-binding agents. Within CP/LF complex LF seems to have an advantage as a chelator of copper due to its proximity to the sites in CP molecule from which copper ions are released. Therefore, formation of CP/LF complex is likely to be physiologically expedient, since it allows for efficient scavenging of pro-oxidant copper ions that may originate in the foci of inflammation upon CP destruction under conditions of oxidative stress.

CONCLUSION

The ability of LF to form a complex with CP may have physiological implication, since it facilitates the scavenging of pro-oxidant copper ions released in the foci of inflammation from CP where stimulated leukocytes generate hydrogen peroxide and other reactive oxygen species. Low pH decreases the capacity of LF to protect CP.

AKNOWLEDGEMENTS.

This work was partly supported by RFBR grants: 00-15-97931; 00-04-49001; 00-04-06146 and 02-04-49698.

REFERENCES

1. Zakharova E.T., Shavlovski M.M., Bass M.G., Gridasova A.A., Pulina M.O., De Filippis V., Beltramini M., Di Muro P., Salvato B., Fontana A., Vasilyev V.B. and Gaitskhoki V.S. Interaction of Lactoferrin with Ceruloplasmin. *Arch. Biochem. Biophys.*, 2000, 374, 222-228.
2. Ryden L. Ceruloplasmin. In: Lontie, R., ed. Copper proteins and copper enzymes, volume III. Boca Raton, FL: CRC Press, 1984, 37-100.
3. Vorland L. Lactoferrin: A multifunctional glycoprotein. *APMIS*, 1999, 107, 971-81.
4. Swan J.A., Darley-Usmar V., Gutteridge J.M.C. Peroxynitrite releases copper from caeruloplasmin: implications for atherosclerosis. *FEBS Letters*, 1994, 342, 49-52.
5. Choi S.Y., Kwon H. Y., Kwon O.B., Eum W. S., Kang J.H. Fragmentation of human ceruloplasmin induced by hydrogen peroxide. *Biochimie*, 2000, 82, 175-180.
6. Yankovski O. Yu. Oxygen toxicity and biological systems. 2000, IGRA Publ., Saint-Petersburg.
7. Halliwell B. and Gutteridge J.M.C. Role of Free Radicals and Catalytic Metal Ions in Human Desease: An Overview. *Methods in enzymology*, 1999, 186, 1-86.
8. Iyer S. and Lonnerdal B. Lactoferrin, lactoferrin receptor and iron metabolizm. *Eur. J.Clin. Nutr.*, 1993, 47, 232-41.
9. Prozorovski V.N., Rashkovetski L.G., Vasiliev V.B., Shavlovski M.M., Neifakh S.A. Evidence that human ceruloplasmin molecule consists of gomologous parts. *Int. J. Pept. Prot. Res.*, 1982, 19, 40-53.

10. Laemmli U.K. Cleavage of structural proteins during the assembly of the head of bacteriophage T_4. *Nature*, 1970, 227, 680-685.
11. Davis D. J. Disc electrophoresis. Method and application to human serum proteins. *Ann.N.Y. Acad. Sci.*, 1964, 121, 404-427.
12. Ravin H.A. Rapid test for hepatolenticular degeneration. *Lancet*, 1956, 1, 6925, 726-727.
13. Ainscough E.W., Brodie A.M., McLachlan S.J., Ritchie V.S. Spectroscopic studies on copper (II) complexes of human lactoferrin. *J. Inorg. Biochem.*, 1983, 18, 103-112.
14. Klebanoff S. and Waltersdorf A.M. Prooxidant Activity of Transferrin and Lactoferrin. *J.Exp.Med.*, 1990, 172, 1293-1303.

*Metal Ions in Biology and Medicine: vol. 7.*Eds L. Khassanova, Ph. Collery, I. Maymard, Z. Khassanova, J.-C. Étienne.
John Libbey Eurotext, Paris © 2002 pp. 523-527.

Selenium, zinc and magnesium - essential components of the antioxidant system

Hlúbik, P. - Opltová, L. - Chaloupka, J. - Vejvodová, M.

Purkyne Military Medical Academy, Třebešská 1575, 500 01 Hradec Králové, Czech Republic

The Czech Republic ranks among the countries with highest prevalence of dyslipoproteinaemia and cardiovascular diseases (CVD). As a valid tool in the primary prevention of the CVD the authors consider the monitoring of the selected components (including metal ions and vitamins) of antioxidant system in the organism.

The study was carried out focused on monitoring of the health state in personnel of the Czech Republic Rescue Fire Brigade. In the project there were involved 933 healthy volunteers. Selected anthropometrical and biochemical parameters (including Se, Zn and Mg serum concentrations) were observed.

The results contribute to information about normal values of serum magnesium, zinc and selenium concentrations in Czech population. Mean serum concentration of Mg was 0,816 (0,057mmol/l, that of Zn was 18,25(2,54µmol/l and mean Se serum concentration was 0,802(0,14µmol/l. The distribution of Mg, Zn and Se serum levels in examined population group approached normal distribution.

In the monitored group, the average serum concentrations of Mg and Se showed a tendency towards lower values of so-called physiological range, however, they corresponded with other findings within the Czech population. The concentrations of Zn corresponded with commonly used normal values. The concept of the study made it possible to reveal relations between the serum Mg, Zn, Se levels, and the age or biochemical and anthropometrical parameters which are generally used as risk indices of cardiovascular disease. No statistically significant relations among the age of examined subjects and their serum Mg, Zn and Se concentrations were proved. Statistically significant negative correlations were revealed for relations among Zn serum concentration and cholesterolemia ($p \leq 0,05$) and triacylglycerolemia ($p \leq 0,01$). High incidence of the obesity and overweight in examined population group prompted to evaluation of Mg, Zn and Se saturation in subjects classified to 4 BMI (body mass index) categories. Mg concentrations showed the tendency to decrease with higher BMI, while by Zn and Se serum concentrations the tendency to increase in higher BMI categories was found. No statistical significance of these changes was proved.

Knowledge of normal levels of concentrations of selenium (Se), zinc (Zn) and magnesium (Mg) in the population serves, among others, in design of regulations concerning health protection, determination of exposure limits and prevention of diseases caused by deficiency of these elements.

Selenium is one of the essential trace elements, its intake in insufficient quantities or none can lead to the development of deficiency symptoms. The role of selenium in human nutrition remained unclear until the late 1970s. This situation changed dramatically when studies conducted in China received worldwide attention, reporting that severe nutritional Se deficiency in discrete regions of China was associated with an endemic juvenile cardiomyopathy (Keshan disease). Further observations implicating Se as a factor in the etiology of chondrodystrophic disease of children have provided the firm evidence of an importatnt role of Se in human health.

Investigators in several countries have assessed the nutritional Se status of their resident populations on the basis of tissue concentrations and/or activities of glutathion peroxidase (SeGSHpx). These results demonstrate a strong geographic variation in Se contents of food supplies. Populations with lowest estimated intakes of Se (among them CzR) also have the lowest concentrations of Se in the accessible tissues.

Two above mentioned human diseases were presented as being associated with severe nutritional Se-deficiency, but several investigators have assessed the Se status of patients with a variety of diseases to determine whether altered Se nutrition and/or metabolism may be involved in these diseases, either as contributing agent or a consequence. Altered Se status has been found consistently in association with relatively few human diseases: alcoholic liver disease, cirrhosis, acrodermatitis enteropathica, kwashiorkor, chronic renal failure and multiple sclerosis. In Finland [7] they were able to correlate the level of serum Se and the frequency of cardiovascular illness and myocardial infarction: low serum Se levels were associated with increased risk to cardiovascular disease. Se seems also to play a role in protecting against certain types of cancer, a lower incidence for cancer is found in areas with high Se content.

Groups of residents having a very low nutritional selenium intake may be expected to occur in Czech population. The retrospective study of concentration levels of Se in 1433 serum samples of the Czech population was performed in time period 1970-1999 [2]. Levels of Se in the interval 1970-1985 were equal (58,1-54,4ug/l).In the interval 1986-1990 the level of Se sharp declined (38,3 ug/l) and since 1990 levels of Se in serum increased again, up to 67,1 ug/l in 1999. During 1996-1998 the concentration of Se in whole blood of 1216 blood donors, average age 33 years, was assessed [1], the median value for adults was 76 ug/l. To investigate the Se status of subpopulation in nordeastern Bohemia [6] 253 serum samples from randomly selected volunteers between 6 and 65 years of age have been analyzed for Se concentrations (mean 55 ± 11 ug/l). Se serum levels in adults proved maximal frequencies between 55-70 ug/l and in children in the range 45-55 ug/l. The results confirmed mild to severe Se deficiency in the population of the region. The same author [5] presented in 1999 the evaluation of Se serum concentrations in the Czech population *(table 1)*. The frequency of deficient Se levels was higher than 65%, the optimal serum concentrations were found in 0,7% inhabitants only.

Table 1. Evaluation of serum Se concentrations in inhabitants of the Czech Republic

evaluation	Se in serum [ug/l]	frequency [%]
pharmacol. Level	> 140	0.1
optimal level	100 - 140	0.7
marginal deficiency	70 - 100	8.6
mild deficiency	55 - 70	25.6
deficiency	45 - 55	28.0
severe deficiency	< 45	37.0

Zinc (Zn) is one of the trace elements essential to animal life: it has been established as being indispensable for the functioning of numerous enzymes common to animals and man. Total amount of zinc in the adult human body is between 1,4 and 2,3 g. The richest organs are prostate, muscles, liver, kidneys and bones. In the serum, in which it is transported, zinc is for 2/3 linked to albumin: this is exchangeable zinc. The remainder is strongly bound to macro-globulin and is not exchangeable. The serum content is of 890 ± 100 ug Zn/l.

Zinc takes part in the formation of numerous enzymes (oxido-reductases, transferases, hydrolases lyases), and acts on a large number of biological phenomena. It has been demonstrated that zinc deficiency couses a decrease in tolerance to carbohydrate: a pre-diabetes condition then appears with decrease in the synthesis and storage of insulin in the?-cells of the Langheren's islets. Different studies tend to show the hypocholesterolemia effect of zinc.

To investigate the Zn status of inhabitants of the Czech Republic [4], 1155 serum and 132 hair samples were analyzed for zinc content. The results (mean 910 ± 276 ug Zn/l serum and 189 ± ug Zn /g hair) demonstrate satisfactory zinc status of the searched population. The frequency of individual serum Zn concentrations proved that the population of the Czech Republic does not suffer from severe zinc deficiency. On the other hand, about one-third of inhabitants has their serum Zn concentrations below the cutoff value of 800 ug Zn/l serum, which means a marginal or mild Zn deficit of the organism. During 1996-1998 the concentration of Zn in whole blood of 1216 blood donors, average age 33 years, was assessed [1], the median value for adults was 5800 ug/l.

Magnesium (Mg) is present in all human tissues and 60% is in the skeleton, where it probably provides the main reserve of the element in the body. Among other tissues the muscles (heart muscle) have the highest Mg content. The adult human body contains 21-28 g of Mg. The intracellular concentration of Mg is more than 10 times higher than that outside the cell, and this is maintained even when the extracellular concentration falls very low. Next to potassium, Mg is the predominant cation in the cells and is an essential part of many enzymes involved especially in carbohydrate metabolism. It is also importatnt for maintaining electrical potential in nerves and muscle membrane.

Magnesium requirement is between 200-300 mg/day. It is widespread in animal and particularly in those of vegetable origin foodstuffs and Mg deficiency is of endogenous origin. The principle cases of disturbances in Mg metabolism are failure of absorptive processes for various reasons, parenteral feeding for prolonged periods, renal vastage from drugs, chemicals or kidney diseases. Mg deficiency in man leads to neuromuscular dysfunction as manifested by hyperexcitability with tremor, convulsion and diarrhoea and is sometimes accompanied by behavioural disturbances. The metabolic relationship to other elements, such as calcium and potassium, as well as its possible significance in cardiovascular disease, makes it important to know the daily intake of Mg and the levels of Mg in serum of monitored population. Jägerstad [3] found a high, significant correlation between death rate from ischemic heart disease and the estimated calcium to magnesium ratios of the average diets in different OECD countries.

The serum Mg level is the simplest and quickest approach to the evaluation of Mg status, even if erythrocyte Mg level is much better indicator of it. The normal Mg serum content in adults is 0,66-0,94 mmol/l. When the dietary intake of Mg is reduced, the urinary excretion of Mg is rapidly lowered. The status of Mg saturation in the Czech Republic population was evaluated in 1999 [8] on the basis of results of magnesiuria determinated in 3587 probands aged 6-65 years. The lowest magnesiuria 3,78 mmol ± 0,08 was found in adults, the highest 6,59 ± 0,16 mmol was found in children.

RESULTS OF THE STUDY

We conducted an epidemiological study focused on monitoring topical health state in the members of Fire Rescue Service from selected areas of the Czech Republic. Altogether, there were 812 healthy volunteers chosen at random. The examination was focused on the evaluation of selected anthropometrical (body weight, body mass index, fat tissue percentage, waist circumference) and biochemical (glycemia, total- and HDL-cholesterol, triacylglycerols, uric acid and selected vitamins serum concentrations, serum AST, ALT and GMT activities) parameters of the nutritional status. The levels of Mg, Zn and Se in serum were determined by atomic-absorption spectrophotometric method (AAS Unicam, GB). Serum which was mineralized in microwave processor (Milestone, Italy) provided selenium analysis.

In the monitored group, mean serum concentration of Mg was 0,816 (0,057mmol/l (19,83 (1,4 ug/l), that of Zn was 18,25 (2,54µmol/l (1193 (166 ug/l) and mean Se serum concentration was 0,802 (0,14µmol/l (63 (11 ug/l). The results of the study contribute to information about normal values of serum Mg, Zn and Se concentrations in Czech population. Their statistical evaluation is documented in *table 2:*

Table 2. Mg, Zn and Se serum concentrations

	mean	**sd**	**median**	**Q_3-Q_1/2**	percentiles			
					5th	**10th**	**90th**	**95th**
Mg [mmol/l]	0,82	0,057	0,81	0,04	0,72	0,74	0,88	0,92
Zn [µmol/l]	18,25	2,54	18,00	1,56	14,74	15,25	21,00	21,75
Se [µmol/l]	0,80	0,14	0,80	0,11	0,56	0,60	1,01	1,10

The distribution of Mg, Zn and Se serum concentrations in examined population group approached normal distribution. No statistically significant difference between serum concentrations of followed elements in men and women was found.

Considering the wide range of monitored characteristics of topical health, the statement is focused on the evaluation of serum levels of Mg, Zn and Se, and their links to selected biochemical and anthropometrical parameters. The concept of the study made it possible to reveal relations between the serum Mg, Zn, Se levels and the age or biochemical and anthropometrical parameters which are generally used as risk indices of cardiovascular disease. The results of that statistical evaluation are documented in the form of correlation matrix *(table 3):*

Table 3. Correlation matrix

	age	**BMI**	**waist**	**Mg**	**Zn**	**Se**	**TCH**	**TAG**	**vit. C**
age	1,00								
BMI	0,265	1,00							
waist	0,356	0,844	1,00						
Mg	ns	ns	ns	1,00					
Zn	ns	ns	ns	ns	1,00				
Se	ns	ns	ns	ns	ns	1,00			
TCH	0,446	0,248	0,282	ns	-0,234	ns	1,00		
TAG	0,204	0,343	0,399	ns	-0,317	ns	0,419	1,00	
vit.C	-0,279	ns	ns	ns	ns	-0,271	-0,212	ns	1,00

Statistical significance of r: $p \leq 0{,}01$, $p \leq 0{,}05$ ns = no significance.

No statistically significant relations among the age of examined subjects and their serum Mg, Zn and Se concencentrations were proved. Statistically significant negative correlations were revealed for relations among Zn serum concentration and cholesterolemia ($p \leq 0{,}05$) and triacylglycerolemia ($p \leq 0{,}01$). Statistically significant ($p \leq 0{,}01$) negative correlation was proved among Se serum concentration and ascorbemia.

High incidence of the obesity and overweight in examined population group prompted to evaluation of Mg, Zn and Se saturation in subjects classified to 4 BMI (body mass index) categories: normal BMI (20-25 kg/m^2), overweight 1st grade (BMI 25,1-28 kg/m^2), overweight 2nd grade (28,1-30 kg/m^2) and the category of obesity (BMI over 30 kg/m^2). The results of that evaluation are documented in the *table 4:*

Table 4. Serum concentrations in BMI categories

	normal	1stgrade overw.	2ndgrade overw.	obesity
Mg [mmol/l]	1,12	0,81	0,83	0,82
Zn [μmol/l]	18,0	18,4	18,4	18,3
Se [μmol/l]	0,81	0,80	0,84	0,88

Mg concentrations showed the tendency to decrease with higher BMI, while by Zn and Se serum concentrations only slight tendency to increase in higher BMI categories was found. No statistical significance of these changes by analysis of variance was proved.

CONCLUSIONS

In the monitored population group, the average serum concentrations of Se showed a tendency towards values of marginal deficiency, however, they corresponded with other findings within the Czech population. The concentrations of Mg and Zn corresponded with commonly used normal values.

No statistically significant relations among the age of examined subjects and their serum Mg, Zn and Se concencentrations were proved. Statistically significant negative correlations were revealed for relations among Zn serum concentration and cholesterolemia ($p \leq 0,05$) and triacylglycerolemia ($p \leq 0,01$).

No statistical significance of relation among serum Se, Zn and Mg concentrations and body mass index of examined subjects was proved by analysis of variance.

REFERENCES

1. Beneš, B.- Spěváčková, V.- Šmíd, J.- Čejchanová, M.- Černá, M.- Šubrt, P.- Mareček, J.: The concentration levels od Cd, Pb, Hg, Cu, Zn and Se in blood of the population in the Czech Republic. Cent. Eur. J. Publ. Health, 8, 2000, 2:117-119.
2. Beneš, B.- Spěváčková, V.- Čejchanová, M.- Šmíd, J.- Švandová, E.: Retrospective study of concentration levels of Pb, Cd, Cu and Se in serum of the Czech population in time period 1970-1999. Cent. Eur. J. Publ. Health, 9, 2001, 4: 190-195.
3. Jägerstad, M.- Abdulla, M.- Svenson, S.- Norden, A.: Magnesium. Scand. J. Gastroent., 14, 1979 (suppl.52): 153-157.
4. Kvíčala, J.- Jiránek, V.: INAA of serum zinc of inhabitants in five regions of the Czech Republic. Biol. Trace Elem. Res., 71-72, 1999: 21-30.
5. Kvíčala, J.: Selen a organismus (Selenium and the Organism). Čas. Lék. Čes., 138, 1999, 4: 99-106.
6. Kvíčala, J.- Zamrazil, V.- Jiránek, V.: Characterization of selenium status of inhabitants in the region Ústí nad Orlicí, Czech Republic by INAA of blood serum and hair and fluorometric analysis of urine. Biol. Trace Elem. Res., 71-72,1999: 31-39.
7. Salonen, J.T.: Association between cardiovascular death and myocardia infarction and serum selenium in a matched pair longitudinal study. Lancet, 1982, 8291: 175-178.
8. Šimečková, A.- Zamrazil, V.- Čeřovská, J.- Kvíčala, J.- Bílek, R.: Status of magnesium, selenium and iodine saturation in the population of seven regions in the Czech Republic. Cent. Eur. J. Publ. Health, 7, 1999, 4: 176-181.

XII PHARMACOLOGY AND THERAPY

Metal Ions in Biology and Medicine: vol. 7. Eds L. Khassanova, Ph. Collery, I. Maymard, Z. Khassanova, J.-C. Étienne.
John Libbey Eurotext, Paris © 2002 pp. 531-535.

Chromium, glucocorticoids, insulin sensitivity and weight control

Richard A Anderson

Nutrient Requirements and Functions Laboratory, Beltsville Human Nutrition Research Center, U.S. Department of Agriculture, ARS, Beltsville, Maryland 20705, USA

ABSTRACT

The demonstration of the effects of chromium on weight control and lean body mass originated from two placebo controlled studies involving young men who were training to become competitive college athletes. There were highly significant effects of supplemental Cr, 200 µg/d, on lean body mass and body fat within 6 weeks. The results appeared to be greater than would be expected and have been difficult to reproduce. However, others studies of longer duration, more that 12 weeks, and higher levels of Cr, at least 400 µg/d, have also reported significant effects of Cr on body weight and composition. In addition, several animal studies involving primarily pigs and goats have documented the effects of Cr on lean body mass and composition. There were significant increases in weight gain and feed consumption in goats consuming a low Cr diet compared with those of the controls with corresponding doubling in circulating insulin. The increases in weight gain were attributed to the antilipolytic effects of insulin leading to accumulation of triglycerides in the adipose tissue. Elevated insulin levels in the low Cr animals would also lead to decreased glucagon. Since glucagon stimulates lipolysis, decreased glucagon may lead to decreased lipolysis and subsequent accumulation of body fat and weight gain. Glucocorticoid administration also increases insulin secretion leading to increased body fat and decreased lean body mass and also increased Cr losses. Supplemental Cr leads to a reversal of the negative effects of glucocorticoid administration and Cr has also been shown to reverse steroid-induced diabetes. In summary, the effects of Cr on lean body mass and weight control can be explained in part by the effects of Cr on insulin sensitivity. This also helps to explain why the effects of chromium on weight control are variable since the effects on weight and lean body mass would be regulated by insulin sensitivity which varies greatly.

Weight control and body composition are related to more than simple caloric or fat intake. Genetics play a role but certainly are only a small part of the overall increases in the incidence of obesity. The incidence of obesity has been increasing yearly rather than over generations. While the incidence of obesity has increased since the sixties, the fat intake in the United States has decreased from 42% to 34% suggesting that dietary factors other than fat are important in the prevention of obesity. One emerging factor that is thought to play a major role is the level of insulin. Foods that lead to higher levels of insulin lead to greater increases in per cent body fat. This concept has been termed the glycemic index (Ludwig, 2000). It is not based simply on the simple sugar content of the diet but other factors including carbohydrate type, fiber, protein, fat and type of fat and method of preparation. All of these variables relate to rises in blood glucose and insulin. Increased weight gain and obesity can result from insulin resistance and hyperinsulinemia without increased energy intake. Therefore, factors that alter blood glucose and(or) insulin may be involved in weight control.

One nutrient that has been shown to alter not only blood glucose and insulin but also body weight and composition is chromium (Cr) (Anderson, 1998a) (Anderson, 2000) (Anderson, 2001). Early experiments demonstrated that Cr increased insulin-dependent amino acid incorporation into heart proteins and cell transport of an amino acid analogue (Schwarz and Mertz, 1959). The first report of significant improvements in lean body mass (LBM) were observed in 10 male college students and 31 college football players following 200 μg of Cr as Cr picolinate daily for 6 weeks (Evans, 1989). These studies appear to have involved a group of subjects very responsive to resistive exercise and Cr. In a study involving 20 M and 20 F swimmers receiving 400 μg daily of Cr as Cr picolinate, Cr significantly increased LBM (3.3%), decreased fat mass (-4.6%) and decreased percent body fat (-6.4%) compared with the placebo group (Bulbulian *et al.*, 1996). Females had a greater change for% fat compared with males (-8.2% and -4.7%, respectively). Effects were not significant after 12 but only after 24 weeks. This study demonstrates that studies involving Cr supplementation and LBM should be longer than 12 weeks and involve 400 μg of supplemental Cr daily or more (Anderson, 1998b).

In a study involving very low calorie diets (3.34 MJ/d), diets were supplemented daily with placebo, 200 μg of Cr as Cr picolinate or 200 μg of Cr as chromium-yeast for 6 months (Bahadori *et al.*, 1997). Subjects were on the 3.34 MJ/d diet for the first 8 weeks. Weight loss in all groups following the initial 8 weeks was similar. After an additional 16 weeks, LBM was lower except in the group consuming Cr picolinate with an increase of 1.81 ± 2.7 kg ($p<0.0001$). Therefore, Cr consumed during and after weight reduction induced by a low calorie diet increased lean body mass.

Kaats *et al.* (Kaats *et al.*, 1996) enrolled 154 patients who consumed a protein/carbohydrate drink that contained either no added Cr, 200 or 400 μg of Cr as Cr picolinate daily. Subjects were free-living and were not provided with weight loss, dietary or exercise guidance. Body composition was measured by using underwater testing with residual lung volumes determined by helium dilution. After 72 days, both groups consuming the Cr supplements had significantly improved changes in body composition. In an earlier study by the same group, subjects consuming both levels of supplemental Cr daily as Cr picolinate displayed significant improvements in body composition. The average increase in fat-free mass was 0.64 kg and decrease in fat was 1.91 kg. There were no significant changes in the placebo group. Body composition changes tended to be greater in the older subjects and in those consuming the higher level of Cr.

These studies involving improved LBM due to supplemental Cr in humans, which have been substantiated in animal studies (Lindemann *et al.*, 1995), have led to exaggerated claims of weight loss in humans. While there is ample evidence that Cr has an effect on body composition, decreases in body weight due to supplemental Cr alone are likely to be small and several well-controlled studies have failed to substantiate effects of supplemental chromium on body composition and weight reduction (Kobla and Volpe, 2000).

Recent studies involving goats have helped to elucidate and substantiate the role of Cr in weight control. Goats fed a high refined carbohydrate low-Cr diet also show elevated blood glucose and insulin (Frank *et al.*, 2000a). The increases in blood glucose following 20 months of a low Cr diet were 33% and almost 200% in circulating insulin in comparison with the control group. There were also large increases in weight gain in the animals consuming the low Cr diet compared with those of the controls (*fig. 1*, lower panel) with corresponding increases in feed consumption (*fig. 1*, upper panel) (Frank *et al.*, 2000b). The increases in weight gain were attributed to the antilipolytic effects of insulin leading to accumulation of triglycerides in the adipose tissue. Elevated insulin levels in the low Cr animals would also lead to decreased glucagon. Since glucagon stimulates lipolysis, decreased glucagon may lead to decreased lipolysis and subsequent accumulation of body fat and weight gain.

There were no effects until after 28 weeks on the low Cr diet marginal in nutritional quality. If it takes more than 28 weeks to detect significant changes in body weight in rapidly growing goats it is not surprising that most of the human studies, which are usually 12 weeks or less in duration, also are unable to detect significant changes in people eating conventional diets. This is consistent with the study of Bulbulian *et al.* (Bulbulian *et al.*, 1996) who were unable to detect significant decreases in body weight after 12 weeks but significant changes were detected in 24 weeks.

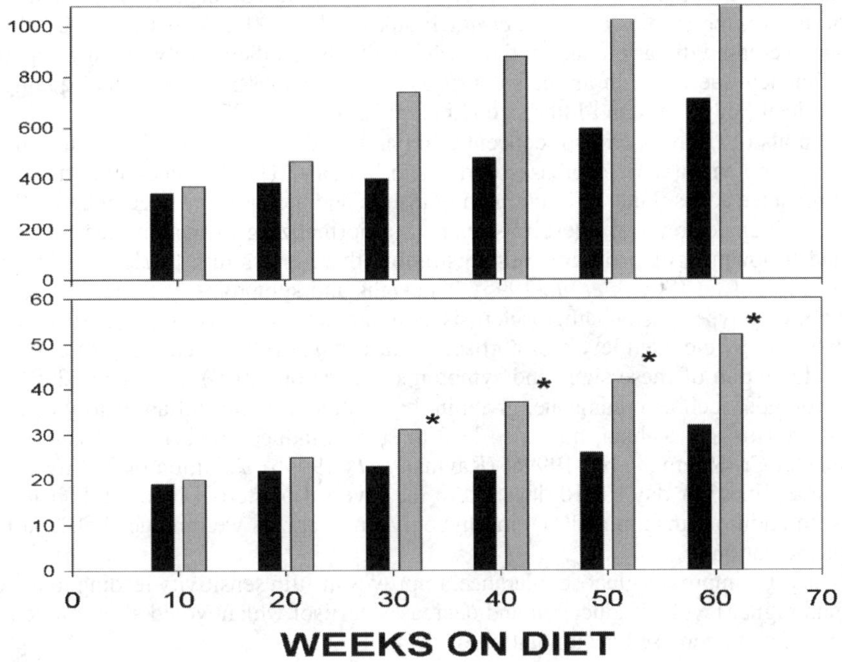

Fig. 1. Feed intake and body weight of goats raised on a low chromium diet, grey bars, and diet supplemented with chromium, black bars. *Significant effect of Cr at P< 0.05; significance not given in original source for food intake. Source: Adapted from Frank *et al.* (Frank *et al.*, 2000b)

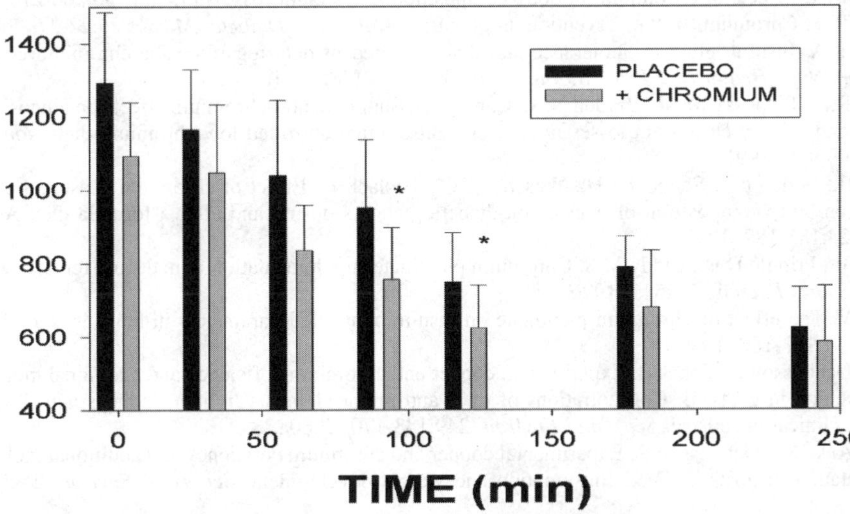

Fig. 2. Supplemental chromium decreases cortisol of subjects consuming a diet low in chromium. Eight adult subjects (90-min glucose greater than 5.56 mmol/L after an oral glucose challenge of 1 g per kg body weight) consumed a diet containing less than 20 µg of Cr per day in a double-blind crossover design. Subjects took either a placebo tablet or a tablet containing 200 µg of Cr as Cr chloride daily for 5 weeks (Anderson *et al.*, 1991). *Significant effect of Cr at p<0.05.

While there are numerous anecdotal reports of Cr changing cravings for sugar and effects on total caloric intake, the studies of Frank *et al.* (Frank *et al.*, 2000a) (Frank *et al.*, 2000b) are the first to report increased dietary intake in the low Cr animals. Studies involving pigs, report increased feed efficiency due to Cr in animals consuming diets of marginal nutritional quality as well as effects on lean body mass and litter size (Lindemann *et al.*, 1995).

Chromium also has alters cortisol concentration and glucocorticoid administration. In *figure 2* is shown that supplemental Cr decreases cortisol in humans. This becomes important regarding weight control since cortisol increases circulating insulin and increases fat accumulation (Freedman *et al.*, 1986). Adrenalectomy of obese rats leads to a normalizing of insulin, decreased fat accumulation and following glucocorticoid administration, there is a return to elevated insulin levels and accumulation of fat (Strack *et al.*, 1995). The signs and symptoms of Syndrome X including abdominal obesity, hypertension, atherosclerosis, coronary artery disease and hyperinsulinemia can all be made worse by elevated levels of cortisol (Strack *et al.*, 1995). Reduction of cortisol due to Cr leads to alleviation of these signs and symptoms (Anderson, 2000) (Anderson, 2001).

Glucocorticoids, such as prednisone, given in the treatment of anti-inflammatory diseases such as allergies, arthritis and asthma, may also lead to steroid-induced diabetes that can be reversed by supplemental Cr (Ravina *et al.*, 1999a) (Ravina *et al.*, 1999b). Addition of 200 µg of Cr as Cr picolinate three times per day for 10 days led to the reversal of steroid-induced diabetes in 47 of 50 patients. In addition, insulin and(or) oral hypoglycemic agents were reduced 50% at the onset of Cr supplementation.

In summary, Cr improves glucose tolerance, improves insulin sensitivity leading to lower levels of insulin and higher levels of glucagon and decreases cortisol which would all promote increased lean body mass and decreased accumulation of fat.

REFERENCES

Anderson R.A. Chromium, glucose intolerance and diabetes. *J. Am. Coll. Nutr.* 17:548-555, 1998a.

Anderson R.A. Effects of chromium on body composition and weight loss. *Nutr. Rev.* 56:266-270, 1998b.

Anderson R.A. Chromium in the prevention and control of diabetes. *Diabetes Metab.* 26:22-27, 2000.

Anderson R.A. Insulin, glucose intolerance and diabetes: Recent data regarding the chromium connection. *Trace Elem. Nutr. Health Dis. Proc. Bio-minerals Symp.* 1:79-86, 2001.

Anderson R.A., Polansky M.M., Bryden N.A., Canary J.J. Supplemental-chromium effects on glucose, insulin, glucagon, and urinary chromium losses in subjects consuming controlled low-chromium diets. *Am. J. Clin. Nutr.* 54:909-916, 1991.

Bahadori B., Wallner S., Schneider H., Wascher T.C., Toplack H. Effect of chromium yeast and chromium picolinate on body composition of obese, non-diabetic patients during and after a formula diet. Acta Med. Austriaca 24:185-187, 1997.

Bulbulian R., Pringle D.D., Liddy M.S. Chromium picolinate supplementation in male and female swimmers. *Med. Sci. Sports Exercise* 28:511, 1996.

Evans G.W. The effect of chromium picolinate on insulin controlled parameters in humans. *Intl. J. Biosoc. Med. Res.* 11:163-180, 1989.

Frank A., Danielsson R., Jones B. Experimental copper and chromium deficiency and additional molybdenum supplementation in goats. II. Concentrations of trace and minor elements in liver, kidneys and ribs: haematology and clinical chemistry. *Sci. Total Environ.* 249:143-170, 2000a.

Frank A., Anke M., Danielsson R. Experimental copper and chromium deficiency and additional molybdenum supplementation in goats. I. Feed consumption and weight development. *Sci. Total Environ.* 249:133-142, 2000b.

Freedman M.R., Horwitz B.A., Stern J.S. Effect of adrenalectomy and glucocorticoid replacement on development of obesity. *Am. J. Physiol.* 250:R595-R607, 1986.

Kaats G.R., Blum K., Fisher J.A., Adelman J.A. Effects of chromium picolinate supplementation on body composition: a randomized double-masked placebo-controlled study. *Curr. Ther. Res.* 57:747-756, 1996.

Kobla H.V., Volpe S.L. Chromium, exercise, and body composition. *Crit. Rev. Food Sci. Nutr.* 40:291-308, 2000.

Lindemann M.D., Wood C.M., Harper A.F., Kornegay E.T., Anderson R.A. Dietary chromium picolinate additions improve gain:feed and carcass characteristics in growing-finishing pigs and increase litter size in reproducing sows. *J. Anim. Sci.* 73:457-465, 1995.
Ludwig D.S. Dietary glycemic index and obesity. *J. Nutr.* 130:280S-283S, 2000.
Ravina A., Slezak L., Mirsky N., Anderson R.A. Control of steroid-induced diabetes with supplemental chromium. *J. Trace Elem. Exptl. Med.* 12:375-378, 1999a.
Ravina A., Slezak L., Mirsky N., Bryden N.A., Anderson R.A. Reversal of corticosteroid-induced diabetes mellitus with supplemental chromium. *Diabet. Med.* 16:164-167, 1999b.
Schwarz K., Mertz W. Chromium (III) and the glucose tolerance factor. *Arch.Biochem.Biophys.* 85:292-295, 1959.
Strack A.M., Sebastian R.J., Schwartz M.W., Dallman M.F. Glucocorticoids and insulin: reciprocal signals for energy balance. *Am. J. Physiol.* 268:R142-R149, 1995.

*Metal Ions in Biology and Medicine: vol. 7.*Eds L. Khassanova, Ph. Collery, I. Maymard, Z. Khassanova, J.-C. Étienne.
John Libbey Eurotext, Paris © 2002 pp. 536-539.

Effects of cadmium, a metallothionein inducer, on cardiac ischemia - reperfusion

Bobillier-Chaumont S.[1], Maupoil V.[1], and Berthelot A.[1]

[1]*Laboratoire de Physiologie et de Pharmacologie - Nutrition Préventive Expérimentale, Faculté de Médecine et de Pharmacie, Place Saint-Jacques, 25030 Besançon cedex, France*

ABSTRACT

Metallothionein (MT), a rich cysteine residue protein, is induced in different organs in response to heavy metals (such as cadmium) and oxidative conditions. The antioxidant properties of MT and the possible implication of oxidative stress in cardiovascular diseases suggests to study the role of MT in the cardiovascular physio-pathology. In the present study, we investigated whether increased level of MT by cadmium (Cd) administration affords protection against cardiac ischemia-reperfusion injury in isolated rat heart.

Male Wistar rats (250-300 g) received an i.p injection of 0, 1, 1.5 or 2 mg/kg Cd 48 h prior to heart perfusion according to the Langerdorff technique. After 10 min of perfusion, a global total ischemia (30 min)-reperfusion (30 min) sequence was performed. Heart rate (HR), coronary flow (CF), left ventricular systolic pressure (LVSP) and left ventricular end diastolic pressure (LVEDP) were measured throughout the experiment. Left ventricular developed pressure and functional recovery were calculated as follow: LVDP = LVSP-LVEDP; FR = $[(LVDP_{TRx} \times HR_{TRx}) \times 100]/(LVDP_{TP8} \times HR_{TP8})$.

During reperfusion, a significant increase of CF (16 ± 1 vs 11 ± 1 ml/min at the end of 30 min reperfusion) and LVDP (82 ± 14 vs 50 ± 5 mmHg at the end of 30 min reperfusion) was observed in the heart of rat pretreatment by 2 mg/kg of Cd. HR was not modified. Therefore the FR was significantly improved by 2 mg/kg-Cd administration (115 ± 23 vs 52 ± 6% at the end of 30 min reperfusion). In conclusion, Cd improve the FR of post-ischemic rat heart. However, the mechanisms underlying this beneficial effect remain to be determined.

INTRODUCTION

Metallothionein (MT), a low molecular weight (6 kDa) metal-binding protein containing a high proportion (30%) of cysteine residues, is known to bind and detoxify heavy metal cations through its numerous thiols (Markant and Pallauf, 1996). Cysteinyl residues also seem to be responsible for the free radical scavenging properties of MT. *In vitro* experimental studies have provided evidence to support the antioxidant function of MT (Min *et al.*, 1999). In addition, enhanced MT gene expression by cadmium (Cd) pretreatment provided protection against DNA damages in cultured rat liver cells (Coogan *et al*, 1994). However, the *in vivo* antioxidant function of MT has not get been clearly demonstrated.

With regard to physiological protection against oxidative injury, the interesting aspect of MT is its inductibility by a variety of physiological and chemical stressors. It has been shown that MT is efficiently induced in many human and animal organs, particularly in liver tissue, in response to heavy metals like Zn or Cd (Norberg, 1979). The possible implication of oxidative stress in the cardiovascular diseases (Laursen *et al*, 1997) and the antioxidant properties of MT, particularly the effect of MT in preventing free radical damage in cultured cardiomyocytes (Wang *et al*, 1999) suggests to study the relation between MT and the cardiovascular function. Ischemia-reperfusion injury is due to an imbalance between the formation of oxidants and the availability of endogenous antioxidants in the heart. In this context, MT could be protective against ischemia-reperfusion injury in the heart.

The present study was designed to identify the protective role of induction of MT against ischemia-reperfusion injury in the isolated rat heart. Cd was used to induce *in vivo* MT in the heart.

MATERIALS AND METHODS

Chemicals

All chemicals were purchased from Sigma (France).

Animals and Diets

Male Wistar rats (280-300 g) were kept in a temperature-controlled room (22 ± 2 °C), at a humidity of 50-60%, with a reverse 12:12h light-dark cycle. Food and water were given *ad libitum*. All animal experiments were conducted according to the guidelines formulated by the European community for the care and use of animals (L 358-86/609/EEC).

Metallothionein induction

Cadmium chloride ($CdCl_2$) was dissolved in sterile 0.9% saline. Rats were injected intraperitoneally with $CdCl_2$(1; 1.5 or 2 mg/kg body weight). Controls were intraperitoneally injected with normal saline. All injections were at a volume of 1 ml/kg body weight.

Ischemia-reperfusion protocols

Fourty eight hours after $CdCl_2$ treatment, rats were anesthetized with nesdonal (60 mg/kg, i.p.) and treated with heparin (250 UI/kg, i.v.). The animals were thoracotomized and their hearts quickly removed and placed in an ice-cold Krebs-Henseleit solution. Hearts were immediately perfused retrogradely using the Langendorff technique at a constant pressure with oxygenated Krebs-Henseleit buffer pH 7.4 prepared in distilled water. Myocardial temperature was measured and maintened constant at 37°C. Left ventricular end-diastolic pressure (LVEDP) was adjusted between 4-10 mmHg. Left ventricular systolic pressure (LVSP) and coronary flow (CF) were measured throughout the procedure. Heart rate (HR), and the contractile force, represented by LVDP (left ventricular developed pressure LVDP = difference between LVSP and LVEDP) were continuously recorded on a polygraph (Gould TA 240). The functional recovery (FR) was calculated as follow $FR = [(LVDP_{TRx} \times HR_{TRx}) \times 100]/(LVDP_{TP8} \times HR_{TP8})$, where $_{TPX}$ represent the time of reperfusion at times x and and $_{TP8}$ the time of perfusion at times 8 minutes.

After 10 min of stabilisation, a global total ischemia (30 min)-reperfusion (30 min) sequence was performed in order to assess the influence of an induction of MT by Cd. At the end of the ischemia-reperfusion protocol, the heart was weighed, immediately frozen in liquid nitrogen and stored at - 80°C for subsequent analysis.

Determination of Metallothionein content

The concentration of MT in the heart was estimated indirectly as the maximal Cd-binding capacity of the samples according to the method of Eaton and Cherian (1991). The concentration of Cd was determined with a flame atomic absorption spectrophotometer (HITACHI Z-9000). The amount of MT in heart was calculated assuming 6 g-atoms of Cd to be bound per mole MT, and the molecular weight of Cd-MT was considered to be 6000 Da.

Statistics

Data were expressed as mean ± SEM (standard error of the mean). Statistical analyses were carried out using a one-way analysis of variance (ANOVA 1). Differences between groups were considered significant at P (0.05 levels.

RESULTS

As shown in *table 1*, LVDP was increased by pretreatment with 2 mg/kg of Cd (not significant). Heart rate was not modified (data not shown).

Table 1. LVEDP, LVSP and LVDP (mmHg) during ischemia-reperfusion protocol of rat injected with 0, 1, 1.5 or 2 mg/kg Cd

	LVEDP (mmHg)		LVSP (mmHg)		LVDP (mmHg)	
	before ischemia	end of reperfusion	before ischemia * vs 1.5	end of reperfusion	before ischemia * vs 1.5	end of reperfusion
Control	5 ± 0.7	77 ± 4.5	83 ± 2*	128 ± 4	78 ± 2*	50 ± 5
Cd 1 mg	6 ± 0.8	62 ± 11.3	84 ± 5*	110 ± 10	79 ± 5*	47 ± 11
Cd 1.5 mg	6 ± 0.6	76 ± 4.4	60 ± 4	125 ± 5	54 ± 5	49 ± 7
Cd 2 mg	6 ± 1	49 ± 13	70 ± 6	131 ± 7	64 ± 6	82 ± 14

During reperfusion, a significant increase of CF (16 ± 1 vs 11 ± 1 ml/min at the end of 30 min reperfusion) was observed in the heart of rat pretreatment by 2 mg/kg of Cd *(fig.1)*.

Therefore the FR was significantly improved by 2 mg/kg -Cd administration (115 ± 23 vs 52 ± 6% at the end of 30 min reperfusion) *(fig.1)*.

Cardiac MT level was gradually increased in the different groups of Cd-injected rats (0.070, 0.091, 0.096 and 0.51 nmol/g tissue for control, Cd 1, Cd 1.5 and Cd 2, respectively).

DISCUSSION/CONCLUSIONS

Different studies have shown that thiol-compounds may be one of the potent defense systems against ischemia-reperfusion (I-R) in the heart; indeed, I-R injury is due to an imbalance between the formation of oxidants and the availability of endogenous antioxidants, such as thiol groups, in the heart (Dhalla, 2000). MT, due to its free radical scavenging capability, may play an important role in the cardiac protection. In the present study, we demonstrated that a significant induction of cardiac MT level (with the dose of 2 mg/kg of Cd), may act as a defense against I-R injury because functional recovery was significantly improved. These results are in accordance with those of Dhalla (2000) which demonstrated that endogenous antioxidants prevent changes in the ischemic-reperfused hearts. However, Cd is not specific for cardiac MT induction, and therefore, cytoprotection may have been due to other effects of the inducing agent. In conclusion, cadmium, a well-known inducer, improve the fonctional recovery of post-ischemic rat heart. However, the mechanisms underlying this beneficial effect remain to be determined.

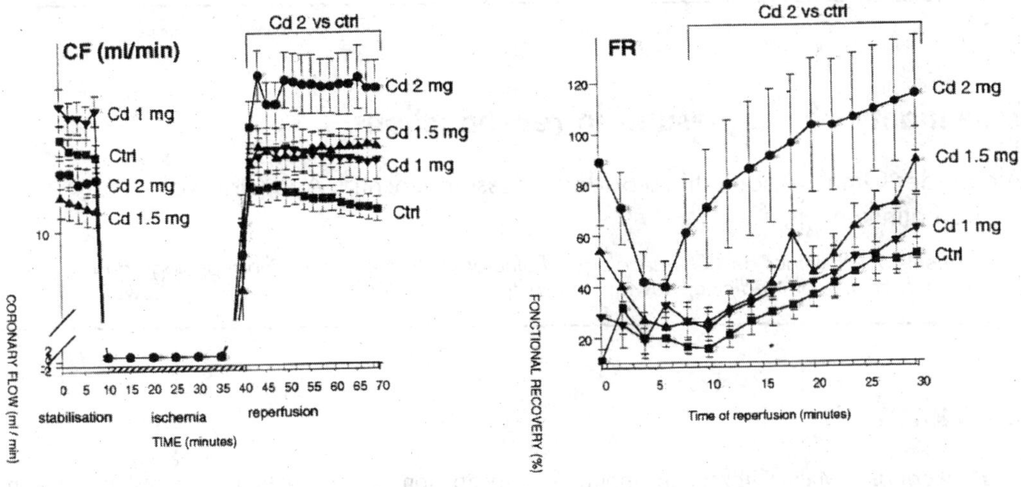

Fig. 1. CF (ml/min) and FR during ischemia-reperfusion protocol of rat injected with 0, 1, 1.5 or 2 mg/kg Cd.

REFERENCES

Coogan T.P., Bare R.M., Bjornson E.J. Waalkes M.P. Enhanced metallothionein gene expression is associated with protection from cadmium-induced genotoxicity in cultured rat liver cells. *J Toxicol Environ Health*, 1994, 41(2), 233-245.

Dhalla N.S., Elmoselhi A.B., Hata T., Makino N. Status of myocardial antioxidants in ischemia-reperfusion injury. *Cardiovasc Res*, 2000, 47(3), 446-456.

Eaton D., Cherian G. Determination of metallothionein in tissues by cadmium-hemoglobin affinity assay. *Methods Enzymol*, 1991, 205, 83-88.

Norberg G.F. Renal effects of mercury- and cadium- metallothionein. *Experientia,*1979, 34supp, 347-350.

Laursen J., Rajagopalan B.S., Galis Z., Tarpey M., Freeman B.A., Harrison D.G. Role of superoxide in angiotensin II-induced but not catecholamine-induced hypertension. *Circulation,* 1997, 95, 588-593.

Markant A., Pallauf J., Metallothionein and zinc as potential antioxidants in radical-induced lipid peroxidation in cultured hepatocytes. *J Trace Elements Med Biol*, 1996, 10, 88-95.

Min K.S., Nishida K., Onosaka S. Protective effect of metallothionein to ras DNA damage induced by hydrogen peroxide and ferric ion-nitriacetic acid. *Chemico-Biological Interactions*, 1999, 122, 137-152.

Wang G.W., Schuschke D.A., Kang Y.J. Metallothionein-overexpresing neonatal mouse cardiomyocytes are resistant to H_2O_2 toxicity. *Am J Physiol,* 1999, 44, 167-175.

*Metal Ions in Biology and Medicine: vol. 7.*Eds L. Khassanova, Ph. Collery, I. Maymard, Z. Khassanova, J.-C. Étienne.
John Libbey Eurotext, Paris © 2002 pp. 540-544.

Essentiality of magnesium in reproduction

Alexander Omu[1], Abdullatif Al-Bader[2], Hussein Dashti[3], Ayo Oriowo[4], Tunde Fatinikun[1]

Departments of Obstetrics and Gynaecology[1], Pathology[2], Surgery[3] and Pharmacology and Toxicology[4], Faculty of Medicine, Kuwait University, P O Box 24923 Safat, 13110 Kuwait

ABSTRACT

Background: Magnesium is an important activator ion in the function of many enzymes involved in phosphate transfer. It is therefore indispensable in cell function. Mg2+ blocks many conditions like muscle contractility and nerve conduction, which are associated with the overloading of intracellular Ca2+ and cAMP. No role has been assigned to Magnesium in male reproductive tract.

Objective of study: A series of studies were carried out to determine the essentiality of Magnesium in human and animal reproduction.

Methodology: in a series of studies, we determined the role of hypomagnesaemia in sexual and sperm dysfunction, like premature ejaculation, impotence, during sperm transport and capacitation of sperm. We tested the hypothesis that changes in Mg^{2+} content of the seminal fluid would affect sperm transport in the vas deferens and the Mg2+ content of the seminal fluid modulates sperm motility in the vas deferens. Fifteen men with genuine premature ejaculation were evaluated in comparison with men with normal sexual satisfaction and normozoospermia and impotence, Zinc, Copper, Selenium and Magnesium were determined by AAS in both serum and semen of groups. Intervention involved treatment with Magnesium for 4-6 weeks and sexual function was evaluated. The influence of external Mg^{2+} concentration on reactivity of the vas deferens was studied in the Sprawly-Drew (SD) rat, with different concentrations of Ca^{2+} and Mg^{2+} in the tissue bath. The possible role of Mg^{2+} and IL-4 as decapacitators of sperm was investigated in 45 men with normal sperm quality.

Results: magnesium Chloride therapy was associated with significant improvement ($p < 0.001$). Induced hypomagnesaemia caused increased electrical twitch contraction of vas deferens. As a decapacitator, Mg2+ inhibited Ca2+ induced increased sperm motility by 50 per cent while initial incubation led to 20 per cent reduction in sperm motility. Malonialdehyde and TNF were increased after sperm incubation with Ca2+ and pentoxifylline. Extracellular increase of Mg2+ was associate with increase in intracellular Mg2+.

Conclusion: Magnesium is essential in the maintenance of erectile function, sperm transport in the vas deferens and in the female reproductive tract and it is an important sperm decapacitator.

INTRODUCTION

Intracellular Mg2+ is known to affect all functions dependent on this ion including glycolysis, protein synthesis, respiration and reproduction. The role of Mg2+ in male reproductive physiology and sperm quality is not fully understood. Magnesium has a four-fold increase in the semen (116 µg /L) compared to the blood (29µg/L), but there is no correlation with semen parameters [1] and no clear role has been assigned to magnesium in semen. The male genital tract has two main related functions. The erectile function is associated with processes of emission and ejaculation in which the sperm is deposited in the female genital tract. The other function is sperm transport in the vas deferens and female genital tract. After sperm capacitation the sperm becomes actively motile as they transcend the female genital tract, to the site of fertilization in the ampullary portion of the fallopian tube. A coat of decapacitors that is deposited on the sperm surface from the epididymal and seminal plasm [2] play a role in suppressing capacitation of the sperm until they have embarked on their migration through female genital tract. Intracellular Mg2+ concentration unlike Ca2+ is very dependent on extracellular conditions [3]. We demonstrated a significant reduction of seminal plasma Magnesium concentration (87 ?g/L) compared to men with azoospermia (122 ?g/L), impotence (121 ?g/L) and control normal men (119 ?g/L) [4]. In another study in rat vas deferens, in a series of in-vitro experiments that external magnesium depletion enhanced the contractile response to electrical stimulation while increasing external Mg2+ inhibited the response [5].

The main objective of this study is to evaluate the central hypothesis that Magnesium is involved in the modulation of erectile function and Sperm motility in the vas deferens and the female genital tract.

MATERIALS AND METHODS

In a series of clinical observations and experiments (a) we investigated the role of magnesium supplementation in 15 men with genuine premature ejaculation and 10 with impotence. Both groups were treated with Magnesium Chloride 500 mg twice daily for a week, and a maintenance dose of 120 mg for one month.

(b) The role of Magnesium in the decapacitation of sperm was investigated among 45 men with normal sperm quality. The following components were evaluated (i) modulators of sperm motility by Ca2+, Mg2+, Pentoxifylline, and Interleukin-4 in a dose- and time-dependent manner. Endpoint markers evaluated included sperm motility, malonialdehyde by a colorimetric method, TNF-? and interleukin-4 by Enzyme Linked Immuno-Sorbent Assay (ELISA).

(c) Sperm membrane transfer of Magnesium was investigated in 7 men with normal sperm parameters. Each semen sample was adjusted to 20 million/ml.

(d) The effect of hypomagnesiaemia was studied in three groups of

Sprawley-Drew adult male rats. (a) In five rats, hypomagnesaemia was induced by diuresis by frusemide (Lasix) for over a week. (b) Five other rats that had no induced diuresis had magnesium supplementation. (c) Five SD rats served as controls. The influence of external Mg2+ concentration on reactivity of the rat vas deferens to electrical stimulation *in vitro* was evaluated. Vasa deferentia isolated from adult male rats were set up in tissue baths containing physiological salt solution at 37°C and were stimulated electrically. Increasing or decreasing concentrations of Mg2+ were added to the bath and their effect on electrically evoked contractions were recorded. Different concentrations of Ca2+ were added to the tissue bath and the effect on the electrical contraction recorded.

RESULTS

Of the 15 men with premature ejaculation treated with Magnesium Chloride, 12(80%) improved significantly after treatment with Magnesium, compared with 2 out of 10 men with impotence ($p < 0.01$). there was improvement in sexual drive ($p<0.05$), sustained erection ($p<0.01$, decrease in occasional impotence ($p<0.05$) and general sexual satisfaction ($p<0.001$). Among men with impotence, only one man achieved sustained erection, orgasm and sexual satisfaction. It must be pointed out however, that six of the 10 men were diabetics on insulin therapy.

In the role of Magnesium in the inhibition of contraction of the male genital tract, the isolated vas deferens of SD rat was used, as shown in *figure 1*. Magnesium induced deprivation was

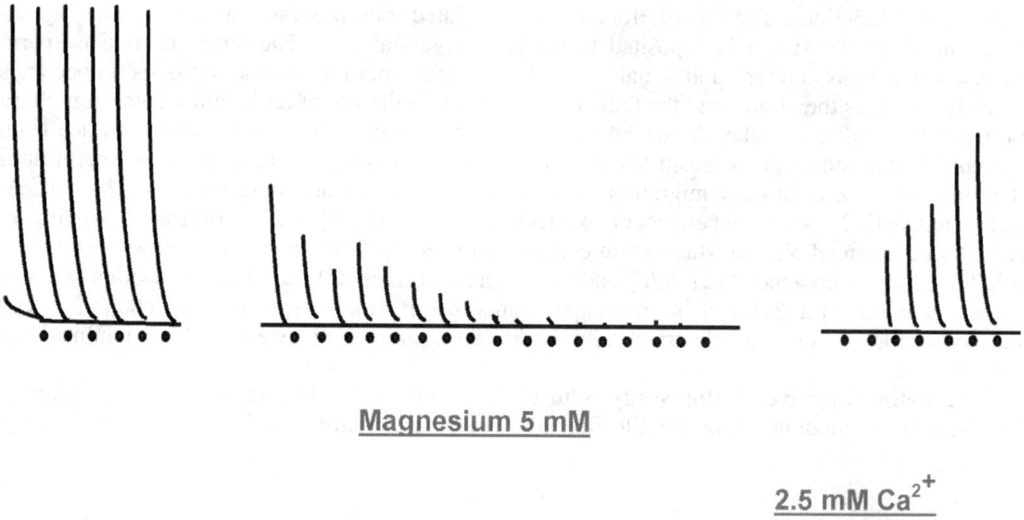

Fig. 1. Inhibition of electrical contractions by magnesium.
A typical original trace showing the inhibitory effect of magnesium?? electrically induced contractions of the rat vas deferens (Ca^{2+} = 0.625 ?? reversal by raising extracellular Ca^{2+} to 2.5 mmol/L. The dots?? stimulation (70 V, 10 Hz in 0.5-strains)

associated with significantly lower tissue level of Magnesium ($p < 0.05$) as in *table 1.* There was increased twitch contractions of the vas deferens compared magnesium supplementation and controls.

Incubation of Ca2+ with normal sperm was associated with a dose-dependent increase of 25%(19-31%) in sperm motility. The increase in sperm motility was inhibited by about 50% by addition of Mg2+, which was reversed by Ca2+. Pentoxifylline incubation with sperm resulted in a 20% in sperm motility, but unlike with Ca2+, it was not associated with inhibition with mg2+. Conversely, initial incubation sperm with Mg 2+ tended to reduction of sperm motility by 20%, reversed by Ca2+. After incubation of sperm with Ca2+ or pentoxifylline, both malonialdehyde and tumour necrosis factor revealed a dose-dependent increases ($P<0.002$, $p< 0.003$) but not with Magnesium *(table 1).*

The possible mechanism of the transmembrane transfer of Magnesium into the spermatozoa, the effect of increasing seminal plasma concentration of Mg2+ on intracellular concentration was investigated. Increase in extracellular Mg2+ levels was associated with a dose related increase in intracellular Mg2+ ($p<0.01$) *(table 2)* and *figure 2.*

DISCUSSION

Magnesium may have important functions in human and animal reproduction. One of the functions of Mg2+ in serum in semen may be to regulate the contractility of the vas deferens. Magnesium is probably involved in semen and sperm transport, by the contraction of smooth muscles of the vas deferens in the propulsion of sperm into the pelvic urethra. This may have important physiological implication on the time of emission and ejaculation [4]. Extracellular hypomagnesaemia in semen may accelerate emission phase of ejaculatory mechanism, thus predisposing to premature ejaculation. Since Mg2+ inhibits phosphodiesterases, it may also be useful in non-medical cases of impotence, since most of the cases included in the present study were diabetic. The present study has demonstrated that intracellular Mg2+ concentration is dependent

Table 1. Comparison of TNF, IL-4 and malonialdehyde after incubation of sperm with modulators of sperm motility

	Calcium		Magnesium		Pentoxifylline	
	(a)	(b)	(a)	(b)	(a)	(b)
Malonialdehyde nmol/ml	1.5±0.8	3.5±1.2	1.6±0.8	1.4±0.8	1.5±0.9	4.1±1.4
TNF (pg/ml)	24±3	41±6	24±3	16±4	23±4	56±11
IL-4	12±4	6±2	12±4	14±6	11±5	7±4
Hypoosmotic Swelling test	64±16	65±16	66±16	65±16	64±16	64±16

(a) Level before incubation and (b) Level after incubation
No difference in the fertilizing capacity of the sperm by the three agents after incubation.

Table 2. Comparison Of Magnesium and Calcium in the seminal plasma, sperm intracellular compartment and sperm membrane

	Mg2+	Ca2+	Ca2+:Mg2+
Seminal plasma Mg/L	58.2±14.4	141.1±24.6	2.4
Sperm Membranes /20million/ml	9.4±2.3	28.6±8.4	3.0
Sperm Intracellular /20million/ml	7.4±2.7	17.3±2.7	2.3

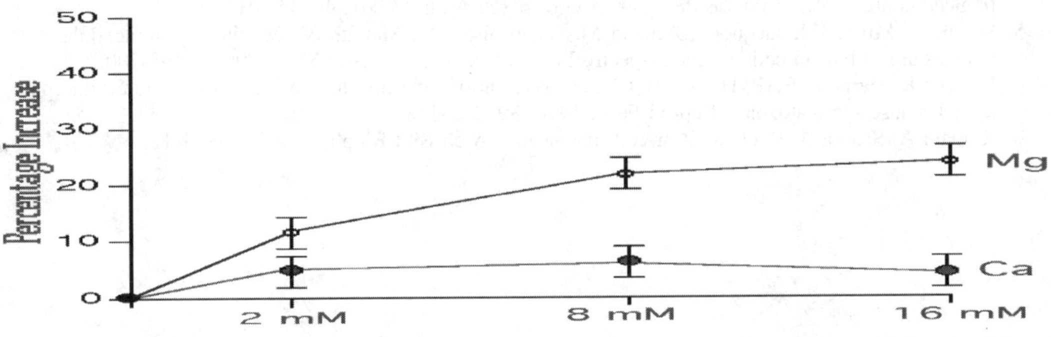

Fig. 2. Effect of extracellular magnesium on intracellular concentarations of magnesium (mg) and calcium (Ca)

on extracellular concentration as previously demonstrated by Sebille and colleagues [3]. Any condition like keto-acidosis and diuretic therapy may therefore lead to hypomagnesaemia with its sequelae. As a decapacitator of sperm, Magnesium is important in the preservation of sperm by delaying capacitation until the sperm under normal physiological conditions, the sperm are deposited in the female genital tract. Mg 2+ could delay capacitation by inhibiting Phosphodiesterase and increase the cyclic nucleotide in the sperm. The cyclic nucleotide cAMP and cGMP are able to stimulate Ca2+ extrusion when present with their dependent protein Kinases. Seminal plasma contains calmodulin-like proteins that are thought to stimulate Ca2+Mg2+ATPase activity in the sperm membrane and prevent the influx of Ca2+ that promotes capacitation process [6]. The plasma membrane Ca2+ pump (PMCA) and the Na+-Ca2+ exchanger are systems that extrude Ca2+ from the cytosol to the extracellular space. This system of Ca2+ extrusion needs more energy than the Ca2+Mg2+ ATPase activity [7]. By delaying capacitation, Magnesium may also be useful in sperm cryopreservation for assisted reproductive technology (ART).

Another possible role of Magnesium in semen is that of preventing lipid

peroxidation that is protective of sperm. As shown in the present study, unlike Ca2+ and Pentoxifylline, Mg2+ did not lead to production of malonialdehyde and tumour necrosis factor, markers of oxidative stress in semen.

CONCLUSION

The present study has demonstrated Magnesium has a role in human and animal reproduction, like modulation of the contractility of the vas deferens to prevent premature ejaculation by preventing premature emission. As a sperm, decapacitator it makes important contribution to reproductive physiology and in assisted reproductive technology. Further research is necessary.

REFERENCES

1. Omu AE, Dashti H, Mohammed AT, Mattappallil AB. Significance of Trace Elements in Seminal Plasma of Infertile Men. Nutrition 1995; 11: 502-505.
2. Chavarria ME, Reyes A, Huacuja L, Rosado A. Capacitation and acrosome reaction of human spermatozoa: an assisted reproduction approach. Arch Med Res 1992; 23: 83-92.
3. Omu AE, AA Al-Bader, H Dashti, Oriowo MA. Magnesium in Human semen: possible role in premature ejaculation. Arch Androl 2001; 46: 59-66.
4. Omu AE, AA Al-Bader, H Dashti, Oriowo MA. Effect of extracellular Mg2+ concentration on electrically induced contractions of rat vas deferens in vitro. Arch Androl 2001; 46: 159-167.
5. Sebille S, Millot JM, Jacquot J, Arnaud M, Delabroise AM, Manfair M: Monitoring intracellular free magnesium in human cells by microspectroflurimentry. Met Ions Biol Med 1996; 4: 198-200.
6. Fraser LR, Harrison RAP, Herod JE. Characterisation of a decapacitation factor associated with epididymal mouse spermatozoa. J.Reprod Fertil 1990; 89: 135-148.
7. Romani A, Scarpa A. Regulation of cell magnesium Arch Biol Biophys 1992; 298: 1-12.

*Metal Ions in Biology and Medicine: vol. 7.*Eds L. Khassanova, Ph. Collery, I. Maymard, Z. Khassanova, J.-C. Étienne.
John Libbey Eurotext, Paris © 2002 pp. 545-549.

Fe, Cu, Co, Vitamin C and folate intake together with basic food ration used in alimentation of soldiers doing military service in polish army and blood hematological parameters of soldiers-blood donors

K. Klos[1], J. Bertrandt[2], A. Klos[2], D. Blinowski[3]

[1] *Department of Infectious Diseases and Allergology, Central Clinical Hospital, Military School of Medicine, 128 Szaserów Street, 00-909 Warsaw, Poland*
[2] *Military Institute of Hygiene and Epidemiology, 4 Kozielska Street, 01-163 Warsaw, Poland*
[3] *Department of Transfuzjology and Transplantology, CBB, CCH, MMA, 128 Szaserów Street, 00-909 Warsaw, Poland*

ABSTRACT

The aim of the work was estimation of iron, copper, cobalt, vitamin C and folate (as factors influencing on hematological parameters of human organism), in basic food ration planned for alimentation for young men doing military service. Values of hemoglobin, haematocrit, number of erythrocytes, leucocytes and thrombocytes in blood soldiers fed this diet and being blood donors were estimated as well. Based on obtained parameters rates of MCV (Mean Corpuscular Volume), MCH (Mean Corpuscular Hemoglobin) and MCHC (Mean Corpuscular Hemoglobin Concentration) were calculated.

Estimation of iron, copper, cobalt, vitamin C and folate content was done using FOOD 2 calculation software and "Tables of Content and Nutritive Value of Food Products". Hematological parameters were determined according to methodology obligatory in Poland.

It was found that iron met the requirements in 123%, copper from 76.8% to 96.0%, vitamin C 232.8% and folate 206.7% of obligatory in Poland recommended norm for young men working hard, including technological losses. Planned food ration contained 76.36 µg cobalt. Examined hematological indicators of 400 donors who gave their blood just after joining the army and among 400 who gave their blood after few months of being fed m/a diet were similar. They were included in proper values limits.

Performed researches allow to state that planned in food ration content of examined elements, except copper, influencing on hematological parameters were included in amounts meeting the requirements for young men working hard. Copper amount should be taken into consideration during alimentation planning.

Key words: blood donors, food ration, sex, homeostasis

INTRODUCTION

Mineral elements including metals make a group of components treated as essential because human organism cannot synthesize them. They must be delivered together with food in adequate amounts and proportions. Mineral elements carry out different functions in living organism. Some of the elements are necessary during erythrocytes and hemoglobin creation processes. This group consist of iron, copper and cobalt [1]. Bioassimilability of mineral elements from the food depends on internal (sex, age etc.) and external factors i.e. form they exist inside food [2]. Relation between unbalanced mineral elements intake and incidence of different chronic illnesses is very often considered. Utilization of mineral elements contained in food depends on many factors that are related both to food products and human organism. Factors influencing on mineral elements utilization are shown below - *fig. 1* [3].

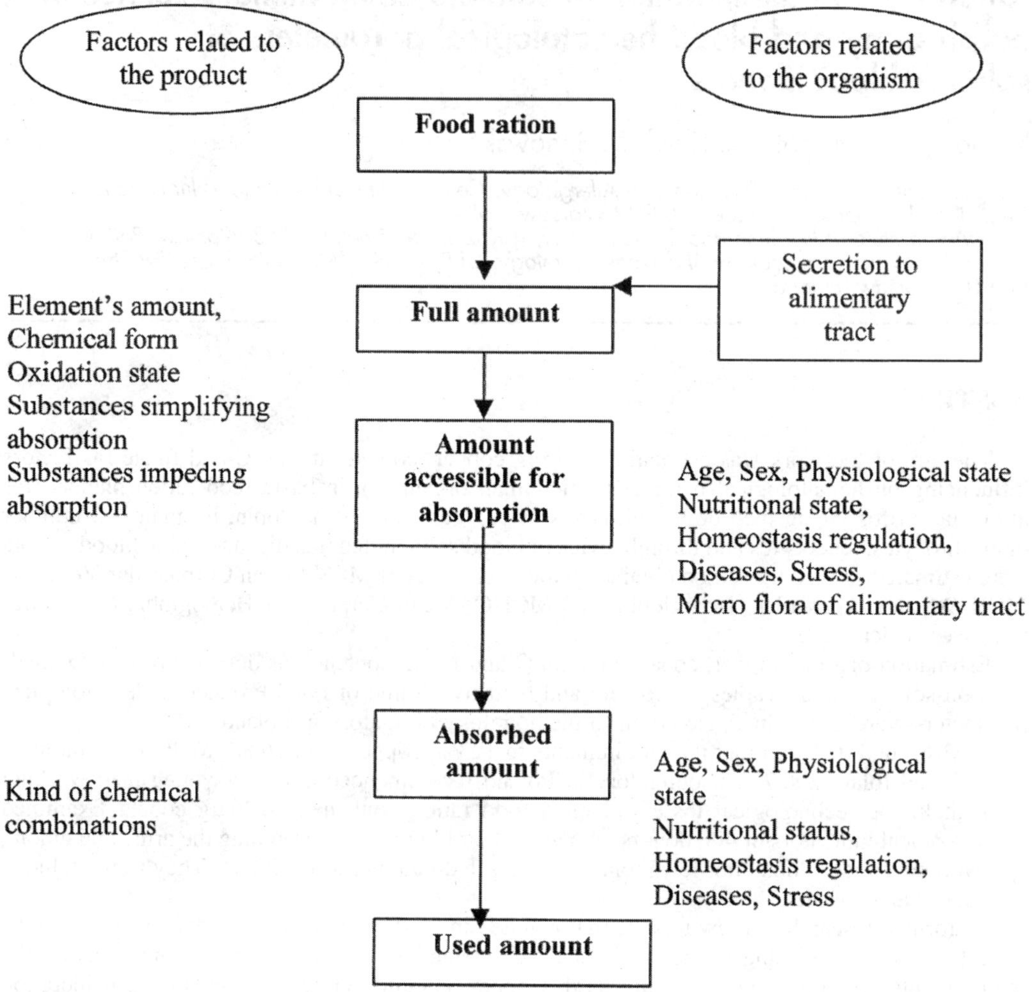

Fig. 1. Factors influencing on mineral elements utilization in organism

It is known that human organism may work even during considerable variations in mineral elements consumption thanks to intestinal absorption processes regulation mechanisms, storing and excretion that keep homeostasis. But long lasting both deficiency and excess of elements in the diet may have serious consequences in the shape of disturbances leading to many diseases occurrence [4]. Anemia is one of the diseases caused by food deficiency. Deficit of such hematopoietic elements as iron, folates, copper or cobalt deficit may be one of the main reasons causing this disease [5].

Total iron amount in human organism amounts 3-5 g, from that 65-67% is contained in hemoglobin, the rest is in myoglobin, ferritin and enzymes. This element participates in oxygen transportation and in cellular respiration. Assimilability of the iron depends on its form in the diet [6].

Copper plays very important role in erythrocytes creation. Copper is a part of respiratory enzymes and allows activation of the iron reserves during hemoglobin synthesis. Copper is an element widely spread in food products. Keeping proper proportions between copper and iron influences on proper absorption of both elements.

To keep proper organism's homeostasis cobalt contained in vitamin B_{12} (cyanocobalamin) molecule is required as well. Inorganic cobalt compounds occurring in food are well water-soluble and following well absorbed.

Folates are necessary to keep proper functioning of hematopoietic system. They are contained in almost all animal and vegetable origin food products. Folates deficit resulting from insufficient consumption, wrong absorption or increased requirements may lead to megaloblastic anemia occurrence.

AIM OF WORK

The aim of work was estimation of iron, copper, cobalt, and vitamin C and folate content in basic food ration used in nutrition of soldiers doing military service. Hemoglobin, hematocrite, number of erythrocytes, leucocytes and thrombocytes value in blood of soldiers fed this diet was assessed as well.

MATERIAL AND METHODS

The source for research was basic food ration used in soldiers alimentation in Polish Army. Based on it iron, copper, cobalt, vitamin C and folate content was calculated. Calculation was done based on calculation software FOOD 2 [7] and "Tables of Food Products Nutritive Value" [8]. Hematological tests were done in two groups of blood donors, 400 persons in each group (according to methodology obligatory in Poland). First group consisted of blood donors giving blood for the first time, just after joining the army; the other group consisted of soldiers serving for few months. Obtained results were compared to norms for young men, obligatory in Poland.

RESULTS AND DISCUSSION

Analyzed food ration is currently a nutrition base of approx 90% young men doing military service in Polish Army.

It was found that average iron content in planned food ration was 20.5 mg/person/day and made 136.6% of obligatory norm for young men heavily exercised. Copper met the recommended requirements in amount ranging from 85.6% to 107.0%, and cobalt amounted 76.4 μg/person/day. Folates and vitamin C occurred in amounts exceeding norm in 200% what made 229.6% and 258.7% of recommended norm respectively. Considering technological losses of about 10% of calculated value, iron met the requirements in 123%, copper 76.8% to 96.%, vitamin C 232.8% and folates 206.7% *(fig. 1)*.

Researches performed by Olejnikat at all [9] in National Fire Department College and group of agriculture schools showed that planned food rations met the requirements in 132% and 139% of norm for iron in 101% and 108% for copper respectively. Klicka at all. [10] in similar researches done in US Military Academy showed that iron met the requirements in 133-161%, folates 115,7-144% and vitamin C 244-382%. Mean copper content in an average German diet met the

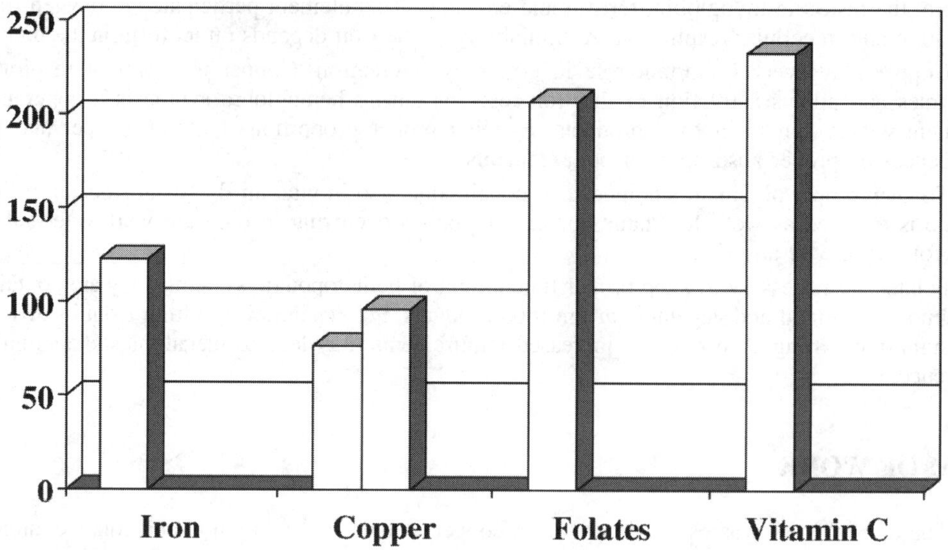

Fig. 2. Percentage of norm of Fe, Cu, folate and vit. C contained in food ration used for nutrition of young men doing military service

requirements in 109% and iron in 122% [11]. Folate mean content in American diet for men amounted 277±192 μg (149-357 μg) [12].

Examined hematological indexes *(tab. 1)* were on the similar level between both groups of blood donors. Average hemoglobin content among beginners and advanced donors amounted 14.6±0.60, hematocrite value amounted 43.8±1.59 and 43.9±1.73 respectively. Number of erythrocytes and leucocytes between both groups were on the similar level and amounted: $4.76 \pm 0.19 \times 10^{12}$ /l and $4.80 \pm 0.18 \times 10^{12}$ /l and $7.7 \pm 1.16 \times 10^{9}$/l and $7.4 \pm 1.28 \times 10^{9}$ /l respectively. These values were included in norms obligatory in Poland. Thrombocytes number in first group of soldiers was $273.4 \pm 47.9 \times 10^{9}$l, and in the other group $280.2 \pm 44.1 \times 10^{9}$l.

Table 1. Young men's, doing military service, blood hematological parameters

Donors	Hemoglobin g/100 ml	Hematocrite value%	Erythrocytes content $\times 10^{12}$/l	Leucocytes content $\times 10^{9}$/l	Thrombocytes content $\times 10^{9}$/l
Group I	14.6±0.60	3.8±1.59	4.76±0.19	7.7±1.16	273.4±47.9
Group II	14.6±0.61	43.9±1.73	4.80±0.18	7.4±1.28	280.2±44.1
NORM	**13.5**	**40.0**	**4.15**	**4.0-7.0**	**150-300**

Indexes of mean erythrocyte volume (MCV) (92.3±2.8 and 91.4± 3.08 $\mu m^{3)}$, mean hemoglobin mass in erythrocyte (MCH) (30.7±1.03 and 30.5±1.14 pg) as well as mean hemoglobin concentration in erythrocyte (MCHC) (33.3±0.61 and 33.4±0.84%) calculated from obtained values were contained in obligatory norms.

Radulski at all. [13] in previously performed researches on nutritive state biochemical estimation of soldiers from selected military unit, observed increase of erythrocyte number, hemoglobin concentration, mean hemoglobin mass and its concentration in erythrocytes during military service. Most of obtained results were contained in obligatory norms. Ziemla?ski [14] got similar values of hematological indexes of young men staying in identical living condition. Authors did not observed significant influence of the diet with controlled content of essential unsaturated fatty acids on hematological indexes values.

Radulski at all [13] found that 100% of examined men doing military service revealed proper iron level in blood serum both at the beginning and in the end of military duty.

Table 2. Calculated erythrocytes indexes

MCH - mean hemoglobin mass in erythrocyte (pg)	MCHC - mean hemoglobin concentration in erythrocyte (%)	MCV-mean erythrocyte volume (μm^3)
30.7±1.03	33.3±0.61	92.3±2.8
30.5±1.14	33.4±0.84	91.4±3.08
27-31	**32-36**	**82-92**

CONCLUSIONS

Performed researches show that in used for nutrition of young men, doing military service, food ration iron, cobalt, folates and vitamin C occurred in amounts higher than recommended norm while copper content differentiated and ranged from 84.6 to 107%.

Significant nutrition influence on changes in blood hematological parameters of young men doing military service, blood donors, was not found.

REFERENCES

1. Anke M., Glej M., Muller R., Dorn W., Vormann J., Anke S.: Macro, trace and ultratrace element intake of adults in Europe problems and dangers. J. Commodity Sci. 2000. 39. 119-139.
2. Babicz-Zielińska E.: Propedeutykażywienia. WSM Gdynia 2000
3. Gawęcki J., Hryniewiecki L.: Żywienie Człowieka. Podstawy Nauki o?ywieniu. PWN Warszawa 2000.
4. Imaeda N., Tokudoma Y., Ikeda M., Kitagawa I., Fujiwaria N., Tokudome S.: Foods contributing to absolute intake and variance in intake of selected vitamins, minerals and dietary fiber in middle aged Japanese. J.Nutr. Sci. Vitaminol. 1999, 45, 519-532.
5. Rafalski H.: Masowe zapobieganie niedoborom zelaza w ochronie zdrowia ludności. I Krajowy Kongres Żywność, Żywienie a Zdrowie, Warszawa 1994, 58.
6. Sempos M.T., Looker A.C.: Iron status and the risk of coronary heart disease: an example of the use of nutritional epidemiology in chronic disease research. J. Nutr. Bioch. 2001, 12, 170-182.
7. Pakiet Programowy FOOD 2. IZZ, Warszawa 1998.
8. Kunachowicz H., Nadolna I., Przygoda B., Iwanow K.: Tabele Wartości Odżywczej Produktów Spożywczych. Warszawa 1998.
9. Olejnik D., Krejpcio Z., Śmigiel-Papińska D., Wójciak R., Gawęcki J., Wiśniewska J.: Zawartość wybranych składników mineralnych (Ca, Mg, Zn, Cu, Fe) w całodziennych racjach pokarmowych młodzieży starszej. Roczniki PZH 1999, 50, 4, 361-367.
10. Klicka V.M., King N., Lawin P.T., Asker E.W.: Assessment of dietary intakes of cadets at the US Military Academy at West Point. J. Am. Coll. Nutr. 1996, 15, 3, 273-282.
11. Betts N.M.: Zinc and copper intakes and their major foods sources for older adults in the 1994-96 continuing survey of food intakes by individuals. Am. Soc. Nutr. Science 200o, 2838-2843.
12. Subar A.F., Block G., James D.: Folate intake and food sources in the US population. Am. J. Clin. Nutr. 1989, 50, 508-516.
13. Radulski J., Rozmysł E., Kłos A., Kołodziejczyk E.: Biochemiczna ocena stanu odżywienia żołnierzy wybranej jednostki wojskowej. Lek. Wojsk. 1997,9-10, 462-467.
14. Ziemlański Ś.: Zalecenia Żywieniowe dla Ludności w Polsce. Warszawa 1998 I.

*Metal Ions in Biology and Medicine: vol. 7.*Eds L. Khassanova, Ph. Collery, I. Maymard, Z. Khassanova, J.-C. Étienne.
John Libbey Eurotext, Paris © 2002 pp. 550-553.

Calcium and vitamin D_3 intake with diet and osteoporosis problem among inmates of military pensioner house

J. Bertrandt[1], A. Klos[1], E. Rozmys[1], M. Schlegel-Zawadzka[2]

[1]*Military Institute of Hygiene and Epidemiology, 4 Kozielska Street, 01-163 Warsaw, Poland*
[2] *Department of Food Chemistry and Nutrition, Collegium Medicum, Jagiellonian University, 9 Medyczna Street, 30-688 Kraków, Poland*

ABSTRACT

Decade menus including two diets, general and light one, were the bases for calcium and vitamin D_3 content estimation. Calculations were done using calculation software FOOD 2 and Tables of Nutritive Value of Food Products. Obtained results were compared to obligatory in Poland norms for these elements for men and women aged over 60. Osteoporosis occurrence was estimated by measuring bones density on the left forearm of examined persons, using densitometry method.

It was found that average planned calcium content in general diet was 1214.9±204.9 mg and in light one 1142.4±178.0 mg. General diet met the requirements of Polish norm in 110.4% for women and 135.0% for men and light diet in 103.8% for women and 126.9% for men. Taking into consideration technological losses during meals preparation calcium content in general diet met the requirements in 99.4%, and in light one in 93.5% of recommended norm for women and 121.5% and 114.2% for men respectively. Average vitamin D_3 content in general diet was 3.53±1.21 µg what made 70.6% of its safe level and 35.3% of recommended norm for both men and women. Average vitamin D_3 content in light diet was 2.86±0.89 µg what made 57.2% of its safe level and 28.6% of recommended norm.

Osteoporosis occurrence was found among 71.0% examined women and 42.3% examined men. Osteopenia was found among 25.8% women and 34.6% men.

Discovered deficiencies of calcium for women and vitamin D_3 for men and women may be one of the main reasons of osteoporosis and osteopenia occurrence among inmates of Military Pensioner House.

Key words: calcium, vitamina D_3 osteoporosis, osteopenia

INTRODUCTION

Calcium is the macro element necessary for proper organism functioning. Over 99.8% of this element is contained in bones and teeth in form of calcic-phosphatic salts. Small calcium amounts occur in blood serum and inside cells playing very important biological role. Calcium activates many enzymes and participates in blood coagulation. Calcium ions influences on proper excitability of neuromuscular system correct heart working and cell membranes permeability as well [1]. Amount of ionized calcium contained in healthy adult human organism comes from consumed and absorbed in intestine calcium, calcium exchanged with bone tissue and calcium resorbed by kidneys. Positive calcium balance conditions osteogenesis predominance over resorption [2].

Decreased osteoblasts activity leading to reduced calcium absorption from gastrointestinal tract and drop of calcium reabsorption in renal tubules is observed among people at advanced age.

Calcium is mostly absorbed in duodenum and jejunum. Absorption degree depends on organism demands and vitamin D_3 saturation. Long lasting calcium and vitamin D_3 deficit in meals may lead to decrease of calcium concentration in blood. Osteoporosis *i.e.* loss of bone mass is a result of organism aging [3]. Mean calcium absorption from food ration amounts 30-40% and decreases at advanced age. The richest calcium sources in Polish diet are dairy products, mainly milk in different forms, cheese and cottage cheese [4, 5]. Calcium assimilability from dairy products is the highest and amounts approx. 32%.

Calcium metabolism consisting of intestine absorption and calcium accumulation in organism is closely related to bone tissue metabolism and undergo dynamic changes during entire human life. Proper calcium balance favors not only accomplishment of peak bone mass but also most of all protects bone tissue from excessive calcium reserves loss and prevents osteoporosis development [6]. Improper nutrition and particularly low consumption of mineral elements with daily meals are the most important factors influencing on osteoporosis threat and development. Calcium deficiency and improper ratio between insufficient calcium intake and phosphates surplus in the diet is the most significant [3, 7].

MATERIALS AND METHODS

The aim of the work was estimation of calcium and vitamin D_3 content in meals planned for consumption for inmates of Military Pensioner House. Osteoporosis occurrence among 26 men and 31 women, inmates of that house, was estimated at the same time.

Decade menus including two diets, general and light one, were the source for estimation of calcium and vitamin D_3 content. Calculations were done based on calculation software FOOD 2 and "Tables of Food Products' Nutritive Value". Obtained results were compared to norms for that elements for men and women aged over 60, obligatory in Poland. Osteoporosis occurrence was estimated by measuring bone density on the left forearm of examined persons using densitometry. Tests were done with a great help of TIMCO Company.

RESULTS

Analyse of 40 menus of the general diet and 40 menus of the light diet showed that mean calcium content amounted 1214.9±204.9 mg and 1122.4±178.0 mg respectively. General diet met the requirements in 110.4% for women and in 135.0% for men and light diet in 103.8% for women and in 126.9% for men.

Considering unavoidable technological losses occuring during meals preparation calcium content in general diet met the requirements for women in 99.4% and in light one in 93.5% and for men in 121.5% and 114.2% respectively.

For better calcium utilization administration of small vitamin D_3 doses is recommended specially for people at advanced age. It is especially important in case of peoples staying in insufficiently insolated rooms. Vitamin D_3 deficit plays fundamental role in osteomalacia and osteoporosis it means porosity and fragility of osseous system [7, 8].

Mean vitamin D_3 content in general diet amounted 3.53±1.21 µg what made 70.6% of the norm on the safe level of 5.0 µg/day and 35.3% of the recommended norm of 10 µg/day. Mean vitamin D_3 content in the light diet amounted 2.86±0.89 µg what made 67.2% of the norm on the safe level and 28.6% of recommended norm.

The main calcium source in general diet was milk and dairy products. M/a products delivered 824.1 mg of calcium what made 67.8% of entire calcium amount in the diet. Light diet consisted of milk and dairy products delivered 806.2 mg what made 70.6% of entire calcium amount in the diet *(table 1, table 2)*.

Table 1. Calcium content in general and light diet

	General diet	Light diet
Mean calcium content in daily diet (mg)	1214.9	1142.4
Amount of calcium coming from milk and dairy products (mg)	824.1	806.2

Table 2. Input of calcium coming from milk and dairy products in this element delivery in examined diets (in%)

	General diet	Light diet
Milk	54.1	65.7
Cottage cheese	1.9	-
Cheese	5.5	-
Melted cheese	2.3	-
Kefir	1.5	4.7
Cream	2.5	0.2
Other products	32.2	29.4

Inquiry researches performed in Kraków within group of people aged in an average 71, showed that milk delivered 55% of calcium, cottage cheese 21% and 16% cheese. Insufficient calcium intake (66% norm) was found while phosphorus was exceeded (145% norm) at the same time, what caused decreased calcium absorption from gastrointestinal tract [1].

Specific changes in osseous system take place during aging process. Spongious substance of bone increases compared to cortical part that become thinner. It comes to enlargement of space between trabecules in spongious part of the bone. It leads to porosity of the osseous system i.e. osteoporosis [9].

Osteoporosis occurrence among inmates of pensioner house was found among 71.0% of examined women and 42.3% examined men. Osteopenia was found among 25.8% women and 34.6% men. Proper bones density was stated among 23.1 men and only 3.2% women.

Researches performed by Sawicki at all [10] showed that in group of women over 50 osteoporosis occurred among 54.2% examined. Milk in daily menu was taken into consideration by 44.0% of female patient while only 35.0% healthy ones did it. The reason of increased milk consumption among patients was early osteoporosis threat suspicion or its diagnosis. Dawson-Hughes at all [11] performed researches on prolonged enrichment of calcium and vitamin D_3 in diets for men and women aged over 68. The following amounts of the elements were found in examined diets: 688±331 mg calcium and 5.0±3.4 μg vitamin D_3 for men and for women 821±478 mg calcium and 4.7±3.8 μg vitamin D_3. Diets of examined groups contained 769±443 mg and 686±293 mg calcium and 5.1±2.9 μg, 4.4±2.3 μg vitamin D_3 respectively. All examined groups got additionally 500 mg calcium and 17.5 μg vitamin D_3 during three following years. Changes in bones density were observed among men and women after two years of enriched diet consumption [12]. It was found that Austrians aged over 65 diets contained 734±213 mg calcium and 2.5±1.8 μg vitamin D_3 [13].

Conclusions

Found calcium deficiencies for women and vitamin D_3 for women and men may be one of the main reasons of osteoporosis and osteopenia occurrence among inmates of Military Pensioner House.

REFERENCES

1. Kostogrys R., Pisulewski P., Cieślik A., Filipiak-Florkiewicz A.: Poziom spożycia mleka i jego przetworów jako źródło wapnia przez ludzi starszych mieszkających samotnie. Żyw. Człow. Metab. 2000, supl. 27, 164-167.
2. Szkop I: Czynniki żywieniowe a metabolizm tkanki kostnej. Żyw. Człow. Metab. 2001, 28, 1, 71-85.
3. Gawęki J., Hryniewiecki L.: Żywienie Człowieka. Podstawy nauki o żywieniu. PWN Warszawa 2000.
4. Marcinkowska-Suchowiecka E.: Wybrane aspekty patofizjologiczne i leczenie osteoporozy starczej. Żyw. Człow. Metab. 1987, 14, 2, 109-116.
5. Lorenc R.S.: Znaczenie spożycia wapnia w fizjologii i patologii człowieka. I Krajowy Kongres Żywność, Żywienie a Zdrowie, Warszawa 1994, 46.
6. Kapuściński P., Marcinkowska-Suchowierska E.: Zastosowanie wapnia w zapobieganiu i leczeniu osteoporozy. Żyw. Człow. Metab. 1998, 25, 4, 403-407.
7. Ziemlański Ś.: Podstawy prawidłowego żywienia człowieka. Warszawa, 1998.
8. Red M., Macdonald H.M.: Nutrition and bone: is there more to it than just calcium and vitamin D? J. Med. 2001, 94, 51-56.
9. Ilich J.Z., Kerstetter J.E.: Nutrition in Bone Health Revisited. A story beyond calcium. J. Amer. Coll. Nutr. 2000, 19, 6, 715-737.
10. Sawicki A., Rutkowska U., Zdrójkowska B.: Spożycie wapnia z mleka i jego przetworów w powiązaniu z występowaniem osteoporozy u kobiet. Żyw. Człow. Metab. 1997, 24, 1, 63-74.
11. Sawicki A., Dębiński A., Skwarek M.: Osteoporoza jako problem żywieniowy i epidemiologiczny. I Krajowy Kongres: Żywność, Żywienie a Zdrowie, Warszawa 1994, 59.
12. Dawson-Hughes B., Harris S.S., Krall E.A., Dall G.E.: Effect of withdrawal of calcium and witamin D supplements on bone mass in elderly men and women. Am. J. Clin. Nutr. 2000. 72, 745-500.
13. Koenig J., Elmadfa I.: Status of calcium and vitamin of different population groups in Austria. Int. J. Vitam. Nutr. Res. 2000, 70, 5, 214-220.

Metal Ions in Biology and Medicine: vol. 7.Eds L. Khassanova, Ph. Collery, I. Maymard, Z. Khassanova, J.-C. Étienne. John Libbey Eurotext, Paris © 2002 pp. 554-557.

Biological activity of sodium salts D,L,-aminoacids, containing tetrahydropyrane ring

Svetlana A. Vasilyeva[1], Liliya Kh. Kalimullina[1], Inna M. Pecherskaya[1], Vladimir A. Sazhin[2], Mars G. Safarov[1]

[1]Department of Organic Chemistry of Bashkir State University, 32, Frunze Street, 450074, Ufa, Russia; [2]Volgograd's Medical Institute, 400087, Volgograd, Russia

ABSTRACT

In continuation of our work on the application of substituted epoxytetrahydropyranes (I) in the synthesis of biological active compounds we about a new approach to the fused sodium salts of the N-tetrahydropuranil-D,L-amino acids (III a-g) via a reaction of the 4-Metyl-3,4-epoxytetrahydropyrane (I) with neuractive D,L-amino acids (II a-f: γ-aminobutyric (a); glicine (b); β-alanine (c); taurine (d); glutaminic (e); aminosuccinic (f) and methionine (g)).

On the purpose to create new psychotropic means a pharmachological screening of sodium salts of compounds III (a-f) has been carried out.

INTRODUCTION

The introduction of aromatic and heterocyclic radicals in the structure of mono- and dicarboxyclic amino acids leads to the emergence of different types of psychotropic activity with compounds, which in characteristic of the derivatives of the γ-aminobutyric acid [5], the glicine [2] and the asparagic acid [5].

The addition of oxygen containing crown-heterocycles to amino acids in particular allowed to reveal compounds with antiamnesty and anticonvulsant activity [1, 6].

In this connection the aim of our research is the synthesis and psychopharmacological study of the N (4-hydroxy-4-metyl-3-tetrahydropyranyl) - derivatives of the five basic neuractive amino acids.

MATERIAL AND METHODS

The structure of the products has been proved by means of spectroscopic methods (^{1}HNMR, ^{13}CNMR). The 4-methyl-3,4-epoxy-tetrahydropyrane (I) was received by the method [4].

The N-(4-Hydroxy-4-Metyl-3-Tetrahydropyranil derivatives of amino acids (III a-g).

Amino acid (0.018 mole) was added to the solution NaOH (0.02 mole) in 20 ml H_2O and 2 ml epoxide (I) (0.18 mole) is poured drop by drop into the prepared sodium salt solution. The mixture is heated for 3 hour at 80°C, then neutralized by 4 N HCl to pH=7.

The fallen out sediment is crystallized from the water-alchoholic solution. The pharmacological screening of sodium salts of compounds (IV a-f) was conducted on the 350 white rats (bote male and female, 180-220 g each) and on 350 white male-mice, 18-20 g each.

The influence of the new compounds on the spontaneous motive activity research reaction and emotionality of the rats was studied with the help of the "open field" test [10].

RESULTS AND DISCUSSION

The compounds (IIIa-g) synthesized with the nucleophylic substitution reaction followed by the determination of their spectral and phisico-chemical epoxide (I) with sodium salts of the D,L-a-amino acids (II a-d) represent smooth procedures by aminogroups which is attained with 10% excess of NaOH that promotes the increase of pH and the transformation of the zwitter-ion into conjugated base, hence compounds (IIIa-g) in 65-95% yields.

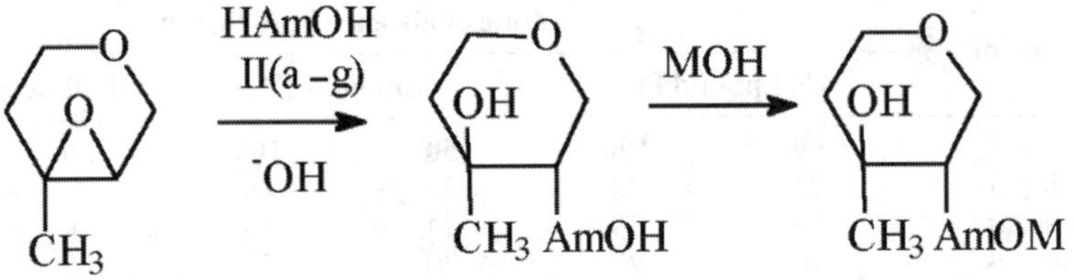

I III(a-g) IV(a-g)

M=Na, K

HAmOH: γ-aminobutyric acid (a); glicine (b); β-alanine (c); taurine (d); glutaminic (e); aminosuccinic (f) and methionine (g).

The screening was conducted on the sodium salts (IV a-g) well dissolved in water in contrast to amino acids (IIIa-g). As a result a number of compounds having obvious psychotropic effects and low toxicality (LD_{50} 1000 mg/kg) have been discovered. The substitution of aminogroup in the structure of neuroactive amino acids results in the emergence with new compounds of psychotropic properties, which the neuroactive amino acids themselves in small doses (less than 100 mg/kg) do not possess [5, 7].

Moreover, pyranil analogues of the inhibitory amino acids - glicine, β-alanine, γ-aminobutyric display anticonvulsant and analgetic activities, diminish the emotional reaction of fear. The sodium salts of N-(4-Hydroxy-4-metyl-3-tetrahydropuranyl) derivatives of glutamate and aspartate potentiate tonic and clonic phases of electroshock convulsions.

Taking into account the considerable structural difference between the N-pyranyl derivatives of midiator amino acids and certain agonists of amino acidic inhibitory and stimulating receptors [8, 9, 11] one can assume that when coming into the brain the pyranyl radical is chipped off the amino-acidic structure which leads to the subsequent inclusion of amino acids into synapsical and metabolic processes.

Besides it has been ascertained that sodium salts of compounds (IV a and IV f) suppress the growth of sulphate restoring bacteria and Pseudomonas. The property can be applied in oil industry for processing water pumped into oil layers.

The comperative date solubility, of sodium salts of IV a and IV g are known compound (4,4-Dimetyl-1,3-Dioxan), suppress the growth SRB was shown in *table 1.*

Table 1

Reagent	**Solubility in water g/l temperature, °C**			
	5	**20**	**79.4**	**91.5**
Compound IV a	50	60	75	80
Compound IV g	51	59	74	79
4,4-Dimetyl-1,3-Dioxan	25.6	18.0	10.7	absent

Solubility of 4,4-Dimetyl-1,3-Dioxan insignificant, at 91.5°C is absent, that it concedes to compounds IV a and IV g *(table 1)*.

Definition of bactericide activity carried out on Pseudomonas aeruginosa, Pseudomonas fluorescens, Pseudomonas aurfacienes, which are the comparisons (SRB) in oil layers, partisipate in oxidation of hydrocarbon oil and worsens its quality.

Table 2

Microorganism	Zone of absence growth, mm					
	Compound IV a		Compound IV g		4,4-Dimetyl-1,3-Dioxan	
	50	**100**	**50**	**100**	**50**	**100**
Ps. aeroginosa	22	28	23	28	5	10
Ps. fluoresceus	25	30	24	30	4	9
Ps. aurefaciens	20	28	20	28	2	7

It can be observed in *table 2* compounds IV a and IV g on comparison with 4,4-Dimetyl-1,3-Dioxan suppress the growth all tried bacterias genus Pseudomonas (zone of absence of growth 28-30 mm), a known compound suppress poorly.

Table 3

Bactericide	Time	Concentration, mg/l			
		50	**100**	**300**	**500**
Compound IV a	**0**	-	-	-	-
	5	+	-	-	-
	10	++	-	-	-
Compound IV g	**0**	-	-	-	-
	5	+	-	-	-
	10	++	-	-	-
	15	++	+	-	
4,4-Dimetyl-1,3-Dioxan	**0**	-	+	-	-
	5	+++	-	-	-
	10	+++	+++	-	-
	15	+++	+++	-	-

Annotation: (+++) intensive growth (+) week growth
(++) moderate growth (-) absent growth

Table 4

Concentration reagen mas. %	Degree of suppression height (SRB), %			
	Compound IV a	Compound IV g	4,4-Dimetyl-1,3-Dioxan	4-Metyl-5,6-Dihydro-2H-pyrane*
0.005	70	80	30	-
0.01	100	100	80	-
0.025	-	-	-	80
0.03	100	100	100	-
0.05	100	100	100	98

* Structural analog

The given results shows that Sodium Salts of N(4-hydroxy-4-methyl-tetrahydropyranil-3) glicine (IV a) and methionine (IV g) have more higher degree of suppression of growth sulphate restoring bacteria (SRB). That, mean concentration bactericide (0.01 mas. %) degree of suppression makes up 100%.

REFERENCES

1. Voronina T.A., Karaseva T.L., Golovenko N.A. Chem. Pharm J. USSR, 1988, 6, 672-682.
2. Voronina T.A., Rudenko O.P., and other, Chem. Pharm J. USSR, 1989, 1, 35-37.
3. Gatsura V.V. Methods of the primary pharmacological research of biological active compounds (Moscow), 1974, 13.
4. Ibatullin U.G., Vasilyeva S.A., Safarov M.G. Izv. Akad. Nauk USSR, Ser. Khim, 1982, 9, 2114-2118.
5. Kopelevich V.M. Gunar V.J., Chem Pharm J, 1981, 5, 27-39.
6. Lukyanenko H.G., Bogatsky A.V., Chem. Pharm. J, 1985, 6, 691-693.
7. Raevskiy K.S., Georgiev V.P. Mediathor aminoacids: neuropharmacological and neuchemical aspects (Sofia), 1986, 240 p.
8. Krogsgaard-Larsen P, Falch E. Mol. And Cell. Biochem., 1981, 38, 129-146.
9. Nielsen E.O., Modsen U, Schauburg K. et. al. Eur. J. Med. Chem. 1986, 21, 433-437.
10. Van Ree I.M., de Wied D. Discuss. Neurosci., 1988, 5, 1-60.
11. Shinozaki H. Prog. Neurobiol., 1988, 30, 399-435.

Metal Ions in Biology and Medicine: vol. 7. Eds L. Khassanova, Ph. Collery, I. Maymard, Z. Khassanova, J.-C. Étienne.
John Libbey Eurotext, Paris © 2002 pp. 558-561.

Lyposome forms of rhenium cluster compounds in models of haemolytic anemia

Shtemenko A.V., Shtemenko N.I.,* Oliyník S.A., Zelenuk M.A.

Ukrainian State Chemical-Technological University, 8 Gagarin avenue, 49005 Dnieprpetrovsk
**Dniepropetrovsk National University, 13 Naukoviy by-street, 49050 Dniepropetrovsk*

To continue our investigations of biological activity of cluster compounds of rhenium, lyposome forms were necessary to obtain and to investigate their structure as many of potentially active compounds had low solubility. UV-spectra of lyposome forms of the substances with organic acids, amino acids and adamantanic acids were investigated, some results about coordination of the compounds with lipid components were obtained. Among the range of cluster compounds of rhenium with organic ligands two ones were chosen for following experiments in vivo according to their ability to interact with biological membranes *in vitro*: 1-dichlorotetra-μ-(i-butirato) dirhenium(III), 2-pentachlorodi-μ-(γ-aminobutirato) dirhenium(III) chloride. Two models of haemolytic anemia were used: A - on rabbits by introducing of $PbAc_2$ - solutions; this model permits to investigate dynamics of anemia in one experimental animal; B - on rats by introducing of phenylhydrazine chloride. 1 and 2 were administrated as in solution as in lyposome forms. Administration of 1 and 2 led to: increase of haemoglobine and resistance of erythrocytes and to prolonging of life for haemolytic animals; significant decrease in quantities of MDA and increase in quantities of reduced glutathion (GSH), glutathionreductase (GSR) and glutathionperoxidase (GSP) in tissues of anemic animals. The most effective was 1 in lyposome form. Comparison of two modes of administration of 2 showed better results for lyposome form.

ABSTRACT

Investigation of interaction of cluster rhenium compounds with phosphatidylcholine was made and spectral characteristics of solutions and lyposomes are presented. Two models of haemolytic anemia were studied and two cluster rhenium compounds were examined in solutions and in lyposome forms. It was shown that rhenium cluster compounds acted as antioxidants in miocardium in these models.

INTRODUCTION

Our previous investigations showed some biological activity of newly synthesized cluster compounds of rhenium such as antitumor activity, erythrocyte stabilizing activity against osmotic haemolysis, changing of morphology of cells, etc. In some experiments it was shown that lyposome forms of metal-organic substances were more effective than their solutions. Necessity to study possibilities to prepare lyposomes is of especially importance for rhenium cluster substances with adamantanic acids as ligands due to their week solubility and potentially high biological activity (adamantanic radical presents in psychotropic, antiviral, immunomodulative medicines). From the other hand there exists some information about stabilizing effects of some metal-organic substances with antitumor properties on the isolated ischaemic-reperfused rat heart [1] throught decrease of malonaldehyde (MDA) production. Some new investigations showed the influence of metal-organic substances on apoptotic processes [2, 3]), that are considered now as the main mechanism of such tissue damages as ischaemia, myocardial infarct, etc. The main purpose of this paper is to discuss mechanism of interaction between phosphatidylcholine and cluster rhenium compounds with different ligands and formation of lyposomes. Among the range of cluster compounds of rhenium with organic ligands two ones were chosen for following experiments *in vivo:* 1 - dichlorotetra-μ-(i-butirato)dirhenium(III), 2 - pentachlorodi-μ-(γ-aminobutirato)dirhenium(III) chloride in order to analyze their antioxidant properties.

MATERIALS AND METHODS

Rhenium cluster compounds $Re_2((CH_3)_2CH_2COO)_4Cl_2$- 1; $[Re_2(NH_3^+CH_2CH_2CH_2COO)_2Cl_5(H_2O)_2]Cl\cdot 2H_2O$ - 2; $Re_2(AdCOO)_4Cl_2$- 3; $Re_2(AdCOO)_2Cl_4$- 4 were synthesized according to [4]. Electronic absorption spectra (EAS) of solutions of 1 - 4 in chloroformium (0,1 mM) without or with phosphatidylcholine (0,25%), loaded and unloaded lyposomes were investigated. Lyposomes were prepared according to [5]. Two models of haemolytic anemia were used: A - on rabbits by introducing of $PbAc_2$ - solutions; this model permits to investigate dynamics of anemia in one experimental animal; B - on rats by introducing of phenylhydrazine chloride [6]. 1 and 2 were administrated as in solution as in lyposome (lyp) forms. All measurements: malonic dialdehyde (MDA), reduced glutathion (GSH), glutathionreductase (GSR) and glutathionperoxidase (GSP) were accomplished according to described procedures.

RESULTS

EAS of 1 *(fig. 1)* showed existence of the absorption band in the area 20000 cm^{-1}, that is relevant to $\delta \rightarrow \delta^*$ electronic transition of quadrupol Re-Re bond.

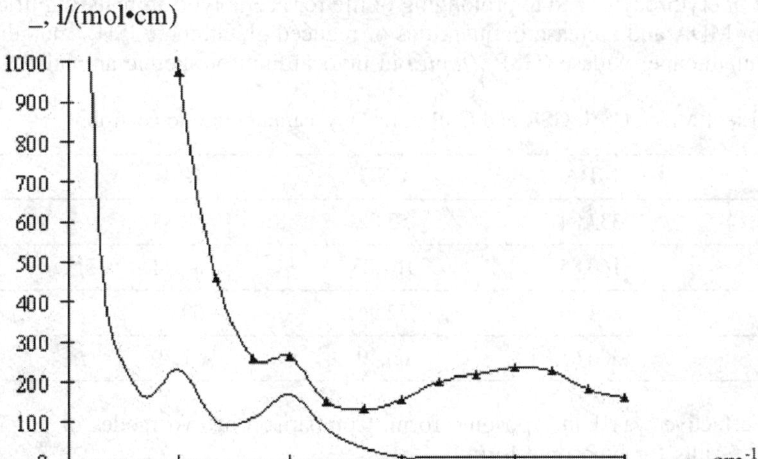

Fig. 1. EAS of 1 in chloroformium (-) and in mixture with phosphatidylcholine (-?-)

In the presence of the lipid there appeared a new band in the area 14500-14000 cm^{-1}, that is the characteristic of $\delta \rightarrow \delta^*$ electronic transition of phosphate groups which are coordinated around dinuclear center Re_2^{6+}. Intensivity of this band was progressing with time, that was one more argument for coordination process that took place in these condition between phosphatidylcholine and Re_2^{6+}. Such bands we discovered in the solutions of 2 (15936 см$^{-1}$), 3 (13937 см$^{-1}$.), and 4 (16287 см$^{-1}$).

In all experiments with loaded lyposomes we discovered that cluster fragment Re_2^{6+} was stable *(fig. 2)* as absorbtion that is relevant to $\delta \rightarrow \delta^*$ electronic transition was observed.

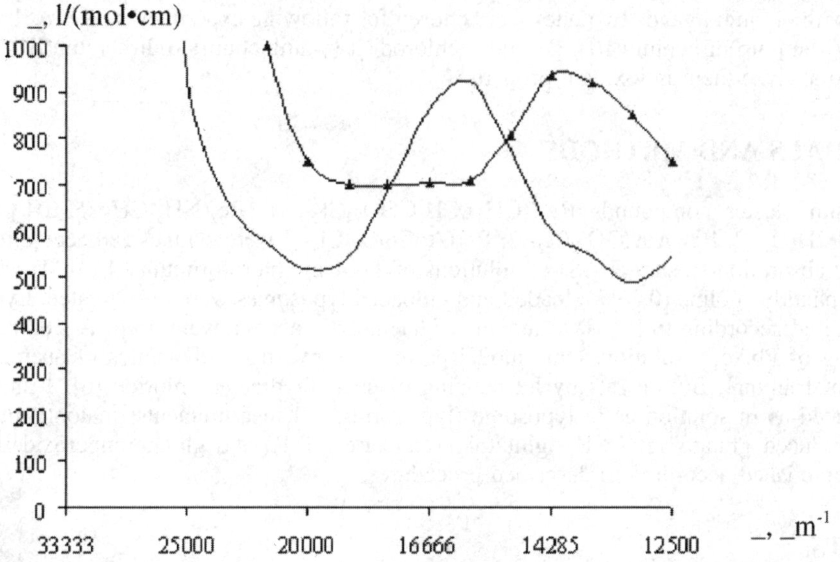

Fig. 2. EAS of lyposomes loaded with 1 (-?-) and 2 (-) against unloaded lyposomes

Administration of 1 and 2 in models of haemolytic anemia led to: increase of haemoglobine and resistance of erythrocytes and to prolonging of life for haemolytic animals; significant decrease in quantities of MDA and increase in quantities of reduced glutathion (GSH), glutathionreductase (GSR) and glutathionperoxidase (GSP) *(table)* in myocardium of anemic animals.

Table. Quantities of MDA, GSH, GSR and GSP in rats myocardium in% to control

Experiment	MDA	GSH	GSR	GSP
Anemia	337,84	53,82	52,85	72,88
Anemia+1 lyp	160,85	102,03	84,24	96,02
Anemia+2 lyp	234,36	72,96	59,05	91,26
Anemia+2	300,65	66,40	57,79	88,65

The most effective was 1 in lyposome form. Comparison of two modes of administration of 2 showed best results for lyposome form.

DISCUSSION

Among the range of cluster rhenium compounds with organic ligands we chose two tetracarboxylates - 1 and 3, and two dicarboxylates - 2 and 4 to demonstrate different mode of coordination in lyposomes. Appearence of a new low-energetic band of absorption in EAS of solutions of 1 and 3 with phosphatidylcholine may be explained by substitution of carboxylic groups by phosphate groups of the lipid with still preservation of the cluster fragment Re_2^{6+}. As for substances 2 and 4, coordination of phosphatidylcholine may take place by transformation of dicarboxylates to mixed dicarboxylate- diphosphate derivatives. It was displayed in EAS of solutions of 1-4 with phosphatidylcholine in comparison with EAS of solutions of 1 - 4 and may be explained by properties of carboxylic groups to formate conugated cyclic system that is impossible for phosphate groups.

Spectral characteristics of loaded lyposomes confirmed these data *(fig. 2)*, as here we have found the bands in the area 14000 CM^{-1} for 1 and 16000 CM^{-1} for 2. Thus, phosphate groups of phosphatidylcholine have possibilities to coordinate around Re_2^{6+} cluster fragment and just this coordination made possible to stabilize halogenocarboxylates of dirhenium against hydrolysis in physiological solutions and to increase their biological activity.

Chemically induced haemolytic anemia is followed by strong process of lipid peroxidation and disbalancing of the system of gluthathion enzymes *(table)*. Both 1 and 2 acted as stabilizers of the peroxidation process. Administration of 1 and 2 led to: increase of haemoglobine and resistance of erythrocytes to acidic haemolysis, and to prolonging of life for experimental animals; significant decrease in quantities of MDA and increase in quantities of reduced glutathion (GSH), glutathion-reductase (GSR) and glutathionperoxidase (GSP) in tissues of anemic animals. The strongest was 1 lyp, as this preparation turned back approximately to the normal state quantity of GSH and activity of GSR and GSP. Comparison of two modes of introduction of 2 showed better results for 2 lyp form, that confirmed above presented considerations about protection 1 and 2 against hydrolysis by lipids.

Mechanism of antioxidant action of rhenium cluster compound is speculated and experiments with some well-known antioxidants to compare with 1 and 2 are working out. Recent model experiments showed strong antiradical properties of 1 and 2, where 1 decreased quantities of artificial radicals as some natural antioxidants did. We guess that antioxidant and antiradical properties of 1 and 2 are realized due to the action of quadrupol metal-metal bond in the first place and due to stabilizing properties of ligands in the second place.

CONCLUSIONS

Phosphate groups of phosphatidylcholine coordinated around rhenium cluster fragment that was demonstrated by electronic absorption spectra data, stabilized halogenocarboxylates of dirhenium in lyposome forms against hydrolysis in physiological solutions and increased their biological activity. Rhenium cluster complexes with organic ligands were active in the models of haemolytic anemia. Their antioxidant properties may be explained by existance of quadrupol metal-metal bond and depends from nature and arrangement of ligands.

REFERENCES

1. Leperre A., Millart H., Prevost A., *et al.* Compared effects of ruthenium red and cis[$Ru(NH_3)_4Cl_2$]Cl on the isolated ischaemic-reperfused rat heart. *Fundam Clin Pharmacol*, 1995, N.9, 545-553.
2. Winter B., Schilling T.,Gey DC., et al. KP46, a new chemotherapeutic drug exerts its antitumour effects by induction of apoptosis. *Exp Clin Endocrinol Diabetes*, 1998, N.106, 130.
3. Syrkin A.B., Zukova O.S., Kikot B.S., *et al.* Teraftal - a new medicine for binar catalytic therapeutic of tumours. *Rus. Chem. J.*, 1998, 42, N.5, 140-146.
4. Shtemenko A.V., Bovykin B.A. Chemistry of Binuclear rhenium Clusters. TMS publication, Pensilvania, 1997. In Book *"Rhenium and Rhenium Alloys"* p.p. 189-197.
5. Munechika L. *United State Patent*, 1997, 5, 662, 931.
6. Meerson F.Z., Evstigneeva M.E., Ustinova E.E. Effect of chronic haemolytic anemia on heart contractile function and increase of its resistance to hipoxia. *Pat. Physiol. and Exp. Therap.*, 1983, N.5, 25-29.

Metal Ions in Biology and Medicine: vol. 7.Eds L. Khassanova, Ph. Collery, I. Maymard, Z. Khassanova, J.-C. Étienne.
John Libbey Eurotext, Paris © 2002 pp. 562-564.

Ferric-adenosine 5'-triphosphate complexes as chelators in thalaseamia

Cristos Pastras, Athanassios Dovas and Jane Anastassopoulou

National Technical University of Athens, Chemical Engineering School, Radiation Chemistry and Biospectroscopy, Zografou Campus, 15780 Zografou, Athens, Greece

ABSTRACT

The interaction of adenosine 5'-triphosphate disodium salt (ATP) with the trivalent $FeCl_3$ iron trichloride salt was investigated in aqueous solutions. Fourier Transform Infrared (FT-IR) spectroscopic studies showed that $FeCl_3$ perturbs the vibrational spectra of the above nucleotide, in particular the phosphate modes and the imidazole ring. The changes in the spectra indicate that Fe(III) cations interact with ATP forming a very stable complex. From the FT-IR spectra it was shown a direct Fe(III)-phosphate bonding at neutral pH values. It was also found from thermal analysis data that the formed complex was a dehydrated compound.

INTRODUCTION

Adenosine triphosphate (ATP) is the most important of the activated carriers in cells [1]. ATP serves as a convenient and versatile storage of energy to drive a variety of chemical reactions in cells. On the other hand, thalaseamia is one of the most common genetic diseases and is posing an increasingly important lethal problem in many countries. The iron chelation therapy started in young people in late 1960s. Without accompanying chelation therapy the iron damages the liver, endocrine organs and heart. Iron has an important role in cell metabolism and growth. However, iron catalyses the generation of oxygen radicals and thus the oxidation of haemoglobin. Thus the iron must be eliminated very quickly. Adenosine monophosphate (AMP) and adenosine triphosphate (ATP) are non-toxic chelators and their affinity for iron is great. In this work we study the chelation of trivalent iron with ATP as a ligand [2].

MATERIAL AND METHODS

Hexaahydrated ferric chloride ($FeCl_3.6H_2O$) and adenosine-5'-disodium salt ($Na_2H_2ATP.3H_2O$) were purchased from Sigma. The ATP ferric complexes were synthesized by dissolving in water hexaahydrated ferric chloride ($FeCl_3.6H_2O$) and $Na_2H_2ATP.3H_2O$ in ratios of ATP:Fe(III): 1:1, 2:1, 3:1, 4:1, 1:2, 1: 3 and 1:4 at a pH of 6-7 and temperature of 36-37°C, and at a concentration of 2.5 mM. The neutralization was done using NaOH 0,1 N solution. FT-IR spectra were recorded with a Perkin Elmer spectrometer, with a resolution of 4 cm^{-1}. The pellets were prepared using KCl as dispersing medium.

RESULTS AND DISCUSSION

The region of the FT-IR spectrum that is most interesting for analytical studies is that between 1700 and 700 cm^{-1}. In *table* are shown the bands of free ATP and of the complex ATP-Fe(III), as well as their possible assignments. The free ATP shows a strong sharp band at 1712 cm^{-1}, which is assigned to the bending mode of NH_2 vibration. This band is greatly changed upon complexation leading to the result that Fe(III) ion interacts may be indirectly through hydrogen bond of coordinated water molecules with the NH_2 group of ATP. This result is in agreement with the thermal analysis data, which indicated the presence of two water molecules of coordination.

Table. FT-IR transmittance bands of free ATP and Fe(III)-ATP. Complex with their possible assignments [3]

ATP cm^{-1}	Fe(III)-ATP cm^{-1}	Assignments
1712	-	δNH_2
1648	**1655**	δH_2O, δNH_2
1551	**1605** **1571**	$\nu C_4{=}C_5$ νN_3- $C_4{=}C_5$,
1495	**1481** **1460**	δN_7-C_8, νC_8-N_9, δC_8-H -
1408	**1415** **1330** **1304**	ν-N_1-C_6-N_6 ν pyrimidine ring
1260	**1251**	νN_7-C_8, νC_2-H, δN_1-C_8
1230	-	νN_7-C_8, δC_8-H,
1103	-	νC-O
1046	**1073**	
1020		
998	**977**	$nPO_3^{2-}deg$
917	-	$n_{sym}PO_3^{2-}$
822	**780**	ν-ribose phosphate

A split of the band at 1550 cm^{-1} occurs due to the metal- N7 binding, since these bands contain C8=N7 stretching vibration as well as the bending C8-H vibration mode. This result is confirmed from the band at 1230 cm^{-1}, that is associated with the stretching and bending vibration of C8=N7 and C8-H bond, and which is disappeared after the reaction.

Considerable changes were also observed after the reaction in the region of 920-1050 cm^{-1}, where the symmetric, degenerate and stretching vibrations of PO_3^{2-} bands absorb. The symmetric band at 917 cm^{-1} is disappeared upon complexation, while the band at 1043 cm^{-1} shifts to higher frequencies.

CONCLUSIONS

From the FT-IR spectra it is shown that ferric ions interact adenosine 5'-triphosphate molecule to produce a very stable complex. The major binding sites are the phosphate oxygens, as it is indicated from the changes of the spectra in the region between 920-1050 cm^{-1}. It was also find that Fe(III) interact indirectly with NH_2 group of ATP through hydrogen bonds of coordinated waters molecules. In conclusion, it seems that ATP can be used as a chelator, since it shows a high affinity for complexation with Fe(III).

REFERENCES

1. Bray A, Lewis J, Walter R R, Essential Cell Biology, An introduction to the molecular biology of the cell, 1997, Gerland Publishing, Inc, New York.
2. Anastassopoulou J, Anifantakis B, Anifantakis Z A, Dovas A, Theophanides T, Thalassaemia and the role of free radicals in our understanding of iron transport and storage in haemoglobin. *Bioinorganic Chem*, 2000: 79, 327-329.
3. Tajmir-Riahi H A, Theophanides T, Adenosine 5'-monophosphate complexes of Pt(II) and Mg(II) metal ions. Synthesis, FT-IR spectra and structural studies, *Inorg Chim Acta*, 1983: 80, 183-190.

*Metal Ions in Biology and Medicine: vol. 7.*Eds L. Khassanova, Ph. Collery, I. Maymard, Z. Khassanova, J.-C. Étienne.
John Libbey Eurotext, Paris © 2002 pp. 565-572.

Experimental study on the effect of metallic copper on fracture healing in chicken

SF Bao*, L Zhao, YJ Liu**, T Cong*, ZJ Cheng*, Z Li*, H Tian***

**Trace Element Research Laboratory, 301 Hospital, Beijing, P R China,*
*** Department of Orthopedics, 301 Hospital, Beijing P R China,*
****No.3 Pharmaceutical Factory, Harbin*

ABSTRACT

Material and methods: 160 growing chickens aging 70 days were fed in 32 cages, five each. The fracture with 1-2 mm defects at bilateral radius of wings was created by osteotomies. The chickens were then divided into four groups, 40 each. Every morning, the first group chickens, as the control, were fed orally with 0.8% Na-CMC solution (sodium carboxymethyl cellulose), the other three groups with copper-Na-CMC suspension (copper powder suspended in 0.8% Na-CMC) of different doses: 20, 40 and 80 mg Cu/kg body weight. All the chickens had free access to water and diet containing Cu 5.4 mg/kg diet. Ten chickens of each group were sacrificed at the 14, 21, 28, and 35thday postoperatively. Liver was taken for analysis of Cu and Zn content. Both radiuses were removed by dissection. Computed tomography scan (CT) was performed for measuring gray values of callus quantitatively. The biomechanical properties of the healing radius were analyzed by a three point bending test. Afterwards, the contents of calcium, magnesium, phosphorus, copper, zinc, iron and hydroxyproline in callus were determined. ***Results:*** The gray values of callus increased along with the increase of copper dose and the observing duration. At the 21 and 35 day after operation, the gray values of callus in high-Cu group were significantly higher than that in the control respectively (909±221 vs. 597±155; 973±100 vs. 763±179 HU p<0.05). The calcium and magnesium in callus 35 days post operation was found to be much higher in middle and high Cu groups than that in the control respectively (Ca: 177±26.7, 176±20.5 vs. 137±34.7 mg/g; Mg: 2.98±0.56, 3.06±0.46, vs. 2.43±0.53 mg/g P(0.05). The contents of hydroxyproline in callus 35 days post operation in high-Cu group were significantly higher than that in the control: 34.23±1.96 vs.32.17±1.93 mg/g, P(0.05. The biomechanical properties of repaired radius had the same tendency. ***Conclusion:*** Copper is helpful in fracture healing. However, the mechanism of the effect of copper on fracture healing is still to be further identified.

Key words: Copper- Fracture healing-Chicken

INTRODUCTION

Copper is one of the essential trace element getting involved in many biological functions in human and animals. Copper deficiency may cause bone disorders (Uauy *et al.* 1998) and the role of copper in osteoporosis has also been reported (Saltman and Strause 1995, Yee *et al.* 1995). The application of copper in fracture healing in China can be traced back to thousands of years ago (Li 1975). Practically, copper powder or copper ore has been widely used as a main composition in prescription for fracture healing in Chinese traditional medicine (Huang and Wu 1990). However, up to now, the report of the experimental research on copper and fracture healing as well as relative mechanism could hardly be found. The aim of the present study is to observe the effect of metal copper powder on bone fracture healing in chicken.

MATERIALS AND METHODS

Animal and diet

160 growing chickens (Shiqiza from Institute of Animal Science, Chinese Academy of Agriculture Sciences) aging 70 days were fed in 32 cages, five each. After eight days adapted feeding, the fracture with 1-2 mm defects of bilateral radius of wings was created by osteotomies. The chickens were then divided into four groups, 40 each, the body weights were 1.07±0.14, 1.04±0.12, 1.06±0.14 and 1.05±0.12 kg (mean±SD) respectively. Every morning the chickens in the control group (group I) were fed orally with 0.8% Na-CMC solution (sodium carboxymethyl cellulose), the other three groups were fed copper-Na-CMC suspension (copper powder suspending in 0.8% Na-CMC) with different doses of 20 (group II), 40 (group III) and 80 mg Cu/kg body weight (group IV). The suspension was fed to chickens by a 5ml syringe with a plastic soft tube instead of syringe needle. All the chickens had free access to water and diet and were weighted weekly.

The diet were prepared as follows: first, the Cu content in main ingredients of mixed diet was measured: corn 1.72, bran 14.32 and dregs 13.77 mg/kg. Based on this result, the composition of diet was decided: corn 84%, bran 5%, bean dregs 6%, ground fish 2.6%, ground bone 1.5%, ground stone. 0.6%, salt 0.3%, methionine 0.05%, and mixed vitamins 0.15%. The main minerals and trace elements requirement for chicken and analytical result of prepared diet is presented in *table 1.*

Tab. 1. Testing and recommended requirement of six elements in diet

Element	Content (mg/kg)	Requirement (mg/kg)
Cu	5.4	6-8
Zn	27.2	35-40
Fe	68.2	60-80
Mn	10.6	30-60
Ca	7680	6000-8000
Mg	1220	350

Ten chickens of each group were sacrificed 14, 21, 28, 35 days postoperatively. Blood were collected for determination of copper, calcium, and alkaline phosphatase (AKP), free calcium and osteocalcin (BGP). The liver was taken out and stored at -30°C until analysis for Cu and Zn. Both radiuses were removed by dissection. The roentgenograms were used for gross evaluation of the fracture healing, and computed tomography scan (CT scan) was conducted for measuring gray of callus quantitatively. The mechanical properties of the healing fracture were analyzed by a destructive three point bending test. The callus was then cut down for determination of calcium, phosphorus, magnesium, copper, zinc, iron as well as hydroxyproline(HyP).

Chemical assay

The liver and callus were digested with HNO_3-$HClO_4$and determined for elements by atomic absorption spectrometry (GGX-2, Geologic Instrument Manufactory Beijing) P was measured photometrically by UV/VIS spectrophotometer (Lambda-7 Perkin- Elmer company). Standard reference material bovine liver (GBW 080193-9207) was used for quality control.

Hydroxyproline measurement: after degreasing, the callus were hydrolyzed by HCL at 124-126°C for several hours and then reacted with chloramine-T and p-dimethylaminobenzal dehyde, determined with spectrophotometer.

CT measurement

CT scan of radius was performed by SELECT/SP (PICKER). The parameters of instrument were: 120kV, 178 mAs, 2.1s and Contiguous 1.5mm thick axial was made through the callus. The gray of callus were determined quantitatively by e-film workstation software and expressed as Hounsfield units (HU).

Biomechanical measurement

Three point bending test was performed with Material Evaluation System (MES)(QTS25 England). The bone was placed on two bars with a distance of 26mm (d) (length of section tested) in the MES and deflected by a bar on the fracture at the opposite side of the bone with a constant speed of 20mm/min. According to the load-deflection curves, the following parameters were calculated automatically by computer: maximum load (P), deflection at maximum load (D), elastic load (L) and elastic deflection (M). Before testing, diameter of callus in the directions of parallel to and perpendicular to the direction of the force (A(mm) and B) were measured with a vernier calliper and the area moment of inertia (J) was calculated, assuming the cross section to be elliptical in shape. The following formula were used in calculation of relative parameters:

Area moment of inertia $(J)(mm^4) = \pi AB^3/64$

Maximum stress $(N/mm^2) = 9.8PBd/8000J$

Elastic stress $(N/mm^2) = 9.8LBd/8000J$

Elastic modulus $(N/mm^2) = 9.8Ld^3/48000MJ$)

Statistical analysis

All the data were submitted to an one-way analysis of variance of MINITAB soft ware. The mean values were tested by a multiple comparison procedure. The ± values shown in the tables represent the standard deviation of the individual values. Significant differences between the means ($p<0.05$) within columns are indicated in the tables by different superscripts.

RESULTS

General

All the chickens in four groups returned to normal activity within 1-2 days postoperatively. No wound infections were found. The chickens grew well during the experiment period. At the end of experiment, the body weight of chickens in the first group was significantly lower than that in other three Cu-fed groups, being 1.49±0.21, 1.62±0.20, 1.68±0.20, and 1.69±0.20 kg respectively.

The contents of copper and zinc in liver were presented in *table 2*. There is no obvious accumulation of both elements in liver in all the groups. In spite of some obvious differences occurring among four groups, there is no practical clinical significance.

Callus density measured by CT scan quantitatively

Callus density increased consistently over the experimental period *(table 3)*. At the 14^{th}day after operation, the fusiform callus surrounding the broken site of the radius was clearly visible on plain radiography. The gray value of callus increased fastest in group 4 from the 14 to 21 day after surgery, 1.93 times as high as that at the 14 day. In contrast, the gray value of callus in the control group increased the lowest, only 1.38 times, and 1.53 times and 1.65 times for group 2 and

3 respectively. At the 28th day, the gray values of callus in three copper fed groups were nearly the same and reached their maximum (922-971HU), but not in the control group (796HU).

Table 2. Cu and Zn level in liver in different groups 14, 21, 28, and 35 days postoperatively

Group	days after surgery			
	14	**21**	**28**	**35**
Cu in liver				
1	3.83±0.70	3.48±0.68	3.93±0.40^{ab}	3.79±0.50^{a}
2	4.87±1.62	4.07±0.35	4.19±0.48^{b}	3.99±0.41^{ab}
3	4.93±0.92	4.25±0.75	3.70±0.44^{a}	4.27±0.71^{b}
4	4.74±0.81	4.48±0.65	3.89±0.36^{ab}	4.02±0.37^{ab}
Zn in liver				
1	32.18±3.23^{b}	32.09±5.72^{b}	28.05±2.80^{b}	25.85±4.25^{a}
2	29.49±2.25^{ab}	27.34±3.97^{a}	27.04±2.22^{ab}	27.98±2.56^{ab}
3	30.08±3.53^{ab}	24.40±1.98^{a}	25.49±2.22^{a}	29.78±3.78^{b}
4	27.74±1.54^{a}	27.33±2.72^{a}	27.25±1.51^{ab}	28.33±1.25^{ab}

^{a,b} means of the same column without a common superscript are significantly different.

Gray values of callus increased with the increase of copper doses. At the 14th day postoperatively, the gray values of callus were nearly the same in four groups. However, at the 21st day, the gray values of callus in middle and high copper doses groups were significantly higher than that in the control (815±93.3, 909±220 vs. 597±155 HU $P<0.05$). At the 35th day, the gray values of callus in three cu-fed groups was higher than that in the control, especially in low and high doses groups (919±172, 973±100 vs. 763±179 HU $P<0.05$).

Table 3. Gray values of callus (HU) in different groups 14, 21, 28, and 35 days postoperatively

Group	Days after operation			
	14	**21**	**28**	35
1	432±94.7	597±155^{a}	796±179	763±179^{a}
2	491±97.8	751±134^{ab}	922±285	919±172^{b}
3	493±113	815±93.3^{b}	971±230	898±157^{ab}
4	471±138	909±220^{b}	968±254	973±100^{b}

^{a,b} means of the same column without a common superscript are significantly different.

Contents of calcium, magnesium, phosphorous, copper, zinc, iron and hydroxypoline in callus

It could be found from *table 4* that at the 14th day after surgery the callus minerals in the control group looked higher, but from the 21st day on, the callus minerals in all of the groups presented an elevated tendency and the calcium in callus in high-Cu group has already been significantly higher than that in the control: 146.7±30.2 and 122.2±24.5 mg/g respectively. At the 28 and 35 day, the calcium in callus in all three of Cu-fed groups was higher than that in the control, especially in low and high dose groups, there was a significant difference compared with the control ($P<0.05$). At the 35th day, the magnesium contents in callus in three Cu-fed groups were significantly higher than that in the control, they were 2.86±037, 2.98±0.57, 3.06±0.46 vs. 2.43±0.53 mg/g respectively.

At the 14th day after operation, the copper in callus of control group was significantly higher than that in other three groups, and then declined. At the 35th day the callus copper in control group was significantly lower than that in other three groups. On the contrary, the iron contents in callus nearly the same at the 14th day after surgery, but from the 21st day the iron in callus in the control group began to increase and at the 28th day it was obviously higher in the control than that in other three Cu-fed groups and still significantly higher than that in middle and high-dose groups up to the 35 day after operation.

In addition, there was a gradually increase in the concentration of hydroxypoline in callus as the duration of healing is prolonged. At the 35th day, the level of hydroxypoline in callus of high-Cu dose group was significantly higher than that in the control (34.23±1.96 vs.32.17 mg/g P<0.05).

Table 4. Contents of calcium, magnesium, phosphorus, copper, zinc iron, and HyP in callus

Group	Days after operation			
	14	21	28	35
Ca (mg/g)				
1	90.6±19.8	122.2±24.5[ab]	147.3±32.6	137.3±34.7[a]
2	86.3±17.5	111.8±28.9[a]	152.5±24.2	160.4±21.7[ab]
3	77.6±34.8	131.3±12.4[ab]	152.1±30.5	177.0±26.7[b]
4	66.7±20.9	146.7±30.2[b]	150.5±33.1	175.6±20.5[b]
Mg (mg/g)				
1	1.89±0.30[b]	1.95±0.48	2.73±0.25	2.43±0.53[a]
2	1.51±0.29[ab]	2.47±1.29	2.64±0.33	2.86±0.37[b]
3	1.35±0.50[a]	2.38±0.52	2.75±0.33	2.98±0.57[b]
4	1.17±0.43[a]	2.47±0.49	2.74±0.51	3.06±0.46[b]
P'(mg/g)				
1	3.35±0.81	4.67±1.07[ab]	6.42±1.00	5.88±1.60
2	3.52±0.56	4.47±0.87[a]	6.09±1.36	6.28±1.01
3	3.42±1.34	4.70±0.98[a]	6.26±1.11	6.88±1.17
4	2.98±0.92	5.70±1.31[b]	6.06±1.48	6.68±1.09
Cu (?g/g)				
1	2.42±1.20[b]	2.15±0.30	2.12±0.44[b]	1.51±0.27[a]
2	1.85±0.31[a]	2.33±0.50	2.41±0.49[b]	2.28±0.29[c]
3	1.80±0.11[a]	2.10±0.31	1.61±0.21[a]	1.95±0.29[b]
4	1.63±0.25[a]	2.43±0.55	2.00±0.58[ab]	2.06±0.42[bc]
Zn (?g/g)				
1	84.15±20.49[b]	101.0±24.95	117.3±14.26	107.4±25.50
2	76.11±6.86[ab]	117.4±13.66	117.5±14.20	119.5±13.14
3	67.43±25.50[ab]	103.3±2.38	121.4±20.94	106.4±24.58
4	54.00±20.55[a]	118.7±18.7	113.0±22.25	117.0±21.13
Fe (?g/g)				
1	24.94±5.79	42.27±5.87	47.99±19.81[b]	47.94±12.75[b]
2	24.73±3.85	43.28±12.81	34.18±10.99[a]	37.45±17.49[ab]
3	29.23±9.39	42.96±12.93	31.66±7.14[a]	27.17±9.06[a]
4	22.66±4.01	35.87±12.29	31.91±7.42[a]	32.97±8.94[a]
HyP (mg/g)				
1	19.37±6.62	21.82±6.32[ab]	31.89±3.24	32.17±1.93[a]
2	21.30±8.49	18.81±2.97[a]	31.96±2.03	33.81±1.92[ab]
3	20.96±6.99	21.66±2.51[ab]	31.77±1.82	33.49±2.14[ab]
4	22.64±6.88	23.33±3.91[b]	29.56±4.55	34.23±1.96[b]

[a,b] means of the same column without a common superscript are significantly different.

Biomechanical observations

Maximum stress, elastic stress and elastic modulus of healing radius in different groups at the 14, 21, 28 and 35 day postoperatively are given in *table 5.*

Table 5. Maximum stress, elastic stress and elastic modulus of healing radius in four groups at the 14, 21. 28 and 35 day postoperatively

Group	Days after operation			
	14	**21**	**28**	**35**
Maximum stress (N/mm^2)				
1	3.04±0.53	12.33±5.98^a	35.53±8.75	49.52±36.98
2	3.26±1.07	11.53±4.37^a	41.91±26.54	63.49±21.00
3	2.93±1.92	14.38±3.44^{ab}	59.76±41.84	78.40±49.03
4	2.89±1.43	18.84±8.26^b	58.71±56.72	73.01±25.84
Elastic stress (N/mm^2)				
1	1.46±0.43	7.62±3.62	22.67±7.68	33.54±29.50
2	1.86±0.50	6.68±2.44	25.09±13.73	38.83±11.54
3	1.53±1.01	8.65±3.24	39.88±28.51	54.18±32.89
4	1.83±0.83	10.05±6.08	39.37±42.30	45.77±15.82
Elastic modulus (N/mm^2)				
1	24.4±8.3	224±135	755±575	1106±1070
2	27.7±17.0	150±110	863±761	1558±862
3	36.3±55.4	168±142	1602±1596	2482±2401
4	17.3±10.7	282±289	1225±1744	1938±1275

^{a,b} means of the same column without a common superscript are significantly different.

In spite there was no significant difference, which was expected and might be exciting, two results could be found obviously from *table 5* with *fig. 1-3.* First, with the going of time and the healing of fracture, the mechanical parameters of radius in four groups were getting improved. Second, during different stages postoperatively, most of the parameters in three Cu-fed groups were superior to that in the control. For instance, at the 35th day after surgery, the maximum stress in the control group was 49.52±36.98, while in three Cu-fed groups were 63.49±21.00, 78.40±49.03 and 73.01±25.84 N/mm^2 respectively. The elastic modulus, which reflects the internal hardness of materials, was also higher in three Cu-fed groups as compared with the control at the 28th and 35th day. The reason why there was no significant difference, may lie in that many factors influenced the biochemical parameters such as the size, the volume, and the shape of the callus as well as the dissimilarity of injury in different degrees, leading to a rather big standard deviation among relative smaller specimen.

DISCUSSION

It demonstrated that metal copper powder had rather lower toxicity for chickens. When being fed copper powder with a higher dose of 80 mg/kg b.w. everyday for 5 weeks, the chickens in the fourth group didn't show any symptom of toxicosis, such as suffering from illness and decrease of body weight. Moreover, the chickens in this group grew well. Meanwhile, no obvious accumulation of copper and zinc in liver showed that chicken had a good tolerance for metal copper. A large quantities of copper was excreted from the feces, which could be seen when the cages were washed everyday, indicated that the homeostasis exerted a very important role in maintaining copper balance in organism (Kirchgessner 1993).

Since many years, the evaluation of fracture healing in animal experiments has been primarily based on radiographic changes of bone, histological and histochemical observation of callus (Nunamaker.1998). However, the histological examination itself is subjective and the result is limited as expressed as enumerative data statistically. Therefore, more reliable quantitative methods for an objective assessing fracture healing are clearly necessary (Markel and Chao 1993).

In this study some biochemical parameters in blood of chickens such as alkaline phospharase, (AKP), free calcium, BGP were determined, but no clear and definite relationship with fracture healing was found.

The use of computed tomography (CT) for determination of gray values in small region of interest (ROI) of callus, which is able to provide mineral density of callus and quantitative assessment of fracture healing, was a new outgrowth technology for CT imaging in recent decade (Cann 1988, Schnarkowski et al 1999). It has been found that the callus density determined in ROI was related to mechanical properties of the healing bone (Boer et al 1998).

In this study three methods were selected mainly for evaluation of fracture healing: 1. CT scan and determination of gray value of callus; 2.determination of minerals and Hyp in callus; and 3.three point bending test for measuring of relative biomechanical parameters of healing bone. It showed that these three methods could give a more comprehensive and accurate evaluation for fracture healing from different angles.

Postoperatively, the contents of calcium, phosphorous and magnesium in callus ascended continuously. Increment was the highest from the 14th to 21th day, and reached the peak value at the 35th day. The change of gray value of callus had the same tendency. Regressive analysis showed a good positive correlation between the contents of calcium, magnesium and gray value of callus 21, 28 and 35 days after surgery, the coefficient was Ca: 0.59, 0.66, 0.77 ($P<0.001$), Mg: 0.47, 0.66, 0.74 ($P<0.001$). Results indicated that gray value represented the degree of mineralization. In addition, there was also a significant positive correlation between the content of calcium in callus and biomechanical parameters such as elastic modulus 14. 21, 28 and 35 days postoperatively, the correlation coefficient was 0.58-060 ($P<0.001$).

It comes the conclusion that the gray value, the contents of minerals in callus and biomechanical properties of healing bone have good positive correlation in assessment of fracture healing in animal experiment. Results from the aforementioned three methods indicated that copper is helpful in fracture healing.

The fracture healing process includes clot, hematoma, inflammation, periosteum reaction, callus formation and remodeling of bone. However, the fracture healing itself is a very comprehensive procedure and is one of the most remarkable of all the repair processes in the body since it results not in a scar, but in the actual reconstitution of the injured in something very like its original form (Frost 1989). A number of factors influence the healing. It is not to be expected therefore, that the mechanism controlling such a process will be easily elucidated.

Many metalloenzymes contain copper. Lysyl oxidase, a copper dependent enzyme, functions in the normal maturation of collagen, particularly in steps important to the formation of lysine-derived cross-links (Rucker and Murray 1978). Being a trace element essential for the synthesis of collagen and elastic protein, copper may exert very important role in growth of bone. In addition, copper is a composition of some antioxidase such as CuZn-SOD, involving in promoting blood circulation, removing blood stasis, and relieving inflammation, which may have benefit effect on the fracture healing. Moreover, it is worth to be further investigated whether copper supplement could increase the absorption of calcium and magnesium in the case of similar calcium intake because it was found in this study that copper supplement increased the deposition of Ca and Mg in callus. Therefore, the role and mechanism of copper in fracture healing should be further identified.

In Chinese ancient medical books it was recorded: "red copper powder could be used in treatment of fracture,'welding'both man's or livestock's broken bone." (Li 1975) and: "a wing broken wild goose was treated by feeding Native Copper, then the goose recovered and flew away" (Tang

1957). To verify these records, we use chicken as our animal model and metal copper powder as copper supplement form. According to the modern medicine knowledge, copper ion should be the form absorbed in intestine, so the next experiment in our laboratory should be focus in the effect of copper compound on the fracture healing in other animal species.

ACKNOWLEDGEMENT

This study was supported by the International Copper Association grant TPT-0601-00.

REFERENCES

1. Boer FC, Bramer JAM, Patka P *et al.* 1998. Quantification of fracture healing with three dimensional computed tomography. Arch Orthop Trauma Surg 117:345-350.
2. Cann CE 1988, Quantitative CT for determination of bone mineral density: A review Radiology. 166:509-522.
3. Frost HM. 1989. The biology of fracture healing. An overviewing for clinician. Part I Clin Orthop 248:283-293.
4. Huang RZ, Wu DZ. 1990. Explanation of ancient prescription for bone wound (Chinese) The Chinese Ancient Traditional Medical Books Publishing House. Beijing.
5. Kirchgessner M. 1993 Homeostasis and homeorhesis in trace elements metabolism. in Proceeding of the eighth international symposium on trace elements in man and animals. Verlag Media Touristik 4-21.
6. Li SZ (1518-1593). 1975 Compendium of Materia Medica Revising medical book (Chinese). The people's Medical Publishing House. Beijing. Book One Vol 4: 345,Vol 8: 465.
7. Markel MD, Chao EYS. 1993. Noninvasive monitoring techniques for quantitative description of callus mineral content and mechanical properties. Clin Orthop 293:37-45.
8. Nunamaker DM. 1998 Experimental models of fracture repair. Clin Orthop 355s: s56-s65.
9. Rucker R and Murray J. 1978. Cross-linking amino acid in collagen and elastin Am J Clin Nutr 31:1221-1236.
10. Saltman PD and Strause LG. 1995. The role of trace elements in osteoporosis J Am Coll Nutr 12: 384-389.
11. Schnarkowski P, Redei J, Peterfy CG *et al.* 1999 Tibial shaft fracture: assesment of fracture healing with computed tomography. J of computer assisted tomography 5: 777-781.
12. Tang SW. 1957 Classic Classified Materia Medica for Emmergencies (Chinese) Photo-offset copy. The people's Medical Publishing House. Beijing. Vol 5:133.
13. Uauy R, Olivares M, Gonzalez M. 1998 Essentiality of copper in humans. Am J Clin Nutr (suppl) 67:952s-959s.
14. Yee CD, Karen SK, Walker M *et al.* 1995 The relationship of nutritional copper to the development of postmenopausal osteoporosis in rats. Biol Trace Elem Res 48, 1-11.
15. Zhao XM (1719-1805) 1983. A supplement to the Compendium of Meteria Medica. Second edision. Beijing. The people's Medical Publishing House. 38.

XIII ONCOLOGY

*Metal Ions in Biology and Medicine: vol. 7.*Eds L. Khassanova, Ph. Collery, I. Maymard, Z. Khassanova, J.-C. Étienne.
John Libbey Eurotext, Paris © 2002 pp. 575-579.

Cisplatin-mediated biochemical changes in mitochondria in tumor-bearing mice

S. B. Prasad* and A. Kharbangar

**Cell and Tumor Biology Laboratory, Department of Zoology, North-Eastern Hill University, Shillong 793022, India*

ABSTRACT

Mitochondrial protein decreased in liver, kidney and tumor cells after cisplatin treatment of ascites Dalton's lymphoma tumor-bearing mice while mt-glutathione level decreased in liver and kidney but increased in tumor cells after the treatment. Cisplatin treatment caused an increase in lipid peroxidation but a decrease in L-malate dehydrogenase activity in the mitochondria of these tissues. It is suggested that the development these biochemical damages in mitochondria could be the important critical events contributing towards cisplatin-induced toxicity/cytotoxicity in the host.

INTRODUCTION

Cis-Diamminedichloroplatinum(II), commonly known as cisplatin, is a potent anticancer agent being used widely in cancer chemotherapy. Various biological properties and effects of cisplatin have been well explored [1, 2], with many reports indicating that therapeutic efficacy of the drug involves its ability to interact with cellular DNA [2-4]. In intact DNA, cisplatin reacts with N-7 position of guanine residues to form monofunctional adducts which close further into bifunctional intrastrand and interstrand cross-links [3]. However, studies with DNA repair-deficient and -proficient cell lines failed to reveal any correlation between concentrations of cisplatin that inhibit DNA synthesis and cell death [5]. Thus, the effect that cisplatin has on DNA replication may not be regarded as a major contributor to its cytotoxicity. Besides its interaction with cellular DNA, changes in the cell surface [6], tissue calcium, potassium concentrations [7], various enzymes (lactate dehydrogenase, cathepsins, arginase, 5'-nuceleotidase) [8] have also been observed and it has been suggested that these changes could also be involved as the additional factors in the anticancer activity of cisplatin.

Mitochondria play a pivotal role in cellular metabolism and are the sites of Krebs' cycle and electron transport. Recently various reports have indicated that mitochondria may be involved in tumorigenesis, Ca^{++} signaling and apoptosis [9-11]. Cisplatin has been shown to bind preferentially to mitochondrial DNA (mtDNA) as compared to nuclear DNA (nDNA) [12]. The increased susceptibility of mtDNA to damaging agents could be due to existence of nonnucleoprotein structure, the lipophilic nature of the mitochondrial membrane, and prevalent oxidative environment in the mitochondria. Mitochondrial DNA encodes 13 proteins which participate in the electron transport chain and a damage to mtDNA will diminish electron transport function resulting a fall in cell bioenergetics and subsequent cellular dysfunction [12]. Mitochondrial dysfunction have been shown in hepatotoxicity [13] and cisplatin-induced nephrotoxicity [14]. Earlier we have reported that cisplatin treatment causes a decrease of succinate dehydrogenase activity in the mitochondria of kidney and tumor cells [15].

The biochemical events in mitochondria that may eventuate cisplatin-induced effects in the hosts remains to be investigated in detail. Thus, present studies were undertaken to look into the changes in mitochondrial protein, glutathione, lipid peroxidation and L-malate dehydrogenase (MDH, EC 1.1.1.37) activity in relation to tumorous condition and cisplatin treatment. This may help to understand further the significance of mitochondria in the effectiveness of cisplatin in toxicity/cytotoxicity in tumor-bearing mice.

MATERIAL AND METHODS

Chemicals

All biochemicals were purchased from Sigma Chemical Company, St. Louis, USA. Other chemicals used in the experiments were of analytical grade. Cisplatin solution (1 mg/ml of 0.9% NaCl) was obtained from Biochem Pharmaceutical Industries, Mumbai, India.

Tumor system and cisplatin treatment

Ascites Dalton's lymphoma is being maintained *in vivo* in 10-12 weeks old inbred Swiss albino mice by serial intraperitoneal (i.p) transplantations of about 10 million tumor cells per animal (0.25 ml vol, in phosphate buffered saline, pH 7.4). Tumor-transplanted animals usually survived for 19-21 days. As per the dose and treatment schedule used earlier by us [8, 15], single therapeutic dose of cisplatin (8 mg/Kg body weight, i.p.) was administered to tumor-bearing mice on the 10th day post-tumor transplantation. After 24, 48, 72 and 96 h of treatment liver, kidney and tumor were collected, mitochondrial fractions were isolated [15] and used for different biochemical determinations. The determinations were also done in the mitochondrial fractions of tissues of normal (untreated, non-tumorous) and control (untreated tumor-bearing mice injected with normal saline only) mice.

Glutathione Estimation

Glutathione was determined as nonprotein (NPSH) and total (TSH) sulfhydryl contents in the mitochondrial fractions under different experimental conditions using the method of Sedlak and Lindsay [16].

Lipid Peroxidation

The concentration of thiobarbituric acid-reacting substances, mainly malondialdehyde, was determined in the isolated mitochondrial fractions using the method of Buege and Aust [17]. The malondialdehyde concentration in the samples was calculated using an extinction coefficient of 1.56×10^5 L/mol and is expressed as nmol/mg protein.

Assay of MDH

Malate dehydrogenase was assayed using the method of Kitto [18]. Mitochondrial homogenate (5%) was prepared in 0.25mol/L sucrose solution by sonication. The reaction mixture contained 0.03ml of NADH (14.3mM), 0.05ml of oxaloacetate (20mM), enzyme and buffer (Potassium phosphate buffer, 0.1M, pH 7.5) to a final volume of 3.0 ml. The reaction was started by addition of either oxaloacetate or enzyme. The optical density was read at 340nm against a blank containing all components of the assay mixture except NADH at one-minute intervals. Enzyme activity was calculated from the initial rate of oxidation of NADH. The amount of enzyme used is adjusted to give a decrease in the optical density of approximately 0.04 per min.

Protein concentration in the mitochondrial homogenates was determined using the method of Lowry et al. [19].

RESULTS

Mitochondrial protein decreased significantly following cisplatin treatment of tumor-bearing mice. However, mt-GSH level decreased in liver, kidney but elevated in DL cells after the treatment (Table). As compared to normal mice, mt-lipid peroxidation (LPO) increased in the liver of tumor-bearing mice. Cisplatin treatment caused an increase in LPO but a decrease in the units of MDH activity in the mitochondria of all the three tissues. Enzyme specific activity decreased in kidney whereas it increased in DL cells after the treatment *(table)*.

DISCUSSION

Two distinct genomes, one located in nucleus and other in mitochondria, are present in mammalian cells. As compared to nuclear DNA (nDNA), cisplatin adduction in mitochondrial.

Table. Total protein (mg/g), total glutathione (TSH), nonprotein thiol (NPSH, (moles/g), lipid peroxidation (LPO, malondialdehyde levels, nmoles/mg protein) and L-malate dehydrogenase (MDH) activity in the mitochondria of liver, kidney and DL cells of mice under different experimental conditions

Tissue	Treatment	Protein	TSH	NPSH	LPO	MDH Units	MDH Specific Activity
Liver	Normal	138.02 ± 7.73	2.65 ± 0.26	0.071 ± 0.002*	0.67 ± 0.02*	2.52 ± 0.17*	0.73 ± 0.05
	Control (TB)	107.42 ± 9.13	3.80 ± 0.27	0.160 ± 0.012	1.19 ± 0.07	2.19 ± 0.13	0.82 ± 0.06
	Cisplatin (24h)	76.73 ± 7.20*	2.61 ± 0.18*	0.050 ± 0.003*	0.88 ± 0.05*	1.26 ± 0.07*	0.71 ± 0.02
	Cisplatin (48h)	76.06 ± 4.72*	1.35 ± 0.11*	0.041 ± 0.002*	1.68 ± 0.11*	1.67 ± 0.10*	0.88 ± 0.05
	Cisplatin (72h)	77.12 ± 7.0*	1.32 ± 0.24*	0.058 ± 0.003*	1.35 ± 0.06	1.86 ± 0.08	0.94 ± 0.04
	Cisplatin (96h)	74.60 ± 6.87*	1.41 ± 0.31*	0.069 ± 0.005*	1.48 ± 0.12*	1.59 ± 0.11*	0.91 ± 0.06
Kidney	Normal	77.86 ± 1.16	2.28 ± 0.37	0.210 ± 0.031	1.31 ± 0.08	1.84 ± 0.14	0.95 ± 0.07
	Control (TB)	82.73 ± 5.84	2.20 ± 0.40	0.130 ± 0.010	1.21 ± 0.15	2.46 ± 0.09	1.19 ± 0.04
	Cisplatin (24h)	68.16 ± 2.80*	1.97 ± 0.12	0.120 ± 0.004*	1.16 ± 0.12	1.39 ± 0.05*	0.81 ± 0.03*
	Cisplatin (48h)	62.90 ± 3.42*	1.45 ± 0.28	0.110 ± 0.011	1.71 ± 0.09*	1.31 ± 0.08*	0.83 ± 0.05*
	Cisplatin (72h)	57.26 ± 3.11*	1.47 ± 0.06*	0.166 ± 0.013	1.76 ± 0.15*	1.55 ± 0.09*	1.09 ± 0.07
	Cisplatin (96h)	51.34 ± 4.57*	1.61 ± 0.09	0.173 ± 0.014*	1.69 ± 0.16*	1.42 ± 0.13*	0.97 ± 0.10
DL cells	Control (TB)	59.26 ± 5.53	2.27 ± 0.18	0.084 ± 0.005	0.60 ± 0.04	1.63 ± 0.18	1.10 ± 0.12
	Cisplatin (24h)	38.84 ± 4.87*	3.00 ± 0.15*	0.140 ± 0.011*	1.08 ± 0.02*	1.32 ± 0.21*	1.36 ± 0.22
	Cisplatin (48h)	38.62 ± 5.20*	2.57 ± 0.16	0.160 ± 0.023*	1.15 ± 0.03*	1.73 ± 0.10	1.79 ± 0.10*
	Cisplatin (72h)	33.14 ± 1.64*	2.85 ± 0.05*	0.140 ± 0.020*	0.97 ± 0.08*	1.41 ± 0.11*	1.84 ± 0.12*
	Cisplatin (96h)	29.86 ± 3.93*	2.97 ± 0.07*	0.133 ± 0.012*	0.94 ± 0.07	1.33 ± 0.15*	1.61 ± 0.23*

Normal = untreated, non-tumorous mice; TB = Tumor-bearing, untreated mice. **A** unit of MDH activity is defined as the amount of enzyme catalyzing the reduction of 1nmole of oxaloacetate per minute and the specific activity as units/mg protein. Data are expressed as means ± SD. Student's t-test; n = 3-4, as compared to respective control, $*P \leq 0.05$.

DNA (mtDNA) has been shown to be 4- to 8-fold higher [12]. Present findings provide an indication that mitochondria could be one of the main targets in the cells during cisplatin-mediated cancer chemotherapy and toxicity. Cisplatin treatment resulted in a significant decrease in mt-protein in all the thrre tissues studied *(table)*. This may involve inhibited protein transport from the cytosol to mitochondria and/or decreased mitochondrial protein synthesis. It has been reported that cisplatin arrests elongation during protein synthesis and this may contribute to the toxic/cytotoxic effects of cisplatin during therapy [20]. Glutathione (a tripeptide; L-γ-glutamyl-L-cystienyl-glycine), an important cellular antioxidant, is maintained mainly (>98%) in the reduced form

(GSH) and plays a key role in the detoxification of reactive oxygen species and xenobiotics [21]. GSH level is critical to various mitochondrial functions including membrane structure and integrity, ion homeostasis, intramitochondrial redox status and activities of sulfhydryl-dependent enzymes [22]. Cisplatin treatment of mice showed a decrease in mt-GSH in liver and kidney (Table). This decrease in mt-GSH may result in an impairment in various mitochondrial functions and may also cause increased mtDNA oxidation and oxidative damage to mitochondria. To authenticate this suggestion, lipid peroxidation (LPO) was determined. Peroxides are formed naturally during mitochondrial respiration and it may be responsible for mitochondrial damage. It was noticed that mt-LPO increased significantly in these tissues after cisplatin treatment and supports the view of cisplatin-mediated increase in oxidative damage to mitochondria. It may, therefore, be suggested that a decrease in mt-GSH and concomitant increase in mt-LPO could be an early and critical factor in cisplatin-induced toxicity in liver and kidney. It has been reported that mitochondrial dysfunction could be a major mechanism during drug-induced hepatotoxicity and nephrotoxicity [15, 16]. In contrast to liver and kidney, in DL cells an increase in mt-GSH was noticed after cisplatin treatment (Table). This may be due to the fact of being fewer and/or structurally altered mitochondria in cancer cells and is perhaps being reflected here in DL cells. It may also be inferred that in DL cells, instead of changes/decrease in mt-GSH, the decrease in cytosolic GSH may play more important role in cisplatin-mediated cytotoxicity. This proposition may be supported by the fact that a decrease in cellular GSH has been noticed in DL cells after cisplatin treatment [23]. In fact, diminished mitochondrially-mediated oxidative phosphorylation has been suggested in cancer cells, with glycolysis being the main source of energy production [24].

L-Malate dehydrogenase (MDH; EC 1.1.1.37) is an important enzyme of Krebs's cycle in mitochondria and catalyzes the oxidation of L-malate to oxaloacetate. The equilibrium of this enzyme catalyzed reaction lies more towards the formation of L-malate, so the MDH activity was assayed by oxaloacetate reduction. Cisplatin treatment of mice showed a decrease in enzyme activity in all the cases *(table)*. This decrease in MDH activity may assist the development of biochemical injury, mitochondrial dysfunction and contribute towards toxicity/death of cells. It should also be mentioned that the observed increase in enzyme specific activity in DL cells after cisplatin treatment may be due to a larger decrease (34-50%, Table) of other proteins rather than increase in MDH. Thus, present findings propose that in cisplatin-mediated chemotherapy, mitochondria are the main targets in cells and support the view of its multilevel/multistep effects in the host.

CONCLUSION

The present studies provide an indication that cisplatin treatment causes definite mitochondrial biochemical injury, which could be involved in ensuing toxicity / cytotoxicity in hosts.

ACKNOWLEDGEMENTS

The partial financial supports were provided by North-Eastern Hill University, shillong and DST, INSA, New Delhi.

REFERENCES

1. Go R.S., Adjei A.A. Review of the comparative pharmacology and clinical activity of cisplatin and carboplatin. *J. Clin. Oncol.*, 1999, 17, 409-422.
2. Rosenberg B. Fundamental studies with cisplatin. *Cancer*, 1985, 55, 2303-2316.
3. Coste F., Malinge J., Serre L., Shepard W., Roth M., Leng M., Zelwer C. Crystal structure of a double-

stranded DNA containing a cisplatin interstrand cross-link at 1.63 Å resolution: hydration at the platinated site. *Nucleic Acid Res.*, 1999, 27, 1837-1846.
4. Pinto A.L., Lippard S.J. Binding of the antitumor drug *cis*-diamminedichloroplatinum (II) (cisplatin) to DNA. *Biochim. et Biophys. Acta*, 1985, 780, 167-180.
5. Sorenson C.M., Eastman A. Influence of *cis*-diamminedichloroplatinum(II) on DNA synthesis and cell cycle progression in excision repair proficient and deficient Chinese hamster ovary cells. *Cancer Res.*, 1988, 48, 6703-6707.
6. Prasad S.B., Sodhi A. Effect of *cis*-dichlorodiammineplatinum(II) on surface of tumor & normal cells: biochemical, fluorescence and electron microscopical studies. *Ind. J. Exp. Biol.*, 1982, 20, 559-571.
7. Prasad S.B., Giri A. Cisplatin-induced changes in tissue calcium and potassium concentrations in tumor-bearing mice. *Med. Sci. Res.*, 1999, 27, 459-462.
8. Prasad S.B., Giri A., Khynriam D., Kharbangar A., Nicol B.M., Lotha C. Cisplatin-mediated enzymatic changes in mice bearing ascites Dalton's lymphoma. *Med. Sci. Res.*, 1999, 27, 723-730.
9. Cavalli L.R., Liang B.C. Mutagenesis, tumorigenicity, and apoptosis: are the mitochondria involved? *Mutation Res.*, 1998, 398, 19-26.
10. Smaili S.S., Hsu Y., Youle R.J., Russell J.T. Mitochondria in Ca^{2+} signaling and Apoptosis. *J. Bioenergetics and Biomembranes*, 2000, 32, 35-46.
11. Mignotte B., Vayssiere J. Mitochondria and apoptosis. *Eur. J. Biochem.*, 1998, 252, 1-15.
12. Olivero O.A., Chang P.K., Lopez-Larraza D.M., Semino-Mora M.C., Poirier M.C. Preferential formation and decreased removal of cisplatin-DNA adducts in Chinese hamster ovary cell mitochondrial DNA as compared to nuclear DNA. *Mutation Res.*, 1997, 391, 79-86.
13. Pessayre D., Mansouri A., Haouzi D., Fromenty B. Hepatotoxicity due to mitochondrial dysfunction. *Cell Biol. Toxicology*, 1999, 15, 367-373.
14. Kruidering M., Van de Water B., De Heer E. *et al.* Cisplatin-induced nephrotoxicity in procine proximal tubular cells: mitochondrial dysfunction by inhibition of complexes I to II of the respiratory chain. *J. Pharmacol. Exp. Ther.*, 1997, 280, 638-649.
15. Kharbangar A., Khynriam D., Prasad S.B. Effect of cisplatinon mitochondrial protein, glutathione, and succinate dehydrogenase in Dalton's lymphoma-bearing mice. *Cell Biol. Toxicology*, 2000, 16, 363-373.
16. Sedlak J., Lindsay R.H. Estimation of total, protein-bound, and nonprotein sulfhydryl groups in tissue with Ellman's reagent. *Anal. Biochem.*, 1968, 25, 192-205.
17. Buege J.A., Aust S.D. Microsomal lipid peroxidation. *Methods Enzymol*, 1978, 52, 302-310.
18. Kitto G.B. Intra- and extramitochondrial malate dehydrogenases from chicken and tuna heart. *Methods Enzymol.*, 1967, 13, 106-107.
19. Lowry O.H., Rosebrough N.J., Farr A.L., Randall R.J. Protein measurement with the Folin phenol reagent. *J. Biol. Chem.*, 1951, 193, 265-275.
20. Heminger K.A., Hartson S.D., Rogers J., Matta R.L. Cisplatin inhibits protein synthesis in rabbit reticulocyte lysate by causing an arrest in elongation. *Arch. Biochem. Biophys.*, 1997, 344, 200-207.
21. Wang W., Ballatori N. Endogenous glutathione conjugates: occurrence and biological functions. *Pharmacol. Rev.*, 1998, 50, 335-355.
22. Chen Z., Putt D.A., Lash L.H. Enrichment and functional reconstitution of glutathione transport activity from rabbit kidney mitochondria: further evidence for the role of dicarboxylate and 2-oxoglutarate carriers in mitochondrial glutathione transport. *Arch. Biochem. Biophys.*, 2000, 373, 193-202.
23. Prasad S.B., Khynriam D. Mutagenicity and endogenous glutathione levels in tumor-bearing mice after cisplatin treatment. Proc. "7th International symposium on metal ions in biology and medicine", May 5-9, 2002, Saint Petersburg, Russia.
24. Warburg O. On the origin of cancer cells. *Science*, 1956, 123, 309-314.

Metal Ions in Biology and Medicine: vol. 7. Eds L. Khassanova, Ph. Collery, I. Maymard, Z. Khassanova, J.-C. Étienne.
John Libbey Eurotext, Paris © 2002 pp. 580-585.

Mutagenicity and endogenous glutathione levels in tumor-bearing mice after cisplatin treatment

S. B. Prasad and D. Khynriam

Cell and Tumor Biology Laboratory, Department of Zoology, North-Eastern Hill University, Shillong 793022, India

ABSTRACT

It is suggested that cisplatin brings about differential effects on the chromosomes of bone marrow and tumor cells of the same tumorous host. Cisplatin-induced chromosomal aberrations (CA) in Dalton's lymphoma (DL) tumor cells were much higher than that in bone marrow cells. At 24h of cisplatin treatment CA as well as aberrant metaphases were maximum which at later periods of treatment (48-96h) decreased appreciably in bone marrow cells while it remained almost unchanged in the tumor cells. Cisplatin treatment caused a decrease of glutathione (GSH) levels in bone marrow as well as tumor cells. Evaluation of mutagenic parameters i.e. CA, micronuclei and sperm head abnormality under the conditions of decreased or increased GSH levels suggested that the induction of cisplatin-mediated mutagenic effects and changes in endogenous GSH level in the host may be inversely related.

INTRODUCTION

Cisplatin (*cis*-diamminedichloroplatinum-II) has been established to be a potent anticancer drug widely used against various malignancies in animals and humans [1-3]. Although, the ability of cisplatin to interact with cellular DNA appears to be the main step in its anticancer activity [1, 4], the involvement of multi-step and multilevel effects of cisplatin in the tumor cell/host has also been suggested as it also affects many other biochemical and mitochondrial changes [5-7]. However, full therapeutic efficacy of this drug is limited by the emergence of various side effects which mainly include nephrotoxicity [8], hematological toxicity [9], and also to the development of acquired drug resistance [10]. Furthermore, its mutagenic potential has also been shown in various test systems [11, 12]. The changes in glutathione levels has been suggested to be of much importance in developing these side effects [10, 12, 13].

Glutathione (L-γ-glutamyl-L-cystienyl-glycine), an important cellular antioxidant, is maintained mainly (>98%) in the reduced form(GSH) and plays a key role in various bio-reductive reactions and protection against harmful free radicals [14]. GSH is also known to be involved in the metabolism of cisplatin causing alterations in the rate of drug uptake and elimination [15]. L-buthionine(S,R)-sulfoximine (BSO), a specific inhibitor of γ-glutamylcysteine synthetase, is commonly used to deplete cellular GSH level [16]. Cysteine is the rate-limiting and predominantly plasma substrate for the synthesis of GSH [17].

The present study was undertaken to investigate the changes in mutagenic effects of cisplatin *in vivo* in relation to endogenous GSH levels, using chromosomal aberrations (CA), micronuclei and sperm head abnormality as mutagenic bioassay parameters. The differential pattern of CA in Dalton's lymphoma (DL) cells and bone marrow cells following cisplatin treatment has also been evaluated.

MATERIALS AND METHODS

Chemicals

Reduced glutathione, L-buthionine-(S,R)-Sulfoximine(BSO), 5,5'-dithiobis-2-nitrobenzoic acid (DTNB) were purchased from Sigma Chemical Company St. Louis, MO, U.S.A. Cisplatin was obtained from Biochem Pharmaceutical Industries, Mumbai, India. Cysteine and other chemicals used in the experiments were of analytical grade and purchased within the country.

Tumor maintenance and drug treatment

Swiss albino mice of about 10-12 weeks were used to maintain the tumor ascites Dalton's lymphoma (DL) by serial intraperitoneal (i.p.) transplantation of ~1×10^7 viable tumor cells per mouse (0.25 volume in phosphate buffer saline, PBS, pH 7.4). Transplanted animals usually survived for 19-21 days. Only tumorous mice were used for the experiments.

Transplanted animals were randomly divided into four groups consisting of 5-6 mice each. On the 10th day following transplantation (~ mid phase of tumor growth), mice in group-I were administered with (i.p.) a single therapeutic dose (8mg/kg body wt) of cisplatin [1]. Since the maximum GSH depletion in DL cells has been noted at 8h of BSO (5mM/kg body wt) treatment [9], mice in group-II were administered with BSO 8h prior to cisplatin treatment. Group-III mice were administered with cysteine (200 mg/kg body wt.) 30 min prior to cisplatin treatment [9]. Control groups were given the same volume of 0.89% NaCl.

Chromosomal analysis

After 24-96h of cisplatin treatment mitotic arrest was initiated 1.5h prior to sacrificing the animal by injection of colchicine (i.p., 4 mg/kg body wt). Metaphase plates of bone marrow and DL cells were prepared by flame-drying technique as described earlier [12]. Chromosomal aberrations were analyzed from 100 good metaphase spreads per animal, classified into the general categories of breaks/gaps, exchanges and sister chromatid unions.

Micronucleus assay

After 30 h of cisplatin treatment micronuclei (MN) from bone marrow cells were assayed following the method of Schmid [18] as described in detail elsewhere [12].

Sperm head abnormality assay

The cauda epididymis from the testis of male mice under different treatment conditions were removed after 10 days of cisplatin treatment. About one thousand sperms from each mouse were examined for the abnormalities in sperm heads (amorphous, hooked, triangular, dwarf, double headed/tailed etc) following the criteria established by Wyrobeck and Bruce [19] and described in detail earlier [12].

Glutathione estimation

GSH in bone marrow cells and DL cells of control or cisplatin treated mice were determined following the method of Sedlak and Lindsay [20] as described in detail [6].

RESULTS

Cisplatin treatment of tumor-bearing mice *in vivo* caused the development of various chromosomal aberrations (chromatid breaks, gaps, isochromatid breaks, exchanges, chromosomal fragments etc) and a decrease in GSH level in bone marrow cells and DL cells *(table 1)*. Total number of aberrations as well as abnormal metaphases decreased appreciably in bone marrow cells with the time of treatment (24-96 h), but in DL cells it remained almost unchanged *(fig. 1A)*. Development of micronuclei (MN) in bone marrow cells and abnormalities in sperm heads were also noticed following cisplatin treatment *(table 2)*. As compared to cisplatin alone, combination treatment of BSO and cisplatin caused an increase in CA, MN and sperm head abnormalities while cysteine plus cisplatin treatment resulted a decrease in these mutagenic parameters *(table 1, 2; fig. 1B)*.

DISCUSSION

Anticancer activity of cisplatin has been attributed to its ability to interact with cellular DNA [1, 4]. Development of CA, MN and sperm head abnormality have been commonly used as a reliable biological indicators in the mutagenic bioassay of different drugs [12, 21]. In present study, the development of these mutagenic parameters were noticed in bone marrow cells and sperms following cisplatin treatment of tumor-bearing mice *(table 1, 2)*, and it supports the earlier findings of its genotoxic properties [11, 22]. As compared to bone marrow cells, total number of aberrant metaphases as well as chromosomal aberrations were much higher in DL cells. These cisplatin-induced aberrant metaphases were maximum at 24h of treatment and then it decreased effectively at later periods (48-96h) in bone marrow cells but in DL cells it remained almost unchanged *(fig. 1A)*. It indicates an effective recovery by bone marrow cells at later periods which could be probably due to clearance of drug from the body, post-replication repair process etc. Cisplatin-induced chromosomal aberrations in DL cells were always higher than that of bone marrow cells which may hint at its differential effect on the chromosomes of DL cells and bone marrow cells of the same host. Tumor cells in general have a deficient DNA-repairing ability than normal cells [1]. This may be reflected here in DL cells in which quite high frequency of chromosomal aberrations/aberrant metaphases were maintained *(table 1)*, and their inability to repair the same could be an important factor leading to cytotoxicity/cell death. It has been suggested that in cancer cells because of deficiency in the repair process, cisplatin lesions are not removed and it may lead to cell death [1]. However, it may be repaired in bone marrow cells which represent normal cells. GSH, a strong intracellular antioxidant, is involved in various protective functions in the cell [14, 23], in the synthesis of nucleic acids [23] and also in the metabolism of cisplatin [15]. Cisplatin treatment caused a significant decrease of GSH in DL cells and bone marrow cells *(table 1)* which should be involved to develop less protective function in the cells and thereby resulting higher mutagenic effects. In an attempt to understand further on the significance of GSH in the mutagenic effects, these mutagenic parameters were also analyzed under the specific condition of decreased or increased GSH level using BSO or cysteine respectively. It showed that as compared to cisplatin alone, all the three mutagenic parameters were significantly enhanced in BSO plus cisplatin treated mice but lowered in cysteine plus cisplatin treated mice. It is known that cisplatin-GSH conjugates could be formed and ejected from the cells in an ATP dependent fashion [24]. It is, therefore, expected that a reduction in intracellular GSH concentrations would favour increased accumulation of cisplatin in the cells and may cause more DNA damage resulting to higher mutagenicity. This suggestion is supported by the observation of a reduction in the frequency of mutagenic parameters when cysteine, a precursor for GSH synthesis, was given in combination with cisplatin *(fig. 1B)*. Cisplatin-mediated decrease in intracellular GSH levels was observed to be prominent at 24h of treatment *(table 1)* which recovered later at 72-96h, but any cisplatin-induced damage exerted on the DNA initially (24h) and retained, as in DL cells, may lead to cell damage and tumor regression.

Table 1. Frequency of chromosomal aberrations and GSH levels in bone marrow cells and Dalton's lymphoma cells of mice under different treatment conditions

	Bone marrow cells				Dalton's lymphoma Cells			
Treatment	**No. of metaphases scored**	**Number of aberrant metaphases**	**Aberrations per cell ± S.D.**	**GSH level (μmoles/g) ± S.D.**	**No. of metaphases scored**	**Number of aberrant metaphases**	**Aberrations per cell ± S.D.**	**GSH level (μmoles/g) ± S.D.**
Control	600	3	0.004 ± 0.001	3.28 ± 0.17	400	24	0.02 ± 0.00	4.43 ± 0.26
CP (24 h)	300	181	1.890 ± 0.140*	2.17 ± 0.14*	300	256	5.07 ± 0.34*	3.51 ± 0.32*
CP (48 h)	400	104	0.320 ± 0.036*	2.20 ± 0.15*	300	268	4.73 ± 0.20*	3.82 ± 0.21*
CP (72 h)	500	86	0.300 ± 0.018*	2.61 ± 0.12*	300	264	4.21 ± 0.30*	4.02 ± 0.32
CP (96 h)	400	43	0.180 ± 0.012*	3.01 ± 0.26	300	241	3.72 ± 0.13*	4.21 ± 0.45
BSO+CP (24 h)	300	219	2.93 ± 0.22^{*a}	2.00 ± 0.18^{a}	300	263	6.03 ± 0.43^{*a}	3.22 ± 0.30
BSO+CP (48 h)	300	192	1.07 ± 0.07^{*a}	ND	300	277	5.17 ± 0.59*	ND
BSO+CP (72 h)	300	88	0.58 ± 0.02^{*a}	ND	300	270	4.57 ± 0.38*	ND
BSO+CP (96 h)	300	50	0.38 ± 0.01^{*a}	ND	300	261	4.17 ± 0.17^{*a}	ND
Cys+CP (24 h)	400	128	0.12 ± 0.01^{*a}	2.89 ± 0.25	300	198	3.01 ± 0.07^{*a}	4.18 ± 0.25^{a}

Table 2. Development of micronuclei in bone marrow cells and morphological abnormalities in sperm heads after cisplatin treatment *in vivo*

	Treatment			MicronucleiSperm head abnormalities		
	Polychromatic Erythrocytes (Mean% ± SD)	**Normochromatic Erythrocytes (Mean% ± SD)**	**Nucleated cells (Mean% ± SD)**	**No. of sperms observed**	**No. of abnormal sperms**	**Mean% of abnormal sperms ± SD**
Control	0.13 ± 0.05	0.06 ± 0.04	0.16 ± 0.06	3000	66	2.20 ± 0.36
CP	1.26 ± 0.12*	0.53 ± 0.13*	0.50 ± 0.10*	3000	228	7.59 ± 0.04*
BSO + CP	2.68 ± 0.09^{*a}	0.92 ± 0.11^{*a}	0.66 ± 0.14*	3000	290	9.58 ± 0.18^{*a}
Cys + CP	0.43 ± 0.04^{*a}	0.19 ± 0.05^{*a}	0.15 ± 0.03^{a}	3000	168	5.53 ± 0.03^{*a}

For *table 1 and 2*: CP = Cisplatin, was given as a single dose (8 mg /Kg body wt, i.p.). BSO = L- Buthionine-[S,R]-sulfoximine was administered i.p. at a single dose of 5 mM/Kg body wt., 8 h prior to cisplatin treatment. Cys = Cysteine, was given (200 mg/Kg body wt) i.p. 30 min prior to cisplatin treatment. Gaps have not been included. GSH = Reduced glutathione; ND = Not determined. Results are expressed as means ± S.D. Student's t-test, n = 3-4; as compared to respective control (*) and cisplatin alone (^{a}). $^{*,a}P \leq 0.05$.

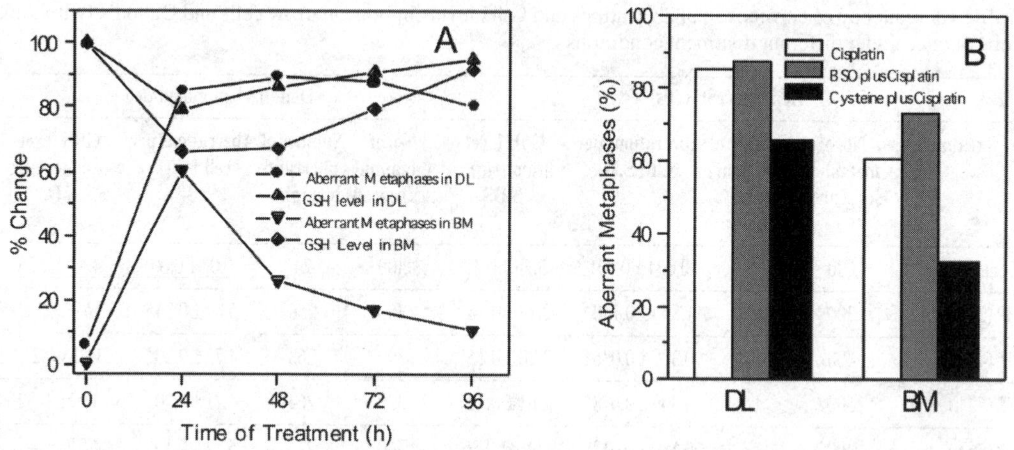

Figure 1. A: Changes in aberrant metaphases and GSH levels in Dalton's lymphoma (DL) cells and bone marrow (BM) cells after cisplatin treatment. B: As compared to cisplatin treatment alone (24 h), aberrant metaphases increase in BSO plus cisplatin- and decrease in cysteine plus cisplatin-treated (24 h) mice.

CONCLUSION

Our findings reveal the mutagenic potential of cisplatin in the host and it is affected by the changes in endogenous GSH levels in the cells. Differential effect of cisplatin on the chromosomes of normal and tumor cells was noted which may possibly involve unlike Pt-DNA interactions and/or repair process.

ACKNOWLEDGEMENTS

The financial support was provided by North-Eastern Hill University, Shillong and DST, INSA, New Delhi. We gratefully thank Prof. S.K. Mishra for his help in the scanning.

REFERENCES

1. Rosenberg B. Fundamental studies with cisplatin. *Cancer*, 1985, 55, 2303-2316.
2. Prasad S.B., Giri A. Antitumour effect of cisplatin against murine ascites Dalton's lymphoma. *Ind. J. Exp. Biol.*, 1994, 32, 155-162.
3. Go R.S., Adjei A. A. Review of comparative pharmacology and clinical activity of cisplatin and carboplatin. *J. Clin. Oncol.*, 1999, 17, 409-422.
4. Coste F., Malinge J., Serra L., Shephard W., Roth M., Leng M., Zelwer C. Crystal structure of a double-stranded DNA containing a cisplatin intrastrand cross-link at 1.63 Å resolution-hydration at the platinated site. *Nucleic Acid Res.*, 1999, 27, 1837-1846.
5. Prasad S.B., Giri A., Khynriam D., Kharbangar A., Nicol B.M., Lotha C. Cisplatin-mediated enzymatic changes in mice bearing ascites Dalton's lymphoma. *Medical Sci. Res.*, 1999, 27, 723-730.
6. Kharbangar A., Khynriam D., Prasad S.B. Effect of cisplatin on mitochondrial protein, glutathione and succinate dehydrogenase in Dalton's lymphoma-bearing mice. *Cell Biol. Toxicol.*, 2000, 16, 363-373.
7. Nicol B.M., Prasad S.B. Sialic acid changes in Dalton's lymphoma-bearing mice after cyclophosphamide and cisplatin treatment. *Brazilian J. Med. Biol. Res.*, 2002, (in press).
8. Krakoff I.H. Nephrotoxicity of *cis*-dichlorodiammineplatinum. *Cancer Treat. Rep.*, 1979, 63, 1523-1525.

9. Khynriam D., Prasad S.B. Hematotoxicity and blood glutathione levels after cisplatin treatment of tumor-bearing mice. *Cell Biol Toxicol.*, 2001, 17, 357-370.
10. Kartalou M., Essigmann J.M. Mechanisms of resistance to cisplatin. *Mutation Res.*, 2001, 478, 23-48.
11. Overbeck TL, Knight JM, Beck DJ. A comparison of the genotoxic effects of carboplatin and cisplatin in Escherichia coli. *Mutation Res.* 1996, 362, 249-259.
12. Giri A., Khynriam D., Prasad S.B. Vitamin C mediated protection cisplatin induced mutagenicity in mice. *Mutation Res.*, 1998, 421, 139-148.
13. Giri A., Khynriam D., Prasad S.B. Use of vitamin C against cisplatin induced mutagenicity and nephrotoxicity. In: *Trends in Radiation and Cancer Biology.* (Sharan R.N, ed.), Forschungszentrum Julich Gmbh, Germany. 1998, pp. 166-176.
14. Wang W., Ballatori N. Endogenous glutathione conjugates: occurrence and biological functions. *Pharmacol. Rev.*, 1998, 50, 335-355.
15. Suzuki C.A.M., Cherian M.G. The interaction of *cis*-diamminedichloroplatinum with metallothionein and glutathione in rat liver and kidney. *Toxicology*, 1990, 64, 113-127.
16. Meister A. Glutathione metabolism and its selective modification. *J. Biol. Chem.*, 1988, 263, 17205-17208.
17. Bannai S., Tateishi N. Role of membrane transport in metabolism and function of glutathione in mammals. *J. Membrane Biol.*, 1986, 89, 1-8.
18. Schmid W. The micronucleus test for cytogenetic analysis. In: *Chemical mutagens: Principles and methods for their detection*, (Hollaender, A. ed), Plenum, New York. 1976, 4, pp. 31-53.
19. Wyrobeck A.J., Bruce W.R. Chemical induction of sperm abnormalities in mice. *Proc. Natl. Acad. Sci. U.S.A.*, 1975, 72, 4425-4429.
20. Sedlak J., Lindsay R.H. Estimation of total, protein-bound and nonprotein sulfhydryl groups in tissue with Ellman's reagent. *Anal. Biochem.*, 1968, 25, 192-205.
21. Giri S., Prasad S.B., Giri A., Sharma G.D. Genotoxic effects of malathion: an organophosphorus insecticide, using three mammalian bioassays in vivo. *Mutation Res.*, 2002, 514, 223-231.
22. Pillaire M.J., Margot A., Villani G., Sarasin A., Defais M., Gentil A. Mutagenesis in monkey cells of a vector containing a single d(GPG) *cis*-diamminedichloroplatinum (II) adduct placed on codon 13 of the human H-*ras* proto-oncogene. *Nuclic Acids Res.*, 1994, 22, 2519-2524.
23. Arrick B.A., Nathan C.F. Glutathione metabolism as a determinant of therapeutic efficacy: A review. *Cancer Res.*, 1984, 44, 4224-4232.
24. Ishikawa T., Ali-Osman F. Glutathione-associated *cis*-diamminedichloroplatinum(II) metabolism and ATP-dependent efflux from leukemia cells: Molecular characterization of glutathione platinum complex and its biological significance. *J. Biol. Chem.*, 1993, 268, 20116-20125.

Metal Ions in Biology and Medicine: vol. 7. Eds L. Khassanova, Ph. Collery, I. Maymard, Z. Khassanova, J.-C. Étienne.
John Libbey Eurotext, Paris © 2002 pp. 586-588.

Selenium and prostate

RM. Pérez-Beriain, A. García de Jalón Comet, J. Castrillo, ML. Calvo Ruata, J.F. Escanero Marcén*, A. García de Jalón Martínez, A. Borque, C. Allepuz**

Biochemistry department. Miguel Servet Universitary Hospital. Pº Isabel la Católica 1-3. 50009 Zaragoza, Spain
**Phisiology department of University. C/ Domingo Miral s/n. 50006 Zaragoza, Spain*
***Urology department. Miguel Servet Universitary Hospital. Pº Isabel la Católica 1-3. Zaragoza, Spain*
The mail will be remitted to: Dr. Ángel García de Jalón Comet.
Jefe de Servicio de Bioquímica Clínica
Hospital Universitario Miguel Servet - Paseo Isabel la Católica 1-3
50009 Zaragoza (Spain) - E-mail: jcomet@teleline.es

ABSTRACT

The aim of the study was to investigate the possible relationship between selenium levels in prostate cancer patients and the healthy population.

Methods : We included three groups in the study. The first group, used as a control, consisted of 107 healthy subjets. In the second group, we considered 197 patients with benign prostatic hyperplasia. In the third group we included 116 patients with positive prostatic biopsy (prostate cancer).

The serum selenium levels were measured by atomic absorption spectrophotometry (AAS) with a graphite furnace and Zeeman background corrector.

Results and conclusions : According to age, the differences are not statistically significant. The analysis of variance shows a statistically not significant difference between the control subjets and the patients with benign hyperplasia, but between this two groups and the group of prostate cancer, the difference was significant with a value for $p<0.05$.

The findings of this study suggest that the individuals with prostate cancer show lower serum concentrations of selenium than the normal population.

INTRODUCTION

Cancer is a major cause of mortality and morbidity throughout the world. Prostate cancer is the second leading cause of cancer deaths in men, therefore it is increasingly important to understand its biology and epidemiology [1, 2].

Subjets with high selenium and vitamin E intake have a lower risk of prostate cancer [3]. The risk of cancer for patients with low serum selenium has been estimated to be up to twice that of subjets with higher levels [4].

Selenium supplementation has been shown for many years to work as an anticarcinogenic agent both in epidemiology and in vitro studies [5]. Selenium supplementation has recently been shown to decrease total cancer incidence [2, 6].

The evidence now available indicates that substantial increases in the consumption of selenium by men taking 80-90 µg a day or more may have a striking impact on prostate cancer rates [7].

However, the mechanism of action of selenium as an anticarcinogenic agent has yet to be elucidated.

AIMS

The aim of the study was to investigate the possible relationship between selenium levels in prostate cancer patients and the healthy population.

MATERIAL AND METHODS

Subjets

We included three groups in the study:

a) The first group, used as a control, consisted of 107 healthy subjets (mean aged= 66.19 ± 9.85 years old).

b) In the second group, we considered 197 patients with benign prostatic hyperplasia, and 66.25 ± 5.11 years old.

c) In the third group we included 116 patients with positive prostatic biopsy (prostate cancer) and 69.19 ± 8.08 years old.

Instrumentation

The serum selenium levels were measured by atomic absorption spectrophotometry (AAS) with a graphite furnace and Zeeman background corrector (*Perkin Elmer 4110 ZL*), using $Pd(NO_3)_2$ solution as matrix modifier.

Statistical analysis

The statistical calculations were carried out using SPSS statistics program. The statistical test used has been the analysis of variance (ANOVA). The Multiple Range Test used was the LSD test (*least significance differences*) with significance level of 0.05.

RESULTS

According to age, the differences are not statistically significant.

The distribution of the selenium levels (probability interval, or mean ± 2 times Standard Deviation (S.D.)) in the differents groups, are defined to the following illustration *(fig. 1)*:

The results of the serum concentrations of selenium in the groups *(table I)* were as follows:

Table 1.

	N	Selenium (μg/l)	C.I. (95%)
Control group	107	74.88	72.00-77.76
Benign prostatic hyperplasia	197	75.54	74.03-77.06
Prostate cancer	116	69.90	67.74-72.07

The analysis of variance shows a statistically not significant difference between the control subjets and the patients with benign hyperplasia, but between this two groups and the group of prostate cancer, the difference was significant with a value for $p<0.05$.

The error bars represent confidence interval for the mean, with a confidence level of 95% *(fig. 2)*.

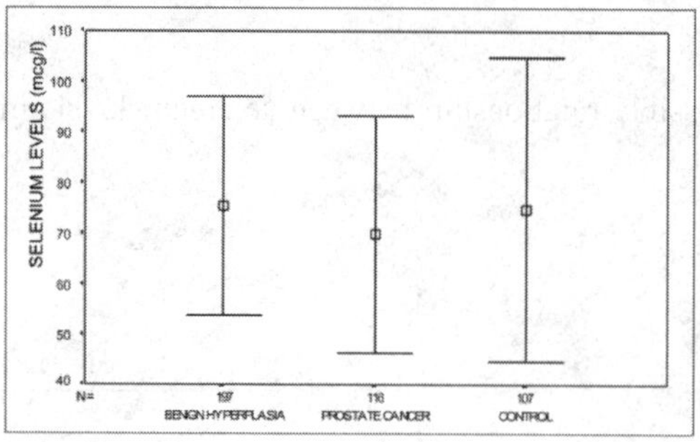

Fig. 1.

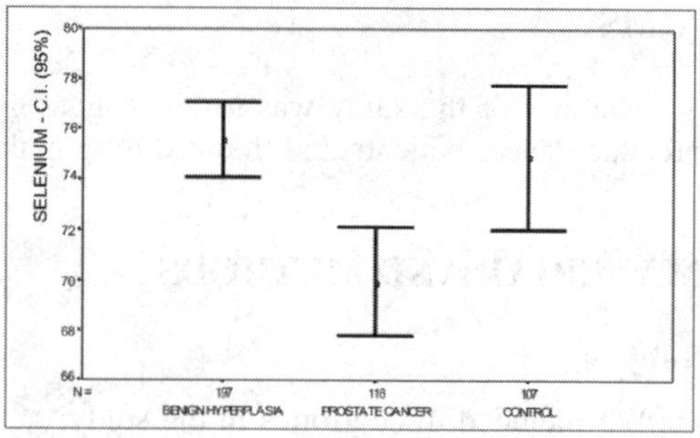

Fig. 2.

CONCLUSIONS

The findings of this study suggest that the individuals with prostate cancer show lower serum concentrations of selenium than the normal population.

On the other hand, we conclude that the selenium doesn't diminish when benign hyperplasia exists, but it decreases when cancer exists.

Even if the other potential benefits of improved selenium status were disregarded, the evidence available for prostate cancer seems to ustify the further assessment of increasing the selenium intake in the population as a priority for public health.

REFERENCES

1. Ashish M. Kamat, Donald L. Lamm. Chemoprevention of urological cancer. The Journal of Urology. 1999, June. Vol. 161, 1748-1760.
2. Clark LC, Dalkin B, Combs GF et al. Decreased incidence of prostate cancer with selenium supplementation: results of a double-blind cancer prevention trial. Br J Urol. 1998 May. 81(5). P 730-4.
3. Brawley OW, Pames H. Prostate cancer prevention trials in the USA. Eur J Cancer. 2000 Jun. 36(10). P 1312-5.
4. Willett WC, Polk BF, Morris JS et al. Prediagnostic serum selenium and risk of cancer. Lancet, 2: 130, 1983.
5. Oberley TD, Zhong W, Szweda LI, Oberley LW. Localization of antioxidant enzimes and oxidative damage products in normal and malignant prostate epithelium. Prostate. 2000 Jul 1. 44(2). P 144-55.
6. Nelson MA, Porterfield BW, Jacobs ET, Clark LC. Selenium and prostate cancer prevention. Semin Urol Oncol. 1999 May. 17(2). P 91-6.
7. E. Giovannucci. Selenium and risk of prostate cancer. The Lancet. Vol 352. September 5, 1998. P 755-6.

*Metal Ions in Biology and Medicine: vol. 7.*Eds L. Khassanova, Ph. Collery, I. Maymard, Z. Khassanova, J.-C. Étienne.
John Libbey Eurotext, Paris © 2002 pp. 589-594.

Metal ions and cancer

Bernard Desoize*, and Philippe Collery**

**Laboratoire de Biochimie, Faculté de Pharmacie, 51 rue Cognacq-Jay, 51000 Reims, France*

***Service de Cancérologie, Polyclinique Maymard, rue Marcel Paul, 20200 Bastia, France*

The existence of relationships between cancer and metals is known by all oncologists. However, various aspects about these relationships are ignored by many. The purpose of this short paper is to propose a brief overview of the relationship between several metals and carcinogenicity and cancer treatment. More details on this subject will be found in two special issues of *Critical reviews in Oncology and Hematology* that will be published in April and June 2002.

METAL AND CARCINOGENESIS

Arsenic

Several studies have indicated that arsenic exposure in certain animal models and in humans contributes to skin neoplasia by chronic stimulation of growth factors. Arsenate (As^{5+}) is structurally similar to phosphate and is probably an inhibitor of reactions involving this anion, such as kinases and phosphatases, thus modulating signal transduction pathways.

Arsenic acts at the level of tumour promotion by modulating the signalling pathways which regulate cell growth and apoptosis. In particular, arsenic has been reported to stimulate activator protein-1 (AP-1) and to either activate or inhibit nuclear factor kappa B (NFκB). AP-1 regulates the transcription of various genes that govern cellular processes such as inflammation, proliferation and apoptosis. The components of AP-1 are activated by MAPKinases, which are activated by translocation to the nucleus, where they phosphorylate target transcription factors. NFκB is a rapidly induced stress-responsive transcription factor that functions to increase the transcription of a variety of genes including cytokines, growth factors and acute response proteins and its activation has been shown to be linked to MAPKinase signalling pathways. Therefore, because of its potential dual role in inhibiting or promoting apoptosis, NFκB may be an important key in determining whether arsenic acts as a carcinogenic agent or as an effective therapeutic agent. Arsenic induces chromosomal abnormalities and disruption of DNA methylation and of repair systems. Arsenic-induced oxidative stress with subsequent DNA damage could explain the toxicity of arsenic. The role of oxidative stress in carcinogenesis will be detailed below.

Cadmium

Cadmium was not believed to generate free radicals by itself, but it has been reported that its administration increased lipid peroxidation. Furthermore, the reducing potential of the cells protects them from cadmium-induced toxicity. A single dose of cadmium chloride was carcinogenic in rats. Cadmium is able to block apoptosis, allowing mutated cells to survive, thereby increasing the probability of cancer.

Chromium

Cr^{6+} has been considered as the most carcinogenic form of this metal, possibly due its high bioavailability. Cr^{3+} decreases the fidelity of DNA polymerase which contributes to its mutagenic and carcinogenic potential. Chromium induces apoptosis through a mitochondrial pathway and through activation of p53Noxa. Chromium also produces reactive oxygen species (ROS).

Manganese

Manganese plays an important role in antioxidant defence mechanisms and forms part of a superoxide dismutase. Manganese is relatively non-toxic to adults. Mutagenicity of Mn^{2+} is low, nevertheless, by modifying the activity of DNA polymerase it decreases fidelity of DNA replication. Arranged in decreasing order of activity we find $MnCl_2 > KMnO_4 > MnSO_4$, but they are far below the mutagenic activity of nickel and chromium. Nevertheless, when injected I.M. to rats, manganese can produce fibrosarcomas. In conclusion, manganese does not represent a significant carcinogenic risk to the population or to workers.

Zinc

Zinc prevents oxidative stress. It stabilizes cell membranes, inhibits sulfhydryl group oxidation and interestingly, it seems to displace redox active metal ions, such as copper and iron, from site-specific loci where damage occurs.

Nickel

According to the International Agency for Research on Cancer (IARC) all nickel compounds, except for metallic nickel, were classified as carcinogenic to humans in 1990. Nickel displays both transforming capability and tumour promoter effects. Its effects could be epigenetic rather than genetic: it increases DNA methylation and histone deacetylation, leading to the inactivation of gene expression. Nickel also activates NF-κB transcription, which in turn induces expression of adhesion molecules. Nickel induces wild type p53 synthesis and then decreases its functional activity. Rb protein, another tumour suppressor, is hypophosphorylated in nickel-transformed cells. Administration of Ni resulted in increased lipid peroxidation products and ROS formation.

Copper

The major functions of copper within metallo-enzymes are in oxidation-reduction reactions, in which it reacts directly with molecular oxygen to produce ROS. Copper binds to DNA, with higher affinity than any other divalent cation, and then promotes DNA oxidation. The oxidation of DNA by hydrogen peroxide using copper as catalyst, which induces single-strand breaks, is 50 times faster than that with iron. Copper induces formation of single-strand breaks, double-strand breaks and chromosomal aberrations. High concentrations of copper can induce growth proliferation and cancer by damaging DNA with ROS. Inadequate dietary copper increases the spontaneous tumorigenesis in the Min mouse.

Vanadium

Probably due to its structural analogy to phosphate, vanadium is able to inhibit cellular tyrosine phosphatases and to activate tyrosine phosphorylases. Both effects activate signal transduction pathways leading either to apoptosis or cell cycle arrest. Vanadium exerts more potent antitumour effects via the initiation rather than the promotion stage of hepatocarcinogenesis. This is mediated mainly through inactivation of carcinogen-derived active metabolites, via induction of liver xenobiotic metabolizing enzymes. Mutagenic and carcinogenic effects and plasma membrane lipoperoxidation could be a result of ROS generation. Vanadium might cause apoptosis in some types of cells, whereas it may be anti-apoptotic in other types.

Iron

Cancer cells have higher numbers of transferrin receptors (TfR1) probably because they take up iron to a larger degree. TfR2, another transferrin receptor has been detected in a wide variety of neoplastic cell lines, and may further increase iron uptake. A membrane-bound transferrin homologue, melanotransferrin, has been reported. and is found in larger amounts in tumour cells. Transferrin secreted by cells may act as an autocrine growth factor, as well as a mitogen. Ferritin expression is up-regulated and secreted in some tumours. Ferritin may also act as an autocrine growth factor. Iron is also able to produce ROS.

Magnesium

Magnesium intake often falls short of the recommended dietary amounts, leading to chronic marginal deficiency that may yield cardiovascular heart diseases, accelerated aging and cancer. Magnesium acts as an intracellular regulator of the cell cycle and controls apoptosis, it plays a significant role in the stabilization of the tridimensional structure of DNA. It also prevents DNA from alkylation. Mortality due to stomach cancer was correlated to the ratio of Mg^{2+}/Ca^{2+}, rather than to magnesium deficiency itself. In magnesium deficient animals there is an increased production of free radicals.

Antimony

Antimony is probably carcinogenic but not genotoxic. The low incidence of toxicity of antimony amongst the general population may be related to the fact that the environmental distribution of antimony is low.

Role of oxidative stress in carcinogenesis

Some metals are necessary for the normal functioning of cells, however, amounts exceeding the physiological levels may be highly toxic. A number of metal ions are able to generate free radicals. Usually there is, within each cell, a fine balance between antioxidant systems and free radicals. When ROS overwhelm the capacity of the defence, the consequences are notably ageing and cancer. Free radicals are chemical compounds possessing one or more unpaired electrons in their outer orbitals, in contrast to non radical species that have pairs of electrons with antiparallel spin. The mechanism of formation of free radicals has been proposed by Haber-Weiss seventy years ago:

$$O^{\bullet -}_2 + H_2O_2 \rightarrow {}^{\bullet}OH + OH^- + O_2$$

The most important factor in the maintenance of cellular redox homeostasis is the thiol group. Under certain conditions thiols forms of thiyl radicals ($-S^{\bullet}$) which can interact with molecular oxygen, generating ROS. The most important molecule containing a thiol group is glutathione. Glutathione-metal conjugates are still redox active, and may induce DNA damage. Most DNA damage seem to be mediated by metal-induced free radicals; in addition, metals such as Cr, Cd, Ni, Co and As, can also inhibit the repair of DNA. Several mechanisms can result in inhibition of repair: directly by free radicals; substitution of zinc in Zn-finger domains, or indirectly by lowering the level of reduced glutathione. Since glutathione is able to reverse most of this inhibition, it is probable that such DNA repair inhibition could be redox-dependent.

METALS AND CANCER TREATMENT

The importance of metal compounds in medicine is undisputed as can be judged by the use of, for example, antimony (anti-protozoal), bismuth (anti-ulcer), gold (anti-arthritic), iron (anti-malarial), silver (anti-microbial) and platinum (anti-cancer) compounds in the treatment of various diseases. Superoxide dismutase inhibition leads to selective killing of cancer cells *in vitro* and *in vivo*, as a consequence, ROS, as well as causing a cancer, could be the solution for the treatment of the tumour.

Antimony

Several series of derivatives have been synthesised and tested, such as antimony(III) polydentate carboxylic acids, diphenylantimony(III) thiolates, Triphenylantimony(V) polyamines, Triorganoantimony(V) disalicylates, Sb_2O_3 (normally regarded as a poison, Sb_2O_3 is less toxic than As_2O_3, which is used against promyelocytic leukaemia) and potassium antimony tartrate (although this compound showed a cross-resistance with cisplatin). All these compounds displayed cytotoxicity in several cell lines, but more studies are in progress to find an efficient derivative.

Bismuth

Bismuth complexes of 6-mercaptopurine were the first antitumour compounds tested yielding interesting results. The emission of α-particles by bismuth isotopes shows potential in radiotherapeutics. The efficacy is increased when the compounds are attached to a monoclonal antibody which can target tumour cells.

Gold

The use of gold in medicine dates back to Arabic and Chinese physicians. Studies of the anti-tumour activity of gold compounds were stimulated notably because patients treated with gold (chrysotherapy) for rheumatoid arthritis had lower malignancy rates than other patients. A large number of gold(I) compounds have been studied, the most potent class was that containing both phosphine and thioglucose ligands. Some compounds combined with ferrocene derivatives showed *in vitro* activity superior to cisplatin. Several Gold(III) compounds, when combined with biologically active molecules, demonstrated interesting activities against cisplatin-resistant cell lines. Some compounds, involving imine donors, are more potent than carboplatin. Other derivatives demonstrated more cytotoxicity than cisplatin in cisplatin-resistant and cisplatin-sensitive cell lines. Several showed a similar cytotoxicity profile to cisplatin. Cross-resistance with cisplatin was found for some of the compounds.

Vanadium

Since it has few adverse effects, vanadium possesses a good therapeutic potential. Anticancer effects of vanadium compounds have been reported with a large variety of malignant cell lines. Antiproliferative effects of vanadium compounds on normal and malignant cell lines appear to be exerted mainly through cell cycle arrest. The concentrations of vanadium were found to be superior in cancerous breast tissue in comparison to normal breast tissue. The organometallic complexes, with vanadium(IV) exhibit antitumor properties *in vitro* and *in vivo*. One of the most promising among these metallocenes, is vanadocene dichloride. Protein tyrosine phosphorylation, induced by vanadium, changes the invasive and metastatic potential of tumor cells, by modulating cell-substrate adhesion, cell-to-cell contact and actin cytoskeleton.

Iron

Iron deprivation is an excellent target since it results in a decreased tumour growth. Amongst iron-containing enzymes, ribonucleotide reductase is one of the most sensitive to iron depletion, as this enzyme contributes to DNA synthesis. Desferrioxamine (DFO) is currently the best drug for iron overload diseases. Once it is bound, iron is metabolically inactive and then does not produce ROS. DFO is effective at inhibiting the growth of a number of tumours. The thiosemicarbazone Triapine markedly inhibits several xenograft tumours, and it crosses the blood-brain barrier. It is synergistic with a number of anticancer drugs. Aroylhydrazones demonstrated distinct activity against mammary tumours and certain leukemias in mice. Tachpyridine inhibits ferritin synthesis and the proliferation of bladder cancer cells in culture, it induces apoptosis. O-Trensox reduces DNA synthesis with greater efficiency than DFO and induces apoptosis.

Rhodium

Rhodium belongs to the same group as platinum. Recent structural studies suggest that the antitumor activity of dirhodium(II) carboxylates may be due to the binding to adjacent guanines on DNA, as with cisplatin. Neutral Rhodium(I) organometallic complexes have activity on Ehrlich and Landschutz ascitic tumors [102-103]. Cationic complexes of rhodium (III) were studied for their antitumor effects, the most promising was the complex including lepidine [116]. Rhodium(III) polypyridyl complexes were studied also for their potential use in phototherapy [118]. Brachytherapy with (-ray $^{106}Ru/^{106}Rh$ plaques can be recommended for choroidal melanomas with good results [121-123]. The most common side-effect of rhodium is nephrotoxicity. Several rhodium compounds have entered phase I clinical trials.

Titanium

The first non-platinum complex tested in clinical trials was a titanium complexe, titanocene dichloride which is now in clinical trials. Bis(β-diketonato)titanium complexes exhibited antitumour properties against various animal tumours; they are more efficient against colon tumours than 5-fluorouracil. The limiting toxicity concerns liver and kidney [7]. Titanocene complexes have strong antiproliferative activity against various cell lines *in vitro* and *in vivo*, even against resistant carcinoma cells. It showed higher activity than conventional anticancer drugs. The ionic titanocene acetonitrile complex exhibited comparable activities. The mechanism of action has not been investigated, the target is unknown and the exact chemical nature of the formulated solutions is still unknown, probably because the complexes are subject to extensive hydrolysis in water, at pH > 5, to yield oligomeric $[Ti(bzac)_2O]_2$ which is insoluble in water. Since transferrin forms a strong complex with titanium(IV), it has been proposed to use transferrin as a mediator for the delivery of the compound to tumour cells.

Gallium

Anticancer properties were described for the first time in 1971. The anionic component of the metal salt has no influence on cytotoxicity. Gallium forms a transferrin-gallium complex that inhibits DNA synthesis by acting on ribonucleotide reductase. Gallium acts as an antagonist to several divalent ions including Mg^{2+}, Fe^{2+}, Zn^{2+} and Ca^{2+}. The tris(8)quinolinolatoGa(III) compound appeared very active in an experimental model, with a reduction of more than 50% of the volume of the tumour, without any significant toxicity. Doxorubicin-gallium-transferrin conjugate is able to reverse resistance to doxorubicin. Protracted infusion of gallium is effective against cancer hypercalcemia. Gallium uptake is selective for the tumour and even more so for metastases, after oral administration when compared to intravenous injection. It was reported that gallium potentiates cisplatin and etoposide; dose adjustment of the three drugs allowed a decrease in toxicity and thus permitted an increase in the number of courses administered. Preclinical studies have demonstrated synergy of gallium with paclitaxel, gemcitabine, vinorelbine, hydroxyurea, fludarabine and interferon-α.

Platinum

The first platinum-containing complex to be used in cancer treatment was cisplatin. Three thousand platinum derivatives have been synthesised and tested, thirty reaching clinical trials and more than half of those have already been rejected. We will focus on the four compounds that are currently available on the market. Efficacy is related to the number of platinum adducts on DNA. *Trans* adducts are more easily repaired than *cis* adducts. Inhibition of DNA synthesis and repair could result from a modification of the three dimensional structure of DNA, induced by the metal adducts. One of the attractions of cisplatin, beside its efficacy, is that its toxicity is different from that of other anticancer agents, thus favouring its combination with other drugs. In testis, bladder, head and neck, small cell lung cancer (SCLC) and in several paediatric malignancies cisplatin remains superior to carboplatin, whereas in other cancers carboplatin has tended to replace cisplatin.

Carboplatin was selected mainly because of its lower non-haematological toxicity compared with cisplatin. Unfortunately, it exhibits cross-resistance with its parent compound. The dose limiting toxicity is myelosuppression, chiefly thrombocytopenia. Its antitumour activity is not superior to that of cisplatin, but it has lower toxicity. Cisplatin and carboplatin sensitize tumour cells to radiotherapy.

Oxaliplatin was selected because it has a higher efficacy and a lower toxicity than cisplatin, it has no cross-resistance with cisplatin. The dose limiting toxicity is sensory neuropathy, which cannot be predicted. It gave interesting results in a number of cancers, especially in colorectal cancer; and its efficacy is remarkable against cancer resistant to other platinum derivatives.

Nedaplatin was selected because it produced better results than cisplatin in preclinical studies, unfortunately, it is cross-resistant with cisplatin. Its main toxicity in humans is myelosuppression, with a delayed nadir and recovery. The official indications in Japan are head and neck, testicular, lung (NSCLC and SCLC), oesophageal, ovarian, and cervical cancer.

In conclusion the second half of the past century gave us chemotherapy, what will happen in the 3rd millennium? We now know that approximately $2/3^{rd}$ of cancers could be avoided by modifying our everyday life (cigarettes, alcohol, nutrition, sexual behaviour, etc.). Thus it can be concluded that it is statistically easier to avoid a cancer than to be cured of it. Could it be more efficient to protect our cells from mutations rather than curing cancers?

*Metal Ions in Biology and Medicine: vol. 7.*Eds L. Khassanova, Ph. Collery, I. Maymard, Z. Khassanova, J.-C. Étienne.
John Libbey Eurotext, Paris © 2002 pp. 595-599.

Soluble cell adhesion molecule-1, selenium and diet in renal and urinary bladder cancers

Anna M. Witkowska[1], Barbara Darewicz[2], Renata Markiewicz[3], Maria H. Borawska[3], Katarzyna Hukalowicz[3]

[1]Department of Food Commodities Science and Technology, Medical Academy, ul. Mieszka I 4 b, 15-054 Bialystok, Poland, [2]Urology Clinic, Medical Academy, ul. Sklodowskiej 24 a, 15-276 Bialystok, Poland, [3]Department of Bromatology, Medical Academy, ul. Kilinskiego 1, 15-230 Bialystok, Poland

ABSTRACT

Soluble cell adhesion molecule-1 (sICAM-1) may play an important role in the cancer progression. Its elevated serum levels were observed in renal and bladder cancers. There is evidence that selenium is an anti-tumor factor. Though, in most papers low serum selenium has been reported.

The purpose of the study was to analyze the influence of dietary preferences on serum sICAM-1 and selenium in renal and urinary bladder cancers.

The serum sICAM-1 and selenium levels in 43 patients of both sexes (mean age 64) with urinary bladder or renal cell carcinoma, and in 15 controls (mean age 63) were measured. Serum selenium concentration was determined using electrothermal atomic absorption spectrometry. Serum sICAM-1 was determined with enzyme-linked immunosorbent assay (ELISA). Food-frequency questionnaires were implemented to collect the dietary data.

The mean selenium concentrations in carcinoma of kidney and urinary bladder were lower than in controls ($p<0.005$ and $p=0.000001$, respectively). No difference was found between the two groups and controls with regard to sICAM-1. However, when men were excluded from the study, significantly elevated levels of sICAM-1 in women were observed at $p<0.002$. There was negative, but not significant, correlation between serum selenium and sICAM-1 levels in cancer patients ($r=-0.26$, $p<0.2$) and controls ($r=-0.34$, $p<0.4$). Significantly lower sICAM-1 level was observed in studied cancers, when alcohol was consumed more frequently. Frequent consumption of wholemeal bread was associated with statistically elevated serum selenium in subjects with renal and urinary bladder carcinoma.

Our results suggest possible association between selenium, sICAM-1 and the diet in renal and urinary bladder cancers.

INTRODUCTION

Selenium is an element with diverse impact on urinary system. On the one side, it prevents oxidative stress leading to major injury of renal cells [1], but on the other, inorganic selenium may induce an excess incidence of renal carcinoma, when drunk with water [2]. Substantially human receive selenium from food as selenoamino acids, selenomethionine and selenocysteine, that demonstrate higher bioavailability than from inorganic species. Ingested selenomethionine is absorbed in the small intestine, and when not metabolized instantly, it is incorporated into organs with high protein retention, such as kidneys [3]. Recent reports reveal high renal selenium concentrations after selenomethionine administration in contrast to sodium selenite. That happens because selenium is possibly involved in the proximal tubule metabolism [4]. In rats, selenium-deficient diet increases the incidence of albuminuria and glomerular sclerosis [1]. The results of the surveys mentioned demonstrate the importance of the dietary selenium in the pathophysiology of kidney. There is not known enough on the significance of dietary selenium in the bladder pathology. However, in bladder cancer lower selenium levels were observed [5].

Selenium supplementation is reported to enhance immune system activity and tumor toxicity. The mechanism involved into the immunostimulant activity of selenium is probably connected to the up-regulated expression of receptors for interleukin-2 on the surface of lymphocytes and natural-killer cells [6]. Intercellular cell adhesion molecule-1 (ICAM-1) is also present on stimulated lymphocytes, however the possible role of selenium in ICAM-1 expression is not known. ICAM-1 is a transmembrane protein, which expression has been reported in renal cancer [7]. Consequently serum ICAM-1 (sICAM-1) levels were found to be significantly higher in renal carcinoma in contrast to non-malignant renal diseases [8]. It is also suggested, that this is sICAM-1 to play an important role in the progression of renal cell carcinoma and urinary bladder cancer [9, 10].

In the present study, we surveyed patients with renal cancers and urinary bladder cancers on their dietary preferences to determine the influence of the dietary factors on serum levels and the reciprocal relationship between selenium and sICAM-1.

MATERIAL AND METHODS

Subjects

Forty three patients of both sexes aged 41-88 years (median age 64 years) with primary urinary bladder or renal cell carcinoma, and fifteen healthy subjects aged 39-87 years (median age 63 years) serving as controls participated in the study. The study was approved by the Local Ethical Committee. All subjects signed informed consent documents.

Table 1. Cancer subjects characteristics

	Urinary bladder cancer (n=27)	**Renal cell cancer (n=16)**	**Total (n=43)**
Number of male subjects	26	7	33
Number of female subjects	1	9	10

Sample storage. Blood samples were collected in VACUTAINER Systems test tubes containing clot activator, Becton Dickinson, France. The samples were allowed to clot within 30 minutes with subsequent centrifugation for 10 minutes at approximately 1000 × g. Then serum was removed and kept frozen at -70 degrees C.

Selenium determination. The selenium contents of the serum samples were assayed by the electrothermal absorption spectrometry (ETAAS) technique on Z-5000 spectrometer, Hitachi, Japan. The samples were diluted in 1M nitric acid and 1% Triton X-100. Certified reference material Seronorm Trace Elements, Serum (704121), SERO AS, Norway, was used to test the accuracy of this method. The results of the quality control analyses were in good agreement with the reference values.

sICAM-1 determination. Serum levels of sICAM-1 were measured with the quantitative sandwich enzyme-linked immunosorbent assay (ELISA), Biomedica GmbH, Austria.

Dietary data. Food-frequency questionnaires were implemented to collect the dietary data. Participation was voluntary and patient confidentiality was insured. Cancer subjects were asked to complete a questionnaire concerning the consumption frequency of twenty five food products: milk, white bread, wholemeal bread, noodles and macaroni, fresh meat (pork or beef), luncheon meat, offal, poultry, fish, cottage cheese, cheese, eggs, butter, pork lard and bacon, vegetable oil, margarine, legumes, vegetables, potatoes, fruits, honey, sweets and pastry, coffee, tea, alcohol. The consumption frequency of foods was estimated according to the following criteria: frequent consumption was defined as an intake of certain food twelve to thirty days monthly, except fish, that were eaten four to twelve days a month. Food products eaten less frequently were classified into "sporadic consumption" group. After the questionnaires had been completed, vegetables and fruits were excluded from the further analysis considering their frequent consumption.

Statistical analysis. The data analysis was performed with parametric tests: unpaired Student's *t*-test for group comparisons and Pearson's correlation test for the correlations between serum selenium and serum sICAM-1.

RESULTS

The mean selenium concentrations in carcinoma of kidney and urinary bladder were significantly higher than in controls *(table II)*.

Table 2. Comparison of the serum selenium and sICAM-1 levels in cancer subjects and in controls

	Bladder cancer	Renal cancer	Whole group[a]	Control
Serum selenium (μg/l)	45.09 ± 16[b]	54.31 ± 13[c]	47.45 ± 15[d]	70.30 ± 13
Serum sICAM-1 (ng/ml)	301.92 ± 110	335.13 ± 110[e]	323.90 ± 134	251.21 ± 40

All results are expressed as means ± SDs. [a]The whole group consisting both of bladder and renal cancer subjects, [b]p=0.000001, [c]p<0.005, [d]p=0.000002, [e]p<0.002.

No difference was found between the whole group of patients and controls with regard to sICAM-1. However, when the urinary bladder cancer subjects were excluded from the study, significantly elevated levels of sICAM-1 in renal patients were observed at p<0.002 *(table 2)*.

There was negative, but not significant, correlation between serum selenium and sICAM-1 levels in cancer patients (r=-0.26, p<0.2) and controls (r=-0.34, p<0.4).

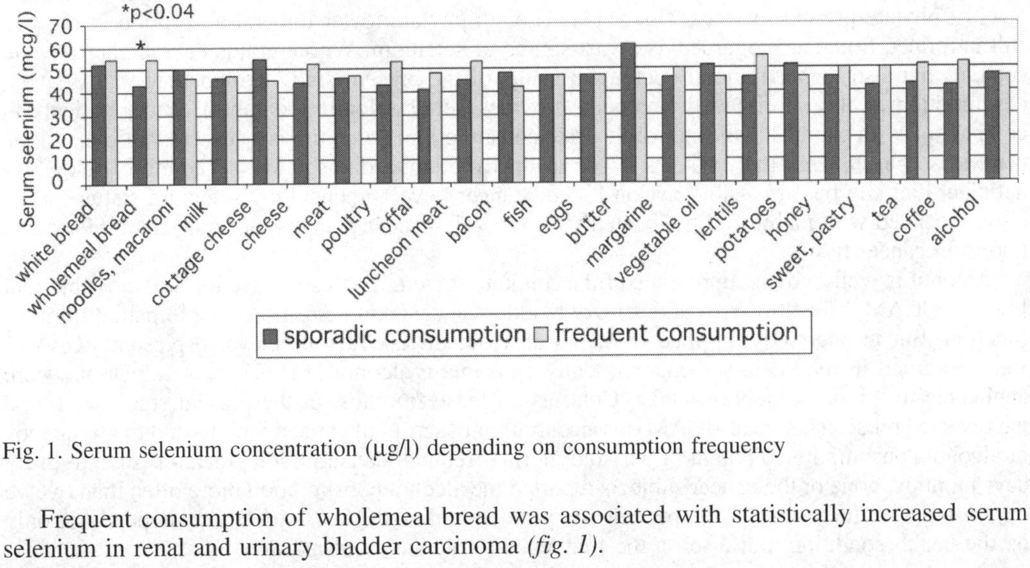

Fig. 1. Serum selenium concentration (μg/l) depending on consumption frequency

Frequent consumption of wholemeal bread was associated with statistically increased serum selenium in renal and urinary bladder carcinoma *(fig. 1)*.

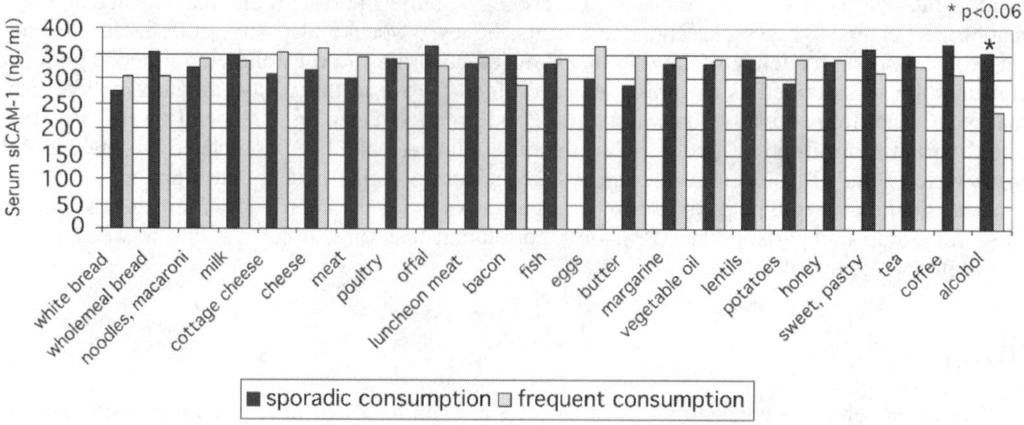

Fig. 2. Serum sICAM-1 concentration (ng/ml) depending on consumption frequency

Significantly lower sICAM-1 levels were observed in studied cancers, when alcohol was consumed more frequently *(fig. 2)*.

DISCUSSION

In the present study we discuss the influence of dietary factors on serum selenium and sICAM-1 in renal and urinary bladder cancers. Selenium is an element absorbed from organic and inorganic compounds. Dietary selenium comes mainly from selenoamino acids. Selenomethionine, present mostly in cereals, wheat and vegetables, shows several times higher bioavailability than that in milk or meat. Whole grains are rich in many components, including minerals, that can be found in the germ and bran, which are reduced in the grain-refining process. However, among cereals relatively lower mineral bioavailability was demonstrated for whole wheat flour, than in refined cereal grains [11]. That lower absorption of minerals should be contributed to high or medium level of phytate present in wheat flour [12]. Though phytate lower mineral absorption in the gut, still unrefined flours are regarded as a vast source of selenium. Whole grains have higher phytate contents but also higher mineral contents. Results of the present study demonstrate that frequent consumption of the unrefined grain foods provides better selenium status in cancer patients in comparison with patients avoiding whole grain consumption. Selenium is a cofactor for glutathione peroxidase, an enzyme that protects tissue against oxidative damage [13]. Therefore, selenium-deficient diet can be a plausible reason for the cancer development. Low selenium status can be also associated with malnutrition repeatedly observed in cancer patients, or with selenium sequestration in cancer tissues.

Alcohol is well-known suppressor of the immune system. In this connection it is possible that lowered sICAM-1 level in renal and urinary bladder cancers can be result of the impaired immune function. But in one study, applied to the cardiovascular risk factors, elevated plasma sICAM-1 were observed in the healthy men, frequently consuming alcohol [14]. These observations were applicable to the daily alcohol intake. Contrary to these findings, in the present study we found the reversed relation between sICAM-1 concentration in serum of cancer patients and the frequency of alcohol consumption. Though in our study term "frequent consumption" meant twelve to thirty days monthly, none of the cancer subjects reported the alcohol consumption more often than twelve days a month. The results of the present study indicate that sICAM-1 level may depend not only on the health condition, but also on the frequency of alcohol consumption.

CONCLUSIONS

Results of our study demonstrate the nutritional impact of the consumption of wholemeal bread and alcohol on selenium and sICAM-1 serum concentration in subjects with renal and urinary bladder cancers. Although selenium level in cancer subjects is statistically low as compared with healthy control, nevertheless the frequent consumption of wholemeal bread may possibly provide better selenium status in cancer subjects than the sporadic one. Frequent alcohol consumption seems to depress immune response in cancer subjects. The study issue proves that the renal and urinary bladder cancer subjects should been given a dietary advise in regard to the more frequent wholemeal products consumption and the alcohol avoidance.

REFERENCES

1. Reddi A.S., Bollineni J.S. Selenium-deficient diet induces renal oxidative stress and injury via TGF-beta 1 in normal and diabetic rats. *Kidney Int.*, 2001, 59, 1342-1353.
2. Vincenti M., Nacci G., Rocchi E., Cassinadri T., Vivoli R., Marchesi C., Bergomi M. Mortality in a population with long-term exposure to inorganic selenium via drinking water. *J. Clin. Epidemiol.*, 2000, 53, 1062-1068.
3. Hansson E., Jacobsson S.O. Uptake of (^{75}Se)selenomethionine in the tissues of the mouse studied by whole-body autoradiography. *Biochim. Biophys. Acta*, 1966, 115, 285-293.
4. Larras-Regard E., Mony M.C. Renal bioavailability of selenium after supplementation with different forms of selenium: ion probe and mass spectrometry study. *J. Trace Elem. Exp. Med.*, 2000, 13, 367-380.
5. Torun M., Aldemir H., Yardim S. Serum selenium levels in various cancer types. *Trace Elem. Electrolytes*, 1995, 12, 186-190.
6. Kiremidjian-Schumacher L., Roy M., Wishe H.I., Cohen M.W., Stotzky G. Supplementation with selenium and human immune cell functions. *Biol. Trace Elem. Res.*, 1994, 41, 115-127.
7. Santarosa M., Favaro D., Quaia M., Spada A., Sacco C., Talamini R., Galligioni E. Expression and release of intercellular adhesion molecule-1 in renal-cancer patients. *Int. J. Cancer*, 1995, 62, 271-275.
8. Heicappell R., Podlinski J., Buszello H., Ackerman R. Cell surface expression and serum levels of intercellular adhesion molecule-1 in renal cell carcinoma. *Urol. Res.*, 1994, 22, 9-15.
9. Shi B.B., Goya N., Okuda H., Ryoji O., Nakazawa H., Toma H. Detection and quantification of soluble intercellular adhesion molecule-1 (sICAM-1) in the serum and urine of patients with bladder cancer. *Int. J. Urol.*, 1998, 5, 324-328.
10. Hoffman R., Franzke A., Buer J., Sel S., Oevermann K., Duensing A., Prost M., Duensing S., Kirchner H., Ganser A., Atzpodien J. Prognostic impact of in vivo soluble cell adhesion molecules in metastatic renal cell carcinoma. *Br. J. Cancer*, 1999, 79, 1742-1745.
11. Franz K.B., Kennedy B.M., Fellers D.A. Relative bioavailability of zinc from selected cereals and legumes using rat growth. *J. Nutr.*, 1980, 110, 2272-2283.
12. Saha P.R., Weaver C.M., Mason A.C. Mineral bioavailability in rats from intrinsically labeled whole wheat flour of various phytate levels. *J. Agric. Food Chem.*, 1994, 42, 2531-2535.
13. Allan C.B., Lacourciere G.M., Stadtman T.C. Responsiveness of selenoproteins to dietary selenium. *Ann. Rev. Nutr.*, 1999, 19, 1-16.
14. Rohde L.E.P., Hennekens C.H., Ridker P.M. Cross-sectional study of soluble intercellular adhesion molecule-1 and cardiovascular risk factors in apparently healthy men. *Arterioscler. Thromb. Vasc. Biol.*, 1999, 19, 1595-1599.

*Metal Ions in Biology and Medicine: vol. 7.*Eds L. Khassanova, Ph. Collery, I. Maymard, Z. Khassanova, J.-C. Étienne.
John Libbey Eurotext, Paris © 2002 pp. 600-603.

Cancer prevention by selenium

G.F. Combs Jr[1], L. C. Clark[2] and B.W. Turnbull[3]

[1]Division of Nutritional Sciences and [3]School of Operations Research, Cornell University, Ithaca, NY, USA, and [2]deceased; formerly of the Arizona Cancer Center, Tucson, AZ, USA

SUMMARY

The Nutritional Prevention of Cancer Trial found that the regular use of a selenium (Se) supplement (200 mcg/day, as a Se-enriched yeast) by older Americans was associated with significant reductions in the risks of total, non-skin carcinomas with strongest effects apparent for cancers of the prostate and colon-rectum. These protective effects were greatest among subjects with pretreatment plasma Se concentrations below ca. 120 ng/ml (1.5 nmoles/ml) and with plasma tocopherol concentrations less than the median for the study population, i.e., 11 mcg/ml. These results suggest that a cancer-protective intake of Se may be ca. 1.5 mcg/kg body weight/day.

Background: The nutritional functions of the trace element selenium (Se) are thought to be discharged by a fairly small number of selenocyteine-containing proteins. These include some that function in antioxidant protection (glutathione peroxidases) and redox regulation (thioredoxin reductases), and others involved in the thyroid hormone production (iodothyronine 5'-deiodinases). That Se supplements can reduce tumor yields in animal models and cancer risk in humans whose dietary intakes of Se are apparently adequate to support the full expression of these selenoproteins, suggests that Se affects carcinogenesis via mechanisms independent of those proteins, i.e., involving specific Se-metabolites. While it cannot be excluded that cancer protection may involve selenoenzymes, such effects might be expected to function in anti-initiation. However, the limited number of human trials completed to date indicate late-stage, cancer-protective effects of Se involving such mechanisms as enhanced immune surveillance, altered cell cycle regulation and enhanced apoptosis. We conducted The Nutritional Prevention of Cancer Trial to determine whether the regular use of supplemental Se (200 ug/day, as a Se-enriched yeast product) by free-living older Americans could reduce cancer risk. The results of the first 10 yrs of this Trial were reported in 1996.

Aims: This report presents the results of the complete Trial.

Methods: We performed analyses of all 13 yrs of the intervention phase of the Trial, and conducted analyses of baseline Se and vitamin E status as covariates of the effect of supplemental Se on prostate cancer incidence.

Results: We found that the use of supplemental Se by Americans with plasma Se levels averaging ca. 120 ug/ml (which suggested an average dietary intake of at least 85 ug Se/day) showed significant reductions in total carcinomas and cancers of the prostate and colon. In the case of prostate cancer, for which Se treatment was associated with a 67% reduction in cases diagnosed, apparent protection was observed within the first 2-3 yrs of the trial. Protection by Se appeared limited to subjects who entered the trial with plasma Se concentrations below ca. 120 ng/ml and with plasma vitamin E levels below the median, 11 ng/ml for the population.

Conclusion: Because most healthy people worldwide have plasma Se levels less than that level, it is likely that increasing Se intakes, either through supplementation, food fortification or the production of Se-enriched crops, may be effective in reducing cancer rates in most countries.

INTRODUCTION

There is substantial evidence that the essential trace element Se can be anti-carcinogenic. The positive results of recent clinical trials [1-4] support this conclusion, particularly when considered in the light of supporting evidence from experimental animal, mechanistic and several epidemiologic studies [5-7]. Thus, it appears that can affect cancer risk by functioning as an essential nutrient that provides the catalytic centers of a number of Se-enzymes with antioxidant and redox-regulatory functions as well as by serving as a source of anti-tumorigenic Se-metabolites. It would appear that the first mechanism might be most relevant to anti-initiation, while the second to anti-progression. That short-term cancer-preventive effects have been observed in humans that were not Se-deficient suggests that supranutritional doses of Se may suppress transformed cells mostly, if not exclusively, through putative anti-carcinogenic Se-metabolites.

Evidence suggests that anti-carcinogenic effects associated with dietary Se may be mediated by certain Se-metabolites: hydrogen selenide (H_2Se) [8-10], methylselenol (CH_3SeH) [11-17] and, in individuals exposed to selenite or selenate, selenodiglutathione (GSSeSG) [18-22]. In animal models these effects have been consistently associated with Se intakes at least 10-fold those required to prevent clinical signs of Se deficiency. On a unit body weight basis (~100 mcg/kg body weight for rodents), they are also much greater than those experienced by most people worldwide, which tend to be much less than 100 mcg/day (or 1-4 mcg/kg body weight for adult humans). That the known Se-proteins appear to be expressed maximally in animal tissues at dietary levels no greater than 0.5 mcg Se/kg has led to the current belief that the anti-carcinogenic effects of such supranutritional levels of Se are not likely to be related to the Se-proteins. In order to test the hypothesis that suprnutritional doses of Se could reduce cancer risk in humans, we conducted the Nutritional Prevention of Cancer (NPC) Trial the preliminary results of which we published in 1996 [2]. This report presents the results of the complete, 13-yr, supplementation phase of that trial with specific attention to the effects of baseline status with respect to Se and vitamin E on cancer responses to supplemental Se.

MATERIALS AND METHODS

We have previously described the design and methods used in the NPC Trial [2]. In short, this was a randomized, placebo-controlled, double-blind trial designed to determine whether the regular use of a Se-supplement (200 mcg/day in the form of Se-enriched yeast) could affect cancer rates in a group of 1312 non-Se-deficient patients with non-melanoma skin cancer in the eastern United States.

RESULTS

The results of the NPC Trial showed that supplemental Se did not affect risks to recurrent basal or squamous cell carcinomas of the skin, but reduced risks to other cancers. Over the first 10 yrs of the Trial, the protective effects associated with Se treatment were: 50% fewer total cancers, 63% fewer cancers of the prostate, 58% fewer cancers of the colon-rectum, and 46% fewer cancers of the lung. These protective effects persisted with the continuation of Se supplementation through 13 yrs: 25% fewer total cancers, 49% fewer cancers of the prostate, 54% fewer cancers of the colon-rectum, and 30% (N.S.) fewer cancers of the lung. These effects were limited to subjects who entered the Trial with plasma Se status in the lowest tertile of the population (i.e., <106 ng Se/ml), to those whose baseline tocopherol concentrations were below the median for the cohort (i.e., <11 mcg/ml), and to males which comprised nearly three quarters of the study cohort.

DISCUSSION

With very few exceptions, patients entering the NPC Trial had apparently adequate plasma Se levels; only 6 of the 1312 subjects had initial plasma Se levels less than 80 ng/ml, a level corresponding to the amount of Se contained in maximally expressed plasma selenoproteins [23] and to the upper limit of glutathione peroxidase responses to Se-supplements in healthy people [24]. This means that few, if any, subjects in the NPC study can be considered to have been of sub-optimal nutritional Se status, at least by traditional standards. In fact, the mean baseline Se level, of ca. 120 mcg/ml, suggests on the basis of the data of Yang et al [25] an average dietary intake of at least 85 mcg Se/day, some 20% greater than the new American RDA [26].

Subjects entering the NPC trial with plasma Se levels less than 106 ng/ml showed not only the highest rates of subsequent cancer but also the strongest apparent protective effects of Se supplementation. In fact, subjects entering with plasma Se levels >121 ng/ml showed no cancer-protective benefits of taking the Se supplement. This would suggest that plasma level, i.e., ca. 120 ng/ml (1.5 nmol/ml), as optimal for cancer protection, and that strategies to reduce cancer risk using a selenomethionine-rich supplement such as Se-enriched yeast might only need to raise Se status above that level. Assuming similar fractional intakes of selenomethionine and related compounds in the diet, it would appear that dietary Se intakes of ca. 1.5 mcg Se/kg body weight/day. (e.g, ?96 mcg/day for American women and 120 mcg/day for American men) are required to support the plasma Se concentrations at the 120 mcg/ml level. Such intakes comprise about twice the newly revised RDA [26].

While the relevance of these findings to other, more divers, populations requires further investigation, it is clear that these apparently efficacious Se-exposures are safe. We found no evidence of adverse effects to Se supplementation at either of two levels: 200 or 400 mcg/day. After 8271 person-years of observation in 1983-1993, only 203 of 1312 NPC Trial subjects had voluntarily withdrawn from the study. That number was distributed in both the placebo (158) and Se-treatment (145) groups, and included only 39 subjects (placebo: 14; Se:25) complaining of side effects. Of the latter number, only 5 subjects in each treatment group had complaints consistent with selenosis (brittle hair, white-patched/brittle nails, skin rash).

CONCLUSIONS

It is clear that Se can be anti-tumorigenic. To the extent that such involve the enhanced expression of selenoproteins (e.g., in Se-supplementation of individuals with Se intakes less than ca. 40 mcg/day [27]), they would appear to support antioxidant defense in the inhibition of tumor initiation. Supranutritional Se exposures would appear to be mediated by other mechanisms, e.g., Se-metabolites produced from relatively high doses of several forms of the element acting in initiation stages. The NPC Trial results suggest that, with intakes of mixed organic forms of Se, cancer risk reduction may be realized when plasma Se levels are raised above ca. 120 ng (1.5 nmol)/ml which may be achieved with Se-intakes of about twice the RDA level.

REFERENCES

1. Li J., Taylor P., Li B., Blot W., Guo W., Dawsey S., Wang G., Yang C., Zheng S., Gail M., Li G., Liu B., Tangera J., Sun Y., Liu F., Fraumeni Jr. F., Zhang Y., Nutrition intervention trials in Linxian, China, *J. Nat. Cancer Inst.* 1993, 85, 1492-1498.
2. Clark L., Combs G., Turnbull B., Slate E., Alberts D., Abele D., Allison R., Bradshaw J., Chalker D., Chow J., Curtis D., Dalen J., Davis L., Deal R., Dellasega M., Glover R., Graham G., Gross E., Hendrix J., Herlong J., Knight F., Krongrad A., Lesher J., Moore J., Park K., Rice J., Rogers A., Sanders B.,

Schurman B., Smith C., Smith E., Taylor J., Woodward J. The Nutritional Prevention of Cancer with Selenium 1983-1993: a Randomized Clinical Trial, *J. Amer. Med. Assoc.* 1996, 276, 1957-1963.
3. Clark L., Dalkin B., Krongrad A., Combs G.., Turnbull B., Slate E., Witherington R., Herlong J.H., Janosko E., Carpenter D., Borosso C., Falk S., Rounder J. Decreased incidence of prostate cancer with selenium supplementation: Results of a double-blind cancer prevention trial, *Br. J. Urol.* 1998, 81, 730-734.
4. Yu S.Y., Zhu Y.J., Li W.G. Protective role of selenium against hepatitis B virus and primary lever cancer in Qidong, *Biol. Trace Elem. Res.* 1997, 56, 117-124.
5. Combs Jr. G.F., Gray W.P., Chemopreventive Agents: Selenium, *Pharmacol. Therapeut.* 1998, 79, 179-192.
6. Ip C., Lessons from Basic Research in Selenium and Cancer Prevention. *J. Nutr.* 1998, 128, 1845-1854.
7. Combs Jr. G.F., Clark L.C., Selenium and cancer: *Nutritional Oncology*, Heber D., Blackburn G.L., Go V.L.W., eds., Academic Press, New York, 1999, 215-222.
8. Wilson A.C., Thompson H.J., Schedin P.J. Effect of methylated forms of selenium on cell viability and the induction of DNA strand breakage. *Biochem. Pharmacol.* 1992, 43, 1137-1141.
9. Lü J., Kaeck M., Jiang C. Selenite induction of DNA strand breaks and apoptosis in mouse leukemia L1210 cells. *Biochem. Pharmacol.* 1994, 47, 1531-1534.
10. Lü J., Jiang C., Kaeck M. Dissociation of the genotoxic and growth inhibitory effects of selenium. *Biochem. Pharmacol.* 1995, 252, 7392-7394.
11. Ganther H.E.. Selenium Metabolism, selenoproteins and mechanisms of cancer prevention: complexities with thioredoxin reductase. *Carcinogensis.* 1999, 20, 1657-1666.
12. Ip C., Ganther H.E. Activity of methylated forms of selenium in cancer prevention. *Cancer Res.* 1990, 50, 1206-1211.
13. Ip C., Ganther H.E. Combination of blocking agents and suppressing agents in cancer prevention. *Carcinogenesis*, 1991, 12, 365-372.
14. Ip C., Ganther H.E. Comparison of selenium and sulfur analogs in cancer prevention. *Carcinogenesis.* 1992, 13, 1167-1170.
15. Ip C., Ganther H.E. Biological activities of trimethylselenonium as influenced by arsenite. *J. Inorg. Biochem.* 1992, 46, 215-222.
16. Ip C., Hayes C., Budnick R.M.. Chemical form of selenium, critical metabolites, and cancer prevention. *Cancer Res.* 1991, 51, 595-600.
17. Vanhanavikit S., Ip C., Ganther H.E. Metabolites of sodium selenite and methylated selenium compounds administered at cancer chemoprevention levels in the rat. *Xenobiotica.* 1993, 23, 731-745.
18. Lanfear J., Flemming J., Wu L. The selenium metabalite selenodiglutathione induces p53 and apoptosis: relevance to the chemopreventive efforts of selenium. *Carcinogenesis.* 1994, 15, 1387-1392.
19. Wu L., Lanfear J., Harrison P.R. The selenium metabolite selenodiglutathione induces cell death by a mechanism distinct from H2O2 toxicity. *Carcinogenesis* 1995, 16, 1579-1584.
20. Bjornstedt M., Kumar S., Bjorkhem L. Selenium and thioredoxin and glutaredoxin systems. *Biomed. Environ. Sci.* 1997, 10, 271-279.
21. Vernie L.N., Hamburg C.J., Bont W.S. Inhibition of the growth of malignant mouse lymphoid cells by selenodiglutathione and selenocystine. *Cancer Lett.* 1981, 14, 303-308.
22. Poirer K.A., Milner J.A. Factors influencing the antitumorigenic properties of selenium in mice. *J. Nutr.* 1983, 113, 2147-2154.
23. Hill K.E., Xia Y., Åkesson B., Boeglin M.E., Burk R.F. Selenoprotein P concentration in plasma as an index of selenium status in selenium-deficient and selenium supplemented Chinese subjects. *J. Nutr.* 1996, 126, 138-145.
24. Nève J. Human selenium supplementation as assessed by changes in blood selenium concen-tration and glutathione peroxidase activity, *J. Trace Elements Med. Biol.* 1995, 9, 65-73.
25. Yang G.Q., Yin S., Zhou R., Gu L., Yan B., Liu Y., Liu Y. Studies of safe maximal daily dietary Se-intake in a seleniferous area in China. Part II. Relation between Se-intake and the manifestations of clinical signs and certain biochemical alterations in blood and urine, *J. Trace Elements Electrol. Health Dis.* 1989, 3, 123-130.
26. Panel on Dietary Antioxidants and Related Compounds. Dietary Reference Intakes for Vitamin C, Vitamin E, Selenium and Beta-Carotene and other Carotenoids. National Academy Press, Washington, D.C., 2000, 600 pp.

Metal Ions in Biology and Medicine: vol. 7. Eds L. Khassanova, Ph. Collery, I. Maymard, Z. Khassanova, J.-C. Étienne.
John Libbey Eurotext, Paris © 2002 pp. 604-608.

Trace metals in neuroblastoma cells and tumors

C. Sergeant[1], B. Gouget[2], M.H. Vesvres[1], G. Deves[1], M. Simonoff[1], G. Raguenez[3], J. Benard[3]

[1]Laboratoire de Chimie Nucléaire Analytique et Bioenvironnementale, CNRS UMR 5084, Le Haut Vigneau, BP 120, 33175 Gradignan cedex, France; [2]Laboratoire Pierre Süe, CEA/CNRS UMR9956, CE de Saclay, 91191 Gif sur Yvette, France; [3]Laboratoire de Pharmacologie Clinique et Moléculaire, Institut Gustave Roussy, 39 rue Camille Desmoulins, 94800 Villejuif, France

ABSTRACT

Neuroblastoma is a severe cancer form in early childhood. N-*myc* amplification is causally involved in the progression of the disease and associated with advanced stages of malignancy. Available metals, including iron, copper and zinc, may have some role in promoting tumor cell growth. In previous studies, we showed that a relationship exists between intracellular trace metal concentrations of cultured neuroblasts and the N-*myc* oncogene amplification rate. We already published preliminary results on determinations of trace metal contents in tumors xenografts developed after injection of cells from the same cell lines in athymic nude mice. In order to confirm our previous results, Fe, Cu and Zn concentrations have been measured in two other human neuroblastoma cell lines characterized by different degrees of N-*myc* amplification as well as in the corresponding tumors. One of them is a non-amplified cell line transfected with N-*myc* gene. This complete study allows a direct comparison of intracellular trace metal contents *in vitro* and *in vivo* and confirms the relationship already established between Fe, Cu and Zn contents and the degree of N-*myc* oncogene amplification in neuroblastoma.

INTRODUCTION

Neuroblastoma (NB), a pediatric cancer, is a complex disease divided into different stages, according to the invasion of the solid tumor. N-*myc* oncogene amplification and subsequent oncoprotein N-Myc overexpression is widely accepted as playing a major role in the biology of NB and contributing to an aggressive phenotype. High degree of N-*myc* oncogene amplification is associated with advanced disease stage and rapid tumor progression. To date, many human NB cell lines have been established, eliciting various rates of N-*myc*. NB cells grow actively *in vitro* and present tumorigenic activity in athymic nude mice. This experimental model constitutes a tool for investigation of NB *in vivo* [1].

A deregulation of intracellular trace metals contents (iron, copper and zinc) can affect cell proliferation and tumor growth. Iron is an essential element required for cell division and increased intracellular iron concentrations may promote malignant cell growth. Fe chelators have been shown to be useful in NB treatment [2]. Iron-rich serum ferritin, an iron-storage protein, is characteristic of patients with advanced stage disease [3]. N-*myc* amplified cells can synthesize zinc metalloenzymes implicated in tumor invasion and metastases formation [4].

In a previous study led on three human NB cell lines, a relationship between trace metal contents and the degree of the N-*myc* oncogene amplification has been underlined [5]. IGR-N-91 [1] and IMR-32 [6] cell lines, 60 and 25-folds amplified respectively were compared to the SK-N-SH [7] cell line presenting no amplification with only one copy of the N-*myc* oncogene per haploid genome. These results were confirmed by the comparison of trace metal concentrations with the degree of N-Myc expression: Fe, Cu and Zn contents were measured in amplified cells treated with retinoic acid, an anti-tumor agent known to down-regulate the N-Myc expression [8].

First results of the comparison of trace metal contents in monolayers and sections of tumor xenografts developed from the same cell lines (SK-N-SH and IGR-N-91) in nude mice have been published in [9]. In the present study, we have investigated two other cell lines and the corresponding tumor xenografts: the SK-N-BE(2) cell line, characterized by 150 copies of the N-*myc* oncogene [10], and for the first time, measures on the non-amplified SK-N-SH cell line transfected (trf) with the N-*myc* oncogene are presented. This latest cell line allows to compare Fe, Cu and Zn concentrations in N-*myc* amplified and non-amplified neuroblasts issued from the same cell line.

MATERIALS AND METHODS

Cell culture and xenografting: NB cells were grown in DMEM at 4.5 g/l D-glucose, supplemented with 10% fetal calf serum and antibiotics, as described in [5]. They were maintained at 37°C in 8% CO_2. At confluence, they were harvested with trypsine/EDTA 0.05% to dilution 4-5. For xenografting, suspensions of 2.10^7 cells in 0.3 ml of antibiotic-free culture medium were prepared immediately prior to injection in each nude mice. Cell suspension was injected sub-cutaneously into the flank of female nude mice 6-8 weeks of age. Within two months later, palpable tumors developed in nude mice and tumor xenografts of a few millimeters diameter were collected in the exponential growth phase. Tumor xenografts were carefully dissected immediately after sacrifice, quick frozen by immersion in liquid nitrogen and stored at -80°C.

A macroscopic heterogeneity of the tumors was observed: it could be distinguished necrotic areas, vascularized or haemorrhagic regions displaying blood vessels and non-vascularized white areas. For determination of trace metal concentrations in NB tumor tissue, non-vascularized white areas were chosen, in order to avoid the analysis of iron-containing haem. Even in these white areas, nuclear microprobe analyses performed on large areas of the tumor xenografts display iron localizations probably due to blood vessels or iron-rich ferritin clusters, as suggested by histological examination. Mean iron concentrations in NB tumors have thus been obtained by selecting other areas.

Nuclear microprobe analyses: Sample preparation for analysis of monolayers of cultured cells by nuclear microprobe was succeeded as already described [5]. Briefly, a cell suspension was seeded onto a very thin Formvar film precoated with attachment factors (polylysin and fibronectin). In the course of 48 h of incubation, a monolayer of tumor cells developed on the polymer film. At the exponential growth phase, the targets were rapidly rinsed in Earle's salt solution and water and rapidly frozen in liquid isopentan chilled at about -165°C by liquid nitrogen. After cryofixation, lyophilisation was achieved in 2 h at -80°C under vacuum. Sample preparation for microprobe analysis of tumors was different. Frozen tumors were embedded in Tissue Tek OCT, frozen in liquid nitrogen and cut with a Reichert-Jung cryomicrotome at -25°C. A 5 µm-section was first collected for histological study and 10 µm-sections deposited on Formvar films were dedicated to nuclear microprobe analysis. Analyses were carried out with a proton beam of 2.5 MeV energy, using the Bordeaux CENG microprobe facility, as already described. Scans were performed on large areas with a spot size of 5 µm diameter and a beam current of 0.8 nA. From the backscattered protons (RBS technique), the total mass of the analyzed sample together with its thickness were calculated [11]. Trace metal concentrations were deduced from X-ray intensities and normalized to the total mass of the sample. Results are thus expressed in µg/g dry weight (or ppm).

RESULTS

We investigated intracellular trace metals contents *in vitro*, with analysis of cultured neuroblasts issued from 5 different cell lines, in comparison with their concentrations *in vivo*, with analysis of tumors xenografted from the same cell lines (except for IMR-32, no tumor having being obtained from this cell line). In this report, a compilation of previous results (for SK-N-SH, IMR-32 and IGR-N-91 cultured cells and for SK-N-SH and IGR-N-91 tumor sections) with new results (for SK-N-SH transfected and SK-N-BE(2) *in vitro* and *in vivo*) is presented. The results, expressed for trace metals in µg/g dry weight, are summarized in *table 1* for cultured cells and *table 2* for tumor sections. The degree of the N-*myc* oncogene amplification or N-Myc expression for each cell line is indicated in *table 1*. For SK-N-SH transfected with the N-*myc* oncogene, the amplification rate is not known but the discussion is based on its N-Myc expression rate compared to that of SK-N-SH and IGR-N-91 [12 and unpublished data].

Table 1. Elemental rates in neuroblastoma cultured cells (means and SD of n values in µg/g dry weight of cells)

Cell line	SK-N-SH	IMR-32	IGR-N-91	SK-N-SH trf	SK-N-BE(2)
Copies of N-*myc*	1	25	60		150
N-Myc expression (fluorescence index)	1		2,5	20	
Fe	85 ± 22	83 ± 21	65 ± 19	52 ± 18	22 ± 12
Cu	8,4 ± 2,6	11,0 ± 2,8	15,6 ± 5,1	3,6 ± 1,7	4,1 ± 1,1
Zn	174 ± 34	277 ± 39	169 ± 19	80 ± 9	82 ± 14
n	38	12	36	6	10

Concerning the cell lines, iron content decreases from 85 to 22 µg/g while the N-*myc* amplification increases from 1 to 150. SK-N-SH transfected cells present N-Myc overexpression 20 times higher than SK-N-SH and the iron content is 40% lower than in SK-N-SH. Copper concentrations have a tendency to increase with N-*myc* amplification from SK-N-SH to IGR-N-91, but the values are very low for the two other lines characterized by very high N-*myc* amplification or expression rate. Zinc values are inhomogenate but high for the three first lines and far much lower for the most amplified/overexpressed ones.

Table 2. Elemental rates in neuroblastoma tumor xenografts (means and SD of n values in µg/g dry weight of tissue)

Tumor	SK-N-SH	IGR-N-91	SK-N-SH trf	SK-N-BE(2)
Fe	721 ± 63	535 ± 187	177 ± 78	56 ± 18
Cu	9 ± 0,3	21 ± 8	6 ± 4	12 ± 6
Zn	159 ± 4	122 ± 40	76 ± 30	93 ± 15
n	3	12	5	7

Concerning the tumors, iron content decreases from 721 to 56 µg/g while N-*myc* amplification of the original cells increases from 1 to 150. Iron content of the tumors developed by injection of SK-N-SH transfected cells is 4 times lower than in the SK-N-SH ones. We can emphasize the fact that iron content in non-vascularized parts of tumors is very high compared to cultured cells (from 2,5 to 8,5 higher). Copper values are less dispersed than iron contents and of the same order than in cultured cells. Zinc values are very close to the concentrations found in the corresponding cultured cells.

DISCUSSION

Results presented here confirm the relationship between Fe concentrations within NB cells and the degree of N-*myc* oncogene amplification: the comparison between SK-N-SH and SK-N-SH transfected by N-*myc* proves the importance of trace metals for different expression rates of the N-Myc protein.

This study confirms as well the first results we obtained on comparison of trace metals content in cultured cells and tumors of two cell lines [9]. Cu and Zn contents are not different in the tumors compared to monolayers of cultured neuroblasts. On the contrary, iron concentrations are far much higher in tumors sections than in monolayers of NB cultured cells for all the studied cell lines. In a previous study [5], we showed that there is no iron-stock in ferritin of cultured neuroblasts, and that iron content in cultured cells is much lower that in well-known tissues. We noticed that when adding 50 µM iron citrate to the culture medium of SK-N-SH cells, intracellular copper and zinc concentrations were not modified but intracellular iron concentrations were multiplied by a factor 3 [13]: NB cells are able to incorporate high contents of iron. By the analysis of trace metal contents in cell culture media, we showed that the content of iron in culture medium was not the limiting factor for intracellular iron accumulation within NB cultured cells.

Other authors have already established that tumor cells, and especially neuroblasts, can accumulate iron in ferritin and hemosiderin when exposed to an iron-enriched medium.

Patients with advanced-stage NB show abnormally high levels of serum ferritin, very likely secreted by NB tumor. We emitted the hypothesis that the iron-rich ferritin, which could have been accumulated in NB tumors (confirmed by the higher iron concentrations found in tumors) before establishing the cell lines, may have been subsequently secreted in the extracellular culture medium *in vitro* or in the patients' serum *in vivo*. In the course of many passages, cells were harvested and cultured in fresh medium which could have progressively impoverished cells in iron. This could explain low levels of Fe that we measured in all the cultured cells. Low iron rates in NB cells could be due to a work on permanent lines: cultured cells could have been adapted to a progressive decrease in their iron-ferritin stocks.

It has been shown that Fe from the extracellular transferrin is more rapidly available for the cell than Fe drawn out of intern sources of Fe storage. And cell culture may influence Fe transport and storage, since the number of transferrin receptors on the cell membrane decreases when cells are cultured in presence of Fe in the culture medium.

Fe present in the culture medium of NB cell lines could be easily bioavailable, reducing the number of transferrin receptors. So cultured cells don't need to accumulate an Fe excess or storage in ferritin or hemosiderin. On the opposite, *in vivo*, Fe might be less bioavailable: the number of transferrin receptors at the cell surface may increase, implying accumulation of Fe in ferritin or hemosiderin, and significant differences in Fe concentrations in NB cells *in vitro* and *in vivo*.

CONCLUSION

This study, by the trace metals measurements in five cell lines and four types of tumors developed in nude mice by injection of the same cells, shows the importance of variations of iron content in function of N-*myc* amplification or N-Myc surexpression. The parallel measurements of trace metals in cultured cells and tumors point out the difference in iron concentrations between these two models and extend the interest of work on three dimensional models like these tumors. For the first time, we had the opportunity to analyze trace metals contents in a non-amplified cell line transfected with the N-*myc* oncogene increasing the N-Myc expression comparatively to the non-transfected cell line. Highly significant differences between trace metal contents in these two cell lines and in the corresponding tumors have been established, confirming the importance of trace metals in oncogenesis.

REFERENCES

1. Ferrandis E., Da Silva J., Riou G., Bénard J. Coactivation of the MDR1 and MYCN genes in human neuroblastoma cells during the metastatic process in the nude mice. *Cancer Res.*, 1994, 54, 2256-2261.
2. Donfrancesco A., Deb G., De Sio L., Cozza R., Castello A. Role of deferoxamine in tumor therapy. *Acta Haematol.*, 1996, 95(1), 66-69.
3. Hann H.W.L., Evans A.E., Siegel S.E., Wong K.Y., Sather H., Dalton A., Hammond D., Seeger R.C. Prognostic importance of serum ferritin in patients with stages III and IV neuroblastoma: the childrens cancer study group experience. *Cancer res.*, 1985, 45(6), 2843-2848.
4. Ara T., Kusafuka M., Inoue M., Kuroda S., Fukuzawa M., Okada A. Determination of imbalance between MMP-2 and TIMP-2 in human neuroblastoma by reverse-transcription polymerase chain reaction and its correlation with tumor progression. *J Pediatr Surg.*, 2000, 35(3), 432-437.
5. Gouget B., Sergeant C., Bénard J., Llabador Y., Simonoff M. N-*myc* oncogene amplification is correlated to trace metal concentrations in neuroblastoma cultured cells. *Nucl. Instrum. Meth. Phy. Res.*, 2000, 170, 432-442.
6. Tumilowicz J.J., Nichols W.W., Cholon J.J., Greene A.E. Definition of a continuous human cell line derived from neuroblastoma. *Cancer Res.*, 1970, 30, 2110-2118.
7. Biedler J.L., Helson L., Spengler B.A. Morphology and growth, tumorigenicity and cytogenetics of human neuroblastoma cells in continuous culture. *Cancer Res.*, 1973, 33, 2643-2652.
8. Gouget B., Sergeant C., Hamon Ch., Llabador Y., Benard J., Simonoff M. Trace metals in differentiated neuroblasts treated with retinoic acid, an anti-cancer drug, *Metal Ions in Biology and Medicine*, vol. 6, Eds J.A. Centeno et al., John Libbey Eurotext, Paris, 2000, 215-217.
9. Gouget B., Sergeant C., Llabador Y., Deves G., Vesvres M.H., Simonoff M., Bénard J. Trace metal and cancer: The case of neuroblastoma. *Nucl. Instrum. Meth. Phy. Res.*, 2001, 181, 465-469.
10. Barnes E.N., Biedler J.L., Spengler B.A., Lyser K.M. The fine structure of continuous neuroblastoma lines SK-N-SH, SK-N-BE(2) and SK-N-MC. *Oncogene*, 1981, 17, 619-631.
11. Moretto Ph., Razafindrabe L., Simulation of RBS spectra for quantitative mapping of inhomogeneous biological tissue. *Nucl. Instrum. Meth. Phy. Res.*, 1995, 104, 171-175.
12. Chassevent A., Benard J., Oncogen N-*Myc* expression and measurement of DNA ploidy in neuroblastoma: a double staining flow cytometric analysis, *Bulletin du Cancer*, 1997, 84, 29-34.
13. Gouget B., Sergeant C., Llabador Y., Simonoff M., Bio-normalyzer: an anti-cancer drug for neuroblastoma? *Nucl. Instrum. Meth. Phy. Res.*, 2001, 181, 470-474.

Metal Ions in Biology and Medicine: vol. 7. Eds L. Khassanova, Ph. Collery, I. Maymard, Z. Khassanova, J.-C. Étienne.
John Libbey Eurotext, Paris © 2002 pp. 609-613.

Ruthenium semicarbazone complexes as potential antitumoral agents

Lucía Otero[1], Pabla Noblia[1], Dinorah Gambino[1], Hugo Cerecetto[2], Mercedes González[2], Antonio Monge-Vega[3], Adela López de Ceráin[3], Olga Ezpeleta[3], Beatriz Parajón-Costa[4]

[1]Cátedra de Química Inorgánica and [2]Departamento de Química Orgánica, Facultad de Química, Universidad de la República, CC 1157, 11800 Montevideo, Uruguay; [3]CIFA, Universidad de Navarra, Pamplona, España, [4]CEQUINOR, UNLP, La Plata, Argentina

ABSTRACT

New complexes of general formulae $[Ru^{II}Cl_2(dmso)_2L]$, where dmso=dimethylsulfoxide and L = 5-nitro-2-furaldehyde semicarbazone (Nitrofurazone®) or N^4-n-butyl-5-nitro-2-furaldehyde semicarbazone, were prepared as an effort to combine free ligand and metal potential antitumoral activity. Complexes were characterized by elemental analyses and FTIR and ^{1}H and ^{13}C NMR spectroscopies. Chemical and structural results have been complemented with an evaluation of some physicochemical parameters (lipophilicity, redox potential) and biological activity of the new complexes.

These last decades have seen a growing interest in coordination compounds of transition metals as potential antineoplastic agents. Some promising results have been obtained with derivatives of different metals. In particular, several Ru complexes have exhibited good to excellent antitumoral activity in some tumor screens. On the other hand, 5-nitro-2-furaldehyde semicarbazone (Nitrofurazone®) has shown increased mammalian cell killing in hypoxia compared to aerobic conditions [1]. In this work, we have developed new Ru(II) complexes with Nitrofurazone® and its derivatives as an effort to combine free ligand and metal potential antitumoral activity. Complexes of general formulae $[Ru^{II}Cl_2(dmso)_2L]$, where dmso = dimethylsulfoxide and L = 5-nitro-2-furaldehyde semicarbazone and N^4-n-butyl-5-nitro-2-furaldehyde semicarbazone, were prepared in good yields by reaction of $[Ru^{II}Cl_2(dmso)_4]$ with L in ethanol or toluene solutions. FTIR and ^{1}H- and ^{13}C-NMR results were analyzed in detail. Chemical and structural results have been complemented with an evaluation of some physicochemical parameters (liposolubility, redox potential) and biological activity of the new complexes. The electrochemical behaviour was studied using cyclic voltammetry at different scan rates. Signals associated with the metal center and the ligand were observed. Compounds were tested in aerobic conditions against MCF-7 (human mammary adenocarcinoma) and TK10 (human kidney adenocarcinoma) tumor lines. Results showed no antitumoral activity in these conditions. The lack of activity could be related with the high hidrophilicity of the complexes. Selective citotoxicity studies in hypoxic conditions against V79 (chinese hamster fibroblastes) are still in progress.

INTRODUCTION

These last decades have seen a growing interest in coordination compounds of transition metals as potential antineoplastic agents. Some promising results have been obtained with derivatives of different metals. Several Ru complexes have exhibited good to excellent antitumoral activity in some tumor screens. In particular, Ru(II)-dimethylsulfoxide (dmso) complexes have shown to possess antimetastatic activity against some murine tumor models [1]. On the other hand, 5-nitro-2-furaldehyde semicarbazone (Nitrofurazone®) **(1)** has shown increased mammalian cell killing in hypoxia compared to aerobic conditions [2]. In this work, we have developed new Ru(II)-dmso complexes with Nitrofurazone® and its derivative N^4-n-butyl-5-nitro-2-furaldehyde semicarbazone as coligands, in an effort to combine free ligand and metal potential antitumoral activity. Ligand formula are depicted in *figure 1.*

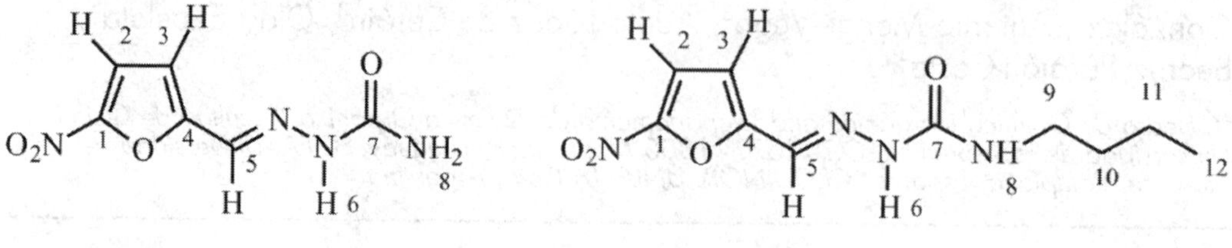

Fig. 1. Formula of the semicarbazone derivatives used as ligands

MATERIALS AND METHODS

All common laboratory chemicals were purchased from commercial sources and used without further purification. 5-nitro-2-furaldehyde semicarbazone (Nitrofurazone®) was commercially available. N^4-n-butyl-5-nitro-2-furaldehyde semicarbazone and $[Ru^{II}Cl_2(dmso)_4]$ were prepared according to literature procedures [3, 4].

Syntheses of the complexes

The investigated complexes of general formulae $[Ru^{II}Cl_2(dmso)_2L]$, where L = 5-nitro-2-furaldehyde semicarbazone (Nitrofurazone®) or N^4-n-butyl-5-nitro-2-furaldehyde semicarbazone, were prepared by reaction of $[Ru^{II}Cl_2(dmso)_4]$ with L in ethanol or toluene solutions. Nitrofurazone® and N^4-n-butyl-5-nitro-2-furaldehyde semicarbazone complexes were recrystallized from methanol and acetone, respectively. Anal. (%) Calcd for $C_{10}H_{18}Cl_2N_4O_6S_2Ru$: C, 22.9; H, 3.43; N 10.7; S, 12.2. Found: C, 23.6; H, 3.41; N 11.1; S, 11.9. Anal. (%) Calcd for $C_{14}H_{27}Cl_2N_4O_6S_2Ru$: C, 28.9; H, 4.64; N 9.62; S, 11.0. Found: C, 28.6; H, 4.44; N 8.81; S, 11.1.

Physicochemical characterization

C, H, N and S analyses were performed with a Carlo Erba Model EA1108 elemental analyzer. FTIR spectra (4000-400cm^{-1}) were measured as KBr pellets on a Bomen M102 instrument. Electronic spectra were recorded on a Spectronic 3000 spectophotometer. ^1H NMR and ^{13}C NMR spectra of the free ligands and of the complexes were recorded at 30°C in dmso-d_6 or acetone-d_6 on a Bruker DPX-400 instrument (at 400MHz and 100MHz, respectively). Heteronuclear correlation experiments (HETCOR), HMQC (multiple quantum) and HMBC (multiple bond), were performed with the same instrument. The NOE experiments were performed using the equipment's NOE-diff pulse sequences at 30°C. The mixing time (t_{mix}) was varied from 250 ms to 500 ms and the irradiated protons were H_6 (both ligands and their Ru complexes) and H_3 (N^4-n-butyl-5-nitro-2-furaldehyde semicarbazone and its Ru complex). Cyclic voltammetric experiments were carried out with a computer controlled PAR (Princeton Applied Research), Potentiostat/Galvanostat model 263A. A standard three-electrode cell was used with a glassy carbon working electrode, a Pt counterelectrode and a Ag/(10^{-3} M) $AgNO_3$ in CH_3CN reference electrode. Measurements were performed in oxygen purged 10^{-3} M dmso solutions with 0,1 M $TBAPF_6$ as supporting electrolyte.

Cyclic voltammograms (CV) were recorded at different scan rates (v) (0.030 to 3 V/s) and at 25ºC. During the measurements a continuos gas stream was passed over the solutions. Lipophilicity tests were performed determining the partition coefficient of the complexes in physiological solution/ n-octanol [5]. Concentration of the complexes was determined spectrophotometrically in the visible absorption maximun of each complex.

Biological activity studies

Compounds were tested at 10^{-4} M doses in aerobic conditions against MCF-7 (human mammary adenocarcinoma) and TK10 (human kidney adenocarcinoma) tumor lines (ATCC American Type Cell Culture Collection), according to previously described procedures [6].

RESULTS AND DISCUSSION

Complexes of general formulae [$Ru^{II}Cl_2(dmso)_2L$], where L = 5-nitro-2-furaldehyde semicarbazone **(3)** and N^4-n-butyl-5-nitro-2-furaldehyde semicarbazone **(4)**, were prepared in good yields and high purity. Analytical data of the complexes are in agreement to the proposed formula.

Selected vibration bands of the ligands and their metal complexes, useful for determining the ligands' mode of coordination are given in *table 1*.

Table 1. Selected vibration bands of the ligands and their Ru complexes (cm^{-1})

Compound	n(CO)	n(C=N)	$n_s(NO_2)$	$n(SO)_{dmso}$
1	1716	1583	1350	-
2	1696	1576	1355	-
3	1665	1566	1347	1078
4	1651	1576	1351	1105

After coordination, the n(CO) bands of the semicarbazone free ligands at *ca.* 1660-1700 cm^{-1} shift to lower frequencies. Simultaneously, the n(C=N) bands of the semicarbazones at *ca.* 1580 cm^{-1} shift. These modifications are consistent with bidentated coordination of the semicarbazone ligands through the carbonyl oxygen and the azomethinic nitrogen. The n(SO) band, observed at *ca.* 1100 cm^{-1} in both complexes, confirms the presence of dmso, sulfur bonded, in the coordination sphere.

The NMR experiments show narrow signals, typical for a Ru(II) diamagnetic complex. The ^1H and ^{13}C NMR chemical shift values of both free ligands and their complexes are listed in *tables 2 and 3. Figure 1* shows the numbering scheme of both free ligands, mentioned in the *tables* and in the text.

HETCOR experiments allowed to assign all signals in the free ligands and in the investigated complexes. Both complexes show similar ^1H and ^{13}C chemical shifts in the nitrofurylsemicarbazone common portion of their molecules. When the ligand is coordinated, the deshielding effect of the metal is apparent at all the protons, causing a downfield shifting of all the ^1H NMR peaks, including those of the dmso ligand (free dmso δ-H = 2.50 ppm). After coordination, the most distinguishing feature of the ^{13}C NMR spectra is the change in the chemical shift of the carbons numbered 1, 3, 5 and 7. The signals of these carbons are significantly displaced downfield, owing to an electron attractive effect of the metal. The NOE-diff experiments show spacial coupling of the ligand protons with those of the coordinated dmso molecules and absence of coupling between H6 and H3 and between H6 and H5. So, although the free ligands show the isomeric form *E*, the NOE-diff experiments would suggest that upon coordination the ligands adopt the isomeric form *Z*.

The redox behavior of Nitrofurazone® and [$Ru^{II}Cl_2(dmso)_2$Nitrofurazone] was investigated using cyclic voltammetry and the results will be briefly discussed. In both cases, as the potential is scanned in a negative direction, at all scan rates investigated, a peak corresponding to the reduction of the nitro group is observed (peak potentials: -0.95/-0.83V for the free ligand and -1.01/-0.9V for the complex *vs.* NHE). So, this peak shifts to more negative potentials after coordination. In addition, the complex shows an oxidation contribution corresponding to the Ru(III)/Ru(II) process (peak potentials: +0.57/+0.46V *vs.* NHE).

Partition coefficient (K) of the complexes was determined as the ratio of the complex concentration in the organic phase over its concentration in the aqueous phase. As expected, K-values obtained (0.067 for **3** and 0.321 for **4**) show an increase of lipophilicity as the complexity of the ligand increases. Nevertheless, both complexes show a quite hidrophylic behaviour.

The results of the biological activity tests performed with MCF-7 (human mammary adenocarcinoma) and TK10 (human kidney adenocarcinoma) tumor lines show very low citotoxicity in aerobic conditions. The lack of activity could be related with the high hidrophilicity of the complexes.

Table 2. ^{1}H and ^{13}C NMR chemical shift values (d) in ppm of L_1 and $[Ru^{II}Cl_2(dmso)_2L_1]$, with L_1 = Nitrofurazone®, in dmso-d_6

L_1	1	2	3	4	5	6	7	8
δ - H	-	7.75	7.21	-	7.80	10.76	-	6.57
δ - C	152.00	115.95	112.80	154.00	128.45	-	157.00	-

Ru^{II}-L_1	1	2	3	4	5	6	7	8	dmso
δ - H	-	7.95	7.70	-	9.34	11.40	-	7.85	3.28, 3.29
δ - C	152.36	114.87	120.28	146.40	138.84	-	163.91	-	45.73, 45.85

Table III. ^{1}H and ^{13}C NMR chemical shift values (d) in ppm of L_2 and $[Ru^{II}Cl_2(dmso)_2L_2]$, with L_2 = N^4-n-butyl-5-nitro-2-furaldehyde semicarbazone, in acetone-d_6

L_2	1	2	3	4	5	6	7	8	9	10	11	12
δ - H	-	7.59	7.09	-	7.92	9.93	-	6.79	3.30	1.54	1.38	0.94
δ - C	151.10	114.23	112.22	153.14	127.79	-	155.71	-	39.54	32.70	20.13	13.57

$Ru^{II}L_2$	1	2	3	4	5	6	7	8	9	10	11	12
δ - H	-	7.77	7.58	-	9.52	10.74	-	7.77	3.49	1.64	1.43	0.94
δ - C	153.00	113.97	120.26	147.14	137.42	-	161.82	-	41.33	32.31	20.49	13.93

dmso: δ - H = 3.38 ppm, δ - C = 45.43, 45.91 ppm

CONCLUSIONS

Two new Ru(II) nitrofurylsemicarbazone complexes were synthesized and characterized. Both complexes showed poor citotoxicity in aerobic conditions against the tested tumor lines. Otherwise, since most tumors are hypoxic tissues, selective citotoxicity studies in hypoxic conditions against V79 (chinese hamster fibroblastes) are now in progress.

REFERENCES

1. Maestroni G., Alessio E., Sava G., Pacor S., Coluccia M., The development of tumor-inhibiting ruthenium dimethylsulfoxide complexes. *Metal complexes in Cancer Chemotherapy*. Ed. B. Kepler, VCH, Weinheim, 1993, 159-185.
2. Mohindra J.K., Routh A.M., Increased cell killing by Metronidazole and Nitrofurazone of hypoxic compared to aerobic mammalian cells. *Cancer Res.*, 1976, 36, 930-936.
3. Cerecetto H., Di Maio R., Ibarruri G., Seoane G., Denicola A., Quijano C., Peluffo G., Paulino M., Synthesis and anti-trypanosomal activity of novel 5-Nitro-2-furaldehyde and 5-Nitrothiophene-2-carboxaldehyde semicarbazones derivatives. *Il Farmaco* 1998, 53 (2), 89-94.
4. Evans I., Spencer A., Wilkinson G., Dichlorotetrakis(dimethyl sulfoxide) ruthenium(II) and its use as source material for some new Ruthenium(II) complexes. *J. Chem. Soc. Dalton Trans.*, 1973, 204-209.
5. Leo A., Hansch C. Elkins D., Partition coefficients and their uses. *Chem. Rev.*, 1971, 71, 525-554.
6. Skehan P.A., et al., New colorimetric cytotoxicity assay for anticancer drug screening. *J. Natl. Cancer Inst*. 1990, 82, 1107-1112.

*Metal Ions in Biology and Medicine: vol. 7.*Eds L. Khassanova, Ph. Collery, I. Maymard, Z. Khassanova, J.-C. Étienne.
John Libbey Eurotext, Paris © 2002 pp. 614-616.

Bone mineral density and the subsequent risk of uterine cancer in the NHANES I follow-up cohort

Richard L., Nelson, Mary Turyk, Jane Kim, and Victoria Persky

Department of Surgery and Department of Epidemiology and Biostatistics, University of Illinois at Chicago
Correspondence:
R. Nelson, Department of Surgery, Room 2204, mail code 957
University of Illinois Hospital
1740 West Taylor Street
Chicago, Illinois 60612, USA
Fax + 1312 996 2704

ABSTRACT

Bone Mineral Density (BMD), measured in 1975, significantly predicted the subsequent incidence of uterine cancer in the NHANES Epidemiologic Follow Up Cohort. This finding further supports BMDas a measure of estrogen exposure and also, inlight of negative data related to breast cancer, weakens the association of estrogen with breast cancer etiology.

INTRODUCTION

Three recent reports, from the Study of Osteoporotic Fractures [1], Framingham [2], and the Fracture Intervention Trial [3] demonstrated a positive association between bone mineral density measurements (BMD) in women and subsequent onset of breast cancer. All of these reports were regarded as significant because estrogen has been thought to increase breast cancer risk [4] and BMD may be a more accurate measure of long term estrogen exposure than recall of hormone supplementation, measurements of endogenous estrogen, parity, or obesity [5,6]. In light of strong associations previously noted between estrogens and corpus uteri cancer, BMD should predict the occurrence of endometrial cancer [4], though this has not yet been directly assessed. In this report we use the NHANES I Epidemiological Followup Study to pursue the hypothesis: that BMD will correlate directly with subsequent uterus cancer incidence.

METHODS

The First National Health and Nutrition Examination Survey (NHANES I) collected data from a national probability sample of the United States civilian noninstitutionalized population between the ages of one and seventy-four years [7]. In addition to the emphasis on nutrition in NHANES I, a subset sample of persons age 25-74 received a more detailed health examination through October 1975. This additional exam included a hand/wrist x-ray. Records for the 6, 260 cases that have been reread are contained in the Public Use File (Radiography Absorptiometry Bone Density).

Outcome date was obtained on the individuals from the NHANES I Epidemiologic Followup Studies (NHEFS), which was conducted in four waves of data collection through 1993. Data collected included in depth interviews with subjects or their proxies, hospital records, including pathology reports, and collection of death certificates for deceased subjects Tracing is complete for 90% of the cohort through 1993. Those with at least one follow-up record in NHEFS were included in the analysis, resulting in a final sample size of 6,046. Disease outcomes were obtained using ICD-9 codes (International Classification of Diseases, 9th Revision) from death certificates and hospital record diagnostic codes in NHEFS mortality and health care facility stay data bases, respectively. They were: uterine cancer (179, 182.0, 182.1, 182.8).

Person-years of follow-up were computed for each cohort as the amount of time since the NHANES I examination to the date of the first of the following events: date of ICD-9 disease code of interest from the health care facility stay files, date of death from the NHEFS mortality file, or the last day of contact from the 1992 NHEFS vital status file. Thus, subjects were right-censored at either death or last date of follow-up. For persons with more than one hospital admission listing a particular disease outcome of interest, the date of the earliest admission for that disease was used.

Cohorts were divided into 4 groups according to their RA BMD: <95, 95-105, 105-115, >115 (mass/volume units established by Compu-Med). Cox proportional-hazards models were fitted for disease outcomes with BMD groups using the PHREG procedure from the SAS System for Windows Version 8.01. Covariates in the analysis were age at NHANES I examination (<35, 35-50, 50-60, and >60 years), body mass index (BMI; weight in kg/height in m^2) at NHANES I examination (<22, 22-25, 25-28, >28), and race (Caucasian versus other). Since we would expect that age would be exponentially related to cancer, and for uterine cancer the slope of the exponential line would change after menopause, the use of age categories did not constrain age to be linearly related to the outcome.

RESULTS

The BMD sample of NHANES I included 6, 046 individuals who had bone density read by RA and follow-up data in NHEFS. This group included 5, 252 Caucasians, 742 African Americans, and 52 individuals of Hispanic origin. There were 2, 818 men and 3, 228 women. The median age at the time of bone density reading was slightly less than 50 years. By 1993 26 cases of uterine cancer had been reported within this subset of the NHANES I cohort. After age, race and BMI adjustment there was a significant ($p = 0.005$) trend to increasing risk with increasing BMD.

DISCUSSION

Uterine Cancer

This is the first report that examines the relationship of BMD to uterine cancer. Because the association of estrogen exposure and uterine cancer development is far less controversial than estrogen and breast cancer risk [4], the veracity of BMD as a measure of estrogen exposure is supported by the significant positive association of BMD with uterine cancer incidence.

Breast Cancer

The relationship of estrogen exposure to subsequent breast cancer risk has been the subject of much debate and investigation [4]. The debate is all the more significant because of estrogen's role in the prevention of coronary heart disease mortality, hip fracture and osteoporosis [5]. We have previously reported to this group that no association between BMD and subsequent breast cancer incidence was found. This was true for all women in the cohort and those over 55 years of age at the time of BMD measurement, i.e. post menopause.

An indirect method has been employed to assess the relationship of estrogen exposure to uterine cancer risk through bone density assessment: the association of osteoporotic fracture to the prior or subsequent diagnosis of uterine cancer *(table 2)*. One recent large case/control study presents compelling evidence for diminished risk of previous recent fracture in women with both breast and uterine cancer [8]. Four earlier reports of fracture and uterine cancer were divided in their results, though the negative studies included fracture after the diagnosis of cancer [9,10,11].

BMD and estrogen

BMD is affected by many factors other than estrogen, including principally age, calcium, body mass index, cigarette smoking, physical activity, and prior bone injury [12-14]. As such it is a non-specific indicator of risk. BMD was also measured in NHANES I at a time when post-menopausal use of estrogen was rare and its increasing use in recent decades may have altered the subsequent risk of cancer in these women. Nevertheless, the results seen in uterus suggest that estrogen exposure up to the point at which BMD was measured was a significant mediator of risk for cancer in NHANES I. In light of these findings, the negative results for breast cancer are all the more important.

ACKNOWLEDGMENTS AND DISCLAIMERS

NHANES I and NHEFS data were provided from the National Center for Health Statistics (NCHS). All analyses, interpretations and conclusions based upon those data are made by the authors only and not the NCHS.

REFERENCES

1. Cauley JA, Lucas FL, Kuller LH, Vogt MT, et al. Bone mineral density and risk of breast cancer in older women: the Study of Osteoporotic Fractures. JAMA. 1996; 276: 1404-1408.
2. Zhang Y, Kiel DP, Kreger BE, Cupples LA, et al. Bone mass and the risk of breast cancer among post menopausal women. N Eng J Med. 1997; 336: 611-617.
3. Buist DSM, LaCroix AZ, Barlow WE, White E, Weiss NS. Bone mineral density and breast cancer risk in postmenopausal women. J Clin Epid. 2001; 54: 417-422.
4. Beral V, Banks E, Reeves G, Appleby P. Use of HRT and the subsequent risk of cancer. J Epidemiol Biostat. 1999; 4: 191-210.
5. Clinical Synthesis Panel on HRT. Hormone replacement therapy. Lancet. 1999; 354: 152-155.
6. Willett WC, Colditz G, Stampfer M. Postmenopausal estrogens - opposed, unopposed or none of the above. JAMA. 2000; 283: 534-535.
7. National Center for Health Statistics. Plan and initial program of the health examination survey. Vital Health Stat.s 1. 1973; 10: 1-3.
8. Newcomb PA, Trantham-Dietz A, Egan KM, Titus-Ernstoff L, Baron JA, Storer BE, et al. Fracture history and risk of breast and endometrial cancer. Am J Epid. 2001; 153: 1071-1078.
9. Persson I, Adami HO, McLaughlin JK, Naessen T, Fraumeni JR jr. Reduced risk of breast and endometrial cancer among women with hip fractures. Cancer Causes & Control. 1994; 5: 523-528.
10. Persson I, Naessen T, Adami HO, Bergstrom R, Lagrelius A, Mollerstrom G, Pettersson B, von Hamos K. Reduced risk of hip fracture in women with endometrial cancer. Int. J epid. 1992; 21: 636-642.
11. Olsson H, Hagglend G. Reduced cancer morbidity and mortality in a prospective cohort of women with distal forearm fractures. Am J Epid. 1992; 136: 422-427.
12. Brot C, Jorgensen N, Madsen OR, Jensen LB, Sorensen OH. Relationships between bone mineral density, serum vitamin D metabolites and calcium: phosphorus intake in healthy perimenopausal women. J Intern Med 1999 May; 245(5): 509-16.
13. Branca F. Physical activity, diet and skeletal health. Public Health Nutr 1999 Sep; 2(3A): 391-6.
14. New SA, Robins SP, Campbell MK, Martin JC, Garton MJ, Bolton-Smith C, Grubb DA, Lee SJ, Reid DM. Dietary influences on bone mass and bone metabolism: further evidence of a positive link between fruit and vegetable consumption and bone health? Am J Clin Nutr 2000 Jan; 71.

*Metal Ions in Biology and Medicine: vol. 7.*Eds L. Khassanova, Ph. Collery, I. Maymard, Z. Khassanova, J.-C. Étienne.
John Libbey Eurotext, Paris © 2002 pp. 617-620.

The effect of iron on prostate and breast cancer cell invasion

André A. Kajdacsy-Balla, Sushma Kaul, Christopher Chitambar, Paul F. Lindholm

Departments of Pathology and Internal Medicine, Medical College of Wisconsin, Milwaukee WI 53226, USA

ABSTRACT

Whereas it has been known for sometime that malignant and nonmalignant cells require iron for their viability and proliferation, it has never been previously reported that iron may be a factor in tumor invasiveness and metastasis. In order to test this hypothesis, PC-3, a prostate cancer cell line, and MDA-MB-231, a breast cancer cell line were pre-exposed to various concentrations of iron salts, either ferric chloride or ferric ammonium citrate (FAC), and tested by assays for cell motility, invasion and adhesion. When compared to untreated cells in Matrigel® invasion assays, PC-3 cells treated with 100 μM FAC or 200 μM $FeCl_3$ had a 52 ± 18% and 122 ± 24% increase in invasion respectively (n=8, $p < 0.05$). Similar results were obtained for invasion of MDA-MB-231 cells treated with 200 μM FAC and above (205 ± 32%, n = 5, $p < 0.05$), but not with various concentrations of $FeCl_3$. Modified Boyden chamber cell motility assays showed 40 ± 5% increase in migration after cells were treated with 100 μM FAC for PC-3 cells (n = 4) and 105 ± 7% for MDA-MB-231 cells (n = 4). A 23 ± 6% increase in motility occurred with 200 μM $FeCl_3$ for MDA-MB-231 cells (n = 4). Cell adhesion to Matrigel® substrate was not changed by FAC or $FeCl_3$ when adhesion was measured at either 15, 30 or 45 min. Concentrations of iron salts that were used in our studies are physiologically relevant, and at least 100-fold less than concentrations that caused loss of viability for these two cell lines. These results suggest that excess iron may have an enhancing effect on invasion and metastasis. This finding may have clinical relevance since epidemiological studies show an association between increased iron levels and iron saturation and increased risk of cancer occurrence and mortality. Postulated mechanisms by which increased dietary iron may enhance the risk of cancer development include the generation of DNA damaging oxygen free radicals and the promotion of cancer cell growth. In addition, we show that iron could possibly cause acceleration of cancer progression by enhancement of cell motility and invasion. We are now studying possible mechanisms for this effect of iron. This work was supported in part by the US Department of Defense grant PCRP/CDMRP-PC010201 and the USA Environmental Protection Agency grant R827152.

INTRODUCTION

There is compelling evidence that environmental agents and dietary factors may act on established cancers to make them more aggressive. One of the best examples is the effect of dietary animal fat on prostate cancer survival [1]. There is also evidence that cigarette smoking is associated with shorter survival after diagnosis, even when corrections are made for tumor differentiation (grade) and spread (stage) at diagnosis [2, 3]. It is however largely unknown whether exposure to less obviously harmful compounds such as iron supplements could make cancer cells behave more aggressively. Iron supplements are of particular concern in view of evidence that primary neoplasms develop more frequently at body sites of excessive iron deposition in both humans and experimental animals. It is thought that iron exerts its carcinogenic effect by catalyzing formation of hydroxyl radicals, suppressing activity of host defense cells and promoting cancer cell multiplication [4, 5, 6]. To our knowledge there is nothing published on the possibility that iron could cause cancer cells to become more invasive. The best evidence in this direction is the work published by Nobel laureate Baruch Blumberg and his group. They showed that mice fed with a low-iron diet had smaller tumors than those fed with a normal-iron diet when injected subcutaneously with several different cancer cell types [7]. They did not however study metastatic or invasive potential modulation by iron depletion or supplementation diet.

MATERIALS AND METHODS

Cell cultures: Human PC-3 prostate cancer cell line and MDA-MB-231 breast cancer cell lines, both highly positive for transferrin receptors, were acquired from American Type Culture Collection (Manassas, VA). Cells were cultured with 10% fetal bovine serum that was not depleted of transferrin.

Iron salts and other reagents: Reagents were acquired from Sigma Chemical Co. (St. Louis, MO) unless otherwise indicated.

Cell culture pre-exposure to iron salts: Cells were exposed to various concentrations of iron salts for 4 hours. Excess iron was washed after cell exposure so that iron effect on the Matrigel® substrate was not a confounding factor. Small amounts of ferritin and other iron-binding serum components were present in the fetal bovine serum used to supplement culture medium.

Invasion and motility assays: Invasion assays were performed as previously described [8]. Chemotactic gradients were not used. Serial ten-fold dilutions of the agent being tested were used for invasion assays. Transwel® chambers (Becton Dickinson, Bedford, MA) are modified Boyden chambers. Matrigel® (Becton, Dickinson) is a gelatinous material extracted from a murine sarcoma cell line that produces large amount of basement membrane. This reconstituted basement membrane was poured over the Transwell® inserts and served as a barrier for cell invasion, similar to the basement membrane and extracellular matrix that cancer cells have to invade *in vivo* in order to metastasize. Transwell® inserts are miniature cups that have a porous membrane. The cells were labeled with tritiated thymidine overnight, detached, washed off excess label, exposed to iron salts, washed off excess added iron, placed on top of the Matrigel®. The cells that migrated through the Matrigel® barrier and the Transwell® membrane to the other side were counted with a scintillation counter (Packard, Downers Grove, IL). Even coating by Matrigel® was confirmed by post-experiment staining of membrane with eosin. For motility assays, the same methods were used, but without Matrigel®. Experiments were done in triplicates and results that showed >20% difference between replicates were discarded.

Cell adhesion assays: A modification of the method of Santoro et al was used [9]. Cells were labeled with ^{3}H-thymidine (Dupont, Boston, MA), washed, pre-incubated with either FAC or $FeCl_3$ for 4 hours, detached from the culture flask, washed off excess iron salts, counted and added to culture plates coated with Matrigel®. Cells were retrieved at either 15, 30 or 45 minutes, and the percent cells that firmly adhered to Matrigel® was estimated with the use of a scintillation counter.

RESULTS

Cell invasion: When compared to untreated cells in Matrigel® invasion assays, PC-3 cells treated with 100 μM ferric ammonium citrate (FAC) or 200 μM $FeCl_3$ had a 52 ± 18% and 122 ± 24% increase in invasion respectively ($p < 0.05$). Similar results were obtained for invasion of MDA-MB-231 cells treated with 200 μM FAC (205 ± 32%, n = 5, $p < 0.05$, see *fig. 1)*, but not with various concentrations of $FeCl_3$ (result not shown).

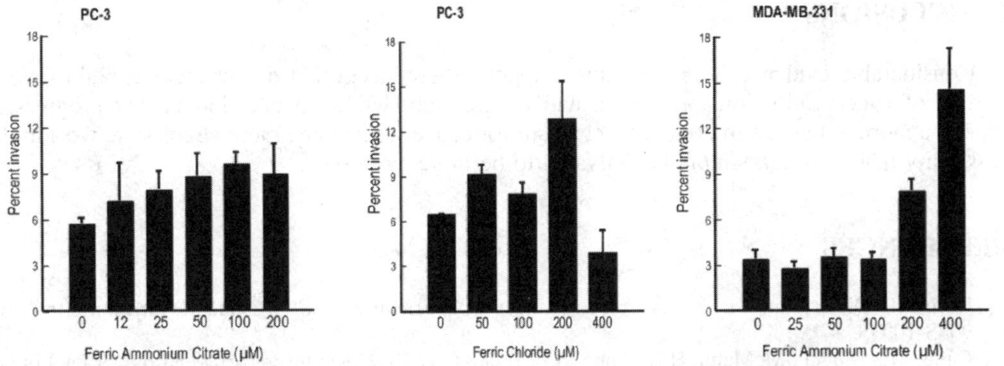

Fig. 1. Effect of ferric salts exposure on cell invasion through reconstituted basement membrane (Matrigel®). PC-3 prostate cancer and MDA-MB-231 breast cancer cells were pre- exposed to various concentrations of ferric ammonium citrate or ferric chloride for 4 hours (n =8 for PC-3; n =5 for MDA-MB-231 cells). The concentrations of iron present in the cell cultures are within the range found in humans, and concentrations of 1000 μM and above are not physiologically relevant. Possible confounding effect of proliferation was automatically corrected since the tritiated thymidine labeling was used to count cells.

Cell motility: Modified Boyden chamber cell motility assays showed increase in migration after MDA-MB-231 cells were treated with either 100 μM FAC or 200 μM $FeCl_3$ ($p<0.05$; see *table I*). For PC-3 cells an increase in motility was seen with FAC ($p<0.05$; see *table I*). We have not yet tested the effect of $FeCl_3$ on PC-3 cells.

Table I. The effect of iron salts on cell motility (n=4)

	100 μM FAC	**200 μM $FeCl_3$**
MDA-MB-231 cells	105 ± 7%	**23 ± 6%**
PC-3 cells	40 ± 5%	**N.D.**

Cell adhesion: Cell adhesion to Matrigel® substrate was not changed by FAC or $FeCl_3$ when adhesion was measured at either 15, 30 or 45 min.

DISCUSSION

Concentrations of iron salts that were used in our studies are physiologically relevant, and at least 100-fold less than concentrations that cause loss of viability for these two cell lines. These results suggest that excess iron may have an enhancing effect on invasion and metastasis. The time frame of the invasion assays (40 hours) was not enough for significant propagation of any iron-induced mutational events, making DNA damage an unlikely explanation for the increase in invasion.

These findings may have clinical relevance since epidemiological studies show an association between increased iron levels and iron saturation and increased risk of cancer occurrence and mortality. Postulated mechanisms by which increased dietary iron may enhance the risk of cancer development include the generation of DNA damaging oxygen free radicals and the promotion of cancer cell growth. In addition, the data suggests that iron could possibly cause acceleration of cancer progression by enhancement of cell motility and invasion. We are now studying the possible mechanisms of these effects of iron on cancer cells.

CONCLUSIONS

Considerable evidence in the literature supports the concept that dna changes explain several aspects of cancer initiation, promotion and progression. while no one doubts that progressive genetic abnormalities are important mechanism for cancer to become more aggressive, we add the possibility that cells can simply be activated to be more aggressive.

REFERENCES

1. Rose D.P. Dietary fatty acids and cancer. [Review]. *American Journal of Clinical Nutrition*, 1997, 66, 998S-1003S.
2. Calle E.E., Miracle-McMahill H.L., Thun M.J., Heath C.W. Jr. Cigarette smoking and risk of fatal breast cancer. *American Journal of Epidemiology*, 1994, 139, 1001-1007.
3. Daniell H.W. A worse prognosis for smokers with prostate cancer. *Journal of Urology*, 1995, 154, 153-157.
4. Wang F., Elliot R.L., Head J.F. Inhibitory effect of deferoxamine mesylate and low iron diet on the 13762NF mammary carcinoma. *Anticancer Research*, 1999, 19, 445-450.
5. Babbs C.F. Free radicals and the etiology of colon cancer. *Free Radical Biology and Medicine*, 1990, 8, 191-200.
6. Nelson R.L. Dietary iron and colorectal cancer risks. *Free Radical Biology and Medicine*, 1992, 12, 161-168.
7. Hann H.W, Stahlhut M.W., Blumberg B.S. Iron nutrition and tumor growth: decreased tumor growth in iron-deficient mice. *Cancer Research*, 1988, 48, 4168-4170.
8. Lindholm P.F., Bub J., Kaul S., Shidham V.B., Kajdacsy-Balla A. The role of constitutive NF-κB activity in PC-3 human prostate cancer cell invasive behavior. *Clinical and Experimental Metastasis*, 2001,18, 471-479.
9. Santoro S.A., Zutter M.A., Wu J.E., Staatz W.D., Suelman E.V.M., Keely P.J. Analysis of collagen receptors. *Methods in enzymology*, 1994, 245, 147-183.

Metal Ions in Biology and Medicine: vol. 7. Eds L. Khassanova, Ph. Collery, I. Maymard, Z. Khassanova, J.-C. Étienne.
John Libbey Eurotext, Paris © 2002 pp. 621-624.

Assessment of some heavy metals in the maternal body, risk in cancer disease

Carmen Hura[1], I. Palamaru[1], B.A. Hura[2]

[1]*Food Toxicology Laboratory, Institute of Public Health, Iassy, Romania*
[2]*Technical University "Gh.Asachi", Iassy, Romania*

ABSTRACT

Ever since humans have become aware that health is inseparably linked to an intact and healthy environment, the control and reduction of pollution have become the focus of worldwide concern. The studies on heavy metals concentrations in human placenta and human milk have been performed, as a continuation of the research concerning heavy metals concentrations in biological and in food samples by atomic absorption spectrophotometer method. The aim of this study was the evaluation of the metals contents in the maternal body.

The study presents the results obtained in 2000-2001 period of some metals [Pb,Cd,Zn,Cu,Mn] in maternal body, in Iassy district, it was investigated if heavy metals levels from human placenta and human milk reflect the body burden of these chemicals. A total of 50 women (18-30 years old) participated in the study voluntarily.

In all analyzed samples these metals were found. Generally, a wide variation between individual samples was observed.

Placenta

The mean metals levels in the maternal placenta varied between 0.2 µg/kg Cd and 8.18 µg/kg Zn.

Human milk

The results of the investigations showed a variation of heavy metals between 0.3 µg/kg Cd and 12.46 µg/kg Zn.

Determinations of these pollutants in human body are important in environmental monitoring for the prevention, control and reduction of pollution as well as for occupational health and epidemiological studies.

INTRODUCTION

In the past years pollution of the environment by heavy metals of antropogenic origin has become a major threat to all living organisms including the man. This threat is aggravated by the ability of heavy metals to accumulate at the top of the ecological pyramid and their high mobility and persistence in ecosystems. Nowadays, the increased environmental levels of metals have been implicated in elevations in the concentrations of several elements in human body. Although current environmental levels of toxic metals rarely produce morbidity or death in the general population, the interest in biological effects from metal has increased during the last decades [1, 2].

The aim of this study was the evaluation of the metals contents in the maternal body [3, 4, 5, 6].

MATERIAL AND METHODS

The study presents the results obtained in 2000-2001 period of some metals [Pb,Cd,Zn,Cu,Mn] in maternal body, in Iassy district, it was investigated if heavy metals levels from human placenta and human milk reflect the body burden of these chemicals. A total of 50 women (18-30 years old) participated in the study voluntarily. The women were giving birth to their first or second child at "Cuza Voda" Hospital of Iassy in 2000-2001 (April - May) period. The milk samples had been collected manually preferentially before the infant was nursed in the morning. Trace elements concentrations were analyzed by atomic absorption spectrophotometer, using a Carl Zeiss Jena, Model AAS-1N, with flame air-acetylene. Atomic absorption spectrophotometer method was used for the determination of heavy metals (Pb, Cd, Cu, Zn, Mn).

The study presents the results of some heavy metals concentrations in maternal body (milk, placenta) during the 2000-2001, in the Iassy area.

RESULTS

The concentrations of toxic and essential metals measured in the placenta and in the human milk are shown in *table 1*, *table 2* and *fig. 1*.

Placenta

Table 1 shows the heavy metals levels in placenta from 28 women. The mean heavy metals in the placenta varied between 8.18 mg/kg Zn and 0.2μg/kg Cd. The mean Zn levels in placenta were 8.18 mg/kg with a variation between 1.67-16.54 mg/kg. The mean Cu levels were 0.96 μg/kg with higher concentration 1.56 μg/kg. The mean lead were 3.75 μg/kg with a maximum level at 20.0 μg/kg and the mean Cd levels were 0.2 μg/kg with a maximum level at 1.0 μg/kg. The mean Mn levels were 0.26 μg/kg with a maximum level at 0.91 μg/kg *(fig. 1)*.

Table 1. The mean levels of heavy metals in placenta

Samples	Metals	Freqv.	Mean	Range
	Zn (mg/kg)	28	**8.18**	**1.67-16.54**
	Cu (μg/kg)	28	0.96	n.d.-8.67
Placenta	Cd (μg/kg)	28	0.2	n.d.-1.0
	Pb (μg/kg)	28	3.75	n.d.-20.0
	Mn (μg/kg)	28	0.26	n.d.-0.91

Human milk

Table 2 shows the heavy metals in human milk from 14 women. The mean heavy metals in the human milk varied between 12.46 mg/kg Zn and 0.3 μg/kg Cd *(fig. 1)*. The mean Zn levels in human milk were 12.46 mg/kg with a variation between 6.17-27.23 mg/kg. The mean Cu levels were 0,81 μg/kg with higher level at 1.56 μg/kg. The mean Cd levels were 0.3 μg/kg with higher concentration at 1.46 μg/kg. The mean lead concentrations were 3.47 μg/kg with a variations between 0.5-5.0 μg/kg. The mean Mn concentrations were 0.7 μg/kg with a variation between 0.07-1.0 μg/kg.

Table 2. The mean levels of heavy metals in human milk

Samples	Metals	Freqv.	Mean	Range
Human milk	Zn (mg/kg)	14	**12.46**	**6.17-27,23**
	Cu (μg/kg)	14	0.81	n.d.-1.56
	Cd (μg/kg)	14	0.30	n.d.-1.46
	Pb (μg/kg)	14	3.47	0.5-5.00
	Mn (μg/kg)	14	0.7	0,07-1.0

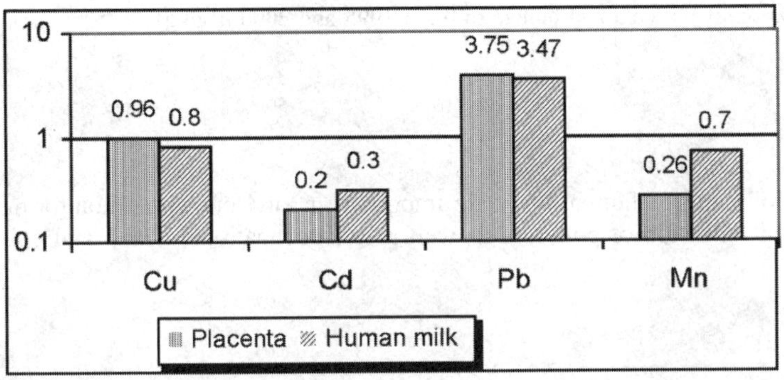

Fig. 1. The mean Cu,Cd,Pb, Mn levels in maternal body (μg/kg)

Comparing of results obtained in 2000 with the preceding years

Table 3 shows the mean levels of the heavy metals in the placenta and the human milk determined in 1995 and in 2000, in Iassy area, by atomic absorbtion spectrophotometer method.

Table 3. The mean heavy metals levels in the maternal body

Samples	1995			2000				
	Pb, μg/kg	Ni, μg/kg	Zn, mg/kg	Pb, μg/kg	Cd, μg/kg	Cu, μg/kg	Mn, μg/kg	Zn, mg/kg
Placenta	5.34	4.5	17.0	3.75	0.2	0.96	0.26	8.18
Human milk	27.8	22.9	31.88	3.47	0.3	0.81	0.70	12.46

The mean Pb and Zn levels in placenta and human milk decreased in 2000 in comparative with 1995.

The mean Pb levels in the placenta decreased from 5.34 μg/kg (1995) to 3.75 μg/kg (2000) and in the human milk from 27.8 μg/kg (1995) to 3.47 μg/kg (2000), *fig. 2*.

In the placenta, the mean Zn levels decreased from 17.0 mg/kg (1995) to 8.18 mg/kg (2000); and in the human milk from 31.88 mg/kg (1995) to 12.46 mg/kg (2000).

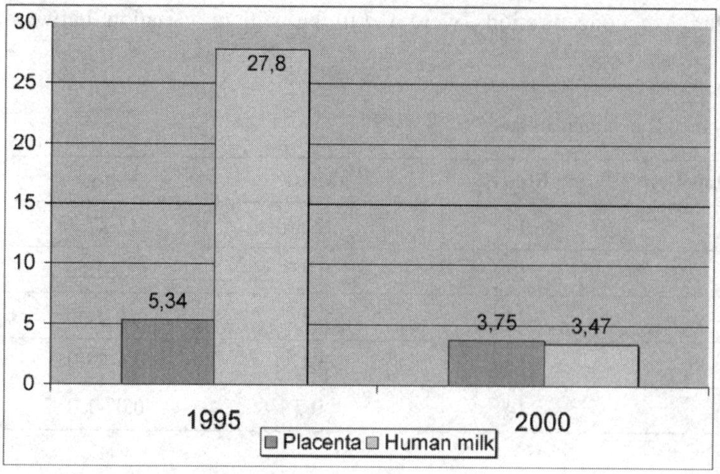

Fig. 2. The mean lead concentration of placenta and human milk, in 1995 and 2000 μ(g/kg)

CONCLUSION

Determinations of these pollutants in human body are important in environmental monitoring for the prevention, control and reduction of pollution as well as for occupational health and epidemiological studies.

REFERENCES

1. Folin M, Contiero E, Vaselli G.M. - Trace element determination in humans. The use of blood and hair, *Biol.Trace Elem.Res.* 1991, 31; 147-158.
2. WHO - Biological Monitoring of Chemical Exposure in the Work Place, vol. 1 and vol. 2.
3. Hura C, Leanca M, Palamaru I - Chemical pollutants in daily diets, a risk for cancer disease. *Rev. Hygiene and Public Health*, 1995, nr.1, vol. 45; 13-16.
4. Hura C. - Chemical Pollution of Food and Health, Iassy, Ed. Socom Hermes, 1997.
5. Hura C.- Chemical contaminants in food, 1980-2000, Ed. CERMI, Iassy, 2001.
6. Hura C. - Chemical contaminants in food and human body (in English, in press).

*Metal Ions in Biology and Medicine: vol. 7.*Eds L. Khassanova, Ph. Collery, I. Maymard, Z. Khassanova, J.-C. Étienne.
John Libbey Eurotext, Paris © 2002 pp. 625-628.

Effect of gallium on the growth rate of U937 cell culture and the activity of tyrosine kinase

Bijan Farzami, Ahmadzadeh Pournaki, Philippe Collery* and Bahram Goliaei#

*Department of Clinical Biochemistry,Tehran University of Medical Sciences P.O.Box 14155-5399 Tehran, Iran, *Service de Cancérologie, Polyclinique Maymard 20200 Bastia, France and #Institute of Biophysics and Biochemistry, Tehran University, Tehran, Iran*

ABSTRACT

In our study **U937** cell, a human myeloid leukemic cell line, has been used to show the effect of Gallium on cellular growth and Tyrosine kinase activity. These cells were cultured in a medium composed of RPMI 1640 (50%), FCS (10%) and deionized water (40%). The study was performed in two steps, and in each step two cultures containing 5×10^5/ ml cells as controls and samples containing 200 and 400 μM Gallium Chloride were used. The cultures were counted and assessed for viability at hours 24, 48, 72, 144, 168. IC50 of 72 hrs and 24 hrs were obtained with the concentrations of 200 and 400 μM respectively. Tyrosine kinase assay was performed by ELISA on the cell free system in control and treated cells and it was showed a concentration dependent increase in the level of tyrosine kinase in Gallium treated cells.

INTRODUCTION

Tyrosine Kinases are active in the signal transduction processes that control cellular differentiation and proliferations [1] This process is performed by phosphorylation of amino acid residues of specific amino acids. Tyrosine kinases are dependent in their activity on metal ions such as Mg(II) and Mn(II). Therefore the key role of the enzyme and the effects of metal ions in activation and repression of the enzyme and consequently its role in tumorogenesis is evident. A significant anti tumor activity of Gallium was first reported by Adamson et al. [2-4]. Several other studies indicate the role of Ga (III) in regression of malignant tumor growth [5-7]. There are two Gallium radio isotopes ^{67}Ga and ^{68}Ga. Possessing nuclear properties that make them attractive for use in nuclear medicine ^{67}Ga citrate has long been established as a diagnostic tool for the localization of neoplastic lesions in animals and humans. In 1973, the toxicity and antitumor activity of salts of the group IIIa metals Aluminium, Gallium, Indium, and Thallium were reported for the first time. All four of the metal salts tested were cytotoxic in tissue culture against cancer cells (Adamson et al.).

During the last two decades, several studies of Gallium have been carried out with different results. Warrel et al [8] published a study indicating that Gallium nitrate administered as a continuous infusion for seven days at 300 mg/ m^2/ day was well tolerated and was an effective treatment for patients with advanced malignant lymphoma. In reported clinical trials, renal toxicity of Gallium has hindered the use of efficient doses.

Consequently in recent years, oral administration of Gallium has been proposed in order to avoid nephrotoxicity derived from parental use of the metal. According to previous experimental results, which showed that Gallium administered orally as Gallium chloride ($GaCl_3$), daily during months, could result in a selective uptake by tumor cells without tumor toxicity, a study was performed that showed that Gallium permits an increase in length of survival with a regression of tumor growth after continuous treatment, but without any diminution of the tumor size (Collery et al, 1986). Due to the lack of renal and hematological toxicity, together with the significant uptake by the tumor, it has been suggested that Gallium chloride can be used orally in conjunction with other cytotoxic agents. In our studies U937, a human myeloid leukemia cell line has been used to show the effect of Gallium on cellular growth as well as on Tyrosine kinase, an enzyme that is known to be involved in signal transduction.

OBJECTIVE

On the basis of the importance of Tyrosine kinase in the mechanism of cancer, it seemed worthwhile to consider the effect of Ga(III) on possible changes on the level of Tyrosine kinase in the cell culture, as well as on the growth and viability of cells.

MATERIAL AND METHODS

Culture medium was chosen to be RPMI 1640 obtained from Gibco.

Gallium chloride was provided by Aguettant Laboratory, France and Tyrosine Kinase Assay Kit was purchased from Boehringer Mannheim (non radioactive test).

U937 Cell line was donated by the Institute of Biophysics and Biochemistry of the Tehran University.

Different concentrations of $GaCl_3$ (100 μM and 200 μM) were used in the logarithmic phase of the growth. Tyrosine kinase was assayed by Elisa using total proteins obtained from culture cells and after complete cell lysis.

RESULTS AND DISCUSSION

Both concentrations of Ga(III) induced the growth inhibition in U937 cell culture.

IC50 for 200 μM concentration of GaCl3 was 72 hrs.

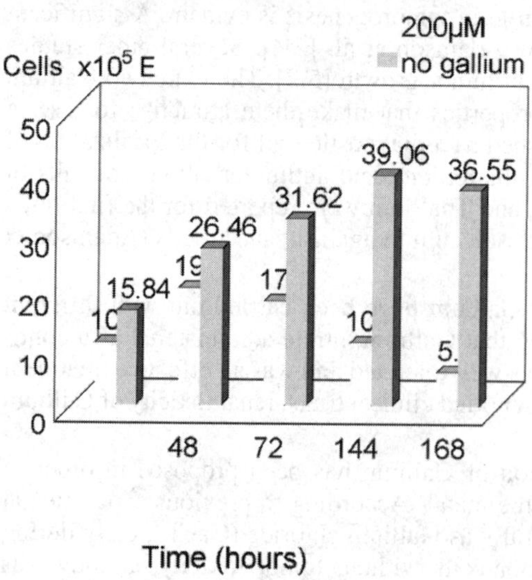

Fig. 1. Cell count after treatement with Ga(III) 200 μmol. as a function of incubation time

The growth inhibition was markedly enhanced when using 400 μM concentrations of GaCl3.

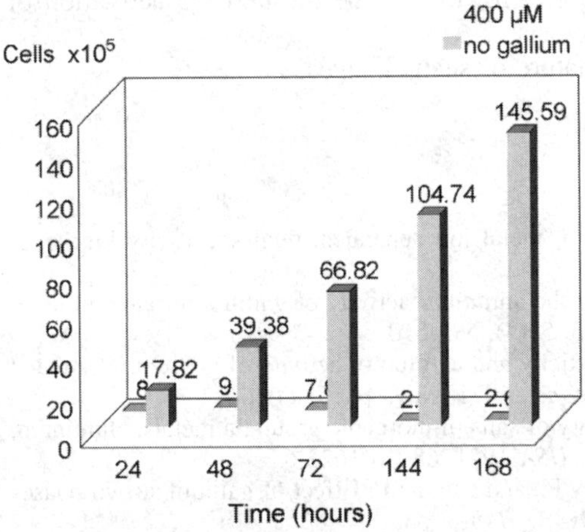

Fig. 2 Cell count after treatment with Ga(III) 400 μmol, as a function of incubation time

It has been suggested that the effect of Ga(III) could be through the interference with mitochondrial membrane potential, but the mechanism has not yet been proven. ***Tyrosine kinase*** activation is through a cascade of events in the process of signal transduction. which is one of the important processes in the mechanism of cancerous tumor production. Our results indicate that Ga(III) could activate the enzyme Tyrosine kinase in the medium of cell free system. In conditions where 400 μM concentration of Ga(III) were used, the enhancement in the activity of tyrosine kinase was observed.

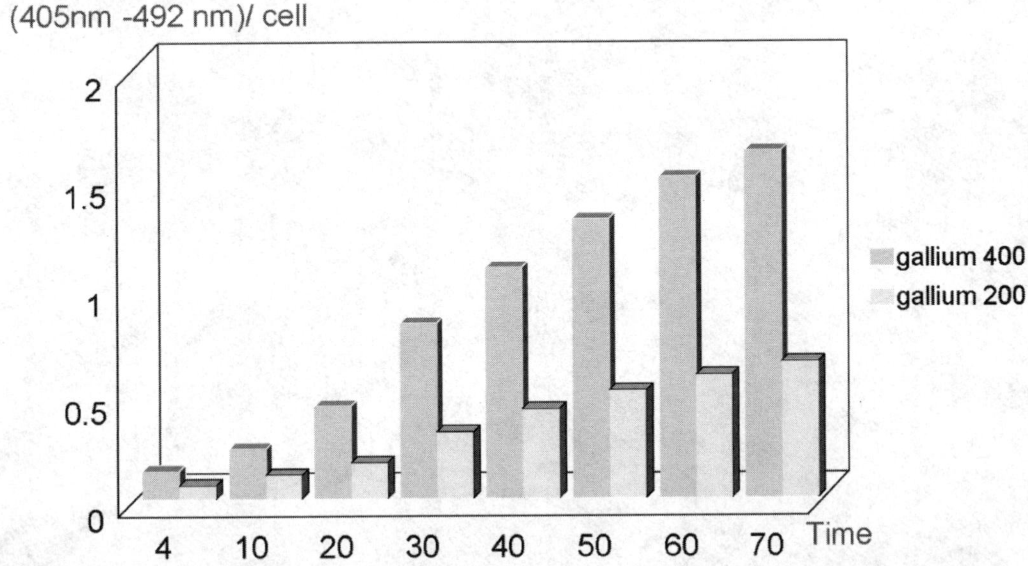

Fig. 3. Tyrosine kinase activity in U937 cells treated with gallium 200 μM and 400 μM

CONCLUSION

The hypothesis that could be assumed for the activation of tyrosine kinase due to the effect of Ga(III) and its relationship with the inhibition of cell growth may be through the activation of Tyrosine kinase biosynthesis in the process of apoptosis.

Further studies are in progress to evaluate the nature of such changes.

REFERENCES

1. Chiun-Jeyeyuan, Chi et al Phosphorylation Kinases a, metal ion dependent dual specificity kinase. *J. Biol. Chem.* 1993, 268, 24, 17683.
2. Adamson R.H., Cannels G.P., Sieber S.M. Studies on the antitumor activity of gallium nitrate and other groups IIIa metal salts. *Cancer Chemother. Rep* 1975, 59(3), 599-610.
3. Hart M., Smith C.F., Yancey S.T., Adamson R.H. Toxicity and anti tumor activity of gallium nitrate and periodically related metal salts 1971. *J. Natl. Cancer Inst.* 1971, Nov. 47 (5) 1121-7.
4. Hart M., Adamson R.H. Antitumor activity and toxicity of salts of inorganic group 3a metals, aluminum, gallium, indium, and thallium. *Proc. Natl. Acad. Sci. USA* 1971, 68 (7) 1623-6.
5. Carpentier Y., Liautaud-Roger F., Loirette M., Collery P., and Coninx P. Effect of gallium at two phases of the CA 755 tumor growth. *Anticancer Res* 1987 7(4B), 745-8.
6. Collery P. et al. Inhibitory effect of gallium chloride and tris 8- quinolinolato gallium III on A549 human malignant cell line. *Anticancer Res.* 2000, 20 (2A), 955.
7. Perchellet E.M,. Ladesich JB., Collery P., Perchelet J.P. Microtubule -disrupting effect of gallium chloride in vitro. *Anticancer drugs* 1999, 10(5), 477-78.

Metal Ions in Biology and Medicine: vol. 7. Eds L. Khassanova, Ph. Collery, I. Maymard, Z. Khassanova, J.-C. Étienne.
John Libbey Eurotext, Paris © 2002 pp. 629-632.

Two new vanadyl(IV) complexes with potential antineoplastic effect on osteoblasts in culture

S. B. Etcheverry[1,2], D. A. Barrio[1], M. S. Molinuevo[1], A. M. Cortizo[1]

[1]*Bioquímica Patológica;* [2]*CEQUINOR, Facultad de Ciencias Exactas, Universidad Nacional de La Plata, 47 y 115 (1900) La Plata, Argentina*

ABSTRACT

Vanadium compounds have been shown to possess pronounced antineoplastic activity both in vivo and in vitro. As part of a project devoted to the development of new vanadium derivatives with potential therapeutical applications, two vanadyl(IV) complexes, one with glucose (GluVO) and the other with naproxen, an anti-inflammatory drug (NapVO) were synthesized and thoroughly tested on osteoblasts in culture. Rat osteosarcoma (UMR106) and mouse calvaria (MC3T3E1)-derived cells were incubated with different concentrations of GluVO and NapVO. Both compounds inhibited cell proliferation (crystal violet bioassay) with higher potency on UMR106 tumoral cells than on the non-transformed osteoblasts. GluVO behaved as a more potent inhibitory agent than NapVO. The differentiation of MC3T3E1 cells (alkaline phosphatase activity) was not significantly affected by these vanadium derivatives. On the contrary, both vanadium compounds inhibited UMR106 differentiation with similar potency, in a dose response manner. In addition, both compounds caused morphological alterations and lipid peroxidation (thiobarbituric reactive substances, TBARS). The levels of TBARS increased in a manner dependent on the concentration of vanadium. Both vanadyl(IV) complexes strongly enhanced the phosphorylation of external regulated kinases (P-Erks). In conclusion, NapVO and GluVO caused osteoblast cytotoxicity and morphological alterations with stronger effects on tumoral than on non-transformed osteoblast-like cells, being potentially useful compounds for antitumoral therapy.

INTRODUCTION

Vanadium, a transition metal, is normally present at very low concentrations (10^{-8}M) in virtually all cells in plants and animals. Vanadium compounds show relevant biological properties, for example they regulate cell proliferation and differentiation as well as several metabolic processes such as glucose uptake and oxidation, lipid metabolism, etc [1]. Furthermore, vanadium compounds are interesting from a pharmacological point of view due to their insulin mimetic actions [2, 3] as well as their antineoplastic activity. Vanadium could be considered a representative of a new class of non-platinum group metal antitumor agents. The chemotherapeutic applications of vanadium derivatives have been shown both in vivo and in vitro studies [4, 5]. We have recently shown, by using a controlled delivery system, that the sustained release of a vanadyl(IV) complex with Aspirin from a poly(beta-propiolactone) film, has antineoplasitc effect on an osteosarcoma cell line [6].

The aim of the present work was the development of new vanadium derivatives with potential therapeutical applications. Two vanadyl(IV) complexes, one with glucose (GluVO) and the other with Naproxen, an anti-inflammatory drug (NapVO), were synthesized and thoroughly tested on the UMR106 osteosarcoma cells and compared with the non-transformed MC3T3E1 osteoblast-like cells in culture.

MATERIALS AND METHODS

GluVO and NapVO were synthesized and physicochemically characterized as previously described [7, 8].

Cell culture of MC3T3E1 and UMR106 osteoblast-like cells, cellular proliferation (crystal violet assay) and differentiation (alkaline phosphatase activity, ALP), morphological studies (light microscopy), lipid peroxidation (thiobarbituric acid reactive substances, TBARS) and phosphorylation of the extracellular regulated kinases (P-Erks) by Western blot, were carried out as previously reported [8, 9].

RESULTS

Effect of vanadium compounds on cell proliferation. UMR106 and MC3T3E1 osteoblast-like cells were cultured in serum free-media with the addition of different concentrations of GluVO or NapVO *(fig. 1 and 2).*

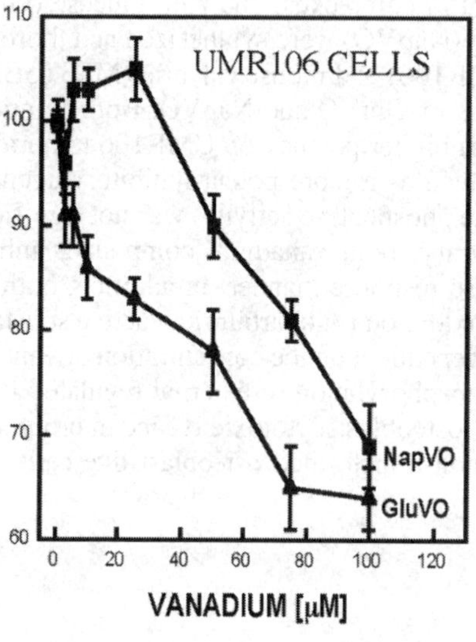

Fig. 1.

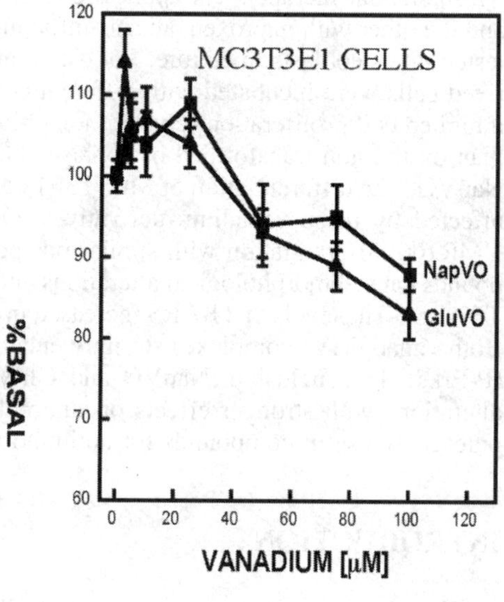

Fig. 2.

Both compounds inhibited cell proliferation in a dose-response manner *(fig. 1 and 2).* In addition, these compounds exerted stronger effects on the tumoral cells (UMR106) than on the non-transformed osteoblasts, being GluVO a more potent inhibitory agent than NapVO. In the MC3T3E1 cells, vanadium-induced cell inhibition in a narrow and high concentration range (75-100 μM). On the other hand, low concentrations of GluVO were effective to decrease UMR106 cell growth *(fig. 1).*

Vanadium compounds selectively inhibited osteoblastic differentiation. The effect of GluVO and NapVO on cellular differentiation was assessed by the alkaline phosphatase activity in UMR106 cells (24 hour-cultures) and in 15 day-cultures of MC3T3E1 cells. Under these conditions both cell express significant levels of the bone-ALP. As can be seen in *fig. 3*, the differentiation of the non-transformed cells was not significantly affected by these vanadium compounds. On the contrary, both vanadium derivatives inhibited UMR106 differentiation in a dose response manner.

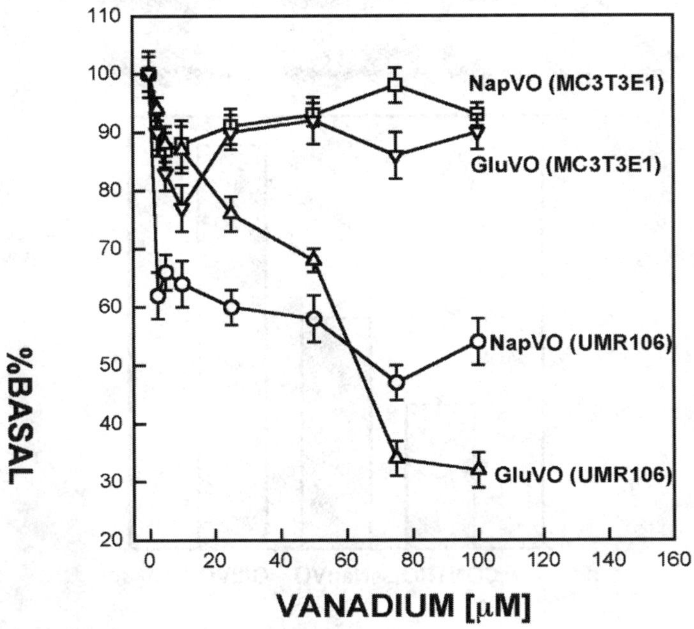

Fig. 3.

Cytotoxic effects

Lipid peroxidation. Both vanadium compounds caused lipid peroxidation on UMR106 cells in culture. TBARS levels increased as a function of vanadium concentration.

Morphological changes. UMR106 and MC3T3E1 cells growing in the presence of these vanadium derivatives became fusiform in shape with loss of processes and cytoplasmic condensation.

Activation of Erks pathway. The effect of vanadium compounds on the activation of Erks in the osteosarcoma cell line can be seen in *fig. 4*. Both vanadyl(IV) derivatives strongly induced the phosphorylation of Erks, as it was determined by Western blot *(fig. 4A)*. *Fig. 4B* shows the relative intensities of the densitometric detection of de bands and indicates the fold of activation over the respective control (P-Erks/Erks).

As can be seen, GluVO was about twice more potent than NapVO in the activation of Erks pathway. This effect correlates with the vanadium-induced growth inhibition in the osteosarcoma cell line.

CONCLUSIONS

NapVO and GluVO caused osteoblast cytotoxicity and morphological alterations with stronger effects on the tumoral than on non-transformed osteoblast-like cells. Induction of lipid peroxidation and specially the activation of Erks through their phosphorylation, may be involved in the antiproliferative and antidifferentiative effects of these compounds on tumoral osteoblasts. These results suggest that NapVO and GluVO are potential candidates for antineoplastic therapy.

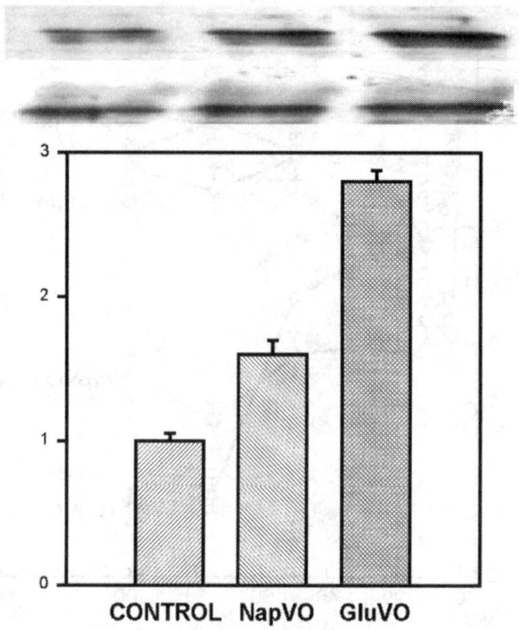

Fig. 4.

ACKNOWLEDGEMENTS

SBE is a member of the Carrera del Investigador, CONICET (Argentina). AMC is a member of the Carrera del Investigador, CICPBA. DAB is a fellowship from CICPBA. MSM is a fellowship from CICPBA. This work was supported by grants from UNLP and CONICET (PIP 1044/98).

REFERENCES

1. Etcheverry SB, Cortizo AM. Bioactivity of vanadium compounds on cells in culture. In: Nriagu JO, ed. *Vanadium in the Environment*. New York: John Wiley & Sons, Inc. 1998, Vol I: 359-394.
2. Schechter Y. Perspectives in Diabetes: Insulin mimetic effects of Vanadate, possible implications for future treatment of Diabetes. *Diabetes* 1990: 39: 1-5.
3. Thompson KH, McNeill JH, Orvig C. Vanadium Compounds as Insulin Mimics. *Chem Rev.* 1999: 99: 2561-2571.
4. Djordjevic C. Antitumor activity of vanadium compounds. In: Sigel H. And Sigel A. (eds) *Metal Ions Biol Syst.* New York: Marcel Dekker Inc. 1995: 31: 595-616.
5. Chatterjee M, Bishayee A. Vanadium-A new tool for cancer prevention. In: Nriagu JO, ed. *Vanadium in the Environment.* New York: John Wiley & Sons, Inc. 1998, Vol II: 347-390.
6. Cortizo MS, Alessandrini JL, Etcheverry SB, Cortizo AM. A vanadium/aspirin complex controlled release using a poly(beta-propiolactone) film. Effects on osteosacoma cells. *J Biomater Sci Polymer* Ed 2001: 12: 945-59.
7. Etcheverry SB, Williams PAM, Baran EJ. Synthesis and characterization of oxovanadium(IV) complexes with saccharides. *Carbohydr Res* 1997: 302: 131-138.
8. Etcheverry SB, Barrio DA, Cortizo AM, Williams PAM. Three new vanadyl(IV) complexes with non-steroideal anti-inflammatory drugs (Ibuprofen, Naproxen and Tolmetin). Bioactivity on osteoblast-like cells in culture. *J Inorg Biochem* 2002: 88: 94-100.
9. Etcheverry SB, Williams PAM, Sálice VC, Barrio DA, Ferrer EG, Cortizo AM: Biochemical properties and mechanism of action of a vanadyl(IV)-Aspirin complex on bone cell lines in culture. *Biometals* 2002: 15: 37-49.

XIV MISCELLANEOUS

*Metal Ions in Biology and Medicine: vol. 7.*Eds L. Khassanova, Ph. Collery, I. Maymard, Z. Khassanova, J.-C. Étienne.
John Libbey Eurotext, Paris © 2002 pp. 635-638.

Zinc intervention in the gene expression of IL-1 and TNF alpha of macrophages from mice in perinatal stages

AE Aguilar, MD Lastra, R Hernández, MA Cabañas, R Pastelin

Laboratorio de Investigación en Inmunología, Departamento de Biología, Facultad de Química, UNAM. Circuito Escolar, Ciudad Universitaria, México, D.F. CP 04510, tel/fax 56 22 37 40

INTRODUCTION

The role played by zinc in the immune system and the mechanisms through which it modulates diverse functions have been subjects of study for decades. Zinc status directly affects the monocyte-macrophage cells, that secrete interleukin-1 (IL-1) and tumor necrosis factor alpha (TNF alpha), among others cytokines. There is evidence that lipopolysaccharide (LPS) stimulated peripheral blood monocytes from preterm infants have a decreased secretion of IL-1 and TNF alpha.

We have previously shown that both phagocytosis and phagocytic metabolism increase significantly in animals supplemented with zinc.

IL-1 has multiple biological activities and is a regulator of the host response to infection and injury. Mononuclear phagocytes are the major source of IL-1, but all other nucleated cells are capable of producing it. IL-1 is a pleiotropic mediator of the host response to infections and injurious insults and it coordinates the activities of other cells and cytokines. TNF alpha with IL-1 set in motion an array of parallel signaling pathways, so exposure of cells and tissues to TNF results in a multitude of biological actions. Thus zinc intervention at the macrophage level could have a direct action of the mentioned cytokines production.

To further study zinc effects on the macrophages functions, we studied the interleukin-1 and TNF alpha gene expression and protein production in peritoneal macrophages from BALB/c mice supplemented with zinc from gestation to weaning.

MATERIALS AND METHODS

BALB/c mice were divided into groups according to zinc oral supplementation (500 mg/L) and periods of administration: from gestation to lactation (6 wk, group I), and from gestation to weaning (9 wk, group II), as formerly reported.

Semicuantitative RT-PCR was applied, cells being prepared as previously described. Total RNA was extracted by the Chomczynski & Sacchi method with a guanidinethiocyanate/phenol solution. After isolation the RNAs were reverse transcribed and amplifications were performed with the rTTh DNA polymerase enzyme and a commercial kit (Gene Amp PCR - Perkin-Elmer).Thirty cycles for IL-1 amplification and thirty five cycles for TNF-alpha amplification were performed in a Perkin-Elmer 2400 thermocycler. PCR products were separated by gel electrophoresis in 1.5% agarose with ethydium bromide. Polaroid photographs were taken.

The following primers were used:
Primer 5' IL-1 alpha: 5'-AAGTTTGTCATGAATGATTCCCTC-3'
Primer 3' IL-1 alpha: 5'- GTCTCACTACCTGTGATGAGT-3'
(Stratagene, La Jolla CA, USA)

Primer 5'TNF alpha:5'-TTCTGTCTACTGAACTTCGGGGTGATCGGTCC-3'Primer 3'TNF alpha:5'-GTATGAGATAGCAAATCGGCTGACGGTGTGGG-3'(Clontech Laboratories Inc,Palo Alto CA USA)

Macrophages were drawn from mice peritoneal cavity in sterile conditions after a 5 mL RPMI intraperitoneal injection for IL-1and TNF alpha evaluation. Macrophages (5×10^6 per well) were stimulated with 1 microgram LPS in polystirene microplates; microplates were incubated in a 5% CO_2 atmosphere at 37 C; supernatants were harvested at 24 h for IL-1, and at 48 h for TNF alpha analysis was performed by ELISA (Mouse ELISA kit Interleukin 1-alpha and TNF-alpha Genzyme).Statistical analysis was performed by ANOVA.

RESULTS

IL-1 Gene expression: RT-PCR for IL-1 showed a four fold increase in IL-1 gene expression (mRNA) at four hours post lipopolysaccharide (LPS) stimulation from group II mice *(fig. 1)*.

IL-1 in supernatant: IL-1alpha in macrophages supernatant. IL-1 concentrations in LPS activated supernatant presented an important increase (2 fold) in zinc supplemented mice from group I (6 wks zinc supplementation). The lengthening of zinc treatment to 9 wks, mice group II, brought a drastic descent on IL-1 concentrations well below the controls, going from 174 to 47 (deltas) *(fig. 2)*.

TNF-alpha gene expression: RT-PCR for TNF-alpha showed a four fold increase in mRNA, relative to controls, especially in mice group I, and a five fold increase relative to controls in mice from group II *(fig. 3)*.

TNF alpha in macrophages supernatant: TNF-alpha concentration assessed in LPS stimulated macrophages presented an increase that went from 818 pg/mL to 1275 pg/mL in mice group I; this tendency to increase persisted with the extension of the zinc treatment in mice group II, where the TNF alpha supernatant concentrations went from 2140 pg/mL to 2574 pg/mL, although in this instance the increase was not so pronounced *(fig. 4)*.

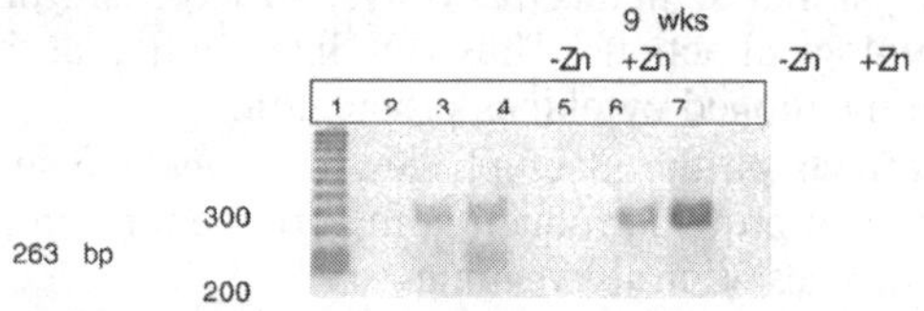

Fig. 1. Zinc (500 mg/L) effect on the IL-1 alpha gene expression in LPS stimulated macrophages from BALB/c mice; mRNA (263 bp cDNA) from mice macrophages (9 wks zinc treatment)
Lanes 3: -Zn -LPS, 4: +Zn -LPS, 7: -Zn +LPS, 8: +Zn +LPS; 1: MW markers

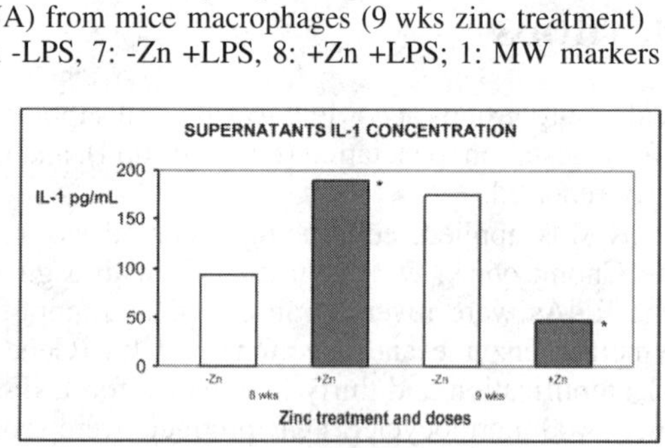

Fig. 2. Zinc supplementation (500 mg/L) effect on IL-1 secretion in LPS stimulated macrophages supernatant. Zinc supplementation provoked a significant increase in the IL-1 alpha concentration in mice group I, 6 wks zinc treatment. Zinc supplement provoked a significant decrease in the IL-1 concentration in mice group II, 9 wks zinc treatment. Results are shown as the media ± standard error in cytokines pg/mL, n=10
*Statistic significance ($p < 0.05$)

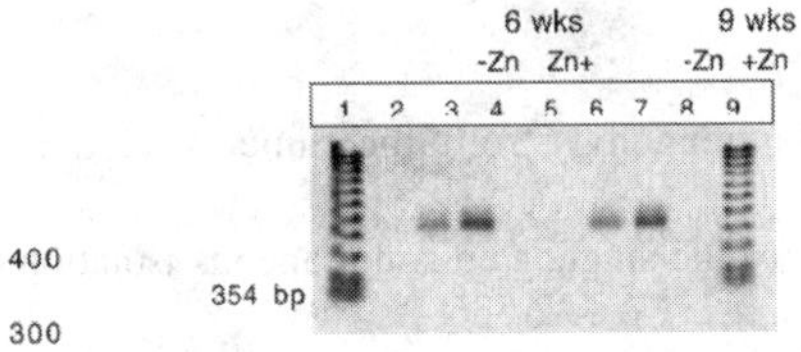

Fig. 3. Zinc (500 mg/L) effect on the TNF alpha gene expression in LPS stimulated macrophages from BALB/c mice; mRNA (354 bp cDNA) from mice macrophages with 6 (group I) and 9 wks (group II) zinc treatment
Lanes 3: -Zn group I; 4: +Zn group I; 7: -Zn group II; 8:+Zn group II;1 and 10: MW markers

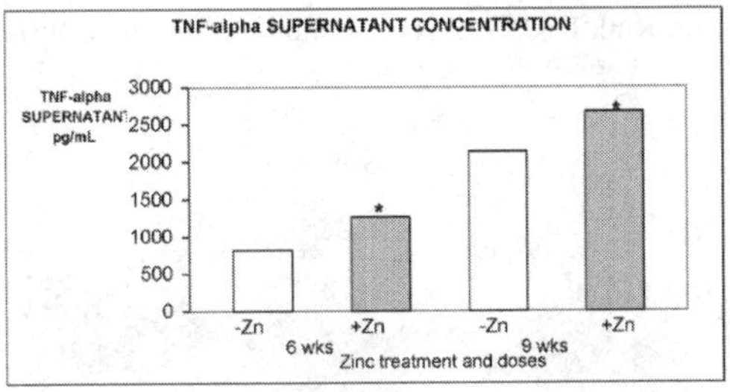

Fig. 4. Zinc supplementation (500 mg/L) on TNF alpha secretion in LPS stimulated macrophages supernatant. Zinc supplementation provoked significant increase in TNF alpha concentration both in mice group I, 6 wks zinc treatment and in mice group II, 9 wks treatment. Results are shown as the media ± standard error in cytokine pg/mL; n=10
*Statistic significance ($p<0.05$)

DISCUSSION AND CONCLUSIONS

In previous work we have reported low levels of IL-1 secretion in newborn animals, and its elevation in zinc treated animals. There is evidence that LPS stimulated peripheral blood monocytes from preterm and term infants, have a decreased secretion of IL-1 and when compared with cells from adults.

The zinc in vivo effect on the macrophages that we observed, resulted in a significant IL-1 increase (20%) in animals supplemented with zinc during the gestation and lactation stages and could determine other IL-1 dependent immune actions that upgrade the immune response.

The fact that zinc *in vivo* administration is evaluated in vitro by macrophages gene expression is of paramount importance since the zinc stimulation is provoking changes at this level and therefore, in the pro-inflammatory cytokines production as an important first line of defense.

On the other hand, the beneficial effects of TNF in resistance to infection are thought to be the result of modest amounts of TNF production at local sites of infection.

In our work TNF alpha production as well as the other cytokines were stimulated The increase was 35% which is indeed the largest we have observed, and with one peculiarity: the stimulation effect of zinc in vivo persisted trough the weaning period; this being the only situation in which zinc administration extension did not diminish dramatically the cytokine production.

The immunological activities dependent on IL-1 and TNF alpha elevate the mice immunological capacity. The fact that no secondary conditions that could be detrimental to mice physiology were observed in the zinc supplemented animals, is fundamental to the experiment..

The zinc mediated stimulation of the immune response by different mechanisms and particularly by the induction of monokines such as IL-1 and TNF-alpha, could favor the growth of the naturally low immune responses of newborns.

REFERENCES

Chandra RK (1999) Nutrition and immunology: from the clinic to cellular biology and back again *Proc Nutr Soc* 58: 681-683.

Chomczynski P (1993) A reagent for the single-step simultaneous isolation of RNA, DNA and proteins from cell and tissue samples *Bio Techniques* 15: 532-536.

Fraker PJ, King LE, Laakko T, Vollmer TL (2000) The dynamic link between the integrity of the immune system and zinc status *J Nutr* 130: 5S Suppl 1399S-1406S.

Frankenberger M, Sternsdorsf T, Pechumer H, Pforte A, Ziengler Heitrock HW (1996) Differential cytokine expression in human blood monocyte subpopulations: a polymerase chain reaction analysis *Blood* 87: 373-377.

Lastra MD, Pastelin R, Camacho A, Monroy B, Aguilar AE (2001) Zinc intervention on macrophages and lymphocytes response *J Trace Elem Med Biol* 15: 5-10.

Wellinghausen N, Kircher H and Rink L (1997) The Immunobiology of zinc *Immunol Today* 11: 80-82.

Metal Ions in Biology and Medicine: vol. 7. Eds L. Khassanova, Ph. Collery, I. Maymard, Z. Khassanova, J.-C. Étienne.
John Libbey Eurotext, Paris © 2002 pp. 639-644.

Sex and age related Ca/P ratio in trabecular bone of femoral neck of healthy humans

M. Tzaphlidou[1], V. Zaichick[2]

[1]*Medical Physics Laboratory, Medical School, Ioannina University, P.O. Box 1186, 45110 Ioannina, Greece;* [2]*Medical Radiological Research Center, 249020 Obninsk, Kaluga Region, Russia*

ABSTRACT

Concentrations of Ca and P as well as the Ca/P ratio were estimated in intact trabecular bone samples from the femoral neck of healthy humans, 34 women and 44 men, aged from 15 to 55 years using instrumental neutron activation analysis. Mean values (M ± S.D.) for the investigated parameters (on dry weight basis) for female and male were: 12.1 ± 3.0%, 5.94 ± 1.71%, 2.07 ± 0.25 and 10.9 ± 2.5%, 5.30 ± 1.23%, 2.07 ± 0.22 respectively. A statistically significant ($p \leq 0.05$) decrease of calcium concentration with age was found for male, but not for female. Concentration of phosphorus and the Ca/P ratio did not change with age. No statistically significant differences of the above parameters were observed related to sex. The mean values for Ca, P and Ca/P ratio were within a very wide range of published data and close to their median. The individual variation for Ca/P ratio in trabecular bone from the healthy human femoral neck was lower than those for Ca and P separately. This means that specificity of Ca/P ratio is better than those of Ca and P concentrations and may be more reliable for diagnosis of bone disorders.

INTRODUCTION

The incidence of bone diseases such as osteoporosis increases with advancing age and the fractures are usually the result of only relatively minor trauma. They represent a significant social and medical problem in terms of treatment and rehabilitation (Lewinnek et al 1980). Under these aspects, it is of particular importance the reliable early diagnosis of a bone disorder.

The most common way to evaluate the risk fracture is based on the assessment of bone mineral density by dual X-ray absorptiometry (DEXA). However, a recent report (Bolotin et al 2001) indicates that the data obtained by this method are not sufficient accurate.

A new method was demonstrated recently for assessing the skeletal Ca/P *in vivo*, using photon absorptiometry (Fountos et al 1997, 1998, 1999). Like dual X-ray absorptiometry (DEXA) for *in vivo* measurement of bone mineral density, the new system uses two photon energies. However, unlike (DEXA), this instrument is optimized for measuring the Ca/P ratio at a fixed site in the distal third of the right radius (Fountos et al 1999). A similar method has also been used for measuring the Ca/P ratio in a finger (Fountos et al 1997, 1998).

In order to evaluate the value of the new method for the *in vivo* determination of the skeletal Ca/P ratio in monitoring bone disorders and aging changes, the precision and accuracy of the technique must be sufficient enough to satisfy the requirement of being able to distinguish differences between study populations. The accuracy of this method is restricted mainly by the presence of fat, marrow and collagen in the measurement field. As judged by experiments, marrow and fat can be a source of considerable error (Fountos et al 1997, Bolotin et al 2001). To determine the accuracy of *in vivo* Ca/P ratio measurements we need to compare them with: 1) the real data for

Ca/P ratio deriving from intact bones, 2) the coefficient of variation (CV) of such data and 3) Ca/P ratio and the CV dependence on age.

The range of Ca/P values on normal bones in response to age and sex can be calculated with better accuracy by *in vitro* studies. There are a great number of reports on this point using chemical analysis techniques, atomic absorption spectrometry, neutron activation analysis etc. However, the majority of these data are based upon non intact bones. In most cases, bone samples are treated with solvents in order to remove collagen, fat, marrow and are then ashed and acid digested. There is evidence that by these methods some amount of Ca and P is lost and Ca/P ratio is disturbed by the treatment (Grynpas et al 1987, Edward 1990, Zaichick 1997, Zaichick et al 1998, Kuo et al 2000). In addition, by this way the measured concentrations of Ca and P are referred to the mineral part of the bone and not to the whole intact bone. Thus, data from untreated samples should be considered to be the ideal data for comparison with those derived from *in vivo* measurements.

In the present study trabecular bone, femoral neck samples from 15 to 55 years old people were analyzed with two objectives. One of the objectives was to use intact bones for Ca/P ratio measurements and to use these data as basis for *in vivo* studies. The second objective was to perform measurements on bone from regions mostly affected by bone disease. It is known that the turnover rate in trabecular bone is about four times greater than that in cortical bone (Hyvonen-Dabek, 1981). Thus any metabolic disturbance in Ca should have its earliest manifestation in trabecular bone.

MATERIALS AND METHODS

Samples of human femoral neck were obtained from cadavers (34 women and 44 men aged from 15 to 55 years) with intact bodies at postmortems within 24 hours of death. Each death was traumatic due to automobile accidents, falls, shootings, knifings, hanging, acute alcohol poisoning, and freezing. The samples were taken as a whole cross section of the femoral neck mainly from the right side. A tool made of titanium and plastic was used to clean samples of soft tissues and blood. Samples were freeze dried until constant mass was obtained. A titanium scalpel was used to cut trabecular bone samples. Then they were weighed and sealed in thin polyethylene films washed with acetone and rectified alcohol. The sample mass varied from about 50 mg to 200 mg. The sealed samples were placed in labeled polyethylene. The determination of Ca and P concentrations by neutron activation analysis as well as the statistics of results are described elsewhere (Zaichick 1997).

RESULTS

Our means for the Ca and P concentrations (Mean ± SEM) in the reference material IAEA H-5, 20.8±0.4% and 9.43±0.25%, were in good agreement with the means of the certified values, 21.2% and 10.2%, and within a 95% confidence intervals, 20.3-22.0% and 7.4-11.8%, respectively. This indicates an acceptable accuracy of the results obtained in our study of these elements in trabecular bone of the femoral neck.

Table 1 depicts our data of Ca, P concentrations as well as Ca/P ratio in the trabecular bone of femoral neck of healthy women and men in different age groups.

A comparison of published data with our results for Ca/P ratio is shown in *table 2*. The data in this Table were collected either from healthy people or from patients without bone disorders. Some values for the Ca/P ratios shown in the above Table were not given by the authors but were calculated by us using the mean concentration values given for Ca and P separately. Among the references in Table 2, data for Ca/P ratio were reported by Aitken (1976) and Grynpas et al (1987) only.

DISCUSSION

A tendency for reduced Ca and P concentrations by age in female and male was observed. However, statistically significant changes ($p \leq 0.05$) were found only in male Ca concentration *(table 1)*. In the same analysis, no statistically significant differences ($p > 0.05$) were detected in Ca and P between women and men. No significant differences ($p > 0.05$) were detected in Ca/P ratio related either to age or sex. The standard deviations (SD) and coefficients of variation (CV) found for Ca and P concentrations were respectively large *(table 1)*. This is due to the very wide individual variation of these values in intact trabecular bone.

The lower SD and CV obtained for Ca/P ratio *(table 1)* than for Ca and P separately is due to the strong correlation between Ca and P in bone. This correlation for healthy women and men aged 15 to 55 years is $r=0.829$ ($p<0.001$) and $r=0.896$ ($p<0.001$) respectively. This shows that the individual variation for Ca/P ratio in femoral trabecular bone of healthy humans is lower than that for Ca and P separately.

The majority of data shown in *table 2* are based upon techniques in which the sample is subjected to a number of treatments in order to remove fat and organic matter. Thus, the concentrations determined for Ca and P are referred to the mineral part of the bone and not to the intact bone. In spite of this, the means of Ca/P ratios range from 1.74 to 4.78, however, the median value, 2.14, is close to the stoichiometric value for hydroxyapatite (2.16). Our mean values for Ca/P ratio in femoral trabecular bone of healthy women and men (2.07 and 2.07) are on a level of 3.3% lower than the median value of published data and of 4.2% lower than the stoichiometric value for hydroxyapatite. To our opinion, the low value for Ca/P ratio obtained by us is due to the influence of the organic matter and particularly of red marrow. It is known (Hongve et al., 1996) that in proximal part of femur the organic part constitutes more than 90% of the whole fresh bone mass. The major component of this organic part of trabecular bone is red marrow. In this part of femur, red marrow constitutes up to 75% of the whole marrow (Brandt, 1977). The concentration of P in red marrow is high enough and about 20 times higher than in yellow marrow (Woodard et al., 1982). Thus, the whole trabecular bone of femoral neck contains more P than the inorganic part of this bone only. We reason that this is the factor that accounts for the lower Ca/P ratio obtained in trabecular bone of femoral neck in comparison with the characteristic value, 2.16, established for the inorganic part only.

We have shown that Ca/P ratio in trabecular bone of femoral neck is not affected by age. This is in agreement with data of Aitken (1976) for trabecular bone of whole femur. To our knowledge there is no published data referring to Ca/P ratio related to age in femoral trabecular bone.

CONCLUSIONS

A tendency for reduced Ca and P concentrations by age in female and male was observed. However, statistically significant changes ($p > 0.05$) were found only in male Ca concentration. No statistically significant differences ($p > 0.05$) were detected in Ca, P and Ca/P ratio related to sex. Also no significant differences were found in Ca/P ratio related to age. The mean values for Ca/P ratio in intact femoral trabecular bone of healthy women and men (2.07 and 2.07) were close to the median value of published data (2.14), some lower than the stoichiometric value for hydroxyapatite (2.16), and within a very wide range of published data (1.74-4.78). The individual variation for Ca/P ratio in trabecular bone from the healthy human femoral neck was lower than those for Ca and P separately. This means that specificity of Ca/P ratio is better than those of Ca and P concentrations and may be more reliable for diagnosis of bone disorders.

Table 1. Effect of age and sex on Ca and P concentrations as well as Ca/P ratio in trabecular bone of femoral neck of healthy women and men aged from 15 to 55 years

Sex	Age Years	n	Ca, % of dry bone weight			P, % of dry bone weight			Ca/P		
			Mean	SD	CV, %	Mean	SD	CV, %	Mean	SD	CV, %
Female	15-35	16	12.8	3.3	25.8	6.09	1.90	31.2	2.12	0.27	12.7
	36-55	18	11.4	2.7	23.7	5.80	1.57	27.1	2.01	0.22	10.9
	P(t-test)			Non-significant			Non-significant			Non-significant	
Male	15-35	21	11.8	2.8	23.7	5.63	1.37	24.3	2.12	0.21	9.9
	36-55	23	10.1	2.1	20.8	5.02	1.04	20.7	2.03	0.23	11.3
	p(t-test)		≤ 0.05		Non-significant			Non-significant			
Female	15-55	34	12.1	3.0	24.8	5.94	1.71	28.8	2.07	0.25	12.1
Male	15-55	44	10.9	2.5	22.9	5.30	1.23	23.2	2.07	0.22	10.6
	p(t-test)			Non-significant			Non-significant			Non-significant	

Table 2. Published data of Ca/P ratio in femoral trabecular bone of women and men

Reference	Method		n, Sex	Age years	Treatment of bone samples	Bone	Ca/P	
	Ca	P					Mean±SD	Range*
Aitken, 1976	AAS	Color	12?16?	30-90	Ach., AD	F	2.27±0.05	-
			12?	-			2.24±0.07	-
			16?	-			2.29±0.08	-
Gasenko et al., 1977	FES	Color	3?	20-40	Without F; Ach.; AD	EF	2.36	-
Grynpas et al., 1987	INAA	INAA	8	-	Intact	FH	2.17±0.09	-
			8	-	Without F&M		2.14±0.11	-
Akesson et al., 1994	EDX	EDX	2?,3?	40-76	Without F&M	FH	1.90	-
	INAA	INAA			Without F&M		1.74	-
	ICPES	ICPES			Without F&M; Ash.; AD		1.77	-
Zhang et al., 1996	PIXE	PIXE	5?	18-30	AD	FH	4.78	-
Our data	INAA	INAA	78	15-55	Intact	FN	2.07±0.23	1.55-2.72
			34?	15-55			2.07±0.25	1.55-2.45
			44?	15-55			2.07±0.22	1.57-2.72
All references (n=6)							Median 2.14	Range** 1.74-4.78

"-" - no information
AAS - atomic absorption spectrophotometry, Color - colorimetric method, FES - flame emission spectrometry, INAA - instrumental neutron activation analysis, EDX - energy-dispersive X-ray microanalysis, PIXE - proton induced X-ray emission, ICPES - inductively coupled plasma atomic emission spectrometry
F - fat, F&M fat and marrow, Ash. - ashing, AD - acid digestion
F - femur, EF - epiphysis of femur, FH - femoral head, FN - femoral neck
* - range of individual data
** - range of means

REFERENCES

Aitken, J.M. Factors affecting the distribution of zinc in the human skeleton. *Calcified Tissue Research*, 1976, 20, 23-30.

Akesson, K., Grynpas, M.D., Hancock, R.G.V., Odelius, R., Obrant, K.J. Energy-dispersive X-ray microanalysis of the bone mineral content in human trabecular bone: A comparison with ICPES and neutron activation analysis. *Calcified Tissue International*, 1994, 55, 236-239.

Bolotin, H.H., Sievanen, H. Inaccuracies inherent in dual-energy X-ray absorptiometry in vivo bone mineral density can seriously mislead diagnostic/prognostic interpretations of patient-specific bone fragility. *Journal of Bone and Mineral Research*, 2001, 16, 799-805.

Brandt, G. Mineral analytical and histometric studies on human bone marrow. *Med Welt*, 1977, 28, 1307-1311.

Edward, J. Ion exchange behavior of fresh human bone. *Journal of Radioanalytical Nuclear Chemistry, Letters*, 1990, 144, 317-322.

Fountos, G., Yasumura, S., Glaros, D. The skeletal calcium/phosphorus ratio: A new in vivo method of determination. *Medical Physics*, 1997, 24, 1303-1310.

Fountos, G., Kounadi, E., Tzaphlidou, M., Yasumura, S., Glaros, D. The effect of inflammation-mediated osteoporosis (IMO) on the skeletal Ca/P ratio and on the structure of rabbit bone and skin collagen. *Applied Radiation and Isotopes*, 1998, 49, 657-679.

Fountos, G., Tzaphlidou, M., Kounadi, E., Glaros, D. In vivo measurement of radius calcium/phosphorus ratio by X-ray absorptiometry. *Applied Radiation and Isotopes*, 1999, 51, 273-278.

Gasenko, O.G., Prohonchukov, A.A., Panikarovsky, B.B., Tigranyan, R.A., Kolesnik, A.G., Pahomov, G.N., Grigoryan, A.C., Antipova, Z.P., Remezov, S.M., Komissarova, N.A.. Condition of microscopic and crystal structures, microhardness, and minerals of human bone after long space flight. *Kosmicheskaya biologiya and aviakosmicheskaya medicina*, 1977, 11, 11-20.

Grynpas, M.D., Pritzker, K.P., Hancock, R.G.V. Neutron activation analysis of bulk and selected trace elements in bone using low flux SLOWPOKE reactor. *Biological Trace Element Research*, 1987, 13, 333-344.

Hongve, D., Johansen, S., Andruchow, E., Bjertness, E., Becher, G., Alexander, J. Determination of aluminium in samples from bone and liver of elderly norwegians. *Journal of Trace Elements in Medicine & Biology*, 1996, 10, 6-11.

Hyvonen-Dabek, M. Proton-induced prompt gamma-ray emission for determination of light elements in human bone. *Journal of Radioanalytical Chemistry*, 1981, 63, 367-378.

Kuo, H.W., Kuo, S.M., Chou, C.H., Lee, T.C. Determination of 14 elements in Taiwanese bones. *The Science of the Total Environment*, 2000, 255, 45-54.

Lewinnek, G.E., Kelsey, J., White, A.A., Krieger, N.J. The significance and a comparative analysis of the epidemiology of hip fractures. *Clinical Orthopaedics and Related Research*, 1980, 152, 35-43.

Woodard H.Q., White D.R. Bone models for use in radiotherapy dosimetry. *British Journal of Radiology*, 1982, 55, 277-282.

Zaichick, V. Sampling, sample storage and preparation of biomaterials for INAA in clinical medicine, occupational and environmental health. In: Harmonization of Health-Related Environmental Measurements Using Nuclear and Isotopic Techniques. *International Atomic Energy Agency, Vienna*, 1997, pp. 123-133.

Zaichick, V., Zaichick, S. INAA application for the assessment of chemical element losses under dry ashing of biological materials. In: International Conference on Nuclear Analytical Methods in the Life Sciences, *Beijing, China*, 1998, pp. 95-97.

Zhang, Y., Zhang, Y., Tong, Y., Qiu, S., Wu, X., Dai, K. Multi-element determination in cancellous bone of human femoral head by PIXE. *Journal Radioanalytical Nuclear Chemistry, Letters*, 1996, 212, 341-351.

*Metal Ions in Biology and Medicine: vol. 7.*Eds L. Khassanova, Ph. Collery, I. Maymard, Z. Khassanova, J.-C. Étienne.
John Libbey Eurotext, Paris © 2002 pp. 645-646.

Influence of low doses irradiation on accumulation of heavy metals in the organism and biological effects in rats

Ivanov S.D., Semenov V.V.*, Kovanko E.G., Yamshanov V.A.

Central Research Institute of Roentgenology and Radiology, Leningradskaya st., 70/4, Pesochny, 197758 St.-Petersburg, Russia
**FGUNPP "Geologorazvedka", Knipovich st. 11/2, 193019 St.-Petersburg, Russia*

ABSTRACT

Recent publications consider mainly results of investigations after combined actions of heavy metal salts and irradiation with doses significantly higher than the permissible ones, however the data absent practically after action of these environmental injuring factors with low doses, that can be met in everyday life.

The purpose of the present investigation is to reveal the consequences of the low doses ionising irradiation exposures on the lead and mercury content in blood and kidney of animals in case of the metal salt prolonged administration with drinking water in comparison with biotesting data of the early and delayed consequences after action of the same injuring factors.

MATERIALS AND METHODS

The investigation was performed on the 225 male rats receiving with drinking water $Hg_2(NO_3)_2$ or $PbCl_2$ in concentrations - 0.7 mcg/l or 2.0 mg/l in account to metal, respectively, during 3 weeks before and 1 month after single total gamma-irradiation with 25 or 50 cGy doses on the "IGUR-1" unit (^{137}Cs, dose rate = 44 cGy/min). The metal concentration measurements in blood, kidney and fodder were performed for mercury - on the atomic-absorption analyser of mercury "RA-915+" with "RP-91" block (made in Russia), and for lead - by an inversed volt-ampere measurement on the ion-scanning system "ISS-82" (made by firm "Radiometer", Denmark).

There were studied the early (24 hrs after irradiation) hematological reactions and genotoxic effects on the base of the blood leukocyte DNA alterations, results of their restorations (30 days after irradiation), as well as delayed consequences - life span shortening. DNA content in leukocytes was determined occuringly during one's lifetime by use of 4',6-diamidine-2-phenylindole (DAPI), and the DNA structure alterations by means of two-parameters fluorescent analysis, including DAPI and ethidium bromide application, as described earlier [Ivanov S.D. et al., 1999].

Besides, the indexes of the natural behaviour activity were studied as a time alterations in search of food in the T-imagery two-branch labyrinth by experimental rats in comparison with the same index of control animals in 1 or 3 months after irradiation. The rats starved for 2 days before this testing to give reasons for food.

The animals were under observation during all life.

RESULTS

Mercury concentration measurements in blood in 24 hrs, 2 weeks and 30 days after irradiation have revealed no significant alterations in the toxicant content in experimental groups of animals in comparison with intact control after both isolated and combined action of injuring factors. At the same time mercury content determinations in kidney of rats in 30 days after irradiation have shown a significant increase of toxicant concentrations only in rats exposed to the radiation-mercury influence (approximately in equal degree after various doses - by 20% if compared with intact control). Such increase of the mercury content was not observed in the animal groups receiving mercury without irradiation.

Neither significant hematotoxic nor genotoxic reactions were revealed in early term after administration of the used mercury salt with low doses in drinking water without irradiation. The early genotoxic effect was revealed in case of combined action only after the mercury with low concentration and irradiation with 25 cGy dose. The significant genotoxic reactions were not observed in 30 days after isolated irradiation or toxicant action. However, the blood leukocyte DNA content reduced by 45% - 34% after combined action of these agents (mercury + 25 cGy) or (mercury + 50 cGy), respectively. The biotesting index alterations did not correlate with the mercury concentration measurement data in blood and kidney.

In 3 months after combined action of mercury with low concentration and 25 cGy irradiation more than doubling of the time in search of food in labyrinth by experimental rats after radio-mercury exposure was observed in comparison with the index of animals in control group ($p<0.01$).

The observation of rat death dynamics has shown that mercury salt administration with permissible concentration resulted in no significant changes of the mortality in comparison with intact control. At the same time, acceleration of animal death was observed after irradiation with 25 cGy dose in rats receiving mercury salt in drinking water both in comparison with intact control group, and with the irradiated group (mean life span was reduced approximately by 15%). In case of irradiation with 50 cGy dose life span of animals, receiving mercury salt, reduced if compared only with the irradiated rats, but not with the intact control group.

In case of lead administration, the irradiation with 50 cGy dose resulted in additional accumulation of lead in kidney, but not in blood of rats. There were observed adaptive reactions as a result of prolonged lead administration with 2.0 mg/l dose in drinking water.

Assessment of natural behaviour activity revealed, that there was a belt of hypersensitivity (corresponding to irradiation with 25 cGy dose in case of lead administration) within the range of radiation-chemical exposures with low doses. The biological effect of combined exposures with low doses can be only approximately appreciated on the base of blood lead measurement, since only the tendency (but not significant correlation) was revealed in relationship between alterations of these indexes.

Significant life span shortening was not observed in rats after combined radiation-lead exposures in comparison with the same indexes of control group animals.

CONCLUSION

So, ionising radiation with low doses (25 and 50 cGy) allows to potentiate the accumulation of mercury in kidney, but not in blood of rats. The prolonged exposures of heavy metal salts with low concentrations combined with the single irradiation with 25 cGy dose resulted in alterations of the natural behaviour activity, and life span shortening in case of the mercury administration and irradiation with 25 cGy dose. Moreover, there was revealed the belt of hypersensitivity, corresponding to irradiation with 25 cGy dose combined with chemical toxicant administration, since the combined exposure in higher dose (heavy metal + 50 cGy) resulted in the lesser damage of the organism. On the base of only alone metal biomonitoring in blood of animals it is impossible to prognosticate the early and delayed biological effects of radio-chemical exposures with low doses. At the same time, the genotoxicity biotesting allowed to evaluate the results of postradiation restoration, that may be used for prediction of delayed consequences after combined action of mercury and irradiation.

REFERENCES

1. Ivanov S.D., Kovanko E.G., Popovich I.G., Zabezhinski M.A. *Radiation Biology. Radioecology.* 1999.-v.39, No 4.-p. 418-424 (in Russian).

*Metal Ions in Biology and Medicine: vol. 7.*Eds L. Khassanova, Ph. Collery, I. Maymard, Z. Khassanova, J.-C. Étienne.
John Libbey Eurotext, Paris © 2002 pp. 647-650.

Copper geoecology in the lake superior province, North America

M. Harthill[1] and E.I. Robbins[2]

[1]*U.S. Geological Survey, Reston VA 20192*
[2]*San Diego State University, Department of Geological Sciences, San Diego, CA 92182*

ABSTRACT

Fossil occurrences, dating from 2.6-2.75 billion years ago (Ga), within the Lake Superior Copper Province, initiated speculation on the historic role of Cu in biological systems and how that essential micronutrient is acquired from the geologic environment. Precambrian rocks, organic tissues, and mine pool water microbiology, electrochemistry, and isotopes were analyzed to attempt to trace copper pathways through time. Contemporary research by others indicates that Cu is bioavailable now, as presumably in the past.

Background

The North American Lake Superior region contains a world-class copper (Cu) province nearly 300,000 km^2. Geologically, the area hosts a variety of life forms, including 2.6-2.75 billion year old microbial fossils.

Aim

Although Cu origins in this region were metagenic and epigenic, palynological and microbial research (Robbins, E.I., Tectonophysics, 94:633, 1983; Robbins, E.I. et al. Geol. Assoc. Canada/ Mineralogical Assoc. Canada Annual Mtg., p. A94, 1994) prompts discussion about the concurrence of Cu and biota in the Province. Cu, an essential miconutrient, provides catalytic and electron transfer functions in almost every known group of organisms. Excess biological Cu is normally excreted, and some bacteria can precipitate excess Cu externally. As an organism enters the fossil record, its metal components accompany it to deposition (Honjo, S., Oceanographic Research Papers, 29:609, 1982).

Methods and results

Interpreting relationships between fossil organisms and their associated metals relies on a variety of microtechniques, including palynological analysis, scanning electron microscopy, and isotopic analysis. Black opaque Cu sulfide mineral(s) were shown enmeshed in organic tissue fragments within regional black shale, and Cu mineralization accompanies microfossils of bacteria and fossilized algal filaments and probable fecal pellets of simple multicellular animals (Robbins, E.I. et al., Proc. Nat. Acad. Sci., 82:5809, 1985), fossil cyanobacterial sheaths, fungal hyphae-sized tissues, and aquatic algae. Despite multiple physical sources of Cu, Zhu (Zhu, X.K., Chem. Geol., 163:139, 2000) interprets wide isotopic variability as biological fractionation at low temperatures. Underground, brine pools are coated with films of Cu^{2+} minerals and petroleum; the bacterial role in forming these oxidized Cu chloride minerals is presently under study.

Conclusions

Research thus suggests biological interactions with Cu over geologic time. Current redox conditions and microbial action may contribute to elevated Cu in food stuffs and has raised concern that additional bioavailable Cu may react synergistically with other metal ions to exceed toxicity thresholds.

INTRODUCTION

The North American Lake Superior region contains a world-class copper province nearly 300,000 km^2. Literature (Robbins and Harthill, in press) indicates that crustal and sedimentary rocks range in Cu concentration from 24-45 mg/kg world-wide, although marine black shales have concentrations up to 300 mg/kg. Important copper minerals in this region are native copper (Cu^0), chalcopyrite ($CuFeS_2$), chalcocite (Cu_2S), bornite (Cu_5S_4), domeykite (Cu_3As), and malachite [$Cu_2(CO_3)(OH)_2$]. Soils from Cambrian sandstones in the Pictured Rocks Lakeshore area have assayed 150 ppm (Shacklette, 1967).

Dissolved bioavailable Cu is cuprous (Cu^+) in dilute waters (Cu<1 ppm) or anoxic water; cupric (Cu^{++}) in acid or neutral oxygenated water; and $HCuO_2^-$ in oxygenated alkaline water (pH >8.5) (Pourbaix (1996). Estimates of worldwide concentrations of copper vary from 0.2 to 30 µg/L in fresh water and 0.05 to 12 µg/L in seawater. Concentrations of total dissolved Cu in Lake Superior (pH 8) range from 1.4 to 6.9 µg/L (Weiler, 1978). At depths below 200 m today, the ground water of the Upper Peninsula of Michigan is a saline brine with Cl^- concentrations >15% (Kelly et al., 1986) and Cu concentrations as much as 7 ppm (Robbins et al., 1994). This paper focuses on rocks that couple Cu with carbonaceous remains or microfossils.

MATERIALS AND METHODS

Interpreting relationships between fossil organisms and metals enclosed within their tissues relies on various microtechniques, including palynological analysis using light microscopy, scanning electron microscopy (SEM), and isotopic analysis. We used the base of data in the literature, including our prior and current work, to trace the flow of Cu, however imperfectly, through time.

RESULTS

Earliest organic carbon within the Lake Superior region appears in the Early Proterozoic (2.6-2.75 Ga); metasedimentary marine rocks were found to contain organic carbon that may have been bacteria or cyanobacteria (Darby, 1996). Microtechniques have yet to be applied to determine if Cu mineralization is related to these *in situ* organic carbon layers. In the Wabigoon Subprovince, the Ontario Atikokan area has chalcopyrite and spectacular domed dolomitic stromatolites created by sediment-coated cyanobacteria, possibly assisted by other bacterial groups (Nisbet and Wilkes, 1987). Pyrite with biogenic sulfur values occurs in horizons above these stromatolites (Strauss, 1986). Copper sulfide mineralization associated with the Kona, Negaunee, and Michigamme Formations has organic carbon in the form of stromatolites (Nordeng, 1963) and algal filaments (Han and Runnegar, 1992). In the Animikie Basin of Minnesota and Ontario, the Rove Formation has Cu mineralization and microfossils of cyanobacteria, pellet-shaped structures that may represent the oldest fecal pellets of simple multicellular animals (Robbins et al., 1985), and possible a worm (Edhorn, 1973).

Tectonic activity of the Middle Proterozoic (0.9-1.6 Ga) (Cannon, 1992), resulted in Cu deposits along with high biological productivity in a rift valley lake. The black shale of the Nonesuch Formation lakebeds contains petroleum, and fossil cyanobacterial sheaths, fungal hyphae-sized tissues, and aquatic algae (Robbins, 1983). The variety of pellet-shaped microfossils (Robbins et al., 1985) and spore triads (Strother, 1986) implies microscopic animals may have lived and reproduced in the water column. Chalcocite enmeshed within organic tissues in the shale (Alyanak and Vogel, 1974; Robbins, 1983) indicates phytoplankton and/or their microbial degraders may have interacted with Cu in the lake water.

Organic matter in the Nonesuch deposits has been buried to petroleum generating temperatures, and Robbins (1985) suggested that catalytic metals such as Cu released from metallo-proteins may have entered into petroleum-generating reactions. Cu concentrations in heavy petroleum and bitumen there assay 2-4 times that of surrounding reservoir rocks and sediments, indicating either higher bioaccumulation (Vasconcelos et al., 2001) and/or greater retention of Cu given local environmental conditions.

Isotopic variations (as δCu^{65} in parts per thousand, ‰) provide evidence of biological fractionation in rocks of the Lake Superior province. Fractionation has been measured on native copper (-1.8 ‰), on chalcocite and on domeykite (0.0 to +1.4 ‰), and in host rock shales/ siltstones/ sandstones (-3.6 to +1.4 ‰) (Zhu et al., 2000). Although multiple physical sources of Cu probably existed, Zhu et al. (2000) interpreted this wide isotopic variability as ancient biological fractionation at low temperatures.

Beginning about 2 million years ago, continental glaciers, perhaps 1000m thick, covered this northern landscape, scraped the land, redeposited boulders containing native copper, and removed evidence of any interactions between Pleistocene organisms and copper.

Pre-European vegetation in the Lake Superior basin was characterized as a conifer-hardwood forest. Where Cu-bearing rocks weather at the surface, conifers can concentrate Cu in tissues (to 700 ppm) and exudates (to 1,500 ppm) above background soil (70 ppm) (Curtin and King, 1986). Shacklette (1967) measured 1.5 weight % Cu in the ash of a copper-accumulating moss that grew on a soil derived from a Cambrian sandstone (150 ppm Cu) in the area. Wild rice in non-contaminated lakes in the Lake Superior region naturally concentrates copper in edible seeds (5.3 ppm), in stems (1.4 ppm) and roots (4.8 ppm) (Bennett et al., 2000).

Chalcocite ore (~1.1% Cu) within the Nonesuch shale was mined at the White Pine Mine, Michigan, and brine pools in the mine were coated with films of blue-green copper minerals and petroleum that drips along roof bolts 820 m underground. The green color of the films was due to the copper chloride minerals, atacamite and paratacamite [$Cu_2Cl(OH)_3$], which surround Gram negative bacterial rods and filaments, leading Robbins et al. (1994) to suggest that modern bacteria participate in the formation of these floating copper minerals from saline mine water. These minerals are enriched in C^{13} relative to the local petroleum, the likely carbon source for heterotrophic microbial activity. Electrochemical studies by Sato (in Robbins et al., 1994) determined that paratacamite group was the stable phase given the oxidizing conditions in the mine water. The data show no redox boundaries, so energy is not available to support autotrophic life. The bacterial role in forming the Cu chloride minerals is presently under study.

Toxicological studies undertaken in Lake Superior showed elevated copper in fish and other aquatic natural resources (Lucas et al., 1970). Fish and wild rice are staples of the Ojibwa Indian subsistence populations, and there is the fear that copper accumulations might reach concentrations toxic to these Native Americans (Joseph M. Rose, Sr., pers. commun., 1993).

DISCUSSION AND CONCLUSIONS

The co-occurrence of copper and fossil biota through time raises interesting questions about evolution. Copper is an essential micronutrient and cofactor contributing to copper-associated polypeptides that provide catalytic and electron transfer functions in almost every known group of organisms alive today from bacteria to humans (Linder and Hazegh-Azam, 1996). As in the geosphere, biological copper is associated with other metals, such as As, Co, Fe, Mo, Ni, Pd, Pt and Zn, each with toxicity thresholds and synergisms of their own. A major concern is that additional bioavailable concentrations of copper may increase above toxicity thresholds to the biota living in the area, some of which are food staples of indigenous Americans.

REFERENCES

Alyanak, N., Vogel, T.A. Framboidal chalcocite from White Pine, Michigan. *Econ. Geol.*, 1974, 69, 697-703.

Bennett, J.P., Chiriboga, E., Coleman, J., Walter, D.M. Heavy metals in wild rice from northern Wisconsin. *Science Total Environment*, 2000, 246, 261-269.

Cannon, W.F. The Midcontinent rift in the Lake Superior region with emphasis on its geodynamic evolution. *Tectonophysics*, 1992, 213, 41-48.

Curtin, G.C., King, H.D. Utility of mull in geochemical exploration, in Carlisle, D., Berry, W.L., Kaplan, I.R., Watterson, J.R., eds. *Mineral Exploration: Biological Systems and Organic Matter*. Englewood Cliffs, New Jersey, Prentice-Hall, 1986, p. 356-376.

Darby, D.G. Evidences of Precambrian life in Minnesota, in Sims, P.K., Morey, G.B. eds., *Geology of Minnesota, A Centennial Volume*. St. Paul, Minnesota Geological Survey, 1996, 264-271.

Edhorn, A.S. Further investigations of fossils from the Animikie, Thunder Bay, Ontario. *Proceedings Geological Association Canada*, 1973, 25, 37-66.

Han, T.M., Runnegar, B. Megasopic eukaryotic algae from the 2.1-billion-year-old Negaunee Iron-Formation, Michigan. *Science*, 1992, 257, 232-235.

Linder, M.C., Hazegh-Azam, M. Copper biochemistry and molecular biology. *American Journal Clinical Nutrition*, 1996, 63, 797S-811S.

Lucas, H.F., Edgington, D.N., Colby, P.J. Concentrations of trace elements in Great Lakes fishes. *Journal Fisheries Research Board Canada*, 1970, 27, 677-684.

Nisbet, E.G., Wilks, M.E. Archaean stromatolite reef at Steep Rock Lake, Atikokan, northwestern Ontario. *Canadian Society Petroleum Geology Memoir*, 1987, 13, 89-92.

Nordeng, S.C. Precambrian stromatolites as indicators of polar shift, in Polar Wandering and Continental Drift. *Society Economic Paleontologists Mineralogists, Special Publication*, 1963, 131-139.

Pourbaix, M. Atlas of electrochemical equilibria in aqueous solutions. New York, Pergamon, 1966, 384-392.

Robbins, E.I. Accumulation of fossil fuels and metallic minerals in active and ancient rifts. *Tectonophysics*, 1983, 94, 633-658.

Robbins, E.I. Petroleum as an ore-bearing fluid: A hypothesis. American Association of Petroleum Geologists, NE Section meeting, Williamsburg, VA, *American Association Petroleum Geologists Bulletin*, 1985, 69(9), 1446.

Robbins, E.I., Harthill, M. Life in a Copper Province, in Selinus, O., ed. *Medical Geology*, Academic, in press.

Robbins, E.I., Porter, K.G., Haberyan, K.A. Pellet microfossils: Possible evidence for metazoan life in Early Proterozoic time. *Proc. National Academy Science*, 1985, 82, 5809-5813.

Robbins, E.I., Stanton, M.R., Tilk, J.E., Congdon, R.E., Evans, H.T., Jr., Gullett, C.D., Sanders, M.B., Sato, M., Schaef, H.T., Seal, R.R., II Association of microbes with authigenic copper-chloride mineral films (atacamite and paratacamite) and petroleum residue at depth in the White Pine copper mine. *Waterloo'94, Program with Abstracts*, Geological Assoc. Canada and Mineralogical Assoc. Canada, Annual Meeting, 1994, A94.

Shacklette, H.T. Copper mosses as indicators of metal concentrations. *U.S. Geological Survey Bulletin B-1198-G*, 1967, G1-G18.

Strauss, H. Carbon and sulfur isotopes in Precambrian sediments from the Canadian Shield. *Geochimica Cosmochimica Acta*, 1986, 50, 2653-2662.

Strother, P.K. Palynomorphs from the copper-bearing Nonesuch Formation. *Palynology*, 1986, 19, 35.

Vasconcelos, T.M., Leal, F.M. Adsorption and uptake of Cu by *Emiliania huxleyi* in natural seawater. *Environmental Science Technology*, 2001, 35, 508-15.

Weiler, R.R. Chemistry of Lake Superior. *Journal Great Lakes Research*, 1978, 4, 70-385.

Zhu, X.K., O'Nions, R.K., Guo, Y., Belshaw, N.S., Rickard, D. Determination of natural Cu-isotope variation by plasma-source mass spectrometry; implications for use as geochemical tracers. *Chemical Geology*, 2000, 163, 139-149.

*Metal Ions in Biology and Medicine: vol. 7.*Eds L. Khassanova, Ph. Collery, I. Maymard, Z. Khassanova, J.-C. Étienne.
John Libbey Eurotext, Paris © 2002 pp. 651-654.

Sensitivity of bacterial biofilms settled on copper tubing to toxic shocks induced by copper ions, with or without previous exposure

L.F. Jacquelin[2], L.A. Khassanova[3], F. Arsac[4], M.P. Gelle [2], C. Choisy[1]

[1]*Microbiologie UFR Pharmacie, 51, rue Cognacq-Jay, 51096 REIMS-CEDEX, France;* [2]*UFR Odontologie, 2, rue Général Koenig, 51100 Reims, Cedex, France;* [3]*Bashkir University, 32 Frunze Street, 450025 UFA, Russia;* [4]*Toxicologie, UFR Pharmacie, 51, rue Cognacq-Jay, 51096 Reims, Cedex, France*

INTRODUCTION

Copper has been known for long for its famous antiseptic properties. As of today, it is incorporated in many pharmaceutical formulations,used as biocid or oligo-element.

Copper now replaces lead in most tubing. Fight against nosocomial infections commitees recommend the use of copper instead of PVC tubing. Yet, the question remains as to whether Cu_0 can stop its own colonization by nosocomial bacteria.

Moreover, maintenance work or biocidal ionization of the water network, involving acids or electrolysis, produces a release of copper salts. Are biofilms settled on copper tubes sensitive or resistant to the toxic shock induced by copper salts (Cu^{++})?

Lastly, water is often acidic. This property may be responsible for a slow yet continuous release of Cu^{++} ions which could induce either an adaptation or a selection. Will copper biofilm bacteria become more sensitive or more resistant to future toxic shocks?

To study the efficacy of Cu^{++} tubing as an inhibitor of biofilms, we compared: 1. The attachment of biofilm on Tygon^R and on copper tubing: monospecific biofilms were made with 2 gram-negative and 2 gram-positive bacterial species. 2. The behaviour of these biofilms with and whithout previous adaptation to a low concentration of copper ions in conditions of high and low toxic shock.

MATERIALS AND METHODS

Strains : *Staphylococcus aureus* ATCC 6538. Slime + with coagulase-negative *Staphylococcus* (CNS), *Pseudomonas aeruginosa* ATCC 15442, *Escherichia coli* ATCC 10536. 3 strains are representative of nosocomial strains. *E. coli* was chosen because of its past use as a test strain in the antiseptic activity assay as defined by the standards methods of AFNOR. *P. aeruginosa* is the test strain according to european standards (CEN).

Pipes : Tygon^R pipe, alimentary PVC, inner diameter: 6,5 mm, outer: 9,5 mm. Copper pipes, inner diameter: 5 mm, outer: 8 mm.

Micro-organisms cultures [1] **:** Planctonic bacterria, grown in Brain Heart Infusion (BHI) Monospecific biofilms were prepared as follows: each strain was subcultured twice in Brain Heart Agar (BHA) then inoculated in adhesion medium: BHI 20 mL, Biofilm medium 20 mL, $MgSO_4,7H_2O$ 20g/L 1%, glucose 1%, water q.s.p. 1000mL. Biofilm medium: yeast extract 1g, casaminoacids 1 g, KH_2PO_4 5g, Na_2HPO_4 12,5 g, water q.s.p. 1000 mL. After 30 min. at +37°C adhesion medium was removed and replaced by culture medium, diluted one fourth.

For Cu $^{++}$ adaptation, culture medium was added with Cu SO_4 to obtain a final concentration of 20 µM/L of Cu^{++} in the medium. Biofilms were used after 24 H at +37°C.

Toxic shocks were performed on one-day-old biofilms with or without copper adaptation. High toxic shock (HTS), final concentration 150 µM/L Cu^{++}, Low toxic shock (LTS) 50 µM/L.

Bacterial count: 3 rinsing in water, followed by transfer into 5mL of water. Removal of bacteria cells by immersion of test-tube in an ultrasonic bath for 1 min., then Vortex agitation, 10 pulses. Serial dilutions and inoculation with Spiral^R System on BHA in Petri dishes. Incubation, 18 to 24 h at +37°C. Calculation of log CFU/mm^2.

Cu^{++} titration: Atomic absorption Spectrometry.

RESULTS

• *S. aureus* biofilms grown onTygon^R or copper, with or without Cu^{++} ons were comparable. This strain is poorly sensitive to HTS, but previous exposure to Cu^{++} sensitized biofilms to toxic shocks.

• Slime + CNS was very sensitive to Cu_0, but poorly to HTS. However, previous exposure to Cu^{++} sensitized biofilms to toxic shocks.

• *P. aeruginosa* appeared to be the most sensitive to HTS, but adaptation to Cu^{++} increase its resistance to toxic shock.

• *E. coli* was as sensitive as *P. aeruginosa* to HTS, but previous exposure to Cu^{++} increased its sensitivity instead of resistance *(fig. 1)*.

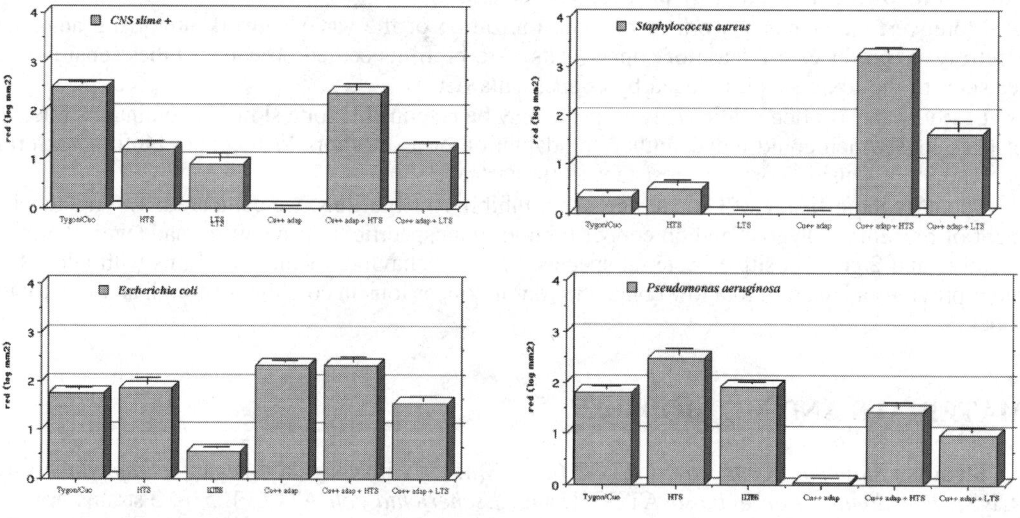

Fig. 1. Reduction, in log/mm^2 of bacteria number in biofilms
Column 1: Tygon® versus copper - other columns: assay versus control biofilm on copper
CNS: Coagulase Negative Staphylococcus HTS: High Toxic Shock LTS: Low Toxic Shock

Table 1 compares sensitivity to Cu^{++} bacteria and biofilms bacteria in different conditions.

Table 2 allows to see the different abilities to concentrate Cu^{++} ions by *S. aureus* and *E. coli.*

DISCUSSION

All strains colonized less copper than Tygon^R. This phenomenon may be due to other reasons than the plain Cu_0 toxicity: it is possible that some strains are able to hydrolyse Tygon^R, finding nutriments in numerous substances included in PVC. If this hypothesis is true, colonization of foreign bodies (prothesis or invasive devices) may lead to a double risk, toxic and infectious.

Table 1. Sensitivity to copper ions (in log/mm^2)

	Planctonic bacteria		Tygon® biofilm		Copper biofilm	
Control	9,20	-	5,40	-	3,65	-
HTS	7,80	1,40	3,39	2,01	1,80	1,85
LTS	9,00	0,20	3,40	2,00	3,10	0,55
Cu^{++} adaptation	7,80	1,40	4,79	1,61	1,32	2,33

Table 2. Intrabacterial Cu content in *S. aureus* and *E. coli*

S. aureus			E. coli	
µg/cell	Increase/control		µg/cell	Increase/control
$2,68.10^{-10}$	-	Control	$8,23.10^{-11}$	-
$7,40.10^{-5}$	2.10^5	Cu^{++} adaptation	$1,22.10^{-1}$	10^{10}
$1,50.10^{-2}$	2.10^8	HTS	6,56	10^{11}
$6,90.10^{-1}$	2.10^9	Cu^{++} adapt + HTS	$7,05.10^{-2}$	10^9
$6,85.10^{-5}$	2.10^5	LTS	$1,40.10^{-1}$	10^{10}
$5,18.10^{-3}$	2.10^7	Cu^{++} adapt + LTS	$5,90.10^{-1}$	10^{10}

Biofilms specialists are numerous to estimate that establishment of a biofilm ceates a stress in bacteria; this stress protects them against other agressions, such as toxic activity, disinfectant activity [5], dehydration, increase of salinity [6, 9] etc. This may be proved by visualisation of new proteins in SDS-Page electrophoresis [2, 4].

In these experiments, planctonic seemed less sensitive to toxic stress than bacteria settled in biofilms. It may be an effect of number [3].

For *E. coli*, copper adaptation increases by a factor of 10^{10} the intrabacterial copper concentration, 10^{11} in HTS and 10^{10} in LTS. In contrast, previous exposure to copper decreases copper accumulation after a high toxic shock (10^9).

For copper adapted *S. aureus*, the intrabacterial copper concentration is multiplied by a bacteria settled in biofilms. It may be an effect of number [3].

For *E. coli*, copper adaptation increases by a factor of 10^{10} the intrabacterial copper concentration, 10^{11} in HTS and 10^{10} in LTS. In contrast, previous exposure to copper decreases copper accumulation after a high toxic shock (10^9).

For copper adapted *S. aureus*, the intrabacterial copper concentration is multiplied by a factor of 2.10^5, in HTS: 10^8, in LTS: 2.10^5. In contrast, the accumulation of copper is stronger after Cu^{++} adaptation followed by toxic shocks.

This implies that mechanisms of defence against copper toxicity are different in these 2 strains.

CONCLUSION

Biofilms behaviours facing toxic shock were different for each strain. Ther was no rule in sensitivity or resistance to copper ionsi in the different conditions of these assays.

In all cases, it appears that copper pipes are very poorly efficient to fight nosocomial bacteria development, except if Cu^{++} ions are periodically released in the water distribution system.

Some authors have proposed to use pipes coated with silver, a metal more efficient than copper in its antibacterial effect [7, 8, 10]. This effect was magnified by intermittent ionization Cu/Ag, releasing Ag ions with a good biocidal effect.

REFERENCES

1. Brisset L., Jacquelin L.F., Le Magrex E., Carquin J., Choisy C. Quantification des biofilms bactériens *in vitro*, intérêt et limites. *Adhésion des Micro-organismes aux Ssurfaces*. Lavoisier Tec-Doc Ed., Paris. 1995, 220-235.
2. Gellé M.P., Jacquelin L.F., Doubeikovskaïa Z., Le Magrex E., Brisset L., Carquin J., Choisy C. Exploration du métabolisme des bactéries au cœur d'un biofilm. *Adhésion des Micro-organismes aux Surfaces*. Lavoisier Tec-Doc Ed. 1995, 236-247.
3. Khassanova L.A. Modifications électrophysiques, electrophysiologiques et biochimiques des bactéries sous l'effet des métaux lourds. *Thèse Doctorat Université de Reims-Champagne-Ardenne*. (URCA) Pharmacie Microbiologie, 1998.
4. Le Magrex E., Brisset L, Jacquelin L.F., Carquin J., Choisy C. Susceptibility to antibacterials and compared metabolism of suspended bacteria versus embedded bacteria in biofilms. *Colloïds and surfaces B: Biointerfaces*, 1994, 2, 89-95.
5. Le Magrex-Debar E., lemoine J., Gellé M.P., Jacquelin L.F., Choisy C. Evaluation of biohazards in dehydrated biofilms on foodstuff' packaging. *Int. J. Food Microbiol.*, 2000, 55, 230-243.
6. Liu Z., Stout J.E., Boldin M., Rugh J., Diven W.F., Yu V.L. Intermittent use of copper-silver ionization for *legionella* control in water distribution systems: a potential option in buildings housing individuals at low risk of infections. *Clin. Infect. Dis.* 1999, 26, 138-140.
7. Mietzner S., Schwille R.C., Farley A., Walder G.E., States S.J., Libert T., Wadowsky R.M. Efficacy of thermal treatment and copper-silver ionization for controlling *Legionella pneumophila* in high volume hot water plumbing systems in hospitals. *Am. J. Inf. Control*, 1997, 25, 452-457.
8. Nyström T., Olson R.M., Kjellerberg S. Survival, stress resistance and alterations in protein expression in the marine *Vibrio* sp. Strain S14 during starvation for different individual nutriments. *Appl. Environ. Microbiol.*, 1992, 58, 55-65.
9. Rohr U., Senger M., Selenka F., Turley R., Wilhelm M. Four years of experience with Silver-Copper ionization for control of *legionella* in a german university hot water plumbing system. *Clin. Inf. Dis.* 1999, 29, 1507-1511.

Metal Ions in Biology and Medicine: vol. 7. Eds L. Khassanova, Ph. Collery, I. Maymard, Z. Khassanova, J.-C. Étienne.
John Libbey Eurotext, Paris © 2002 pp. 655-658.

Complexation of three *N*-carboxyalkyl derivatives of 3-hydroxy-4-pyridinones with Fe(III), Al(III), Ga(III) AND In(III). STUDIES *in vitro* and *in vivo*

M. Amélia Santos[1], Marco Gil[1], Sérgio Marques[1], M. Lurdes Gano[2] and Sílvia Chaves[1]

[1]*Centro de Química Estrutural, Complexo I, Instituto Superior Técnico, Av. Rovisco Pais, 1049-001 Lisbon, Portugal;* [2]*Instituto Tecnológico e Nuclear, Estrada Nacional N° 10, 2686-953 Sacavém, Portugal*

ABSTRACT

The study of chelating agents for Fe(III) and the group 13 (IIIA) metal ions is of biomedical interest. The 3-hydroxy-4-pyridinones (3,4-HP) are very effective chelators in the neutral range of pH and one of those compounds (Deferiprone) has even been used in clinical trials. Aimed at developing new potential orally active drugs we have studied a set of *N*-carboxyalkyl derivatives of 3,4-HP (alkyl = ethyl, propyl and butyl). We describe herein a summary of those results, namely the effect of changing the alkylic chain length of the *N*-carboxyalkyl substituents on both the *in vitro* and the *in vivo* properties. Thus, the lipohydrophilic character of the ligands as well as their interactions with Fe(III), Ga(III), Al(III) and In(III) were evaluated, in aqueous solution. They all present high affinity for these metal ions and, in some cases, even higher than Deferiprone (DMHP). The formation constants of the complexes follow the order ((FeL_3) > ((GaL_3) > ((InL_3). The lipophilic character of the ligands and the stability of their complexes decrease with the increasing size of the alkylic chain. Results from the *in vivo* studies clearly evidence that the administration of any of the three chelators interfere in the normal biological distribution profile of the tracer enhancing the excretion of ^{67}Ga, leading to lower bone uptake and higher clearance with increasing size of the chain.

INTRODUCTION

The development of hydroxypyridinones, namely 3-hydroxy-4-pyridinones, has been attracting considerable attention due to the number of potential applications in medicinal chemistry. In fact, this class of compounds are very good chelating agents for a variety of hard M(III) metal ions which have been associated to disease states due to gradual accumulation in man [ex: Fe ((-thalassemia) [1], Al (Alzheimer's disease and dialysis encephalopathy sindrome) [2]] or to radiodiagnostic/radiotherapy purposes (^{67,68}Ga, ^{111}In) [3].

The great interest of the 3-hydroxy-4-pyridinones started from their potential use as clinical iron chelating agents for the replacement of the most usual drug, the desferrioxamine (DFO) [4, 5]. The interaction of hydroxypyridinones with Al^{3+} has also been the object of recent research [6], namely *in vitro* studies [7, 8] and biological assays as Al-mobilizers from animal models of aluminium intoxication [9, 10]. The interaction of hydroxypyridinones with gallium [11, 12] and indium [13] has also been studied in relation with their potential use in radiodiagnosis.

As part of an ongoing project aimed at developing ligands for this set of highly charged metal ions with potential clinical use as oral drugs, we have studied a set of hydrophilic *N*-carboxyalkyl

3,4-HP derivatives (1-carboxyalkyl-2-methyl-3-hydroxy-4-pyridinone, alkyl = ethyl, propyl and butyl) [14]. A summary of the obtained results is included herein, namely the lipohydrophilic characterization of the ligands and their interactions with Fe(III), Ga(III), Al(III) and In(III) in aqueous solution. The evaluation of their *in vivo* behaviour was also performed, including the study of the effect of each ligand administration on the biodistribution of the ^{67}Ga-loaded mice, taking into account potential pharmaceutical applications.

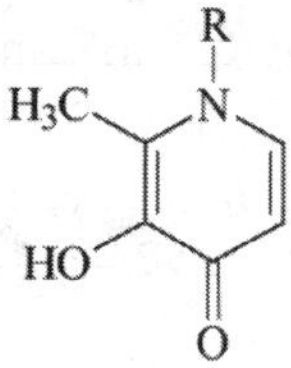

$R = -(CH_2)_nCO_2H$

n = 2	**H_2L^1**
n = 3	**H_2L^2**
n = 4	**H_2L^3**
$-CH_3$	**DMHP**

Scheme

MATERIALS AND METHODS

The materials and methods used are according to the literature [12, 14]. The ligands were obtained by reacting 3-benxyloxy-2-methyl-4-pyrone with the aminoalkylcarboxylic acids followed by a standard hydrogenolysis. The octanol-water partition coefficients (log D) are basically concentration ratios (D) of each compound between the octanol phase and the aqueous phase (*Tris*, pH = 7.4) at 25 °C. The stability constants of the complexes were calculated by fitting the potentiometric and spectrophotometric data. The biodistribution studies were performed in mice intravenously injected with ^{67}Ga-citrate. A separated group of animals was simultaneously injected intraperitoneally with ligand in saline solution. The radioactive dosage administered and the radioactivity in the sacrificed animal (main organs) were determined by gamma counting. Biodistribution results were expressed as percent of injected dose per total organ (% ID/ organ).

RESULTS AND DISCUSSION

The results of the equilibrium studies performed in aqueous solution are summarized on *table 1*. Analysis of the protonation constants of these *N*-carboxyalkyl derivatives shows comparable sets of values, which are attributed to the hydroxypyridinone hydroxyl, carboxyl and pyridyl groups, respectively, according to evidences from the literature (*e.g.* 1,2-dimethyl-3-hydroxy-4-pyridinone, DMHP) [8]. There is a small increase on the basicity of the hydroxypyridone hydroxyl and carboxyl groups with the size of the alkylic chain, which can be attributed to the concomitant increasing of the electron donation effect of the alkylic group. According to the calculated log K_i values, for a pH wide range centred around the physiological pH [5-8], the major ligand species is $[HL^{-1}]$, thus anticipating the hydrophilic character of these pyridinone derivatives.

The complexation behaviour of these ligands towards this set of M^{3+} metal ions in aqueous solution was evaluated through the global stability constants of the complexes ($b_i = [ML_i]/[M][L]^i$) or, better, the corresponding pM values (which are the negative log of the free metal ion concentration in presence of ligand excess, $C_L/C_M = 10$, $C_M = 10^{-6}$ M, at physiological pH 7.4). Analysis of *table 1* shows that the stability constants of these complexes are quite high, in agreement to previous studies [15]. Some of these ligands,, namely the one with the shortest alkylic chain (H_2L^1), have higher affinity for these metal ions than the drug Deferiprone [8] or even transferrin (Tf)

[16-19]. The calculated stabilities follow the order Fe(III) > Ga(III) > In(III), in agreement with predictions of HSAB theory. The same trend has been reported for DMHP and also Tf, for which the co-ordination environment is known to be dominated by hard oxygen atoms. Concerning the Al(III) complexes, although their stability constants have usually values between those of Ga(III) and In(III) species, in some cases, they are less stable than the In(III) complexes, as happens with Tf.

The lipo-hydrophilic characterization of the ligands, through their partition coefficients, shows that they are quite hydrophilic (-1.2>log D>-1.9) and there is a slight increase of the lipophilicity with the size of the alkylic chain.

Table 1. Stepwise protonation constants (log K_i) and partition coefficients (log D) of a set of 1,2-substituted 3-hydroxy-4-pyridinones and transferrin as well as the global formation constants (log b_{ML3}) and pM* values at pH = 7.4 of the corresponding M(III) complexes (M = Fe, Al, Ga, In)

Ligand				Complex			
	Log D	**Log K_i**		**Fe**	**Ga**	**Al**	**In**
H_2L^1	-1.89	9.83 4.13 3.34	log b_3	37.97[a]	36.84[b]	32.81	32.98
			pM	**21.5**	**20.1**	**16.1**	**16.8**
H_2L^2	-1.38	9.91 4.38 3.53	log b_3	37.04	36.19	31.27	31.95
			pM	**20.1**	**19.6**	**14.4**	**16.4**
H_2L^3	-1.22	10.04 4.46 3.25	log b_3	37.16	36.41	31.60	31.35
			pM	**19.9**	**19.4**	**14.4**	**16.4**
DMHP (Deferiprone)	-1.03[c]	9.77[d] 3.62	log b_3	35.88	35.76	32.62	31.71
			pM	**19.3**	**17.8**	**16.0**	**16.5**
Transferrin (Tf)	-	-	log K_{2M}[e]	20.34	18.80	12.72	16.44
			pM	**20.3**[f]	**20.3**[g]	**14.5**[h]	-

*at $C_M = 10^{-6}$ M, $C_L = 10^{-5}$ M.[a] ref 15; [b] ref 12; [c] ref. 9; [d] ref 8; [e] ref. 16; [f] ref. 17; [g] ref 18; [h] ref. 19.

The *in vivo* studies (see summary presented in *fig. 1*) show that these chelators can interfer inthe normal biodistribution of the tracer, enhancing the ^{67}Ga excretion and leading to higher bone uptake. There is also a slightly lower blood clearance and slightly higher hepatic uptake. ITLC analyses of mice urine samples, collected at sacrifice time, indicated that the radioactivity is mostly excreted as ^{67}Ga-complexes. Although the mechanism involved is not clear, these complexes may be the main species retained in bone, where the free carboxyl group could play some important role.

Although the transferrin is known to compete with citrate for the ^{67}Ga, the ligands in this study present a very high affinity for gallium, so that some of them can even compete with transferrin.

The size of the alkylic chain has also some effect on the biodistribution. Increasing the chain size seems to diminish the bone fixation and to enhance the excretion. On the other hand, there is a slight effect on decreasing the blood clearance and increasing the hepatic retention. The bone fixation effects may be related with the increase of ligand affinity for the gallium, while the other effects seem to be mostly related with the lipophilicity/molecular weight of the ligands/complexes and, probably, some subsequently interference in the excretion mode (renal/hepatobiliar).

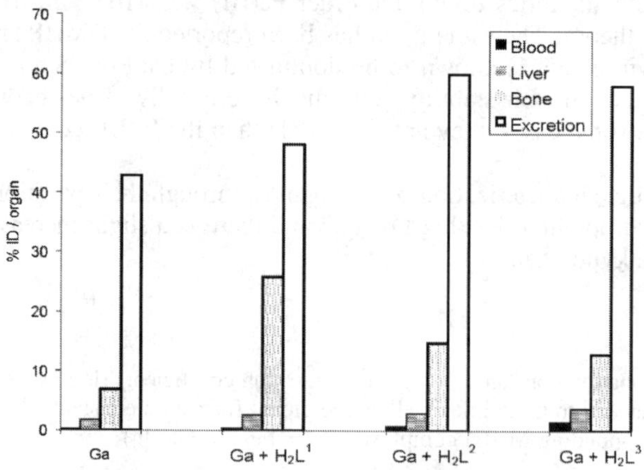

Fig. 1. Histogram of the biodistribution data of the most representative tissues and urinary excretion for ^{67}Ga-citrate and ^{67}Ga-citrate with simultaneous intraperitoneal injection of each ligand, 48 h after intravenous administration in female mice.

ACKNOWLEDGMENT

The authors are thankful to the Portuguese *Fundação para a Ciência e Tecnologia* (FCT) for financial support (project POCTI 35344/99).

REFERENCES

1. R.C. Hider, A.D. Hall, Clinical Useful Chelators of Tripositive Elements. In Progress in Medicinal Chemistry; G.P. Ellis, G.B. West, eds.; Elsevier, New York, 28 (1991) 41-173.
2. R.A.. Yokel, Neurotoxicology, 21 (2000) 813-828.
3. M.A. Green and M.J. Welch, Nucl. Med. Biol. 16 (1989) 16 435.
4. G.J. Kontoghiorghes, The Lancet (1985) 817; *Ibid*, Inorg. Chim. Acta 135 (1987) 145-150.
5. P.S. Dobbin, R.C. Hider, A.D. Hall, P.D. Taylor, P. Sarpong, J.B. Porter, G. Xiao, D. van der Helm, J. Med. Chem. 36 (1993) 2448-2458. B.L. Rai, L.S. Dekhordi, H. Khodr, Y. Jin, Z. Liu, R.C. Hider, J. Med. Chem. 41 (1998) 3347-3359.
6. M.A. Santos, Coord. Chem. Rev., in press.
7. D.J. Clevette, W.O. Nelson, A. Nordin, C. Orvig, S. Sjoberg, Inorg. Chem., 28 (1989) 2079-2081.
8. E.T. Clarke and A.E. Martell, Inorg. Chim. Acta, 196 (1992) 185-194.
9. R.A. Yokel, A. K. Datta, E.G. Jackson, J. Pharmacol. Exp. Ther., 257 (1991) 100-106.
10. R.A. Yokel, K.A. Meurier, T.L. Skinner and A.M. Fredenburg, Drug Metabolism and Disposition, 24 (1996) 105-111.
11. D.J. Clevette, D. M. Lyster, W.O. Nelson, T. Rihela, G.A. Webb; C. Orvig, Inorg. Chem. 29 (1990) 667-672.
12. M.A. Santos, R. Grazina, A.Q. Neto, G. Cantino, L. Gano and L. Patricio, J. Inorg. Biochem. 78 (2000) 303-311.
13. B.L. Ellis, A.K. Duhme, R.C. Hider, M.B. Hossain, S. Rizvi and D. van der Helm, J. Med. Chem. 39 (1996) 3659-3670.
14. M.A. Santos, M. Gil, S. Marques, L. Gano G. Cantino and S. Chaves, (submitted).
15. M.A. Santos, R. Grazina, A.Q. Neto, G. Cantino, L. Gano and L. Patrício, Metal Ions in Biology and Medicine, John Libbey Eurotext, vol.6, Paris (2000), 613-615.
16. W.R. Harris, Y. Chen, K. Wein, Inorg. Chem. 33 (1994) 4991-4998.
17. R.J. Motekaitis and A.E. Martell, Inorg. Chim. Acta 183 (1991) 71-80.
18. W.R. Harris, V.L. Pecoraro, Biochemistry 22 (1983) 292-299.
19. W.R. Harris, J. Sheldon, Inorg. Chem. 29 (1990) 119-124.

Index
Authors

T

U

V

W

Y

Z

Achevé d'imprimer par Corlet, Imprimeur, S.A. - 14110 Condé-sur-Noireau (France)
N° d'Imprimeur : 57831 - Dépôt légal : avril 2002 - *Imprimé en U.E.*